T0189158

Lecture Notes in Artificial Intelligence 604

Subseries of Lecture Notes in Computer Science
Edited by J. Siekmann

Lecture Notes in Computer Science

Edited by G. Goos and J. Hartmanis

F. Belli   F. J. Radermacher (Eds.)

Industrial and Engineering Applications of Artificial Intelligence and Expert Systems

5th International Conference, IEA/AIE - 92
Paderborn, Germany, June 9-12, 1992
Proceedings

Springer-Verlag

Berlin Heidelberg New York
London Paris Tokyo
Hong Kong Barcelona
Budapest

Series Editor

Jörg Siekmann
University of Saarland
German Research Center for Artificial Intelligence (DFKI)
Stuhlsatzenhausweg 3, W-6600 Saarbrücken 11, FRG

Volume Editors

Fevzi Belli
University of Paderborn, Dept. of Electrical Engineering
P. O. Box. 1621, W-4790 Paderborn, FRG

Franz Josef Radermacher
Forschungsinstitut für anwendungsorientierte Wissensverarbeitung (FAW)
P. O. Box 2060, W-7900 Ulm, FRG

CR Subject Classification (1991): I.2, J.6

ISBN 3-540-55601-X Springer-Verlag Berlin Heidelberg New York
ISBN 0-387-55601-X Springer-Verlag New York Berlin Heidelberg

© Springer-Verlag Berlin Heidelberg 1992
Printed in Germany

Typesetting: Camera ready by author/editor
Printing and binding: Druckhaus Beltz, Hemsbach/Bergstr.
45/3140-543210 - Printed on acid-free paper

Address by the Federal Minister for Research and Technology

The general conditions which will characterize industrial production in the next century have already become visible; they include increasing complexity, greater demands for flexibility and quality, growing division of labour, and even shortening cycles of development and production.

Good provision for the next century is the preparation for future systems that allow rapid and independent access to general knowledge whenever required for solving particular problems while being capable of handling even incomplete and inaccurate knowledge, in other words: preparation for systems of artificial intelligence.

This is a complex goal, however, which undoubtedly involves enormous difficulties. The approach should be via reasonable and reachable subgoals in order to avoid setbacks and frustration.

The Federal Minister for Research and Technology has been supporting research and development projects in the field of artificial intelligence since 1984; up to now about DM 225 million have been earmarked predominantly for expert systems as well as for image and natural-language processing systems.

This support has contributed substantially to the dissemination and further development of the scientific methods for artificial intelligence in universities and research institutions throughout the Federal Republic of Germany; cooperative projects provide a direct link with the industry. Germany's efforts in this field are recognized by the international community.

In Germany today a number of enterprises offer tools for expert systems, and various system houses using the relevant additional knowledge market complete expert systems. Mainly in industry as well as in the banking and insurance business such systems have so far been applied above all for diagnosis, counselling, configuration and planning.

When supporting artificial intelligence research in the years to come the Federal Minister for Research and Technology will attach increasing importance to the subsequent application of research results. This does not mean, however, that the need for further basic research is questioned.

All the best for the success of the Fifth International Conference on Industrial and Engineering Applications of Artificial Intelligence and Expert Systems in Paderborn.

Dr. Heinz Riesenhuber

Address by the General Chairman and the Program Chairmen

Welcome to The Fifth International Conference on Industrial and Engineering Applications of Artificial Intelligence and Expert Systems (IEA/AIE-92). Through this annual international conference we have been achieving our objective of providing a forum for the international community to discuss recent successes in Artificial Intelligence, Expert Systems, Neural Nets, CAD/CAM, and many other related areas in the industrial and engineering environment. The 72 reviewed papers and 5 invited papers presented at this conference cover these areas from various perspectives and multilateral points of view.

This is the first time that an IEA/AIE conference is taking place outside USA. We have received more than 120 papers from 23 countries, clearly indicating the international character of this conference series. Each paper has been reviewed by at least three referees. The organization of the IEA/AIE-92 was made possible only because of the intensive and diligent efforts of many people, and because of the sponsorship of several companies and of the University of Paderborn, Southwest Texas State University, and the FAW Ulm/Germany. Our thanks are due also to the members of the program committee, to the external referees, and - last but not least - to the authors who submitted their papers. Our special thanks go to the Springer-Verlag for the excellent cooperation.

We hope you will gain deeper insight into the multidimensional research concerning applied Artificial Intelligence topics in industry, interact with other participants in addressing your own challenging research problems, and enjoy the exciting settings of Paderborn and Central Germany.

Moonis Ali   Fevzi Belli   Franz Josef Radermacher

Program and Organization Committees

General Chairman	Program Chair	Program Co-Chairs	Ex Officio
Moonis Ali	Fevzi Belli	Gr. Forsyth Edward Grant F. J. Radermacher	Jim Bedzek Manton Matthews

Program Committee

O. Abeln	A.G. Cohn	P. Jedrzejowicz	I. Plander
J.H. Andreae	A. Costes	M. Klein	M.M. Richter
W. Bibel	R. Freedman	J. Liebowitz	E. Sandewall
G. Biswas	T. Fukuda	R. Lòpez de	J. Shapiro
H. Bonin	O. Günter	Màntaras	P. Sydenham
I. Bratko	U.L. Haass	B. Moulin	L. Thomas
L. Carlucci Aiello	G. Hartmann	B. Neumann	M. Trivedi
T. Christaller	K.M. van Hee	S. Ohsuga	I. Witten
P. Chung	A. Herold	D. Peacocke	T. Yamakawa

External Referees

Y. Ali	R. Inder	P. Mesequer	M. Shirvaikar
D. Barschdorff	S. Isenmann	A. Miola	P. Simmet
R.A. Boswell	H. Johnson	T. Mowforth	C. Souter
M. Burgois	V. Junker	D. Nardi	M. Sprenger
C. Chen	H. Kleine-Büning	G. Neugebauer	M. Valtorta
R. Cunis	M. Knick	T. Niblett	A. Vellino
J. Davy	S. Kockskämper	H. Reichgelt	Th. Vietze
J. Diederich	M. Kopisch	W. Riekert	A. Voss
F.M. Domini	N. Kratz	H. Ritter	M. Wallace
S. Dzeroski	J. Kreys	G. Sagerer	L. Wieske
K. Echtle	V. Küchenhoff	T. Schaub	S. Wrobel
U. Egly	J. Lamberts	K. Scheuer	T. Zrimec
N. Eisinger	M. Linster	C. Schlegel	
I. Ferguson	O. Ludwig	J. Schneeberger	
H.P. Gadaghew	J. Meads	B. Shepherd	

Acknowledgments to

Universität Paderborn
Southwest Texas State University
Forschungsinstitut für anwendungs-
orientierte Wissensverarbeitung (FAW)
Dornier - Deutsche Aerospace
VW-Gedas
Daimler Benz
Digital Equipment GmbH
Siemens-Nixdorf Informationssysteme

Table of Contents

Finance (Chair: H. Bonin, FH Nordostniedersachsen)

Knowledge-Based Systems I (Chair: L. Thomas, SNI/Paderborn)

Knowledge Representation (Chair: A.B. Cremers, Univ. Bonn)

Knowledge Acquisition and Language Processing (Chair: Th. Christaller, GMD/Bonn)

Planning and Scheduling (Chair: R.V. Rodriguez, Univ. W. Florida)

Data/Sensor Fusion (Chair: U.L. Haass, FORWISS/Erlangen)

CAD II (Chair: R. Aiken, Temple Univ.)

Gaining Strategic Advantage with Real-Time Distributed Artificial Intelligence

Yuval Lirov

Salomon Brothers
Rutherford, NJ 07070

Abstract. Synergy of communications technology with artificial intelligence results in increased distribution, specialization, and generalization of knowledge. Generalization continues migration of the technology to the main-stream programming practice. Distribution further personalizes automated decision-making support. Specialization scales up domain expertise by architecting in real-time a series of limited-domain expert systems that work in concert. These important trends resolve a list of current shells inefficiencies revealed via analysis of a distributed systems management platform.

1 Introduction

Widespread deployment of strategic advantage systems that include knowledge-based components guides the progress of artificial intelligence. Strategic advantage systems are those that fundamentally change the ways to conduct business and provide firms and individuals with sustained competitive advantage. As we are quickly approaching the era of having tens and even hundreds of computers in every room communicating via microwaves (Weiser, 1991), will artificial intelligence continue to provide leadership in architecting strategic advantage systems?

1.1 First-order Effects

Communications technology is another major contributor in the recent strategic advantage systems that profoundly impact our society. Telephone network, for example, is the infrastructure that contributes to diversification of capital power by supporting such an important distributed economic system as the OTC market - the largest distributed national US stock market. A "first-order cost reduction effect" is achieved by substituting information technology for human coordination to eliminate hundreds of clerks from conventional stock markets (Malone and Rockart, 1991). Telephone technology alone, however, is unable to support truly mass economic and social activities because of its narrow information bandwidth. The phone cuts distance and time only. An expert must accumulate necessary information first and then make the right decisions.

1.2 Second-order Effects

Wider information bandwidths, introduced by distributed computing, enable both broadcasting and soliciting vast amounts of information. Local and wide area networks enable new strategic services as well as pose new technology challenges. For instance, a real-time reflection of remotely located transactions (e.g., an IBM-PC-based trade capture in Tokyo to a SUN-based database in New York) has become a routine service. Thus, we witness growing economic interaction between private investors and a reduced role for centralized markets, i.e., a "second-order cost reduction effect", facilitated by increased coordination because of distributed computing.

1.3 Third-order Effects

Although private accumulation of information is simplified by personal computers and networking, it still takes an expert to make the right decision. Today we can only speculate on the effects of dissemination of expertise, or, in other words, on the effects of a "third-order cost reduction effect". The third-order effect will improve market efficiency and, judging from the experience of London International Stock Exchange, will reduce the profits of brokers and trading specialists (Clemons, 1991). For example, would an integration of Black-Scholes modeling tools with data feeds and news understanding on a personal computer at home result in massive options trading by private investors? Will the personal short-term investment sector outgrow its institutional investment counterpart? On the other hand, would a distributed case-based reasoning capability operating across a wide area network be the right strategic advantage tool to retain the current activity volume in the institutional investment sector? Distributed expertise, resulting in a third-order effect, will affect our society in more ways than any of the technological advances mentioned above.

1.4 What is Distributed Artificial Intelligence?

Distributed artificially intelligent systems address the following topics (Bond and Gaser, 1988):

1) formulate, decompose, and allocate problems, and synthesize results among a group of intelligent problem-solvers (agents);
2) enable agents to interact (provide language and protocol);
3) prevent harmful interaction between the agents and accommodate global effects in local decisions;
4) reason about coordinated processes and actions of other agents.

Typical advantages of distributed artificially intelligent systems include:

1) adaptability: temporal and spatial distribution of processing elements provides alternative perspectives on emerging situations;
2) cost-efficiency: large number of low-cost workstations achieve tremendous computational speed-ups, outperforming in many cases the supercomputers;
3) improved development/management: separate development/management of different subsystems enabled by separating specialists;
4) isolation/autonomy: parts of a system may be isolated;
5) reliability: redundancy and cross-checking;
6) specialization: better control provided by a specialized computation.

This paper uses a real-life strategic advantage system as a working example to expose a list of important problems that require immediate attention. For example, a robust and fast distributed reasoning architecture is needed to enable vendor-independent data base access via automated query creation. Moreover, the creation of real-time distributed systems management platforms requires rigorous methodology (Melamed, 1991).

1.5 Paper Outline

Knowledge-based data interpretation is a critical part of such platforms. After discussing the working example in Section 2, we enumerate the shortcomings of conventional shell-based systems in Section 3. Knowledge based components need better interfaces, wider domain expertise, and quicker development. CASE techniques (e.g., knowledge compilation, presented in Section 4) quicken system development and ease maintenance. Distributed reasoning and integrated quantitative and qualitative computations open new opportunities to resolve previously intractable problems. Section 5 broadly outlines current state of the enabling technology.

2 Case Analysis: Distributed Systems Management

Information technology is commonly used to leverage existing corporate resource advantage. Deployment of a distributed multi-vendor platform in which local processors are assigned to each business unit in a brokerage firm is a case in point. Such a platform integrates market data services, trade execution, clearance and settlement, and analytical/risk management applications. Its purpose is to achieve a sustainable strategic advantage in aiding the real-time analysis by traders of complex trading strategies, in helping management to assess and control trading risks, in selling effectively to increasingly sophisticated institutional investors, and in settling large volumes of trade. It is an example of decentralization of resources that currently takes place in many firms.

The new platform must increase data-processing productivity, must be fail-safe around the clock, and must be responsive to change, enabling simple assimilation of new techniques and products. The last requirement is especially important for sustaining strategic advantage as competitors who started later should not gain competitive advantage because of reduced technical risks.

Development of such a system today requires the design and implementation of a sophisticated real-time systems management facility that includes network and software monitoring and alarm interpretation in real-time. Such a facility, although under development by several vendors, does not yet exist. Therefore, the development of a strategic advantage system may require managing technical risk associated with providing technology that is not yet available.

The next sub-section elaborates on the challenges of monitoring and diagnosis. Sub-section 2.2 summarizes the difficulties that human operators face when using traditional network monitoring systems. Sub-section 2.3 overviews the architecture of our intelligent diagnostic system.

2.1 Monitoring and Diagnosis

Automated real-time systems management is important because it speeds up troubleshooting, reduces system down time, and ensures service quality. Systems management includes non-obtrusive monitoring of system components to identify service disruptions and requires the quick analysis of evidence to indict the perpetrator of a disruption. Systems management is difficult because of a quantity versus expediency tradeoff: the more evidence gathered the more certain the indictment but the longer the assimilation period (Figure 1). The conflict is settled by an automated alarm filtering system that both processes the evidence in real-time and reduces the flow of insignificant fault alarms using knowledge-based clustering and averaging techniques. The speed-up achieved by the improved troubleshooting process positively affects system operators by re-allocating their time from mundane message interpretation tasks to more demanding and productive endeavors. Therefore, once such an intelligent tool becomes operational, overall improvements are achieved in service quality.

the more evidence gathered
the more certain the indictment

the longer the assimilation period

Figure 1.   Quantity versus Expediency

2.2 Difficulties of Traditional Systems Management

Systems monitoring task requires automation even for mildly complex systems. Advanced systems monitoring tools record systems fault data automatically. Paradoxically, improved data acquisition only complicates subsequent data interpretation tasks as there is more data to analyze in a shorter time.

The current troubleshooting practice of computer systems follows the Monitor-Compare-Alarm-Interpret paradigm depicted in Figure 2. Here, a monitoring platform monitors various system behavior indices, identifies extraordinary situations, and generates alarm messages to operators. The operator must then interpret all alarms.

Systems management is difficult for the following typical key reasons:

1) complexity: a network is usually a very large system, containing several thousand different interacting network components;
2) cascading: a primary fault causes a number of secondary faults throughout the system (e.g., a router going down causes all services traversing that router to go down too); this phenomenon is exacerbated by an avalanche of secondary alarms to the operators, further obstructing correct and timely problem resolution;
3) low observability: some network components are difficult to monitor (e.g., cables) and their state must be deduced by observing other components;
4) non-linearity: the number of messages grows faster than linearly with time (e.g., the number of retransmission requests grows just before a network storm);
5) inherent disguise - often only the symptoms of secondary faults are observed (e.g., most network protocols incorporate retry mechanisms to recover from errors and increase reliability);

such protocols substitute for numerous real faults a single obvious symptom - slow response time;

6) myopic monitoring mechanisms - each mechanism measures only a single system aspect - no global view;

7) synergistic alarm interpretation - alarms must be interpreted collectively, using chronological, causal, structural, and textual information simultaneously;

8) transience - generation of alarms can be characterized by a random process that manifests itself in short-lived bursts of alarms and their negations (e.g., "process down, process up.")

Increasing complexity of systems is outstripping the capability of human beings to control them

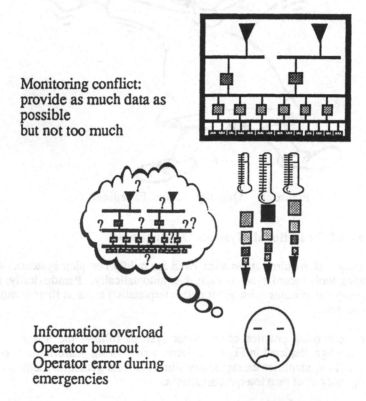

Monitoring conflict:
provide as much data as
possible
but not too much

Information overload
Operator burnout
Operator error during
emergencies

Figure 2. Traditional Systems Management

Synergistic alarm interpretation is particularly important and difficult as it requires simultaneous processing of causal, structural, and chronological information and results in a combinatorial explosion of possibilities.

2.3 Knowledge-based Component

A knowledge based diagnostic tool resolves the problem by automating data interpretation tasks and providing both numerical averaging and clustering effects together with symbolic rule and text processing capabilities. Our knowledge based alarm filtering system (AFS) quickly simulates the typical message interpretation activities of an experienced operator by cycling through the following loop: observe symptoms, hypothesize, test and evaluate, and conclude.

Its design follows the Monitor-Compare-Interpret-Alarm paradigm depicted in Figure 3.

Monitoring conflict
resolution:
provide only
interpretations

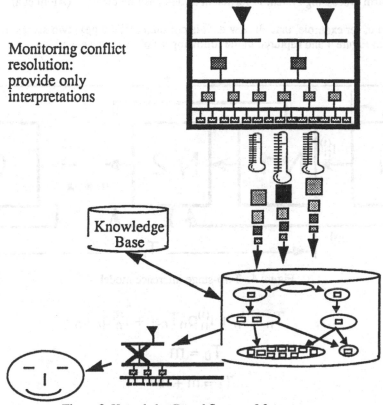

Figure 3. Knowledge Based Systems Management

According to this paradigm, the network monitoring platform observes the network and sends messages to AFS, which, in turn, interprets received messages and generates only an aggregate summary message to the operator. The advantage of such an approach versus the Monitor-Compare-Alarm-Interpret approach consists in shifting from the user-processing intensive task to the computer-processing intensive one. Note, that AFS itself can be implemented as a distributed system (Factor et al, 1991).

AFS performs the following synergistic message interpretation tasks:

1) integration of structural and chronological data from various system components
2) reasoning with integrated data using statistical and causal inferencing rules

AFS achieves the desired message interpretation by retaining an internal model of the network and by relating the incoming stream of messages to that model. AFS relates messages to parts of the model and makes message clustering decisions in real-time using a rule-based system. Thus, AFS integrates rule-based (shallow) and model-based (deep) reasoning approaches.

2.4 Expert Systems Performance Analysis

Interestingly, distributed knowledge based systems not only help troubleshooting, but also enhance system monitoring capacity by reducing the amount of traffic between the fault data collection and message display components. Surprisingly, there is almost no published work on quantitative expert system capacity evaluation methods, even though the design of a simple real-time alarm processing system is sufficient to demonstrate the need (Aloni et al, 1991).

Let us assume, for example, that all new data is consumed at the next two stages of reasoning as depicted in Figure 4 and captured in the following set of recursive equations.

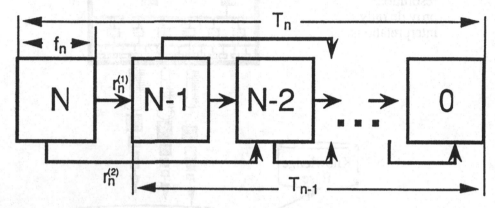

Figure 4. Two-stage inference model

$$T_n = f_n + r_n^{(1)} p_n T_{n-1} + r_n^{(2)} q_n T_{n-2}$$

$$T_0 = m$$

$$T_1 = m + rf$$

Here f_n is the time required to process messages at the n-th phase. p_n is the probability to transfer $r_n^{(1)}$-th portion of the initial message load to the (n-1)-st stage and q_n is the probability to transfer $r_n^{(2)}$-th portion of the initial message load to the (n-2)-st stage. T_n is the total processing time for the initial load for n stages. Since a closed-form solution of above system is difficult, we only demonstrate numerical solutions graphically, viewing the impact of filtering on total system performance from three different angles, in Figure 5.

The horizontal axes denote the changes in r_1 and r_2, while the vertical axis corresponds to the total reasoning time over the several reasoning stages. We elected to depict the dependency of T_n on r_1 and r_2 for $n = 4$ and for $n = 3$. We assume

$$f_n = f, \quad r_n^{(1)} = r_1, \quad r_n^{(2)} = r_2, \text{ and}$$

$$p_n = p, \quad q_n = q$$

Notice also that the scope of expertise of the overall expert system is magnified because of the required development a series of interacting limited-domain expert subsystems.

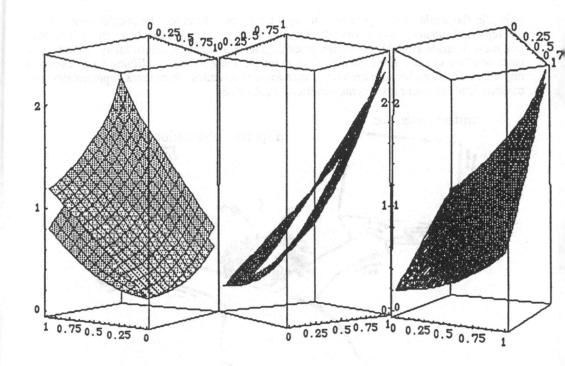

Figure 5. Numerical solution of two-stage inference model

3 Why Aren't Shells Good Enough?

This section lists the main shortcomings of expert systems shells as perceived from distributed computing viewpoint.

Knowledge-base reasoning shells represent the marketing legacy of AI hype of the early 80's: make one knowledge-based product and sell it a thousand times. Current knowledge-base reasoning shells fall short from enabling real-time distributed reasoning because of three overheads: communications, representation, and callback. Each overhead hurts real-time, distributed, and process control aspects of reasoning respectively. Collectively, these overheads impair the most important AI feature - learning, which is a cardinal feature to achieve scalability of the software to production standards.

Communications overhead hurts the real-time aspect of the system. The overhead is incurred when communicating data between a real-time data source (e.g., alarms) and an expert system working memory. Usually, a fact assertion mechanism makes the new data available for the inference engine. This mechanism is the undesirable communications overhead.

Representation overhead hurts the distributed aspect. It is incurred when some conventional software module must interact with a knowledge-based module. The shell must resolve the conflict of a uniform interface to disparate information sources (Figure 6). Conventional inference engines assume some standard knowledge representation scheme and enforce every object in the world to be represented using the scheme. However, the problem arises when conventional modules have a sophisticated non-standard data structure (e.g., an esoteric tree) that must be made available to the inference engine. For example, homogeneity of production rules can lead to convoluted rules of obscure intent. Furthermore, modifying a rule can have unpredictable effects because dynamic interaction between rules. Sometimes, representation and communications overheads become inextricably entangled.

uniform interface

disparate information sources

Figure 6. Representation Overhead

Finally, the callback overhead hurts the process control aspect and it results from the common way of passing the execution control to the reasoning mechanism. The foreign module regains control (is called back from the knowledge-based shell, hence the name) only after the reasoning module completes its processing. Popular graphical interface situations (e.g., Motif) exacerbate this phenomenon because they have a separate callback mechanism. It is very difficult to implement two callback mechanisms (one, for reasoning and another, for graphics interface) in the same application.

In summary, modern expert systems are often constructed using the tools developed according to the computational paradigms of the previous decade. These archaic tools are unable to adjust their architecture for solving different types of computational problems. Knowledge compilation, discussed in the subsequent Section 4, offers a viable way to overcome aforementioned difficulties while retaining the advantages of knowledge and inference separation.

4. Further Examples of Distributed Reasoning

Distributed reasoning also empowers such important areas as knowledge compilation, synergistic quantitative and qualitative computations, fast case-based reasoning, and efficient multi-objective optimization.

4.1 Synergistic Computations

Distributed processing systems open new avenues for merging quantitative and qualitative reasoning approaches, with separate machines performing separate kinds of reasoning. Such synergistic systems performing, for example, heuristic search and discrete event or qualitative simulations have been surveyed in (Lirov and Melamed, 1991). Previously intractable applications, such as discovery of correlation patterns in price data, may be now solved routinely on networks of workstations. Speeding combinatorial search problems by using distributed processing was also discussed in (Monien and Vornberger, 1987), where it is shown that sending messages between asynchronous processors with local memory is well suited to handle many problems from operations research and artificial intelligence.

Distributed computing also enables efficient merging of multi-objective optimization and expert systems technology, which, in turn, results in reduced modeling efforts and enhanced problem-solving tools. Search is one of the ways to combine multi-objective optimization and knowledge-intensive computational schemes. Search is usually associated with prohibitive computational costs and heuristics are often used to alleviate computational burden. Algorithmic heuristic construction is an important facet of knowledge compilation technology (Lirov (B), 1991). Distributed computing solves a multi-objective optimization problem by assembling a solution from its parts obtained from independent computations of multiple objectives.

4.2 Knowledge Compilation

Knowledge compilation provides new directions for distributed reasoning implementation. For example, one can create an array of knowledge-based skeletons, each operating as a separate process, possibly on a separate processor. The knowledge base skeletons can be created off-line before populating the knowledge base with detailed advice. A skeleton contains the rules to generate and evaluate hypotheses except for actual technical details of troubleshooting advice. The advice may include proper outputs of the correctly functioning components, though values themselves will be acquired at a later knowledge acquisition stage. Pre-compiling a skeleton for the knowledge base speeds up execution because of compiler optimization, and enables better code portability because of improved modularity (Lirov (A), 1991).

The divide-and-conquer approach is effective for hypothesis evaluation - the cost of the hypothesis equals the minimum over the set of sums of the tests associated with the hypothesis plus the cost of the subsequent hypothesis.

$$C(T) = \min\{C_t + pC(T^+) + qC(T^-)\}$$

Here p denotes the probability of passing the test t, and q - of failing. The bad news with this equation is that its computational cost is prohibitive for all practical purposes (NP-hard). Thus, we face a new conflict: the equation dictates the diagnostic tree that wastes the fewest resources, but solving the equation is impossible for any reasonably large system (Figure 7).

compute the diagnostic tree that
wastes the fewest resources

solving the equation is impossible
for any reasonably large system

Figure 7. Computational Explosion

This kind of incompatibility is common in Artificial Intelligence, a heuristic being the standard
remedy. Indeed, an easily computed heuristic, based on a modification of the Huffman coding
scheme, substitutes here for the recursive computation to generate a workable estimate (Lirov
and Yue, 1991).

4.3 Case-Based Reasoning

Distributed computing also introduces a qualitative change in traditional case base reasoning
(Figure 8). A choice of an appropriate case base must be made prior to or during the reasoning
process. One possibility for providing such a function is to create a uniform method to access
disparate databases. Significant advances in contents-based data base search will enable progress
in this area. Additionally, a search in a single case base may produce only an interim lead
towards a final problem resolution that may belong to some other case in a distant case base.
Thus, an enabling instrument, such as blackboard, is required to provide chaining of cases. In
summary, distributed case base reasoning systems will integrate several AI paradigms.

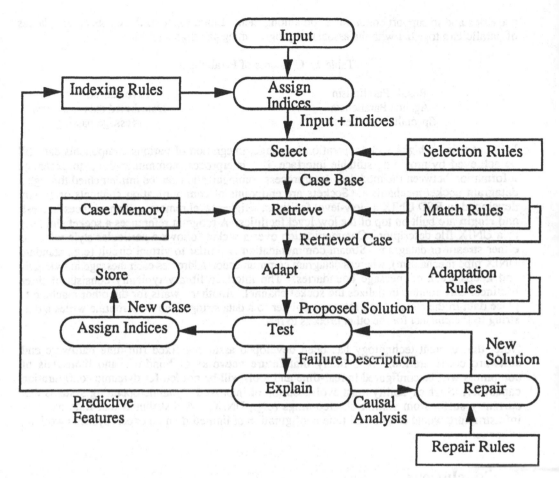

Figure 8. Distributed Case Base Reasoning Flow

5 Enabling Technology

Distributed reasoning is facilitated by choosing either a language-independent approach or a special-purpose language. A special-purpose language typically describes data and procedural abstractions in objects and allow for object interaction using message communication. These languages differ by the level of concurrency that is defined by the degree of granularity. Languages, like ACTORS (Agha, 1986) create fine-grain programs, where every object is independent. Languages like ABCL/1 (Yonesawa, 1986) introduce large granularity.

Language-independent methodology can be further classified into high and low level approaches. A coordination language, e.g., Linda (Gelernter and Carriero, 1992) and blackboard systems are examples of a high level language independent approach. Blackboard systems enable a collection of independent processes read and write under the supervision of a control system (Corkill, 1987). Unlike the production rules, knowledge sources in blackboard systems are not homogeneous. It is still hard to predict the effects of modifying knowledge source. Additionally, the dynamic interaction of knowledge sources precludes systems's performance analysis. A coordination language provides operations to create computational activities and to support communication among them. Linda supports three conceptual classes of parallelism together with the associated programming paradigms (Table 1).

Table 1. Concepts of Parallelism

Result Parallelism	Live data structures
Agenda Parallelism	Distributed data structures
Specialist Parallelism	Message passing

Without using a high-level cooperation language, integration of various components can still be achieved by providing suitable interface, i.e., interprocess communication, to exchange information between the modules. Interprocess communication can be implemented through a datagram socket mechanism. Sockets are endpoints of communication channels originally developed in BSD UNIX as low-level facilities. Higher level remote system services like rsh and rlogin are built on top of the low level facilities. A program references a socket similarly to a UNIX file descriptor. Communication over a socket follows a particular style which is either stream or datagram. Stream communication is similar to virtual circuit (e.g., standard UNIX pipe mechanism), whereas datagram communication addresses each message individually and keeps track of message boundaries. The interface library typically consists of three routines. One routine initializes the socket channel. Another - waits for the other machine to place data in the channel and returns a pointer to a data string. The third routine writes a data string to the channel for the other process to read.

Note, that current technology has been developed assuming fixed run-time hardware and software configuration. It is likely that future networks of hundreds and thousands of computers will be configured in real-time and tools will be needed for dynamic configuration capability. Such capability must evolve on top of improved dynamic operating systems that currently suffer from similar shortcomings (e.g., UNIX). A distributed real-time network infrastructure would enable real-time configuration of limited-domain expert systems working in concert.

6 Conclusions

In summary, synergistic distribution of various processes on a multitude of machines results in dramatic and previously unattainable increases of computing power. The changes in turn provide for implementing new strategic applications. Therefore, the nineties will see continued growth of deployed strategic advantage systems that will include knowledge-based components. These components will experience three major trends. First, is a distribution trend that individualizes research and decision making by disseminating expertise. Second, is a specialization trend, whereby specialized knowledge base systems are becoming available for future systems integration (e.g., specialized Help Desk applications for PC's). Third, is a generalization trend that manifests itself via a continued migration of the technology to the main-stream programming practice (e.g., distributed case based reasoning will become standard distributed data base management technique). Future distributed computing environments will need dynamic configuration capability.

In particular, future expert systems will have better distributed real-time processing mechanisms that will eliminate current communications, representation, and callback overheads. Real-time processing performance evaluation will adopt formal methods of classical performance analysis and queueing theory. Additionally, expert system design environment will provide standard mechanisms for knowledge compilation. Knowledge compilation will migrate to general practice of software development and will become a standard Computer Aided Software Engineering (CASE) practice. Both distributed reasoning and synergistic quantitative/qualitative reasoning schemes scale up domain expertise by architecting in real time a series of limited-domain expert subsystems that work in concert.

References

G. Agha, An Overview of Actor Languages, SIGPLAN Notices, 21(10):58:67, 1986.

O. Aloni, Y. Lirov, A. Melamed, and F. Wadelton, Performance Analysis of A Real-Time Distributed Knowledge-Based Message Filtering System, World Congress on Expert Systems, pp. 2346-2354, Orlando, Florida, 1991.

A. Bond and L. Gaser, eds., Readings in Distributed Artificial Intelligence, Morgan Kaufmann, 1988.

E. Clemons, Evaluation of Strategic Investments in Information Technology, Communications of the ACM, 34(1), 22-36, January 1991.

D. Corkill, K. Gallagher, and P. Johnson, Achieving Flexibility, Efficiency, and Generality in Blackboard Architectures, Proceedings of 1987 Conference of American Association for Artificial Intelligence, 18-23, 1987.

D. Gelernter and N. Carriero, Coordination Languages and Their Significance, Communications of the ACM, 35(2), pp. 97-107, February 1992.

Y. Lirov (A), Computer-Aided Software Engineering of Expert Systems, Expert Systems with Applications, 2:133-144, 1991.

Y. Lirov (B), Algorithmic Multi-Objective Heuristics Construction in the A* Search, Decision Support Systems, 7(2):159-167, May 1991.

Y. Lirov and O. Yue, Automated Troubleshooting Knowledge Acquisition for Network Management Systems, Applied Intelligence, 1:121-132, 1991.

Y. Lirov and B. Melamed, Expert Design Systems for Telecommunications, Expert Systems with Applications, 2 (2/3):219-228, 1991.

T. Malone and J. Rockart, Computers, Networks, and the Corporation, Scientific American, 128-136, September 1991.

A. Melamed, Distributed Systems Management on Wall Street - AI Technology Needs, The First International Conference on Artificial Intelligence Applications on Wall Street, 206-212, 1991.

B. Monien and O. Vornberger, Parallel Processing of Combinatorial Search Trees, Invited talk in Parallel Algorithms and Architectures, Sahl (SDR), 1987.

M. Weiser, The Computer for the 21-st Century, Scientific American, 94-104, September 1991.

A. Yonezawa, J-P. Breit, and E. Shibayama, Object-Oriented Concurrent Programming in ABCL/1, Proceedings of 1986 Conference on Object-Oriented Programming Systems and Languages, 258-268, 1986.

Intelligent Databases and Interoperability

Armin B. CREMERS, Günter KNIESEL, Thomas LEMKE, Lutz PLÜMER

Rheinische Friedrich-Wilhelms-Universität
Institut für Informatik III
Römerstr. 164; W-5300 Bonn 1; Germany
E-Mail: {abc,gk,tl,lutz}@uran.informatik.uni-bonn.de

Abstract

Reuse of existing applications for building new applications requires the integration of heterogeneous and potentially distributed autonomous systems. Therefore interoperability will be one main issue in the design of future intelligent information systems. In the framework of a layered architecture we suggest a cyclic model of interoperable systems in order to allow reuse of *new* applications as interoperable information sources. We disuss some possible contributions of AI techniques to the solution of interoperability problems. A short description of an information system for planning, surveying and engineering applications in the mining industry substantiates the necessity of interoperability in a real life setting.

1   Introduction

Today's Database Management Systems (DBMSs) are not powerful enough to cover the needs of future information systems. Discussions about the directions of DBMS research agree in the increase of complexity, quantity and distribution of relevant data ([Silberschatz et al. 91]). For the efficient management of very large data bases whose size will grow rapidly, new mechanisms will have to be developed. "Intelligent" database systems, incorporating advanced integrity checking mechanisms ([Cremers & Domann 83], [Lipeck 89]), deductive capabilities, object orientation and hypermedia representation formalisms, are expected to cope with the increasing complexity of data and its evaluation ([Parsaye et al. 89]).

However, in many applications it will be impossible to move all relevant data into one new intelligent information system. Thus, the information management of the future will be *distributed*, and, regarding the variety of data models in use, *heterogeneous*. Since economic and organizational constraints often require that existing systems still have to be used in the future and it is crucial to access them without restricting their functionality, the systems which are used as information sources have to retain their *autonomy* while simultaneously cooperating with the global system[1].

Distribution, heterogeneity and autonomy are the most important dimensions of *Federated Database Systems* (FDBS). Following the taxonomy of [Sheth & Larson 90], FDBSs may be categorized by the way they control the access to their component DBMSs: In *tightly coupled* FDBSs the access to component DBMSs is controlled by the federated system (using predefined integrated schemata), whereas there is no central control enforced in a *loosely coupled* FDBS. In the latter kind of systems, which are also called *interoperable database systems* ([Litwin & Abdellatif 86]), it lies in the application programmer's responsibility to create and maintain his own federations of component systems.

---

[1]   The need for such systems has early been recognized (e.g., [Heimbigner & McLeod 85]).

While the former approach appears to be adequate in a closed environment with a relatively small and fixed number of component DBMSs (e.g., in a corporation), the latter seems to be more appropriate in an open environment where the number of component systems is frequently increased by the inclusion of new DBMSs. Since future information systems need to support the dynamic integration of new information sources, long term research should emphasize the integration of interoperability in intelligent database systems.

2 Layers of Interoperable Systems

An interoperable system incorporating many independent subsystems needs to be extensible, allowing the integration of new subsystems. Since extensibility has to be fully transparent for an interoperating application, an architecture of interoperable systems requires the definition of standardized, application independent interfaces. As the experience in other areas (e.g. databases, communication protocols) has shown, the conceptual organization of systems into distinct layers is a good starting point for the structuration of the tasks which have to be solved and the definition of standardized interfaces.

Several layered architectures have been proposed for federated database systems (for an overview, see [Sheth & Larson 90]). In the following, we will abstract from a specific model and use a simplified three-level model that allows to discuss our proposals for future research. When talking about interoperable systems, at least the following layers can be distinguished (fig. 1):

- The *Information Source Layer* consists of a variety of autonomous component systems using different data models and languages. These systems do not support cooperation. They are used as information servers for the next layer.

- The *Federation Layer* accesses local information sources and provides the application layer with a uniform view of the information available at the lower level. This uniform view comprises a common data model (not necessarily a common schema) and a standardized programming interface, abstracting from the distribution of information on several information sources.

- The *Application Layer* consists of autonomous application programs accessing the data relevant to them via the federation layer.

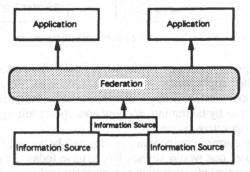

Fig. 1: Three-layer model of interoperable systems

The integration of non-cooperative, pre-existing systems into global applications raises a lot of theoretical and practical problems E.g., *transaction management* may no longer rely on 2-phase-locking since autonomy of information sources hinders the global cooperation layer from effecting transaction management in the local systems, and, vice versa, transaction managers of information sources do not even realize that their transaction is part of a global one. Also, few results have been achieved on *query optimization* because global and local optimization strategies are not allowed to make any assumption about each other due to the autonomy of local optimizers[2]..

We concentrate on two questions which have received less attention yet. In the next chapter, we extend our three-layer model, taking into account the interdependency of application and information source layer. After this, possible contributions of AI techniques to the solution of problems occurring in the federation layer are discussed.

3   A Cyclic Model of Interoperable Systems

The federation layer accesses information sources and provides (part of) their contents to the application layer. The general model makes no assumptions about the complexity of information sources. They may be simple file systems but also intelligent data or knowledge base systems, expert systems or any other system providing information for its users. [Sheth & Larson 90] suggest the term *Federated Knowledge Base Systems* for FDBSs incorporating intelligent component systems.

The increase of "intelligence" at the information source layer clearly enhances the power of the federated system, but the overall concept still remains restricted to the integration of pre-existing systems into a higher federation. Application programs built upon the federation layer can themselves be regarded as highly specialized information sources. It is thus sensible to request that they should be reusable, in the sense of making their functionality available to other applications.

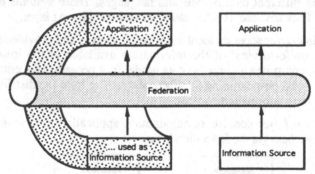

Fig. 2: Cyclic model of interoperable systems

This leads to the cyclic extension of the layer model illustrated in figure 2, where applications access other applications via the federation layer. In order to support their intended mutual interoperation, new applications should be designed for their future use in an interoperable system by taking into account concepts resulting from the integration of pre-existing, non-cooperative systems.

Problems raised by current component systems can be avoided by providing future systems with the facilities one would already like to have today. Examples include better protocols for the arrangement of transaction management and query optimization between

---

2   Recent surveys cover the spectrum of current research on distributed heterogeneous systems ([Hsiao & Kamel 89], [Breitbart 90], [Elmagamird & Pu 90], [Bukhres & Elmagamird 91], [Ram 91], [Sheth 91] and others).

a component system and the federated system, support for negotiations at various system levels and the use of more improved data models aiding the recognition of the data's semantics.

4   The Federation Layer

Even if the consideration of reusability requirements will lead to a higher degree of interoperability in future systems, at least in the near future it is unrealistic to expect that research will converge towards one generally accepted data model. Therefore the *heterogeneity* of data models used by component systems will continue to be one of the main problems[3]. In order to support management of heterogeneity, a powerful meta-information model is desirable. Metainformation is a prerequisite for accessing information sources and for comparing and relating the data of different sources.

One result of such comparisons may be that the data extracted from different sources represents information about the same object (e.g. "John's computer" versus "The computer on John's desk"). The problem of determining object identity across multiple information sources is only one topic that poses the question of the role of *object orientation* in interoperable systems ([Scheuermann & Yu 90], [Eliassen & Karlsen 91]). Aiming at extensibility and reusability, the object oriented paradigm seems to be a suitable basis for the development of a data model for the representation of heterogeneous information sources in an open environment.

However, object orientation alone is not a solution for all problems. Due to the autonomy of component systems, it is possible that data describing the same object is

- semantically identical   ("... costs 100$"   versus   "... costs 178DM")
- contradictory   ("... is damaged"   versus   "... is ok")
- complementary   ("... is red"   versus   "... is heavy")
- imprecise   ("... is near X"   versus   "... is 3 cm from X").

Moreover, the metainformation base may be incomplete or even invalid because the autonomous information sources cannot be forced to regularly update the metainformation base. Approaches for handling such fuzzy and inconsistent knowledge are better known in *logics*. In distributed AI, different kinds of logics have been proposed for solving problems like, for instance, representing and handling the metaknowledge that an agent has about himself and about other agents and information sources, their abilities, cooperation strategies, intentions and believes ([Boman & Johannesson 90], [Galliers 90]).

The integration of both, the object-oriented and the logic paradigm, which is advocated by many researchers ([Monteiro & Porto 89], [McCabe 89]), appears to be a promising approach for interoperable systems. Moreover, the *agent paradigm* itself is a good framework for the discussion of problems occurring at the federation layer, which result from the essential features of interoperable systems: distribution, heterogeneity, autonomy and the absence of central control. Multi agent systems are especially designed for such environments with a high need for cooperation of autonomous entities.

We propose the use of an agent model for a more detailed description of the federation layer as it is illustrated in figure 3. In this model, interface agents have the task to translate information between the information sources' resp. the applications' data models

---

[3]   [Kim & Seo 91] present a classification of schematic and data heterogeneity in interoperable systems. A recent survey is given in [Sheth 91].

and the federation's common data model, which is then used throughout the cooperation layer as the basic model for data and metadata representation.

The cooperation layer may be arbitrarily structured consisting of simple as well as highly specialized agents. Many different forms of interaction between agents can be thought of. However, all agents must offer a repertoire of basic cooperation mechanisms. Then, the functionality of the federated system can be enhanced by adding new agents which will automatically be able to cooperate with the existing ones. Thus, describing the structure of the federation layer in terms of agents instead of traditional system components has the advantage of increased flexibility and extensibility.

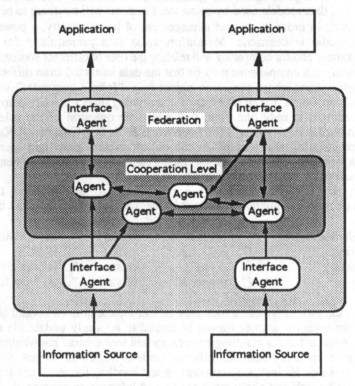

Fig. 3: An agent model of the federation layer

5   Steps Towards an Integrated Approach: EPSILON

A system which integrates object oriented features, logic programming and databases in a uniform framework is EPSILON[4], a distributed knowledge base management system ([Kniesel & Cremers 90], [Kniesel et al. 91]). The development of EPSILON offered the opportunity to study some aspects of interoperable systems. In this chapter, we will sketch relevant features of EPSILON and give a prospect of their relationship to interoperability.

An EPSILON knowledge base (KB), corresponding to one application, is a collection of theories and links between theories. *Theories* represent entities of the real world. They

---

4   EPSILON is the acronym for the ESPRIT I Project 530, "Advanced KBMS Based on the Integration of Logic Programming and Databases".

are the basic units of a knowledge base and can contain knowledge in different representation formalisms (e.g. logic programs, facts from a database). Each theory has an associated *inference engine*, which implements the corresponding knowledge representation features (e.g. certainty factors) and inference control mechanisms (e.g. forward reasoning or planning). One of the most important features of EPSILON is that an inference engine can itself be defined in a theory. This allows users to extend the system (e.g. to add new representation formalisms) simply by writing a theory that implements a new inference engine.

Links specify relationships among theories. For instance, a *dict link* from theory T1 to theory T2 expresses that T2 is a data dictionary, containing metainformation about T1. In the case that T1 is a database theory, T2 simply contains its schema definition. Dictionaries associated to other types of theories may contain other types of metainformation, depending on the power of the corresponding inference engines. The user can define his own types of links by writing corresponding inference engines. Predefined inference engines that implement new types of links are, among others, constraint and inheritance. A *constraint link* between theories T1 and T2 specifies that T2 contains the integrity constraints of T1. These are automatically checked every time T1 is updated. Different types of inheritance links allow the combination of existing theories in a way that goes beyond the ability of many current object-oriented programming systems[5].

In the distributed environment a KB may be decomposed into a set of sub-KBs which are allocated to different nodes in a network. Information about the structure of a knowledge base (theories, inference engines, links) and about the allocation of theories is stored in the distributed knowledge base dictionary (DKBD). The actual allocation of sub-KBs is fully transparent to users who work with one logically integrated application. Therefore, it is possible to distribute a KB that has been designed and developed at one node without changing its behaviour towards the user. Vice versa, a KB whose parts have been developed and tested on different nodes can be integrated simply by creating suitable links between its theories. Due to the cooperation facilities of EPSILON instances on different nodes, every theory has access to the linked theories stored on other nodes.

In the development of EPSILON problems closely related to interoperability have been treated. Theories offer a framework for integrating heterogeneous knowledge sources. The logic foundation of theories provides a clear declarative semantics that is an indispensable prerequisite for a target model used for the unified representation of different source models. The theory and link model provides an inherently extensible mechanism which allows the application programmer to increase the system's representational power by integrating new representation formalisms and formerly unaccessible (sub) systems. In order to integrate a new system only a specific inference engine has to be developed. After this, data managed by the system may be represented as an EPSILON theory. Such theories can fully be integrated within an EPSILON application by creating suitable links, in the same way as it is done when linking different parts of a distributed knowledge base to form a logically integrated application.

---

[5] Since the introduction of inheritance, many different variants of inheritance have been developed. Nevertheless, most current object-oriented programming systems support only *one fixed notion* of inheritance, while in the framework of EPSILON different forms of inheritance are just different relationships modeled by different types of links.

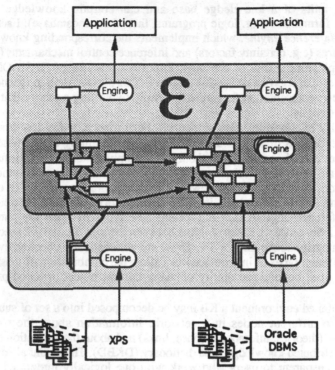

Fig. 4: Accessing independent systems using EPSILON.
The cooperation between the engines is not shown explicitly.

The different instances of the EPSILON kernel on different nodes and the user defined inference engines form a community of cooperating agents. As it is illustrated in figure 4, these engines administer theories which are connected by a variety of links, reflecting the relationships between the chunks of knowledge imported from independent systems. Inference engines do not only serve as interfaces to external systems. They can accomplish a variety of tasks which are essential in an interoperable system, e.g.:

- global integrity constraint checking[6]
- extraction and verification of metainformation (based on theory dictionaries)
- comparison of data from different sources
- support for integration of schemata
- management of uncertain information

These tasks may be tackled by single specialized inference engines as well as by a community of interacting engines that are specialized on different aspects of a problem. Such communities of engines and the corresponding theories may be structured in a suitable way, for instance, in order to allow integrity engines to enforce the validity of global constraints, a network of linked engines can be used.

The design and implementation of inference engines solving these problems is subject of ongoing research. Another aspect, the suitability of EPSILON for building agent communities has already been explored in [Kniesel et al. 91].

---

6   Efficient integrity constraint checking exploiting the theory and link features is described in [Levi et al. 90].

6 Interoperability in Real Life – An Example from the Mining Industry

We face a typical example of the problems described previously in a project where we are involved in designing and developing an information system for planning, surveying and engineering applications in the mining industry ([Baumewerd-Ahlmann et al. 91]).

Mine planning is a task of high responsibility. There are risks of dislocations, rock bursts, inbursts of water, toxic gases, fire and explosions. Some of them can only be identified from a global perspective based on reliable information. An information system for the mining industry must therefore provide complete, correct, consistent and actual information. On the other hand, today there is a great variety of heterogeneous, special purpose application programs (mostly written in FORTRAN), each using its own data, a tailored data structuring and a special file structure. The resulting redundancy and semantic overlapping causes serious problems regarding data availability, compatibility and consistency. For instance the basic data for planning the ventilation of a mine and designing maps of a mine are similar, but incompatible.

Economic and organizational constraints require that existing data and systems have to be used also in the future. It is not possible to integrate all existing data in one new centralized data management system, replacing the existing ones. Moreover, different types of workstations are in use at several sites. On each of them, a subset of the existing programs is running, and a subset of the existing data is provided. The engineers using these workstations are working on the same object (the mine), but from different perspectives and with different models. A solution in this situation is the development of an interoperable information system, able to access all pre-existing, distributed systems, to coordinate all data management tasks and to offer a standardized interface and data model to the various planning, surveying and engineering applications.

A basic precondition is the definition of a common data model, which integrates the different viewpoints of several users like surveyors, mining engineers, electrical engineers and others and which allows them to refer to the same objects with the same data. It is for instance interesting to see how overall geological attributes of the rock or the coal seam and their modifications in the space result in technical problems and influence the costs. In order to do that one has to make sure that an object in the geological database can be identified as an object or a part of an object in the databases that represent the mine from a commercial point of view. A common, integrated data model for all applications must be flexible enough to allow the definition of various refined models on top of it.

The integration of existing data is less a problem of syntactic conversions but of semantic compatibility. Basic problems are different representation formalisms and incomplete, inconsistent or imprecise information. The interpretation of existing data quite often requires to understand the context in which it has been used in the past. Transformation of data in a form such that they can be reused in a modern information system in a reliable way forms a subproblem of high complexity.

Summarizing our experiences from this project we feel that there is a high need for well founded concepts for the integration of independent information sources. We believe that a system like EPSILON can be enhanced to fill such need.

7 Conclusion

Distributed heterogeneous federations of autonomous information sources will be an integral part of future intelligent information systems. In this paper, after a short introduction to interoperable information systems we focussed on two areas that we would like to recommend as directions for future interoperability research.

One area is concerned with the reuse of *new* applications in interoperable systems. We introduced a cyclic model allowing future applications to serve as information source as well. Having this in mind, the development of new applications should be directed towards a direct support of interoperability requirements as they are identified in current research.

The second issue is the integration of (distributed) AI concepts and techniques, object orientation and logics in order to provide a framework for the description of problems and their solution and realization. We think that the concepts already developed in agent models (cooperation and coordination of heterogeneous and autonomous agents etc.), object-oriented systems (encapsulation, object identity, inheritance, among others) and logics (e.g., belief logics and non-monotonic reasoning) can be very useful for the further discussion of interoperability aspects.

As an example following the approach of coupling the logic and the object-oriented paradigm, we presented EPSILON which of course is just one first step towards the development of interoperable intelligent information systems. The need for enhanced mechanisms for the integration of pre-existing information sources has been substantiated in an application example covering planning, surveying and engineering tasks in the mining industry.

References

[Baumewerd-Ahlmann et al. 91]
A. Baumewerd-Ahlmann, A.B. Cremers, G. Krüger, J. Leonhardt, L. Plümer, R. Waschkowski: An Information System for the Mining Industry. In: *Database and Expert System Applications (DEXA '91). Proceedings of the International Conference*. Berlin 1991. Berlin: Springer-Verlag 1991.

[Boman & Johanesson 90]
M. Boman, P. Johannesson: Epistemic Logic as a Framework for Federated Information Systems. In: S.M. Deen (ed.): *Proceedings of the International Working Conference on Cooperating Knowledge Based Systems*, October 1990, University of Keele, UK. London: Springer-Verlag 1990.

[Bukhres & Elmagamird 91]
O.A. Bukhres, A.K. Elmagamird: Interoperability in Multidatabase Systems. *Encyclopedia of Microcomputers* 9 (October 1991).

[Breitbart 90]
Y. Breitbart: Multidatabase Interoperability. In: *SIGMOD record* 19, 3 (September 1990), pp. 53-60.

[Cremers & Domann 83]
A.B. Cremers, G. Domann: AIM: An Integrity Monitor for the Database System INGRES. In: M. Schkolnick, C. Thanos (eds.): *Proceedings of the Ninth International Conference on Very Large Data Bases*. Florence, Italy, November 1983. pp. 167-171.

[Eliassen & Karlsen 91]
F. Eliassen, R. Karlsen: Interoperability and Object Identity. In: [Sheth 91], pp. 25-29.

[Elmagamird & Pu 90]
A.K. Elmagamird, C. Pu (eds.): Special Issue on Heterogeneous Databases. *ACM Computing Survey* 22, 3 (September 1990).

[Galliers 90]
J.R. Galliers: Cooperative Interaction as Strategic Belief Revision. In: S.M. Deen (ed.): *Proceedings of the International Working Conference on Cooperating Knowledge Based Systems*, October 1990, University of Keele, UK. London: Springer-Verlag 1990.

[Heimbigner & McLeod 85]
D. Heimbigner, D. McLeod: A federated architecture for information management. In: *ACM Transactions on Office Information Systems* **3**, 3 (July 1985), pp. 253-278.

[Hsiao & Kamel 89]
D.K. Hsiao, M.N. Kamel: Heterogeneous Databases: Proliferations, Issues and Solutions. *IEEE Transactions on Knowledge and Data Engineering* **1**, 1 (March 1989).

[Kim & Seo 91]
W. Kim, J. Seo: Classifying Schematic and Data Heterogeneity in Multidatabase Systems. In: [Ram 91], pp. 12-18.

[Kniesel & Cremers 90]
G. Kniesel, A.B. Cremers: Cooperative Distributive Problem Solving in EPSILON. In: Commision of the European Communities (ed.): *Proceedings of the 7th Annual ESPRIT Conference*. Brussels, September 90. Dordrecht: Kluwer Academic Publishers, 1990.

[Kniesel et al. 91]
G. Kniesel, M. Rohen, A.B. Cremers: A Management System for Distributed Knowledge Base Applications. In: W. Brauer, D. Hernandez (eds.): *Verteilte Künstliche Intelligenz und kooperatives Arbeiten*. 4. Internationaler GI-Kongreß Wissensbasierte Systeme, München 1991. Berlin: Springer-Verlag 1991. pp. 65-76.

[Levi et al. 90]
G. Levi, U. Griefahn, S. Lüttringhaus: Integrity Constraint Checking. In: *Final Report of the ESPRIT project EPSILON*. 1990.

[Lipeck 89]
U. Lipeck: *Dynamische Integrität von Datenbanken*. Berlin: Springer-Verlag 1989. (Informatik-Fachberichte, Vol. 209).

[Litwin & Abdellatif 86]
W. Litwin, A. Abdellatif: Multidatabase Interoperability. *IEEE Computer* **19**, 12 (December 1986).

[Monteiro & Porto 89]
L. Monteiro, A. Porto: Contextual Logic Programming. In: *Proceedings of the Sixth International Conference on Logic Programming*. MIT Press 1989, pp. 284-299.

[McCabe 89]
F.G. McCabe: *Logic and Objects – Language, Application and Implementation*. PhD thesis, Imperial College, London 1989, UK.

[Parsaye et al. 89]
K. Parsaye, M. Chignell, S. Khoshafian, H. Wong: *Intelligent Databases*. New York: Wiley 1989.

[Ram 91]
S. Ram (ed.): Special Issue on Heterogeneous Distributed Database Systems. *IEEE Computer* **24**, 2 (December 1991).

[Scheuermann & Yu 90]
P. Scheuermann, C. Yu: Report on the Workshop on Heterogeneous Database Systems held at Northwestern University, Evanston, Illinois, December 1989. In: *SIGMOD record* **19**, 4 (December 1990), pp. 23-31. Also appeared in: *Data Engineering* **13**, 4 (December 1990), pp. 3-11.

[Sheth & Larson 90]
A.P. Sheth, J.A. Larson: Federated Database Systems for Managing Distributed, Heterogeneous, and Autonomous Databases. In: [Elmagamird & Pu 90], pp. 183-236.

[Sheth 91]
A.P. Sheth: Special Issue on Semantic Issues in Multidatabase Systems. *SIGMOD record* **20**, 4 (December 1991).

[Silberschatz et al. 91]
A. Silberschatz, M. Stonebraker, J. Ullman (eds.): Database Systems: Achievements and Opportunities. In: *Communications of the ACM* **34**, 10 (October 1991), pp. 110-111.

The TOVE Project
Towards a Common-Sense Model of the Enterprise

Mark S. Fox

Department of Industrial Engineering, University of Toronto
4 Taddle Creek Road, Toronto, Ontario M5S 1A4 Canada

tel: +1-416-978-6823, fax: +1-416-971-1373, internet: msf@ie.utoronto.ca

Abstract. The goal of the TOVE project is fourfold: 1) to create a shared representation (aka ontology) of the enterprise that each agent in the distributed enterprise can jointly understand and use, 2) define the meaning of each description (aka semantics), 3) implement the semantics in a set of axioms that will enable TOVE to automatically deduce the answer to many "common sense" questions about the enterprise, and 4) define a symbology for depicting a concept in a graphical context. The model is multi-level spanning conceptual, generic and application layers. The generic and application layers all also stratified and composed of micro theories spanning, for example, activities, time, resources, constraints, etc. at the generic level. Critical to the TOVE effort is enabling the easy instantiation of the model for a particular enterprise TOVE models will be automatically created as a by product of the enterprise design function. TOVE is currently being built to model a computer manufacturer and an aerospace engineering firm.

1. Introduction

Within the last 10 years there has been a paradigm shift with which we view the operations of an enterprise. Rather than view the enterprise as being hierarchical in both structure and control, a distributed view where organizational units communicate and cooperate in both problem solving and action has evolved [12]. Consequently, enterprise integration focuses on the communication of information and the coordination and optimization of enterprise decisions and processes in order to achieve higher levels of productivity, flexibility and quality. To achieve integration it is necessary that units of the enterprise, be they human or machine based, be able to understand each other. Therefore the requirement exists for a language in which enterprise knowledge can be expressed. Minimally the language provides a means of communicating among units, such as design, manufacturing, marketing, field service, etc. Maximally the language provides a means for storing knowledge and employing it within the enterprise, such as in computer-aided design, production control, etc.

We distinguish between a language and a representation. A language is commonly used to refer to means of communication among people in the enterprise. Whereas a representation refers to the means of storing information in a computer (e.g., database). A representation is a set of syntactic conventions that specify the form of the notation used to express descriptions, and a set of semantic conventions that specify how expressions in the notation correspond to things described. With the advent of distributed systems, we are seeing the need for processes (aka agents) to communicate directly with each other. As a result, the representation has become the language of communication. For example, in an object oriented system, we both store and communicate objects without distinction.

The problem that we face today, is that the computer systems to support enterprise functions were created independent of each other; they do not share the same representations. This has led to different representations of the same enterprise knowledge and as a consequence, the inability of these functions to share knowledge. Secondly, these representations are defined without an adequate specification of what the terminology means (aka semantics). This leads to inconsistent interpretations and uses of the knowledge. Lastly, current representations are passive; they do not have the capability to automatically deduce the obvious from what it is representing. For example, if the representation contains a 'works-for' relation and it is explicitly represented that Joe 'works-for' Fred, and that Fred 'works-for' John, then the obvious deduction that Joe 'works-for' John (indirectly) cannot be made within the representation system. The lack of a 'common-sense' deductive capability forces users to spend significant resources on programming each new report or function that is required.

The advent of object-oriented systems does not necessarily resolve any of these concerns. Being object oriented has two different interpretations. The more common interpretation is from the programming language perspective: an object is an abstract data type which supports polymorphic invocation of procedures. Consequently the programming paradigm changes from procedure invocation to message sending. The second interpretation is representational. An object represents both classes and instances of things, and they have properties that can be inherited along type hierarchies. Either interpretation does not directly solve the problems that we have raised.

The goal of the TOVE project is fourfold: 1) to create a shared representation (aka ontology) of the enterprise that each agent can jointly understand and use, 2) define the meaning of each description (aka semantics) in a precise and as unambiguous manner as possible, 3) implement the semantics in a set of axioms that will enable TOVE to automatically deduce the answer to many "common sense" questions about the enterprise, and 4) define a symbology for depicting a concept in a graphical context.

In the following, we review representation efforts of relevance, describe the TOVE project and discuss measurement criteria and limitations of the approach.

2. Enterprise Modeling Efforts

In trying to construct an ontology that spans enterprise knowledge, the first question is where to start. Brachman provides a stratification of representations [5]:

- **Implementation:** how to represent nodes and links.

- **Logical:** nodes are predicates and propositions. Links are relations and quantifiers.

- **Conceptual (aka Epistemological):** units, inheritance, intension, extension, knowledge structuring primitives.

- **Generic:** small sets of domain independent elements.

- **Application (aka Lexical):** primitives are application dependent and may change meaning as knowledge grows.

The conceptual level received much attention in the 1970s, with the development of knowledge representation languages such as FRL [19], KRL [3], SRL [11], KLONE [4] and NETL [10]. More recently, there has been a resurgence in interest in conceptual level representations both from a logic perspective, i.e., terminlogical logics, and a standards perspective. In the 1980s, attention turned to Generic level representations, such as Time [1], Causality [18, 2], Activity [20], and Constraints [13, 7]. CYC represents a seminal effort in codifying, extending and integrating generic level concepts [15].

At the application level, various efforts exist in standardizing representations. For example, since the 1960's IBM's COPIC's Manufacturing Resource Planning (MRP) system has had a shared database with a single representation of corporate knowledge. In fact, any MRP product contains a standard representation. Recently, several efforts have been underway to create more comprehensive, standard representations of industrial knowledge, including:

CAMI: A US-based non-profit group of industrial organizations for creating manufacturing software and modelling standards.

CIM-OSA: A reference model being developed by the ACIME group of ESPRIT in Europe [23] [9].

ICAM: A project run by the Materials Lab. of the US Air Force [8] [17] [16, 22].

IWI: A reference model developed at the Institut fur Wirtschaftsinformatik, Universitat des Saarlandes, Germany [21].

PDES: Product Data Exchange Standard. Defined by a standards group initially to cover geometric information but then extended to cover additional product data. The model provides a deep view of product descriptions but does not address enterprise modeling.

Though all of these efforts seek to create a sharable representation of enterprise knowledge, there has neither been a well defined set of criteria that these efforts should satisfy, nor has a formal underlying ontology and semantics been created to enable common-sense reasoning. Consequently, their interpretation varies from user to user.

3. The TOVE Project

As stated above, the goal of the TOVE project is fourfold: 1) to create a shared representation (aka ontology) of the enterprise that each agent can jointly understand and use, 2) define the meaning of each description (aka semantics), 3) implement the semantics in a set of axioms that will enable TOVE to automatically deduce the answer to many "common sense" questions about the enterprise, and 4) define a concept symbology.

We are approaching the first goal by defining a reference model for the enterprise. A reference model provides a data dictionary of concepts that are common across various enterprises, such a products, materials, personnel, orders, departments, etc. It provides a common model that represents a starting point for the creation of an enterprise specific model. Our reference model will incorporate standard models, where available, e.g., CIM-OSA, IWI, ICAM, CAMI, but deviate from standards where research dictates.

We approach the second goal by defining a generic level representation in which the application representations are defined in terms of. Generic concepts include representations of Time [1], Causality [18, 2], Activity [20], and Constraints [13, 7]. The generic level is, in turn, defined in terms of an conceptual level based on the 'terminological logic' of KLONE [6].

We approach the third goal by defining at each level of the representation, generic and application, a set of axioms (aka rules) that define common-sense meanings for the ontological terms. We view definitions as being mostly circuitous, as opposed to be reducible to a single set of conceptual primitives. The axioms can be used to deduce the answers to many questions that will be posed by users.

4. A Microtheory for Resources

An example of a generic level representation is a "microtheory" for resources. A microtheory is a locally consistent syntax and semantics for the representation of some portion of knowledge[1].

We view that "being a resource" is not innate property of an object, but is a property that is derived from the role an entity plays with respect to an activity. Consider the role of a steel bar in the activity of machining it into a 3D shape. Properties that derive from an object's role as a resource in this activity may include:

- **Consumption:** A resource is "used" or "used up" by an activity. The former indicates that the resource, once used, is no longer available in its original form once the activity is completed. In fact, its former self may no longer exist. "Using" a resource indicates the original resource exists after the completion of the activity.

- **Divisibility:** Stuff, like water, is still stuff no matter how you divide it - to some limit. Divisibility can occur along a physical or temporal dimension.

    - **Physical Structure:** Resources may be randomly, physically divisible, such as fluids, or in a structured manner, such as an oven. The nature of the structuring my be imposed by its role.

    - **Temporal Structure:** Resources may be temporally divided either randomly, such as a pizza oven, or in a structured manner, such as a communication line or autoclave. Again it depends on its role.

- **Resource Availability:** The availability of a resource for usage is a characteristic of both consumable and reusable resources. Given a role, a resource may have a maximum capacity.

A set of axioms have been defined that relate and operationalize the meaning of these properties.

5. Measurement Criteria

The success of this project can be measured in two ways. The first measure is the extent to which the representation models successfully two or more enterprises. The second approach focuses on the intrinsic characteristics of the representation:

- *Generality:* To what degree is the representation shared between diverse activities such as design and troubleshooting, or even design and marketing? What concepts does it span?

- *Competence:* How well does it support problem solving? That is, what questions can the representation answer or what tasks can it support?

- *Efficiency:* Space and inference. Does the representation support efficient reasoning, or does it require some type of transformation?

- *Perspicuity:* Is the representation easily understood by the users? Does the representation "document itself?"

- *Transformability:* Can the representation be easily transformed into another more appropriate for a particular decision problem?

---

[1]Micro-theories have been used extensively in the CYC project at MCC [15].

- *Extensibility:* Is there a core set of ontological primitives that are partitionable or do they overlap in denotation? Can the representation be extended to encompass new concepts?

- *Granularity:* Does the representation support reasoning at various levels of abstraction and detail?

- *Scalability:* Does the representation scale to support large applications?

- *Integration:* Can the representation be used directly or transformed so that its content can be used by existing analysis and support tools of the enterprise?

Satisfaction of these criteria directly affect its acceptability within the enterprise and ultimately its ability to increase the productivity and quality of decisions and actions.

These criteria bring to light a number of important issues and risks. For example, where does the representation end and inference begin? Consider the competence criterion. The obvious way to demonstrate competence is to define a set of questions that can be answered by the representation. If no inference capability is to be assumed, then question answering is strictly reducible to "looking up" an answer that is represented explicitly. In contrast, Artificial Intelligence representations have assumed at least inheritance as a deduction mechanism. In defining a shared representation, a key question then becomes: should we be restricted to just an ontology? Should the ontology assume an inheritance mechanism at the conceptual level, or some type of theorem proving capability as provided, say, in a logic programming language with axioms restricted to Horne clauses (i.e., Prolog)? What is the *deductive capability* that is to be assumed by a reusable representation?

The efficiency criterion is also problematic. Experience has demonstrated clearly that there is more than one way to represent the same knowledge, and they do not have the same complexity when answering a specific class of questions. Consequently, we cannot assume that the representation will partition the space of concepts, but there will exist overlapping representations that are more efficient in answering certain questions. Secondly, the deductive capability provided with the representation affects the store vs compute tradeoff. If the deduction mechanisms are taken advantage of, certain concepts can be computed on demand rather than stored explicitly.

The ability to validate a proposed representation is critical to this effort. The question is: how are the criteria described above operationalized? The *competence* of a representation is concerned with the span of questions that it can answer. We propose that for each category of knowledge within a partition and for each partition, a set of questions be defined that the representation can answer. Given a conceptual level representation and an accompanying theorem prover (perhaps prolog), questions can be posed in the form of queries to be answered by the theorem prover. Given that a theorem prover is the deduction mechanism used to answer questions, the *efficiency* of a representation can be defined by the number of LIPS (Logical Inferences Per Second) required to answer a query. Validating *generality* is more problematic. This can be determined only by a representation's consistent use in a variety of applications. Obviously, at the generic level we strive for wide use across many distinct applications, whereas at the application level, we are striving for wide use within an application.

6. Problems in Usage

The effort in creating an Enterprise model is fraught with problems. The identification of measurement criteria is one step towards being able to compare alternatives. But there are other problems that are not addressed by these criteria. One is the **Correspondence Problem**. What is the relationship among concepts that denote the same thing but have different terminological descriptions? It is common for enterprises, especially those that cross country boundaries to use different names to refer to the same concept. No matter how rationale the idea of renaming them is, organizational barriers impede it.

Another problem is the sheer size of the model. Consider the following basic relations and objects in their range defined for the part concept in the ICAM model form the design perspective [17] [16]:

- **IS CHANGED BY:** Part Change (105) (also shown as "is modified by")
- **APPEARS AS:** Next Assmbly usage item (119) (also shown as "is referenced as").
- **HAS:** Replacement part (143).
- **HAS SUBTYPE (IS):** Parts list item (118), Replacement part (143).
- **IS USED AS:** Next Assembly Usage (40), Advance material notice item part (144), Configuration list item (170).
- **IS TOTALLY DEFINED BY:** Drawing (1).
- **IS LISTED BY (LISTS):** Configuration list (84).
- **IS USED IN:** Effectivity (125).
- **IS FRABRICATED FROM:** Authorized material (145).

and from a manufacturing perspective:

- **HAS:** N.C. Program (318), Material issue (89), Component part (299), Alternative part (301), Part/process specification use (255), Material receipt (87), Work package (380), Part tool requirement (340), Part requirement for material (397), Standard routing use (254), Image part (300), Part drawing (181).
- **IS ASSIGNED TO (HAS ASSIGNED TO IT):** Index (351).
- **IS DEFINED BY (DEFINES):** Released engineering drawing (12).
- **IS SUBJECT OF:** Quote request (90), Supplier quote (91).
- **IS TRANSPORTED BY:** Approved part carrier (180).
- **IS RECEIVED AS:** Supplier del lot (309).
- **APPEARS AS:** Part lot (93), Ordered part (188), Serialized part instance (147), Scheduled part (409), Requested purchase part (175).
- **CONFORMS TO:** Part specification (120).
- **IS INVERSE:** Component part (299), Alternate part (301), Section (363), End item (5), Configured item (367), Image part (300).
- **IS USED AS:** Component part callout (230), Process plan material callout (74).
- **IS SUPPLIED BY:** Approved part source (177).
- **MANUFACTURE IS DESCRIBED BY:** Process plan (415).
- **SATIFIES:** End item requirement for part (227).
- **IS REQUESTED BY:** Manufacturing request (88).
- **IS STORED AT:** Stock location use for part (227).
- **IS SPECIFIED BY:** BOM Item (68).

We expect that the size of an an enterprise model to be beyond the abilities of any database manager or knowledge engineer to understand and effectively use. Consequently, the instantiation of a enterprise model for a particular firm may have to be performed in another way.

Our recommendation is that the instantiation of a firm's enterprise model be a byproduct of the the enterprise design function. Our view is similar to that of the IDEF family of modeling languages in that it is the design of the firm's activities that will entail a subset of enterprise modeling classes to be instantiating. The result of enterprise goals and activity specfications should be an automatically instantiated enterprise model. But in order to successfuly generate a model, the activity modeling methods most be more explicit in the specification of goals, activities, constraints, resources, etc. than is currently found in IDEF-like modeling tools.

7. TOVE Testbed

TOVE is not only a research project but an environment in which to perform research. The TOronto Virtual Enterprise (TOVE) is a virtual company whose purpose is to provide a testbed for research into enterprise integration. TOVE grew out of need to provide a single testbed that would integrate our research efforts. Our short term goal for TOVE is to define a company, existing solely in the computer, to support the exploration of issues in planning, and scheduling with fully specified models for both flowshop and jobshop experiments. Consequently, the criteria for selecting a product that TOVE produces includes:

- It would provide a testbed for primarily mechanical design, with the opportunity for electrical and electronic design. A domain that involves description and manipulation of 3D objects which have interesting but not too detailed design features.

- It could be designed to be as simple or complex as desired.

- Components would have to be fabricated and assembled so that planning and scheduling research could explore both.

- Components could be made out of a variety of materials, both mundane and exotic.

- A variety of resources and processes which provide complex challenges for process planning, facility layout, and scheduling systems.

- Components could actually be fabricated at CMU or purchased externally.

- Students and faculty at CMU would want to purchase it.

Desk lamps were selected with these criteria in mind. Lamp components fit the design criterion quite well, as many are relatively simple, but all have at least a few interesting and unique features. For example, some arm components are straightforward hollow cylinders, while some base and head components are irregular polygons in 3D. With respect to the materials criterion, lamp components can be metal, hard plastic, soft plastic, wire, and foam. Some components can actually be either metal or plastic. With respect to the material and process variety criteria, lamp manufacturing requires purchasing, fabrication, assembly, sub-contracting, non-destructive testing, packing, and distribution, as well as front end marketing and sales operations. The resources for these processes are large in number and type, as well as diverse in their operational and maintenance needs. Parts can be produced either in batches or on an individual basis. Major lamp components are heads, arms, and bases. Three styles of each are produced, with a standard interface between base/arm and arm/head components. The parts mix is achieved with a mix-n-match of these major components.

An earlier version of TOVE, called CARMEMCO, was developed in LISP and Knowledge Craft[R] at Carnegie Mellon University by Lin Chase. We have adapted this to a C++ environment using the ROCK[TM] knowledge representation tool from Carnegie Group. TOVE operates "virtually" by means of knowledge-based simulation [14]. Future versions of TOVE will extend it to be multi-plant and multi-region situations.

8. Conclusion

In conclusion, the TOVE project's goals are 1) to create a shared representation (aka ontology) of the enterprise that each agent can jointly understand and use, 2) define the meaning of each description (aka semantics), 3) implement the semantics in a set of axioms that will enable TOVE to automatically deduce the answer to many "common sense" questions about the enterprise, and 4) define a symbology for depicting a concept in a graphical context. We are approaching these goals by defining a three level representation: application, generic and conceptual. Each level will have a well-defined terminology which will be defined in terms of lower level terms. Each term and each level will have an axiomatic definition of it terms, comprising a micro theory for a subset of terms at that level, and enabling the deduction of answers to common-sense questions. The instantiation of a TOVE model will be the by product of the activity model of the enterprise. TOVE is currently being built to model two enterprises:a computer manufacturer and an aerospace engineering firm.

9. Ackknowledgements

This research is supported in part by an NSERC Industrial Research Chair in Enterprise Integration, Carnegie Group Inc., Digital Equipment Corp., Micro Electronics and Computer Research Corp., Quintus Corp., and Spar Aerospace Ltd.

Lin Chase designed and implemented the precurser to TOVE, called CARMEMCO. Donald Kosy and Marty Tenenbaum have contributed both ideas and inspiration to our modelling effort.

References

[1]   Allen, J.F.
       Towards a General Theory of Action and Time.
       Artificial Intelligence 23(2):123-154, 1984.

[2]   Bobrow, D.G.
       Qualitative Reasoning About Physical Systems.
       MIT Press, 1985.

[3]   Bobrow, D., and Winograd, T.
       KRL: Knowledge Representation Language.
       Cognitive Science 1(1), 1977.

[4]   Brachman, R.J.
       A Structural Paradigm for Representing Knowledge.
       PhD thesis, Harvard University, 1977.

[5]   Brachman, R.J.
       On the Epistemological Status of Semantic Networks.
       Associative Networks: Representation and Use of Knowledge by Computers.
       In Findler, N.V.,
       Academic Press, 1979, pages 3-50.

[6]   Brachman, R.J., and Schmolze, J.G.
        An Overview of the KL-ONE Knowledge Representation Systems.
        Cognitive Science 9(2), 1985.

[7]   Davis, E.
        Constraint Propagation with Interval Labels.
        Artificial Intelligence 32:281-331, 1987.

[8]   Davis, B.R., Smith, S., Davies, M., and St. John, W.
        Integrated Computer-aided Manufacturing (ICAM) Architecture Part III/Volume III:
            Composite Function Model of "Design Product" (DES0).
        Technical Report AFWAL-TR-82-4063 Volume III, Materials Laboratory, Air Force
            Wright Aeronautical Laboratories, Air Force Systems Command, Wright-
            Patterson Air Force Base, Ohio 45433, 1983.

[9]   ESPRIT-AMICE.
        CIM-OSA - A Vendor Independent CIM Architecture.
        In *Proceedings of CINCOM 90*, pages 177-196. National Institute for Standards and
            Technology, 1990.

[10]   Fahlman, S.E.
        A System for Representing and Using Real-World Knowledge.
        PhD thesis, Massachusetts Institute of Technology, 1977.

[11]   Fox, M.S.
        On Inheritance in Knowledge Representation.
        In *Proceedings of the International Joint Conference on Artificial Intelligence.* 95
            First St., Los Altos, CA 94022, 1979.

[12]   Fox, M.S.
        An Organizational View of Distributed Systems.
        IEEE Transactions on Systems, Man, and Cybernetics SMC-11(1):70-80, 1981.

[13]   Fox, M.S.
        Constraint-Directed Search: A Case Study of Job-Shop Scheduling.
        PhD thesis, Carnegie Mellon University, 1983.
        CMU-RI-TR-85-7, Intelligent Systems Laboratory, The Robotics Institute,
            Pittsburgh,PA.

[14]   Fox, M.S., Reddy, Y.V., Husain, N., McRoberts, M.
        Knowledge Based Simulation: An Artificial Intelligence Approach to System Model-
            ing and Automating the Simulation Life Cycle.
        Artificial Intelligence, Simulation and Modeling.
        In Widman, L.E.,
        John Wiley & Sons, 1989.

[15]   Lenat, D., and Guha, R.V.
        Building Large Knowledge Based Systems: Representation and Inference in the CYC
            Project.
        Addison Wesley Pub. Co., 1990.

[16] Martin, C., and Smith, S.
Integrated Computer-aided Manufacturing (ICAM) Architecture Part III/Volume IV: Composite Information Model of "Design Product" (DES1).
Technical Report AFWAL-TR-82-4063 Volume IV, Materials Laboratory, Air Force Wright Aeronautical Laboratories, Air Force Systems Command, Wright-Patterson Air Force Base, Ohio 45433, 1983.

[17] Martin, C., Nowlin, A., St. John, W., Smith, S., Ruegsegger, T., and Small, A.
Integrated Computer-aided Manufacturing (ICAM) Architecture Part III/Volume VI: Composite Information Model of "Manufacture Product" (MFG1).
Technical Report AFWAL-TR-82-4063 Voluem VI, Materials Laboratory, Air Force Wright Aeronautical Laboratories, Air Force Systems Command, Wright-Patterson Air Force Base, Ohio 45433, 1983.

[18] Rieger, C., and Grinberg, M.
The Causal Representation and Simulation of Physical Mechanisms.
Technical Report TR-495, Dept. of Computer Science, University of Maryland, 1977.

[19] Roberts, R.B., and Goldstein, I.P.
The FRL Manual.
Technical Report MIT AI Lab Memo 409, Massachusetts Institute of Technology, 1977.

[20] Sathi, A., Fox, M.S., and Greenberg, M.
Representation of Activity Knowledge for Project Management.
IEEE Transactions on Pattern Analysis and Machine Intelligence
PAMI-7(5):531-552, September, 1985.

[21] Scheer, A-W.
Enterprise-Wide Data Modelling: Information Systems in Industry.
Springer-Verlag, 1989.

[22] Smith, S., Ruegsegger, T., and St. John, W.
Integrated Computer-aided Manufacturing (ICAM) Architecture Part III/Volume V: Composite Function Model of "Manufacture Product" (MFG0).
Technical Report AFWAL-TR-82-4063 Volume V, Materials Laboratory, Air Force Wright Aeronautical Laboratories, Air Force Systems Command, Wright-Patterson Air Force Base, Ohio 45433, 1983.

[23] Yeomans, R.W., Choudry, A., and ten Hagen, P.J.W.
Design Rules for a CIM System.
Elsevier Science Publishing Company, 1985.

Automatization in the Design of Image Understanding Systems

Bernd Radig

W. Eckstein[2], K. Klotz[2], T. Messer[1], J. Pauli[2]

[1] Bayerisches Forschungszentrum für Wissensbasierte Systeme
[2] Institut für Informatik IX, Technische Universität München
Orleansstraße 34, D-8000 München 80

Abstract. To understand the meaning of an image or image sequence, to reduce the effort in the design process and increase the reliability and the reusability of image understanding systems, a wide spectrum of AI techniques is applied. Solving an image understanding *problem* corresponds to specifying an image understanding *system* which implements the solution to the given problem. We describe an image understanding toolbox which supports the design of such systems. The toolbox includes help and tutor modules, an interactive user interface, interfaces to common procedural and AI languages, and an automatic configuration module.

1 Introduction

Machine Vision, and in general the interpretation of sensor signals, is an important field of Artificial Intelligence. A machine vision system coupled with a manipulator is able to act in our real world environment without the human being in the loop. Typical tasks of understanding the semantics of an image involve such applications as medical diagnosis of CT-images, traffic monitoring, visual inspection of surfaces etc. Given an image understanding problem to be solved for a specific application, usually a long process of incremental design begins. A huge space of parameters and decisions has to be mastered to achieve satisfying results. Illumination conditions, if they are under control, have to be experimentally set. Operators acting on input images and intermediate results have to be selected. The procedures implementing them are usually further specified by individual sets of parameters. Appropriate data structures have to be invented which somehow correspond to the image structures which have to be computed. The search space of interpretation hypotheses has to be managed and hypotheses have to be evaluated until a final result is obtained.

No theory of Machine Vision exists to guide the design of an image understanding system and the methodological framework to support the design is not sufficiently developed. Therefore, to improve the efficiency of the design process and the quality of the result, tools have to be utilized which offer support for the different phases of the design process of an image understanding system. Most of those which are available now concentrate on fast prototyping of low-level image processing. Older tools merely consist of FORTRAN or C libraries (e.g. SPIDER [Tamura et al. 83]) of modules which implement typical signal and image processing operators, e.g. filters. Newer ones supply front ends which use some window-mouse based interaction, e.g. [Weymouth et al. 89]. A CASE-tool for constructing a knowledge based machine vision system will not be available in the near future. Nevertheless, some of the limitations of existing tools can be overcome by directing research into those phases of the design process in which parts can be operationalized and therefore incorporated into a next generation of those tools [Matsuyama 89], [Risk, Börner 89], [Vernon, Sandini 88].

2 Computer Aided Vision Engineering

In solving a machine vision problem, an engineer follows more or less a general life-cycle model. He has to analyse the problem, collect a sample of typical images for testing, specify a coarse design on a conceptual level, match this design with available hardware and software modules, specify and implement some missing functionality, realize a prototype, perform tests using the collected set of images, improve the prototype, observe more and more constraints which accompany the transfer into real use, specify and implement the final system, validate the system using a different sample of images, monitor the system during operation, and do final modifications. In the first part of the life-cycle, to perform a feasibility study, fast prototyping is the usual strategy. Experience tells that in reality even for not to complex problems prototyping is slow. This was our motivation to start building a toolbox which supports machine vision engineering, especially suited for feasibility studies. As a consequence of this focus, we omitted those parts which could handle real-time and process control requirements but concentrated on solving the image understanding problem.

2.1 Toolbox Properties

A toolbox supporting fast prototyping should have some of the following properties:

Sensor Input and Documentation Output Control. A wide variety of sensors may be utilized to take images and image sequences, e.g. CCD-camera, satellite, scanner etc., which deliver various image formats. Also a variety of display and hardcopy devices is in use, e.g. film recorder, colour-displays, black-and-white laser printers and the like. The toolbox should accept input from usual sources and produce output for usual display and hardcopy devices as well as for desktop publishing systems.

Repertoire of Operators. The core of a toolbox are efficient operators which manipulate and analyse images and structural descriptions of images, e.g. affine transformations, linear filters, classifiers, neuronal networks, morphological operators, symbolic matching methods etc. If the toolbox provides uniform data structures and their management together with preformed interfaces, new operators can be easily incorporated. They should be written with portability in mind so that they can be transferred later into the target system and are reusable in other toolbox versions.

Data Structures. The concept of abstract data structures has to be implemented to hide implementation details and to allow programming in different languages, e.g. C, C++, Pascal, PROLOG, LISP, and even AI shells to access these structures.

System Architecture. Typical modules in a toolbox should be the internal data management system, an external image database management system, device drivers, operator base, network communication subsystem to allow for remote access and load sharing within a distributed environment, procedural interface with extensions specialized for each host language, graphical user interface, command history with undo, redo, replay facilities, online help, advice and guidance module, automatic operator configuration module, knowledge acquisition tool including editors for different forms of knowledge description and a consistency checker, and a code export manager which generates source or binary code for the operators and data structures, used in the prototype for compilation and linking in the target environment.

2.2 Goals

Portability. To save investments into software, as well into the target systems as into the toolbox itself, such a system must be highly portable and has to generate reusable code for the target system. A UNIX environment for the toolbox and eventually for the runtime system, too, is the current choice. The user should not been forced - as far as possible - to invest time in integrating different hardware and software concepts and products before being sure that his idea of a problem solution will work.

Uniform Structures. If on all levels of an image understanding development system - signal processing through symbolic description - functions and data structure are presented in an uniform way which hides most implementation details, the user is motivated to explore even apparently complex constructs of operator sequences. He can easily combine and switch between different levels of abstraction in his program, focussed on functionality, on what-to-do and not on how-to-do it.

Interaction. This motivation has then to be supported by an highly interactive, "user friendly" toolbox interface which stimulates the engineer to test alternatives and not to think from the very beginning in terms of technical details and implementation effort.

Tutorial, Tactical and Strategical Support. Of the user, especially a novice one, it cannot be expected that he is able to keep the functionality, applicability, and parameters of all image analysis methods, cast into operators and sequences of them, which the toolbox can offer in his mind. Therefore a tutorial, tactical and strategical support is a must. He needs advice what operators to select, what alternatives exists, how to determine parameter values for their arguments, time and space complexity to be expected, and how to optimize if the prototype does not produce acceptable results.

Automatic Configuration. If the toolbox has some knowledge about the methods included in its operator base, automatic configuration becomes available. If at least a partial image understanding problem can be described in a formal way or even better by demonstrating examples and eventually counterexamples of what kind of image feature has to be detected, a configuration module could be able to choose an optimal sequence of optimal parameterized operators automatically from the operator base. First results in this direction are reported by [Ender 85], [Haas 87], [Hasegawa et al. 86], [Hesse, Klette 88], [Ikeuchi, Kanade 88], [Messer, Schubert 91], [Liedtke, Ender 89].

Automatic Adaption. If (a part of) an image understanding problem is given by supplying a generic model e.g. of the real world object to be detected or by showing representative examples it becomes essential to guarantee the transfer the problem solution based on this description to the real operation of the final system. To do this with a minimum of the engineer's intervention, an automatic adaption or specialization of the generic description to the varying situations during operation has to be included into the design of the toolbox which in turn has to include this capability into the runtime system [Ender, Liedtke 86], [Pauli et al. 92].

Competence Awareness. If a machine vision system is able to adapt itself to (slightly) varying conditions it should be able to detect when its limited capabilities to follow changing situations is exceeded. To be able to report this is a prerequisite for controllable reliability. Then it can ask for human intervention, from manual parameter tuning through a complete redesign, to analyse and handle such situations.

We do not believe that it is an easy task realizing a machine vision toolbox those properties striving at these goals. Automatic design and programming of problem solutions is a dream. Nevertheless, in such a special environment as image understanding is, methods of knowledge engineering, software engineering, and life-cycle support may be – with greater success than in general data processing – combined into a form of computer aided vision engineering, CAVE.

3   The HORUS Project

About five years ago, we started building two alternatives of tools intended to support research and education in image processing and understanding. One based on special hardware (digitizer, display system, Transputer board, ...) and special software (OCCAM) which required high competence in hardware and software engineering to become operational and to be maintained. This concept survived in a spin-off company of the author. The weak point with this approach is the intensive use of hardware dependant implementation and therefore the costly adaptation to technical innovations.

The other alternative aimed at realizing some of the properties and goals as described in Chapter 2.

Portability: As hardware platform for the HORUS toolbox a generic UNIX workstation environment was chosen, using standard C as implementation

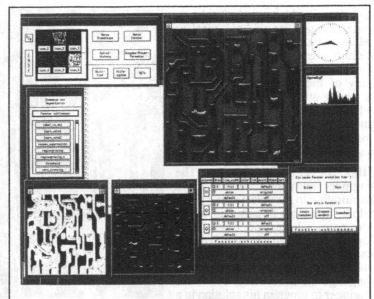

Fig. 1. HORUS window

language, TCP/IP protocols for communication, and X-Windows and OSF/Motif to interact with the user [Eckstein 88a], [Eckstein 90]. The interfaces to the operating system are so well defined and localized that it has been transferred – with an effort of two hours through one day – to different platforms such as DECStation Ultrix, even DECVAX VMS, HP9000 family, SUN Sparc, Silicon Graphics family, and even multiprocessor machines such as Convex and Alliant.

Load Sharing. If more than one processor is available in a workstation network or a multiprocessor machine, execution of operators can automatically be directed to a processor which has capacity left or which is especially suited, e.g. a signal or vector processor. The interprocess communication uses standard socket mechanisms, the load situation is monitored using Unix system commands [Langer, Eckstein 90].

Interaction. A typical screen using the interactive version of HORUS looks as in Fig. 1. The user starts with the main menu (sub window 2 in Fig. 2) where he may open sub window 8 to create or modify windows to display pictures or text, sub window 4 (see Fig. 3) to see a list of operators (organized in chapters) which are contained in HORUS, sub window 7 (see Fig. 4) to set parameters, sub window 1 to see a directory of images or image objects he has used and produces in his session, or other windows to call the online help and manual, select input and output devices etc. Fig. 4 shows an example of the variety of parameters which control the display of

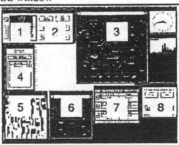

Fig. 2. Sub windows

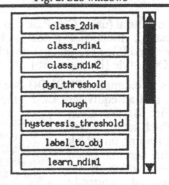

Fig. 4. Default window parameters for image display

colored	draw	line_width	color	lut	paint	shape	part
	☐ 3	fill	1		default		
◇	☐	white			original		
		default			off		

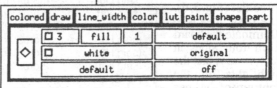

an image, in this example 3 colours, raster image, line width 1 pixel in white colour, default lookup table, default presentation (optimized for the screen), original shape, and display of the whole picture without zooming etc. Since HORUS knows which screen is in use, this menu automatically offers only such parameter values which are needed and applicable.

High Level Programming. The engineer may choose the interactive exploration to get some feeling about which operators to apply in which sequence. A history of commands is logged which can be modified and executed again. Furthermore, HORUS offers a comfortable host language interface which allows the engineer to program his solution in a

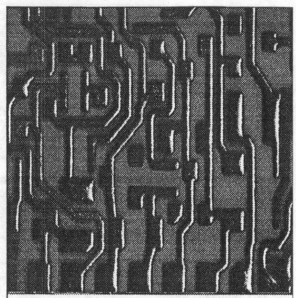

Fig. 5. Thick film IC, illuminated from right

```
seg(L,R,O,U,Solder,Fault) :-
   threshold([L,R,O,U],Dark,0,70,1),
   unionl(Dark,Line),
   mean([L,R,0,U],Mean,21,21,1,2),
   dyn threshold(Mean,Light,20,1,2),
   unionl(Light,Reflex),
   fillup(Reflex,Solder),
   complement(Line,Solder,H1),
   connection(H1,H2),
   select shape(H2,Fault,area,20,max).
```

Fig. 7. PROLOG program

procedural (using C or Pascal as language), functional (LISP), object oriented (C++) or declarative (PROLOG) style [Eckstein 88b].

As an example a PROLOG (see Fig. 7) and an equivalent LISP program (Fig. 8) are given which solve the task of finding on a thick film ceramic board areas which are not covered correctly by solder (Fig. 5, 6). The basic idea is to illuminate the board from four sides in such a way, that the highlights form the contours of the lines where these are

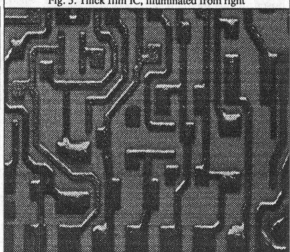

Fig. 6. Illumination from top

```
(defun seg (L R O U)
   (let (Dark Line Mean Light Reflex Solder H1 H2 Fault)
   (setq Dark      (horus 'threshold      (list (list L R O U) 0 70 1)))
   (setq Line      (horus 'union1         Dark))
   (setq Mean      (horus 'mean           (list (list L R O U) 21 21 1 2)))
   (setq Light     (horus 'dyn threshold  (list Mean 20 1 2)))
   (setq Reflex    (horus 'union1         Light))
   (setq Solder    (horus 'fill up        Reflex))
   (setq H1        (horus 'complement     (list Line Solder)))
   (setq H2        (horus 'connection     H1))
   (setq Fault     (horus 'select shape   (list H2 "area" 20 "max")))
   (list Solder Fault)))
```

Fig. 8. LISP program

covered by solder. The solution is then straightforward.

By thresholding all pixels with an intensity value lower than 70 (this can be determined interactively) in each image L, R, O, U, an object Dark is obtained which contains four components of dark areas. A union of these four pixel sets representing the lines is formed. Low pass filtering, dynamic thresholding, and set union produces the contour image of Fig. 9. The

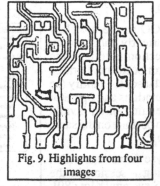

Fig. 9. Highlights from four images

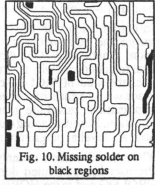

Fig. 10. Missing solder on black regions

threshold parameter is also selected by an interactively controlled test. The regions which are enclosed by contours are filled, forming a cover of all lines except where solder is missing. Subtracting this result from the object Line uncovers those areas. The remaining pixels are collected and connected regions are formed. Regions with an area of less than 20 pixels are excluded and the result is stored in the object Fault which is displayed in Fig.10.

The LISP program is a transcription of the PROLOG program. The data structures are objects which have hidden components created and selected automatically by operators. This kind of abstraction allows the engineer to write compact programs without being involved in implementation details.

Help Modes. Operators perform exception handling to inform the user. A window is opened which displays related information (Fig. 11). Here the name of the operator where the error occurred is displayed. The bottom part of the window explains which result is generated and which errors arise in what situation, in this case dilation on an empty image. In the upper part of the window, a menu offers related information. Pressing the help button gives an explanation how to use these menus. The buttons keywords, see_also, and alternatives open access to related topics, e.g. description of operators with similar functionality or explanation of the meaning of the word threshold. The button parameter tells how to choose parameter values for this operator, the button abstract delivers a short description and the button manual gives a long description of the operator (Fig. 12). The manual is written in LAT$_E$X, therefore a standard UNIX tool has been used to provide the functionality of the manual management and a LAT$_E$X preview program generates the window. A new version will use a Postscript previewer which allows the inclusion of pictorial examples and

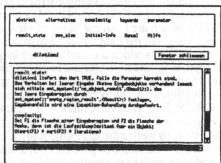

Fig. 11. Error related window: information about operator

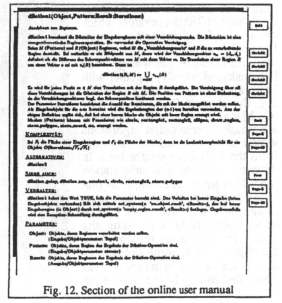

Fig. 12. Section of the online user manual

explanations into the manual.

In combination with PROLOG as the host language, the help module is currently enhanced by more functionality. One direction is to help the user to understand why an error occurred. It traces back the chain of PROLOG clauses to identify the place where a chain of computations and operator applications caused the generation of an invalid situation. Here debugging can be done on the same abstract level as programming.

The other direction is to analyse the actions which the user has performed during a session in the PROLOG and in the HORUS system [Klotz 89]. Here the help systems monitors the user input and is therefore able to give more precise advice in case the user needs help or an error message comes up. A first step in the realisation of this concept is the module operated by the window of Fig. 13. It provides the user with a complete list of predicates he has defined in field 2 and HORUS operators in field 1 which are from the point of view of PROLOG built-in predicates. Field 3 displays the history of actions which are available for inspection, modification and redo application. Field 4 accepts as input PROLOG and HORUS commands which are performed in the current context of execution. Similar concepts of debugging are well known from several LISP systems and applications. The innovation here is the uniform handling of actions in PROLOG as well as in HORUS. Further development will include the automatic analysis of interaction protocols to guide the user in understanding why an error occurred. Traces of rule activations and procedure calls as well as the assignment of variables at the moment when the error occurred will be available. The idea is to give the user a meaningful access to all those objects which might be associated with an exception, and to suggest alternatives for control structure and parameter values.

Fig. 13. HORUS / PROLOG debuging

Automatic Configuration. The online manual contains all information about HORUS operators which are needed for advice and guidance. Using this information, represented in a frame-based knowledge base [Polensky, Messer 89] together with basic knowledge of image processing, the module FIGURE [Messer 92a,b] is able to automatically generate image interpretation algorithms which are composed of operators from the HORUS operator base. In the current version, the configuration module is given an example of that kind of objects it should detect in an image or a sequence. This can be done interactively, e.g. using the mouse to indicate the region where the object is, or by some simple segmentation method. In the application illustrated in Fig. 14 the pins of the

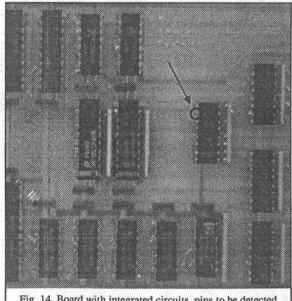

Fig. 14. Board with integrated circuits, pins to be detected

integrated circuits have to be detected. FIGURE analyses the boundary, the background surrounding the object, and its interior. It then constructs a search space of all reasonable operators which might be applied. Various rules and constraints restrict the search space to a manageable size. Static restrictions help to determine the order of operators within the algorithm to be constructed. An example of such a rule is: *If dynamic thresholding is selected then it should be preceded by smoothing.* The even more difficult task is to supply the modules with values for their parameters. Elementary knowledge about computer vision is included in the rule base as well as operator specific knowledge. No domain specific knowledge is incorporated in the rule base. These rules are rather simple, telling the system such elementary statements as *for edge preserving smoothing a median filter is better than a low pass filter.* Constraints between operators forbid e.g. *applying a threshold operator on a binary image.* But from the operator specific knowledge the system knows that *a threshold operator needs a threshold value between 1 and 255.*

Fig. 15. Small bright spots detected by operator sequence first configured

The quality of the generated proposals of operator sequences depends not only on the knowledge base but even more on the precision with which a vision problem can be described and the sequences can be evaluated. Of course, there is a correlation between both. Since at the moment a problem is described by indicating the boundary of a region which has to be extracted, two aspects are used for evaluation. The evaluation function takes into account how good the area of the found object matches the area of the given object and the distance between the boundaries of both objects. To force the configuration system to generate alternative sequences which include different operators in their sequences and not only different parameter values for essentially the same sequence, two templates are generated from the indicated boundary, namely a boundary oriented and a region oriented one. The best of both alternatives generated for these two classes survives.

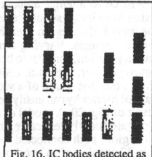

Fig. 16. IC bodies detected as dark regions by second operator sequence

As an example consider the problem of detecting the pins of integrated circuits on a printed board as in Fig. 14. One pin's boundary is drawn, e.g. the one indicated by an arrow in Fig. 14. The configuration system generates an operator sequence

```
findpins(IC,SPOTS,PINS) :-
    convex(IC,BODY),
    contour1(BODY,CONTOUR),
    circle(CIRCLE,0,0,5),
    dilation(CONTOUR,CIRCLE,STRIPE,1),
    intersection(STRIPE,SPOTS,PINS).
```

Fig. 17. PROLOG program, combining outputs to final result

which detects correctly most of the pins but also a lot of other small regions. The simple remedy is to filter out all regions which are not close to the body of the IC. Therefore, the detection of IC bodies is given as a second task to the configuration system. Fig. 16 shows the result. The closeness constraint cannot be implemented in the current version. The vision engineer has to do some programming, formulating such code as in Fig. 17. The arguments to the simple PROLOG rule are both images as inputs and the resulting image as output. For the IC bodies the contours of their convex hulls are computed. A circle of radius 5 pixels is defined and used to widen the contour, implementing the predicate *close*. A simple intersection of this image with that of the small regions eliminates most of the unwanted spots.

It is obvious from Fig. 18 that the result is not completely correct.

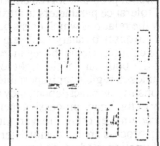

Fig. 18. Spots close to IC boundary, combined from both results, mostly pins

Anyhow, the automatic configuration is utilized here in the context of rapid prototyping. The result gives a good starting point to improve on. To prepare this example for this paper took less than one hour. It could be done on a very abstract but natural level in terms of design decisions such as *find small bright regions close to IC bodies*. Two mouse interactions and one PROLOG rule implemented it.

Model Adaptation. The models for describing objects, used so far, are simple closed contours. This is, of course, not sufficient for the analysis of more complex situations [Liedtke, Ender 86] where scenes are to be composed of objects constructed as a part-of hierarchy and related by constraints

Fig. 19. First frame of an image sequence

[Fickenscher 91]. The more complex such models are the more difficult it will be to attempt a generalizable description. A typical situation exists in an image sequence with moving objects [Nitzl, Pauli 91]. Here the model of an region in each image has to follow the expected or predicted changes. Varying attributes are area,boundary, shape, grey-value, position, contrast to background etc. as well as relations such as below, darker, between, inside etc. In the example of Fig. 19 which shows the first frame of a sequence, a man is moving his right leg. A part-of hierarchy can easily be constructed, guided by the segmentation result of Fig. 20. A model of the right leg is obtained by specifying an elongated nearly vertical region whose centre of gravity is left of a similar region, both connected to a more compact region which is above. Fig. 22 consists of an overlay of Fig. 20 and the isolated right leg in different positions.

Fig. 20. Initial segmentation

For matching the model with the image structure, maximal cliques (totally connected subgraphs) in an assignment graph are computed [Radig 84]. The nodes of the assignment graph are formed from potentially corresponding pairs of image and model elements. Edges in this graph link relational consistent assignments. This approach is able to establish correspondence even in the case of deviations of relation and attribute values. To match image and model structure is a NP-hard problem. Application of heuristics using some kind of A*-search technique reduces complexity [Pauli 90]. Other techniques of matching structures are presented in [Pfleger, Radig 90]. The problem of how to evaluate the quality of the match has received much attention in the last decade. A recent workshop on Robust Computer Vision [Förstner, Ruwiedel 92] discussed different approaches.

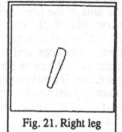

Fig. 21. Right leg

A major problem, solved only for simple situations, is the specification of tolerance parameters to attribute values given their interdependence by the relations which exists between different elements. It is impossible to describe analytically e.g. how the height and width of the rectangle circumscribing the leg in Fig. 22 varies with the motion. The length of the boundary of the right leg is somehow related to the area of the enclosed region but impossible to state exactly. Even if for some rela-

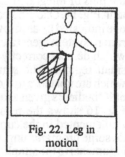

Fig. 22. Leg in motion

tionships an exact dependency might be found, it usually becomes corrupted by the unreliability of the image processing methods, by noise, or by the effects of digital geometry on the rastered image. Therefore, variations of attribute values which should be tolerated by the matching process are not easily determined and need time consuming experimentation. The model adaptation module, to be enclosed in the HORUS toolbox, contains methods of qualitative reasoning to help the engineer determining trends of parameter values and to follow difficult interrelationships without violating consistency between those parameter tolerances.

4. Conclusion

After more than twenty years of image understanding research the situation with respect to the methodology of designing image understanding systems is disappointing. A theory of computational vision still does not exists which guides the implementation of computer vision algorithms. Our approach originated from an engineering point of view. We identified some of the problems which decrease quality of the results and productivity of the design process in the area of image understanding. Obviously, we could not address all aspects and could not offer solutions to all the problems we are faced with.

Our advantages in the problem areas of portability, user interfacing, parallelisation, multi host language interfacing, tutorial support, high level debugging, model adaptation, reusability, and automatic configuration are sufficient to start integrating all related software modules in a toolbox for Computer Aided Vision Engineering. In the HORUS system the availability of an interactive and a high level programming interface, the online help and debugging system, and the automatic configuration module have been in use for some period of time. We use it extensively in a practical course on Computer Vision as part of our Computer Science curriculum. We observed a drastic increase of creative productivity of our students working on their exercises. Other Computer Vision Labs testing our system reported a similar experience.

In the near future the model adaptation will be more closely integrated into HORUS. The tutorial module will be able to give recommendations to the designer which module and parameter values to choose – an interactive complement of the automatic configuration module. During the process of integration some new challenges will appear. One is the extension of the internal object data structure – which is effective and efficient for low- and medium-level processing – to structures needed by high-level image understanding. A second problem is the description of operators in such the way that all modules are able to extract automatically that part of the information which it needs. To describe more than 400 operators ranging from a simple linear filter up to a complete Neural Network simulator is a time consuming task. To specify formally for an author of a new module how he has to encode for HORUS the knowledge about his operator is not solved in general.

Nevertheless, we could demonstrate that some parts in the design process of image understanding systems can be automatized successfully.

References

[Eckstein 88a]: W. Eckstein: Das ganzheitliche Bildverarbeitungssystem HORUS, Proceedings 10th DAGM Symposium 1988, H. Bunke, O. Kübler, P. Stucki (Eds.), Springer-Verlag, Berlin 1988

[Eckstein 88b]: W. Eckstein: Prologschnittstelle zur Bildverarbeitung. Proceedings 1st IF/Prolog User Day, Chapter 5, München, June 10, 1988

[Eckstein 90]: W. Eckstein: Report on the HORUS-System. INTER, Revue International de L'Industrie et du Commerce, No 634, Oct. 1990, pp. 22

[Ender 85] M. Ender: Design and Implementation of an Auto-Configuring Knowledge Based Vision System, in: 2nd International Technical Symposium on Optical and Electro-Optical Applied Sciences and Engineering, Conference Computer Vision for Robots, Dec. 1985, Cannes

[Ender, Liedtke 86]
M. Ender, C.-E. Liedtke: Repräsentation der relevanten Wissensinhalte in einem selbstadaptierenden regelbasierten Bilddeutungssystem, Proceedings 8th DAGM-Symposium 1986, G. Hartmann (Ed.), Springer-Verlag, Berlin 1986

[Fickenscher 91] H. Fickenscher: Konstruktion von 2D-Modellsequenzen und -episoden aus 3D-Modellen zur Analyse von Bildfolgen, Technische Universität München, Institut für Informatik IX, Diplomarbeit, 1991

[Förstner, Ruwiedel 92] W. Förstner, R. Ruwiedel (Eds.): Robust Computer Vision, Proceedings of the 2nd

International Workshop, March 1992 in Bonn, Herbert Wichmann Verlag, Karlsruhe 1992

[Haas 87] L.J. de Haas: Automatic Programming of Machine Vision Systems, Proceedings 13th International Joint Conference on Artificial Intelligence 1987, Milano, 790 - 792

[Hasegawa et al. 86] J. Hasegawa, H. Kubota, J. Toriwaki: Automated Construction of Image Processing Procedures by Sample-Figure Presentation, Proceedings 8th International Conference on Pattern Recognition 1986, Paris, 586 - 588

[Hesse, Klette 88] R. Hesse, R. Klette: Knowledge-Based Program Synthesis for Computer Vision, Journal of New Generation Computer Systems 1 (1), 1988, 63 - 85

[Ikeuchi, Kanade 88] Katsushi Ikeuchi, Takeo Kanade: Automatic Generation of Object Recognition Programs, Proceedings of the IEEE, Vol. 76, No. 8, August 1988, 1016 - 1035

[Klotz 89] K. Klotz: Überwachte Ausführung von Prolog-Programmen. Proceedings 2nd IF/Prolog User Day, Chapter 9, München, June 16, 1989

[Langer, Eckstein 90]: W. Langer, W. Eckstein: Konzept und Realisierung des netzwerkfähigen Bildverarbeitungssystems HORUS, Proceedings 12th DAGM-Symposium, R. E.. Großkopf (Ed.), Springer-Verlag, Berlin 1990

[Liedtke, Ender 86] C.-E. Liedtke, M. Ender: A Knowledge Based Vision System for the Automated Adaption to New Scene Contents, Proceedings 8th International Conference on Pattern Recognition 1986, Paris, 795 - 797

[Liedtke, Ender 89] C.-E. Liedtke, M. Ender: Wissensbasierte Bildverarbeitung, Springer-Verlag, Berlin, 1989

[Matsuyama 89] T. Matsuyama: Expert Systems for Image Processing: Composition of Image Analysis Processes, in: Computer Vision, Graphics, and Image Processing 48, 1989, 22 - 49

[Messer 92a] T. Messer: Model-Based Synthesis of Vision Routines, in: Advances in Vision - Strategies and Applications, C. Archibald (ed.), Singapore, World Scientific Press, 1992, to appear

[Messer 92b] Tilo Messer: Acquiring Object Models Using Vision Operations, Proceedings Vision Interface '92, Vancouver, to appear

[Messer, Schubert 91] T. Messer, M. Schubert: Automatic Configuration of Medium-Level Vision Routines Using Domain Knowledge, Proceedings Vision Interface '91, Calgary, 56 - 63

[Nitzl, Pauli 91] F. Nitzl, J. Pauli: Steuerung von Segmentierungsverfahren in Bildfolgen menschlicher Bewegungen, Proceedings 13th DAGM-Symposium, München, B. Radig (Ed.), Springer-Verlag, Berlin, 1991

[Pauli 90] J. Pauli: Knowledge based adaptive identification of 2D image structures; Symposium of the International Society for Photogrammetry and Remote Sensing, SPIE Proceeding Series, Band 1395, S. 646 - 653, Washington, USA, 1990

[Pauli et al. 92] J. Pauli, B. Radig, A. Blömer, C.-E. Liedtke: Integrierte, adaptive Bildanalyse, Report I9204, Institut für Informatik IX, Technische Universität München, 1992

[Pfleger, Radig 90] S. Pfleger, B. Radig (Eds.): Advanced Matching in Vision and Artificial Intelligence, Proceedings of an ESPRIT workshop, June 1990, Report TUM-I9019, Technische Universität München; to be published by Springer-Verlag, Berlin 1992

[Polensky, Messer 89] G. Polensky, T. Messer: Ein Expertensystem zur frame-basierten Steuerung der Low- und Medium-Level-Bildverarbeitung, Proceedings 11th DAGM-Symposium 89, Hamburg, H. Burkhardt, K. H. Höhne, B. Neumann (Eds.), Springer-Verlag, 406 - 410

[Radig 84] B. Radig: Image sequence analysis using relational structures; Pattern Recognition, 17, 1984, 161 - 167

[Risk, Börner 89] C. Risk, H. Borner: VIDIMUS: A Vision System Development Environment for Industrial Applications, in: W. Brauer, C. Freksa (Eds.): Wissensbasierte Systeme, München Oct. 1989, Proceedings, Springer-Verlag Berlin, 477 - 486

[Vernon, Sandini 88] D. Vernon, G. Sandini: VIS: A Virtual Image System for Image-Understanding Research, in: Software - Practice and Experience 18 (5), 1988, 395 - 414

[Weymouth et al. 89] T. E. Weymouth, A. A. Amini, S. Tehrani: TVS: An Environment for Building Knowledge-Based Vision Systems, in: SPIE Vol. 1095 Applications of Artificial Intelligence VII, 1989, 706 - 716

Constraint Programming -
an Alternative to Expert Systems

Mehmet Dincbas

COSYTEC S.A.
Parc Club Orsay Université, 4 rue Jean Rostand
91893 Orsay Cedex / France

Extended Abstract

Constraint Programming is a new emerging and very promising technology. It belongs
to the class of *declarative programming* which comprises other paradigms like rule-based
programming or logic programming. It tends to generalise conventional Artificial
Intelligence programming by introducing, besides basic logical inference mechanisms,
more mathematical and algorithmic techniques.

A *constraint* expresses a relationship among different objects defined on a computation
domain. Constraints can be of different types :
  • arithmetic : e.g. $4*X + 3*Y >= 6*Z + 5$
  • boolean : e.g. A = not (B or C)
  • membership : e.g. Color . [blue, white, red]
  • functional dependencies : e.g. a cost matrix

Most of the work in Constraint Programming (CP) has been done in the Constraint
Logic Programming (CLP) framework during the last 6-7 years. CLP schema generalises
very well the Logic Programming schema by keeping its fundamental caracteristics, e.g.
the equivalence of its declarative, procedural and fixpoint semantics. Logic
Programming, as examplified by Prolog, can be seen as a special case of CLP where the
computation domain is restricted to the Herbrand universe (i.e. interpretation-free terms)
and the unique constraint is the equality constraint between the objects of this universe.
Several CLP languages have been developed around the world, the most important ones
being Prolog III, CLP(R) and CHIP.

The introduction of the *constraint* concept brings two advantages : on one hand it allows
to describe problems in a more natural way, on the other hand, it allows to benefit from
efficient specific algorithms developed in different areas like Artificial Intelligence,
Operations Research and Mathematics.

Concerning the expressiveness of the language, Constraint Programming provides the user to manipulate directly :
• computation domains much closer to the user's domain of discourse (e.g. natural numbers, rational terms, boolean formulas)
• different types of numerical and symbolic constraints (e.g. equality, disequality, inequalities, functional dependencies)

Concerning the efficiency, Constraint Programming systems use more mathematical techniques, namely constraint solving techniques on some algebraic structures, than usual declarative systems like Expert Systems or Logic Programming. The use of specialized algorithms and methods developed in different domains can increase drastically the efficiency for solving several complex problems.

The most important constraint handling techniques used in the current Constraint Programming systems are coming :
• From Artificial Intelligence : Consistency checking techniques, constraint propagation, constraint satisfaction, theorem proving in propositional calculus
• From Operations Research: Linear programming (simplex), Integer Programming, Branch & Bound search
• From Mathematics : Equation solving (e.g. in Boolean or Finite Algebra), Decision procedures for polynomials (e.g. Gröbner Bases)

The main difference between Expert Systems (and more generally Artificial Intelligence approaches) and Constraint Programming can be summarized by :
• Expert Systems are aimed to solve "ill-defined" problems by using deductive methods, symbolic computations and heuristic approaches
• Constraint Programming aims to solve "well-defined" problems by combining the above-mentioned techniques with more mathematical ones (i.e. numerical computation, algorithmic approach).

This combination offers new perspectives to declarative programming languages for solving complex decision making problems where we need both :
• *qualitative* and *quantitative* knowledge
• *deductive* and *mathematical* methods
• *symbolic* and *numerical* computations
• *heuristic* and *algorithmic* approaches

While Expert Systems are well adapted to solve problems where domain expertise (together with some logical inference mechanisms) is enough, the use of Constraint Programming is especially required for complex "constraint search problems" where heuristics alone are not enough. The main application domains of current Constraint Programming systems are the following :
• Production & Manufacturing (e.g. scheduling, sequencing, planning)
• Logistics and Services (e.g. ressource allocation, warehouse location)
• Circuit Design (e.g. formal verification, test pattern generation, routing, placement)

• Finance (e.g. logical spreadsheets, portfolio management, investment planning)

Compared to the more classical methods used for solving these complex problems and coming mainly from Operations Research (e.g. integer or linear programming) Constraint Programming offers :
• short development time
• flexibility
• interactivity
without sacrifying efficiency.

Another important issue about Constraint Programming is the *modelisation* of problems. While this is an important step when using Expert Systems it becomes the crucial point when using Constraint Programming for solving complex search problems. The conciseness and especially the efficiency of the program depends directly on the modelisation. Different choices exist for :
• computation domains (e.g. discret versus continuous variables),
• variables and values (often can be exchanged)
• constraints (more or less active)
A good modelisation requires a very good knowledge of the problem as well as of different methods.

Constraint Programming brings together flexibility and versatility of Artificial Intelligence tools with the efficiency of conventional algorithmic approaches. This combination allows to solve problems which are hard to tackle by other means. The results obtained so far with Constraint Programming are rather impressive. As an example we can mention the CHIP system which has permitted to solve several real life problems in the areas of scheduling, planning or ressource allocation in a much more efficient way than with conventional approaches while requiring much less programming effort and providing more flexibility in the solution.

References

1. M.S.Fox : "Constraint-Directed Search : A Case Study of Job-Shop Scheduling". PhD. Thesis, Carnegie-Mellon University, 1983
2. J.Jaffar and J-L.Lassez : "Constraint Logic Programming". Proc. ACM POPL, Munich, pp.111-119, 1987
3. M.Dincbas et al. : "The Constraint Logic Programming Language CHIP". Proc. FGCS'88, Tokyo, pp.693-702, 1988
4. M.Dincbas et al. : "Applications of CHIP to Industrial and Engineering Problems". Proc. IEA/AIE-88, Tullahoma, Tennessee, 1988
5. A.Colmerauer : "An Introduction to Prolog III". Communications of ACM, 33(7), pp. 69-90, 1990

CASE-BASED REASONING IN EXPERT SYSTEM ASSISTING PRODUCTION LINE DESIGN

Hidehiko YAMAMOTO

Toyoda Automatic Loom Works, Ltd.   (presently with
Faculty of Education,
Wakayama University
930, Sakaedani, Wakayama-shi, Wakayama-ken, 640, Japan)

and

Hideo FUJIMOTO

Dept. of Mechanical Engineering,
Nagoya Institute of Technology,
Gokiso-cho, Showa-ku, Nagoya-shi, Aichi-ken, 466, Japan

Abstract. This paper proposes the knowledge acquisition idea to create new machine tool specifications by case-based reasoning and using the expert system assisting production line design. This proposal is the algorithm proceeding case-based reasoning under the frame work of hypothetical reasoning. In this algorithm, both the knowledge concerning machine tools and the meta knowledge with frame knowledge representations are expressed. A new machine tool knowledge is acquired by analogy as a frame knowledge expressions. This algorithm is programmed with LISP.

Key Words. Case-based reasoning, Hypothetical Reasoning, Expert System, Production Line Design, Computer-aided Design

1 Introduction

When we develop expert systems whose assisting domains are design, diagnosis or control, we hope that in the future, the goal will be for these expert systems to take the place of experienced planners. However, this does not only depend upon the domains. In order to obtain this goal, we have developed many expert systems which efficiently expressed experienced planners knowledge without contradiction. We have also constructed a reasoning engine which conforms to an experienced planners' consideration process. These kinds of systems have been applied to certain restricted domains[1][2], and the capabilities of such systems were proven.

However, most of the systems' reasoning methods corresponded to selecting suitable solutions from among a knowledge data base, so-called, the method of solving a search problem. Therefore, the solutions acquired by carrying out the expert systems capabilities in the design and diagnostic domains, were selected as a result of combinations satisfying the problem conditions. The solutions represent a combination of some of the data which was described beforehand in a data base. In this case, there are some experienced people who will discover that the solutions can be easily forecasted by anyone. When the system obtains solutions by utilizing reasoning and calculations which are extremely detailed and high leveled compared to experienced people, the experienced people appraise that the solutions are indeed correct. A strong advantage in this system is that inexperienced people can easily operate the systems and obtain the same level solutions as experienced people are capable of performing. There are some experienced people who have come to the conclusion that the systems are not worth developing because a tremendous amount of time and developers are necessary to further develop the system.

As a result of extensive research we consider it extremely important that in the future development of expert systems to create and use the knowledge which has not been described beforehand in a knowledge data base. This function is a type of learning function. In this research, based on the expert system we have mentioned above, we propose the knowledge acquisition idea to create new machine tool specifications by case-based reasoning [3] .

2 Expert System Assisting Production Line Design [4][5]

The expert system we have developed is capable of examining the following contents by using multi-objective evaluation reasoning.

Multi-objective evaluation reasoning can narrow down the single most suitable machine tool type from among the machine tool types which are selected with the basis of machine precision alone because of proceeding the two stage reasoning process. The two stage reasoning process includes elimination reasoning and the comparison of multiplicative utility function values. Elimination reasoning eliminates unsuitable machine tool types based on the aspect of machine tool mechanism and function aspect. The comparison of multiplicative utility function values determines a single suitable machine tool type by simultaneously evaluating five objectives of machine tool purchasing, such as cost, the machine tool maker's technological level and machine tool maker service system. In the comparison method, forward reasoning and frame data search are performed by using hybrid knowledge representation from semantic network representation and if-then rules in order to determine multiplicative utility function values such as the certainty equivalent. Thanks to the comparison method it's unnecessary for the system to make extensive interviews with users to determine the values concerning attributes and it also prohibits users from inputting incorrect knowledge.

The expert system determines machine tool types, process sequence, machine set number and cutting conditions.

3 Knowledge Acquisition In Production Line Design

Planners in production line design develop production lines by using machine tools which they are either familiar with or have used before.   The objectives of production line design are, for example, to determine machine set number and machine sequence.   In this thinking process they take all restrictive conditions into consideration.

However,   there are instances when the situation cannot be satisfied with the restricted conditions even if they use all machine tools they've used before or are familiar with.   When this situation occurs they will acquire resolutions by either loosening the restricted conditions or create new machine tools which will satisfy the restricted conditions without loosening the restricted conditions.   Planners that are capable of performing the latter method are looked upon as excellent planners.   As a result, computer systems assisting production line design which are unable to carry out an excellent planners' thinking consequently cannot become excellent systems.

The method used to acquire resolutions from among a knowledge data base by loosening the restricted conditions corresponds to the reasoning which proceeds a design process by changing the hypothesis which was presumed to be true into false and then reconsidering the hypothesis which now has broader   true restrictive conditions as true.   This reasoning method can be expressed as hypothetical reasoning based on Truth Maintenance System [6].

The method used to create new machine tools by being persistent in keeping the restricted conditions is regarded as knowledge learning which begins from the true-false judgment in hypothetical reasoning.   It is because of the basis in an exsisting knowledge data base that the reversal process from true to false in hypothetical reasoning can be carried out.   Hypothetical reasoning cannot help abandoning the hypothesis which has been presumed true owing to the result that a satisfying resolution was not acquired when all the knowledge in a knowledge data base was used. On the contrary, if the hypothesis manages to remain true as it was before and the knowledge acquisition used to create new machine tools which conforms to the hypothesis can be carried out,   it's possible for computer systems to create new machine tool knowledge. Excellent planners usually get the knowledge acquisition to hit upon machine tools that have new mechanisms by inferring on the analogy of machine tools they are familiar with or they have used before. This kind of thinking process corresponds to a case-based reasoning problem solution.

4 Case-based Reasoning In Production Line Design

4.1 Algorithm of Case-based Reasoning

Reasoning used to create new machine tools which excellent planners also create is described below.   Here we propose the algorithm to accomplish case-based reasoning within the framework of hypothetical reasoning.

In general ,   the difficulty in applying case-based reasoning to design support systems is in establishing the goal to which the system will carry out case-based reasoning.   For example,   if we develop data search systems which include case-based reasoning, we only need to consider it to an analogous degree by comparing it with a target specification which a user inputs. A problem with a design support system such as ours is that it consists of many data search problems and it's difficult for a user to input goal informations every time. We have to have an automatic decision method as to when and how the goal information should be established. We therefore propose to establish goal specifications whithin the framework of hypothetical reasoning.

Among competitive hypotheses in the hypothetical reasoning process, a corresponding hypothesis exists that enables us to decide beforehand which hypothesis should be chosen first as a true hypothesis. This kind of hypotheses which is the competitive hypotheses and includes a select priority order, is selected by applying the experienced planners' concept. The concept is that experienced planners select a true hypothesis because it is the most ideal situation among other hypotheses. When a true hypothesis selected from among the hypotheses which includes select priority order is considered false, highly experienced planners will take a characteristic action. The planners are inclined to take a creative action toward an ideal fact in order to revive the true hypothesis which was considered false. This creative action which highly experienced planners used to develop new machine tools or production lines is taken because the true hypothesis which was judged false is indeed the most ideal. Highly experienced planners want to hold fast to the most ideal hypothesis and take it through the entire process in order to make the best production line. This kind of ideal hypothesis can be started as the goal for case-based reasoning.

When we consider knowledge acquisition in production line design, we conclude that our goal is to develop a method capable of creating new machine tool knowledge which was not described beforehand in a knowledge data base, or in other words, acquiring a new machine tool specification.

New machine tool specifications are acquired by the following algorithm. When the hypothesis, which is one of the priority order hypothses, is considered false, case-based reasoning is automatically initiated. Case-based reasoning is carried out with the objective that the false hypothesis will become true. Case-based reasoning consists of six specific steps.

1) To select knowledge elements for machine tool specifications from among meta knowledge in case-based reasoning. The elements must correspond to the contra-dictory causes which created the violations when a certain fact contradicts restricted conditions. The elements are described in the knowledge data base for machine tools as an object knowledge which indicates which machine tool construction or function caused the violations. In the elements, important degrees of violations for each element are also described.

2) To pay close attention to a machine tool which is being considered as the most suitable, and to then examine the similarity between this machine tool and all other machine tools among the knowledge data base. This examination is performed in order to compare an object knowledge which is selected from meta knowledge.

3) To select the machine tool whose analogous degree has a minimum of one as a temporary ideal machine tool type.

4) To take out the object knowledge of the temporary machine tool type knowledge,

and insert this object knowledge, which is being considered as the most suitable described in step 2, into the temporary machine tool type knowledge.

5) To examine the adjustments of this new machine tool knowledge construction.

6) To repeat the design process by using the new machine tool if the adjustments are proven to be true. If the adjustments are not suitable or if a contradiction occurs again when repeating the design process, it will systematically return to step 3 and select another machine tool whose analogous has a next minimum of one as a temporary ideal machine tool type, and continue to step 4.

4.2 Machine Tool Knowledge Base

Machine tool knowledge which is required in the algorithm described in section 4.1 contains principle expressions which will be further defined in this section.

Knowledge concerning machine tools is described with frame knowledge expressions whose frame name is a machine tool type name, slot name is a construction part name, and whose facet name is a construction part value. This kind of frame knowledge corresponds to a type of knowledge relating to machine tools or a machine tool specification [7].

A machine tool type name is the name of the machine tool type itself, for example, in a case where the machine tool type is one which is on the market, the type name " by the company " is described. A construction part name is the name given to each construction element of a machine tool, for example, in a case where a machine tool is a lathe turning machine, the construction elements correspond to a head stock, a bed and a tool rest. Characteristic knowledge taken from these kinds of construction elements is described as a facet name, for example, in a case where the construction element is a head stock a vertical type is described.

We have found through our research that when we acquire machine tool knowledge by using this type of frame description, it corresponds by creating a new machine tool which is not described beforehand in a knowledge data base.

4.3 Meta Knowledge Base

In order to carry out the case-based reasoning algorithm described in section 4.1 it is important for computer systems to have the knowledge described as to which kind of knowledge among machine tool knowledge should be selected as the object of analogy or meta knowledge.

Meta knowledge corresponds to the knowledge group that describes the connection as to which construction part name of a machine tool specification caused a contradiction when the priority order connection hypothesis selected from competing hypotheses becomes a contradictory hypothesis. We express this meta knowledge as constructed in Figure 1. The frame name describes the names of competing hypotheses, the slot name describes contradictory phenomena and the facet name describes construction part name of machine tools.

For example, we consider the contradiction frame name "select the minimum process order of a process number", the slot name may be described as "cycle time is over", resulting in contradictory phenomenon. The facet name may be described as " Loading mechanism" and "NC feed mechanism" where experienced

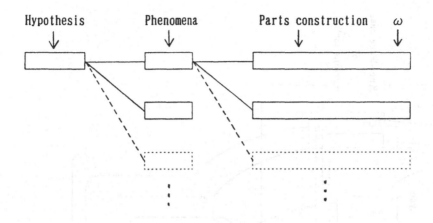

Fig. 1 Meta Knowledge

planners are required to guess the construction part name of machine tool specifications which will also cause this phenomenan.   As a result,   several values exist in a facet name and therefore ω, an important degree value to contradictory phenomena between each value will also be describd in a facet name. This important degree value is expressed as a percentage value or crisp value which planners with extensiveexperience possess.

4.4 Analogy Management

By using both the knowledge base concerning machine tools and the meta knowledge base mentioned in section 4.3 above, we can now proceed to the analogy process as shown in Figure 2.
　　When a priority order relation hypothesis among competing hypotheses becomes contradictory, we establish the machine tool type which has been considered true untill this point as the most suitable target type. By using a combination of the contradiction and contradictory phenomenon,   the construction part names which cause the contradiction and the important degree value ω can be reasoned from the meta knowledge base. Accessing the knowledge concerning the machine tool re-garded as most suitable target type,   the characteristic value of this machine tool construction part can also be reasoned.   In the same manner,   the characteristic values of another machine tool types which are the object of analogy are reasoned.
　　Next,   we compare characteristic values of the construction parts of these two kinds of machine tools. We call this comparison value λ_{ij}. The "i" indicates the most suitable target type and "j" represents another machine tool type which is the object of analogy.   The variable factor λ_{ij} takes the value "0" when the

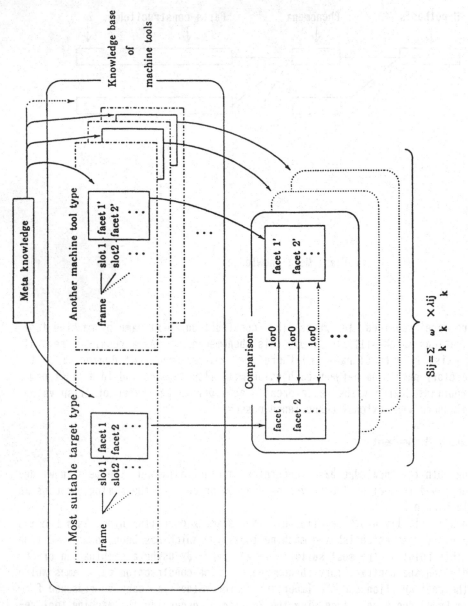

Fig. 2 Analogy Management

$$S_{ij} = \sum_k \omega_k \times \lambda_{ij_k}$$

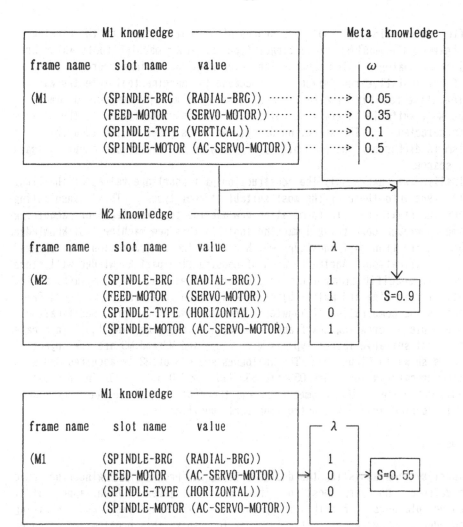

Fig. 3 Analogy Example

characteristic values of i and j are dissimilar and when the characteristics are the same it takes the value "1".

In this method, after acquiring each value of ω and λ, these two values are then substituted with the equation shown below in order to calculate the S_{ij} analogous score which will indicate the analogous degree between the two machine tool types.

$$S_{ij} = \sum_{k} \omega_k \times \lambda_{ijk}$$

ω_k = weighted value concerning construction part name k
λ_{ij} = comparison value concerning construction part name k with the value 1 or 0

After calculating all analogous scores of S_{ij} which represent the analogous degree between the most suitable target type and other machine tools which are the object of analogy, the machine tool type with the minimum score is selected as the first tentative resolution. Because the machine tool selected was the first tentative resolution with the most dissimilar machine mechanism compared with the most suitable target type which caused a contradiction, the minimum score is selected. The selection process which is based on the idea that if a mechanism is different, the probability of the same contradictions occuring again becomes scarce.

Finally, we change only the construction part knowledge values of the first tentative type into those of the most suitable target type. This transplanting indicates the final phase in formulating new machine tool knowledge or acquiring new frame knowledge concerning a machine tool. In this new machine tool knowledge, only the construction part knowledge which created the contradiction is changed to a new specification. Another branch of construction part knowledge will store the remaining specifications which were determined as successful results after hypothetical reasoning and multi-objective evaluation reasoning. New frame knowledge is the combination of knowledge from the two kinds of specifications.

An example of creating new frame knowledge for a new machine tool, in a case where the most suitable target type is M1, and M2 and M3 are the object of analogy, as shown in Figure 3. The analogous score S of M2 is aqcuired 0.9 with the following calculation : $S=0.05\times1+0.35\times1+0.1\times0+0.5\times1=0.9$. In this method, the analogous score of M3 becomes 0.55 and therefore M3, whose score is lower than M2 is adopted as a new construction part specification.

5 Conclusions

Many expert systems assisting the design process in production engineering have been developed, however, most of them are not given a high appraisal by experienced planners. For this reason ordinary expert systems ceased acquiring new knowledge as a matter of fact. In order to resolve this problem, we attended to attack the automatical knowledge acquisition function using case-based reasoning to the expert system we had already developed.

During this research, a new method of creating machine tool knowledge in the expert system used to assist production line design was proposed. This proposal is the algorithm proceeding case-based reasoning under the framework of hypothetical reasoning. In this algorithm, it is important to express both the knowledge concerning machine tools and the meta knowledge with a frame knowledge representation and to acquire a new machine tool knowledge expression by analogy. Systematically examining the adjustments for new machine tool knowledge construction is a problem which must be further developed in the future in order to create efficient machine tools for production line design.

Acknowledgements

This research has been supported by Artificial Intelligence Research Promotion Foundation. We are grateful for their help and encouragement.

References

1. S. C-Y. Lu, C. B. Blattner, and T. J. Lindem: A Knowledge-Based Expert System for Drilling Station Design, Applications of Artificial Intelligence in Engineering Problems, Springer-Verlag, Vol. 1, 1986, pp. 423-443.
2. W. A. Taylor: Development of a Knowledge Based System for Process Planning in Arc Welding, Applications of Artificial Intelligence in Engineering Problems, Springer-Verlag, Vol. 1, 1986, pp. 545-561.
3. K. J. Hammond: SHEF:A Model of Case-Based Planning, Proc. of AAAI-86, 1986, pp. 267-271.
4. H. Fujimoto and H. Yamamoto, A Multiobjective Evaluation Expert System for Production Line Design, The 5th International Conference on Applications of Arttificial Intelligence in Engineering, Springer-Verlag, 1990, pp. 447-466.
5. H. Fujimoto and H. Yamamoto: Development of Design Support System with New Reasoning and its Applications to Production Line Design, Proc. of 1990 ASME International Computers in Engineering Conference, Vol. 1, 1990, pp. 17-24.
6. J. Doyle: A Truth Maintenance System, Artificial Intelligence, Vol. 12, 1988, pp. 231-272.
7. H. Fujimoto and H. Yamamoto: Expert System for Process Planning in Production Facility Design, Proc. of the 3rd IFIP International Conference on Computer Applications in Production and Engineering, North-Holland, 1989, pp. 393-399.

Scaling-up Model-Based Troubleshooting by Exploiting Design Functionalities

Johan Vanwelkenhuysen

Artificial Intelligence Laboratory
Vrije Universiteit Brussel
Pleinlaan 2, B-1050 Brussels, Belgium
E-mail: johan@arti.vub.ac.be

Abstract

To perform model-based troubleshooting, it is commonly suggested to model the structure and correct behavior of the device under focus. Numerous
problems have been observed though when trying to scale-up this approach
to complex real-world devices. We have explored the alternative idea to model knowledge about the intended use of a device (i.e. design functionalities)
as a more global understanding of the behavior of the device. In this paper,
such a functional model for digital processor boards is presented, simplifying the board description considerably. In addition, it is highlighted how this
model is exploited by an implemented diagnostic system to deal with the time-
dependent behavior of processor boards and to exhibit an efficient problem-
solving behavior.

1   Introduction

Several authors have reported the difficulties they have encountered when trying to scale-up the well-known model-based troubleshooting approach [Davis, 1984,
deKleer and Williams, 1987] to complex and real-time problems [Hamscher, 1988,
Gallanti et al., 1989]. These difficulties mainly relate to the impractical or time-
intensive effort to design a detailed behavioral model of complex devices in order to make appropriate outcome predictions. Reasoning on such a model is often too costly to be of practical use. Suggestions have been made to overcome
these problems to some extent by considering additional diagnostic-oriented knowledge [Hamscher, 1988, Struss and Dressler, 1989, Steels, 1989] and by viewing the
device from different perspectives at multiple levels of abstractions [Hamscher, 1988,
Gallanti et al., 1989, Sticklen et al., 1989]. Work discussed in this paper continues
this line of research.

In our attempt to develop a diagnostic system for processor boards, we too have
observed numerous problems with designing a detailed behavioral model as suggested by [Davis, 1984] or even by applying Hamscher's improved methodology

[Hamscher, 1988]. As an alternative, we have exploited the idea of using knowledge of the device's intended use (i.e. design functionalities) to serve as domain knowledge for model-based troubleshooting. In the literature, this approach has been called the *functional approach* [Sticklen et al., 1989].

After characterizing the target troubleshooting environment in which our diagnostic system has to be operational (section 2), the problems which we observed with the methodologies suggested by [Davis, 1984] and [Hamscher, 1988] are presented in section 3. Section 4 discusses how a processor board can be viewed from the functional perspective. The last section highlights how this model has been exploited by TroTelC's diagnostic engine to bypass some problems presented in section 3.

2 The Target Troubleshooting Environment

Since March 1989, we have been involved in a joint project with Alcatel Bell in which a diagnostic knowledge system for processor boards (called *TroTelC*) has been developed.

Alcatel Bell is a telecommunications company producing the System 12 digital switching system for application in public and special networks. System 12 has a distributed architecture consisting of a large number of subscriber line circuit boards and (combined) processor, memory, and terminal interface boards (generally called *Printed Board Assemblies (PBAs)*). Each PBA is manufactured and extensively tested in Alcatel Bell's production environment before it is installed in the final customer's configuration.

After assembly, each PBA undergoes three test stages: the *manufacturing defect analyzer test (MDA), functional board test (FBT)* and *functional unit test (FUT)*. A MDA test inspects the board to identify process faults. After MDA test, each PBA undergoes a FBT. The PBA is subjected to a system functional test under different operating conditions (e.g. the test environment's temperature can be increased). It is verified during FBT whether the PBA realizes its design functionalities correctly. Lastly, the PBA is installed into a rack assembly and the entire system is tested during FUT. When a board fails the FBT, a fault report is generated and both are forwarded to a troubleshooter whose task is to identify the component(s) causing the observed misbehavior.

The goal of the project is to introduce a knowledge system (KS) into this environment. This KS has to serve as a resource for the electronics technician. The technician is not required to be an expert troubleshooter. He is expected to be able to operate the diagnostic station and the available measurement equipment to control and to observe the PBA's behavior. The KS should provide the technician with specific knowledge about the UUT and with problem solving capabilities to assist diagnosis of the UUT. Whenever the KS fails to provide the required knowledge in a particular situation, the UUT is forwarded to an expert troubleshooter.

3 Problems with Model-Based Troubleshooting

Authors have reported difficulties they have encountered when trying to scale-up the model-based reasoning approach (as described in [Davis, 1984]) to complex and real-time problems [Hamscher, 1988, Gallanti et al., 1989]. These problems are mainly due to the impractical or time-intensive effort to design a detailed behavioral model of complex devices in order to make appropriate outcome predictions. Also reasoning on such models is often too costly to be of practical use.

• Model-based reasoning requires a detailed description of the behavior of each component. For complex components, such a detailed behavior description is not always available. If it is available, it is often not feasible to represent the correct behavior such that it can be applied to make the required predictions. One also has to be aware that the designer has specified the component behavior with the assumption that the device is in its normal operation mode. Therefore, that behavior description does not necessarily apply when a component receives inputs which, under normal conditions, cannot be generated by the surrounding components.

• Another problem is related to the observations required from the operator. Queries for observations are in terms of the domain model which is used to predict the expected behavior. Since this model is very detailed, observations also need to be very detailed, many times resulting in costly observation procedures which are prone to human error.

Hamscher reports about his work in which he has tried to scale-up model-based reasoning to build a program that can troubleshoot complex digital circuit boards [Hamscher, 1988]. He views a circuit at multiple levels of abstractions to handle the problem of costly reasoning through a detailed behavioral model. Temporal abstractions are introduced to make efficient predictions in easily observable features (such as change, cycle, frequency). Additionally, his program is given knowledge of how often and in what way some components fail. This knowledge is interpreted as heuristic knowledge which helps the problem solver to generate plausible diagnoses instead of all logically possible ones. In spite of these significant improvements, there still remain some problems with this methodology (as noted in [Hamscher, 1988]). These problems are related with the difficulty and time-consuming effort to design the behavioral rules. Efficiency problems also endanger the practical use of Hamscher's approach. These problems are not only due to implementation issues but also to the inflexible control and costly prediction task using the behavioral model.

4 Taking a Functional Perspective

By considering a device from the perspective of its intended use,* it turns out that many of the problems mentioned in the previous paragraph can be resolved or bypassed (section 5). In this section, a functional model of processor boards is pre-

---

*In the literature, this approach has been called the *functional approach* [Sticklen et al., 1989]

sented. This model has been created by a knowledge engineer who observed the behavior of expert troubleshooters in their troubleshooting environment (the acquisition process is discussed in detail in [Vanwelkenhuysen, 1990]).

4.1 A Functional Model for Processor Boards

The functional model reflects what the UUT is intended to be used for (i.e. its design functionalities). It describes the UUT in terms of *circuit operations* that can be executed on the board and the *hardware module functions* which are elementary functions whose causal interactions make the circuit operation happen. Due to the functional perspective taken of the board, it becomes relevant to augment the model with knowledge of the *test programs* which were executed during FBT. These test programs are described in terms of the circuit operations on the UUT.

4.1.1 Circuit Operations

In principle, four kind of basic operations can be performed to memory or processor boards: One can read or write from/into a memory or I/O component. Mapping these basic operations onto the memory and I/O components of a PBA (called *hardware modules*) results in what we call the *circuit operations* of a particular PBA. Examples of memory hardware modules are the *RAM* and the *EPROM*. Examples of I/O hardware modules are the *Process Control Register* and the *Timer*. Consequently, some of the circuit operations for this control processor board are *Write Value into RAM*, *Read Value from EPROM*, *Write Value into PCR*, and *Read Timer Counter Status*.

A circuit operation is represented in terms of its input/output parameters, causal relations with the involved hardware module functions, and in terms of parameter relations.

• For each circuit operation it is known what its input/output parameters are and on which bus this parameter is put. For example, the *Read Value from EPROM* circuit operation expects as input parameter an address (within the EPROM range) via the external address bus and outputs the data located at that address at the external data bus.

• Instructing a PBA to perform a circuit operation directly causes the activation of one or more hardware module functions. An activation of a hardware module function may, in turn, cause the activation of other hardware module functions at some point in time. The sequence of (correct) causal interactions between all activated hardware module functions make the PBA behave such that the circuit operation is performed as expected.

As an example, consider again the circuit operation *Read Value from EPROM*. Instructing the PBA to perform this operation causes interactions between the hardware modules depicted in figure 1. The *Address Input Buffer* receives the address input parameter and loads this to the address-data bus. The *Address Latch* filters the address from that bus and sends it to the *Address Decoder* and *EPROM*. The *Address Decoder* analyses the address and decides which hardware module to select for activation (i.e. hardware module for which this address is intended), which is the *EPROM*. The *EPROM* retrieves the data located at the address and sends this to the internal bus. The *Internal Address Data Transceiver* takes the data from the internal bus and forwards this, through the address-data bus, to the *Data Buffer* which finally outputs the data. The *Control Buffer* and *IPCU* are responsible for the flow of control, i.e. deciding which hardware modules are to be activated including timing sequence. This causal relation between the circuit operation and the involved hardware module functions is represented by a circuit operation.

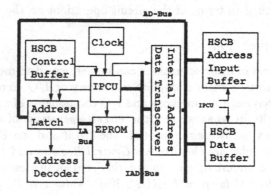

Figure 1: The hardware module functions associated with circuit operation *Read Value From EPROM*.

- The last relation associated with each circuit operation is the parameter relation. The parameter relation specifies the way in which the circuit operation's input/output parameters are transmitted through the causally related hardware module functions.

4.1.2 Hardware Module Functions

Hardware module functions are functions which emerge from the causal interactions between a number of gates. A board designer labels such a group of gates as a hardware module. The different hardware modules of a PBA are provided by the designer through the block diagram. The latter are useful concepts, because by grouping the gates, the combined behavior can be abstracted and represented independent of the underlying implementation or technology.

A hardware module function is represented in terms of its underlying gates, the state controlling enable and control signals, and the transformation relationship between the function's input and output. Figure 2 depicts the representation of the hardware module function *Filter Address from Address Data Bus*.

- A hardware module function is defined by its control and enable signals. The enable signals define when a hardware module is in an active state. The control signals identify which function the hardware module has to realize (since a hardware module often can realize a number of functions). Both signals are defined as the prerequisite conditions of a hardware module function.

For example, the hardware module function *Filter Address from the Address Data Bus* (figure 2) has one enable signal (DBEL-) and one control signal (ITR-). To be activated, the enable signal has to receive a low pulse. A high pulse is expected at the control signal to indicate the transmission direction[†]

- A hardware module is a virtual component grouping a number of gates. The relation between a hardware module function and the underlying gates is represented. It is not necessary that a specific gate only contributes to one hardware module function.

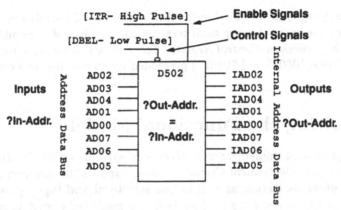

Figure 2: Hardware module function *Filter Address from Address Data Bus.*

- The relation between the inputs and outputs are also represented with a hardware module function. The function *Filter Address from the Address Data Bus*, for example, receives its inputs from the Address Data Bus and the same value (i.e. identity relation) is put onto the Internal Address Data Bus.

4.1.3 Test Programs

The functional model of a PBA is augmented with knowledge of the test programs. These test programs are executed during the FBT and can be selectively executed through a diagnostic test station. A test program consists of a number of test segments, each test segment verifies a particular (sub)function of the UUT. For each available test program, the test segments involved and the type of program is indicated. Three types are distinguished, each type exploiting the specific configuration of the test box.

[†]The *Internal Address Data Transceiver* can realize two functions: Transmitting data from the *Internal Address Data Bus* to the *Address Data Bus* and transmitting data in the other direction.

A test segment is defined as a sequence of test routines. Each test routine is defined as a sequence of instantiated circuit operations.[‡] Each test routine has an associated fault code. When executing a test segment, the test routines are sequentially executed. As soon as one test routine fails, the fault code of the failing test routine is displayed. Figure 3 shows as example the *Processor Control Register Test*.

```
PROCESSOR CONTROL REGISTER TEST        (a test segment)

    Write Value into PCR
        Address: PCR-address                                }
        Input-data: [hexa-value "6000"]                      }  Initialization

             Read Low Byte from PCR
1st Test         Address: (>> low-byte address write-value-into-pcr)
Routine  {       Expected-output-data: (>> low-byte input-data write-value-into-pcr)
                 Fault-code: [code "0601"]

             Read High Byte from PCR
2nd Test         Address: (>> high-byte address write-value-into-pcr)
Routine  {       Expected-output-data: (>> high-byte input-data write-value-into-pcr)
                 Fault-code: [code "0602"]
```

Figure 3: A test segment can be divided into a number of initializing circuit operations and test routines whereby each test routine consists of a number of circuit operations. The *Processor Control Register Test* consists of two test routines (identified by fault code "0601" and "0602") preceeded by an initializing write operation.

5   Exploiting the Functional Model

We have designed and implemented a diagnostic system, called *TroTelC* (a Troubleshooting Tool for Electronical Circuits), which exploits the functional model discussed in the previous section as well as the structural and topological knowledge of the PBA. In this section we explain how the model of circuit operations, the hardware module functions and test programs are exploited by TroTelC.

5.1   The Troubleshooting Process

Circuit operations and hardware module functions are abstractions of component behaviors. Reasoning at this abstract level enhance the troubleshooting process by decreasing and simplifying the number of observations.

An important reason underlying this observation is that by taking a functional view, test programs become a relevant source of knowledge. We have exploited the knowledge embodied in a test program and have incorporated this into the diagnostic system. These test programs allow circuit operations to be verified in a very inexpensively and more easy manner. In addition, they provide the system with knowledge to create effective hardware module function tests and in this way limit the number

---

[‡]By "instantiated" we mean that for each circuit operation, the input/output parameters are defined

of operations to be performed to the board§ An additional advantage is that the tests and observations are always performed with the circuit in its normal mode of operation, i.e. the reasoning is guided primarily by possible circuit operations rather than what is required to test a component.

As soon as the TroTelC system receives a fault listing, it verifies the symptom and suspects the circuit operations executed by the failing test segment's subroutine with code equal to the fault code. More test segments are proposed to be executed to gather more evidence.

When sufficient evidence has been gathered of a suspected circuit operation or if no further discrimination at this level is possible, hardware module functions are then investigated. For example, consider it is decided to investigate the hardware module function *Filter Address from the AD-bus* contributing to the circuit operaton *Read Value from EPROM*. The UUT is instructed to perform the *Read Value from EPROM* circuit operation. This causes a chain of actions and interactions between a number of hardware module functions including the suspect under focus. The troubleshooter is then asked to perform some measurements. The measurements are related to investigating whether the involved hardware modules are in an active state at some point in time during the operation (i.e. whether the enable and control signals arrive simultaneously). While the hardware modules are in the active state (i.e. the enable and control signals serve as triggering actions), it is verified whether the expected inputs are transformed into the expected outputs according to the definition of the hardware module function.

In summary, to emphasize three interesting features. (1) By taking a functional view, test programs and segments become relevant because they can be described in terms of circuit operations which are executed to verify certain functionalities of the UUT. Executing and observing these tests can be easily done through the diagnostic station; (2) Once sufficient evidence has been gathered of a misbehaving circuit operation, the causally related hardware module functions are then investigated. This guides the search towards the entities actively involved in the circuit operation which limits the number of observations considerably rather than those reasoning approaches considering the circuit structure. (3) Once a level is reached at which a particular function may be assumed working, further observations along that path are unnecessary. For most of the faults which appear in practice, this speeds up the troubleshooting process considerably. In some cases however, the cause might be situated outside the normal contributing components.

5.2   Finding causes "outside" the normal contributing entities

In case of bus faults or when the UUT cannot realize any of the intended functions (e.g. the primary communication with the UUT fails), focusing only on the involved functions is unsufficient. We have extended the reasoning process to these problems,

---

§Due to space limitations, this is not further discussed. The reader is referred to the longer version of this paper, published as technical report at the VUB AI-Lab.

fully and only relying on the functional model described in this paper. This extension is clarified by showing how bus faults are treated.

Bus faults cover the set of faults in which a chip is incorrectly activated and consequently sends data to its output lines interfering with the correct data from another source. Consider for example the buffer shown in figure 2. The signals at the output lines of the buffer can be observed to be incorrect although the chip receives the correct input signals and behaves as expected. In the circuit, there may be a number of other chips connected to the IAD00 through IAD07 lines (i.e. the low byte of the internal address data bus). Normally, when the buffer is enabled, the other chips which can output data onto the IAD-bus should be disabled. However, due to spikes or an internal defect, one of these other chips may be (occasionally or continuously) enabled such that it sends interfering data to the IAD-bus. This type of fault causes incorrect signals to be observed at the output of the chip.

When there is high probability that a bus fault exists, the TroTelC system gathers all hardware module functions whose outputs are connected to the interfered bus (no matter whether or not these functions are involved in the misbehaving circuit operation). For all these hardware module functions it is observed whether they are active (i.e. whether the expected enable and control signals of these hardware module functions arrive simultaneously) while the hardware module under focus (e.g. the buffer) is enabled (i.e. the enable signal serves as triggering action). Those hardware module functions which become active become highly suspect and investigated (stepping out of the set of normally contributing entities).

5.3   Dealing with Temporal Dependencies

One of the difficulties when designing a behavioral model of a complex device is related to capturing the time-dependent behavior of that device. When performing a circuit operation, many behaviors are being realized by numerous components, causing the board to be in a continuous state of change. The realization of some of these behaviors might be a prerequisite for activating other behaviors. On the other hand, some behaviors may be realized in parallel. A board is so carefully designed that a delay of a few nanoseconds can cause incorrect interaction between the different behaviors performed by the components.

The functional approach allows one to deal with these temporal dependencies: The functional model indexes component behavior by viewing their abstracted behavior as a temporal entity in the chain of interactions ([Sticklen et al., 1989] and [Abu-Hanna et al., 1991]). Circuit operations identify the hardware module functions whose interaction make the board behave as expected for that operation. A hardware module function abstracts from the underlying component behaviors and is an entity which is temporally active only. By the indexing mechanism provided by the functional model, we mean that this representation allows us to focus on (read "index") a group of components (hardware modules), test whether these components are active some point in time during the execution of the circuit operation and verify whether the appropriate data is provided. When focusing on these components, we are not concerned of the detailed timing information when certain signals must occur. What is only relevant is that the required signals are provided and appear simultaneously at some point in time and consequently put the hardware module into an active state.

6 Acknowledgements

The author acknowledges the influence of the members of the Knowledge Based Systems group at the VUB AI-Lab, and in particular Luc Steels, Viviane Jonckers, Philip Rademakers, Walter Van de Velde, Kris Van Marcke, and Koen De Vroede. Discussions with Jim Jamieson (Alcatel), Robert De Bondt, Robert Laskowski, Jef Schraeyen, and Roger Segers (Alcatel Bell) provided me with important insights in the problem domain. Jim Jamieson, Viviane Jonckers, and Walter Van de Velde gave very useful comments on an earlier draft of this paper.

References

[Abu-Hanna et al., 1991] Abu-Hanna, A., Benjamins, R., and Jansweijer, W. (1991). Device understanding and modeling for diagnosis. *IEEE*, 6(2).

[Davis, 1984] Davis, R. (1984). Diagnostic reasoning based on structure and behavior. *Artificial Intelligence*, 24.

[deKleer and Williams, 1987] deKleer, J. and Williams, B. (1987). Diagnosing multiple faults. *Artificial Intelligence*, 32.

[Gallanti et al., 1989] Gallanti, M., Roncato, M., Stefanini, A., and Tornielli, G. (1989). A diagnostic algorithm based on models at different level of abstraction. In *Proceeding of the Eleventh International Conference on Artificial Intelligence*, pages 1350, 1355.

[Hamscher, 1988] Hamscher, W. (1988). *Model-Based Troubleshooting of Digital Systems*. PhD thesis, MIT Lab.

[Steels, 1989] Steels, L. (1989). Diagnosing with a function fault model. *Applied Artificial Intelligence*, 3(2-3).

[Sticklen et al., 1989] Sticklen, J., Goel, A., Chandrasekaran, B., and Bond, W. (1989). Functional reasoning for design and diagnosis. In *International Workshop Model-Based Diagnosis*. Paris.

[Struss and Dressler, 1989] Struss, P. and Dressler, O. (1989). Physical negation - integrating fault models into the general diagnostic engine. In *Proceeding of the Eleventh International Conference on Artificial Intelligence*, pages 1318, 1323.

[Vanwelkenhuysen, 1990] Vanwelkenhuysen, J. (1990). Modeling and implementing the task of diagnosing pcbs: A task-oriented approach. AI-Memo 90-6, AI-Lab VUB.

Application of Knowledge-based Systems to Optimised Building Maintenance Management.

G Clark P Mehta

Brunel University
UK

T Thomson

Dunwoody & Partners
UK

Abstract. This paper outlines research into the application of knowledge-based systems (KBS) to maintenance management within an integrated building management system (IBMS). Many buildings already have maintenance scheduling programs which schedule breakdown and preventative maintenance work. There is a requirement to 'tune' this scheduling to best fit the exact requirements of the building and as these change with time there is a need for a continuously adaptive system especially in the field of preventative maintenance. The goal is to maximise equipment reliability and minimise maintenance, repair and replacement costs by the optimum scheduling of man power and resources. The concept is illustrated by the use of a case study into lighting systems using data from the Lloyd's 1986 building in London.

Keywords: Intelligent Buildings, Knowledge-based Systems, Maintenance

1 Introduction

Maintenance management can be considered as the direction and organisation of resources in order to control the availability and performance of industrial plant to some specified level. Some of the more technologically advanced buildings such as the Lloyd's building already use computerised maintenance management packages. Their existing system consists of a database which stores data for all plant equipment within the building. Such data includes the location of the equipment and plant description along with the breakdown history including costing information and details of the maintenance staff. This information is updated via a central fault reporting facility called front desk administration which receives details of breakdowns and problems from the occupants of the building. A record of the planned maintenance intervals for all plant equipment is stored in the system and this enables the system to determine the planned maintenance periods for equipment and schedule maintenance at a given time in the future. Dockets are automatically printed and then issued to the relevant personnel on a regular basis. After the work is completed the dockets are returned and the system is updated with the results of the maintenance work written on the dockets by the maintenance staff. This provides the ability to keep track of the status of individual items of work at any time within the building and enables efficient administration of contract maintenance work by accounting for the time of contract staff.

The maintenance management system works well and achieves the requirements set out above but it is very much an isolated entity within the automation system in a building. There is no ability for direct connection to any of the building's control systems for automatic data capture and update. If the maintenance system were packaged as part of the overall IBMS the scheduling of maintenance could take place based on actual breakdown and costing data other than simply at set intervals entered into the system when plant equipment is commissioned.

There has been research into knowledge-based maintenance but more from the diagnosis point of view than scheduling preventative maintenance. Finley [1] discusses the possibility of integrating expert systems into planned maintenance programs and using rules to interpret condition monitoring and provide technical support. Other authors specifically target the building services environment as an area where expert systems can be usefully deployed - particularly in diagnosis and maintenance for complex HVAC systems where the current maintenance costs are specifically high [2]. There have been some prototype systems developed in recent years [3,4]. Arueti [5] reports on a knowledge-based prototype for optimisation of preventative maintenance scheduling developed for a nuclear power plant based on probabilistic judgement and inference rules using data on failure rates, repair times, costs and indirect economic costs such as accidental power loss. Similar approaches have been used for job scheduling within a manufacturing environment [6].

1.1 Advantages of Knowledge-based Enhancements

If the maintenance management system is integrated into the IBMS then the planned maintenance period could become a dynamic variable which could be tuned to achieve the most economic schedule. In the Lloyd's system the maintenance interval for equipment is fixed and it can only be changed by direct input from the user. Ideally the planned maintenance could be tuned to best match the requirements of the building throughout its life cycle. The buildings use and requirements will change with time. Hence the maintenance work and the priorities allocated to such work should change with the building's requirements and its occupants. By using a KBS approach to this 'intelligent tuning', the system can explain why maintenance is scheduled at a particular time by use of engineering rules, actual breakdown data and statistical analysis. This would be of particular use to the maintenance and engineering staff within the building who may not have faith in KBS 'judgements' without knowing the facts behind the decision making. Expert systems in particular lend themselves to explaining the reasons behind decisions.

For finite maintenance resources the KBS can optimise maintenance scheduling on the basis of priorities. For example, if two separate pieces of maintenance work should be undertaken during the same period and there is insufficient resources to undertake both of them the KBS can re-schedule one of them based on economic or other factors and then explain its reasoning. The KBS can also be used for predicting breakdowns, maintenance cost and spare part requirements. This is particularly useful for planning ahead in terms of budget requests and ordering materials. The system can advise on reorder times and stock of spares based on predicted requirements. It can also advise on specific maintenance requirements following a breakdown, highlight common or particularly costly faults and provide decision support for repair or replacement. A KBS can be used for asking 'what if questions' such as setting constraints and conditions or to find the predicted effect of changing equipment operation. For example, to consider if it would be more economic to reduce or increase staffing levels, re-distribute staffing levels between different disciplines or to reduce or increase planned preventative maintenance.

2 Maintenance within an IBMS

It is expected that future IBMS systems will communicate through hierarchical local area network systems. An IBMS which includes maintenance management can enable automatic transfer of monitored information as shown in Fig 1. This data is collected from the building control systems at the sensor level and passed to the maintenance management system on a regular or real time basis. Information can be collected on plant performance in terms of running time, efficiency, flow rates and power consumption. Other parameters such as trends as can be measured or calculated for transmission via the local area network to the BMS historical data base to assist in the operation of planned preventive maintenance. Faults, breakdowns and alarms would also be automatically signalled to the IBMS.

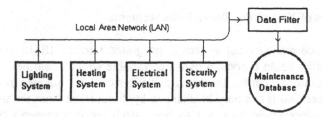

Fig 1 Integrated Maintenance Program

This avoids the need for a manual input and thus ensures that it always has the latest and the most accurate data. Should an alarm occur which requires attention, the appropriate message and work order can be printed out automatically. The urgency of the alarm can be categorised so that the time scale over which corrective action is taken is appropriate to the situation. Archived data can also be called upon to assist the maintenance engineers to plan future strategies.

2.1 Intelligent Archiving of IBMS Data

There is a great deal of information stored within an IBMS's control and management system. Some of this data must be archived either as a statutory requirement or for planning and analysis. The problem is deciding just what to archive as archiving requires some form of computer backing store such as disk or tape streamer and is expensive. For example, an access control system might register people entering the building every few seconds, the temperature in different rooms is continuously changing, many lifts within a building receive calls to floors all the time and the electrical power consumption of the building changes every time plant machinery is turned on or off. Firstly one must decide just what information should be archived and then how frequently this should be done. However if these are fixed decisions the opportunity may be missed for identifying trends and unusual behaviour.

The IBMS should have the flexibility to archive some information when this information is judged to be worth archiving. For example, the electrical power consumption may be recorded and archived every five minutes. However, a sudden peak in consumption at say 150% of normal, which is very unusual, may occur for ten seconds between the five minute archiving period. Under the normal archiving rules this would not be recorded and the fact that this may happen every day at exactly the same time may not be noticed. An intelligent archiving system must be selective in what is archived based on what is deemed to be useful and also it needs to be adaptive when identifying abnormal situations. The knowledge-based approach to data archiving would provide this ability for adaptation.

2.2 Proposed Knowledge-based Approach

There is an opening for an application of KBS to close the feedback loop outlined above by tuning the planned maintenance schedules within the maintenance management package to best fit the requirements of the building. Feedback is the key element in determining whether or not the planning and scheduling of preventative maintenance actually reduces costs. Management by definition implies control, and the scheduling of maintenance is the manager's tool for control. For example, if a chiller system breaks down on average once a month and its planned preventative maintenance take place every two years there is a clear case for carrying out preventative maintenance more frequently. In the case of the Lloyd's building where more than thirty-five thousand maintenance jobs were undertaken in 1990 alone, the task of identifying every area of intelligent feedback into the maintenance program by manual means would not be cost effective. An intelligent IBMS could fine tune itself to best manage the building and adapt as the system requirements change throughout its life by the use of adaptive knowledge-based optimised maintenance scheduling. Rules can be written within an expert system to schedule maintenance based on the following factors.

(a) The probability of a breakdown for a particular piece of plant equipment can be predicted by statistical analysis of actual breakdown data from the same class of equipment. This can also be linked to the use of the equipment within the building.
(b) Conditioning monitoring using sensors at the control level to monitor the system and signal to the management system when a problem or fault condition is detected. For example, pressure sensors on either side of an air filter within a ventilation system can be used to detect the time to change an air filter.
(d) The cumulative cost of a reduction in performance can be considered. In the case of lighting an example of this might be the cost of wasted electricity not being converted into light due to dirt build up on the light fittings.

(e) The cost of the labour for specific trades can be used to decide if there is an advantage in using different maintenance staff to perform planned maintenance. This might favour investing more money into preventative maintenance or conversely to reduce the overall maintenance cost allowing for the cost of servicing breakdowns.

(f) The prediction of actual use of the equipment up to the time of planned preventative maintenance which requires prediction based on the data stored within the control part of the IBMS.

(g) The scheduling of other equipment within a particular class. For example, if a particular piece of equipment is to be serviced at a given time then it would probably be most cost effective to also service other equipment of the same type at the same time.

(h) Scheduling of different classes of equipment with finite resources can take into account the number of maintenance staff and their working hours, the level of spares in stock and the constraints on taking equipment out of use for maintenance..

(i) Consideration of the equipment's location and site specific factors will allow the access time to the plant equipment to be predicted. Knowledge of location will also allow the system to schedule maintenance on plant in close proximity.

(j) Scheduling at economic 'down times' for a parent class. For example, if a piece of equipment breaks down and goes out of service, preventative maintenance of its sub-components can be conducted at the same time. This may avoid having to take it out of service again.

2.3 Proposed Database Structure and Prototype Knowledge Base

The specific classes of data required for the KBS has been identified and the structure of an object oriented database has been determined and a prototype has been built within Microsoft's Excel spreadsheet package. Excel is an ideal tool for a prototype database of this nature due to it's advanced 3 dimensional spreadsheet facility and dynamic cell links between sheets. The spreadsheet provides a gateway between the data acquisition and the knowledge base application. The data has been classified into five main classes. Within these classes some sub classes, objects and object specific properties have been identified for the purpose of defining inheritance.

(a) There is a plant equipment database which has been divided into six sub-classes namely security, transport, HVAC, electrical supply, lighting and fire prevention. Each of these sub-classes inherits eight class properties.

    (i)   A property describing the use of the equipment
    (ii)   A property which can be used to predict failure probability
    (iii)   The date of last maintenance (planned or breakdown)
    (iv)   The cost of planned maintenance
    (v)   The planned maintenance period
    (vi)   The predicted use at the time of maintenance
    (vii) An indication of monthly running cost

(viii) Location within the building

Individual sub classes may have specific properties. For example lighting has the additional specific properties of average tube life, labour rate for a bulk replacement, average light level, minimum light level, and luminaire light depreciation constants, etc.

(b) A building information database storing a description of the building. This enables the KBS to interpret the location data for the plant equipment.

(c) A preventative maintenance database which is used to store summarised data on faults and breakdowns. Every fault and item of repair work can be classified into a finite number of jobs. For each job the average time, frequency and cost can be calculated from feedback from the breakdown dockets (archived data) and then used by the KBS for optimising scheduling in the future. As this data is continuously being updated from actual breakdown data this is one way in which the feedback loop is closed.

(d) Maintenance resources database recording data relating to the available maintenance staff including trade classes and working hours. This enables the KBS to predict what capacity there is available to undertake maintenance work. Other resources would include the level of spares or maintenance machinery.

(e) Scheduled work database; this is a 'calendar' for listing what work is scheduled and for when along with predicted costing and resource requirements.

Almost all the specific classes of data outlined above are already available distributed amidst the existing individual systems within a building such as Lloyd's. However, they are not in the structured form required for a KBS to process the data easily. An object oriented format would reduce the possibility of multiple occurrences of the same data and the problems that may lead to. Nexpert Object is the expert system shell which is being used to prototype the proposed knowledge-based system. Nexpert supports some advanced features including an object oriented data structure and the ability to modularise knowledge bases and define inference priorities within rules. Both Nexpert and Excel run within the Microsoft Windows 3 environment and they both support the windows direct data link feature as well as SYLK database retrieval format which is a derivative of standard SQL. Nexpert can automatically read in an excel file from disk and generate its own object oriented data structure which is then processed by the rule base.

3   Case Study - The Lighting System in the Lloyd's Building

Lighting is one of the importance areas within a building. The Lloyd's 1986 building has a complex lighting control system with 8,900 separate luminaires each under the control of the building's Energy Control System. The lights are grouped with each group of lights having an individual switching pattern switched by 285 autonomous lighting control boxes. For the purposes of this case study one specific floor was selected namely the second floor. This floor contains 760 luminaires, all under computer control.

3.1 Historical Maintenance Data

The computerised maintenance management system at Lloyd's records the service history and costing data on most of the plant equipment within the building. Breakdown data was obtained for every controlled light fitting on floors 2 and 4 from April 1990 to January 1991. The faults were classified into one of five main groups. The cause of the fault was either the starter, the tube, the circuit breaker or some combination of all three. Secondly, the fault data was summarised in the form of a distribution table for each maintenance period since the lamps were last block replaced. The number of faults in each period were totalled along with the number of hours of maintenance time. This enables the frequency distribution of the individual faults to be calculated along with the average repair time for each job. In addition to the breakdown data specific data on the light fittings is also required such as the material cost of each tube, its starter, average life and the labour rate for spot replacements.

3.2 Failure Prediction

For effective optimisation of planned maintenance the condition of the lighting installation must be predicted at a specific time in the future. This prediction is based on the archived data on breakdown history and depreciation of light level. The light level from a light fitting falls with time. This decrease in output is due to a combination of the two independent factors. Firstly the output from the tube itself will fall, very rapidly at the start of its life and then at a rate proportional to the amount of time it is used. The second factor which reduced light output is the build-up of dirt in the light fitting. This increases at a rate proportional to the elapsed time since the fitting was cleaned. In both cases the decrease in output can be approximated to a linear equation with the addition of an initial constant.

The data given on failures only covers the first part of the full graph as the fittings are block replaced well before the 50% failure point, as this is most economical. However, the survival probability can be predicted outside the known data range by extrapolation. Many failure-causing mechanisms give rise to measured distributions of times-to-failure that approximate quite closely to probability density distributions of definite mathematical form, known in statistics theory as p.d.f.s. Such functions can therefore be used to provide mathematical models of failure patterns, which can then be used in performance forecasting calculations. Some p.d.f.s. relating to maintenance studies include the Negative exponential, Hyper exponential, Normal (or Gaussian) distribution, the Weibull, gamma, Erlang and Lognormal distributions [7]. Combinations of these can be used to construct more complex failure rate models such as the 'bath tub curve' where failures are high during a 'running in' period, then after a short time they adopt a random pattern and finally near the end of the equipment's life they begin to increase due to components wearing out. The KBS is used to select the p.d.f. that best fits the failure data and the actual parameters calculated by the use of an iterative algorithm.

In the case of luminaires which exhibit a definite wear-out failure pattern the normal distribution provides the best fit. They tend to fail at some mean operating age, μ, with some failing sooner and some later, thus a dispersion, of standard deviation σ, in the recorded times to failure. To be able to predict the total number of failures at a particular time the cumulative distribution function is required. This probability can be obtained in an exact form from the relevant probability density function by integrating it with respect to time as follows,

$$F(t) = \int_{-\alpha}^{t} f(t)\ dt = \frac{1}{\sigma\sqrt{2\pi}} \int_{-\alpha}^{t} \exp\left(\frac{-(1-\mu)^2}{2\sigma^2}\right) dt$$

One simple way to represent this cumulative failure prediction function is by means of the function shown below,

Failure probability, $P_f = \{\ 1/[\ 1 + (P_{uh}/A_{lh})^i\ ]$

Predicted survival probability, $P_s = (1 - P_f)$

It is not exact but has the advantage that it can be represented by only two numeric constants. The first A_{lh} is some indication of the average life of the tube in hours which is the value around which the curve falls off. This will differ slightly from the manufactures own data as the average life is to some extent a function of the switching pattern. The second constant, i, is a measurement of the rate of this fall off. The term P_{up} is the predicted use for the luminaire in hours. The actual and predicted data is shown plotted together in Fig 2.

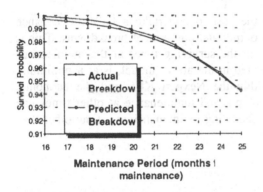

Fig 2. Fitting a Curve to Actual Breakdown Data

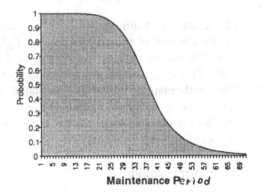

Fig 3. Luminaire Survival Probability

This function can now be used to extrapolate outside the known data range as shown in Fig 3. This enables the cost of breakdowns to be predicted up to and beyond the average life of the luminaires.

3.3 Economic Considerations

If the lighting installation is not maintained a cost is incurred in spot replacements of faulty luminaires and also light is wasted due to inefficient tubes and dirty light fittings; this lost light can be estimated. However, conducting a bulk replacement too early means that a small amount of useful tube life is wasted. The most economic time to schedule maintenance is before the predicted cost of not conducting maintenance exceeds the cost of conducting maintenance. This is calculated for lighting as follows,

C_u -   unit cost of tube + starter
C_{lb} -   labour cost of bulk planned maintenance and cleaning per unit
C_{ls} -   spot labour cost for maintenance per unit
n   -   number of lights on a particular floor

total cost to conduct bulk maintenance at after time T, $C_m = \sum_1^n (C_u + C_{lb})$

total light breakdown cost until time T, $C_b(T) = \sum [P_f(T) \times (C_{ls} + C_u)]$

remaining unit cost of fittings, $C_{ur}(T) = \sum [A_{lh} - AVERAGE(P_{uh}(T))] / A_{lh} \times C_u$

light running cost, $C_r(T) = \sum [P_{uh}(T) \times \text{pwr rating of tube} / 1000 \times \text{cost/kw/h}]$

cost of lost light, $C_l(T) = C_r(T) \times L_1$

the cost of maintenance per unit time as a function of the maintenance interval T is,

$$C(T) = \left(\frac{C_m + C_b(T) + C_l(T) - C_{ur}(T)}{T} \right)$$

The economic maintenance time is when the cost to undertake a bulk maintenance equals the sum of the cost of the lost light output and cumulative breakdowns. This is allowing for the cost of the remaining life in the surviving tubes at the maintenance time. This occurs when the total cost, predicted in the above model, is at a minimum. The mathematical model is calculated within the Nexpert KBS and the economic maintenance period is predicted. The cost of planned maintenance at other times can also be predicted to asses the extra cost of scheduling maintenance at a time other than the optimum.

4   Conclusions

A KBS approach to lighting maintenance within an integrated building management system would improve maintenance management in the following four ways.

- Enable luminaire failures to be predicted based on archived breakdown data which would in turn enable maintenance costs both labour and materials to be predicted.
- Predict lighting levels for an installation at a given time based on actual light level readings. This may reduce the number of spot light readings that have to be made to assess the condition of the lighting installation prior to bulk lamp replacement.
- Advise on bulk maintenance based on economic considerations, occupant comfort and finite maintenance resources.
- Provide cost savings by administering a system of replacing spot failures with used lamps which have sufficient life left to last to the next bulk replacement.

The KBS itself would require minimum overheads as the data required for it's analysis will be contained within the IBMS.

Optimised lighting maintenance must be seen as only a small part of the whole integrated maintenance management system. To make significant savings these ideas must be applied to the areas of maintenance that currently have the highest expenditure. In the case of Lloyd's this is most probably heating, ventilation and air conditioning (HVAC). The next stage is therefore to conduct a similar case study applying the same ideas to HVAC equipment. This would then lead to formulation of a rule base to schedule one class of maintenance against another with finite maintenance resources. When this is complete, the area of 'what if' hypothetical analysis can be explored. It is expected that this will lead to a specification for a general model of the building and its system from the point of view of maintenance. For example, it would be possible to predict the consequences of a lift going out of service in terms of increased waiting times for the other passengers.

Acknowledgements

The work presented here is part of a collaborative project between Brunel University and the insurance corporation Lloyd's of London, consulting engineers Dunwoody and Partners and Thorn Security who are manufacturers of IBMS. The authors are particularly grateful to Lloyd's for their assistance in providing data from their building to be used in the case study.

The project is funded by the UK Department of Trade and Industry, the Science and Engineering Research Council and the industrial collaborators.

References

1. H.F.Finley, 'Reduce Costs with Knowledge-Based Maintenance' Hydrocarbon Processing, Vol. 66 Part 1, p64-66, January 1987
2. C H Culp, Expert Systems in Preventative Maintenance and Diagnostics, ASHRAE Journal -American Society of Heating Refrigerating and Air-Conditioning Engineers, Vol. 31 No.8 p24-27, 1989
3. Pignataro V, Details of the Computerised Maintenance Management-System at Quebec-and Ontario-Paper Co. Ltd, Thorold Ontario, Pulp & Paper-Canada Vol. 90 Issue 1 pp 109-110, 1989
4. Warren T, Pender R, Maintenance Management Control Through Utilisation of a Computerised System, CIM Bulletin, Vol. 81 Issue 911 pp 94-94, 1988
5. S Arueti, D Okrent, 'A Knowledge-Based Prototype for Optimisation of Preventative Maintenance Scheduling', Reliability Engineering and Safety, Vol. 30, Pt 1-3, p93-114, 1990
6. Majstorovic V D, Milacic V R, Expert Systems for Maintenance in the CIM Concept, Computers in Industry, Vol.15 No.1-2 pp.83-93, 1990
7. N A J Hastings, J B Peacock, 'Statistical Distributions', Butterworth, 1974

Application of Model-Based Reasoning to the Maintenance of Telecommunication Networks[1]

Walter Kehl, Heiner Hopfmüller, Traytcho Koussev, Mark Newstead

Alcatel SEL Research Centre
Lorenzstr. 10, W-7000 Stuttgart, Germany

Abstract. This paper describes the application of model-based reasoning techniques to the maintenance of telecommunication networks. Model-based reasoning is used in all major steps of the overall maintenance process. The necessary knowledge is represented as structural knowledge (functional entities and their inter-dependencies) and as behavioural knowledge. The inference engine uses this knowledge for abductive and deductive reasoning. Experiences with a prototypical application (modelling the BERKOM network) are described and evaluated. In the light of these experiences new concepts and improvements are discussed.

Keywords. Model-Based and Qualitative Reasoning, Expert & Diagnostic Systems, Knowledge Representation, Telecommunication Management Network

0 Introduction

This paper discusses the application of expert system techniques to the maintenance of telecommunication networks and describes work done in the RACE project AIM[2]. The maintenance problem tackled here is the on-line corrective maintenance of hardware faults, but in the long run, the techniques and concepts developed in AIM should be an integral part of a TMN (Telecommunication Management Network) and applicable to any kind of faults. Maintenance for telecommunication networks is characterized by a vast number of fault reports which have to be correlated down to a small number of fault hypotheses and by the constraint that maintenance of any part of the network must not affect the overall working of the system. In AIM's approach to solve the maintenance

[1]The work described in this paper is not only that of the authors but bases on previous work done by all partners of the AIM project. Special thanks to J. Bigham for contributions to this paper. This work was supported by the Commision of the European Community under the RACE programme in the project AIM, R1006.

[2]RACE Project R 1006, AIM: AIP (Advanced Information Processing) Application to IBCN Maintenance

problem expert system techniques, especially model-based reasoning, have been applied and implemented in two prototypes.

1 The MAP-B Application

1.1 The Maintenance Task

The overall goal of the AIM project is to develop a Generic Maintenance System (GMS) for IBCNs[3]. Such a highly generic system which has to be independent of vendor- and technology-specific implementations can only be realized by a very clear separation between the generic procedural knowledge (i.e. the inference mechanisms and tools) and the specific declarative knowledge (i.e. the specific and explicit models of the different network elements of an IBCN).

The Generic Maintenance System (GMS) has been used in a prototypical application for the BERKOM network, a Broadband-ISDN field trial network. The overall maintenance task can be divided into four subtasks, which are connected together in the maintenance cycle[4] shown in figure 1:

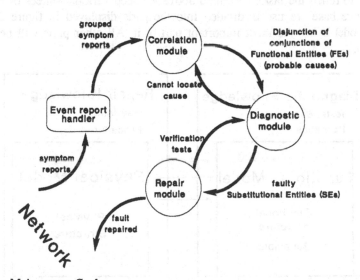

Figure 1. The Maintenance Cycle

Fault Detection: The event report handler accepts the observed symptoms from the telecommunication network, groups them together into batches and forwards these batches to the correlation module.

Fault Correlation: The generation of fault hypotheses. Telecommunication networks can

---

[3]"IBCN" stands for "Integrated Broadband Communication Network"
[4]Unfortunately it is not possible to describe the maintenance cycle with an example. It would require too many details ot the BERKOM system which are company confidential, too.

create a multitude of fault reports caused by the same reason. There must be a function which can cope with these large numbers of symptoms. Correlation performs this function and provides a list of fault candidates which can explain the aforementioned event reports.

Diagnosis (Fault Isolation): The verification of fault hypotheses. When multiple fault candidates exist, steps must be taken to verify whether the candidates supplied by correlation are faulty or not. The primary tool of this task is testing. Appropriate tests have to be determined by applying certain criteria (e.g. preference for automatic tests, minimization of network disruption, minimal costs, resource availability, etc).

Repair (Fault Correction). Once a faulty resource has been identified it has to be repaired. This normally involves replacement of the faulty component. One of the criteria for replacement is whether a component can be hot replaceable or not. In any case protective actions have to be taken to minimize disturbances of the rest of the telecom system. Once replacement has been carried out tests are performed on the network to ensure that the replacement was done successfully.

1.2 The Functional Model

In order to fulfill the tasks described above we need various sources of knowledge. The knowledge base we use is divided into the parts displayed in figure 2, with the functional model being the most important part of it. All other parts will be described further down.

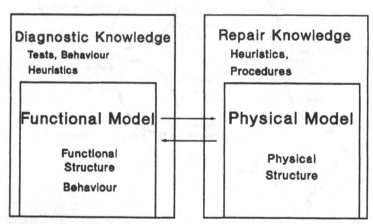

Figure 2. Structure of the knowledge sources

The functional model consists of structural information (the structural model) and behavioural knowledge. We describe here the functional model which we implemented for the BERKOM network. The BERKOM prototype exchange is a field trial network to gain experience with broadband subscriber equipment and broadband switching.

1.2.1 The Structural Model

The structural model is built out of functional entities (FEs) corresponding to specific functionalities of the modelled telecommunication network. A functional entity is for

Subscriber view

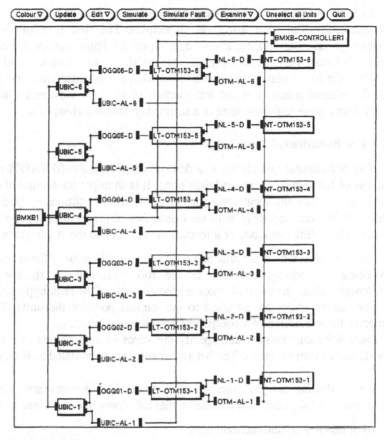

Gabriel CAD graph: Version 3.0

Figure 3. A part of the structural model

example a multiplexing functionality or the functionality of transmitting data between two FEs. There is a mapping between the FEs of the functional model and the elements of the physical model.

A functional entity (a unit) consists out of internal attributes and ports. The ports are connected to the ports of other functional entities, e.g. the power-out port of a modelled converter is connected with the power-in port of a modelled multiplexer-group. The functional entities together with these port-to-port connections are building the structural model. In figure 3 a part of the BERKOM structural model is displayed showing the powering of some multiplexers (UBIC) and their data connections to electrical-optical convertors (LT-OTM).

Several functional entities can be seen together as a higher-order functionality. This concept has been incorporated into the structural model as the organization of the functional entities on four different levels of granularity which are connected via a has-subfunctions/is-subfunction-of relation. The reasoning is done on the fourth level which has the finest granularity.

In order to define the functional entities a classification hierarchy has been set up. This hierarchy was built using two principles. The first principle is the distinction between generic and specific knowledge: while the upper part of the hierarchy reflects generic telecom classifications (e.g. the division into auxiliary and communication devices), the lower part introduces BERKOM-specific classes like special multiplexers, etc. The second principle is the introduction of different viewpoints into the structural model, like a powering viewpoint or a subscriber-oriented viewpoint.

1.2.2 The Behavioural Knowledge

The behavioural knowledge is a description of how the BERKOM network behaves in terms of functional entities and their ports. It is an important feature of the model-based approach that only the local behaviour of the functional entities is specified; the overall behaviour is given implicitly with the interaction of the local behaviour of all functional entities. Three different kinds of information are added to the structural model:

- *internal states* which are attached to the functional entities. These internal states (e.g. working/not-working) are seen as the root causes for a specific behaviour and statements about the internal states are the contents of the fault hypotheses.
- *port states* which are attached to the various ports of the units. These port states specify the interaction of a component with its neighbours.
- *local behaviour rules* which specify the states of output ports as a function of input ports and of internal states. The behaviour rules are formulated as if-then constructs.

The following simple example of a power line demonstrates the definition and interpretation of the rules. Let us assume that the power line is defined as follows:

```
(defclass power-line (generic-line)
    ((source-port :domain '(power no-power))
     (drain-port :domain '(power no-power))
     (myself-internal-state :domain '(working not-working))))
```

Each power-line has the two ports *source* and *drain*, which can be in the states *power* or *no-power*. The source port is connected to the power source (e.g. a converter) and the drain port to a power consumer. The internal state is represented by the attribute *myself* and can be *working* or *not-working*. Now the rules defining the behaviour in terms of these ports and the internal state:

```
(define-rule power-rule-1
    (if (and (working ?myself)
             (power ?source))
        (power ?drain)))
```

```
(define-rule power-rule-2
    (if (or (not-working ?myself)
            (no-power ?source))
        (no-power ?drain)))
```

Rule 1 says that if the line itself is okay and if power is coming in through the source port, then there is also power on the drain port. Rule 2 describes the fault behaviour: if either no power is coming in or the line itself is out of order then no power can be supplied at the drain port. Although these rules are locally very simple, in combination with a complex structural model they can result in very complex overall behaviour.

1.3 Diagnostic Information

The diagnostic information consists mainly of the models of the available tests, i.e. structural and behavioural knowledge. Such tests are e.g. a loop-back test or a reflectometer test for optical fibres. When a test is performed, the test functional element is connected to the "real" model (imitating e.g. the connection to loop-back test hardware) and the test symptoms are interpreted in the same way as the fault symptoms have been interpreted in correlation. Each functional element has knowledge about the tests it can use. Additional information like resource requirements and costs of a certain test is available for cases when a test has to be chosen among several possible tests.

1.4 The Physical Model

The Physical Model is a representation of all physical components of BERKOM. It has been built using the same modelling techniques as were used in the functional model. The main classification criterion is the common distinction between replaceable objects (printed boards, cables) and objects which can't be replaced (exchange, racks). The dependence between the modeled components are represented via the has-part/is-part-of and the is-connected-to relations. The interaction between the physical and the functional components is represented by correspondence links, which can be one-to-many or many-to-one. Additional repair knowledge is associated with the physical model.

1.5 Knowledge Acquisition

Knowledge Acquisition for MAP-B was made mainly using technical documents (functional and physical specifications, circuit diagrams, cabling lists etc.). Only a few exhaustive interviews were made. To arrange the material an informal strategy has been used: The first step was to use the cabling list and its categories of cables and the link information. So a rough physical and partly functional structure was achieved. The nodes to which the cables were connected to had been investigated step by step. By doing so completeness of the physical structure was achieved. This structure then was analysed according to categories such as clocking, powering, flow of data, control, functional decomposition and functional dependence. After all of the structural information had been acquired, the behavioural knowledge could be added.

As maintenance and reliability issues have a very high priority for telecommunication products our modelling could rely on a lot of already built-in maintenance functionality. This means that the granularity of the model does not need to be too detailed and that we don't have any "sensor problems". Instead we can use alarm reports which carry already some useful information. This means that behaviour rules can be formulated over value domains which are a limited set of discrete values.

1.6 Reasoning Techniques

The reasoning technique which is used for fulfilling the maintenance tasks is model-based reasoning. The inference engine consists of two main modules:

The suggestion interpreter. The suggestion interpreter applies the behaviour rules in an abductive[5] way to suggest possible causes for a symptom. In our example in 1.2.2 a symptom could be *(no-power ?drain)* for a power line *pl-1*. The suggestion interpreter uses the behaviour rules in an inverted way and fires the ones in which the consequent matches the symptom. Here power-rule-2 would apply and the first suggestion would be *(or (not-working ?myself) (no-power ?source))*. If in this suggestion there is a statement about an internal state of the unit, like *(not-working ?myself)*, it is kept as a valid hypothesis to explain the symptom. The other clauses of the suggestion describing states of input ports are used to generate a local symptom for the unit which is connected to the source port of *pl-1*. Here the same process starts again with this unit. In this way the suggestion interpreter walks along the connectivity information and collects all possible hypotheses explaining the original symptom.

The Model Interpreter. The model interpreter is used for deductive reasoning. It takes some hypotheses, e.g. *(not-working ?myself)* of *pl-1*, and uses the rules in the way they are written, so that it obtains the result *(no-power ?drain)*. The model interpreter is therefore doing a simulation which can start with any assumption, whether it is about an internal state or about any port state. The model interpreter is based on an ATMS[6] which acts as an internal memory. This is due to the fact that very often the same assumptions have to be considered again and again.

Applications. The two reasoning modules are used in the three major parts of the maintenance cycle, in correlation, diagnosis and repair. *Correlation* is done in two steps. The first one is reasoning from the observed fault symptoms to the assumed causes, done by the Suggestion Interpreter. The second step is the checking of the suggested hypotheses for consistency with the help of the model interpreter. This step is simulating the effects of the hypotheses, asking the question "Is this hypothesis consistent with the current status of the network?". The current status is represented by the in-service/out-of-service information of the "intelligent components". We call these components in this way because they have the ability to send out information about their status; of course we must as well formulate the intelligent components behaviour as part of the model. Only the suggestions which have been found consistent by the model interpreter are given out as a result of correlation.

In *diagnosis* also the two reasoning steps are necessary, but the observed symptoms which have to be explained are now test results instead of fault reports. In *repair* we use only the model interpreter in simulating the effects of a replacement. In this way all the affected components can be determined and it is possible to initiate protective actions.

Note that the inference engine can deal with multiple faults at the same time. For a

---

[5]"Abductive" reasoning means reasoning from effects to causes, as opposed to deductive reasoning going from causes to effects.
[6]ATMS - Assumption Based Truth Maintenance System, see [DEK 86]

more detailed account of the inference engine see [BIG 90a]; the underlying mechanisms are explained in [DEK 86] and [LAS 88].

2 Evaluation of Experiences

2.1 Modelling

The experiences from the prototype have shown that the unit-port representation is a very generic and useful format for modelling network functionalities. There were no difficulties to represent the structural model, the physical model or the test units with unit-port connections. The same holds for the local behaviour rules which are associated with the functional entities. We could formulate fault behaviour as well as working behaviour, behaviour for normal operation and for tests. Realizing the unit-port model in an object-oriented way has brought additional benefits:

- The inheritance mechanism allows a very easy separation of generic and specific knowledge. This holds for the structural as well as for the behavioural information. As we have acquired a corpus of generic telecom knowledge, it should be quite easy to model a new system by adding only new facts to the knowledge base and using inherited generic knowledge wherever possible.
- It is possible to incorporate different viewpoints of the modelled knowledge very easily by separating the whole knowledge into modules which can be seen separately as well as combined with other modules. Thus the model can be looked at at different levels of complexity and abstraction.

2.2 Knowledge Acquisition

We have seen that all relevant information of a telecommunication system like a telephone exchange, even for a prototypical network, is available in a more or less formal shape. If in the design process of a technical system the requirements of knowledge acquisition are taken into account, it is possible to avoid additional knowledge acquisition (which would be a sort of "reverse engineering") and to derive a knowledge base largely automatically during the design process. The main work in this area would be to define the structure of the information base and the different objects and their hierachy.

2.3 Model-based Reasoning

The center of our approach is model-based reasoning, although the design of our system does not preclude the use of heuristic or experiential knowledge. We think that this approach - as opposed to a approach using mainly production rules - is specially geared for the maintenance of telecommunication networks, because of several reasons:

- Telecommunication networks are usually quite large. It is easier to model such a large network with many simple and discrete facts than with a big corpus of compiled knowledge.
- Telecommunication networks are quite often installed in variants of a given basic system. Modelling of these variants is very easy with a deep model based approach and with the strict distinction between generic and specific knowledge. The same holds for changes to the system.

- The knowledge base which is the basis for inference is also accessible to other components of the system. Therefore the knowledge base can reflect always the current state of the telecommunication network (e.g. the actual connection between various subscribers) and the reasoning can use all the dynamic knowledge which is available.
- Another advantage of using deep knowledge is robustness, i.e. the ability to handle faults not explicitly foreseen,

2.4 Problems

One of the main problem areas is efficiency, which is for a maintenance system - which should in a realistic case operate under real-time conditions - an important issue. These problems occur only when calling the model interpreter but they are due to the fact that we use the model interpreter in "full simulation mode". That means that we compute all port states which can be deduced from an assumption. But when using the model interpreter in the consistency checking in correlation, it could stop immediately after the first inconsistency has been found. With this and other efficiency improvements which we are currently implementing we are confident that we won't get into troubles with the exponential complexity of the ATMS (see [SEL 90] and [FOR 88]).

To give an impression of the dimensions we are talking about, here some rough numbers: for simulating typical events with only local consequences the model interpreter has to compute about 20 port states and needs about 1.5 seconds. This increases: for the computation of 200 port states it needs 50 seconds. But if the size of the simulation increases again to about 700 port states, the time goes up only to 60 seconds[7].

3 Further Work

A great deal of work has to be done to integrate our work into an overall Telecommunication Management Network (TMN). To discuss all this is out of the scope of this paper, for a detailed discussion see [NEW 91]. We want to concentrate here more on the topics which are directly connected with model-based reasoning. We see the following important issues for further work:

Other forms of reasoning. It is clear for us that the core of the inference engine should be model-based, but also that it should be complemented by other forms of reasoning. So it should not exclude principally reasoning with heuristic knowledge or reasoning with experiential data, e.g. for transient faults.

Second order maintenance. We speak of second order maintenance when we are reasoning about the built-in maintenance functionality (test circuits, alarm lines, controllers, etc). To take into account that these components, from which we get our symptoms, can fail itself leads to having different modes of reasoning.

Temporal and Uncertain Information. A maintenance system must make use of temporal and uncertain information. There is an uncertainty module integrated into the inference engine, but it has not been used in the prototype (see [BIG 90b]).

---

[7]The background of these numbers is our model which consists of about 1000 instances with (estimated) 8000 connections between them. It is implemented in CLOS in the Harlequin LispWorks environment and runs on SUN Sparcstations under X-Windows.

Development Environment. Seeing the aspect of acquiring behavioural knowledge, the inference engine should be extended to a kind of development environment for behavioural knowledge. Therefore a set of tools should be available, like a stepper, a tracer, etc. These tools could as well be the basis of an explanation component.

Knowledge compilation. One way of achieving more efficiency could be the compilation of the knowledge base. Different approaches to compilation are possible, from the complete off-line compilation to the on-line storage of already executed abductions and deductions in the model.

4 Conclusion

The experiences with MAP-B have shown that the basic concepts and techniques developed within AIM are appropriate to solve the maintenance problem in a generic way. Model-based reasoning together with object-oriented modelling makes it possible to develop a Generic Maintenance System (GMS) which can operate on different specific knowledge bases. Problems which were identified during the implementation of the prototype are mainly in the area of efficiency, but solutions are foreseeable. Whereas a lot of improvements need to be done, we see as the next step the complete integration of our concepts and techniques into the "intelligent MIB" (see [NEW 91]. We are confident that also the other management functions inside the TMN could be realized using model-based techniques. We could arrive at a point where the overall presence of uniform and generic concepts help to cope with the complexity of broadband telecommunication networks.

References

[BIG 90a]   Bigham, J., Pang, D. and T. Chau (1990). Inference in a Generic Maintenance System for Integrated Broadband Communication Networks, Proceedings of the fourth RACE TMN Conference, November 1990, 199 - 206

[BIG 90b]   Bigham, J. (1990). Computing Beliefs according to Dempster-Shafer and Possibilistic Logic. Third International Conference on Information Processing and Management of Uncertainty in Knowledge Based Systems, Paris.

[DEK 86]   de Kleer, J. (1986) An assumption based TMS, Artificial Intelligence, Vol 28, 127 - 162

[FOR 88]   Forbus, K.D. and J. de Kleer, Focussing the ATMS, Proceedings of the AAAI, August 1988, 193 - 198

[NEW 91]   M. Newstead and H. Hopfmueller, G. Schapeler, A Design of the Operation, Maintenance and Construction of an Intelligent MIB, Fifth RACE TMN Conference, November 91

[LAS 88]   Laskey, K.B. and P.E. Lehner, Belief maintenance: an integrated approach to uncertainty management, Proceedings of the AAAI, August 1988, 210 - 214

[SEL 90]   Selman, B. and H.J. Levesque, Abductive and Default Reasoning: A computational core, Proceedings of the AAAI, 1990, 343 - 348

Advanced Information Modelling for Integrated Network Management Applications

Mark A. Newstead, Bernd Stahl, Gottfried Schapeler.

Dept. ZFZ/SW2, Alcatel SEL, Research Centre,
Lorenzstr. 10, D-7000 Stuttgart 40,
Federal Republic of Germany.

Abstract. The current trend in user requirements for network management proceeds towards integrated network management. That is, users of telecommunications services want a uniform management view of, and want to subscribe to, specific sets of network management functionality over the whole network they are using. A prerequisite to this uniform network management view is a uniform information model of the network. The object-oriented approach is currently used and standardized for this information modelling. This paper presents such concepts together with appropriate AIP[1] technology, especially model based reasoning, for carrying out this information modelling effectively and efficiently.

Keywords. RACE[2], AIM[3], Object-oriented Paradigm, Object, Managed Object, Network Management, TMN[4], MIB, Managed Resource, Model based reasoning.

1 Introduction

This paper will attempt to outline the requirements by the telecommunications industry for the application of AIP techniques to TMNs (Telecommunication Management Network). Relevant AIP techniques will then be explained and their applicability to the TMN shown. This will be followed by a description of prototype TMN applications developed at the SEL-Research centre, using the AIP techniques mentioned.

---

[1] Advanced Information Processing. An acronym used in RACE and ESPRIT to describe modern information processing technologies including Distributed Processing and Databases, Real Time and Distributed Knowledge Based Systems, Advanced Processor Architectures, etc.

[2] RACE (Research and development in Advanced Communications technologies in Europe).

[3] RACE TMN Project R1006 AIM: AIP Application to IBCN Maintenance.

[4] Telecommunication Management Network.

Future telecommunications networks will be characterized by the interconnection of a whole range of multi-vendor equipment using a large number of different technologies providing constantly increasing sets of services to telecommunications service users. Besides proper usage of their subscribed services, users will require facilities with which they can manage all hardware / software resources they are using for their telecommunication services. Therefore, a certain defined portion of network management functionality will be offered as an additional service to telecommunication service users. Furthermore, users of these network management services are requiring a uniform view to the equipment and networks they are using and partly managing.

These user requirements to network management leads to the definition of *Integrated Network Management,* which is defined as a the ability to manage all resources contributing to network communications uniformly, in terms of the user interface and the network capabilities, regardless of the architecture of the network management solution. Integrated Network Management builds upon *Interoperable Network Management,* which is defined as the specification and application of the means by which network management products and services from different suppliers can work together to manage communications and computer networks. The key aspect of Integrated Network Management is the uniform user interface, whereas uniform communications is the key aspect of Interoperable Network Management. Lots of standardization is going on in this area, both within international standardization organizations (ISO, CCITT, etc.) [7] and vendor groups (OSI/NMF, etc.) [8].

Essential for Integrated and Interoperable Network Management is a uniform information model for the relevant network management information to be exchanged and presented to the network management service user. Object-oriented and semantic association approaches are used to carry out this uniform information modelling. Model based reasoning techniques are then applied to the models, providing higher level, uniform, management interaction capabilities. AIP technology is currently been developed for the production of tools and products that support these advanced information modelling techniques.

2 Principles of Advanced Information Modelling of Network Management Information

2.1 The Object-oriented Paradigm

The current trend in designing network management systems is that they are to be designed in an object-oriented way, and implemented using object-oriented tools. In an object-oriented system everything is an object in the sense of the object-oriented paradigm: information, applications, man-machine-interface, communication messages. However, object-oriented techniques have up to now mostly been used during the design of network management information. The object-oriented approach encompasses at least the following nine well-known principles for software development:

a) **Abstract Datatypes (ADT)**: an ADT is the combination of a set of data structures plus a corresponding set of operations that can be performed on these data structures.

b) **Information Hiding:** implementation details are not visible to the outside world; the object-oriented paradigm uses this principle in that it makes only the names of the object's operations and public data-structures visible to the outside world, the implementation of the operations and internal data structures are not visible.

c) **Data Encapsulation**: data is protected from direct access, i.e. data structures within an object can only be manipulated via the object's operations.

d) **Communication** between objects is carried out by messages. These messages access the visible operations of the message receiving object, plus if necessary the data structures that are affected by this message.

e) **Classification**:   Similar objects are classified into object classes. By this classification an ontological order is imposed on the objects that occur and are processed in a system. The objects of class are also called instances. The class of an object instance defines the structure of this instance: i.e. its data and its operations.

f) **Meta Class** object protocol allows classes to be treated as instances with respect to their meta class. Thus allowing class level communication and data encapsulation.

g) **Inheritance**: In order to achieve modular system components, class definitions of one class are to be used in the definitions of other classes. This is carried out via inheritance. Inheritance imposes a structure on the construction of object classes. Subclasses can be defined by expansion or modification of existing classes. This process is called specialization, as the derived classes are specifying their instances more specific than the used more general classes. A class can have more than one superclass, this is referred to as *multiple inheritance*.

h) **Polymorphism/Dynamic Binding**: polymorphism is the capability to generate references (during message passing) to objects of which their particular class is not yet known during software definition or translation time, but is computed during runtime.

i) **Instantiation**: Instantiation means the generation of new object instances during runtime of an object-oriented software system. The structure and the behaviour of the generated object instance is defined by the class of this instance.

2.2   Semantic Models

Object oriented models characterize real world systems through the definition of classes and instances, the forming of inter-relationships between instances, and attributing behaviour to classes of objects. Semantic modelling describes the attachment of semantic meaning to instance inter-relationships thereby forming semantic associations. For example, power supplies send power to voltage converters, voltage convertors send power to clock distributors, clock distributors send clock signals to multiplexers, etc.

The attachment of semantics to relationships is achieved by providing a domain of applicability for the relationship. For example, a power relationship can either propagate power or can propagate no power (or even some pre-determined state in-between).

The behaviour of an object can be explained by describing how it interacts with the outside world and its internal states. Thus the behaviour of a clock distributor can be described by stating that if it is receiving power and is working, then it is able to send out clock signals.

2.3   Model based reasoning

Network management systems are highly intricate systems handling complex interactions of network information. There exists a strong desire to formulate complete solutions to network representation and their management interaction. Therefore Advanced Information Processing (AIP) techniques are required to represent and reason about such structures in a declarative, understandable way, supporting the process of design, maintenance, management and usability of such networks.

Model based reasoning is about the construction of models to represent and understand systems whose component interaction is structured, but highly intricate [12]. Thus it is a suitable AIP technique for representing network management systems.

Given an object oriented paradigm, a way to describe the semantic association between objects and a way to describe the behaviour of an object, the principles of model based reasoning [1] can be applied. The main principles of model based reasoning are as follows:

a) **Unit**:   an object which describes the characteristic of a resource in terms of heuristics, behaviours, ports, and internal states.

b) **Port**:   the interface points of a unit to the outside world. A port has a domain describing the semantics of the connection to other ports of other units.

c) **Connection**:   the semantic relationship between two ports of two units.

d) **Behaviour**:   constraints / mappings between input / output ports and internal states of a unit.

e) **Heuristic**:   a description of a units' operation as defined by experts and not by observable characteristics of the unit in question.

f) **Compositional**:   the description of a systems behaviour must be derivable from the structure (connection between units) of the system and associated local unit behaviours.

g) Locality: Behaviours must describe effects (causality) which propagate locally through specified connections.

2.4   Overall Advantages of principles used

The AIP techniques above offer many advantages [9] over conventional software practices and offer much requested capabilities to TMN design, representation, construction, maintenance, and use. Below is a list of some of the many benefits on offer.

- **a) Explicit**   definition of resources modelled.
- **b) Constructive**: can build upon and inherit from, generic (standards based) TMN model libraries.
- **c) General**   domain independent reasoning systems.
- **d) New technology**   resources can be modelled - even if no expert experience is available.
- **e) Multiple**   faults can be handled.
- **f) Novel**, unspecified, overall network behaviours can be realized from the interaction of localized behaviours (reasoning from first principles).
- **g) Natural**   representation of structure and semantic association.
- **h) Competence** is better handled, as no requirement to use unproven expert heuristics.
- **i) Second generation** technology is used, which is a major advancement over first generation rule based expert system technology.
- **j) Modification** of the model is supported by the locality principle, as all changes are local and so do not effect global consistency.

3   Application of Advanced Information Modelling to Network Management

3.1   The Managed Object

Management systems exchange information modelled in terms of managed objects. Managed objects are software representations of telecommunications resources. Besides its "normal telecommunications" functionality a resource provides an interface of some arbitrary kind to management. The managed object suppresses irrelevant detail of the resource and provides a "standardized" interface for the represented resource to a management application function.

The correspondence between the real resource and its managed object(s) can be many-to-many. Furthermore, managed object may provide different views of the same resource or set of resources on different levels of abstraction. An abstraction always suppresses irrelevant details. Managed object classes are organized within an inheritance tree [6]. Managed object instances are organized within a semantic graph, covering a multitude of relationships like containment, power, clocking etc. Essentially, four different qualities are described concerning managed objects:

1. **Attributes** define managed object states or other properties and are represented as single or multiple-valued variables of arbitrary complex type.
2. **Operations** define the management operations that can be issued against the managed object and its corresponding resource(s) from some management application function.
3. **Notifications** define events that are received by some management application function from a managed object and its corresponding resource(s).
4. **Behaviour** defines the dynamic behaviour of a managed object and its corresponding resource(s).

These qualities are used to describe the information that is presented to the management service user, the information that is exchanged between open network management systems, and information that management application functions may work with.

3.2 The Management Information Base

Management information can be grouped into what is known as the Management Information Base (MIB) [2]. This term has been widely accepted by the standards making and similar bodies, viz. ISO with OSI, ANSI in T1M1, OSI Systems Management Forum and ETSI in NA4 [7,8]. This is the central repository of all knowledge and data held in the TMN about the network and the TMN itself. The information base is required to store information on network and system configuration, customers and services, current and historic performance and trouble logs, security parameters and accounting information.

It is proposed that such knowledge is represented in a form which is amenable for processing by model based reasoning systems, the principles of which are mentioned earlier. Such a representation allows reasoning about the MIB itself, [3] thus allowing user / application access to the task specific services provided by such reasoning systems (eg. simulation, inference, etc.).

Knowledge about the network resources is stored as a managed object within the MIB. A managed object can be viewed as the interface between a resource and management application functions which use the resource.

Object oriented model based reasoning is a powerful and highly generic representational and processing technique. It allows the MIB to define and use managed objects of sufficient capability to provide a uniform high level view of network resources to users, applications and distributed parts of the MIB. The MIB can provide for a management application function one or more managed objects which reflect particular views about a resource enabling and supporting the application function to manage the resource without knowing the detailed functionality and architecture of the resource.

4   Applications using Advanced Information Modelling techniques

4.1   Fault Management

In the RACE project AIM we have successfully applied object oriented and model based reasoning techniques to the area of fault management of two networks; a broadband field trial network [4] and a MAN (Metropolitan Area Network). Clear, highly declarative, functional and physical object oriented models of the networks have been defined. Localized network element resource behaviour is captured and used to define the overall behaviour of the networks.

Model based reasoning techniques are applied using the semantic connections between the telecom objects and their associated behaviours, allowing simulation and fault management tasks (alarm correlation, diagnosis, and repair) to be performed on the two networks mentioned above. A much more detailed account of the application of model based reasoning to fault management can be found in a sister paper in these proceedings called "Application of Model-Based Reasoning to the Maintenace of Telecommunication Networks" [10]. Our work in this area expands on the work by De Kleer [13] on multiple fault diagnosis, which is currently a major problem for TMN fault management systems.

Using object oriented model based reasoning techniques greatly facilitates updating of the knowledge base; due primarily to the locality principle (ie. all behaviours must describe effects which are local to the modelled resource). So any change in resource functionality (due to technology upgrade for instance) results in minor, localized changes to the model resource description and no change to any of the reasoning components.

4.2   Gabriel

In order to facilitate the construction of the Fault management models described above in section 4.1 a knowledge acquisition / representation environment was developed, called 'Gabriel' which stands for *G*raphical *a*cquisition, *b*rowsing and *r*easoning of *i*ntelligent *el*ements. It is an experiment to determine the requirements of graphical front ends to MIBs supporting object oriented model based reasoning systems. Hence, it must support all the object oriented principles mentioned in this paper and also the principles of model based reasoning.

Network representations have very complex semantic interactions and must represent large numbers of resources. Thus Gabriel must solve the joint problems of complexity and size. In the following sections some of the more important features of Gabriel are explained. The screen dump below shows Gabriel displaying three viewpoints of the broadband field trail network model, as mentioned earlier. It should be pointed out that this model contains around one hundred classes and about one thousand class instances, with over ten thousand semantic interconnections.

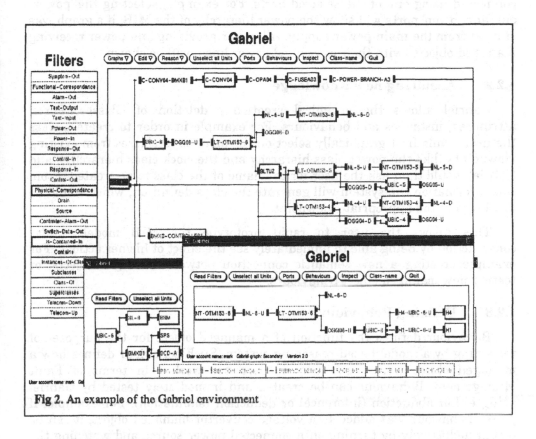

Fig 2. An example of the Gabriel environment

4.2.1 Filtering out unwanted knowledge

Because of the complexity of the MIB and managed objects, users must be given the ability to not view information which is not directly relevant to the type of information they wish to browse. This can be achieved by filtering out unwanted information, and is an important part of defining a viewpoint. A viewpoint is defined as a portion of the MIB which contains managed objects with some common semantic relationship (ie. power, clock, subscriber, containment, etc.).

Managed Objects have many functionalities (power, clocking, subscriber info., etc.), these functionalities are usually visable in the way the managed object semantically interacts with its peers. For example, a multiplexer power receiving functionality will have a power connection to its power source. Such connections are clearly identifiable through the use of ports. A port defines the point of connection between two managed objects.

Ports can be used as a means of filtering out unwanted knowledge, as users can specify what ports to use as filters when a graph is shown. The graph will recursively show all children of a managed object which are connected using one of the selected ports. For example, selecting the 'power-out' and 'drain' ports will show the power hierarchy of the MIB, if a graph was started from the main power supply. All power providing and power receiving managed objects (with their associated power lines) will be shown.

4.2.2 Acquiring new Knowledge

Gabriel allows the graphical creation / deletion of Classes, Ports, attributes, instances and behaviours. For example in order to create a class the user would first graphically select one or more superclasses from existing viewpoints (like the power class hierarchy and the clock class hierarchy), and Gabriel would then ask the user for the name of the class to be created. From this information the system will generate the class definition, which would be displayed in the appropriate viewpoint.

This allows the user to rapid prototype the MIB models under construction by being able to immediately see the effect of his/her actions. For example creating a new semantic connection between managed objects, or testing new managed object behaviours.

4.2.3 Defining Behaviour

Behaviour defines the function of a managed object for the purposes of reasoning by a model based reasoning system. Such behaviour defines how a managed object works, and why it does not work - in terms of Fault Management. Behaviour can be created and immediately tested by Gabriel using either abduction (inference) or deduction (simulation). For example, if power behaviour was added to a voltage convertor managed object, it can be tested deductively by turning off a connected power source and watching the effect on the convertor (ie. it should go out of service and change colour). It can also be tested abductively by asking the reasoning system why it is not sending out power, which it should reply that an upstream power source, or the convertor, is faulty.

4.2.4 Implementation environment

Gabriel is implemented in CLOS (Common Lisp Object System) using the LispWorks programming support environment by Harlequin [11]. Gabriel currently runs on a SUN Sparc machine running unix and the X windows (X11r4) graphics environment.

5   Conclusion

The use of advanced information modelling to allow management application functions to receive a unified view of the modelled network is desirable and possible using object oriented model based reasoning techniques. Also such techniques have been shown to be highly successful in the area of fault management and the definition of high level graphical user interfaces.

The structure of the network lends itself to such advanced information modelling techniques, as both require the representation of complex semantic interactions of resources. Therefore model construction appears as a natural reflection of the network, enhancing, not detracting, from the original network design.

Such techniques are also highly declarative and explicit, enabling a greater degree of reliability and understandability of the model and hence the MIB itself.

Such techniques do not rely on unprovable expert heuristics (rules), but on detailed descriptions on the behaviour and interconnections of the resources themselves, enabling reasoning from first principles, and are thus much more reliable and consistent than first generation rule based techniques.

In this paper, and in our ongoing research, we are using state of the art advanced information technologies to enhance the capabilities of current and future network management systems. Enabling a better understanding of the networks under management.

We maintain that object oriented and model based reasoning techniques are crucial in the development of advanced network management systems. Therefore application specialists should become familiar with object oriented and especially model based reasoning technologies and techniques, as we feel certain that this is the technology of the 1990's. It should be stressed that we have applied only a fraction of the AIP techniques on offer, much more work is required researching into the areas of qualitative and temporal reasoning and applying such model based reasoning techniques to all areas of network management.

Acknowledgements

We wish to thank our partners in the RACE R1006 AIM project, whose help and support enabled us to define a framework for advanced information modelling for integrated network management functions. Especially Dr. John Bigham and his colleagues at Queen Mary and Westfield College, London, who designed and constructed the model based reasoning system [5] mentioned in this paper.

References

1. Daniel G, Bobrow, et al. Artificial Intelligence - Journal volume 24 Numbers 1-3 December 1984, special volume on Qualitative Reasoning about physical systems. Introduction.
2. Williamson G., Callaghan J., Cochrane D., Schapeler G., Stahl B., Galis A. Implementation Architecture for TMN - Intermediate Version, April 1991, RACE Project Guideline.
3. Newstead M., Schapeler G., Hopfmueller H. A design of the Operation, Maintenance and Construction of an Intelligent Management Information Base, Nov. 1991, 5th RACE TMN conference.
4. Kehl W., Newstead M., Schapeler G., Koussev T., Hopfmueller H. Conclusions from the BERKOM Maintenance Prototype and Recommendations for future Maintenance Systems, 5th RACE TMN conference Nov. 1991.
5. Bigham J., Pang D., Chau T. Inference in a Generic Maintenance System for Integrated Broadband Communication Networks, Proceedings of the 4th RACE TMN conference, November 1990.
6. Stahl B., Azmoodeh M., Knowledge Representation of Networks in the RACE Project AIM, Proceedings of the 4th RACE TMN conference, November 1990.
7. Sub committee ISO/IEC JTC/SC21/WG4, document number ISO/IEC DP10164. Information Retrieval, Transfer and Management for OSI, parts 1 to 8.
8. OSI Network Management Forum Object Specification framework, Issue 1 , Sept. 1989.
9. Struss P., Model-Based Diagnosis of Technical Systems, Tutorial Program, European Conference on Artificial Intelligence, August 6-10, 1990.
10. W. Kehl, H. Hopfmüller, T. Koussev, M. Newstead. Application of Model-Based Reasoning to the Maintenance of Telecommunication Networks. The Fifth International Conference on Industrial and Engineering Applications of Artificial Intelligence and Expert Systems (IEA/AIE-92), Paderborn, Germany.
11. LispWorks, a Common Lisp programming environment. Vendor: Harlequin Ltd, Barrington Hall, Barrington, Cambridge CB2 5RG, England.
12. R. Davis. Diagnostic reasoning based on structure and behaviour. Artifical Intelligence, 24:347-410, 1984.
13. J. De Kleer and B.C. Williams. Diagnosing multiple faults. Artificial Intelligence, 32:97-130, 1987.

An Integration of Case-Based and Model-Based Reasoning
and
its Application to Physical System Faults†

Stamos T. Karamouzis*   Stefan Feyock**

Department of Computer Science
College of William & Mary
Williamsburg, VA 23185
U.S.A

Abstract. Current Case-Based Reasoning (CBR) systems have been used in planning, engineering design, and memory organization. There has been only a limited amount of work, however, in the area of reasoning about physical systems. This type of reasoning is a difficult task, and every attempt to automate the process must overcome the problems of modeling normal behavior, diagnosing faults, and predicting future behavior. We maintain that the ability of a CBR program to reason about physical systems can be significantly enhanced by the addition to the CBR program of a model of the physical system to describe the system's structural, functional, and causal behavior. We are in the process of designing and implementing a prototypical CBR/MBR system for dealing with the faults of physical systems. The system is being tested in the domain of in-flight fault diagnosis and prognosis of aviation subsystems, particularly jet engines.

1 The Problem

We consider a physical system to be a set of components connected together in a manner to achieve a certain function. *Components* are the parts that the system consists of, and may themselves be composed of other components. For example, an engine is a component in an airplane and it is composed of other components such as a compressor, a combustor, a fan etc. Components which are composed of other components are called *subsystems*.

Reasoning about physical systems is a difficult process, and every attempt to automate this process must overcome many challenges. Among these are the tasks of generating explanations of normal behavior, fault diagnoses, explanations of the various manifestations of faults, prediction of future behavior, etc. The reasoning process becomes even more difficult when physical systems must remain in operation. During operation, a physical system is changing dynamically by modifying its set of components, the components' pattern of interconnections, and the system's behavior.

---

†   Work Supported by NASA grant NCC-1-159
*   stamos@cs.wm.edu
**   feyock@cs.wm.edu

Explaining normal behavior is the process of elaborating the function of each subsystem and how this function contributes to the overall operation of the system. Explaining the operation of an automobile, for example, would require knowledge of the function of the carburetor, operation of the fuel pump, movement of the wheels, etc., and how all these affect each other and contribute to the final operation of moving the automobile. There are several approaches to explaining the normal behavior of physical systems by means of a model of the system. These approaches include naive physics [9], qualitative physics [4, 8, 10], bond graphs [13, 5], causality models, and others, each of them achieving various degrees of success and various advantages over the others.

Fault diagnosis is the process of explaining why the behavior of a system deviates from the expected behavior. Such diagnoses are the answers to the questions "Why has my watch stopped?" and "Why were the lights flickering after yesterday's storm?" Fault examples include a broken spring, a dead battery, a leak in a fuel line, etc. The task of diagnosis presents particular challenges such as identifying the faulty component, taking into consideration fault propagation, and accounting for multiple faults.

A number of systems have been developed to deal with these problems. Such systems fall into two categories. *Associational* or *shallow-reasoning systems* are systems that do diagnosis based on predefined links between sets of symptoms and pre-existing explanations [2]. These systems are fast but inflexible, since their lack of deep domain knowledge makes them incapable of dealing with problems outside their preset rule bases. *First-principle* or *deep-reasoning systems* use causal reasoning to produce explanations for the set of symptoms [3]. These systems are more flexible, but are slower, since they must derive each new diagnosis from the underlying model.

In maintenance diagnosis, i.e. diagnosis of physical systems not in operation, it is sufficient to identify the source of the problem (faulty component) in order to determine which component(s) need to be repaired. In domains where the system is in continuous operation, however, it is desirable that the system operators be aware of fault consequences in order to facilitate corrective actions. A pilot who observes abnormal behavior in the plane's sensor values needs to know not only what the fault is, but also how the fault will propagate and what its subsequent effects will be.

Automating the process of predicting the future behavior of physical systems is a difficult task because physical faults manifest themselves in various ways and it is difficult to enumerate all possible consequences. Current efforts to incorporate prognostication features in diagnostic systems that reason from physical system models succeed in predicting the expected course events but are limited by the level of detail of their models [6]. For example, a model-based reasoning (MBR) system that has a model of an airplane's functional and physical connections among components may, after establishing that the fan in the left engine is the faulty component, predict that the fault will affect the operation of the compressor since there is a functional link between the two components. Such a system is incapable, however, of deducing that flying fragments from the faulty fan may penetrate the fuselage and damage the right engine. Humans, on the other hand, are good at making such predictions, since their reasoning is based not only on pre-existing models of the world, but also on previous directly or vicariously experienced events which remind them of the current situation.

2 Approach

The approach we are taking is a novel methodology for dealing with physical systems in operation, and involves the use of case-based techniques in conjunction with models that describe the physical system. Case-Based Reasoning (CBR) systems solve new problems by finding solved problems similar to the current problem, and by adapting solutions to the current problem, taking into consideration any differences between the current and previously solved situations. Because CBR systems associate features of a problem with a previously-derived solution to that problem, they are classified as associational-reasoning systems.

We are employing a case-based reasoning methodology for fault diagnosis and prognosis of physical systems in operation. A hybrid reasoning process based on a library of previous cases and a model of the physical system is used as basis for the reasoning process. This arrangement provides the methodology with the flexibility and power of first-principle reasoners, coupled with the speed of associational systems. Although domain independent, the proposed work is being tested in the domain of aircraft systems fault dia- and prognosis.

In contrast to other CBR research efforts, each case in our methodology is not only a set of previously observed symptoms, but also represents sequences of events over a certain time interval. Such temporal information is necessary when reasoning about operating physical systems, since the set of symptoms observed at a particular time may represent improvement or deterioration from a previous observation, or may reveal valuable fault propagation information. In a jet engine, for example, the fact that the fan rotational speed was observed to be abnormal prior to an abnormal observation of the compressor rotational speed is indicative that the faulty component is the fan and that the fault propagated to the compressor, rather than the reverse.

The model represents the reasoner's knowledge of causal relationships between states and observable symptoms, as well as deep domain knowledge such as functional and physical connections among the components of the physical system about which the reasoner must reason. Our research alleviates the knowledge acquisition problem to which current model-based systems are subject by letting each case of the CBR reasoning mechanism contribute its causal explanation, gained from adapting previous incidents, to the formation and maintenance of the causality model. The model can therefore be considered as a general depository of knowledge accumulated through time. In return the model aids the matching and adaptation processes of the CBR reasoning mechanism.

3 Methodology

The described research integrates case-based and model-based reasoning techniques for dealing with physical system faults. In order to demonstrate the challenges and benefits of such work a prototypical system is being designed and implemented in the aircraft domain. The system contains a self-organizing memory, as defined by [14], for storing previously encountered problems. Initially each case has been represented in a memory organization packet (MOP) as implemented in [12], but eventually MOPs will be implemented using LIMAP, a matrix-based knowledge representation tool [7].

Each case represents an actual aircraft accident case and consists of a set of features that identify the particular accident, a set of observable symptoms, and a causal explanation that describes the relationship between various states and observable features. The set of identifying features includes information such as aircraft type, airline, flight number, date of the accident, etc. The set of symptoms includes information about abnormal observations from mechanical sensors or "human sensors" such as the value of the exhaust gas temperature, the value of engine pressure ratio, the sound of an explosion, or the smell of smoke in the passenger cabin. These symptoms are presented in a sequence of symptom groups. The inter-group time intervals are of unknown and uneven length; it is their ordering that it is of importance.

Additionally, the system incorporates a model, called the *world knowledge model*, that consists of deep domain information such as the physical and functional dependencies between the components of the physical system, and causal information describing the transitions between various states of the physical system, as well as dynamical (sub)models of the physical system. Along with the causal information between two states, e.g. "inefficient air flow" and "slowing down of the engine", the model maintains a frequency count of the number of times that the system witnessed that inefficient air flow caused the engine to slow down. The physical and functional connections include information of the type "the Fan is connected to sensor N1 via a functional link", "the Fan is physically connected to the compressor", etc. The causality knowledge of the world model contains information such as "fan-blade separation causes the rotational speed of the fan to fluctuate" and "the rotational speed of the fan causes the engine pressure ratio to fluctuate."

When the system experiences a new set of symptoms it searches its case library for the most similar case. Based on the observation that in most cases similar faults manifest themselves in similar ways only during the first moments of the fault occurrence [1], the system takes advantage of the available temporal information in each case, and will try to establish similarity based on the observable symptoms during the first moments of the fault occurrence. It is likely that input cases do not exactly match any previous cases in memory. To allow for a partial match between an input case and a library case the system consults the world model. The model aids the matching process in dealing with features that appear different on a superficial level, but are accounted for by the same initial cause. For example, if during the current incident (case) it is reported that "engine pressure ratio is fluctuating" then using the causality knowledge of the model the system can match the current case with a library case that includes the related feature "rotational speed of the fan is fluctuating."

If the system finds and retrieves a similar case, the causal explanation of the retrieved case is adapted to fit the current case, and stored in the case library for future usage. The system is provided with a set of adaptation rules which, in addition to adapting the retrieved causal explanation to fit the current case, find possible gaps in the causal explanation and fill in the missing causalities. This causal explanation connects the symptoms to a justifying cause, and thus the system's causal reasoning ability produces a causal analysis of the new case, rather than simply a reference to a previous solution. The new causal analysis is not only stored in the case library as part of the input case, but is used to augment and modify the causality knowledge of the world model. The causal analysis consists of a sequence of pairs of the type "event A causes event B", "event B causes event C" and so on. Each of these pairs is stored in the causality section of the model. In the case that the model already knows about the causal relation between two events from a previously seen case, the system updates the frequency count between the two events. The world model is therefore created based on the previous behavior of the physical system, and is constantly updated based on the current behavior, either by augmenting its previous causal knowledge or "becoming more sure" about causal relations.

When the retrieved case fails to explain some of the observed symptoms the system can stand by for additional symptoms, or it can search the case library again for an additional match that will explain the remaining symptoms. If a new case is found, its causal explanation is used as if these symptoms were the result of another fault. Using the model the system attempts to establish a relation between the two faults by searching for a causal, structural, or functional link between them. If no link is found it is assumed that the system is experiencing multiple faults.

Constant consultation of the model gives the system its prognostication ability. For example, having achieved a match of the current situation with a previous case where the faulty component was a bad fuel controller, the system will hypothesize that the same fault is occurring. By referencing the world model it is able to predict that an engine flameout may occur, although that did not happen in the retrieved case, because the model may have recorded at least one previous instance where this happened. The operator is provided with a list of possible consequences of the fault along with a frequency count of each one. The following figure is a diagram of the various modules involved in the proposed reasoning system along with their interactions.

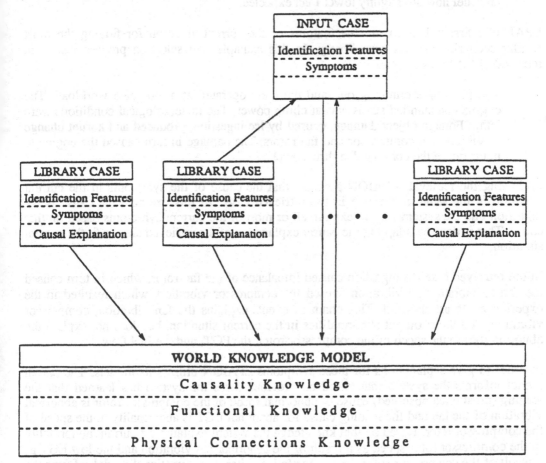

Fig. 1. EPAION's various modules and their interactions

In addition to the use of the described causal, functional and physical dependencies models that describe domain information through associations, an attempt will be made to investigate the required techniques for integrating CBR with other models. For example using a model that describes its domain via equations requires the CBR component to employ different techniques during the matching process. The behavior of the equations contained in such models is predicted from the equations' initial conditions. As soon as the CBR component solves a case the model should be refined by determining the intervals of initial variables for which the model's equations would give the same solution. The matching criterion would not be identical initial conditions, but conditions in the same interval.

4 An Example

The ideas presented here can be demonstrated in terms of the following simple example. We assume that EPAION is given the following data:

> The plane is climbing out, with the crew operating at moderate workload. The engine commanded status is at climb power. The weather is icing. The crew observes a small thrust shortfall and vibration in the compressor and fan rotors. The compressor rotor speed shows a 5% shortfall. The exhaust gas temperature (EGT) and fuel flow are slightly lower than expected.

EPAION's first task is to use the features of the current situation for finding the most similar scenario from its case library. In this example, the selection process results in retrieval of the following case.

> The plane was climbing out, and the crew operated at a moderate workload. The engine commanded status was at climb power. The meteorological conditions were icing. Foreign object damage, caused by ice ingestion, produced and abrupt change of vibration in compressor and fan rotors. The damage in turn caused the engine to not produce the commanded thrust level.

Following the retrieval, EPAION assumes that the cause of the symptoms in the current situation is the same as the one in the retrieved case. In this example the cause is ice ingestion and the system tries to explain all or most of the current symptoms based on that cause. This is done by adapting the causal explanation of the retrieved case to fit the current situation.

In the retrieved case ice ingestion caused imbalance of the fan rotor, which in turn caused the fan to vibrate. Fan vibration caused the compressor vibration, which resulted in the experienced thrust shortfall. This chain of events explains the fan vibration, compressor vibration, and thrust output abnormalities in the current situation, but does not explain the abnormalities in the speed of the compressor rotor, the EGT, and the fuel flow.

In order to give explanations for these symptoms EPAION utilizes its models. The causal model informs the system that based on previous cases the system has learned that the leading (most often observed) cause of abnormal speed of the compressor rotor is abnormal vibration of the fan and the leading cause for abnormal EGT is abnormality in the speed of the compressor rotor. Based on that knowledge the system explains the current abnormality in the compressor rotor speed as a result of the abnormal fan vibration and the low EGT as a result of the compressor rotor speed shortfall. Similarly the functional model informs the system that the fuel flow is functionally linked with the speed of the compressor rotor and therefore the current abnormality in the fuel flow is successful explained. The retrieved scenario does not contain any symptoms that are not experienced in the current situation; therefore no further explanations are needed.

As soon as all symptoms are explained EPAION creates the causal explanation of the current case by connecting each symptom to its cause. This causal explanation is associated with the current situation and is stored in the case library. Figure 2 displays the chain of causal events in the retrieved and current case.

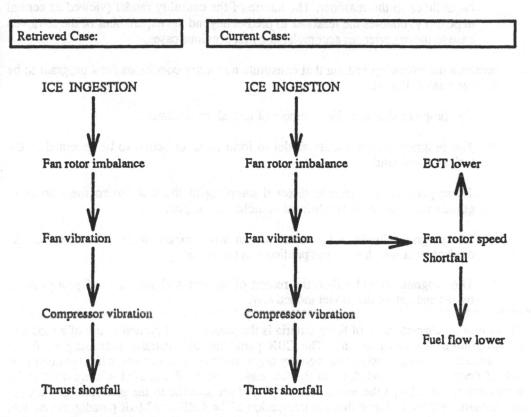

Fig. 2. Causal explanations of retrieved and current case.

5 Results

Experience with integrated CBR/MBR has led to the following conclusions:

- Combining a memory of past cases with models combines the efficiency of associational reasoning with the flexibility of model- based reasoning.

- The integration of CBR and models enhances the ability of the model-based component by the CBR component's capacity to contribute new links into the causality model. The adaptation rules of the CBR component not only adapt the retrieved causal explanation to fit the current case, but also find possible gaps in the causal explanation and fill in the missing causalities. These additional causalities serve in the causal explanation of the current case and to expand the available knowledge to the model.

- The integration of CBR and models enhances the ability of the CBR component by using the model to aid the processes of matching, and adaptation. The model aids matching and adaptation in dealing with features which appear different on a superficial level, but are accounted for by the same initial cause.

- The use of the causality model provides enhanced fault-propagation forecast capabilities to the reasoner. The nature of the causality model (viewed as central depository) enables the reasoner to predict beyond the experiences of the retrieved case to the experiences accumulated by all previous cases.

[11] presents the following criteria that constitute necessary conditions for a program to be characterized as intelligent.

- The program should utilize a model of its task environment.

- The program should use its model to form plans of action to be executed in the task environment.

- These plans should include directed sampling of the task environment so as to guide execution along conditional branches of the plan.

- The program should re-formulate a plan when execution leads to states of the environment which were not predicted in the model.

- The program should utilize the record of failures and successes of past plans to revise and extent the model inductively.

The dominant characteristic of these criteria is the presence and particular use of a model in the program's task environment. The CBR paradigm demonstrates promising results in areas such as planning, design and memory organization but its success is limited due to the lack of deep domain knowledge. In the few cases where CBR is used in conjunction with deep domain knowledge the techniques employed are specific to the particular application and domain. We have shown that the integration of the CBR and MBR paradigms can help overcome these limitations.

References

1. Air Accidents Investigation Branch. (1990). *Report on the accident to Boeing 737-400 G-OBME near Kegworth, Leicestershire on 8 January 1989*. AAIB-AAR-4/90.

2. Buchanan, B. G., and Shortliffe, E. H., editors (1984). *Rule-Based Expert Systems*. Addison-Wesley Publishing Co., Readings, MA.

3. Davis, R. (1984).Diagnostic Reasoning Based on Structure and Behavior. *Artificial Intelligence*, vol. 24, pages 347-410.

4.   De Kleer, J. and Brown, S. J. (1985). A Qualitative Physics Based on Confluences.
      In *Qualitative Reasoning about Physical Systems*. Bobrow, D. G., editor. MIT Press,
      Cambridge, Massachusetts

5.   Feyock, S. (1991). *Automatic Determination of Fault Effects on Aircraft
      Functionality*. Midgrant Report 1990, NASA grant NCC-1-122, February 1991.

6.   Feyock, S., and Karamouzis, S. (1991). Design of an Intelligent Information System
      for In-Flight Emergency Assistance. In *Proceedings of 1991 Goddard Conference on
      Space Applications of Artificial Intelligence*.

7.   Feyock, S., and Karamouzis, S. (1991). *LIMAP*. Technical Report WM-92-1,
      College of William & Mary, Computer Science Department, Williamsburg, Virginia.

8.   Forbus, K. D. (1985). Qualitative Process Theory. In *Qualitative Reasoning about
      Physical Systems*. Bobrow, D. G., editor. MIT Press, Cambridge, Massachusetts

9.   Hayes, P. J. (1979). The naive Physics Manifesto. In *Expert Systems in the
      Microelectronics Age*. Michie, D., editor. Edinburgh University Press, Edinburgh.

10.   Kuipers, B. (1985). Commonsense Reasoning about Causality: Deriving Behavior
      from Structure. In *Qualitative Reasoning about Physical Systems*. Bobrow, D. G.,
      editor. MIT Press, Cambridge, Massachusetts

11.   Michie, D. (1971). Formation and Execution of Plans in Matching. In *Artificial
      Intelligence and Heuristic Programming*. Finoler & Meltzer, editors, American
      Elsevier.

12.   Riesbeck, C. K., and Schank, R. C. (1989). *Inside Case-Based Reasoning*. Lawrence
      Erlbaum Associates, Hillsdale, New Jersey.

13.   Rosenberg, R. C., and Karnopp, D.C. (1983). *Introduction to Physical System
      Dynamics*. McGraw-Hill, New York.

14.   Schank, R. C., (1982). *Dynamic Memory: A Theory of Learning in Computers and
      People*. Cambridge University Press.

Analysing Particle Jets with Artificial Neural Networks

K.-H. Becks, J. Dahm and F. Seidel

Physics Department, University of Wuppertal
Wuppertal, Germany

Abstract. Elementary particle physics includes a number of feature recognition problems for which artificial neural networks can be used. We used a feed-forward neural network to seperate particle jets originating from b-quarks from other jets. Some aspects such as pruning and overfitting have been studied. Furthermore, the influence of modifications in architecture and input space have been examined. In addtition we discuss how self-organizing networks can be applied to high energy physics problems.

1 Introduction

During the last years there has been an increase of interest for "brain style computing" in terms of artificial neural networks (NN). The reason is the power NNs have shown for a wide variety of real-world feature recognition problems. The attractive features are adaptiveness, robustness, inherent parallelism, etc.

In particle physics there are many challenging feature recognition problems such as quark flavour tagging [1]. Furthermore, with increasing luminosity and energy of new particle accelerators efficient extraction procedures will become more and more important. The standard procedure for extracting information from experimental data is performed by various cuts in parameter space. One wants to get the optimal choice of cuts which seperate the different classes from each other. This is really what NN aim to do.

First a feed-forward NN with a backpropagation learning algorithm will be discussed. This architecture is used for classification of data where the features to be recognized are known beforehand. A feed-forward NN is an example of supervised learning. Secondly, we shortly explain how self-organizing networks, which form their own classification of the input patterns as the features to be recognized are unknown, could be used to detect new physical properties. Details about both kinds of networks can be found in [1].

The feed-forward networks are used in the context of high energy physics data analysis, e. g. for the quark flavour tagging problem. To understand this task better we shortly discuss the physics problem: Quark flavour tagging means the reconstruction of the initial quarks, the basic building blocks of nature, created in a particle collision process. The reconstruction process has to be performed from the observed particles in a detector system. In the relatively simple case of an electron-positron

---

[1] This work has been supported by the "KI-Verbund NRW", founded by the Ministry for Science and Research of North Rhine Westfalia

annihilation experiment, a quark/antiquark pair (of the same flavour) might be created. The different quark flavours are up, down, strange, charme, beauty, and the (to now undetected) top quarks. The process of event reconstruction is complicated by the so-called fragmentation process of quarks into hadrons, by decays of short living hadrons into stable particles, and by clustering of single hadrons to particle jets. These complications make a direct way back from observed hadrons to the initial quarks impossible.

2 Quark Flavour Tagging

One main topic of interest for this study is "heavy flavour physics". For precise measurements of b-quark parameters one has to be able to distinguish whether a hadron jet originates from a b-quark or a light quark (u, d, s, c).
To solve this problem, we used two feed-forward network architectures with the standard backpropagation algorithm [2]. A layered feed-forward network with 40 input nodes, 20 hidden nodes, and 1 output node, as well as a network with two hidden layers, 40 input nodes, 20 hidden nodes in the first layer, 10 nodes in the second hidden layer, and 1 output node. The data used in our analysis were generated with a physics generator plus a full detector simulation. The train set consisted of 10000 event patterns from which the following simple input variables have been used:

- the absolute values of the momenta of the ten fastest particles in the two most energetic jets, called jet 1 and jet 2;
- the transversal components of momenta of the ten leading particles in jet 1 and jet 2 with respect to the jet axes.

From a physical point of view the two jets are almost independent. Taken this into consideration we have split the input layer in two halves, each receiving the variables of one jet. The aim was to force the net to deal with seperated jets. It turned out that such a partially connected network provides a better performance than a fully connected one. To allow a comparison of the network performance we defined:

$$\text{b-efficiency} = \frac{\text{number of correctly classified b}}{\text{number of all b events}} \tag{1}$$

$$\text{b-purity} = \frac{\text{number of correctly classified b}}{\text{number of patterns classified as b}} \tag{2}$$

To achieve a further improvement of performance, we have included event shape features to the input patterns for the two hidden layer network. These features are the relative orientations of the three most enrgetic jets to each other.
Fig. 1 shows the result for such partially connected networks: one hidden layer (open squares), two hidden layers without relative orientations (black triangles) and with orientation (open circles).

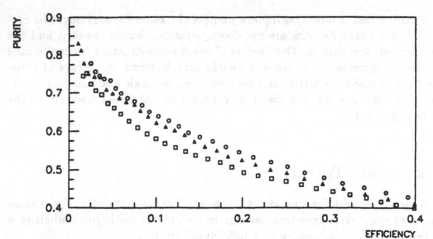

Figure 1: *Purity vs. efficiency for different network architectures*

Overfitting

Generalization means the ability of a network to extract general rules out of a train set. If too many degrees of freedom are available the neural network learns the train pattern by heart. This reduces the capability of generalization. In our application the input space is finite. The input data are normalized to the intervall [0,1]. As our input space is of high dimensionality, the used networks have many degrees of freedom. Therefore the danger of overfitting is always given. We have observed this effect by studying various networks: While the performance on the test set decreases, the performance on the train set increases. It should be remarked that overfitting can occur before the maximal generalization has been archieved.

Pruning

To avoid overfitting and to ease possible hardware implementations, pruning is an important topic. A simple procedure is to cut weights which are lower than a pre-assigned threshold. We have applied this method to different architectures with different thresholds. To summarize: for low thresholds the performance of the nets did not change significantly. When passing over a critical threshold value, the performance breaks down but after training the pruned net again the old performance can be achieved again (Fig. 2).

3 Outlook

For the study of new physical questions in high energy physics, especially if one wants to proceed into a new energy domain, there exist a number of simulation programmes. In order to understand the physics outputs, the simulations have to describe the experimental data very well. But especially the phenomenological part in the simulations gives rise to differences between different programmes. To find

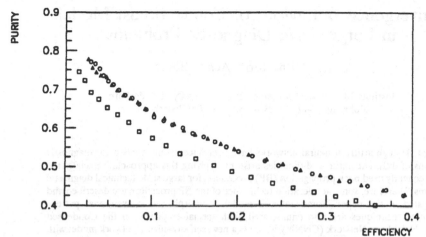

Figure 2: *Performance of a partially connected net (black triangles) and a pruned net before (open squares) and after retraining (open circles)*

out more detailed information about such differences the authors of Ref. [3] used a self-organizing network. They fed the network with data of b-quark jets generated by two different simulation packages. The feature (output) map showed a different clustering behavior of both programmes. The weight vector components enabled them to disentangle the parameters on which the deviations were based.

We are interested in comparing simulated data with real experimental data (from present accelerator experiments). With the help of self-organizing networks it should be possible to extract the physical features which could provide a chance for improved simulation studies. Our next aim is the extrapolation of simulation data to higher energies. After the construction of the next generation of accelerators we will then be able to compare these data with real data. It is our hope to detect new physical properties by applying this method.

References

[1] J. Hertz, A. Krogh, R. G. Palmer
Introduction to the Theory of Neural Computation, Addison-Wesley (1991)

[2] D. E. Rumelhart and J. L. McClelland (eds.)
Parallel Distributed Processing: Explorations in the Microstructure of Cognition (Vol. 1), MIT Press (1986)

[3] L. Lönnblad, C. Peterson, H. Pi, T. Rögnvaldsson
Self-organizing Networks for Extracting Jet Features, Preprint Lund University, LU TP 91-4 (1991)

Convergence Behaviour of Connectionist Models in Large Scale Diagnostic Problems

Laszlo Monostori*, Achim Bothe

Institute of Electrical Measurement, University of Paderborn
Pohlweg 47-49, P.O.Box 1621, D-4790 Paderborn

Abstract. Though artificial neural networks have been successfully applied for diagnostic problems of different natures, difficulties arise by applying this approach, especially the most frequently used back propagation (BP) procedure, for large scale technical diagnostic problems. Therefore, some acceleration techniques of the BP procedure are described and investigated in the paper. Some network models isomorphic to conventional pattern recognition techniques are also enumerated, with special emphasis on the Condensed Nearest Neighbour Network (CNNN), which is a new, self-organizing network model with supervised learning ability. The surveyed techniques are analyzed and compared on a diagnostic problem with more than 300 pattern features.

Keywords: Technical diagnosis, Pattern recognition, Artificial neural networks

1 Introduction

Pattern recognition learning and classification approaches play a significant role in solving technical diagnostic assignments. A great part of monitoring and diagnostic systems actually utilize the pattern recognition technique [1, 17, 19, 27].

Artificial neural networks (ANNs) or *connectionist models* are systems composed of many simple processing elements operating in parallel whose function is determined primarily by the pattern of connectivity. These systems are capable of high-level functions, such as adaptation or learning, and/or lower level functions, such as data preprocessing. Pattern recognition is the field, where artificial neural networks seem to have the most potential benefits for practical applications [2, 9, 12, 18].

It was very straightforward to compare the neural network approach with traditional pattern recognition techniques in the field of technical diagnostics. It is worthy of note, that these investigations were made nearly simultaneously. Dornfeld applied neural networks for cutting tool monitoring [10]. Barschdorff and his colleagues reported on the application of ANN technique for the quality control of electric drive motors [5], and of gear boxes [2]. Monostori and Nacsa investigated ANN techniques for detecting the wear of small diameter drills [18, 20]. The results showed, that artificial neural networks surpassed conventional pattern recognition techniques.

As a logical consequence of these and similar investigations, these authors report on the development of "neuro monitoring systems", using accelerator cards for ANN computations [6, 10, 20].

*) During the period 1.8.1990 - 31.12.1991 with a Humboldt Research Fellowship on the leave from: Computer and Automation Institute, Hungarian Academy of Sciences, Kende u. 13-17, Budapest, H-1518 Hungary

A recently published international survey on ANN applications in computer aided manufacturing (and so in technical diagnostics) pointed out [7], that the majority of applications relied on the back propagation (BP) technique [23]. However, the well known drawbacks of the original BP technique, i.e. inclination for jamming in local minima and slow convergence rates, become real barriers on large scale problems.

Consequently, there is a qualitative leap between small sized and large scale technical diagnostic problems (patterns with some hundreds features or more), if one wants to attack these problems with connectionist techniques. Therefore it is urgently needed, to reconsider the applicability of the ANN approach, especially of the BP procedure for large scale problems.

Accordingly, the aim of this paper is to describe some acceleration techniques of the BP approach proved to be very effective for smaller problems. Moreover the convergence, knowledge representation and classification abilities of the following approaches will be compared:

- original pattern learning BP technique with some accelerations,
- batch learning BP technique with adaptive modification of learning rates dedicated to the network weights and
- CNNN a new, self-organizing network model developed at the University of Paderborn, Germany.

2 ANN Models for Large Scale Technical Diagnostic Problems

2.1 Back Propagation Technique

The BP algorithm is a least mean square (LMS) back-coupled error correction learning procedure for feedforward, multilayered networks with "hidden layers" [23]. However it is a much more general technique, and is applicable for a wide class of complex, interconnected systems where the elementary subsystems are represented by known continuous and differentiable functions [28]. It uses a gradient descent technique, which changes the weights in its original and simplest form by an amount proportional to the partial derivative of an error function E (1) in respect to a given weight (2).

The square of the error between the target value t_p and the output value of the net o_p is computed for the pattern p of the training set:

$$E_p = \frac{1}{2} \sum_k (t_{p,k} - o_{p,k})^2 ,$$
(1)

where k denotes the node number at the network output. Omitting the p subscript the gradient descent technique gives:

$$\Delta w_{j,i} = -\eta \frac{\partial E}{\partial w_{j,i}} .$$
(2)

Here j denotes a node in a layer and i a node in the preceding layer, and $w_{j,i}$ the weight between these two nodes (the nodes' thresholds can be similarly handled [23]). The constant η is named as learning rate, and usually $(0 < \eta < 1)$.

The key of the BP procedure is, that the above partial derivatives can be computed using the chain rule from the actual output of the network, from its difference to the target, from the actual weight values and the partial derivatives of the nodes' activation functions, from layer to layer, also for nodes not in the output layer [23].

For a given pattern-target pair the method is implemented in two stages. During the forward pass, the activity levels (outputs) of all nodes are calculated. Then using a backward pass, starting at the output units, the derivatives required for the weight modification are computed.

It is almost obligatory to distend equation (2) with a momentum factor α:

$$\Delta w_{j,i}(n+1) = -\eta \ \frac{\partial E}{\partial w_{j,i}} + \alpha \ \Delta w_{j,i}(n) , \qquad (3)$$

where α is usually taken to be $0 \leq \alpha < 1$, and n is the number of times a pattern has been presented to the network. The effects of the momentum term are to magnify the learning rate for flat regions of weight space where the gradients are more or less constant, and to prevent oscillations.

2.2 Acceleration Techniques of the BP Procedure Applied in the Investigations

In contrast to the perceptrons' training algorithm [22], one of the predecessor of the BP, this algorithm does not converge always (it can jam in local minima), and its convergence can be very slow. However the procedure was successfully applied for a great number of assignments [9, 12, 23], and it seems the real problem is its slowness which gets worse with scaling up of the problem being treated.

In the following acceleration techniques of the BP procedure applied in the investigations will be shortly described.

Dealing with pattern recognition problems, *target values* different from the usual 0 and 1 (0.1 and 0.9) were used, because 0 and 1 can only asymptotically be approximated by the sigmoidal function, and even small changes at the nodes' outputs in their saturated states, require relatively great modifications of weights.

The *network structure* fundamentally determines the network performance (e.g. classification and generalization abilities) and the learning time. The *number of input nodes* was given by the problem, i.e. the number of considered features. Networks with *three layers* were applied.

Regarding the necessary *number of nodes in the hidden layer* equations (4) and (5) were used as theoretically minimum values. There exists a theorem [13, 16], which states, that in a J-dimensional space, the maximum number of regions that are linearly separable using H hidden nodes is given by

$$M(H,J) = \sum_{j=0}^{J} \binom{H}{j} \qquad if \ H > J , \qquad (4)$$

and

$$M(H,J) = 2^{H} \qquad if \ H \leq J . \qquad (5)$$

In "*batch learning*" the weights are updated only after all patterns have been presented. In contrast to the usual "*pattern learning*" (3), the changes for each weight are summed over all of the input patterns and the sum is applied to modify the weights after each iteration over all the patterns:

$$\Delta w_{j,i}(m+1) = -\eta \sum_P \frac{\partial E_p}{\partial w_{j,i}} + \alpha \, \Delta w_{j,i}(m) \, . \tag{6}$$

Here m is the number of presentations of all patterns.

It was found, that both the rate and the existence of the convergence depend on the *proper selection of the learning rate η and of the momentum factor α*. Moreover it seems, their optimum values can not be determined a priori by some of the recommended rules of thumb, and the optimums change during the training iterations [25].

Vogl and his co-workers use the batch learning technique (6) and *vary η dynamically* according to whether or not an iteration decreases the total error of all patterns [26]. Moreover α is set to zero when, as signified by a failure of a step to reduce the total error, the information inherent in prior steps is more likely to be misleading than beneficial. Only after the network takes a useful step, i.e. one that reduces the total error, α assumes a non-zero value again.

It may hardly be supposed, that in a given moment of the iteration procedure a single learning rate η is optimal for all the weights. More probably, there are weights, which can be modified faster than others, i.e. the weights must we equipped with their own learning rates. In this case the batch learning (6) has the following form:

$$\Delta w_{j,i}(m+1) = -\eta_{j,i}(m) \sum_P \frac{\partial E_p}{\partial w_{j,i}} + \alpha \, \Delta w_{j,i}(m) \, . \tag{7}$$

In our investigations, instead of the total error, which is not appropriate to differentiate between the weights, the sign of weight derivatives was considered. When the derivative of a weight possessed the same sign for consecutive steps, the learning rate for that weight was increased, assuming the convergence can hereby be accelerated. On the contrary, as long as the derivative kept changing sign, the corresponding η was decreased, until a step could be done without causing the given weight derivative to change sign. Both increase and decrease of η were exponential but the decrease is faster than the increase. In contrast to the similar algorithm for learning rate modification of Tollenaere [25], both increase and decrease were limited to ensure convergence. The influence of the previous steps through the momentum factor was similarly switched off as in the previously described algorithm of Vogl [26].

The potential benefits of this acceleration algorithm are the following:

- the initial value of the learning rate η hardly influences the learning procedure,
- faster learning,
- better scaling properties
- adaptation of the individual learning rate for every weight and threshold without the appearance of extreme rates.

These benefits are expected to surpass significantly the inconveniences caused by the enhanced complexity and by the higher memory requirement of the algorithm.

2.3 Network Models Isomorphic to Conventional Pattern Recognition Techniques

Taking into account the shortcomings of the most popular ANN learning approach, the BP technique illustrated in the previous section, and the similarities between the structure of the conventional discriminant function based pattern classifiers and of feed forward ANNs [6], it is not surprising, that there are approaches which map well established pattern recognition principles and techniques into the ANN world.

Two main tendencies can be observed. The first is to convert given pattern recognition techniques into appropriate ANN forms (*probabilistic neural network, PNN* [24], *minimum distance automata, MDA* [29]), while in the second approach the pattern recognition arsenal is used rather for the initialization of the network's weights (*Gaussian isomorphic network, GIN* [30]).

Condensed Nearest Neighbour Network. It is a significant drawback of the PNN and MDA solutions, that a part of the networks (let us say, the number of nodes in hidden layers) linearly increases with the number of references (training patterns).

Barschdorff and Bothe described an ANN model based on nearest neighbour classification [4]. Consequently, their approach has no limitations concerning the statistical properties of the input patterns. Using the *Condensed Nearest Neighbour (CNN) concept* introduced by Hart [11], by which the number of reference pattern to be considered during the classification can be decreased, neural networks with reasonable size can be produced even for problems with a great number of training patterns. The resulted *Condensed Nearest Neighbour Network (CNNN)* model has three layers, nodes with quadratic feature in the hidden layer and simple nodes performing logical OR operation in the output layer [4].

Some results of the application of this self-organizing ANN model also in the field of technical diagnostics and their comparison with conventional techniques are described in [4]. The performance of the CNNN approach on large scale technical diagnostic problems will be treated in the next section.

3 Results of Investigations

A series of tests was accomplished at the Institute of Electrical Measurement, University of Paderborn, in order to investigate the applicability of the ANN approach to large scale technical diagnostic problems. The state of the system to be investigated was monitored by 308 pressure sensors. The patterns composed of 308 features proved to be strongly overlapped and sparse (only some of them took up values different from zero, which related to the normal state of that part of the system). A system model was not available.

Three kinds of tests were made on a 33MHz PC/AT with 80386/387 processors. 95 patterns were available for investigations, which were divided into learning and prediction sets as it is shown in Table 1. The artificially generated patterns aiming at describing the normal state (all of the features are near to zero) were not considered in Test2, but the networks still had 10 output nodes, and one of them was always forced to small values.

	Learning set	Prediction set	No. of classes
Test1	60	35	10
Test2	56	33	9
Test3	95	0	10

Table 1. The tests with the corresponding parameters

All of the three tests described in this section were accomplished with the following ANN learning procedures treated in the first part of this paper:

- original pattern learning BP procedure with some accelerations,
- batch learning BP technique with adaptive modification of learning rates dedicated to network weights,
- Condensed Nearest Neighbour Network.

3.1 Back Propagation Techniques

During the tests with different versions of the BP procedure the elements of the target vectors were set to 0.1 and 0.9 respectively (Sect. 2.2). Three layer networks were applied with $\geq$ 4 nodes in the hidden layer corresponding to the minimum number determined by equation (5). Very hard convergence criteria were set: 100% recognition rate on the learning set, with network outputs $\geq$ 0.8 for the maximum of the outputs and $\leq$ 0.2 for the other output nodes. The trained networks were tested with the prediction sets (Table 1) consisting of elements which were not considered during the corresponding learning phases.

Original Pattern Learning BP Procedure with some Accelerations. The investigations were accomplished with formula (3) which can also be regarded as the original version of the BP. The sequence of patterns presented to the network was randomly determined. Both the learning rate η and the momentum factor α were set to 0.5. The results are summarized in Table 2.

Test	Net struct.	No. of iter.	Conv. time [h]	Total error	Recognition [%]		
					right	altern.	faulty
Test1	308-10-10	18108	64.4	0.0092	57.143	8.571	34.286
Test2	308-10-10	6565	21.9	0.0096	60.606	6.061	33.333
Test2	308- 6-10	12136	25.3	0.0186	60.606	6.061	33.333
Test2	308- 5-10	13514	26.3	0.0305	63.636	3.03	33.333

Table 2. Results of tests with the original version of the BP procedure

Table 2 shows that the pattern learning procedure converged in all the accomplished tests. The results were categorized in three groups, i.e. *right, alternative, faulty*. The term alternative was used for cases, in which the output node with the second largest value indicated the class, where the just classified pattern belonged to. The classification results in Table 2 are not satisfying, and the enormous training times make this original version of the BP procedure under the described circumstances unsuitable for large scale diagnostic problems like the investigated one.

Batch Learning BP Technique with Adaptive Modification of Learning Rates Dedicated to Network Weights. The batch learning techniques of Sect. 2.2 with the adaptive modification of dedicated learning rates for every weight and threshold was also investigated. This technique resulted in an acceleration in convergence with a factor between 4.4 and 11, depending on the network structures. Convergence was reached with the theoretical minimum number of nodes in the hidden layer (5) too, which was in this case 4.

The convergence criteria were mitigated in the subsequent investigations, i.e the criterium was a recognition rate of 100% with patterns in the test set, without any prescriptions on the output values of the network as described at the beginning of Section 3.1. The results are summarized in Table 3.

Test	Net struct.	No. of iter.	Conv. time [h]	Total error	Recognition [%]		
					right	altern.	faulty
Test1	308-10-10	3500	10.0	0.0234	54.286	20.000	25.714
Test2	308-10-10	470	1.3	0.0233	63.636	6.061	30.303
Test2	308- 6-10	465	0.8	0.0454	66.667	3.03	30.303
Test2	308- 5-10	600	0.9	0.0381	60.606	9.091	30.303
Test2	308- 4-10	1192	1.4	0.0622	78.788	0.000	21.212

Table 3. Results obtained by the adaptive modification of learning rates attached to every weight and threshold

Comparing with the results of the pattern learning BP procedure with a unique and constant learning rate, (Table 2) accelerations between 6.4 - 31.6 were reached, and the time needed to convergence for Test2 data seems to be acceptable. With the aid of an appropriate number crunching processor it can be brought down to some minutes. The performance of the structure 308-4-10, with the theoretical minimum number of hidden nodes, merits particular attention, indicating the influence of the number of these nodes on the generalization ability of the network. It must be mentioned, that convergence has been reached by both of the investigated versions of the BP procedure, with the data of Test3, which incorporated all of the patterns available (Table 1).

3.2 Condensed Nearest Neighbour Network

In the case of classification problems there is the opportunity to generate the network or at least to initialize its weights using conventional pattern recognition techniques (Sect. 2.3). As a possible approach to large scale diagnostic problems, the *Condensed Nearest Neighbour Network (CNNN)* shortly described in Section 2.3.1 was also drawn into the investigations. CNNN is a self-organizing three-layer feed-forward network with supervised learning ability, which generates the required number of nodes in the hidden layer. With the modification of its r_{min} parameter, by which the minimal influence regions of subclass centres are prescribed, recognition rate of 100% can be reached by any learning set [4].

This r_{min} parameter of the CNNN approach was determined by the data in Test3 (Table 1), which guaranteed recognition rates of 100% also for the learning sets in Test1 and Test2. The results are summarized in Table 4.

Test	Net struct.	Recognition [%]		
		right	alternative	faulty
Test1	308-47-10	71.428	8.571	20.000
Test2	308-42-10	69.697	9.091	21.212

Table 4. Test results of the Condensed Nearest Neighbour Network (CNNN)

Taking into account the above correspondence, the results of the CNNN approach usually surpass the recognition rates reached by investigated versions of the BP algorithm, except the 308-4-10 structure generated by the batch learning BP technique with adaptive learning rates (Table 3). However, the CNNN approach requires a considerably greater number of nodes and consequently of weights too. This fact can be attributed to the dispersed nature of available patterns. At the same time the CNNN approach converges considerably faster, e.g. during 2.33 minutes with Test3 data. In contrast to the BP versions, a trained CNNN network can easily incorporate new knowledge, without beginning the whole training process nearly from scratch.

The relatively small recognition rates achieved by the investigated ANN models can be attributed to the nature of the given diagnostic problem, i.e. to the sparse and strongly overlapped patterns.

4   Conclusions

The main conclusions of the investigations described in this paper are the following:

- The original version of the BP procedure is not suitable for large scale diagnostic problems, at least what the learning concerns.
- The border of applicability of the BP can significantly be shifted away with properly chosen combinations of various acceleration techniques and with appropriate hardware.
- The introduction of modular, hierarchical, loosely coupled network structures, decreasing the dimension of individual modules, seems to be one of the most promising further opportunities.
- Self-organizing neural network models can significantly contribute to the solution of large scale technical problems, and it can not be precluded, that these models will have some relationship to the conventional pattern recognition techniques, like the described CNNN model.
- The comparison of the applicability of symbolic and subsymbolic (e.g. ANN) AI techniques for large scale diagnostic problems must be initiated, and the integrated use of both approaches must be attempted.

Acknowledgements

The authors would like to express their gratitude to *Prof. D. Barschdorff* head of the Institute of Electrical Measurement, University of Paderborn for his encouragement and valuable comments. Grateful acknowledgments are due to *Prof. G. Warnecke* head of the Institute of Manufacturing and Production Engineering, University of Kaiserslautern, Germany and *Mr. Ch. Schulz* head of the CIM-Centrum Kaiserslautern, University of Kaiserslautern, who called our attention to the problem and provided the investigated data.

Special thanks to the *Humboldt-Foundation* allowing L. Monostori to carry out scientific research work in Germany.

References

1. D. Barschdorff, Th. Dressler, W. Nitsche: Real-time failure detection on complex mechanical structures via parallel data processing, North-Holland, Computers in Industry, Vol. 7., 1986, pp. 23-30.

2. D. Barschdorff, A. Bothe, G. Wöstenkühler: Vergleich lernender Mustererkennungsverfahren und neuronaler Netze zur Prüfung und Beurteilung von Maschinengeräuschen, Proc. Schalltechnik'90, VDI Berichte Nr. 813, pp. 23-41.

3. D. Barschdorff, D. Becker: Neural networks as signal and pattern classificators, Technisches Messen, No. 11, 1990, pp. 437-444.

4. D. Barschdorff, A. Bothe: Signal classification using a new self-organising and fast converging neural network, Noise & Vibration, Vol. 22, No. 9, Oct. 1991, pp. 11-19.

5. D. Barschdorff: Case studies in adaptive fault diagnosis using neural networks, Proc. of the IMACS Annals on Computing and Applied Mathematics, MIM-S2, 3-7, Sept, 1990, Brüssels, pp. III.A.1/1-1/6.

6. D. Barschdorff, L. Monostori, A.F. Ndenge, G. Wöstenkühler: Multiprocessor systems for connectionist diagnosis of technical processes, Computers in Industry, Special Issue on Learning in IMS, 1991, pp. 131-145.

7. D. Barschdorff, L. Monostori: Neural networks, their applications and perspectives in the intelligent machining, A survey paper, Computers in Industry, Special Issue on Learning in IMS, 1991, pp. 101-119.

8. P. Bartal, L. Monostori: A pattern recognition based vibration monitoring module for machine tools, Robotics & Computer-Integrated Manufacturing, Vol. 4, No. 3/4, 1988, pp. 465-469.

9. DARPA neural network study, AFCEA International Press, 1988

10. D.A. Dornfeld: Unconventional sensors and signal conditioning for automatic supervision, AC'90, III. CIRP International Conference on Automatic Supervision, Monitoring and Adaptive Control in Manufacturing, 3-5. Sept, 1990, Rydzyna, Poland, pp. 197-233.

11. P.E. Hart: The condensed nearest neighbour rule, IEEE Transaction on Information Theory, May, 1968, pp. 515-516.

12. R. Lippmann: An introduction to computing with neural nets, IEEE ASSP Magazine, April 1987, pp. 4-21.

13. J. Makhoul, A. El-Jaroudi, R. Schwartz: Formation of disconnected decision regions with a single hidden layer, Proc. of the Int. Joint Conference on Neural Networks, 19-22 June, 1989, Washington, D.C., Vol. I., pp. 455-460.

14. P. Mertens: Expertensysteme in der Produktion, Oldenburg, München, Wien, 1990.

15. M. Minsky, S. Papert: Perceptrons: an introduction computational geometry, MIT Press, Cambridge, 1969.

16. G. Mirchandi, W. Cao: On hidden nodes for neural nets, IEEE Trans. on Circuits and Systems, Vol. 36, No. 5, May 1989, pp. 661-664.

17. L. Monostori: Learning procedures in machine tool monitoring, North-Holland, Computers in Industry, Vol. 7., 1986., pp. 53-64.

18. L. Monostori, J. Nacsa: On the application of neural nets in real-time monitoring of machining processes, Preprints of the 22nd CIRP International Seminar on Manufacturing Systems, 11-12, June, 1990, Enschede, the Netherlands, pp. 6A/15-27.

19. L. Monostori,: Signal processing and decision making in machine tool monitoring systems, The Journal of Condition Monitoring, Vol. 3, No. 1, 1989, pp. 1-20.

20. J. Nacsa, L. Monostori: Real-time monitoring of machining processes, Proc. of the AC'90, III. CIRP International Conference on Automatic Supervision, Monitoring and Adaptive Control in Manufacturing, 2a. Poster Papers, 3-5. Sept, 1990, Rydzyna, Poland, pp. 8-22.

21. N. J. Nilsson: Learning machines, Mc Graw-Hill Book Company, New York, 1965.

22. F. Rosenblatt: The perceptron: A probabilistic model for information storage and organization in the brain, Psychological Review, Vol. 65, 1958, pp.386-408.

23. D.E. Rummelhart, J.L. McClelland: Parallel distributed processing, MIT Press, Cambridge, 1986

24. D.F. Specht: Probabilistic neural networks, Neural Networks, Vol. 3, 1990, pp. 109-118.

25. T. Tollenare: SuperSAB: fast adaptive back propagation with good scaling properties, Neural Networks, Vol. 3, 1990, pp. 561-573.

26. T.P. Vogl, J.K. Mangis, A.K. Rigler, W.T. Zink, D.L. Alkon: Accelerating the convergence of the back propagation method, Biological Cybernetics, Vol. 59, 1988, pp. 257-263.

27. M. Weck, L. Monostori, L. Kühne: Universelles System zur Prozess- und Anlagenüberwachung, Vortrag und Berichtsband der VDI/VDE-GMR Tagung Verfahren und Systeme zur technischen Fehlerdiagnose, Langen, FRG, 2-3 Apr. 1984, pp. 139-154.

28. P.J. Werbos: Backpropagation through time: what it does and how to do it, Proceedings of the IEEE, Vol. 78, No. 10, Oct. 1990, pp. 1550-1560.

29. J.H. Winters, C. Rose: Minimum distance automata in parallel networks for optimum classification, Neural Networks, Vol. 2, 1989, pp. 127-132.

30. H.C. Yau, M.T. Manry: Iterative improvement of a Gaussian classifier, Neural Networks, Vol. 3, 1990, pp. 437- 443.

Pattern Recognition Approach to an Acoustical Quality Test of Burnt Ceramic Products

Dr.-Ing. Benno Kotterba

rte Ges. für Datenverarbeitung in der Technik mbH, Karlsruhe

Abstract. The productivity of a fabrication is determined by the weakest member. There are many test benches in the ceramic industry today, with inspectors, who qualify the manufactured products according to subjective points of view. Ever increasing piece numbers as well as increased quality requirements lead to ever increasing demands that the inspectors are no longer equal to. An important characteristic in quality determination of ceramic products is their being free from cracks. Cracks can only be discovered by acoustical evaluation. Test equipment for acoustical pattern recognition analysis takes over the inspector's task today. It recognizes cracks reliably. The error rate is reduced greater selectivity increases productivity.

1 Quality Testing of Clay Roof Tiles

Everyone who once bought porcellaine in a shop watched the salesman knock against the cup or plate or even insisted on his doing so, to assure himself of the product's quality by its sound. The sound of a ceramic object depends on the material state. Even inner invisible cracks change the sound audibly. The acoustical test as method for quality assurance is used today in the final test by the ceramic industry. Up to now this is the only way to determine the quality. By knocking against ceramic products sounds are produced that depend on the material state as well as on the shape.

Mostly unknown is that quality conscient producers of clay roof tiles perform a 100 % test. When the burnt roof tiles leave the kiln they are first detached from each other and go to the test bench. Every inspector knocks against appr. 16,000 roof tiles per day and evaluates the sound of every single object. This is the only way to guarantee that the delivered goods are free from cracks. Testing requires high concentration by the inspectors. The sound of every single tile must be listened to and evaluated as far as its acoustic pattern is concerned. Unluckily human beings are not able to work with constant concentration for a whole work day. Added to this is the fact that the surroundings themselves are relatively loud and that they distract the inspector from his work from time to time through other events.

The application of objective methods for the 100 % test of roof tiles provides the preconditions to sort with constant test quality. Apart from a productivity increase the producer is now able to prove the delivered quality and thus to reduce the guarantee demands.

2 Subjective Acoustic Analysis

A 100% test through an objective acoustic analysis had not been possible successfully in the past as the methods and technical preconditions did not exist. Up to now the technically possible processes determined which test variables could be used. Following is the description of a method which orientates itself at first by the subjective abilities of human beings and deducts the required test variables from this.

Already v. Helmholtz [1] distinguishes in case of acoustical perception between musical consonance and harmony. He discovered that beats can be percepted as rough and dissonant. Sounds, i.e. periodic, distorted vibrations are percepted as harmonic by human beings, if the frequencies of its partials are in certain relations to each other. v. Helmholtz discovered that as a rule only up to six partials are relevant for the sensation of harmony. Human beings prefer the partial intervals of octave, fifth and fourth. Terhardt [2] and Aures [3] affirm these investigations and prove that not only beats but amplitude variations in general are determinative as roughness for the sensoric consonance. The sensoric harmony is additionally influenced strongly by the sensoric values loudness, sharpness and timbre. Sounds as produced e.g. by knocking against ceramic objects are always short-time sounds. The hearing event consists on the one hand of the knocking itself as well as of the answer on the other hand. The acoustic pattern depends in this case on the material state as well as the shape.

To verify the results of the literature for use for ceramic products [4], test persons evaluated in several acoustic tests the sounds in accordance with the method of pair comparison as well as that of polarity profile. The used sounds resulted from roof tiles which had been evaluated subjectively by inspectors. The test persons classified the sound pairs in accordance with the question which sound of a pair was more harmonic.

Test sound	1	2	3	4	5	6	7	8	9	10	11	12	13	14	15
subjectiv judgements			good					limit					bad		
relative frequency	12	10	13	12	8	7	9	6	9	9	1	0	2	4	3
priority	2	4	1	3	8	9	5	10	6	7	14	15	13	11	12

Table 1: Assignment of the sounds to the subjective pre-evaluations [4]

Table 1 shows the results of a test. The 15 test sounds have been sorted in the table in accordance with the subjective judgements of the inspectors. The third line states the relative frequency of preference by the test persons. The test persons preferred e.g. thirteen times test sound 3 as more harmonic. At the other end of the scale test sound 12 remained without any preference. The last line of the table shows the priority of the test sounds as far as the preference within the pair comparison is concerned. The evaluation of harmony through the test persons thus mainly agreed with the quality evaluation of the inspectors. The results of all acoustical tests prove that quality evaluation of ceramic parts depends on the inspector's sensation of hamony.

The correlations of the subjective polarity judgements show the linear dependency between the sound attributes. The dependency between the sensed roughness and irritation (92.7 %) is very obvious. Inversely proportional are sound height and sharpness. Sharpness decreases with sinking sound height. Sound height and sharpness do however influence the sensed irritation only slightly.

3 Metrological Acoustical Analysis

The sounds of roof tiles have a die rate of appr. 300 ms and a band width of appr. 5 kHz. With metrological methods analyses were performed of time period, spectrum, critical-band function as well as cepstrum.

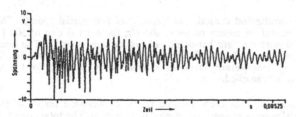

Picture 1: Time domain signal of a roof tile sound

Picture 1 shows the typical periodic gradient of the microphone voltage of a sound. After short rising and reaching of the maximum amplitude, periodically structured vibrations follow which decrease in the course of time. The appertaining power density spectrum (picture 2) shows a large number of dominating spectral lines. Clearly recognizable are lines with the same spacing, mainly in the higher frequency range. The dominating spectral lines are the multiples of a common fundamental frequency.

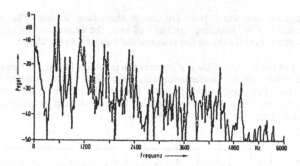

Picture 2: Power spectrum of a roof tile sound

The critical-band function of the sounds (picture 3) do no longer dissolve the spectral interrelationships between the individual spectral parts. The thus determined total loudness does show the sensation of loudness very well, does however not permit a conclusion as far as harmony is concerned. Loudness is of little importance for the sensation of harmony as far as these sounds are concerned. The safest statements on the sensed melodiousness result from the evaluations of the spectrum structure. Of main importance for this is the sensation of harmony as it has been formulated principally by v. Helmholtz [1].

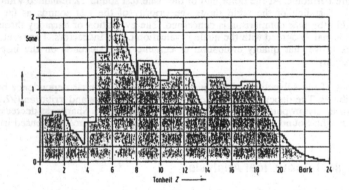

Picture 3: Critical-band function of a sound

The sounds of the investigated ceramic parts consist of two partial sounds. The first partial sound has a fundamental frequency of appr. 400 Hz, the second one of 500 Hz. Added to these fundamental frequencies are harmonic ones of varying intensity. The frequency of the fundamental tone changes with material composition, thickness of roof tile, its shape and above all the burning temperature. Decisive is furthermore whether the tile surface is left natural or whether it is enamelled.

If the roof tile has a visible or even invisible crack, its vibration response changes during knocking. The partial sounds change their frequency position. The total sound is irritating. A detailed evaluation of the octave and fifth relationships of all dominating spectral lines of the sounds confirmed the assumption that the subjective quality evaluation depends on the sensoric harmony [4].

4 Increased Reliability and Productivity

Today it is possible, with the technical facilities of fast signal processors and computers, to practically use these long-known theories and to reproduce the ear's ability in small steps. Starting from the presented test results an automatically working system for sound analysis was developed and installed. The roof tile companies Max Jungmeier, Straubing, were the first to use such test facility which automatically tests the roof tiles. Every 1.5 seconds a roof tile reaches the test bench. Here it is knocked against by a mechanical pestle. A microphone records the emitted sound and transmits it to a test computer. This determines the test variables for the evaluation of the sensoric harmony and compares them with the previously taught acoustical patterns. A hierarchically structured discriminance classifier processes the test patterns and makes the quality decision. If the sound shows characteristic deviations from the acoustical patterns the roof tile is marked as bad and sorted out.

Human beings are however still required as guide for the correct adjustment of the test facility. As no absolute measure exists for harmony and thus a crack-free roof tile, the adjustment of the facility must depend on the subjective pre-evaluation by the inspector. In a learning process the test facility is taught the acoustic differences between pre-evaluated good and bad roof tiles. These acoustical patterns are stored in the test facility as quality references. Has the test facility been taught these acoustical patterns, it can decide automatically whether a tile is free from cracks that is whether the quality evaluation must be good or bad.

Today the test facility already works with high recognition accuracy. In an extensive test phase appr. 20,000 roof tiles have been evaluated by the inspector and the test facility as far as their quality is concerned. This lead to the results shown in table 2.

	test facility	good inspector	normal inspector
clearly bad	0,30%	1,50%	5,00%
clearly bad	1,0 % - 3,0%	2,0% - 5,0 %	5,0 % - 20 %

Table 2: Error rates of the test facility and very good or normal inspectors in the quality evaluation of roof tiles

The test facility has an error rate of 0.3 % in the recognition of definitely bad roof tiles. The inspector does not recognize 1.5 % of definitely bad roof tiles. This result is however valid only for a very good inspector. The error rate of a normal inspector in the recognition of bad tiles is appr. 5 %. The test facility evaluates as bad appr. 1 - 3 % of the roof tiles evaluated as good by the inspector. The error rate of a normal inspector in quality evalution of good roof tiles varies between 5 and 20%. These variations are caused by varying concentration and attention.

The accuracy of the test facility for automatic quality evaluation is very high at appr. 99 %. It extends a human beings determination abilities. By regular additional teaching the recognition performance of the facility and thus the accuracy in qualilty evaluation can be increased in the course of time.

These results demonstrate very clearly that the application of the objective test technology increases the reliability of quality tests. Especially striking however is the decrease in mistaking good roof tiles for waste. If this decrease is set at a rate of only 5 %, these are at a daily production of 16,000 roof tiles already 800 parts which are not sorted out as cracked by mistake. The application of the test facility can thus lead to a clear increase in productivity.

References

1. H. v. Helmholtz: Die Lehre von den Tonempfindungen als physiologische Grundlage für die Theorie der Musik. Vieweg Verlag, Braunschweig 1863

2. E. Terhardt: Ein psychoakustisch begründetes Konzept der musikalischen Konsonanz. Acustica 58 (1985) 121 - 137

3. W. Aures: Berechnungsverfahren für den sensorischen Wohlklang beliebiger Schalle. Acustica 58 (1985) 282 - 289

4. M. Eberhard: Die objektive Ermittlung des Wohlklanges technischer Schalle. Diplomarbeit FH Düsseldorf 1990

Enhancing Software Engineering Capabilities of PROLOG by Object-Oriented Concepts

Bernd Müller

Fachbereich Informatik, Universität Oldenburg

D-2900 Oldenburg

Abstract

This paper presents some parts of the hybrid object-oriented language PPO. PPO extends PROLOG by a type concept and object-oriented features. These extensions are used to improve the software maintenance capabilities of PRO-LOG, which are very rudimentary.

We give a survey of the language and describe in detail the type system and the implementation of multiple inheritance. An example is used to demonstrate this language features and to motivate the estimated improvements of software maintenance capabilities of PPO compared with PROLOG.

1 Introduction

PROLOG as a modern programming language (although introduced in the middle 70th PROLOG is sometimes classified as a 4th or 5th generation language) lacks some primitive software engineering concepts such as a type and a module system. On the one hand the absence of these concepts makes PROLOG a powerfull language for rapid prototyping. On the other hand it is hard to use PROLOG in real life, i.e. large applications.

To overcome this deficiencies many type systems [MO84, DH88, Smo89, Han89, YS91] and tools for the development of PROLOG applications like module systems [Qui90, Sco90] intelligent debuggers [DE90], mode analysers [DW88] and program termination provers [DM79, Plü90a, Plü90b] have been developed.

A new programming paradigm, namely object-oriented programming, has also been applied to PROLOG [Mos90, Zan84]. This new paradigm spreads widely in the last few years because of its power to reduce software complexity. Object-oriented extensions are integrated *into* the language in this approaches, in opposite to most of the above mentioned tools, which are available *outside* the language.

The language PPO (*PRO*LOG *P*lus *O*bjects), developed at the University of Oldenburg, integrates object-oriented concepts like encapsulation and inheritance with a type system into PROLOG. PPO's test environment is the EUREKA project PROTOS (*PRO*LOG *TO*ols for Building Expert *S*ystems) [ACH90] where some research in the area of knowledge based production planning and scheduling is done.

2 Basic Notations

Basic notations and motivations of object-oriented concepts are taken from [Cox86]. Object-oriented programming is mentioned there as a new code packaging technique. The central ideas are *encapsulation* and *inheritance*. Some single data items are gathered and encapsulted by allowing only special procedures to access the data. The data items are called *attributes* and the procedures are called *methods*. The whole thing is called an *object*. Objects are described by *classes* where each object is an *instance* of a class. Programmers are distinguished in *suppliers* and *consumers*. A supplier describes the structure and behaviour of objects, while the consumer uses the objects (possibly to describe other objects). The difference to conventional programming is the place of responsibility to choose type-compatible operators in expressions, depending on the used operands. In object-oriented programming the supplier and the operand are responsible to choose the type-compatible operator while in conventional programming the consumer is. Though the progress is that the consumer has not to specify how to do something by stating operator and operands but only to specify what should be done in terms of a message to an object. The receiving object has to choose a suitable method out of its method pool to do it right. This kind of looking for a suitable method at run time is called *dynamic* or *late binding* in opposite to *static binding*, which is done at compile time in conventional languages when function calls are mapped to addresses.

Another central feature of object-oriented languages is inheritance. *Inheritance* is a mechanism to reuse code which already exists. This is done by scattering code of a class into other classes. Though a new class description has not to be started with a blank sheet but can be built up on existing code. Inheritance hurts the encapsulation principle because internal structures are available outside the class. The arising problems are beyond the scope of this paper and are only discussed in some special cases, but the interested reader can find further information in [Sny86].

Encapsulation and inheritance are used by Cox to define a hybrid language, namely Objective-C, a descendant from C extended with this object-oriented concepts. The class descriptions are compiled into so called *Software-ICs*, which can be used analogously to Hardware-ICs to construct larger systems simply by putting them together. Cox motivates that this new software construction methodology gives software engineering the same efficiency gain as VLSI technique gave to hardware construction and is therefore one step toward the solution of what is known as the software crisis.

Now we want to give a raw sketch of PROLOG type systems. They can be classified by

- many-sorted / order-sorted, and

- monomorph / polymorph,

where the most powerful combination is polymorphically order-sorted. The first type system for PROLOG was introduced by Mycroft and O'Keefe [MO84] in 1984, and developed further, for example by [RL91, DH88]. For order-sorted monomorphic first order logic a calculus was presented in [Wal83] and implemented for PROLOG in [HV87, Mül88]. Polymorphically order-sorted type systems were introduced in [Smo89, YS91] and the former implemented in [BBM91].

The system in [MO84] can determine the well-typedness of a program at compile time, whereas the order-sorted systems have to do some work at run time, because of the possibility of order-sorted variables to be constrained to smaller types by unification.

In the following we assume that the reader is familiar with PROLOG notations like predicate, clause, backtracking etc.

3 PPO

PROLOG as a programming language lacks some software engineering principles, for instance there is no modularization, no data abstraction and no typing mechanism. Furthermore, the programming environment and the language itself do not provide any software maintenance concept. Object-oriented programming offers techniques, especially encapsulation and inheritance, which can help to realize modularization and data abstraction and therefore basically support software maintenance and reusability in keeping interfaces small. These concepts together with a type concept have been integrated into PROLOG resulting in PPO.

PPO follows mainly Cox's model but differs in some details. An object's structure and behaviour is defined by a class definition, where each object belonging to this class (an instance of this class) consists of the attributes and procedures defined there plus the inherited attributes and procedures. There are two different kinds of attributes and two different kinds of procedures in PPO. *Slots* are imperative attributes, *relations* are relational ones, which means that writing to a slot overrides the old value while writing to a relation adds a new tuple. Thus slots are single valued and relations are multi valued attributes. The two kinds of procedures are *methods* and *routines*, where both are syntactically similar to PROLOG predicates. Methods establish the interface of an object, routines are only visible inside the object. So methods may be bound (dynamically) to messages and routines may be called only by methods and routines of the same object.

3.1 Types in PPO

As mentioned earlier the research work on PROLOG type systems was started by [MO84] with a many-sorted polymorphic system. Today the question turns on polymorphically order-sorted type systems because such systems are the strongest ones in a descriptional as well as in a computational view. The reason therefore is, that the sort hierarchy is used to *describe* hierarchical structures of the application domains and is used *computationally* to reduce the SLD search tree. For PPO we have decided not to use a type concept of this class, because of the interrelationship to class hierarchies. Providing two mechanisms to describe hierarchies of the application domain, namely subclass relations originating from the object-orientation of the language and subsort relations originating from the type system would lead to an unmanageable language.

The type system which meets our demands and is close enough to be easily adopted is the one from [MO84]. It is introduced there as a pure compile time system. Types are used as in PASCAL[1] to check if complex terms are well-constructed corresponding to the declaration of the involved function symbol and to check if unification would lead to forbidden variable substitutions (in PASCAL: type mismatch in an assignment). If a program is well-typed there is no need to deal with types at run time, because the resolvent of two well-typed clauses is proven to be well-typed. Therefore the PROLOG computation would never result in a wrong typed goal list if started with a well-typed goal and a well-typed programm.

The advantage of the sketched type system is that it is fairly simple: The types have to be declared, which means that for every type constructor (function symbol) the signature has to be declared. Additionally the predicates have to be augmented with their signatures, too. After the type-checker has verified the well-typedness of the program the augmented declarations were removed and the resulting PROLOG program runs on every standard PROLOG implementation. The type system also handles polymorphic types, which means that PROLOG standard predicates for the polymorphic type list[2], like member/2 and append/3 can be used without recoding them, only by adding the list type and augmenting the type declarations for the predicates. The only drawback of the type system is its inability to deal with order-sorted types, which is compensated, as mentioned earlier, by the class hierarchie.

PPO's type system differs from [MO84] in the following points: There is a new basic type *obj*, denoting any object in the system (object ids), and the type handling between modules (classes) has been changed. To fit into object-oriented philosophy, especially to support Software-ICs, we decide not to export types as proposed by Mycroft and O'Keefe. Exporting types has the drawback that changes in one class can involve changes or at least recompilation of other classes (if non-opaque types are used). This hurts the Software-IC principle and makes reusability much more difficult. In PPO objects themselves are elements of an abstract data type, namely the class, without an explicit defined type identifier. Therefore, in PPO there is a difference between classes and types while some other authors use the terms synonymously. For a discussion of types versus classes see [Weg87]. The operations of the abstract data type (all instances of the class) are the methods defined for the corresponding class. The principle of class independence also influences the computational model of Mycroft and O'Keefe's type checker, which is a compile time one. Since in PPO there are no export-/import declarations it is not possible to guarantee the type compatible usage of a method in a message at compile time, because the typing of the method is not known. To avoid type errors, type information has to be available not only at compile time but also at run time. The only allowed types in messages are the four basic types (because no types are exported) and the unification algorithm which unifies messages with methods has to be changed to work with run time type informations (as it is done in PROLOG systems which have an order-sorted type system). The unification between subgoals and routines remains standard Robinson unification.

---

[1] type compatible usage of identifiers within expressions, with no special run time semantics
[2] and any other polymorphic type

3.2   Multiple Inheritance in PPO

Since multiple inheritance in PPO is an elegant and straight forward continuation of single inheritance we want to describe this first. If a method foo/1 is defined, say in class c1, as

```
method foo : atom.
foo(X) :- write('Atom '),
    write(X),
    write(' written with method from c1').
```

it can be used in messages to instances of class c1 and of all its subclasses. As described earlier, defining a method with the same identifier and arity as a method of a superclass does not override this one but simply adds the new clauses. What is left to define is the sequence of clauses in terms of the PROLOG-style search strategy of PPO: If an object receives a message it first looks for appropriate methods in its own method pool and then, if the search was not successful or backtracking occurs, in the method pools of its superclasses. The search through the superclasses starts from the direct superclass and ends at the root class, called object in PPO.

Going further in the example we now define a method foo/1 in class c2, a subclass of c1:

```
method foo : atom.
foo(X) :- write('Atom '),
    write(X),
    write(' written with method from c2').
```

The query

```
?- new(O, c2), !, send(O, foo(foo_test)).
```

yields

```
Atom foo_test written with method from c2
O = obj798 ;
Atom foo_test written with method from c1
O = obj798 ;
no
```

The first goal of the query in the example above is the PPO builtin predicate new(Obj,Class) which creates Obj as a new instance of class Class. The cut, which behaves as in PROLOG, prevents backtracking into new/2. Since O in the example becomes unified with the id of the new created object of class c2, namely obj798 (an uniquely generated identifier), the foo/1 method of class c2 is bound to the message (send/2 is also an PPO builtin) as expected. The ';' causes backtracking into the inherited method from class c1, generating the second answer. This behaviour is not possible in conventional object-oriented languages because defining a method in a class already defined in a superclass is normally seen as specialization. In PPO it is seen as an extension, which is more appropriate when a relational semantics is prefered. This use of inheritance also opens the door to a novel definition of the semantics of multiple inheritance as described below.

```
defclass color_apparatus.
superclass machine.
relation possible_product : atom.
    /* possible products to produce on this apparatus */
method produce : atom x integer x integer x integer.
produce(Product, Amount, StartDay, EndDay) :-
    /* produce Amount liters of product Product,
       starting at StartDay and ending at EndDay */
  getrel(possible_product, Product),
  ...
endclass color_apparatus.
```

Figure 1: Rules to produce colors

Inheriting the same method (same identifier, same arity, same typing[3]) from two different superclasses is handled in PPO as follows: The two clause sets are joined into one clause set which is the definition of the method for instances of the common subclass. Looking for a model theoretic semantics this is sufficient, but the operational semantics needs a sequence of clauses:

Let $level_c(sc)$, the level of a superclass sc wrt class c, be the number of superclasses from c up to sc. The level of the direct superclass of a class is one. The sequence of clauses of one method is the textual order (as in PROLOG).

The Algorithm to gather clauses for a method of class currentclass is:

```
              Join := empty list
              FOR i := 1 TO level_currentclass(object) DO
                FOR EACH superclass of level i DO
                    append clauses of method to Join
              END
            END
```

where the superclasses of the innermost loop are scanned from left to right following the superclass declarations in the class definition file. We have decided for this arrangement of the two loops, i.e. for breadth-first search, because a inherited method which is closer in the superclass hierarchy has to be noticed stronger than a method far away. This is in opposite to the depth-first strategy of PROLOG, but it seems to fit better into the object-oriented class hierarchy concept. Because of this method definition multiple inheritance in PPO doesn't cause conflicts in opposite to many other object-oriented languages. This is demonstrated by an example from a toy world production planning scenario.

The classes color_apparatus, glaze_apparatus and multi_purpose_apparatus are defined in the figures 1, 2 and 3. Instances of this classes can produce colors, glazes or both. Since the production rules for colors and glazes are different let us assume that producing colors needs the double time than producing glazes and differs in some other facts as well. The smallest time interval represented is one day. An

---

[3] differing only in one of them means a different method

```
defclass glaze_apparatus.
superclass machine.
relation possible_product : atom.
    /* possible products to produce on this apparatus */
method produce : atom x integer x integer x integer.
produce(Product, Amount, StartDay, EndDay) :-
    /* produce Amount liters of product Product,
        starting at StartDay and ending at EndDay */
  getrel(possible_product, Product),
  ...
endclass glaze_apparatus.
```

Figure 2: Different rules to produce glaze

```
defclass multi_purpose_apparatus.
superclass color_apparatus, glaze_apparatus.
endclass multi_purpose_apparatus.
```

Figure 3: An example of multiple inheritance

instance of class **multi_purpose_apparatus** is capable to produce both colors and glazes since it inherits both **produce/4** methods. Every instance of one of the three classes knows the products it is able to produce. This knowledge is stored in the relation **possible_product**, which is used as first subgoal in both definitions of **produce/4**. Both methods can be bound to a corresponding message received by an object of class **multi_purpose_apparatus**. Depending on the instantiation of the first argument both methods can fail in the first subgoal if the apparatus is unable to produce this special product or, if not, compute a production plan considering the different production rules for colors and glazes. If the first three arguments are instantiated but the fourth is a free variable this is unified with the end date of the computed production plan. Depending on the first argument the proper method for colors or glazes is used and computes a different end date because of the different production times.

This kind of programming is standard PROLOG coding technique. However its implication to object-oriented programming is a novelty. Separate development of the classes **color_apparatus** and **glaze_apparatus** is allowed and supported. Separate in PPO means different files with no interface between them. These two definitions can be joined with a new class definition (also a file) by declaring the two classes as common superclasses. Further, the two classes can be supplied as binary Software-ICs as proposed by Cox. There is no editing necessary, which is the case in PROLOG, where you have to join the clause sets manually with an editor. The implications of this advance to software maintenance, i.e. no change in existing code, was mentioned earlier.

We want to close this section with a discussion of specialization versus generalization of inheritance. The usual semantics of a class hierarchy is set inclusion. All the instances of a subclass are a subset of the instances of the superclass. That means that traversing the class hierarchy the sets of instances get smaller and smaller. This kind of specialization is also reflected in object methods in most object-oriented languages. The redefined method in a subclass is a specialization of the method of the superclass, sometimes implemented with a 'send super' construct. In PPO, as it was shown for multiple inheritance, it's just the other way round. Given the usual Herbrand model semantics for PROLOG inheriting the same two methods from two superclasses can be combined in a specializing manner by (logically) ANDing both. Semantically, this means intersection of the two relations (the semantics of the original methods). In the above example this could be implemented this way:

```
producemulti_purpose(P,A,S,E) :-
    producecolor(P,A,S,E), produceglaze(P,A,S,E).
```

which has no solution because of the different production times. PPO doesn't implement specialization but generalization, which results in

```
producemulti_purpose(P,A,S,E) :-
    producecolor(P,A,S,E); produceglaze(P,A,S,E).
```

(However, the prototype is not implemented in that way). This is only possible in languages with relational semantics of methods. It turns out that this is not only one more mechanism to handle multiple inheritance but is really necessary in some applications as shown by the example. Further, a feature of this approach is that there is no difference in defining the semantics of single and multiple inheritance and therefore we yield a uniform building for both.

4 Related Work

PPO can be seen as an object-oriented extension of PROLOG in a similar fashion as C++ [Str84, Str86] and Objective-C [Cox86] are object-oriented extensions of C. The basic entities are encapsulated objects communicating with each other using messages. These messages trigger methods which are PROLOG predicates in PPO in opposite to C functions in C++ or Objective-C. The relational view of methods was proposed in [Gal86] but the introduced language lacks encapsulated objects. The PPO type system has its origin in the work [MO84] and was adapted to the dynamic binding mechanism and to the missing import/export declarations of class definitions.

There are many approaches, and we can only look at some of them, which combine object-orientedness with logic. These can be classified in systems combining

- objects and (pure) logic [AP90b, AP90a, CW89, Con88, KL89], and

- objects and PROLOG [FH86, Mor90, Mos90, PLA89, SB90, Zan84].

Members of the first group are almost always basic research projects. The resulting languges are, from a theoretical point of view, worth to look about, but at this time not a tool to implement real life applications. The second group takes PROLOG as a kernel and looks for some enhancements from the object-oriented world. Commercial systems as well as research systems belonging to this group. We can only sketch a few of them.

PROLOG++ [Mos90] is an object-oriented extension of PROLOG developed at Imperial College. It is implemented as a compiler with target language PROLOG and runs on LPA-PROLOG systems. The language supports multiple inheritance and chooses in a conflict situation the first[4] matching method. This was done because of efficiency reasons. PROLOG++ does not support types. A feature worth to mention is a deamon concept. A program can specify procedures which are executed before and after attribute assignments. This is done to guarantee consistent attribute manipulation. Deamon concepts have their origin in frame-based languages but are superfluous in object-oriented languages, because in a language with fully encapsulated objects only methods of the object itself can manipulate its attributes and *this* methods have to be implemented to do this in a consistent manner. There is no reason to separate action and responsibility.

AMORE [PLA89] (and in the discussed features in a very similar fashion SPOOL [FH86]) uses the Smalltalk- and frame-like style of object-oriented features to enhance PROLOG to be an implementation language for large knowledge based applications. AMORE supports also deamons and furthermore facets for slots. This is heired from frame-based languages, too. The language is type-free but deamons can be used to implement a value-based type-checker at run time. Multiple inheritance is supported and conflicts are resolved by a precedence list associated with each class, as done in many object-oriented languages.

Focusing on the shown multiple inheritance mechanism from a language independent level it is obvious that only relational (to read: nondeterministic and backtrackable) methods can be combined in that way. This is not the same as finding all solutions of all inherited methods and putting them in a list, as it can be done in FLAVORS[5] [Moo86].

5 Conclusions and Future Work

We have introduced parts of the language PPO which is a descendant from PROLOG. It can be summed up as

$$PPO = PROLOG + types + objects + Software\text{-}ICs.$$

The type system and the object part of PPO, especially how multiple inheritance is handled, has been described in detail. PPO enhances PROLOG with a module system, based on object-oriented encapsulation, the possibility of structured knowledge representation, based on the class hierarchie, and a type system. Other object-oriented features which improve reusability and maintenance of software systems, like inheritance and polymorphic methods are also available in PPO.

---

[4] where 'first' is not specified further
[5] to verify, look at a method, which generates infinitly many solutions

We think (but can not prove up to now) that this improvements would lead to a better acceptance and usage of PROLOG, or better PPO, as an implementation language for knowledge based systems.

PPO's first prototypical implementation, which was done by compiling into PRO-LOG, was only tested with toy programs. At present we are working on an implementation in C and, in parallel, on a re-implementation of a knowledge based production planning system [ACH90] in PPO, which was originally done in PROLOG.

The current focus of our work is on PPO's type system. The type system presented here originates from Objective-C. Its capabilities for strong type checking are limited, because all objects are of type *obj*. Therefore every message can be send to every object. This makes code reuse very simple, but does not support *type-safe code reuse* [PS91], which we hope to obtain with the new type system.

References

[ACH90] Hans-Jürgen Appelrath, Armin B. Cremers, and Otthein Herzog, editors. *The EUREKA Project PROTOS*, Zürich, 1990.

[AP90a] Jean-Marc Andreoli and Remo Pareschi. Linear Objects: Logical Processes with Built-in Inheritance. In D.H.D. Warren and P. Szeredi, editors, *7th International Conference on Logic Programmming*, Jerusalem, June 1990.

[AP90b] Jean-Marc Andreoli and Remo Pareschi. LO and Behold! Concurrent Structured Processes. In *OOPSLA'90*, pages 44–56, October 1990.

[BBM91] Christoph Beierle, Stefan Böttcher, and Gregor Meyer. Draft Report of the Logic Programming Language PROTOS-L. IWBS Report 175, IBM German Science Center, Institute for Knowledge Based Systems, June 1991.

[Con88] John S. Conery. Logical Objects. In *Logic Programming*. MIT Press, 1988.

[Cox86] Brad J. Cox. *Object Oriented Programming - An Evolutionary Approach*. Addison-Wesley, 1986.

[CW89] Weidong Chen and David Scott Warren. C-Logic of Complex Objects. In *Eighth ACM SIGACT-SIGMOD-SIGART symposium on Principles of Database Systems*, 1989.

[DE90] Mireille Ducasse and Anna-Maria Emde. OPIUM: A Debugging Environment for Prolog Development and Debugging Research. *ACM Software Engineering Notes*, 16(1):54–59, 1990.

[DH88] Roland Dietrich and Frank Hagl. A Polymorphic Type System with Subtypes for Prolog. In H. Ganzinger, editor, *2nd European Symposium on Programming, LNCS 300*. Springer, 1988.

[DM79] N. Dershowitz and Z. Manna. Proving Termination with Multiset Orderings. *Communications of the ACM*, 22(8):465–476, 1979.

[DW88] Saumya K. Debray and David S. Warren. Automatic Mode Inference for Logic Programs. *The Journal of Logic Programming*, 5(3):207–229, 1988.

[FH86] Koichi Fukunaga and Shin-ichi Hirose. An Experience with a Prolog-based Object-Oriented Language. In Norman Meyrowitz, editor, *Proc. OOPSLA'86, Conference on Object-Oriented Programming Systems, Languages and Applications*, Portland, OR, September 1986. ACM.

[Gal86] Herve Gallaire. Merging Objects and Logic Programming: Relational Semantics. In *Proc. AAAI'86*, 1986.

[Han89] Michael Hanus. Horn Clause Specifications with Polymorphic Types. Dissertation, Universität Dortmund, 1989.

[HV87] Martin Huber and Igor Varsek. EPOS — Extended Prolog with Order-Sorted Resolution. Interner Bericht 4, Universität Karlsruhe, 1987.

[KL89] Michael Kifer and Georg Lausen. F-Logic: A Higer-Order Language for Reasoning about Objects , Inheritance, and Scheme. In *SIGMOD*, 1989.

[MO84] Alan Mycroft and Richard A. O'Keefe. A Polymorphic Type System for Prolog. *Artificial Intelligence*, 23(3), 1984.

[Moo86] David A. Moon. Object-Oriented Programming with Flavors. In Norman Meyrowitz, editor, *Proc. OOPSLA'86, Conference on Object-Oriented Programming Systems, Languages and Applications*. ACM, 1986.

[Mor90] Larry Morlan. *IBM PROLOG Language Workbench/VM (IPW), User's Manual, Release 1.0*, 1990.

[Mos90] Chris Moss. An Introduction to Prolog++. Research Report DOC 90/10, Imperial College, 1990.

[Mül88] Bernd Müller. Design and Implementation of an Abstract Machine for order-sorted Logic Programms. Studienarbeit 711, Universität Stuttgart, Institut für Informatik, 1988. (in german).

[PLA89] Z. Palaskas, P. Loucopoulos, and F. Van Assche. AMORE — Object Oriented Extensions to Prolog. In *TOOLS '89*, 1989.

[Plü90a] Lutz Plümer. *Termination Proofs for Logic Programs*. LNCS (LNAI) 446. Springer, 1990.

[Plü90b] Lutz Plümer. Termination Proofs for Logic Programs based on Predicate Inequalities. In *Seventh International Conference on Logic Programming*, Jerusalem, 1990.

[PS91] Jens Palsberg and Michael I. Schwartzbach. What is Type-Safe Code Reuse ? In Pierre America, editor, *Proc. ECOOP'91, European Conference on Object-Oriented Programming, LNCS 512*, pages 235–341, Geneva, July 1991. Springer.

[Qui90]   Quintus Computer Systems Inc. *Quintus Prolog Development Environ-ment*, 1990. Release 3.0.

[RL91]   Uday S. Reddy and T.K. Lakshman. Typed Prolog: A Semantic Recon-struction of the Mycroft-O'Keefe Type System. In *ILPS*, 1991.

[SB90]   A. Schmidt and F. Belli. An Extension of PROLOG for Object-Oriented Programming in Logic. In *The Third International Conference on Indus-trial & Engineering Applications of Artificial Intelligence & Expert Sys-tems*, 1990.

[Sco90]   Roger S. Scowen. PROLOG Draft for Working Draft 4.0, N64. ISO/IEC JTC1 SC22 WG17, International Organization for Standardization, September 1990.

[Smo89]   Gert Smolka. Logic Programming over Polymorphically Order-Sorted Types. Dissertation 386, Universität Kaiserslautern, Fachbereich Infor-matik, 1989.

[Sny86]   Alan Snyder. Encapsulation and Inheritance in Object-Oriented Program-ming Languages. In Norman Meyrowitz, editor, *Proc. OOPSLA '86, Con-ference on Object-Oriented Programming Systems, Languages and Appli-cations*, Portland, OR, September 1986. ACM.

[ST87]   Ehud Shapiro and Akikazu Takeuchi. *Concurrent Prolog*, chapter Chapter 29, Object Oriented Programming in Concurrent Prolog. MIT Press, 1987.

[Str84]   Bjarne Stroustrup. Data Abstraction in C. *AT&T Bell Laboratories Tech. Journal*, 63(8), 1984.

[Str86]   Bjarne Stroustrup. *The C++ Programming Language*. Addison-Wesley, Reading, 1986.

[Wal83]   Christoph Walther. A Many-Sorted Calculus based on resolution and paramodulation. In *8th Internation Joint Conference on Artificial Intel-ligence*, pages 882–891, Karlsruhe, 1983. Morgan Kaufmann.

[Weg87]   Peter Wegner. Dimensions of Object-Based Language Design. In Norman Meyrowitz, editor, *Proc. OOPSLA'87, Conference on Object-Oriented Programming Systems, Languages and Applications*, Orlando, Florida, Oc-tober 1987.

[YS91]   Eyal Yardena and Ehud Shapiro. A Type System for Logic Programms. *The Journal of Logic Programming*, 10(2), February 1991. also in [ST87].

[Zan84]   Carlo Zaniolo. Object-Oriented Programming in Prolog. In *IEEE Sym-posium on Logic Programming*, 1984.

SPECIFYING DECISION-MAKING PROCESSES

Mr. Andrew Borden
Dr. Unver Cinar

SHAPE Technical Centre, The Hague, The Netherlands

ABSTRACT: An explicit specification for the implementation of any process is important to provide a standard for design, verification and validation. Since the control structure of Decision Making Processes (DMP's) tends to be very complex, it is difficult to apply structured methods to the development of specifications. This paper provides a formal meaning for the term "specification" as it applies to DMP's and which can be used as a design and evaluation standard.

Keywords: Information Theory, Decision-Making Process, Decision Tree, Specification.

1 INTRODUCTION

In this paper, a Decision Making Process (DMP) will be defined as a universe of objects, a set of sensors which can measure the specified attributes of these objects, a knowledge base which can support a classification scheme and an algorithm to perform the classification.

The specification of any process is a formal, implementation-independent, characterization of the process. A tight specification is important as a requirement against which a proposed implementation can be evaluated. The authors are not aware of any existing, formal definition of the term "specification" as it applies to DMP's as defined in this paper. Specs are usually generated using one or more structured analysis methods. Since the flow of control in the context of this paper tends to be very complex, it is difficult to apply these methods. For this reason, it is difficult to verify, validate, benchmark or otherwise evaluate the implementation of a DMP.

The purpose of this paper is to develop a formal definition of "specification" as it applies to DMP's. The DMP spec will be developed in two stages. The first is the logical (L) spec which ensures that the DMP will always make a classification. The second stage is the performance (P) spec which determines how efficient the control strategy of the DMP must be. The division of the spec into these two components means that the designers task can be reduced from one complex effort, to two more manageable tasks which can be performed in tandem. The challenge will be first, to exhibit one logically correct DMP to ensure that the L spec exists, then to select a control strategy which will satisfy the P spec.

The DMP considered here has the following properties:

o Performance is important because time critical action depends on the outcome of the process.
o The relative cost in time or other resources of making sensor measurements varies significantly.

o A result must be produced despite residual uncertainty.

Maintenance diagnostics and fault isolation in a nuclear power plant are an example. Sensors must measure the value of attributes like temperature and radiation levels and the process must select an appropriate action plan.

The use of information theory in designing DMP's is discussed in Reference (1), but resource costs are not considered. Reference (2) emphasized knowledge acquisition, not resource-constrained DMP design.

2 A FORMAL DEFINITION OF A DMP

A DMP consists of the following five elements:

(O, M, S, Po, K) where:

O is a universe of objects.

M is the space of all attribute values for objects in O.

S is a mapping (S: O --> M) which associates a set of attribute values with each object in O.

P_o is a user-specified partition of O.

K is a mapping, (K: M ---> (O/P_o)), by which experts assign objects to the correct element of the partition P_o.

If the mapping K of M into the quotient space O/P_o is single-valued, then the inverse mapping, K^{-1} induces a partition P_m on M and the quotient mapping:

K/P: M/P_m ---> O/P_o

exists and is one to one. Moreover, if K is surjective, then K/P is surjective and every class of objects specified by the end-user will match some subset of sensor indications in the measurement space (M). The partition P_m is defined to be the Logical spec for the DMP. If P_m exists, then any object in O can be identified or classified by assigning its sensor measurements to the correct cell in the partition.

This assignment is done by selecting an algorithm $\hat{S}$ which assigns the attribute values of an object O to the correct class in the partition: $\hat{S}$:M ---> (M/P_m). If the L spec exists, at least one correct implementation of $\hat{S}$ can be generated in the form of a decision tree by backward chaining. The characterization of DMP spec in terms of mappings and partitions is shown in the Figure.

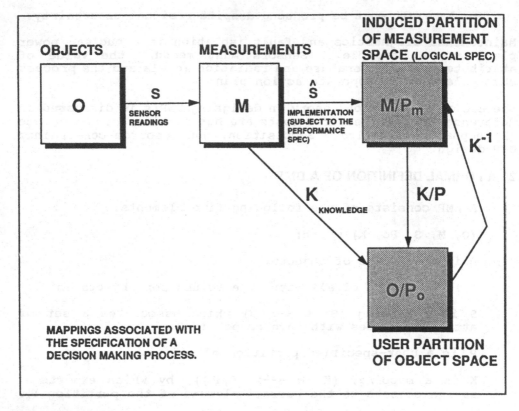

3 PROBLEMS RELATED TO THE LOGICAL SPEC

There are two potential problems with the existence of the L spec. The first is that the mapping K might not be single-valued. This means that one set of sensor indications might correspond to different objects in the environment. This would correspond to indefinite boundaries for the partition P_m in the Figure and would mean that there is residual noise in the system. The second is that the mapping K might not be surjective. This means that at least one class of objects can never be reached by the mapping K, so the knowledge base is incomplete.

The most practical procedure to test for the existence of the L spec may be to attempt to generate one correct decision tree. At least one CASE tool uses this approach (the Knowledge Shaper). (3,4) If Knowledge Shaper fails to produce a decision tree, it will identify the source of the difficulty so that the designer can create one or more ad hoc rules to remove the ambiguity or incompleteness in the knowledge base.

When the designer has modified the knowledge base so that one correct decision tree can be generated, it becomes one (of many) candidates to be the implementation of Ŝ. Unfortunately, there are many possible correct decision trees which cannot be generated by backward chaining. If a "good enough" tree is not found among the trees produced by backward chaining, the selection of the control strategy might have to rely on other tree generation techniques. (5)

4 SELECTING A CONTROL STRATEGY THAT MEETS THE PERFORMANCE SPEC

The design of an efficient control strategy in the form of a decision tree is accomplished by incorporating domain knowledge. Nilsson (6) discusses the relationship between the effort expended in doing this and run-time efficiency. In our application, the designer must consider the ratio between entropy reduction and resource cost at each node of the tree to achieve the best possible performance.

Although, the task of optimizing the control strategy is NP-complete (7), it may be possible to find a sub-optimal solution using CASE tools which rely implicitly on a two-stage definition of specification as presented here.

5   CONCLUSION

One of the conspicuous benefits of the use of a formal spec and compatible CASE tools is maintainability. In many applications, the incorporation of new knowledge must be accomplished with some urgency. If the designer used canonical methods to comply with a formal spec, then the maintainer will be able to modify the process more readily.

REFERENCES

1. Goodman, R.M., "Decision Tree Design Using Information Theory", Knowledge Acquisition, Volume 2, Number 1, March 1990, Pps 1 - 19.

2. Bohanec, M. and Rajkovic, V.; 8th International Workshop, Expert Systems and Their Applications, 30 May to 3 June 1988, Volume 1, Pps. 59 - 78.

3. Cockett, J.R.B and Herrera, J. A., "Prime Rules Based Methodologies Give Inadequate Control", University of Tennessee Knoxville, Department of Computer Science Technical Report, CS-85-60, 1985.

4. Vrba, J. A. and Herrera, J. A., "Knowledge Shaping Provides Expert System Control", Computer Technology Review, Volume VIII, Number 11, September, 1988.

5. Herrera, J. A. and Cockett, J.R.B., "Finding all the Optimal Forms of a Decision Tree," University of Tennessee Knoxville, Department of Computer Science Report, CS-86-66, 1986.

6. Nilsson, N. J., Principles of Artificial Intelligence, Tioga Publishing Company, Palo Alto, CA, 1980.

7. Genesereth, M. R. and Nilsson, N. J., Logical Foundations of Artificial Intelligence, Morgan Kaufmann Publishers, Inc., Los Altos, California, 1987.

Operationalizing Software Reuse as a Problem in Inductive Learning

Robert G. Reynolds, Jonathan I. Maletic, Elena Zannoni

Computer Science Department, Wayne State University
Detroit, Michigan 48202

Abstract. Biggerstaff and Richter suggest that there are four fundamental subtasks associated with operationalizing the reuse process [1]: finding reusable components, understanding these components, modifying these components, and composing components. Each of these subproblems can be re-expressed as a knowledge acquisition problem relative to producing a new representation able to facilitate the reuse process. In the current implementation of the Partial Metrics (PM), the focus is on operationalizing the first two subtasks.

This paper describes how the PM System performs the extraction of reusable procedural knowledge. An explanation of how PM works is carried out thorough the paper using as example the PASCAL system written by Goldberg [4] to implement the Holland's Genetic Algorithm.

Keywords: software reuse, inductive learning, decision trees, chunking, software metrics.

1 Introduction and Related Work

Software reuse is predicated on the idea that problem solving knowledge encoded in the form of computer programs can be reapplied to solve related problems in the future. From this perspective, knowledge must be represented in a framework that facilitates its reuse. In terms of a software system, this corresponds to a new encoding of its structure so as to facilitate its reuse. In order to achieve such a new reformulation, an understanding of the knowledge embodied in the original code is required. Here, achieving such an understanding is viewed as a knowledge acquisition task. This overall task can be broken down into a sequence of knowledge acquisition subtasks.

Biggerstaff and Richter suggest that there are four fundamental subtasks associated with operationalizing the reuse process [1]: finding reusable components, understanding these components, modifying these components, and composing components. Each of these subproblems can be re-expressed as a knowledge acquisition problem relative to producing this new representation.

In the current implementation of the Partial Metrics (PM) system, the focus is on operationalizing the first two subtasks. A component diagram of the PM system is given in figure 1. PM extracts planning knowledge from a software system at three different levels of granularity. Knowledge can be extracted at the system level, procedural level, or code level. For an incoming candidate system, there are three possible passes. The goal of the first pass is to assess the reusability of the system as a whole. If a candidate is accepted for reuse at this level, it can be used as input for a second pass. In this second pass, collections of procedures can be extracted as candidates for reuse. Successful candidates can then be used as input for a third pass. In this pass, code segments can be

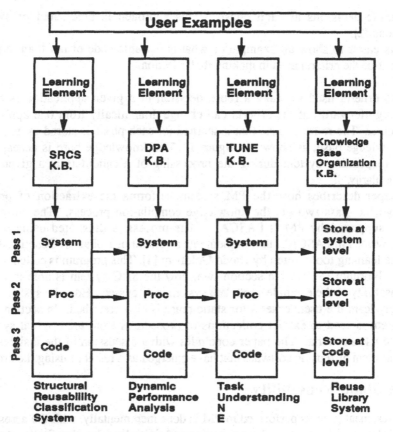

Fig. 1. Component diagram of the PM System

extracted from procedural candidates.

Each pass consists of four basic steps. New information is added at each step to a frame based description of the transaction history of the candidate. The first two steps support the isolation of reusable objects at a given level of granularity, the first phase in Biggerstaff's reuse process. In the first step, the Structural Reusability Classification (SRC) System assesses the static structure of the candidate object relative to user specified criteria for reuse. In the second step the Dynamic Performance Analysis (DPA) System is used to isolate reusable components in the candidate system based upon their performance relative to user specified criteria for reuse. The criteria used for the assessment is a function of the granularity at which the programming knowledge is extracted.

The Task Understanding (TUNE) System supports Biggerstaff's second phase of the reusability process, understanding the component to be reused. The goal here is to produce a generalized description of the context in which the candidate can be reused. In the current version, a faceted classification scheme is used as a basis for automatically classifying and storing candidate modules. The faceted method synthesizes the classification structure from the modules given to the system for storage [7].

Each of the phases described above is supported by a knowledge-base that is used to generate the reuse decisions performed in that phase. The knowledge stored in each is acquired automatically during the configuration phase of the PM system. In the configuration process, the system is provided with examples of acceptable and unacceptable decisions at each phase. This approach is predicated on two basic assumptions [2].

1. It is easier to show an example of what is reusable code or not than to precisely describe the criteria used in making that selection.

2. The criteria used to make a reuse decision in a given application is relatively straightforward but the criteria can change dramatically from one application to another. The reuse criteria that is learned for each phase is stored in its associated knowledge base, as shown in figure 1. That knowledge base is accessed by its corresponding system during the processing of a candidate at a given level of granularity.

This paper describes how the PM system performs the extraction of procedural knowledge, i.e. pass two of the knowledge compilation process. The current target language supported by PM is PASCAL. The process is illustrated using a piece of software, written in PASCAL, that implements a version of Hollands' Genetic Algorithm, a machine learning tool written by David Goldberg [4]. This program is called the Simple Genetic Algorithm (SGA). In section two, how the SRC system is able to assess the static reusability of the example candidate system is discussed. How the system is able to inductively learn the users criteria for static reuse is also described. In section three the DPA system is used to extract collections of procedures from the original system that satisfy the users criteria. The paper concludes with a discussion of the insights into the general problem of reusing software that have emerged as a result of using the prototype.

2 The Static Reusability

The assessment process performed in PM is done incrementally, with the assessment of a candidates static structure being performed initially by the Static Reusability Classification System (SRCS). The assessment of static structure is performed prior to the behavioral assessment for two reasons. Firstly, information for the static analysis can be derived from information generated by a compiler and is less resource intensive than the behavioral analysis. Secondly, ill-structured programs that can complicate behavioral assessment at a given level of granularity are removed early on, since those programs that do not pass the first phase are withdrawn from consideration.

The criteria used to perform the static assessment process will vary with the level of granularity. At the system level the criteria will relate to the complexity of its interface to the environment. At the procedural level, the criteria relates to procedural structure and content. At the code level, the criteria relate to the code structures used in the implementation. In this section the current implementation of the approach used in PM to assess procedural structure will be described.

While the criteria for reuse are provided implicitly by those configuring the system in terms of positive and negative examples, the SRCS system utilizes an explicit domain model to aid in explicitly operationalizing the criteria. The domain model currently in use at the procedural level is given in Figure 2. This is an example of a software quality model, similar to that of [3]. In this model the concept of reuse is decomposed into successively more specific factors, ultimately into metrics that assess reuse in terms of a systems static structure.

In the model, static reuse is broken down into two subsidiary factors, the ease of revision of a system and its portability. Portability is then viewed as a function of two factors, the system machine independence and its modular structure. A system's ease of revision is influenced by factors such as maintainability and testability. Maintainability is, in turn, a function of program documentation, code complexity, and modular complexity. Testability is seen as a function of code complexity and modular complexity. Code complexity is a function of code size, data structure complexity, and control structure complexity. A system's cohesion, coupling and modular structure contribute to its modular complexity. Each of these factors is now measured in terms of one or more structural metrics. This collection of metrics is used to describe the structure of a candidate system. Details concerning the metrics used can be found in [2].

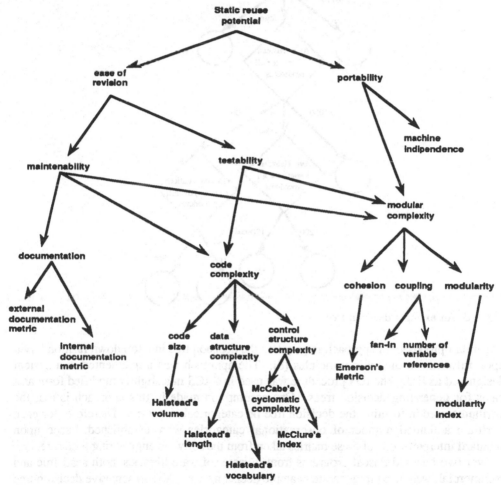

Fig.   2. Quality Model Used to Assess Static Reusability

While the model given above specifies what factors can be important in determining the reusability of software, the acceptable values of and relative importance for each will vary from one application to another. The goal of the learning element associated with the SRCS system is to acquire this information for a specific application. The source of this information is a training set of both positive and negative examples of reusable code. The system attempts to generate a decision procedure expressed in terms of values for selected metrics that is capable of discriminating between the positive and negative examples. This is an example of an inductive concept learning problem in which the concept to be learned is reuse as a function of a program's static structure.

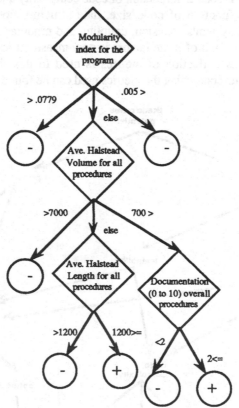

Fig. 3. An example decision tree.

Quinlan proposed an approach to generate the decision tree in a top down fashion based upon information theoretic principles [8]. The approach was implemented in a system designated as ID3. The ID3 procedure was used in SRCS in a slightly modified form as a basis for generating decision trees. One assumption made in this approach is that the attributes used in forming the decision tree be categorical in nature. Therefore, for each attribute a limited number of observational categories were established, based upon standard interpretations of these metrics taken from the software engineering literature.

Over two hundred Pascal programs from diverse software libraries, both academic and commercial, were used to generate example decision trees. A representative decision tree produced from these examples is given in figure 3. Decision nodes that represent the partitioning based upon values for a given attribute are displayed as diamonds. Leaf nodes are represented as circles and correspond to either acceptance (+) or rejection (-) of a module as the result of following the branches dictated by the metric values for the system. Note that this decision tree requires information about each of the major subtrees in the software quality model described earlier; documentation, code complexity, and modularity. Also, based upon fan-in from the higher level concepts one would expect modularity concerns to be most important of the three, followed by code complexity issues, and then documentation. This is exactly what occurs in the example.

In fact, experiments with the same system have led to some general observations about the structure of the generated trees. First for any given application environment, the generated tree is relatively shallow and broad. The average number of decisions needed to classify a system is 3.1, with an average number of 7.4 internal nodes. Second, the decision procedure exhibited performance equal or better than human experiments in making such discriminations, with accuracies ranging from 75% to 98% depending upon the number of training examples used. In general, a training set of 25 or more examples, evenly divided between positive and negative instances, guaranteed accuracies in excess of 90%. Third, attributes associated with the module complexity are most frequently found at or near the root. This reinforces the validity of the quality model used, since in that model, modular structure influences three higher level factors in the underlying quality model: testability, maintainability, and portability. Lastly, although the decision trees for any given application are small , they can change dramatically in terms of the variables used from one application to the next.

Once a decision tree is configured for an application, it is stored in the knowledge base associated with the SRC system. Let us assume that the decision tree just discussed is the currently active one. Each incoming candidate system is then assessed in terms of this decision tree. In order to do this, the metric values required by the currently active decision tree are calculated. Here, the example Pascal program that will be used, implements a simplified version of Holland's Genetic Algorithm, a machine learning tool, that was written by Goldberg. The procedural structure for Goldberg's program is given in figure 4. In order to simplify the diagram, procedures that are called by several others are indexed by integers in the diagram, and the names of the indexed procedures are given.

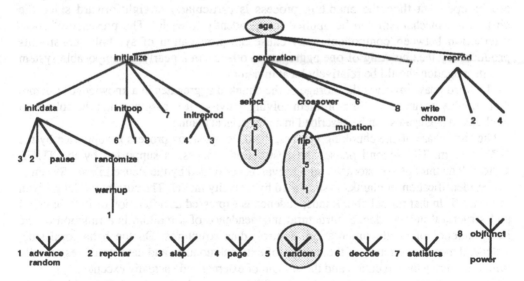

Fig. 4. Structure Chart for the Simple Genetic Algorithm Program.

The modularity index is the first metric used to evaluate the program. This metric is the ratio of the number of procedures to the number of lines of executable code as proposed by Prieto-Diaz [7]. The value here is .07 which falls within the acceptable range. Thus, the next decision is based upon the average Halstead volume for all procedures. The value (~840) is also greater than 700 but much less than 7000. Halstead's volume is defined as

$$(\text{Length}) \, \log_2(\text{vocabulary}) \, ,$$

where length is the total count of operators and operands in the procedure and vocabulary is the total number of unique operators and operands [5]. The next decision is based upon the average length, as defined above, of each procedure. The average length here is ~175 which is again within acceptable limits (>=1200). As a result, the SRC system returns the class associated with the terminal node in the path which is '+'. The program is therefore accepted at this stage and sent on to the DPA subsystem.

3 The Dynamic Performance Analysis Subsystem

The goal of this subsystem is to isolate candidate structures at a given level of granularity. The general model for this process supported here is that of chunking. The chunking process is motivated by the need for humans to associate specific input with more general or abstract patterns. Chunking consists of several distinct phases. The first phase decomposes input into aggregates of low-level objects. These aggregates should be relatively independent in order to simplify the encoding process. Simon observed [11] that many complex systems can be decomposed into a collection of hierarchically structured subsystems, such that the intrasystem interactions are relatively independent of intersystem ones. If there is no dependence between the inter and intra-subsystems, then the system is said to be completely decomposable. Nearly decomposable systems exhibit a few interactions, but still permit efficient encoding to take place [11].

The second phase of the chunking process is encoding or abstraction. If the aggregates are independent then the encoding process is particularly straightforward since the abstraction mechanism can be applied independently to each. The presence of some interaction between components may cause the propagation of symbolic constraints produced by the chunking of one aggregate to others. In a nearly decomposable system this propagation should be relatively local in nature.

The third phase involves the storage of the chunked aggregates in a knowledge structure that enables its future use in a problem solving activity. It is assumed that the collection of chunked aggregates can be described in a hierarchical manner.

The first phase of the chunking process, the decomposition process is supported by the DPA system. The second phase, the abstraction process, is supported by the TUNE system. The third phase, storage and retrieval, is performed by the Reuse Library System.

The identification of chunks is supported by a quality model. The current model is given in figure 5. In that model chunk independence is expressed as a function of its behavioral and structural independence. Structural independence of a module is a function of the independence of its documentation, control, data structures. On the behavioral side, independence is a function of the number of external modules and data structures actually imported during the execution and the amount of external code actually executed.

For a given software system, such as the SGA example here, the set of possible chunks correspond to each of the possible connected subgraphs present in the call graph. The null hypothesis here is that only the whole system itself can be chunked. This corresponds to chunking the whole system with the SGA main module as the root. It is considered the null hypothesis since the system had already been accepted for reuse in the previous phase.

The criteria for selecting a collection of procedures (chunk) reflects the amount of effort

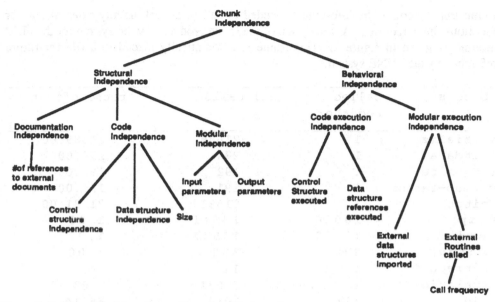

Fig. 5. Software Model to Assess Procedural Independence

required to implement its reuse in a new program. The performance function associated with reuse effort has two components. The first component assesses the call frequency metric for a hypothesized chunk, x, defined as the ratio of the total number of calls made to x (intended as calls to the root of the subgraph corresponding to the chunk), divided by the number of total calls made to the procedures that are members of the hypothesized chunk. This component assesses the cohesiveness of action within the code. The second component reflects the dependency of code on outside information. An increase in the complexity of this component increases the "reuse effort". The metric used is Halstead's Volume, which measure the size of the shared code structures. The ratio of the volume for the shared component relative to the volume for the code component associated with the hypothesis represents the contribution of this second factor.

$$\text{reuse\_effort(H)} = C_1 \times \frac{\text{all calls (members of H)}}{\text{called together (members of H)}} + C_2 \times \frac{\text{volume (code component)}}{\text{volume (shared)}}$$

The reuse effort for a code structure associated with a hypothesis, H, is a weighted linear function of the two components. Currently, the amount of reuse effort for a hypothesized collection to be acceptable is constrained to be close to one. It should take little or no effort to reuse it. The current prototype of the DPA system computes the first component, automatically generating the collections of procedures associated with acceptably low reuse effort values.

The averaged values for the call frequency component for all the runs of the SGA program are given in Figure 6. Each hypothesized chunk is labeled by its root node. The average number of calls made to the chunk per run is then given followed by the total number of calls per run made to all the procedures of the chunk by any procedure of the system. The selected hypotheses are those with a reuse effort lower than 2. The selected collections are circled in Figure 4. Note that these aggregates are all positioned at low levels in the call graph. Once the acceptable aggregates are isolated, the system prepares a frame that describes the information needed to call it, as well as any external data or functions that it requires. A description of the frame produced by the system for the FLIP module is given in figure 7. This frame is automatically placed in a file for future reference by the TUNE system.

objects	called together	all calls	reuse effort
initialize	1	21689	21689.00
initdata	1	274	274.00
randomize	1	192	192.00
warmup-random	1	191	191.00
initpop	1	21581	21581.00
flip	**10050**	**10541**	**1.05**
random	**10350**	**10557**	**1.02**
objfunc	**330**	**330**	**1.00**
initreport	1	14	14.00
generation	10	31031	3103.10
select	300	10541	35.14
crossover	150	29591	197.27
mutation	9000	20591	2.29
report	10	686	68.60

Fig. 6. Selection of Reusable Objects for the Simple Genetic Algorithm Program

```
              Object  :  flip
****** Object Global Variables ******
global  jrand        : integer
global  oldrand      : array[1..55] of real

********* Object Interfaces *********
procedure_type_qualifier : function
procedure_name           : flip
procedure-result-type    : boolean
Formal-Parameter-List    : probability : Real

********** Object Contents **********
function flip(probability :real) : boolean ;
begin
    if probability=1.0 then flip:=true
    else flip:=(random<=probability);
end;

function random : real;
begin
    jrand := jrand + 1;
    if (jrand > 55) then begin
      jrand := 1;
      advance_random;
```

```
      end;
      random := oldrand[jrand];
   end;


   procedure advance_random;
   var   j1          : integer;
         new_random : real;
   begin
      for j1 := 1 to 24 do begin
        new_random:=oldrand[j1]-oldrand[j1+31];
        if (new_random <0.0) then new_random:= new_random+1.0;
        oldrand[j1] := new_random;
      end;
      for j1 := 25 to 25 do begin
        new_random:=oldrand[j1]-oldrand[j1-24];
        if (new_random <0.0) then new_random:= new_random+1.0;
        oldrand[j1] := new_random;
      end;
   end;
```

Fig. 7. The description of a candidate code object produced by the DPA subsystem.

4 The Task Understanding Element

As a result of the above processes, a code object that is given to the Task Understanding Element (TUNE) possesses high structural decomposability and behavioral integrity. The goal of TUNE is to use the structural and behavioral information to insert the object into a hierarchically arranged collection of tasks. The task structure can be predefined by a user and generated by the system. The prototype for this subsystem is under development. A brief overview of its proposed structure will now be given.

Each task possesses a frame description. The contents of the frame for an incoming module are compared with those for leaf node tasks. If an acceptable fit is produced, then the code is stored relative to this task in a Case Base. If an acceptable fit is not produced then a new task is created. The frame description for this new task is the current data for the new individual originating the task.

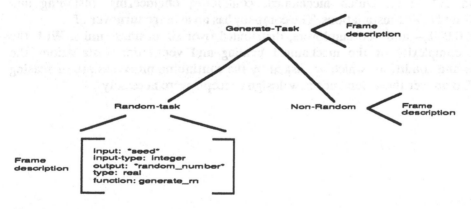

Fig. 8. Task Hierarchy

Towards a Real Cad-system using AI technology

Ir. W. Zeiler

Kropman B.V. Mechanical Contracting
Verrijn Stuartlaan 21, 2288 EK Rijswijk, The Netherlands

Abstract. In this paper the Kropman Intelligent Computer Aided Design system is presented, the development of an Intelligent Computer Aided Design system for integral design, building and application of mechanical HVAC installations. Up to now the use of the computer in the engineering phase is limited to automised drawing, assisted by a number of stand-alone applications. The ICAD-system is able to support the entire engineering process. This can be achieved by providing the system with an amount of knowledge about the design of systems. By means of a hierarchical, functional structuring of the process components, based on a meta model, it is possible to consider the entire systems from different aspective views. The base of the ICAD-concept is an object oriented specification language. With this language the model can be described and processed untill it answers the functional specifications. Up to now not all applications have been implemented but some are already been used in the daily working process. From this experience it can be concluded that the ICAD-concept is a very usefull guide-line for further applications of A.I. The intention of Kropman B.V. is to use the system primarily for the optimalisation of their own design and engineering process.

1   Introduction

1.1   Background

Kropman B.V. is a Dutch mechanical contractor, engineering, installing and maintaining HVAC-installations. The company has an average turnover of
f. 106.000.000.,-- and 480 employers, distributed over six business units. With the growing complexity of the mechanical heating and ventilation installations the demands and conditions which are asked by the continuing processes are increasing rapidly. To answer these demands new design concepts were necessary.

As a new object is associated with a task, the frame information in the current task is generalized or specialized in order to be consistent with the new example. Conceptual clustering, as developed by Michalski, is used to support this activity [6]. An example of a portion of a task hierarchy that might support the FLIP hypothesis is show in figure 8. The asterisks bracket variables that are to be bound with elements in the candidate code object. The other values are constants that represent constraints on the bindings that can occur.

5 Conclusion

In this paper it was demonstrated how inductive learning techniques can support the acquisition of knowledge about procedural reuse, and how this knowledge can be applied to the identification of reusable procedures. In fact, it is suggested that the properties associated with reusable software components are easy to check. The problem is that the criteria can change dramatically from one application environment to another. That is why it is essential to have a system that can efficiently learn this criteria from examples provided by the user. The current PM system is able to support the learning of relevant criteria using appropriate domain knowledge as demonstrated by its processing of the SGA program.

References

1. T. Biggerstaff, C. Richter: Reusability Framework, Assessment, and Directions. IEEE Software, Vol. 4, No. 2. March 1987.

2. J.C. Esteva, R.G. Reynolds: Learning to Recognize Reusable Software by Induction. IJSEKE, Vol. 1 No.3 1991.

3. N. E. Fenton: Software Metrics: A Rigorous Approach. Chapman and Hall Press, London, 1991.

4. D. Goldberg: Genetic Algorithms in Search, Optimization and Machine Learning. Addison-Wesley Publishing, 1989.

5. M.H. Halstead: Elements of Software Science. Elsevier-North Holland, New York, NY, 1977.

6. R.S. Michalski, R.E. Stepp: Conceptual Clustering of structured Objects: A Goal Oriented Approach. Artificial Intelligence, Vol. 28, pp 43-67, 1986.

7. R. Prieto-Diaz: A Software Classification Scheme. Ph.D. Thesis, Dept. ICS, University of California, Irvine, 1985.

8. J.R. Quinlan: Induction of Decision Trees. Machine Learning, Vol. 1, No. 1, pp 81-106, 1986.

9. H. A. Simon: The Sciences of the Artificial. M.I.T. Press, Cambridge, Mass., 1969.

1.2   Approach to intergrated system design

Already in 1985 Kropman has started research on Expert Systems, in 1988 their first process control expert system was ready. This system was built for the monitoring and process control of the installations of the burned people hospital of the Zuiderziekenhuis in Rotterdam. This system worked quite well, but because the knowledge engineering of the system took quite a lot of time, it was decided that this was not the way to go. Also the system, which was built in the Expert system shell Personal Consultant Plus, was rather slow in responce. To solve this problem the original programme was transfered to Turbo Prolog, which gave a considerable advantage in runtime. For the next project there was not a systematically approach for the knowledge engineering but rappid prototyping was used. This resulted much faster in a working system, but as it contained quite a lot of omissions and because of the poor possibilities of maintaining the knowledge software this approach was never used again. The conclusion of this experience was that the re-engineering necessary for knowledge engineering wasn't an effective way of approach. This led to the idea of integrating the building of the expertsystem within the engineering process itself and use it as an intelligent computer aided design tool.

1.3   Objectives

In order to achieve a better integration of the companies activities it is necessary to combine all data concerning one project, from management level, supervision level through process control level. An essential point of this concept is the distribution of information with a multi-user system. Untill this moment the facilities of computer aided design are still used in an inadequate way. By viewing the building services installation and the engineering process from a different angle it is possible to create a design system with a higher degree of support. Starting from this new vision, the Intelligent Computer Aided Design concept, optimal use of the computer- resources during the engineering phase will be realised. This ICAD-concept is not an entirely new approach [2,4,5] but the pragmatic modulair way which Kropman has taken is new. The Kropman approach [7,8,9] is principaly based on Struck's dissertation [3]. This vision requires a central model description of the installation. This model, named metamodel, consists of all the components of the installation and their individual relations. The metamodel gives a description of the design object which is independant of any aspect. From this model it should be possible to execute the different engineering tasks (figure 1). Facts, generated during the execution of these tasks, are added to the model. Aspect models are used to focus on a certain aspect of the design, which is created by executing a scenario. The rules of the scenario, which can be a calculation programm or a knowledge module, represent the declarative knowledge of the aspect, the functions of the scenario denote the procedural knowledge.

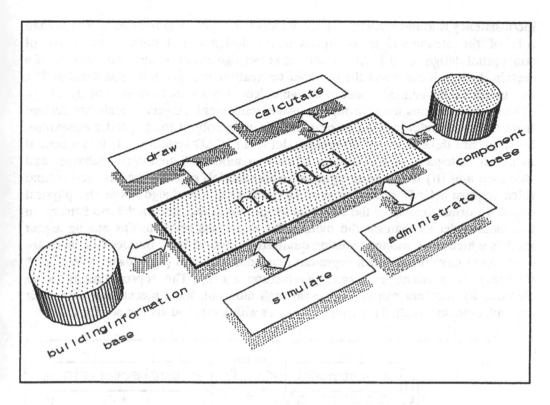

2   Metamodel

The first role of the metamodel e.g. maintaining relationships among models is realized by the metamodel mechanism. In the metamodel mechanism is a model representing qualitative relationships among aspect models. An aspect model assumes a set of definitions of concepts for representing properties of a design object. The definitions are the background theory of the aspect model. For instance algebraic geometry is a background theory of a solid model. It defines concepts such as vertex edge, face, and solid object. An aspect model is a selective representation of design object filtered by its background theory. Aspect models are generated by interpreting a metamodel in the context of its background theory. Thus relationships between aspect models essentially originate from relationships between their background theories. This means that knowledge for integrating aspect models can be represented as relationships among concepts defined in their background theories. Knowledge about relationships among aspect models is represented on the level of background theories, rather than on the level of aspect models. There is an advantage of representing knowledge on the level of background theory. Since the knowledge can be independent of a particular representation of the aspect model we can use a coherent universal definiton of concepts to represent relationships among aspect models. Besides the representation of the aspect model itself the metamodel represents the dependecies among aspect models. When an aspect model is modified, validity of relationships is evaluated by the metamodel mechanism. If an inconsistency is found relevant aspect models are adjusted to remove it. The second role of the metamodel is to represent the design object during the course of conceptual design. If an ICAD sytem has knowledge about physical phenomena of a certain domain it can assist the designer to create new artifacts in that domain. The metamodel is the central model of design object. It is a serie of models of the design object which obtains data through stepwise refinement. Aspect models are derived from the metamodel in order to evaluate the design object from specific viewpoints. So there are three roles of the metamodel in an ICAD system i.e. I) it is a central model to integrate aspect models II) it is a model of qualitative behavior and structure and III) it is a working space where models evolve. The first and second roles require a CAD system to possess ontological knowledge about the physical world. In order to achieve the first role we proposed the metamodel mechanism. In the metamodel mechanism the metamodel represents relationships among aspect models which are reasoned by using qualitative physical laws. Besides this there are other local mechanism which represent depended variables (e.g. Power = Voltage * Current) or constant facts (e.g. maufacture: Stork). The representation of an elementairy problem can be danoe through different, local mechanism with their dependencies as so called dependency graph's with knots and arrows, see fig. 2.

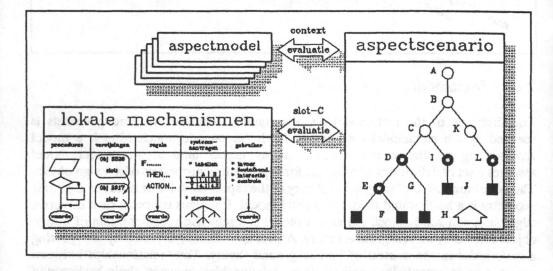

3   The specification language

The implementation of knowledge is done by the so called specification language. By means of a language the design is specified and translated on computer level. After this the model can be reworked. The specification language is based on the object representation. The object-orientation makes it possible to order features in taxomonies of classes or families. The subclasses describes the further specialities and inherit their features form their "parents". Expressions which exists of different features of the parents and determine the features of the "children".Knowledge which is put to the ICAD-system, must be put as less as possible in source-code, but has to have the form of knowledge rules and be present as a stand-alone module. To put all kind of knowledge to the ICAD-system it is necessary to have a flexible system. The introduction of semantics by putting in context of the object data and algoritms allows the user to provide the system with knowledge (figure 3).

To make this possible the objects have got slots. A slot is an object that holds information regarding a particular attribute of a unit. A unit can have as many slots as desired. Unit structure is similar to class structure, except that a unit is like a table storing all data regarding characteristics of specific objects. A unit can have sub-classes. Slots can be filled up with all kinds of things: values, formulas, knowledge rules, programm calls, search tasks for databases etc. Slots makes it possible to activate objects to structures within the programm, with their own knowledge and functionality. This is called local integration of data and algoritms. This forms the base of the specification language. The

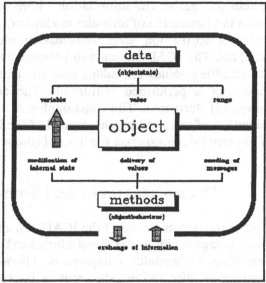

Figure 3: the object.

specification language contains knowledge, which can be used and presented if ask for by the designer. The designer is being assisted by the CAD-system because he can insert the slots with the available rules and knowledge of the system. This object orientated programming is being done in C^{++} and partially in Lisp. The engineering process has the character of a top-down discription annex specification phase, which is being followed by a bottom up engineering phase. The phases have an iterative character, but the changing from top-down to bottom-up phases is clearly visable (figure 4).

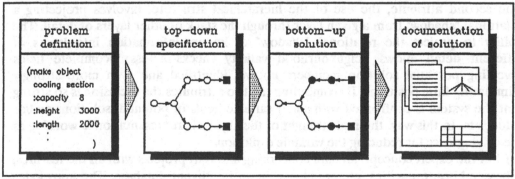

Figure 4: stepwise refinement of the model.

During the top-down phase the designer is working mainly creative. He generates alternative solutions for his designproblem. During the engineering phase the designer works out the globale shape. During this process he can be supported by the knowledge which has been implemented in the design language. He can instruct the ICAD-system to execute the calculation of a powerrange within a slot. The ICAD-system will gather the information within the formula or knowledge rule. This can mean the evaluation of formulas in another slot from the same or another object on a different abstraction level. Information which is lacking will be asked from the designer. The ICAD-system will proceed untill all necessary facts are gathered and the resulting values are added into the slots. The calculations and the knowledge reasoning is performed bottom-up. The designer is always the one making the important decissions. The functionality of the ICAD-system will be limited to the execution of calculations, the control of the input from the designer and the offering of alternative solutions on the level of components to choose from.

4 The modular principle: Hierarchical Abstraction

One of the corner-stones of the ICAD-concept is the modular principle, by which the bases is created for a functional hierachical structuring based on abstraction level, a modelling in modular components. Hierarchical abstraction means the decomposition of information into levels of increasing detail, where each level is used to define the entities in the level above. In this sense each level forms the abstact primitives of the level above. These terms in the upper level, form condensed expressions of a given relational and/or operational combination of primitives from the level below. In any given level of detail the representation is based on a partly repetition of a structural pattern. In this pattern are places for each combinatorial product which can be made of the variables defined. The repetition of the pattern has the property of excluding elements in a search neighbourhood faster than it includes new ones. This property greatly reduces the variable explosion problem when adding new facts at a given level.

The second attribute, the use of the hierarchical structure, involves projecting a relational "shadow" from a given layer through the stack of other layers of detail. The ability to project the relational "shadow" of known information into layers of different detail allows neighbourhood validity checks across incomplete fields revealing potential solution members not yet identified and even more strongly eliminating false positions. By combining the two attributes the exclusion is so strong that the system is confronted with small variable fields of potential solutions and/or value sets. In this way, the real strenght of the algebraic representation is working as the drivingforce for reducing the variable explosion.
Within the expert concept, all data concerning different projects with all the features, subassemblies, capacities, etc are stored in an intelligent database. The knowledge system is able to configure the mainlines by which an optimal specific installation is generated. The execution is performed interactive using the specification language.

As the system grows, it obtains access to more and more data, and it is possible to contribute more, using the modular principle to the final design.

5   Present experience

This concept is first being tested on the control section of the mechanical installations. As the result of this approach the system generates the main- and controlcurrent diagrams using the interactive specification language. This system is already beyond the testing phase and has been used for the engineering of the control section of several building service designs. The functionality of the system has been extented with further modules; such as Kvs-calculations, valve selection, measurementlists, and prefab drawings of hydraulic systems.

Object-oriented programming offers the opportunity of common objets gathered in classes to be put into so called libraries. The objects can be used again and again because the classes are independent of an application and form seperate units with their own semantics. Through this they can be used as building stones for new and different solution structures again and again. Within the AI-CAD project there has been started with an object-library. With an in C++ written programm [1,10] the ICAD-system can innitiate a symbol library by DXF-files transformation. The active symbol library contains only the drawing symbols which are nescecary on that moment. These libraries are put in the different databases. These databases, elements in a central database system, are the integrating elements between user and the knowledge of the experts, which is put into the system as building stones (see fig. 5) the so called object-prototypes which can be filled through teh aspect tasks, tabel 5.1, 5.2 and 5.3.

The design system is acting as an experienced adviser working close with and supporting the human designer e.g. process operator. Consequently the creativity and interpetive capabilities of the human designer are used as good as possible in connection with the possibilities of the automised system to compare and to verify large quantities of connected facts. Through this teamwork of computer and human new possible strategy for process optimalisation will be generated. At present the first applicatons are being used succesfull in daily use. For the present working configuration see fig. 6.

```
if (!strcmp (object_name, "Circulatiepomp"))
{
obj -> SetWidthHeight (2, 2);
asp_list.Append (new ActiveScreenPart (obj, x, y, x + obj -> GetWidth(), y + obj -> GetHeight()));
obj -> AddConnection (WARMW_AANVOER, FLOW_IN, 0, 1, FIXED);
obj -> AddConnection (WARMW_AANVOER, FLOW_OUT, 2, 1, FIXED);
obj -> AddConnection (NO_MEDIUM, FLOW_IN, 1, 0, FIXED); //objecten op gelijk niveau
obj -> AddConnection (NO_MEDIUM, FLOW_OUT, 1, 2, FIXED);

obj -> CreateAndAddSlot ("fabrikaat",SLOT_STRING,"-","?",-1,-1,-1,"");
obj -> CreateAndAddSlot ("type",SLOT_STRING,"-","?",-1,-1,-1,"");
obj -> CreateAndAddSlot ("nominale_doorlaat",SLOT_INT,"mm","?",-1,-1,-1,"");
obj -> CreateAndAddSlot ("drukklasse",SLOT_INT,"PN","?",-1,-1,-1,"");
obj -> CreateAndAddSlot ("flensdiameter",SLOT_INT,"mm","?",-1,-1,-1,"");
obj -> CreateAndAddSlot ("hoogte",SLOT_INT,"mm","?",-1,-1,-1,"");
obj -> CreateAndAddSlot ("breedte",SLOT_INT,"mm","?",-1,-1,-1,"");
obj -> CreateAndAddSlot ("diepte",SLOT_INT,"mm","?",-1,-1,-1,"");
obj -> CreateAndAddSlot ("soort",SLOT_STRING,"-","?",-1,-1,-1,"");
obj -> CreateAndAddSlot ("prijs",SLOT_DOUBLE,"f","?",-1,-1,-1,"");
obj -> CreateAndAddSlot ("stuksnummer",SLOT_INT,"-","?",-1,-1,-1,"");
obj -> CreateAndAddSlot ("x",SLOT_INT,"-","?",-1,-1,-1,"");
obj -> CreateAndAddSlot ("y",SLOT_INT,"-","?",-1,-1,-1,"");
obj -> CreateAndAddSlot ("filenaam",SLOT_STRING,"-","?",-1,-1,-1,"pomp");
obj -> CreateAndAddSlot ("is_a",SLOT_STRING,"-","?",-1,-1,-1,"appendage");

return obj;
}
```

tabel 5.1: object-prototype

Object: Circulatiepomp				
Part of: Warmwaterregeling				
Connections:				
medium	dir	x	y	destination
Warm aanvoer	Flow in	0	1	pipenet
Warm aanvoer	Flow out	2	1	pipenet
No medium	Flow in	1	0	-
No medium	Flow out	1	2	-

Slots:
- fabrikaat: Wilo -
- type: DPN100/200-3/4 -
- nominale_doorlaat: 0 mm
- drukklasse: 0 PN
- flensdiameter: 0 mm
- hoogte: 0 mm
- breedte: 0 mm
- diepte: 0 mm
- soort: none -
- prijs: 0.00 f
- stuksnummer: 0 -
- x: 0 mm
- y: 0 mm
- filenaam: pomp -
- is_a: appendage -

tabel 5.2: object vóór aspecttaak

Object: Circulatiepomp				
Part of: Warmwaterregeling				
Connections:				
medium	dir	x	y	destination
Warm aanvoer	Flow in	0	1	pipenet
Warm aanvoer	Flow out	2	1	pipenet
No medium	Flow in	1	0	pipenet
No medium	Flow out	1	2	pipenet

Slots:
- fabrikaat: Wilo -
- type: DPN100/200-3/4 -
- nominale_doorlaat: 100 mm
- drukklasse: 16 PN
- flensdiameter: 220 mm
- hoogte: 550 mm
- breedte: 368 mm
- diepte: 546 mm
- soort: dubbel -
- prijs: 5885.00 f
- stuksnummer: 17 -
- x: 5500 mm
- y: 2250 mm
- filenaam: pomp -
- is_a: appendage -

tabel 5.3: object ná aspecttaak

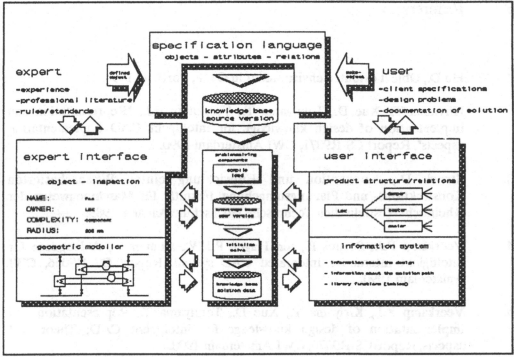

figure 5 system building stones

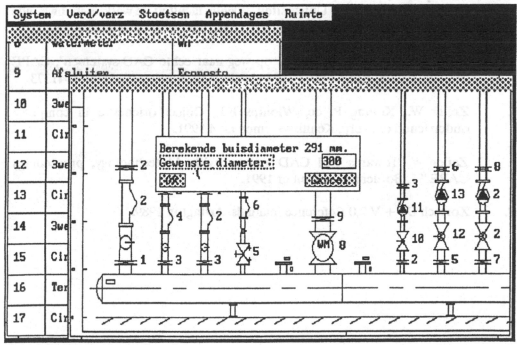

figure 6 Present working configuration

6 References

1. Hu D., Object oriented environment in C++, Portland 1990.

2. Kiriyama T., Xue D., Tomiyama T., Veerkamp P.J., "Representation and implemention of design knowledge for intelligent CAD implementation aspects", Report CS-R9071, CWI Amsterdam 1990.

3. Struck D., Konzeption und Entwicklung einter Wissensbasierten Konstruktions- und Planungsumgebung, Fakultät für Maschinenwesen der Rheinisch-Westfälischen Technische Hochschule Aachen, 1989.

4. Veerkamp P.J., Bernus P., ten Hagen P.J.W., Akman V., A language for intelligient interactive integrated CAD system, Report CS R9056, CWI Amsterdam 1990.

5. Veerkamp P.J., Kiriyama T., Xue D., Tomiyama T., Representation and implementation of design knowledge for intelligent CAD; Theoretical aspects, ReportCS-R9070, CWI Amsterdam 1990.

6. Winnips F.J., Ontwerpen en kennissystemen in de installatietechniek, OC-D-200 afstudeerverslag Universiteit Twente, faculteit de werktuigbouwkunde vakgroep ontwerp- en constructieleer, Enschede 1991, Kropman Rijswijk.

7. Zeiler W., "Met behulp van AI op weg naar echte CAD-systemen", CAPE '91. Rotterdam, April 1991, CIM in Nederland, Samson ISBN 9014041373.

8. Zeiler W., Koning P. de, Winnips F.J., Object orientatie in computer ondersteund ontwerp, Kennissystemen nr. 4 1991.

9. Zeiler W. Towards real CAD systems with AI technology, proceedings CAPE '91, Bordeaux, september 1991.

10. Zortech C++ V 2,0 Reference manuals, Arlington 1989.

WIZ* - A Prototype for Knowledge-Based Drawing Interpretation

B. Pasternak, G. Gabrielides, R. Sprengel

Labor für Künstliche Intelligenz – Universität Hamburg
Bodenstedtstraße 16 – 2000 Hamburg 50 – Germany

Abstract. The WIZ-system is a prototype for interpretation of technical drawings. The paper introduces drawing interpretation as a knowledge processing task. Starting from a discussion about the necessary kinds of knowledge we present our frame-based knowledge structure and explain its implementation using a generic blackboard model, having easy adaptibility to new application domains in mind. The main part of the paper deals with the basic interpretation cycle which exploits a-priori knowledge about the drawing structure and associated geometrical constraints. This is discussed in detail for an example from the area of engineering drawings. The interpretation starts with the vectorization primitives and stops when physical objects (axles, cylinders, etc.) and non-physical dimensioning objects (dimension sets, metric arrows, etc.) are recognized. It is shown that the results of drawing interpretation can be used for domain-specific tasks, e.g. dimension checking or drawing redesign. Finally, a graph-based strategy for dimension checking is described in detail.

Keywords: Drawing Interpretation, Knowledge-Based System, Blackboard Model

1 Introduction

Technical drawings are compound documents conveying a mixture of graphical and textual information. The goal of computer-based drawing interpretation systems is to capture the information contained in technical drawings and to convert it into electronic form representing the physical properties of the drawn objects. The interpretation is based on the results of vectorization, which generates a description of the drawing in form of unstructured collections of vector primitives. An important prerequisite for achieving an interpretation is the use of knowledge about the structure of drawings. The next paragraphs highlight some interesting approaches from the knowledge representation point of view.

An early attempt towards the design of a drawing analysis system was DELPHI [2]. It is characterized by a separation between domain-dependent and domain-independent knowledge, a hierarchically structured knowledge representation in form of frame-like structures, and a blackboard-based administration. Unfortunately, it remains only at the theoretical level.

The AI-MUDAMS Recognizer [6] is an interpretation system for hand-made mechanical drawings. Beginning with a raster image it employs a production system

---

* WIZ is an acronym of the german term "Wissensbasierte Interpretation von Zeichnungen".

that attempts recognition of line types and arrows. The design of the system concentrates on efficiency. Therefore, a procedural knowledge representation is used. This has disadvantages with respect to the systems adaptability to different application domains.

Another interpretation system that is based on procedural knowledge representation is presented in [5]. This system analyzes 3-view mechnical drawings to obtain a 3D representation. Data and intermediate results are kept in a blackboard-fashioned database. An additional feature of this system is the use of the dimensioning of the different views to obtain the size of the drawn object.

In contrast to the above mentioned systems ANON [7] uses a declarative knowledge representation. ANON is a knowledge-based image analysis system intended to extract elements and symbols from mechanical engineering drawings. The system operates directly on the image, combining the extraction of descriptive primitives with their interpretation. However, the interpretation does not go beyond the recognition of line types and dimension symbols.

The WIZ-system concentrates on the high-level interpretation of machine-made drawings, which is based on a set of vector primitives extracted from a scanned image by the AEG Gradas-GES vectorization system. One of the design criteria has been an easy adaptability to new application domains. This requires a primarily declarative knowledge representation and a decomposition of the knowledge base into independent parts. The structuring of the knowledge is described in Section 2, Section 3 describes the corresponding representation techniques. The system's architecture and design are presented in Sections 4 and 5 respectively. The system pursues an interpretation up to the recognition of physical objects (e.g. axles, plates) and their dimensioning. This is the basis for further tasks, like dimension checking and drawing redesign (Section 6).

2 The Role of Knowledge

To obtain an interpretation of a drawing, a-priori knowledge must be embedded in the interpretation system. The interpretation is seen as the process of applying a-priori knowledge to given data.

2.1 Domain Dependency of Knowledge

In general a drawing interpretation system should deal with technical drawings from different application domains. Therefore the knowledge used for interpretation is divided into domain-dependent (specific) knowledge, which is applicable only to a certain class of drawings, and domain-independent (general) knowledge, which is valid in the wider area of technical drawings. In order to perform an adaption to a new application domain with an acceptable amount of effort, the domain-dependent knowledge should be kept separate from the domain-independent knowledge and must be formulated in a well-structured way that allows easy modification. The basic domain-dependent knowledge can be structured into several kinds of knowledge. Despite the fact that the knowledge is domain specific its underlying structure is widely domain independent. The main kinds of knowledge are:

Knowledge about Objects and their Properties. The drawing consists of basic objects (e.g. lines, arrows, arcs, texts, etc.) that are extracted from the bitmap image of the paper drawing during a vectorization process. Knowledge pertains to higher-level objects based on these graphic primitives. These higher-level objects can be physical (like cylinder, axle, etc.) or non-physical (like metric-arrow, dimension-set, etc.). The objects have different properties. Important properties of the graphic primitives are position and orientation. Higher-level objects have additional properties like size, shape, etc. .

Knowledge about Structural Relations. Structural relations between objects are essential for building higher-level objects from lower-level ones. Based on decomposition a hierarchy of objects can be defined. For example, a dimension-set in an engineering drawing consists of a double-headed arrow, a text string and optional subsidiary lines. Although the decomposition hierarchy is the central structural relation, additional structural relations may be required in specific domains.

Knowledge about Geometrical Constraints. In addition to knowledge about structural relations (which determines which objects can be combined to higher-level objects) there is also knowledge about geometrical constraints which must be fulfilled, e.g. collinearity of dimensioning arrows.

3 Representation of Knowledge

The a-priori knowledge, upon which the interpretation process is based, is represented in a frame-based way.

Knowledge about object types is represented in concepts. Concepts are generic, because they do not describe the properties of concrete objects but the properties that are true for all objects of this type. Based on these concepts a taxonomic and decomposition hierarchy is built, according to techniques from object-oriented programming.

In addition to the decomposition through HAS-PARTS relations, the necessary geometrical constraints must be expressed. For example, in the domain of axles the contour lines of each cylinder must be symmetric to a center line. Knowledge about the geometrical constraints is embedded in knowledge sources that are described in more detail later on.

Only the knowledge about the basic concepts that describe the graphic primitives and the knowledge about the very basic geometrical relations (like parallelism, symmetry, etc.) are domain independent. These basic concepts and fundamental geometrical relations can be viewed as built-in.

This kind of knowledge encapsulation into frames and knowledge sources allows a modular definition of knowledge bases. One of the advantages of this modularity is the ease of reuse of already defined knowledge sources in new application domains.

4 The System's Architecture

Figure 1 shows the basic structure of the architecture of the WIZ-system. The a-priori knowledge about concepts is formulated in the generic knowledge structure.

The objects on the blackboard are instances of the concepts, created by knowledge sources, if all necessary parts of the concept are available and the geometrical constraints concerning these objects are fulfilled.

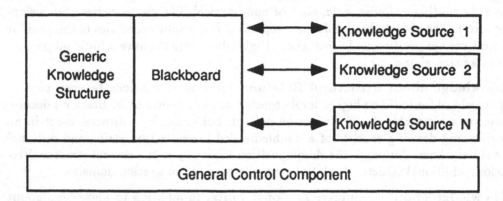

Fig. 1. The basic architecture of WIZ

4.1 Basic Cycle of Processing

The interpretation process is an interaction between the blackboard and the knowledge sources. Each change on the blackboard caused by inserting a new object releases a signal to the knowledge sources called an event. Events carry information about the type of the affected objects. Each knowledge source possesses activation conditions that contain a list of events on which they react and a list of conditions that must be fulfilled by the object that causes the event. When a new event occurs, each corresponding knowledge source specialized for this type of event becomes activated. If the activation conditions (e.g. geometrical constraints) are satisfied an appropriate entry into the agenda of the control component is made.

A scheduler executes the action part of the knowledge source that has been registered in the agenda. The action part usually contains instructions for creating new blackboard objects. The processing cycle starts again until a knowledge source terminates the process or the agenda becomes empty. At the end of interpretation the blackboard contains the results as related objects that constitute an instantiated subtree of the decomposition hierarchy of the a-priori knowledge.

4.2 Efficient Blackboard Operations

A blackboard should manage all interpretation results and hypotheses, while the knowledge sources should have the fastest possible access to the blackboard objects. In this context, fast and direct access is very important for the performance of blackboard systems. Therefore special access functions for blackboard objects are used. These functions are equipped with conditions that reduce the search space. The acceleration of the interpretation process is important for large volumes of data which are expected in the case of technical drawing interpretation.

4.3 Specialized Blackboards

The objects represented on the blackboard have different geometric properties. For example, a circle has position, but no orientation. On the other hand a line has position and orientation. Therefore, specialized blackboards are defined that allow the representation of objects according to their main geometric properties. The WIZ-system defines two blackboards specialized for the representation of position and orientation.

4.4 Geometric Access Functions

Geometric access functions are the result of the combination of geometric specialized blackboards and object access functions with search space reduction. For example, suppose that an arrow is represented on the blackboard for oriented objects. The access to an arrow in the opposite direction can be done simply by using the built-in access functions and reducing the search space to the desired orientation. These geometric access functions are the building blocks for the representation of the geometrical constraints inside the knowledge sources.

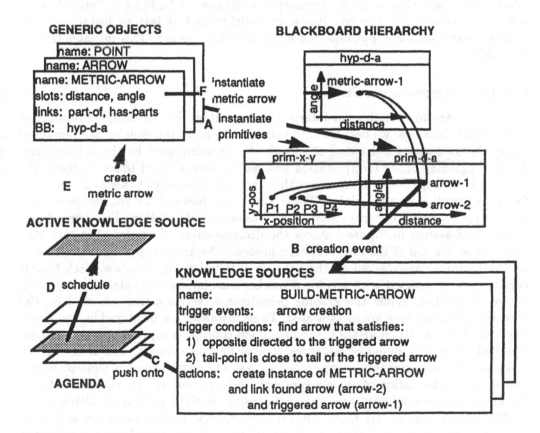

Fig. 2. Interpretation of two arrows as a metric arrow

4.5 Generic Blackboard Developement System

To reduce the implementation effort of the above mentioned architecture the commercially available generic blackboard development system GBB [3, 4] has been used. This tool predefines generic blackboards, blackboard objects, and control shells, that can be specialized to the application needs.

5 Design of the Interpretation System

This section describes the design of the WIZ-system based on the architecture presented in Section 4. The first part of this section explains the interpretation cycle by means of an example. The second part introduces an interpretation strategy in a subdomain of engineering drawings.

5.1 The Interpretation Cycle

Figure 2 illustrates a typical interpretation cycle. The starting point is given by instanting two graphic primitives (*arrow*). These arrows trigger a corresponding knowledge source that tests the geometric conditions for building a double-headed arrow (*metric-arrow*). The conditions are satisfied and at last an instance of the concept *metric-arrow* is put onto the hypotheses blackboard. This initiates further processing.

5.2 The Interpretation Strategy

As preliminary to investing an interpretation strategy in the complex domain of engineering drawings, we decided to do the first steps in the simpler subdomain of axles. Figure 3 shows a drawing of an axle that is interpreted by the WIZ-system. At the beginning of the interpretation process the user is asked to select one of the existing knowledge bases. Momentary there are two knowledge bases available - one for axles, another one for plates. These knowledge bases share knowledge sources for common tasks, like recognition of dimensioning information. The screendump of the WIZ-system in Figure 4 shows the decomposition hierarchy of instantiated concepts at the end of the interpretation process. The knowledge sources that build up the hierarchy are stacked on the left side of the hierarchy window. Each knowledge source is depicted with a colored marker that indicates its status - triggered, active or finished. While the graphic primitives of vectorization are read in, the corresponding concepts are instantiated. These instances are displayed in the 'Primitives' area of the HIERARCHY window. Each concept is represented by a named box including the number of instances created for this concept. The HAS-PARTS relation between instances is represented by a solid line between the concept boxes. During the instantiation of objects various knowledge sources are triggered. At first the knowledge source DRAW-KS displays the instances of the graphic primitives in the PRIMITIVES window. By the use of priority settings the interpretation is divided into two sequential paths. First the physical objects in the drawing are interpreted until reaching the top-level concept *axle*, then the dimensioning is interpreted up to the concept *dimension-set*.

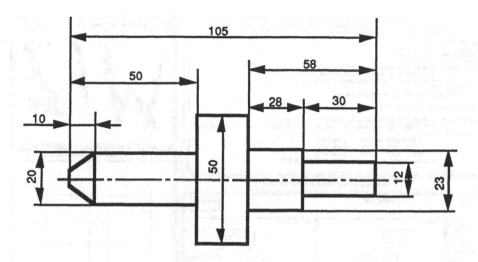

Fig. 3. An engineering drawing of an axle

5.3 Interpretation of Physical Objects

In general the line thickness is a good cue for finding the lines that build the contour of physical objects in the drawing. Unfortunately, the vectorization we used does not guarantee the right thickness in all cases. Therefore, the distinction between contour and non-contour lines has to be done by knowledge about the domain instead of using line thickness. The steps of interpretation are:

- LC-KS tests the connectivity of the lines in the drawing. Lines that are open ended at one side are assumed to be *subsidiary-lines* and extended by SUB-KS through crossings with arrows. Lines with both sides connected to other lines are candidates for *contour-lines*.
- CON-KS checks the proposed *contour-lines* for building a *closed-contour* loop.
- CYL-KS tests the *closed-contours* for rotational symmetry with respect to a *center-line* and instantiates the concept *cylinder* for each of these contours.
- SUBC-KS tests the lines of *closed-contours* that do not constitute a *cylinder* as parts of a *subsidiary-line*. Matching lines are added to already existing *subsidiary-lines*.
- AX-KS combines neighbouring *cylinders* to an *axle* and creates *dimension-nodes* for all *axle* and *cylinder* parts that need dimensioning information. The *dimension-nodes* are connected to the objects via the domain specific HAS-DIMENSIONING relation, which is displayed as a dashed line in the hierarchy.

5.4 Interpretation of Dimensioning

In the second phase the dimensioning information in the drawing is interpreted.

- MA-I-KS and MA-O-KS combine *arrows* to form an instance of one of two types of *metric-arrows*.

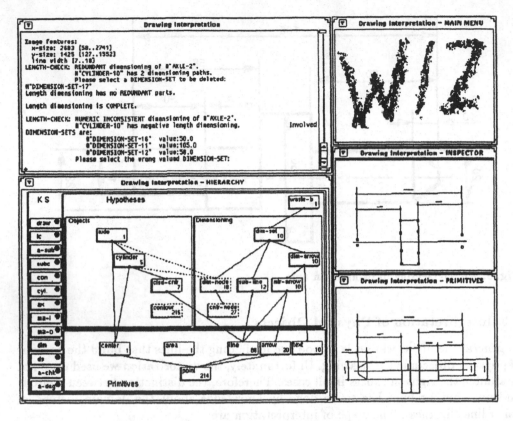

Fig. 4. Screendump of the WIZ-system while interpreting an axle

- DIM-KS builds the object *dimension-arrow* from a *metric-arrow* and *text*.
- DS-KS combines a *dimension-arrow* with optional *subsidiary-lines* and the *dimension-node*s built during physical object interpretation.

6 What can be Obtained from the Interpretation?

During the interpretation process described above two coupled hierarchies of concepts are established. This section explains how the gained information is used to perform a check of the drawing's dimensioning and describes the drawing redesign facilities of the WIZ-system.

6.1 Dimension Checking

In the domain of engineering drawings there are two major types of entities, physical objects and dimensioning information. The WIZ-system connects objects and dimensioning via the concept of a *dimension-node*. *Dimension-node*s specify the points of the objects where dimensioning information is needed. In the WIZ-system the task of checking the dimensioning of an object is reduced to checking the graph that is made of *dimension-nodes* and arcs built from the *dimension-sets*. Figure 5 shows the *dimension-nodes* that are relevant for length checking of the axle and the derived graph that is used for dimension checking.

To allow a clear description of the dimension-checking process we define the graph of *dimension-nodes* in more detail.

- The arc value is the value of the corresponding *dimension-set*.
- Two *dimension-nodes* that are connected by a concatenation of n arcs are defined to have a path of length n between them.
- The value of a path of length one is defined as the signed value of the arc. The sign is determined by the ordered relation between the *dimension-nodes* (startnode < endnode results in a positive value).
- The value of a path of length greater than one is defined as the sum of the values of all subpaths.

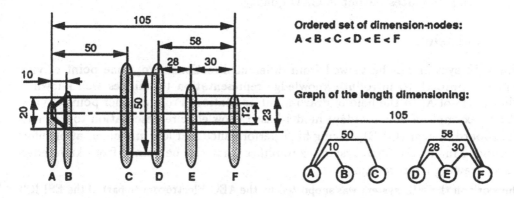

Fig. 5. Axle with marked *dimension-nodes*, ordered set of nodes, and dimensioning graph

Incompleteness means that there are two *dimension-nodes* with no path between them. Assuming that the top-most *dimension-set* (valued 105) in Figure 5 were missing, there would be no path between the nodes A and F. As a result the length dimensioning of the axle would be incomplete.

Redundancy means that there are two or more paths between two *dimension-nodes*. The drawing in Figure 5 has redundant dimensioning between the nodes D, E and F.

Inconsistency type A means that there is redundancy and the different paths do not have the same value.

Inconsistency type B means that there is a path from *dimension-node* X to *dimension-node* Y that is negative valued, but X < Y. Figure 5 has such type of inconsistency. The value of the path from C to D has a negative value (-50 + 105 - 58) inspite of C < D.

The knowledge source A-CHK-KS of the WIZ-system checks the axle for incompleteness, redundancy, and inconsistency. Figure 4 shows an example of a dimension check session. The LISTENER window shows the finished detection of redundancy solved by user-driven deletion of the redundant *dimension-set*. At the moment of the screen-dump the WIZ-system has detected inconsistent (type B) length dimensioning of the cylinder in the middle of the axle. The INSPECTOR window displays this cylinder and the path of *dimension-sets* that is proved to be inconsistent. The user is asked to select the wrong valued *dimension-set* and to correct that value.

6.2 Drawing Redesign

A major problem in using scanned and vectorized images in CAD-systems is the poor quality of the vectorization results: The extracted lines do not meet exactly at one point, the line thickness is not recognized correctly, the lines are slightly slanted, and the size of the drawing parts does not fit perfectly to the values given in the dimensioning. Based on a complete interpretation of the drawing with verified dimensioning a redesign of the axle can be done that satisfies all the mentioned quality requirements. The WIZ-system has a knowledge source A-DSG-KS for redesign of axles that produces output in CAD quality.

7 Conclusion

The WIZ-system can be viewed from different directions. From one point of view it is an approach to introduce knowledge representation techniques well known in other parts of AI to the field of drawing interpretation. From another point of view, it is an example of how to adapt model-based knowledge representation to a generic blackboard system (GBB), having high performance and low implementation effort in mind. Finally, the WIZ-system examplifies what can be gained from knowledge-based drawing interpretation.

The work on the WIZ-system was supported by the AEG Electrocom as part of the ESPRIT Project 2001: SPRITE.

References

1. D. D. Corkill, K. Q. Gallagher, K. E. Murray: GBB – A Generic Blackboard Development System, in Blackboard Systems, Eds. Engelmore & Morgan, pp.503-518, 1988
2. U. Domogalla: Ein Expertensystem für die automatische Erfassung von technischer Graphik, Proc. DAGM/OEAGM, pp.297-303, 1984
3. K. Q. Gallagher, D. D. Corkill, P. M. Johnson: GBB Reference Manual – Version 1.2, COINS Technical Report 88-66, 1988
4. Blackboard Technology Group, Inc.: GBB Refernce Manual – Version 2.0, 1991
5. K. Iwata, M. Yamamoto, M. Iwasaki: M.Recognition System for 3-View Mechanical Drawings, Proc. 4th ICPR (BPRA), pp.240-249, 1988
6. W. Lu, Y. Ohsawa, M. Sakauchi: A Database Capture System for Mechanical Drawings Using an Efficient Multi-dimensional Graphical Data Structure, Proc. 9th ICPR, pp.266-269, 1988
7. T. P. Pridmore, S. H. Joseph: Using Schemata to Interpret Images of Mechanical Engineering Drawings, Proc. 9th ECAI, pp.515-521

This article was processed using the LATEX macro package with CLNCS style

Intelligent System for Feature- Based Modelling of machine parts

Anna Lekova, Dentcho Batanov
Department of Programming and Computer Applications,
Technical University, Sofia 1156, BULGARIA

Nikolay Nikolaev
Artificial Intelligence Laboratory, Department of Computer Science,
Technical University, Sofia 1156, BULGARIA

Abstract. This paper proposes a new system for feature- based modelling- ISFBM with built in strategy for defining details by their features, realizing design process over the computational model of rule- based systems paradigm. The construction includes facts for describing the features and rules for modelling with them. The presented approach gives a flexible and convenient mechanism for modifying the model and semantic analysis of the received product. Specific characteristic is the tool for automatic generation of general features, based on the primitive ones, with which the domain of application is extended.

1 Introduction

Conventional CAD systems realize design process, which does not correspond to the real methods of human- they are based on procedural approach, not on declarative one. The final product is obtained by selecting from preliminary defined constraints on a typesetting geometric primitives. Radical changes in it can't be made[13]- its possible improvement is equivalent to building a new model. These shortcomings have led to new CAD constructions on the principles of the intelligent approach: ICOOPS[7] and FBMS[15,16], simulating real style of design. During their work common information models are preliminary specified from machine part features, which carry information for geometry, technology and manufacturing. They are substantiated with values by applying heuristics.
Another problem in using CAD systems is the low degree of integration with automated manufacturing/CAM/. The CAD product definition lacks engineering data as: tolerances, material properties, surface finishing etc.[5]. For more complete product definition CAM, based on the method of representing details by features are developed, which fulfil this CAD/CAM integration. Underlying purpose in CAD/CAM is object orientation. The aim of this approach is that each machine part is presented as "capsulated" information containing the description of the part, procedures for processing and relations with other objects. Otherwise the term features, pieces of specific geometrical and engineering information for often used machine parts are denoted, that is the use of features can be considered in particular as object-oriented approach. For this reason the approach of modelling with features has become popular and it is a technology for describing machine parts in terms used about

reasoning during the whole detail life cycle. The most researches in the available literature, based on this approach, are for recognizing of the features, or the design process starts at a stage in which model with features has been already done. The existing products: PC-CAPP[11], GCAPPS[10], DLINK[12], FBMS[15], IKOOPS[7] are oriented to concrete types of machine details- to prismatic, to turned, to sheet etc. Besides this insufficient generalization of engineering features and their different descriptions, difficulties in formalization, because of covered and interactive functions, are reasons the path for building product definitions in them not to be systematized.

In current paper a new method and tools for realization are proposed, with built in strategy for defining details by their features, realizing design process over the computational model of rule- based systems paradigm. The construction includes facts for describing the features and rules for modelling with them. The presented approach gives a flexible and convenient mechanism for modifying the model and semantic analysis of the received product. Specific characteristic is the tool for automatic generation of general features, based on the primitive ones, with which the domain of application is extended.

The description of the system is organized in the following sequence: in the second chapter features and their representations are given. In the third chapter the new basic language PLEM for rule- based programming is proposed. After that we examine the functional scheme of ISFBM and implementation of its structural components. In Appendix 1 the graphic model is given.

2 Features and their hierarchical representation

With the term features pieces of specific geometrical and engineering information for often used machine parts are noted [13,15]. For increasing intelligence of our system two new kinds of features are introduced:
- features for semantic analysis- they serve as a basis for carrying out semantic control of the received detail;
- inheritance features- they give information for inheritance of the features from the connected parts of the detail and their positioning.

The most usual way of representing features is the symbolical. In this way the system becomes more general and gives opportunities for using by different designers. They become acquainted in a natural way with the specific problem, because they can work with terms from their field. This is an important condition for easy and convenient modification of the product.

In the current method features are defined with data abstractions at different levels. The features are described hierarchically with an informal apparatus, based on set theory. At the highest level are the basic features groups for every component of the detail. In the common case for an assemble part, the preliminary information model- PIM is:

$PIM:\{(B1,B2,\ldots,Bm)_1,(B1,B2,\ldots,Bp)_2,\ldots(B1,B2,\ldots,Bq)_k\}$
where:
$(B1,B2,\ldots,Bm)_1,$
$(B1,B2,\ldots,Bp)_2,$

```
. . . ,
(B1,B2,...,Bq)k,    (m,p,q,k>=1)
```
- are basic features groups /ordered n-tuples/ for k- parts of the detail.
For a single monolith detail (k=1) the PIM is:
```
PIM:{(B1,B2,...,Bm)1}.
```
Each basic group Bi includes a finite number of groups features: form features
G1, material features G2, precise features G3, semantic features G4,
inheritance features G5, technological features G6, manufacturing features
G7, functional features G8 etc.:
```
B:{G1,G2,...,Gl}  ,   l>=1
```
The groups G_i consist of features subgroups S_i:
```
G:{S1,S2,...,Sp},   p>=1,
```
where S_i are ordered triples object O, attributes A, values V -
```
S:{F1,F2,...,Fr},
```
the facts are:
```
F:(O,{A1,...,Ax},{V1,...,Vy})
```
and r,x,y >= 1.
In the modelling process the initial empty values {V1,...,Vy} are substantiated.
The above symbolic way for representing features is built in the system
presented here. The method is based on the composition of proper feature
raised-primitives and create a copy-primitives. Every machine part of the
detail may be described with a basic features group such that the depth level
of complexity of the groups in it to be consistent with geometrical model of
the details. The degree may be changed by the modeller for automatic
generation of general features groups. The restrictions of generalization come
from basic groups and defined operations for each kind of features, which is
for specific applications.
Here are proposed some of the features groups and subgroups, made in our
method:
-example for basic features group:
```
(fact (meta_description   rot/pris/  ext/int/  subf/
prim/  name))
```
;rotational or prismatic features, etc.
-form features group:
```
(fact (extern_comp_features  {A1,...,Ax}))
```
;subgroup for external compatible basic features group
```
(fact (user_def_param_descr  {A1,...,Ax}))
```
;features subgroup for user defined parameters description
```
(fact (prim_prototype  {A1,...,Ax}))
```
;features subgroup for geometry prototype description
-features semantic group:
```
(fact (descr_semantic_rule  {A1,...,Ax}))
```
;features group of rules description that check the relation between components
in the model
-precision features group:
```
(fact (cylindricity   datum   cutting_tools ))
```
;tolerance features subgroup for cylindrically
-material features group:

```
(fact    (descr_material    {A1,...,Ax}))
```
;description material type subgroup
-technological features group:
```
(fact    (descr_spec_inf    {A1,...,Ax}))
```
;features subgroup for tool and machine descriptions, stress-analyses, GT, CNC, assemble.

Prescribing of the basic features groups is done with a descriptor /module for describing the features/ or automatically during the work of the intelligent system /see chapter 4/.

The most frequently used basic features groups are for: holes, steps, slots and pockets, such that for every specific application standard ones are defined. For example in shafts manufacturing such are step, keyway, holes etc.. Identification and classification of these features is important, because the different types require different networks of operations, machines, cut tools and fixtures.

3 Production Language with Eager Matching- PLEM

The approach, which is proposed, for building a system for modelling by features is implemented with the Production Language with Eager Matching PLEM. It supports the paradigm for rule- based programming [2] and is a convenient tool for design of systems, whose work resembles intelligent human behaviour. For easier understanding is described the language PLEM with the formal apparatus of abstract syntax:

```
P ∈ Prog Program        P ::= F R
F ∈ Fact Initial facts  F ::= (fact (OU)) | F1 F2
U ∈ Uni  Unions         U ::= ^OH | U1 U2
R ∈ Rul  Rules          R ::= (rule O A -> C )|R1R2
A ∈ Ant  Antecedents    A ::= (OT)|~(OT)|(V->(OT)|
                              (test E)|A1 A2
E ∈ Exp  Expressions    E ::= (<I1 I2)|(> I1I2)
                              (=I1 I2)|(and E1E2)|
                              (or E1E2)|(not E1E2)
T ∈ Tem  Templates      T ::= ^OI|^OH |H |I |T1 T2
C ∈ Cons Consequents    C ::= (create (OT))|
                              (delete V) |
                              (print <string>) |
                              (load-facts <name>)
                              C1 C2
O ∈ Obj   Object atoms
H ∈ Hcon  Constant atoms
I ∈ Ide   Identifiers-always begin with the symbol
          ?. After the symbol & in their end may
          be added constraints.
V ∈ Var   Element variables - begin with ? too.
```

The interpreter of PLEM solves a problem task, presented with a rule- based program, performing succesive transformations of current working facts according to the rules till reaching a goal [9]. That is organized with a specially selected version of the chronological backtracking algorithm, which is the best appropriate for rule- based languages, made on incremental matching.

The second distinguishing characteristic of PLEM is the effective algorithm for incremental eager matching. The interpreter performs tests of all the rules with respect to the last changes in the working facts. The results are accumulated in a non-shared structure, which consists of the environment with variable bindings and a table with antecedent statuses. Combinations with consistent bindings of the common variables in the rules are not stored in the memory as in TREAT [8]. Only when all of the antecedents of a production are satisfied checks are made. In this way the main shortcomings of RETE [4] are overcome:

- in PLEM interpreter are not allowed memories, which size may be combinatorially explosive;
- redundant intermediate tests with the variables from inactive antecedents are not performed.

The Production Language with Eager Matching PLEM is implemented in a dialect of Lisp- Scheme [14].

4 Intelligent Method and System for Feature- Based Modelling of machine details- ISFBM

In this method the design process is considered as a kind of intelligent behaviour of the CAD system. Here a general model is found, which can be formalized as a sequence of creating a symbol model with following analyses and modifications to create a final detail model for testing in manufacture. The design activities are defined as a sequence of individual automatically modelling operations. They are integrated in a process by transferring the information between them. The new method for modelling by features is presented as a result of general operations, applied on the set theoretical data description, described above. Originating from a preliminary given structure of the detail, the desired properties are substantiated in it using heuristics. The concrete information model KIM is presented as:

$KIM\{G,S,F\}=\{OPgi,OPsj,OPfk\}\{(B1,B2,\ldots,Bm)_1,(B1,B2,\ldots,Bp)_2,\ldots(B1,B2,\ldots,Bq)_k\}$.

This design sequence is organized with production rules, according to which the current components, described by features in groups and subgroups in PLEM language, are transformed. In the process of work the different properties are conveyed directly to the model and an instance of every feature is received after substantiating the values for the parameters.

The system ISFBM works in two modes- initialization: the base groups of features are defined by specifying their properties and modelling: the properties are fixed and substantiated, put into the model etc..

The construction consists of the following elements:
- descriptor for determining the basic features groups;
- subsystem of modellers / they work with the object model /:
  - for automatic generation of general basic features groups;

- form modeller;
- modeller for semantic control;
- position modeller;
- modification modeller;
- precision and material modellers;
- interfaces with the system for geometric modelling and engineering data base
/ Appendix 1/.

The current state of rule- based language working memory determines operating of the whole system. At start moment the preliminary formed feature description group with the specific part of the detail is loaded in it. The facts for different features are designed preliminary with the descriptor. In its language tools for easy defining of rules and equations exist. Only the attributes Ai are defined by the descriptor as terms from application domain and their values Vi are substantiated in modelling mode.

Besides the above method, the basic features groups can be designed also by the modeller for automatic generation of general groups. It serves for deducing new descriptions from the existed specified models CIM1 and CIM2:

$$CIM1: \{ (B_{11}, B_{12}, \ldots, B_{1m})_{11}, \ldots, (B_{11}, B_{12}, \ldots, B_{1q})_{1k} \}$$
$$CIM2: \{ (B_{21}, B_{22}, \ldots, B_{2m})_{21}, \ldots, (B_{21}, B_{22}, \ldots, B_{2q})_{2k} \}$$

This way the designer is freed from the obligation of initial defining by the descriptor and the system alone generates features descriptions on the basis of the models, made by it, i.e. it can learn. The designer points the name of the already created file, basic groups and the name of the new group features. The new through axial hole description is formed in the working memory as a union of preliminary descriptions of the constituent basic features groups and inheritance information in the model of the detail:

$$IMgener = Saddit \ U \ CIM1 \ U \ CIM2$$
$$IMgener : \{ Saddit \ U$$
$$(B_{11}, B_{12}, \ldots, B_{1m})_{11}, \ldots, (B_{11}, B_{12}, \ldots, B_{1q})_{1k} \} U$$
$$(B_{21}, B_{22}, \ldots, B_{2m})_{21}, \ldots, (B_{21}, B_{22}, \ldots, B_{2q})_{2k} \}$$

At that the features, which do not require substantiating, are deleted and ready facts are appended for them in Saddit. The basic groups of components may lack features groups or subgroups. In order to be pointed to other modellers, that the basic group is general, a fact with object feat_complex and attributes- the names of the constituent groups is added. There are special rules for eliminating errors in using groups with equal names. The facts in working memory are written in a symbolic file as the others.

This is illustrated here by the following example: if we have a model of through axial hole with basic groups for thread and chamfer in existing file, we may generate this general basic group under a given name, so the positioning relations between the volume features are saved with the concrete values. According to the rule sub_feat_posit2 the preliminary facts for positioning-?f1, ?f2, ?h1 and ?h2 are removed and the substantiated instances from them-f-pos are saved in the working memory.

```
(rule  sub_feat_posit2
    (feat-complex  ?feat  ?featr)
    (?f1  ->  (descr_pos  ?feat  ?kind  ?param1))
    (?f2  ->  (descr_pos  ?feat  ?kind  ?param2))
```

```
(?h1  ->  (f-pos  ?feat  ?numb  ?kind  ?param1  ?val1))
(?h2  ->  (f-pos  ?feat  ?numb  ?kind  ?param2  ?val2))
->
    (delete  ?f1  ?f2  ?h1  ?h2)
    (create  (f-pos  ?feat  x  ?k  ?param1  ?val1)
    (create  (f-pos  ?feat  x  ?k  ?param2  ?val2))))
```

In the process of work the form modeller loads all the facts from the general file with the basic group in the working memory. So the rules for creating the initial shape and positioning are triggered. The form modeller designs the initial common structure of the detail with differentiated components and relations between them: the result is instances of features with substantiated properties- these are the attributes in facts for the features. Substantiation is done in two ways:
- it is pointed with a special fact that the parameter value has to be defined by interface from the user /this is the subgroup:user_def_param_descr in the rule user-definition-parameter/;
- the value of the parameter is inherited on the basis of the preliminary given rule / descr_inherit_rule in feature-inheritance-rule / in the basic group features.
For example after substantiating the attributes for through axial hole on the basis of facts from form group and inheritance group features:

```
(fact  (user_def_param_descr  cy-bore  diam))
;user-defined parameters,
(fact  (user_deriv_descr  cy-bore  length))
;inheritance parameters and the rules, which these facts will activate:
(rule  user-definition-parameter
    (file_name  ?nam  ?nu  ?feature&~nil)
    (?re  ->  (user_def_param_descr  ?feature?type&~q))
->
    (create  (f-param1  ?feature  ?nu  ?type  (read))))

(rule  feature-inheritance-rule
    (relation  ?feature  ?numb  ?feat-old&~nil  ?nu)
    (?inh  ->  (descr_inherit_rule  ?feature  ?bpar
              ?feat-old  ?apar))
    (f-param1  ?feat-old  ?nu  ?apar  ?val)
->
    (create  (f-param2  ?feature  ?numb  ?bpar  ?val)))
```

the concrete state of working memory will be changed, hence and the sequence of design process, because of the facts: f-param1 and f-param2.
At the stage of construction changes in the conventional CAD the user himself takes care of the wholeness of the model. In system ISFBM these actions are performed automatically. A modeller is built which checks the proportion between the parameters and proportion between position parameters in the model with the added component.
The space functions in three-dimension space are determined by the position modeller. Using them the coordinates of volume form features of the new component are defined. These are dependences between old coordinates and parameters, with which the volume features will be positioned, fixed in the

basic group. Abstract position parameters are found with which is worked in the same way, as with the form features parameters. They are defined in PIM for rotational, prismatic parts, etc.

```
(fact   (user_def_pos  cyli  rot  penetration-depth))
(fact   (user_def_pos  cyli  rot  distan-betwbasepnts))
(fact   (user_def_pos  cyli  rot  angle-betaxis-front))
(fact   (user_def_pos  cyli  rot  angle-axis/x-left))
(fact   (user_def_pos  cyli  rot  dist-axis-centre))
```

The structure obtained till this design stage is changed by the modification modeller. An original mechanism for modifications is proposed, the so called "design in the past". The designer activities hasn't been protocoled during the creation of the detail models, in order to modify itself. The knowledge of the creation history of KIM and geometric data prototype formats is found implicitly in PIM. Modeller rules define the links between parameters of participating groups after the modification. Specific for the system ISFBM is that it is not necessary to memorize the history of creating the model, because after loading the facts in working memory, the inference process will be started and the deleted information will be updated with new data. Actions, described by rules, look in this way:

```
(rule   successor
;looking for descendants of the modified component
        (modif  ?featr  ?numb)
        (prim_abs  ?fe  ?nu  bols  ?featr  ?numb  ?fe)
 - >
        (create  (child  ?featr  ?numb  ?fe  ?nu)))

(rule   hide-changes
;this rule changes the relation facts and fires the design with new data
        (?rel  ->  (relation  nil  nil  ?feat-old  ?numb))
        (modif  ?featr  ?numb)
        (prim_abs  ?feat-old  ?numb  bols  ?ferel  ?nu2  ?)
 - >
        (delete  ?rel)
        (create  (relat  ?feat-old  ?numb  ?ferel  ?nu2)))

(rule   change
;this rule changes the work memory for update the modifications
        (?child  ->  (child  ?feat-old  ?numb  ?feature  ?nu))
        (change  ?feature)
        (?f  ->  (file_name  ?name))
 - >
        (load-facts  ''name.mod'')
        (create  (relat  ?feature  ?nu  ?feat-old  ?numb)
        (create  (feature  ?feature  ?nu))
        (create  (file_name  ?nam  ?nu  ?feature))) )
```

Precision and material modellers work analogously. The descriptor includes selecting this type features groups. The first modeller loads the facts that regulate the form and dimension tolerances: facts for tolerances, surface finish,etc. Two types of tolerances are defined: dimension tolerances and geometry form tolerances. The former are specified by their upper and lower values, so that this information will be extracted implicitly depending on the type of processing and modified by the designer. The geometry form tolerances as roundness, concentricity, etc may be determined in order to put requirements to machine tools. The material modeller processes existing information for the kind and code of the material, hardness, etc. A strategy for creation a based knowledge system from table data of reference books in specific domains is built in this modeller. There is proceeding from syntactic level - variables, ranges, decisions, estimations and probabilities. With production language these relations are presented as facts with identified attributes from the user point of view.

After the final symbolic formation of the model, on the basis of geometric primitives concrete geometric formats are created by the interface. It uses the facts for inherited or user defined parameters. After possible modification of the values the substantiated model is updated. The general geometric task is solved to find the points coordinates of the new added Bi from the concreted subgroups with object "prim_prototype" and concrete facts f-param1, f-param2 and F-position. This modeller allows the geometric modelling system to work hidden from the user, so that to support high level of graphic language for defining solid primitives and operations. The model is obtained indirectly from the information model and it is not necessary special generations of parameter programs to be designed.

5 Conclusion

A methodology for development of a CAD system was presented in which modelling is done analogously to real methods of humans by applying engineering features. Our method for implementing it with a new production languages proposes new opportunities for:

- defining the properties of designing parts in symbolic way with the concrete terms from the field of application;
- building in different approaches for design of the modelling process.

These advantages as well as the improved mechanism for modifying the product model and the possibility for automatic generation of general features indicate universality of the intelligent features modelling system ISFBM.

The object of future work is test for this method and prescription of the features for applications as GT, CNC, stress-analyses, assemble, etc.

The proposed approach, method and tools can be used as software, witch is enough general, it's built as modules with clear relations with the other specific programs and is open for future application where a specific information, structured as features, is available.

References

1. H. Abelson, G. Sussman: Structure and Interpretation of Computer Programs. MIT Press, Cambridge, MA, 1985.
2. L. S. Brownston, R. G. Farrell, E. Kant, N. Martin: Programming Expert Systems in OPS5: An Introduction to Rule-Based Programming. Addison Wesley, Reading, MA, 1985.
3. B. Choi, M. Barash, D. Anderson: Automatic recognition of machined surfaces from a 3D solid model. CAD, 16, (2), 1984.
4. C. L. Forgy: Rete: A Fast Algorithm for Many Pattern / Many Object Pattern Match Problem. Artificial Intelligence, 9, (1), 1982.
5. D. Gossard, R. Zuffante, H. Sakurai: Representing Dimensions, Tolerances and Features in MCAE Systems. Proc. IEEE Computer Graphics & Applications, March, 1988.
6. S. Joshi, T. Chang: Graph-based heuristics for recognition of machined features from a 3D solid model. CAD, 20 (2), March, 1988.
7. I. Lee, B. S. Lim, A. Y. NEE: IKOOPS: an intelligent knowledge based oriented approach planing system for the manufacture of progressive dies. Expert Systems, 8 (1), Feb, 1991.
8. D. P. Miranker: TREAT : A Better Match Algorithm for AI Production Systems. Proc. AAAI- 87 Sixth National Conference on Artificial Intelligence, Seattle, WA, 1987.
9. N. Nilsson: Principles of Artificial Intelligence. Tioga, Palo Alto, CA, 1980.
10. S. Pande, N. Palsule: GCAPPS - a computer-assisted generative process planning system for turned components. Computer-Aided Engineering Journal, August, 1988.
11. S. Pande, M. Walvekar: PC-CAPP - a computer-assisted process planning system for prismatic components. Computer - Aided Engineering Journal, August, 1989.
12. R. Patel, A. McLeod: The implementation of a mechanical engineering design interface using engineering features. Computer - Aided Engineering Journal, December, 1988.
13. R. Patel, A. McLeod: Engineering feature description in mechanical engineering design. Computer -Aided Engineering Journal, October, 1988.
14. J. Rees, W. Clinger,./ed./: Revised[3] Report on the Algorithmic language Scheme. Sigplan Notices, 21, (12), 1986.
15. J. Shah, M. Rogers: Functional requairements and conseptual design of the Feature-Based Modelling System. Computer Aided Engineering Journal, February, 1988.
16. J. Shah, M. Rogers: Expert form feature modelling shell. CAD, 20, (9), 1988.
17. T. Smithers, A. Conkie, J. Doheny, B. Logan: Design as intelligent behaviour: an AI in design research programme. Artificial Intelligence in Engineering, 5, (2), 1990.
18. A. Lekova, D. Batanov: Inteligent Interface for the interactive design of three-dimensional objects by features in association with geometric modelling systems. Automation and computers, 7-8, Sofia, 1991.

Apendix 1: Structural scheme of the ISFBM

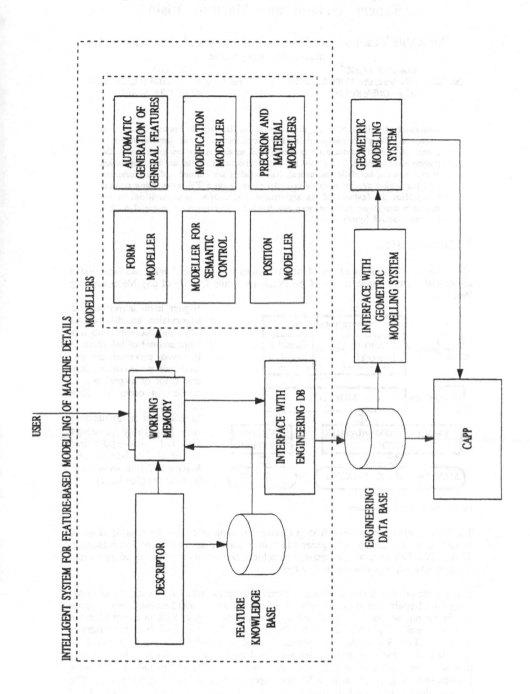

Intelligent Process Control by means of Expert Systems and Machine Vision

Lluís Vila*, Carles Sierra*, Antonio B. Martinez+, Joan Climent+

e-mail contact: vila@ccab.es

IIIA-CEAB-CSIC*
Camí de Sta. Bàrbara, s/n, 17300 BLANES
CATALONIA, SPAIN

FIB-UPC+
Pau Gargallo, 50, 8028 BARCELONA
CATALONIA, SPAIN

Abstract. In this paper an architecture for Intelligent Process Control is presented. The main components of the architecture are an inspection vision system to identify defects on products and an expert system to diagnose the process malfunctions. The expert system architecture is based on the cooperation of two models, heuristic and causal. Both models are defined and its connexion with the vision system is explained. An example on a TV flowcoating process exemplifies the features of the architecture and shows the possibilities of the approach achieving performances not reachable in current commercial on-line Process Control Systems.

1 Introduction

The use and integration of advanced information technologies in Control and Supervision are crucial for the improvement of the Productivity and Quality of any Manufacturing Process.

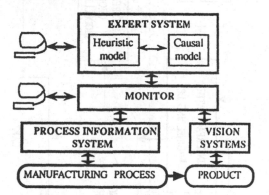

Higher level activities like supervision are difficult to carry out because there is a huge amount of information involved, processes are not well-understood and unsteady and a lot of experience is needed in order to solve problems.

In this paper an architecture for Integrated Process Control incorporating Artificial Intelligence and Machine Vision techniques is presented (See fig. 1).

Fig. 1 The IPCES architecture.

The IPCES project (Intelligent Process Control by means of Expert Systems) is aimed at extending process control with general human characteristics like vision and reasoning. Thus IPCES like systems are intended to achieve performances not reached by current commercial on-line process control systems.

The process control domain includes several intelligent tasks like Vision, Monitoring, Diagnosis, Repair, Prognosis, Generation of control actions and Learning. In this paper we focuss on the part of the Vision System and the Expert System concerning the Diagnosis task. A generic system for Diagnosis in the domain of Process Control is presented. Two knowledge representation models are defined: the first based on *experiential knowledge* and the second based on *causal knowledge*. The way they interact to get improved results is defined. A real application, including a TV tube Flowcoating Diagnosis System, a Screen Inspection Vision System and the interaction between them is presented. In figures 2-5 some illustrating photographs of correct and defective screens are showed. Discussion on this application is made in the last section.

---

[1]This research and development project has been partially supported by the ESPRIT program, project n. 2428.

Fig 2. Correct phosphorus triad lines
appearance in a TV tube screen
(field of view 4.8 mm x 5.4 mm)

Fig 3. Microdefect: Contamination
of red phosphorus on green
and blue phosphorus
(field of view 4.8 mm x 5.4 mm)

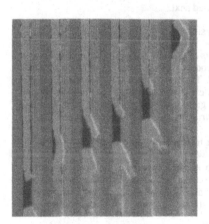

Fig 4. Microdefect: Red missing lines
(field of view 4.8 mm x 5.4 mm)

Fig 5. Macrodefect: water splash
(field of view 36.8 mm x 41.4 mm)

2 Diagnosis in Process Control

The Diagnosis of a system consists of determining those out-of-order elements -*causes*-that account for the observed abnormal behaviour [1].
Different kinds of knowledge can be used as a human expert does. On one hand the so called *shallow* or *heuristic* knowledge that embodies the experience gained through solving problems for a long time. On the other hand the *deep* or *fundamental* knowledge that expresses how the physical process actually works -causal relations-. The IPCES Diagnosis System incorporates both types of knowledge and takes advantage of their cooperation in order to get more accurate and complete solutions, and to increase the system robustness. It is the fundamental thesis of the *2nd Generation Expert Systems* [2, 3, 4, 6, 8]. In this section generic formalisations for both types of knowledge and their interaction are presented.

2.1 The Hypothesis Space

The Hypothesis Space model (HS) is a general approach to *Diagnosis* using heuristic knowledge [1] based on the idea of *refinement* of the set of possible causes. Evidences about them are propagated along a hypothesis hierarchy towards the most plausible causes.

To define this model several requirements of the Diagnosis task in a changing domain -as it is the case in Industrial Process- have been taken into account: *Dealing with Uncertainty*, *Reasoning with Incomplete information*, *Non-monotonicity*, *Efficiency* and *Adequacy for Knowledge Acquisition*. The model definition is inspired on the results of a behavioural analysis of the human expert explained next.

2.1.1 The Expert's Behavioural Analysis

The HS model has been derived from the results of the analysis of the expert's behaviour while diagnosing the product defects due to process malfunctions [9]. It has been made over three flowcoating process control experts. The conclusions of this analysis are that the Diagnosis task can be defined as a flexible process of generation, refinement and pruning of hypothesis while case data is being gathered. Main knowledge components is a *hypothesis-refinement relationships* static structure on which Focussing and Testing knowledge is organised:
i) *Focussing knowledge* is used by the expert to *focus* on more specific causes in the presence of some findings.
ii) *Testing knowledge* is used by the expert to *verify* a particular hypothesis currently considered. Hypothesis are tested when a certain amount of evidence for them has been gathered. This test is based on the evidence about other hypothesis and findings from the problem. In the trivial case these tests are just the results of verifying the hypothesized cause in the reality.

2.1.2 The Hypothesis Space Model

A *Basic Hypothesis* represents an atomic cause of the problem. Given a domain, the set of basic hypothesis is determined by the expert. Also a set of observable data, called *Data*, is defined. The elements of this set represent any information that can be obtained from the domain problem.

Basic Hypothesis and Data are represented as events in the probabilistic sense, i.e. an expression $X = x_i$ where X is a variable and x_i is a value for it. Next the basic definitions of the HS model are presented:

- The *Basic Hypothesis Set* of a given domain is defined as BHS = {bh_i / bh_i is a Basic Hypothesis}.
- We define the *Hypothesis Set* of a given domain as H $\subset$ $\mathcal{P}$(BHS). For any h $\in$ H, h = {bh_1,...,bh_n} means the expression (bh_1 v...v bh_n). P means power set.
- The set of *Hypothesis Refinement Link*, for short L, is L $\subset$ H x H x [0,1], where for

any $(a,b,\alpha) \in L$, $b \subset a$, and $\alpha = p(b|a)$, being p a probabilistic measure.

- A *Hypothesis Space Net* HSN = (H, L) is a directed acyclic graph where nodes, H, represent hypothesis, and links, L, represent refinement relationships, in the sense that $a \rightarrow b$ means $b \subset a$.
- A *Valuation function* on H is a function ψ_i: H $\rightarrow$ [0, 1]. The set of these functions is defined as $\Psi_H = \{ \psi_i / \psi_i$: H $\rightarrow$ [0,1] $\}$. These functions represent the belief on the hypothesis.
- A *Heuristic Rule* is a quadruple $(h_1, cond, h_2, cf)$ where $h_1, h_2 \in$ H, $cf \in$ [0, 1], and $cond \in COND = \{ \wedge_i c_i / c_i \in BHS \cup Data \}$.
- A *Focussing Rule* is a heuristic rule $(h_1, cond, h_2, cf)$ where h_1 is predecessor[1] of h_2.
- A *Testing Rule* is a heuristic rule $(h_1, cond, h_2, cf)$ where $h_1 = \emptyset$.

Now we can define what a Hypothesis Space is.

Def. A *Hypothesis Space* is a quintuple HS = (N, FR, TR, σ, δ) where

N = (H, L) is an HSN

FR = $\{fr_i / fr_i$ is a focussing rule $\}$ σ: FR x $\Psi_H \rightarrow \Psi_H$

TR = $\{tr_i / tr_i$ is a testing rule $\}$ δ: TR x $\Psi_H \rightarrow \Psi_H$

σ and δ are functions that transform the current valuation function by the application of an heuristic rule

Def. A *Hypothesis Space State* is the composite object HSS = (HS,ψ) where $\psi \in \Psi_H$.

2.1.3 The Inference Control

The HS as such is an static structure over which an inference control is defined. The reasoning is performed by applying the knowledge contained in the hypothesis space net on the current case data. The inference control algorithm defines an exploration of the hypothesis space net i.e. the way the hypothesis are visited[2] and also defines what visiting a hypothesis consists of. The algorithm takes an HS state which is transformed to the hypothesis valuation that best reflects the reality according to the diagnosis objectives and the problem at hand.

```
Function Heuristic_Diagnosis (HS, ψ₀) is

    ψ = ψ₀
    Loop
        h = Select_Hypothesis(HS, ψ)
        Exit if h = ∅
        r = Select_Rule(HS, h)
        ψ = Apply_rule(r, ψ)
    Endloop
    return(ψ)
EndFunction
```

The *Select_Hypothesis* function takes a Hypothesis Space structure with a particular valuation function and determines which is the most appropriate hypothesis to visit. The *Select_Rule* function selects one rule from the set of rules related to the hypothesis determined by the Select_ Hypothesis function. The *Apply_Rule* function first *executes* the selected rule and then *Reconfigurates* the Hypothesis Space.

A tailored inference control method is defined by: i) Specifying these three functions, and ii) Defining the evidence management for each of the different types of knowledge and their combinations. It will determine the two steps of the Apply_rule function.

2.1.4 HS Evaluation

The HS model offers a solution to the requirements stated at the beginning of the paper. Not all the relevant information is needed to provide solutions and the accuracy of the outcome suffers a graceful degradation when less information is available. The static hypothesis structure and the organisation of the knowledge spread among the hypothesis it refers to, allows to incorporate new information by revising the evidence valuation function and giving a refined solution.

The clear meaning of what a hypothesis and its links to other hypothesis represent leads to a natural and understandable knowledge representation -even for experts- that allows the knowledge engineer to follow a structured knowledge acquisition and to know at any moment what has been already obtained and what is still lacking.

In [11] it is explained how the HS model is tailored to the domain of Industrial Process Control. Among others the requirement of quick answers is stressed.

2.2 The Qualitative Causal Model

The fundamental knowledge about how the physical process works can be useful for Diagnosis. Basically we are interested in representing the elements that change in the process, the process and product parameters -variables-, and how these changes influence each other i.e. the *causal* knowledge -influences-. It is because i) the process physics are not precisely known and ii) human experts reach acceptable performances exploiting just qualitative information. We will propose a *Qualitative model*:

Def. The relevant *parameters* in a given process is represented by a set of *Variables* defined as $V = \{v_i / v_i$ is a Variable$\}$. Variables take values over a *Quantity Space* [5], a qualitative values set which represent qualitative abstractions of real values of parameters.

Def. $QS = \{qs_i / qs_i$ is a possible qualitative value of a variable$\}$ with a total order relation $<$ and a *central* value denoted by 0.

Def. For a certain process, given a QS, a set of *Influence* functions is defined as $IF = \{If / If: QS \rightarrow QS\}$ (See Fig.7).

Def. Given a certain process, a *Causal Model* is defined as a finite, acyclic, directed and labelled graph where nodes are Variables and links are Causal Relations labelled with influence functions. $CM = (V, QS, C)$ where V is the set of Variables, QS is the Quantity Space over which the variables take values and a set of Causal Relations $C \subset V \times IF \times V$, where $(v_i, If, v_j) \in C$ iff the parameter represented by v_i influences the parameter represented by the variable v_j according to the *If* function.

In a CM the variables in V are classified according to the influences received and sent:
- *Initial*: v is an Initial var. iff there is no influence to it, i.e.
$$\not\exists\,(v_i, If, v) \in C$$
- *End*: v is an End variable iff $\not\exists\,(v, If, v_j) \in C$
- *Intermediate*: v is an Intermediate var. iff it is neither an Initial variable nor an End var.

The state of the process specifies the values of the parameters. In the model this state is described by assigning values to the variables.

Def. An *state* of a CM is a function $\psi : V \rightarrow QS \cup \perp$, where $\perp$ represents the undefined value and $\alpha = \psi(v_i)$ is called the *value* of v_i.

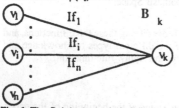

To define how the variable values are propagated through the Causal Model we define the basic building block -called *brick* (See Fig.6). Any Causal Model can be seen as a composition of a set of bricks. The extension of the analysis of a single brick to the whole CM is straightforward.

Fig. 6. The *Brick*, the basic building block of a Causal Model.

Def. The *brick* of a CM related to the variable v_k is defined as $B_k = \{(v_i, If, v_k) \in C\}$. The v_i are called *causing variables* and the v_k is called *explained variable* of the brick.
It is assumed that the influences on a variable from its causing variables are independent and that the final behaviour of the explained variable can be computed from a combination of the effects that would induce each causing variable separately, namely the *additive assumption*. Let's see how the value of the explained variable of a brick is computed from

an state over the causing variables:

First, by the M function, it is computed the marginal influence that every causing variable induces on the explained variable through the influence function:

$$M: QS \times IF \rightarrow QS$$
$$M(v_i, If_i) = mi_i = If_i(\psi(v_i))$$

Then the explained variable value is computed combinning of the marginal influences:

$$F: QS^n \rightarrow \mathcal{P}(QS)$$
$$F(mi_1, mi_i, ..., mi_n) = \psi(v_k)$$

F is defined as a recurrent application of a binary function called F'. So $F(mi_1, mi_i, ..., mi_n) = F'(mi_1, F'(mi_i, ..., mi_n))$. Given that QS is finite, only a finite number of such binary functions exist. The set of possible functions is reduced by determining some properties that the function has to satisfy. Among them, next properties are stressed: *Associativity*, *Commutativity* and *Non-decreasing respect to any variable* (See [12] for further details). To define the functionalities to be provided by the Causal Model and their algorithms it is introduced the condition of consistency of a given valuation over the variables.

Def. A valuation function ψ on a Causal Model CM = (V, QS, C) is said to be *consistent* iff $\forall$ brick = $\{(v_i, if, v) / (v_j, if, v) \in C\}$: $\psi(v) \in F(M(\psi(v_1)), ..., M(\psi(v_n)))$.

Let's now define the functionality of the tasks that we perform upon the Causal Model. A *cause* means a disturbance on an initial variable, a *compound cause* a set of them, an *effect* a disturbance on an end variable and a *compound effect* a set of them. The complexity of the algorithms for these tasks is exponential in the general case but for Causal Models where i) the number of disturbances is limited and ii) the number of different paths between parameters is small[1], the combinatorial explosion is kept under control.

2.2.1 Causal Simulation

Given a set of disturbances on one or more variables $\{(v_i, qs_i)\}$, simulation on a Causal Model CM consists in generating every possible set of effects consistent with it $\{\psi_i\}$:

$$S_{CM} : P(V \times QS) \rightarrow P(\Psi)$$
$$S_{CM}(\{(v_i, qs_i)\}) = \{\psi_i\}$$

The Input variable-value pairs fix values for some variables in the Causal Model i.e. partially define the valuation function. By a *Closed-World Assumption* every undefined predecessor to these defined variables are set to the *undef* value ($\perp$). Then, simulation propagates the variable disturbances forward through the causal links maintaining the consistency condition. If this disturbances are not inconsistent this computation produces a set of partially defined valuation functions $\{\psi_i\}$ corresponding to the different possible behaviours. The projection of these functions on the end variables is the result of the simulation task, a set of sets of *effects*.

2.2.2 Causal Diagnosis

Given a set of disturbances in one or more variables $\{(vi, qsi)\}$, diagnosis on a Causal Model CM consists of generating every possible set of *causes* consistent with it $\{\psi i\}$:

$$D_{CM} : P(V \times QS) \rightarrow P(\Psi)$$
$$D_{CM}(\{(v_i, qs_i)\}) = \{\psi_i\}$$

The diagnoser computes all the possible value combinations of influencing variables affecting explained variables related with the disturbed variables through a forward

---

[1] These conditions are usually fulfilled by real applications.

propagation. If the effects are not themselves inconsistent, this computation produces a set of valuation functions $\{\psi_i\}$. The projection of these functions on the initial variables results in the set of diagnosis, i.e. a set of sets of causes.

2.2.3 Explaining

Given a set of disturbances on one or more initial variables and another set of them over a set of end variables $\{(v_i, qs_i)\}$, explaining on a Causal Model CM consists in generating every possible *explanation -causal chain* from effects to causes-, consistent with the set of observations $\{\psi_i\}$, that justify the effects from the causes, and it can be described by the function:

$$E_{CM}: P(V \times QS) \rightarrow P(\Psi)$$
$$E_{CM}(\{(v_i, qs_i)\}) = \{\psi_i\}$$

2.3 Interaction between the Heuristic and the Causal models

Up to this point two models has been presented based on different types of knowledge. In general heuristic knowledge is used first to generate solutions efficiently. But this efficiency is obtained in return for no guarantee about neither soundness or completeness of the solution. Thus when heuristic solutions are not satisfactory fundamental knowledge is applied. The heuristic results can be improved in several ways using the causal knowledge: i) to complete them, ii) to validate them, and iii) to provide sound explanations of them (See [11]). Besides it, the heuristic knowledge also embodies efficient strategies for determining which is the most interesting knowledge to be obtained. So the subsequent application of causal knowledge will take advantage of the information made available during the heuristic process.

3 An example: The Flowcoating Application

The flowcoating process consists of coating the TV screen glass with a phosphorous suspension in order to get a periodic sequence of triads of colour lines (red, green and blue) on the screen that will emit colour light when an electron beam excite them. It is made by treating several input materials -like the screen glass, the mask, the phosphorous powder,...- through a sequence of physical and chemical processes -like precoating, suspension flowcoating, exposure, developing.

3.1 The Flowcoating Diagnostic System

In the flowcoating hypothesis hierarchy the topmost hypothesis represents that there is a problem -without specifying- in the flowcoating process. Sub-hypothesis of it are defects. A classification of real problems in the screen founded during the production has been made by the flowcoating process experts in a reduced set of classes. A class is called a defect and is identified by a pattern on the screen. For instance in the figure 5 you can see a real water splash defect. The generic defect Water splash represents the set of defective screens with patterns similar to it. Basic Hypothesis are either a deviation on a process parameter (e.g. too high suspension Temperature), deviation on a environment parameter (e.g. too high room Humidity), a problem in the equipment (e.g. broken lamp) or a input material out of specifications (e.g. phosphorus powder to old, glass roughness out of spec). according to our Structured Process Description [10].

The Data set defined in the HS model in this case is compound of information about the product or any intermediate product feature and previous actions taken by the operators on the process (e.g. mill maintenance, change of exposure table).

The general inference control algorithm defined generically for the HS model has been particularised. Once an hypothesis is selected it is selected as many times as rules are applicable till the set or rules is exhausted. The rule selection is simplified to just activate the rules associated to the current hypothesis in the following order: first try the testing rules and afterwards the focussing rules.

A Causal Model has been built that represents the process and product parameters of the flowcoating process and the influence links between them. The leftmost side variables represent input material parameters and process parameters. The rightmost side variables represent product parameters. Intermediate variables represent subproduct parameters. A three-value Quantity Space has been defined, with {-, 0, +} representing a smaller, normal and greater value with respect to the right specified value. Also a five-value QS has been studied [12]. Four Influence functions have been defined (See Fig.7).

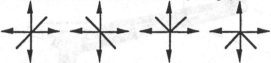

Fig 7. These are the Influence functions defined in our application: Direct, Inverse, Extreme positive and Extreme negative. Axis x represents the domain values and axis y the range values.

The defined functionalities of the Causal Model, their algorithms and the interaction between the heuristic and the causal model has been implemented following the general ideas described previously.

3.2 The Interface SIVS - FDS

The communication between the Screen Inspection Vision System (SIVS) and the Flowcoating Diagnostic System (FDS) is performed through the Monitoring System (See Fig. 1). The SIVS delivers a list of the defects that match with the currently detected defect on the flowcoated screen. It is the initial input of the Flowcoating Diagnostic System (FDS). It includes information about features general to all defects and other specific to each defect that has been specified in the defect definition. Beside it, the SIVS provides data about other features of the product not directly concerning the defects. Also a statistical treatment on these informations is performed in order to asses the pattern of occurrence of the defects and trends of specific product parameters (e.g. the trend of the line width).

3.3 The Screen Inspection Vision System (SIVS)

3.4.1 SIVS Constrains

The aim of the Screen Inspection Vision System (SIVS) is to 100% check each screen to detect any defect and to analyze it, in order to deliver the information to the Diagnosis Expert System. The customer's requirement of checking the whole screen area in 12 seconds with a resolution of 50μm is a major determining factor for the whole visual inspection system. It means an information flow rate of about

$$\frac{(0.6 * 0.45) \text{ m}^2/\text{screen} * 8 \text{ bits/pixel}}{(50.* 10^{-6})^2 \text{ m}^2/\text{pixel} * 12s} = 72 \text{ M bits /s * screen}$$

It is not easy to deal with this enormous amount of data. Solving it by software is prohibitive because of time constraints. To manage all this data it is imperative to pre-process it in order to achieve two objectives: first, to reduce the flow data without loosing information and second, to extract the most relevant characteristics from the raw data.

There are different types of defects. They can appear at macro level, where a general view is necessary to detect them (See figure 5) and at micro level where a closer view is necessary (See figures 3 and 4). They are grouped into similar kinds, the different kinds must be treated with tailored pre-processors which have to work in parallel to fit the time constraints.

3.4.2 Surface Scanning of the Screen

All the needed information about the defects to be detected will be gathered by an image scanner consisting in a galvo-mirror system which has a colour TV camera associated (See

Fig. 8). The image scanner is controlled by a X and an Y signal which deflect the horizontal and vertical mirrors giving the coordinates to centre the optical system. There are also a focussing signal and a zooming signal.

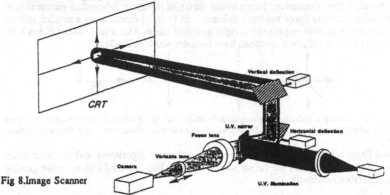

Fig 8.Image Scanner

In order to get the desired resolution,the screen is scanned by small areas. A resolution of 50μm means about 7 pixels per phosphorus line (a phosphorus line is about 280μm wide). For a screen with 1800 phosphorus lines and a colour camera with 512 horizontal pixels, this implies

$$\frac{(1800 \text{ phl} * 7 \text{ pixel/phl})^2 * \frac{3}{4}}{(512 \text{ pixel})^2/area} = 454 \text{ area.}$$

for a screen 600 mm wide, each area will be:

$$\frac{512 \text{ pixel/wide} * 600 \text{ mm.}}{1800 \text{ phl.} * 7 \text{ pixel/phl}} = 24 \text{ mm/wide}$$

and

$$24 \text{ mm /wide} * \frac{3}{4} \text{ wide/high} = 18 \text{ mm. high}$$

3.4.3 Screen Inspection Vision System Architecture

The figure 9 shows the final SIVS Architecture. We kept in mind the currently produced commercial hardware [7]. This can save designing and developing time allowing us to concentrate in the specific problems for which there are no existing solutions because of time constrains. We can say that the choice of either commercial or specific hardware strongly depends on the time constraints. For the detection and location of defects we have decided on specific hardware. For classification and characterisation of defects we have decided on commercial hardware and software.

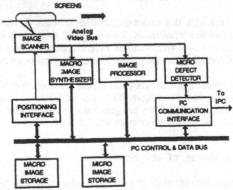

Fig 9. The Screen Inspection Vision System Architecture.

The IVS architecture has four main functions: the microdefect detection, the microdefect analysis and the macrodefect detection and analysis.

The information to be used by the whole SIVS comes from the Image Scanner. So the obtained video signal is send to the micro-defect detector that consists of several preprocessors specialised in each kind of defect. When a microdefect is detected an interrupt signal is given to the controller which captures the x,y coordinates of this defect. Those coordinates are used to zoom-in on the defect. Once the zooming-in has been realised, the new enlarged image of the defect is input to the image processor unit and transferred to the micro image storage board. Later on, the image processor will analyze the defect image with several algorithms which, using the information of the preprocessors allow the characterisation of the defect.

The video signal also goes to the macro image synthesizer unit. In it an averaging system obtains an equivalent, small resolution image of the whole screen built from the scanned microareas and sends it to the macro image storage board. This whole screen image is passed over the image processor unit. This unit performs an analysis over this information which detects, locates and classifies the possible macro-defects. The information about any macro defect is given to the Defect Definition Frame.

4 Conclusions

The integration of advanced information technologies are leading to fully automation and higher performances in Process Control and Supervision. Vision Systems are able to inspect the product and deliver information about it. On this and other informations from the process, Expert Systems reason to identify, diagnose and repair a current problem.

On the hand of the Knowledge-Based system, the combination of two different kinds of knowledge permits to achieve a more correct, robust and reliable system. A general model to represent heuristics from experience has been defined, the *Hypothesis Space* model. It is very well-suited for representing the experiential knowledge for process diagnosis and repair. Among its properties the ability to reason under incomplete knowledge, the management of changing information by revising its conclusions and the facilities for knowledge base development are the more important. A *Causal Model* has also been defined to complement the heuristic one. It has been shown how deep models can refine and complete the heuristic results, but further work is required in order to reduce the computational complexities of the algorithms in general cases.

Acknowledgments. We would like to thank Mr. Rambla, Mr. Kikkert, Mr. Singlas and Mr. Gouraud (from Philips consortium) for their cooperation as process experts, and José C. Ortiz for his implementational work.

References

[1] W. Clancey: *Heuristic Classification*, Artificial Intelligence 27, 1985.
[2] R. Davies: *Diagnostic Reasoning based on Structure and Behaviour*. Artificial Intelligence 24, 1984.
[3] P. Fink and J. Lusth: *Expert Systems and Diagnosis Expertise in the Mechanical and Electrical Domains*. IEEE Trans. SMC, Vol. 17, 1987.
[4] M. Genesereth:*Use of Design Descriptions in Automated Diagnosis*, Artificial Intelligence 24, 1984.
[5] B.J. Kuipers: *Qualitative Simulation*, Artificial Intelligence, 29, 1986.
[6] L. Steels: *The Deepening of Expert Systems*. AI Communications. Vol 1(1), 1987.
[7] SGS. Thomson: *Image Processing*, Databook, Italy, 1990.
[8] W. Van de Welde: *Learning from Experience*. Ph. D. thesis of VUB, 1988.
[9] Ll. Vila: *A Behavioural Analysis of the Process Control Expert*. W.R. IIIA-91, 1991.
[10] Ll. Vila: *Process Description*. Working Report IIIA-91, 1991.
[11] Ll. Vila: *Intelligent Process Control by means of Expert Systems and Machine Vision*. Research Report IIIA-91/21, CEAB-CSIC, 1991.
[12] Ll. Vila, N. Piera: *Qualitative Reasoning by Causal Models*. W.R. IIIA-91, 1991.

Tracking and Grasping of Moving Objects
– a Behaviour-Based Approach –

Alexander Asteroth [†] Mark Sebastian Fischer [†] Knut Möller [†] Uwe Schnepf [‡]

[†] Department of Computer Science, University of Bonn, Bonn, Germany
[‡] AI Research Division, GMD, Sankt Augustin, Germany

Abstract. Behaviour-based robotics (cf. Brooks [2]) has mainly been applied to the domain of autonomous systems and mobile robots. In this paper we show how this approach to robot programming can be used to design a flexible and robust controller for a five degrees of freedom (DOF) robot arm. The implementation of the robot controller to be presented features the sensor and motor patterns necessary to tackle a problem we consider to be hard to solve for traditional controllers. These sensor and motor patterns are linked together forming various behaviours. The global control structure based on Brooks' subsumption architecture will be outlined. It coordinates the individual behaviours into goal-directed behaviour of the robot without the necessity to program this emerging global behaviour explicitly and in advance. To conclude, some shortcomings of the current implementation are discussed and future work, especially in the field of reinforcement learning of individual behaviours, is sketched.

1 Introduction

By the end of the sixties and the early seventies robotics was considered to be the most important field of Artificial Intelligence research. It was expected that a successful robot system would feature the main constituents of intelligent behaviour. Later on, the interest in robotics has declined considerably. In the following, we will shortly discuss why this (from our point of view) seems to be the case.

Knowledge-based techniques were used to develop various functional modules for different aspects of robot-world-interaction such as perception, planning, and control. These modules should enable a robot to act intelligently in the real world. As a result, these modules suffered from the same limitations as any other knowledge-based system (e.g. expert systems)[1].

On the one hand, this is the qualification problem, i.e., the question of what aspect of the real world should be considered when building the knowledge base. On the other hand it is the frame problem, i.e., the question which changes to the knowledge base should be made due to real world interactions. The unpredictability of the real world, which is reflected in these problems, limits the accuracy of the domain model as represented by the knowledge base. Unpredictable events will lead to system failures. In limited (closed world) domains pragmatic considerations of these problems can actually help to build properly working knowledge-based systems. But this is not possible for the field of robotics applications, where the real world has to be dealt with.

---

[1] Knowledge-based systems are mainly characterised by the use of an explicit world model, i.e. the knowledge base, as well as adequate state transition operands. These operands are used to modify the knowledge base in accordance with the implications of system activity in the real world hence maintaining a complete and accurate world model in the knowledge base.

Another important aspect is the geometry-based methodology used for the numerical modelling of the real world in knowledge-based robot controllers. The main problem herein is the vast number of sensor data, which must be converted into geometrical descriptions in order to maintain a precise model of the real world. This approach requires tremendous computing power and is quite error–prone as even slight inaccuracies may cause severe numerical problems [6]. Another dimension is added to the complexity of geometric world modelling by an attempt to incorporate any possible uncertainties into the world model. This finally leads to computational intractability of this approach.

Finally, the dominance of cognitive functionality in the design of an intelligent robot controller appears to be a problem for knowledge-based systems. By this we refer to the emphasis on modelling higher cognitive abilities, such as planning, problem solving, reflection, and reasoning, while disregarding far simpler aspects of every day life, such as perception and action in a dynamically changing environment. We consider this "cognitivism" as one reason why there does not exist a single explicit computational model of a behavioural organisation in robot control. Such a behavioural organisation, which undoubtedly plays an important role in natural biological systems (even for the development of cognitive abilities), is not involved in the design of classical knowledge– or geometry–based robot controllers. In our goal to build more intelligent and more flexible robot controllers, we have started exactly at this point. The studies we have conducted should give insights into what features the control structure should have for the realisation of such a behavioural organisation of a robot arm controller.

In the following section, an example of the behavioural organisation in simple animals taken from Ethology[2] is given. In this example it was studied how complex overall behaviour can be decomposed into simple behavioural patterns [4],[5]. In section 3, the work carried out by Brooks and Connell, which is basic to all our considerations, is presented. In Section 4, the general ideas to fulfill the control task will be outlined, and the individual behaviours used to build the controller presented. Finally, the performance of the system will be discussed pointing to future work such as parallelisation of the control hierarchy, and incorporation of learning techniques to form adaptive behavioural modules.

2   About Coastal Snails

Let us consider an introductive example of animal behaviour which shows how obviously complex behavioural patterns can be suffiently explained in terms of simple, cooperating behavioural sequences. The food seeking behaviour of the coastal snail **Littorina** (cf. Fraenkel [5]) can easily be decomposed into simple behavioural sequences which are triggered by very simple sensory patterns (i.e. light and gravity). Indeed, the following seemingly complicated navigational behaviour of the Littorina can be explained by a simple algorithmic control structure similar to the control structure used in behaviour-based robotics.

Littorina is a snail which mainly lives on algae that have been washed ashore. In order to get to the location of its food, the snail has to behave in different ways when roaming its environment in seek of food. The snail is able to surmount even large obstacles such as rock notches and holes underwater. The basic behaviour of the snail is to move uphill and into the dark (the former guided by gravity changes, the latter by light intensity). This is a useful behaviour since the snail's food is mainly to be found near the water edge, where it is darker than in open sea. As useful this general guiding behaviour might be, it can cause problems. If the snail moves into a notch, this behaviour is not of any use anymore: the snail cannot move any higher nor

---
[2]Ethology is the study of animal and human behaviour.

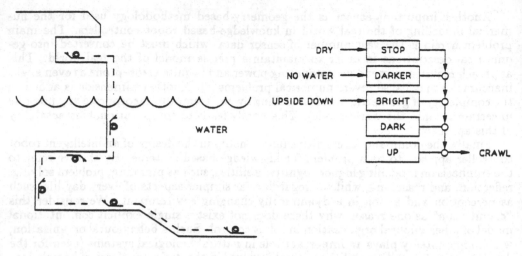

Figure 1: Food seeking behaviour of Littorina [4].

into the dark. Without an additional behaviour, the snail would remain there until it starved. However, when the snail comes to move upside down in the notch (following the surface of the rock notch) the basic behaviour is inverted and the snail moves now towards brighter areas. This way, the snail does not get trapped in a hole. When the snail gets out of the water, further behaviour is encountered: even when the snail is moving upside down, it is heading towards darker areas. This helps the snail to reach a shady environment out of the water. Additionally, the algae can only be found in rock notches near the water surface where they have been washed ashore. A final behaviour saves the snail from crawling too far away from the water: if the surface the snail is moving on becomes dry, the snail stops crawling, preventing it from wandering too far inland. Here, the snail waits until a wave washes it back into the sea, where the whole cycle starts anew.

Fig. 1 shows a schematic description of these five behavioural patterns and their organisation. The arrows indicate actions which are triggered by internal states of the snail, and sensory stimuli coming from the environment. In this behavioural organisation, not only activities depend on each other, they also interact by suppressing each other. This is indicated by arrows pointing to circles, where the output of a behavioural pattern is replaced by the behavioural pattern on top of it. The way of interaction, however, is only unidirectional, i.e. only higher-level modules can influence lower-level modules, not vice versa. The main question remaining, however, is whether these organisational principles present in animal behaviour can be used to build intelligent robot systems. This question will be the major focus of attention for the rest of this paper.

3   Behaviour-Based Robotics

Rodney Brooks [2] was one of the first researchers in AI and Robotics who decided to leave the main roads of knowledge representation and knowledge-based techniques in order to develop intelligent systems (cf. Brooks [3]). He argued that, looking at the evolution of intelligence during the billions of years of biological development, it seems that the (in evolutionary terms) recent emergence of cognitive abilities such as problem solving, language, expert knowledge etc. are relatively simple once the substrate of being and reacting in the real world is provided.

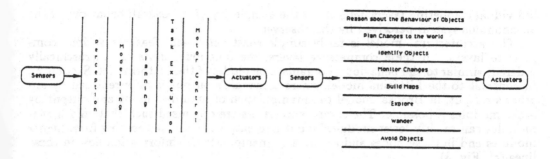

Figure 2: Modular and behaviour-based decomposition of a robot controller [2].

The classical approach in AI to build a robot controller would be to decompose the system to be developed into various subparts according to their functionality. Hence, there would be a system for sensing, a system for situation interpretation, a system for choosing the appropriate actions according to the specified situations and a system to execute the actions chosen (cf. Fig. 2). Each of these subsystems would have been very complex, too. Based on various knowledge-based techniques such as knowledge representation, model-based vision, planning, qualitative reasoning, truth maintenance etc., each subcomponent itself would be built of various complex components.

Brooks suggested, instead of building systems which intend to cover the complete range of human reasoning activities, one should follow an incremental path from very simple, but complete, systems to complex autonomous intelligent systems. The incrementality of this approach is constituted by a horizontal decomposition of tasks according to various "activities" in an autonomous agent rather than a vertical decomposition based on functional entities (cf. Fig. 2). Each activity producing subsystem, or behaviour, is a complete system in itself featuring sensing, reasoning and acting capabilities. Once such a behaviour has been implemented and tested, it remains unchanged and covers a particular aspect of the agent's overall behaviour.

By introducing several layers to enhance the agent's behavioural repertoire, it becomes necessary to define the appropriate organisation and ordering of the layers. Since all layers are running in parallel, they have permanent access to the agent's sensing facilities and try to become active as soon as they receive the appropriate state description through the sensors. This somewhat chaotic behaviour, where multiple activities are competing in gaining control over the agent, has to be organised by means of suppression of less important behaviours in order to achieve a reasonable global behaviour of the agent.

3.1 The Subsumption Architecture

To this end, Brooks opted in favour of a hierarchically fixed organisation of the behaviours building each new layer on top of already existing ones. Additionally, each layer is equipped with the ability to monitor and to influence the behaviour of the next lower layer. Hence, a community of relatively independent, parallel behaviours is created in which each behaviour constitutes a subtask of the agent's cognitive and non-cognitive skills.

The complete control system of the agent consists of various layers and is referred to as the subsumption architecture (the name derives from the fact that the individual layers are combined by suppressing and inhibiting mechanisms). The control flow in the agent emerges from the changing activities of the distributed control flow resident in the various layers, but there is no central control involved. The interaction of the individual competing layers determines the complexity of the overall behaviour of the autonomous system perceived by the observer.

The architectural model is fairly simple constructed and features various competence layers. In these competence layers, the functional units are hierarchically ordered, similar to the behavioural organisation model of the Littorina (cf. Section 2). In contrast to the Littorina model, where a module's output is only replaced by another's output, in Brooks' model the manipulation of modules' input and output by other modules is possible. The competence layers are organised such that only higher modules can modify lower modules' input and output. The lines coming from higher modules end in these nodes and are able to manipulate the information flow in these lines (cf. Fig. 3).

Brooks' approach is mainly based on engineering considerations in the design of his autonomous systems and does not focus on more theoretical aspects of system behaviour. Though, we consider the robots developed at his lab being much more efficient, reliable, robust, and flexible facing the dynamic changes in their environment than robots developed elsewhere following the traditional approach. Nevertheless, the subsumption architecture which represents an implementation of fixed behavioural patterns having a fixed relation to each other lacks some flexibility as it is present in the behavioural organisation of animals.

In his PhD thesis [4] Jonathan Connell described an architectural model for robot control which resembles Brooks' subsumption architecture. Connell used the strictly hierarchical relation between the behaviours only as a general idea and introduced less hierarchical features to improve the performance of his system. In implementations of the subsumption architecture, frequently output of lower control levels is used to modify input of higher control levels. This is done for pragmatic reasons.

Summarising, one can describe Connell's colony architecture as a modular structure organised into layers where influence is *mainly* pointing from modules within higher levels to modules within lower levels.

A concrete example of such an architecture will be explained later by describing our own implementation.

4 The Realisation

In order to study the appropriateness of the behaviour-based approach for robot arm control, we decided to build a robot controller for a particular real-world application — the tracking and grasping of a moving object (i.e. a ball). We believe that this example[3] shows sufficient problem complexity to provide insights into how this approach can be used for a wider range of robot manipulator applications. The robot manipulator is located on a platform on which the ball is rolling. There were no a priori assumptions about speed and direction, i.e. what trajectory the ball would be moving when rolling through the workspace. The final robot controller should be flexible and robust enough to deal with all possible different situations.

4.1 The General Control Ideas

First we need to consider what kind of sensory information and world knowledge should be provided to our robot arm controller. As far as sensory information is concerned, we decided to use a camera which is located on the robot arm wrist. This sensor was considered to provide all necessary informations, apart from the internal state

---

[3]Actually we already implemented a number of other approaches to this task [9] giving us the opportunity to compare the effort needed for development and the resulting performance of the systems.

descriptions of the robot arm joints themselves. We opted in favour of a wrist camera since this sensor provides only local information without the necessity to consider a global world model. This should make the controller more flexible and robust.

The robot arm should be controlled by simple interacting behaviours such as moving up and down, or turning right or left, in a way that a moving object in the robot's workspace could be found in the camera's sensor field. After detecting a moving object, the robot arm should be guided by other behaviours in a way that the object gets centered in the camera image. Then the robot arm should move in a way that the image of the object remains centered though moving, and grows larger and larger in the camera image. When the image of the object in the sensor field has grown to a particular size and nearly fills in the camera image completely, a grasping reflex should perform a secure grasping assuming that the object is actually close enough to grasp it. As one can see, this is the only knowledge about object location or object movement direction necessary to control the robot. The overall scenario is shown in Fig. 3.

4.2 Hardware Description

The robot control program is running on a Sun Sparc-Station 1 operating under SUN-OS 4. The image processing is performed on a general purpose image processing unit with a 68030 prozessor and a graphics card connected to a CCD-camera. For technical reasons we could not fix our camera on the robot arm wrist. So we decided to fix it on the ceiling of our lab focussing on the robot's workspace from above. The images coming from this camera are transformed to a virtual wrist camera (for more information concerning the transformation of the image data please refer to Asteroth and Fischer [1]). The image processing system delivers information about the center point of a moving object relatively to the camera image. By using data reduction[4] we managed to accelerate the image processing cycle from 5 Hz up to 20 Hz, i.e. in case of a moving object within the visual field of the camera, the robot controller receives information on the relative object location 20 times a second. Current sensory information is essential to the behaviour-based approach, since no world model is available to the controller in order to compute future states of the world. We use the world as its own model.

The robot is a Mitsubishi RV-M1, having five DOF (base, shoulder, elbow, pitch and roll) and a simple gripper. The working space radius is 70cm approximately. The robot arm is controlled by a Z80 processor the operating system of which has been reprogrammed to allow absolute and relative joint movements. Additionally, it is possible to replace current robot commands by new commands without interrupting the robot motion.

4.3 The Controller

In this section we describe the individual behaviours we used to build up our subsumption controller. On the basis of the general control ideas outlined in the previous section we divided the control task into several modules and organised them in a hierarchical way (cf. Fig. 3).

The "Avoid" Behaviour The most important task of any robot control is to prevent the robot from crashing into obstacles. In our case, this means preventing the

---

[4]We blur the image and only evaluate 16 areas of the camera image instead of all pixels — this technique is sufficient within our approach to detect moving objects. The image processing routine does not perform any "object recognition" in its very sense. It only looks for intensity changes in the camera image divided into large regions

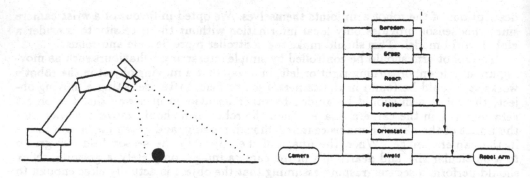

Figure 3: The scenario and our robot controller.

robot from hitting the table. Since we do not have any e.g. force sensors we decided to use geometrical information given by the robot joint and link parameters. By some simple trigonometric computations, we calculate the height of the robot joints above the table surface. This function determines whether control sequences coming from other modules going to the robot joints would make the robot arm to collide with the table. If so, the provided robot joint positions are modified. In our experiments it has shown that keeping five centimeters distance to the table surface guarantees secure operation of the robot. This lowest module does not perform any activity, it just takes care that other modules' activities do not harm the robot.

The "Orientate" Behaviour The next step is to orientate the robot towards a visible moving object. This movement is performed only by the base and wrist joint of the robot. The reactions triggered by the camera output are fairly simple: if there is no moving object visible, the robot does not move. If there is something moving in the right (or left) part of the camera image, the robot arm turns a little to the right (or left). If something is moving in the upper (or lower) part of the image, the robot moves up (or down) a bit. The extent of these basic actions is modulated by the current joint positions, the displacement of the object's picture to the image center, and the *moment*[5] to form new goal configurations to be attained by the manipulator. Since these joint positions are calculated for every new object position computed by the image processing, this behavioural module passes 20 joint changes per second to the RV-M1 controller. Those commands currently executed by the robot are interrupted and immediately replaced by new ones. Hence, a sequence of short trajectories is actually driving the robot. Using these two basic behaviours, it is possible to track a moving object detected in the visual field of the camera.

The "Follow" Behaviour For locating a moving object in the camera image it is sufficient that the robot is only turning the base and moving the wrist up and down. But in order to get closer to the ball, it is necessary to drive the shoulder and elbow joints into the direction of the ball. This is done by calculating joint modifications leading to a movement into the direction pointed to by the wrist. This again can be computed by trigonometric manipulations using the knowledge about the simple kinematic structure of the robot. The risk of colliding with the table does not occur since the **Avoid** behaviour is active. But the robot arm is, in our example task, never able to reach the ball, since the **Avoid** behaviour keeps it at a level five centimeters above the table.

---

[5]By the *moment* of an object we refer to the number of non-black pixels in the corresponding part of the camera image, weighted by the light intensity of these pixels.

The "Reach" Behaviour Now, since we have managed to bring the robot hand safely close to the moving object, we need to enable the robot to finally grasp the ball. Therefore, we have to influence the activity of the **Avoid** behaviour. For this purpose, we add another input line to the **Avoid** module. This input line provides the **Avoid** behaviour with a value for the required distance to the table surface. This value is now allowed to be decreased by higher modules, with a certain delay. If the distance value is decremented for a long enough period, it actually reaches the zero value[6]. If the value is not decreased anymore, it is incremented until it finally reattains the default value. The decrease is triggered by an object visible in the center of the camera image that surmounts a certain moment. Under the influence of this behaviour, the robot is actually able to reach the moving object. If the object is not in the center of the image anymore (which increases the danger of larger compensation movements of the robot to relocate the object in the image center), the distance value is immediately increased until a safe hight above the table is reached.

The "Grasp" Behaviour Now we have to provide the robot controller with the grasping ability. For this purpose, we implemented a grasping reflex which is triggered by an object located in the center of the camera image having a certain large *moment*. If this is the case, the output of all other behaviours below in the hierarchy will be suppressed, the gripper will be closed, and the robot arm will be moved back into its *home-position*.

The "Boredom" Behaviour A special case has to be treated at the end: if the robot arm covers the object, the arm has to be moved into a new position, where the ball can be seen again. Hence, if the object is hidden for some time, this top-level behaviour suppresses the output of all other behavioural modules below and moves the arm *randomly*. This only happens for static cases i.e. if the ball is hidden by the manipulator, and in cases where no balls are in view. The complete control architecture is shown in Fig. 4.

4.4 Discussion

We have tested the robot controller in many situations and found it nearly 100% reliable as far as the tracking and grasping of a ball is concerned.

It is flexible in the sense that the controller is able to handle unexpected events due to the emergent global behaviour which does not need to consider all possible events when implementing the system. Secondly, it is robust as the failure of individual behaviours (apart from the **Avoid** behaviour) does not cause a system crash. Thirdly, it is modular, as parts of the controller can be developed individually, tested and added to existing layers.

Low level behaviours can be run without high-level behaviours. This was shown by operating the system with more and more levels. In our experiments, we started with the two basic behaviours (**Avoid** and **Orientate**). The system performed as expected. Adding the next level did not cause any problems, it just increased the behavioural repertoire of the robot. This way, we easily added more and more levels until the complete range of competence had been reached.

The controller proved to be flexible, robust and modular — interesting features for any robot application. One technical (not a conceptional) shortcoming, however, is that our robot (just 5 DOF) is only able to *grasp* balls once they have stopped.

---

[6] The zero level corresponds to 1cm distance above the table (the minimal distance of the robot to the table for secure operation).

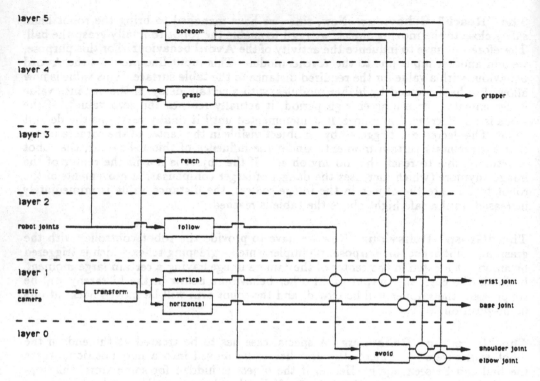

Figure 4: The complete control architecture.

Grasping rolling balls is still not possible. The reason for this is the fairly low communication rate between the host computer, the RV-M1 controller, and the image processing unit.

Due to this, the execution of a robot command takes place only some time after the corresponding image has been evaluated (i.e. too long a time gap between sensor and motor patterns). This causes the robot arm to always be somewhat *behind* the moving object. In order to change this, and to avoid calculating ball trajectories (we do not want to use a priori world models), we introduced so-called *histories*: the changes in robot arm joint positions during previous action sequences are recorded and considered when computing the next movement of the robot.

Using histories, it is possible to get closer to the ball when following it, or, if the ball moves on a straight line, even get ahead of it. Nevertheless, it is still not possible to grasp the ball since the compensation movements during the following of the ball are not stable enough to locate the ball in the image center for a few cycles. But this is essential to lower the distance value as performed by the **Reach** behaviour. We hope to overcome these limitations by speeding up the communication rate between the different processing units. Additionally, we could use adaptive mechanisms to learn the dynamic properties of the robot arm when following a ball. This knowledge could be integrated into the histories and could serve to compute more reliable predictions about the ball's position in motion.

We are currently investigating the use of adaptive mechanisms such as neural nets and classifier systems to learn these dynamic system characteristics. Additionally, we are studying different reinforcement learning algorithms such as temporal-difference methods (cf. Sutton [8]) and Q-learning (cf. Watkins [10]) to be incorporated into the adaptive mechanisms mentioned above.

Finally, we are considering the use of a parallel implementation of the robot controller based on a transputer architecture. This should help us, first, to carefully reconsider our own methodology in terms of modularity and conceptualisation, and second, to speed up computing time in order to at least stay at the same level of performance while including time consuming learning algorithms.

5 Conclusion

In this paper we described a modular and flexible robot controller based on the subsumption architecture. We have shown that it is possible to use the behaviour–based approach for ordinary robot manipulators applied to real–world tasks. We consider this approach of great importance since it allows us to overcome the well-known limitations of knowledge-based systems. Being aware of the shortcomings of a purely behaviour–based robot control technology (i.e. anti-representational), our future work will mainly be in the field of building-up internal representations of the robot-world interaction and learning of world concepts. But these world models will be based on behavioural concepts developed during this interaction rather than predefined knowledge bases (cf. Patel and Schnepf [7]).

Acknowledgements

We would like to thank Thomas Christaller, Michael Contzen, Michael Eichhart, Michael Fassbender, Christian Haider, Frank Smieja, Andreas Thümmel, Sebastian Thrun, and Gerd Veenker and all the members of the Robotics and Neural Networks working group at the University of Bonn and GMD for their support and help.

References

[1] A. Asteroth and M.S.Fischer. *Ballgreifen mit einem Roboter — ein verhaltensbasierter Ansatz*. University of Bonn, Department of Computer Science, Internal Report, July 1991.

[2] R.A. Brooks. *A Robust Layered Control System for a Mobile Robot*. Massachusetts Institute of Technology, AI Memo 864, Cambridge, 1985.

[3] R.A. Brooks. *Achieving Artificial Intelligence Through Building Robots*. Massachusetts Institute of Technology, AI Memo 899, Cambridge, 1986.

[4] J.H. Connell. *A Colony Architecture for an Artificial Creature*. PhD Thesis, Massachusetts Institute of Technology, Cambridge, 1989.

[5] G. Fraenkel. "On Geotaxis and Phototaxis in Littorina", in *The Organization of Action: A New Synthesis*, C.R.Gallistel (ed.), Lawrence Erlbaum, 1980.

[6] Hoffmann, C. M.. *The Problem of Accuracy and Robustness in Geometric Computation*, in Computer, 22,3, 31–41, 1989.

[7] M.J. Patel and U.Schnepf. *Concept Formation as Emergent Phenomena*. To appear in: Proc. of the First International Conference on Artificial Life, Paris, France, 1992.

[8] R.S. Sutton. *Learning to Predict by the Methods of Temporal Differences*. Machine Learning 3:9-44, 1988.

[9] S. Thrun, K. Möller, and A. Linden. *Planning with an adaptive world model*. In D. S. Tourtezky and R. Lippmann, editors, Advances in Neural Information Processing Systems 3 , San Mateo, November 1990. Morgan Kaufmann.

[10] C.J. Watkins. *Learning with Delayed Rewards*. PhD thesis, Cambridge University Psychology Department, 1989.

PANTER -
Knowledge Based Image Analysis System
for Workpiece Recognition

Bärbel Mertsching
University of Paderborn, Department of Electrical Engineering
D-4790 Paderborn

Abstract. This article describes a knowledge based image analysis system that uses the Hierarchical Structure Code (HSC) for the initial segmentation of the image data. The generation of this code is domain independent and allows for an elegant coupling with a knowledge base. The system was successfully tested in the field of workpiece recognition. In the first section of the article, a short introduction of the image transformation is presented, a presentation of methods to extract the size and position invariant features follows. The main emphasis of the article is on the description of a semantic network language for a knowledge based image analysis system using the Hierarchical Structure Code.

1 Introduction

Knowledge based image analysis is the synthesis of pattern recognition methods and of artificial intelligence, in which general problem solving methods are combined with methods for image preprocessing and segmentation. This field is characterized by the use of sensor data which are often superimposed by noise. For the analysis of complex images, explicit knowledge of the task domain is required in order to derive a symbolic description of the input image data automatically. When developing a system for image analysis, the coordination of an image's initial segmentation and its latter knowledge based analysis poses the main problem.

In this article, a system is described that combines the advantages of an hierarchical image segmentation scheme (the Hierarchical Structure Encoding) and a semantic network for knowledge representation.

2 The Hierarchical Structure Code

Each contour and region of objects in an image can be mapped to the so-called *code trees* of the HSC. Invariant features such as shape descriptions or relations between structures and components can be easily extracted from the HSC. This method provides a straightforward, domain independent transition between the signal space of the image and the space of its symbolic representation.

To generate the HSC, first a *Gaussian pyramid* (a sequence of images in which each is a low-pass filtered copy of its predecessor) and then a *Laplacian pyramid* is constructed in a preprocessing step. The result of these procedures is that enhanced features are segregated by size - with fine detail in the highest resolution level and progressively coarser detail in each lower level.

During the encoding procedure the hexagonal digitized image is subdivided into overlapping islands of seven pixels. In this manner, continuous structures such as bright or dark contours or regions are divided into *structure elements* which are represented by *code elements* $<t;m;\varphi|k;n=0;r;c>$. The type of a structure element is encoded by t, its shape and orientation by m and φ; r;c denotes the coordinates of a code element, whereas $|k;n=0;r;c>$ presents its *hierarchical coordinate*. The encoding is carried out at each of the different resolution levels k with a pixel distance of 2^k.

During the linking process, seven islands $I|k;n>$ are described by an island of double size $I|k;n+1>$. At first, the code elements are analyzed for continuity. In the second step continous structures are generalized and represented by one code element $<t;m;\varphi|k;n=1;r;c>$. This procedure results in a hierarchical data structure, code trees of an object which are stored in a HSC data base. This method can be repeated until the whole structure is represented by just one single code element $<t;m;\varphi|k;n;r;c>$ (*root node*). The linking process finally ceases when one island covers the entire structure. A more detailed description of the HSC can be found in [HAR87].

The advantage of such an image transformation method is the resulting easy, rapid to an image's dominant structures. Since only local operations are employed during the generation process, strategies for parallel programming can be easily implemented (see [PRI90]). It can even be executed in real time by using a HSC processor [WES90].

As the HSC code trees are not invariant against modifications of size, position and orientation of objects, explicit knowledge is applied to interpret the results of the segmentation. During image analysis, domain independent operations extract an image's invariant features which are matched to modelled features in the knowledge base. These operations can be categorized in four groups: structure-describing operations search for root nodes or the leaves of code trees; a shape-analyzing operation investigates the form of a code element sequence; topological operations examine neighbourhood relations among image structures and geometric operations are used to determine an object's size and position.

3 Semantic HSC Networks

After comparing different knowledge representation formalisms, i.e. *formal grammers*, *production systems*, *formal logics* and *semantic nets*, the last method was chosen for application in the PANTER system. The main reasons for employing semantic networks in computer vision are the following: Knowledge can be formulated explicitly and transparently. The abundant possibilities of structuring a semantic net permit the integration of different knowledge types and support an epistemologic modular design of a knowledge base. Existing systems prove that complex image analysis tasks can be solved by using this method. (A detailed discussion can be found in [SAG91]). The semantic network language described in the following chapters is designed to analyse hierarchical encoded images which result in constraints for the analysis. The granulation of the knowledge is dependent on the HSC and the domain. The knowledge representation language used in the PANTER system has a well-defined syntax, on which the domain independent, intensional semantic of the employed data structures and inference rules is based.

3.1 Syntax of a HSC Network

Two different basic types of nodes exist in the PANTER system. First, a *concept* consists of an intensional description of an object or an event; secondly, an instance contains an extensional description of a signal and satisfies the definition in a concept. The data structure of a node is frame-like (see fig. 1). Edges to other concepts and attributes are stored in the slots of a concept. Standard relationships between nodes can be built up by three different types of edges and their inverses. They are denoted as *part-of*, *is-a* (*specialization*) and *instance-of* and are written explicitly in the first five slots of a concept. Special relationships (e. g. characterizing the topology or geometry of an object) are described separately in so-called *relational concepts*.

```
CONCEPT                  name of concept
  (GENERALIZATION        [name of concept])
  (SPECIALIZATION        [name of concept]
    [RELATION            definition of relation]
    [ATTRIBUTES
                         name of attribute: definition of attribute
                         {,name of attribute: definition of attribute}])
  INSTANCES              [name of instance]
  (PART_OF               [name of concept])
  (PART                  [name of concept]
    [RELATION            definition of relation]
    [ATTRIBUTES
                         name of attribute: definition of attribute
                         {,name of attribute: definition of attribute}])
  [ATTRIBUTES
                         name of attribute: definition of attribute
                         {,name of attribute: definition of attribute}]
END                      name of concept
```

Fig. 1   Syntactical structure of a concept: On the left, the terminal syntactical categories are written in capital letters; on the right, the non-terminal categories. The expression [x] denotes a zero times or single occurrence of x , (x) represents a single or multiple repetition of x, whereas {x} denotes a multiple including the zero times repetition of x.

An attribute characterizes the typical features of a concept. Fig. 2 shows the syntactical components of an attribute definition. While the first three slots contain declarative knowledge, the fourth is used to address procedural knowledge. In this case, the name and parametes of an HSC operation are stored to extract invariant features out of code trees in order to determine the value of an attribute.

TYPE   type of attribute
RANGE   range of attribute values
NUMBER   number of attribute values
OPERATION   procedure for evaluating the attribute
 OPERAND
 OPERATIONRANGE
 PARAMETER
 FORMAL_RESULT

Fig. 2 Syntactical structure of an attribute's definition

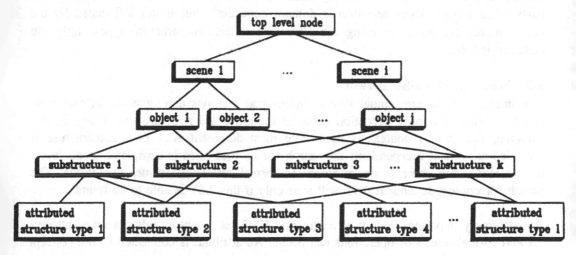

Fig. 3 PART_OF hierarchy of a semantic HSC net

The TYPE of an attribute is generally an elementary or composed data type (e. g. an integer or array), although it can also be a concept. The value of an attribute and its quantity can be restricted by entries in the slots RANGE and NUMBER. The OPERATION is supplemented by three arguments: OPERAND, OPERATIONRAN-GE and PARAMETER. The OPERAND can encompass the entire HSC or particular HSC elements, sequences and groups of HSC elements. The OPERATIONRANGE restricts the working space of an operation to *volumes of interest* (minimal or maximal detector or linking level) or *regions of interest* (window within a level). In the slot PARAMETER additional restrictions can be supplemented. The FORMAL RESULT denotes the HSC elements which are extracted by the operation. As above, formal results of an operation can be HSC elements, sequences, groups etc. Subsequent operations can use formal results as operands.

The granulation of the 'PART-OF' hierarchy (see fig. 3) is adequate for the domains used; if necessary, more conceptual levels can be introduced. The lowest conceptual level is formed by the so-called *attributed structure types* which are derived from the HSC (see below). Described at the second level, *substructures* consist of one or seve-ral attributed structure types. A group of substructures describes an object, while several objects form a scene description. The top-level node is employed to comprise several scene descriptions in one knowledge base.

The elements on the lowest abstraction level of the part-of hierarchy can be generated directly out of the HSC. Due to the hierarchical structure of an HSC data base, an image structure is encoded as a code tree. It can be characterized by its root node's code type (edges, bright and dark lines, regions and vertices). In addition, a sequence of leaf elements can be developed from the root, thus the shape of the sequence can be analyzed. The information on the code type, number of elements and shape are combined with an attributed structure type (AST), a primitive for symbolic image description, and can be generated by three consecutive operations. Substructures are parts of an object. Their selection is domain dependent, but is not influenced by the initial image segmentation using the HSC. For attributed structure types, only the contrary is true.

3.2 Semantic of a HSC network
The introduction of conceptual levels implies that relations among nodes at one level are prohibited. The only exception exists at the attributed structure type level: in this instance, the idea of context is used. The most dominant AST of a substructure is regarded as context independent; the search for its root node is completed at all levels of the HSC, thus there is no reference to formal results of former operations. The search for context dependent AST will start only if this AST is able to be found.

An extremely important feature of a semantic HSC net is that an inheritance of attributes and formal results of operations can occur. An attribute is connected to the concept where it semantically relates. *Vertical inheritance* signifies that an attribute definition can be overridden by the definition in a superclass concept in order to formulate a similar but not identical attribute. In this way, redundance in the knowledge base is reduced because similar concepts can be avoided. *Horizontal inheritance* is used mostly on one abstraction level. The formal result of a context independent attribute can be propagated as an operand to attributes of the same or another concept, yet the range of formal results' validity is restricted. This method reduces the costs of the search in a HSC data base.

3.3 Procedural semantic of a HSC network
The procedural semantic of an HSC net contains inference rules for its analysis. They do not depend on syntactic or semantic reasons, but rather are used to facilitate the interpretation of a network. At first, restrictions for the use of edges are introduced: No cycles are permitted in an HSC network and the relationship SPECIALIZATION is permitted only on one abstraction level. The term 'instance' is a complete structural interpretation of a concept.

4 Example of a semantic HSC network
The semantic network language can be demonstrated by the following example. The scene shown in fig. 4 portrays one or two connecting rods. The network contains five abstraction levels sufficient to model scenes for workpiece recognition. A distinction

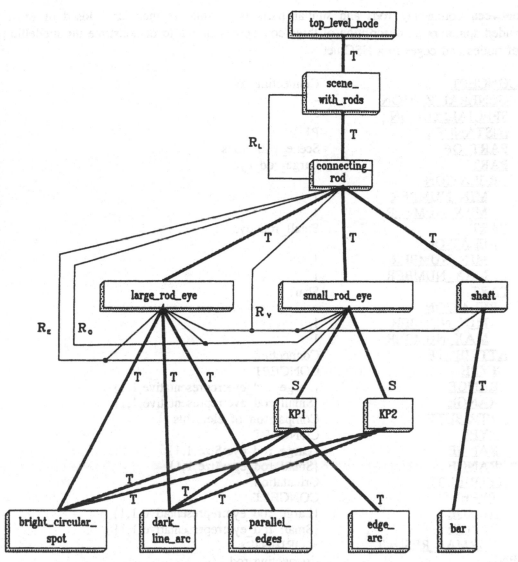

Fig. 4   A semantic HSC net of a connecting rod. The nodes are concepts which
describe the complete rod, its substructures and attributed structure types.
Standardized and special relationships are denoted by edges. (S: SPECIALI-
ZATION/ GENERALIZATION, T: PART/ PART_OF, R_L: position, R_V:
connection, R_E: number of elements, R_O: orientation)

between connecting rods and similar parts is possible as there are closed or open-ended spanners. A description of three concepts is shown to demonstrate the modelling of nodes and edges in a HSC net.

<u>CONCEPT</u>	Connecting rod
<u>GENERALIZATION</u>	-
<u>SPECIALIZATION</u>	-
<u>INSTANCES</u>	P1
<u>PART_OF</u>	Scene_with_rods
<u>PART</u>	Large_rod_eye
<u>RELATION</u>	
<u>MIN_NUMBER</u>	1
<u>MAX_NUMBER</u>	1
<u>PART</u>	Small_rod_eye
<u>RELATION</u>	
<u>MIN_NUMBER</u>	1
<u>MAX_NUMBER</u>	1
<u>PART</u>	Shaft
<u>RELATION</u>	
<u>MIN_NUMBER</u>	1
<u>MAX_NUMBER</u>	1
<u>ATTRIBUTE</u>	Connection
<u>TYPE</u>	CONCEPT
<u>RANGE</u>	{Large_rod_eye:representative,1,1}
<u>RANGE</u>	{Small_rod_eye:representative,1,1}
<u>ATTRIBUTE</u>	Comparison_of_elements
<u>TYPE</u>	CONCEPT
<u>RANGE</u>	{Large_rod_eye:Size,1,1}
<u>RANGE</u>	{Small_rod_eye:Size,1,1}
<u>ATTRIBUTE</u>	Orientation
<u>TYPE</u>	CONCEPT
<u>RANGE</u>	{Large_rod_eye:representative,1,1}
<u>RANGE</u>	{Small_rod_eye:representative,1,1}
<u>FORMAL_RESULT</u>	<ORIENT>
<u>END</u>	Connecting rod

Fig. 5 Textual description of the object node 'Connecting rod'

Fig. 5 depicts the textual description of the object node 'Connecting rod'. Note that there are no entries in the slots GENERALIZATION and SPECIALIZATION. The concept is connected to the concept 'Scene_with_rods' by a PART_OF edge and has three daughter concepts which can be reached using the PART edges: 'Large_rod_eye', 'Small_rod_eye' and 'Shaft'. The attributes of the concept used to model special relationships point to relational concepts. This process is marked by the entry CONCEPT in the TYPE slot. The names of the attributes are identical to the names of the relations. The arguments of the special relations are found in the slots RANGE . The term 'representative' indicates that the first entry of the argument represents a formal result which is computed by the pseudo-operation TRANS. The relation 'Connection' demands that the substructures 'Large_rod_eye'and 'Small_rod_eye' must be interconnected. The digits denote the computation of one value for the relation. The second relation, 'Comparison_of_elements', describes the ratio of the elements in the two edge sequences around the eyes of a connecting rod. The third attribute is used to analyse the position of the connecting rod.

CONCEPT	Bright_circular_spot
GENERALIZATION	-
SPECIALIZATION	-
INSTANCES	Spot
PART_OF	Large_rod_eye
PART_OF	KP1
PART_OF	KP2
PART	-
ATTRIBUTE	Structure_types
TYPE	STRING
RANGE	{bright_closed_region}
RANGE	{bright_closed_vertex}
NUMBER	1,1
OPERATION	ROOT
OPERAND	<HSC>
OPERATIONRANGE	kLOI-1<=k<=kLOI+2, 0<=n<=8
PARAMETER	-
FORMAL_RESULT	<ROOT_NODE 1>
ATTRIBUTE	Number_of_elements
TYPE	INTEGER
RANGE	{3, 10}
NUMBER	1,1
OPERATION	SEQU
OPERAND	<ROOT_NODE 1>
OPERATIONRANGE	-
PARAMETER	-
FORMAL_RESULT	<SEQU 1>
ATTRIBUTE	Shapes
TYPE	STRING
RANGE	{circular}
NUMBER	1,1
OPERATION	SHAPE
OPERAND	<SEQU 1>
OPERATIONRANGE	-
PARAMETER	-
FORMAL_RESULT	<ORSEQU 1>
END	Bright_circular_spot

Fig. 6 Textual description of an AST concept

The use of procedural knowledge - HSC operations - is evident in fig. 6 which shows the concept of an attributed structure type. The concept is interconnected to three concepts at substructure level by PART_OF edges. The slot PART has no entry which indicates that the concept describes an AST. In the first attribute ,'structure_types', the root node of an AST is modelled: Its structure type should be a bright region or vertex code. The operation ROOT should search for exactly one root node. The concept is a context independent AST, because the OPERAND encompasses the entire Hierarchical Structure Code of an image. The OPERATIONRANGE is restricted to four resolution levels in respect to the level-of-interest k_{LOI}, the level k, where the entire connecting rod is encoded in the highest resolution. The formal result < ROOT_NODE 1 > is the OPERAND in the next attribute, 'Number_of_elements'. The operation SEQU should develop the code element sequence < SEQU 1 > at resolution level by climbing top-down through the code tree with this root node. The number of sequence's elements should be 3 to 10. The third attribute, 'Shapes', models the shape of the sequence of code elements < SEQU 1 >; it should be circular.

CONCEPT	Connection
GENERALIZATION	-
SPECIALIZATION	-
INSTANCES	-
PART_OF	-
PART	-
ATTRIBUTE	Argument
TYPE	Boolean
RANGE	{TRUE}
NUMBER	1,1
OPERATION	CONNECT
OPERAND	-
OPERATIONRANGE	-
PARAMETER	straight
FORMAL_RESULT	-
END	Connection

Fig. 7   Textual description of a relational concept

The modelling of the special relation 'Connection' has been explained above. The related relational concept is shown in fig. 7. As previously mentioned, the slots of the standardized relations are empty. The operation for the evaluation of the relation is found in the slot OPERATION of the attribute 'Argument'. If the connection between the given structures is 'straight', the symbolic result of the operation is 'true'.

5 Applications and Results

The image transformation and the semantic network language described in this article were successfully applied in the knowledge based image analysis system PANTER to investigate gray level images of flat industrial parts. These objects may lie separately or may overlap. It is a shortcoming of the system that it is restricted so far to the evaluation of 2D models but plans exist to extend it to 3D models. The entire PANTER system was implemented in PASCAL under VMS; it is discussed in detail in [MER91a]. At the moment, PANTER is reimplemented at T800 transputers and a SUN as a host computer in order to develop parallel analysis strategies for higher efficiency (see [MER91b]). Traffic scenes are studied as a second domain.

References

[HAR87]   Hartmann, Georg: Recognition of Hierarchically Encoded Images by Technical and Biological Systems. In: Biological Cybernetics 56, 1987, 593-604

[MER91a]   Mertsching, Bärbel: Lernfähiges, wissensbasiertes Bildanalysesystem auf der Grundlage des Hierarchischen Strukturcodes. Fortschr.-Ber. VDI Reihe 10. Düsseldorf (VDI-Verlag) 1991

[MER91b]   Mertsching, Bärbel; Büker, Ulrich; Zimmermann, Stephan: Ein Satz von merkmalbestimmenden Operationen zur Auswertung von Bildpyramiden. In: Radig, B. (Ed.): Mustererkennung 1991. Informatik-Fachberichte 290. Berlin (Springer-Verlag) 1991, 180-186

[PRI90]   Priese, Lutz; Rehrmann, Volker; Schwolle, Ursula: An asynchronous parallel algorithm for picture recognition based on a transputer net. In: Journal of Microcomputer Applications 13, (Academic Press), 1990, 57-67

[SAG91]   Sagerer, Gerhard; Niemann, Heinrich; Hartmann, Georg; Kummert, Franz; Mertsching, Bärbel: Semantische Netzwerksysteme in der Musteranalyse. (In Press)

[WES90]   Westfechtel, August: Entwurf und Realisierung eines Prozessors zur hierarchischen Codierung von Flächen, Kanten und Linien. Fortschr.-Ber. VDI Reihe 10, 138. Düsseldorf (VDI-Verlag) 1990

KITSS: Using Knowledge-Based Software Engineering For Functional Testing

Uwe Nonnenmann and John K. Eddy

AT&T Bell Laboratories

Murray Hill, NJ 07974, U.S.A.

Abstract. Automated testing of large embedded systems is one of the most expensive and time-consuming parts of the software life cycle. The *Knowledge-based Interactive Test Script System* (KITSS) automates functional testing in the domain of telephone switching software. KITSS uses novel Artificial Intelligence approaches to improving this Software Engineering activity. KITSS has a statistical parser to support the specification of test cases in English. These tests are converted into a formal representation that is audited for coverage and sanity. To accomplish the conversion, KITSS uses a theorem prover-based inference mechanism and a knowledge base as the domain model with both a static terminological logic and a dynamic temporal logic. Finally, the corrected tests are translated into an automated test language.

Keywords. Software testing, large embedded systems, knowledge-based software engineering, statistical parsing, automatic programming.

1 Introduction

Software Engineering is a knowledge intensive activity that involves defining, designing, developing, and maintaining software systems. In order to build effective systems to support Software Engeneering activities, Artificial Intelligence techniques are needed. The application of Artificial Intelligence technology to Software Engineering is called Knowledge-based Software Engineering (KBSE) [Lowry & Duran, 1989]. The goal of KBSE is to change the software life cycle such that software maintenance and evolution occur by modifying the specifications and then rederiving the implementation rather than by directly modifying the implementation. The use of domain knowledge in developing KBSE systems is crucial. We have taken this philosophy and applied it to the task of functional software testing.

As software grows in size, there is an increase in the effort and cost that is needed to develop software projects. It is generally accepted that the development of large-scale software with zero defects is not possible. A corollary to this is that it is not possible to uncover all defects by testing. This is because of the many inherent problems in the development of large projects [Brooks, 1987]. As just a few examples, a large project provides support for many interacting features, which makes requirements and specifications complex. Also, many people are involved in the project, which makes it difficult to ensure that each person has a common understanding of the meaning and functioning of features. Finally, the project takes a long time to complete, which makes it even harder to maintain a common understanding because the features change through time as people interact and come to undocumented agreements about the real meaning of features.

The consequence of these problems is that programs that do not function as expected are produced and therefore extensive and costly testing is required. Once software development is complete, even more testing is needed for product maintenance. The major cost of maintenance is in re-testing and re-deployment and not in re-coding. Estimates are that at least 50%, and up to as much as 80% [McCartney, 1991], of the cost on the life cycle of a system is spent on maintenance.

To find and eliminate software problems early in the development process, we designed an automated testing system that is well integrated into our existing design and development process [Nonnenmann & Eddy, 1991]. The typical end-to-end software system development process involves sequentially producing the outputs shown in Figure 1. The figure is "V" shaped to emphasize that a good design is a testable design. The shape highlights the inherent connection between the pre-coding design outputs and the post-coding test outputs [Myers, 1979]. Our focus is on bridging the gap across this "V" from the "external design" to the "functional tests". In functional testing, the internal design and structure of the program are ignored. It corresponds directly to uncovering discrepancies in the program's behavior as viewed from the outside world. This type of testing has been called *black box* testing because, like a black box in hardware, one is only interested in how the input relates to the output. The resulting tests are then executed in a simulated customer environment which corresponds to verifying that the system fulfills its intended purpose.

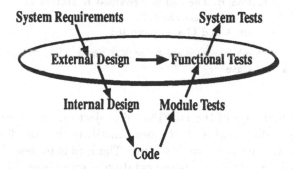

Fig. 1. KITSS Testing Process

In summary, KITSS improves the testing process by generating tests of higher quality, which allows more frequent automated regression testing. Furthermore, tests are generated earlier, *i.e.*, *during* the development phase not *after*, which should allow the earlier detection of problems. The result is higher quality software at a lower cost.

2 Functional Testing with KITSS

The *Knowledge-based Interactive Test Script System* (KITSS) was developed at AT&T Bell Laboratories to reduce the increasing difficulty and cost involved in testing the software of DEFINITY®PBX telephone switches. Although our system is highly domain specific in its knowledge base and inference mechanisms, the approach taken is a general one and applicable to any functional software testing task.

DEFINITY supports hundreds of complex features such as call forwarding, messaging services, and call routing. Additionally, it supports telephone lines, telephone trunks, a variety of telephone sets, and even data lines. At AT&T Bell Laboratories, PBX projects have many frequent and overlapping releases over their multi-year life cycle. It is not uncommon for these projects to have millions of lines of software code.

> **GOAL:** Activate CF[1] using CF Access Code.
> **ACTION:** Set station B without redirect notification[2].
> Station B goes offhook and dials CF Access Code.
> **VERIFY:** Station B receives the second dial tone.
> **ACTION:** Station B dials station C.
> **VERIFY:** Station B receives confirmation tone. The status
> lamp associated with the CF button at B is lit.
> **ACTION:** Station B goes onhook.
> Place a call from station A to B.
> **VERIFY:** No ring-ping (redirect notification) is applied
> to station B. The call is forwarded to station C.
> **ACTION:** Station C answers the call.
> **VERIFY:** Stations A and C are connected.

Fig. 2. Example of a Test Case

The design methodology of the DEFINITY project involves writing *test cases* in English before any coding begins. Test cases constitute the only formal description of the external functioning of a switch feature. The idea is to describe how a feature works without having a particular implementation in mind [Howden, 1985]. Figure 2 shows a typical test case. Test cases are structured in part by a *goal/action/verify* format. The goal statement is a very high-level description of the purpose of the test. It is followed by alternating action/verify statements. An action describes stimuli that the tester has to execute. Each stimulus triggers a switch response that the tester has to verify (*e.g.*, a specific phone rings, a lamp is lit, a display shows a message etc). Overall, there are tens of thousands of test cases for DEFINITY. All these test cases are written using an editor and they are executed manually in a

---

[1] CF is an acronym for the call-forwarding feature, which allows the user to send his/her incoming calls to another designated station.

[2] Redirect notification is a feature to notify the user about an incoming call when he/she has CF activated. Instead of ringing the phone issues a short "ring-ping" tone.

test lab. This is an error prone and slow process that limits test coverage and makes regression test intervals too long.

Some 5% of the above test cases have been converted into *test scripts* using an in-house automation language. Although this automation process is an improvement over the manual testing process, the automation languages themselves have several problems that KITSS addresses. The request in Figure 2 to "set station B without redirect notification" does not contain the actions necessary to accomplish this task. A test case succinctly and directly expresses its intentions independent of a particular switch or configuration in the real world. However, in order to write an automated test script a tester must provide all of the expansions, missing details, and instantiations necessary for a correct automated test script. KITSS automates the conversion from test case to test script.

This meant that, prior to KITSS, the conversion from test case to test script took a long time and required the best domain experts. The process did not support any automatic semantic checking. There were only limited error diagnosis facilities available as well as no automatic update for regression testing. Also, the test scripts became so specific that they lost the generality of test cases, which may be a template for many test scripts covering multiple software releases. In contrast, the test cases are easier to read and maintain since they are not cluttered with language initializations or switch configuration specifics.

KITSS takes English test cases as input and translates them into formal, complete functional test scripts that can be used to test DEFINITY switch software. To make KITSS a practical system required novel approaches in two very difficult and different areas. First, a very informal and expressive language needed to be transformed into formal logic. Second, incomplete test cases needed to be extended since humans have difficulties specifying tests that are both complete and consistent.

To address these two difficulties, KITSS provides a natural language processor, which is trained on example test cases, and a completeness and interaction analyzer, which audits test coverage. However, these two modules have only been feasible due to the *domain-specific* knowledge-based approach taken in KITSS. Therefore, both modules are supported by a hybrid knowledge-base (the "K" in KITSS) that contains a model of the DEFINITY PBX domain. Concepts that are used in telephony and testing are available to both processes to reduce the complexity of their interpretive tasks. KITSS is also interactive (the "I" in KITSS). For example, imagine that a module does not have enough information to settle on only one interpretation for a phrase. KITSS would present the possible choices along with its preferred choice and then ask the test case author to verify or override its choice. Finally, KITSS also provides a translator that generates the actual test scripts (the "TS" in KITSS) from the formal representation derived by the analyzer.

The above requirements led to the architecture as shown in Figure 3. It shows that KITSS consists of four main modules: the domain model, the natural language processor, the completeness and interaction analyzer, and the translator. The domain model is in the center of the system and supports all three reasoning modules. The next section briefly describes each module (for more details see [Nonnenmann & Eddy, 1992]).

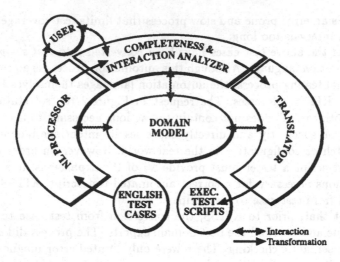

Fig. 3. KITSS Architecture

3 KITSS Modules

3.1 Domain Model

A domain model serves as the knowledge base for an application system. Testing involves experience with the switch hardware and testing equipment as well as an understanding of the switch software with its several hundred features and their interactions. The domain model formalizes this knowledge. For test cases, it enforces a standard terminology and simplifies the maintenance. Additionally, the domain model provides the knowledge that reduces and simplifies the tasks of the natural language processor, the analyzer, and the translator modules. The focus of KITSS and the domain model is on an end-user's point of view, *i.e.*, on (physical and software) objects that the user can manipulate.

From a knowledge representational point of view, we distinguish between static properties of the domain model and dynamic ones [Brodie *et al.*, 1984]. Static properties include the objects of a domain, attributes of objects, and relationships between objects. Dynamic properties represent all operations on objects, their properties, and the relationships between operations. The applicability of operations is constrained by the attributes of objects. Integrity constraints are also included to express the regularities of a domain.

In KITSS, the *static model* represents all telephony objects, data, and conditions that do not have a temporal extent but may have states or histories. It describes major hardware components, processes, logical resources, the current test setup, the dial plan and the current feature assignments. All static parts of the domain model are implemented in CLASSIC [Brachman *et al.*, 1990], which belongs to the class of terminological logics (*e.g.* KL-ONE).

The *dynamic model* defines the dynamic aspects of the switch behavior. These are constraints that have to be fulfilled during testing as well the predicates they are defined upon. Objects include predicates, stimuli which can be either primitive or abstract, and observables. Additionally, the dynamic model includes invariants and rules as integrity constraints. Invariants are assertions which describe only a single state, but are true in all states. These are among the most important pieces of domain knowledge as they describe basic telephony behavior as well as the *look & feel* of the switch. Rules describe low-level behavior in telephony. This is mostly signaling behavior.

Representing the dynamic model we required expressive power beyond CLASSIC or terminological logics, which are not well-suited for representing plan-like knowledge. We therefore used the WATSON Theorem Prover (see Section 3.3), a linear-time first-order resolution theorem prover with a weak temporal logic. This non-standard logic has five modal operators *holds, occurs, issues, begins,* and *ends* which are sufficient to represent all temporal aspects of our domain. The theorem proving is tractable due to the tight integration between knowledge representation and reasoning.

In adding the dynamic model, we were able to increase the expressive power of our domain model and to increase the reasoning capabilities as well. The integration of the hybrid pieces did produce some problems, for example, deciding which components belonged in which piece. However, this decision was facilitated because of our design choice to represent all dynamic aspects of the system in our temporal logic and to keep everything else in CLASSIC.

3.2 Natural Language Processor

The existing testing methodology used English as the language for test cases (see Figure 2) which also is KITSS' input. English is undeniably quite expressive and unobtrusive as a representation medium, yet it is difficult to process into formal descriptions. In practice, however, test cases are written in a style that is a considerably more restrictive technical dialect of English, a naturally occurring *telephonese* language that is less variable and less complex.

Recent research in statistical parsing approaches [Jones & Eisner, 1992] provides some answers to the difficulty of natural language parsing in restricted domains such as testing languages. In the KITSS project, the parser is trained on sample test cases and collects distribution statistics. The statistics are used to prune unlikely syntactic and semantic structures.

The use of statistical likelihoods to limit search during natural language processing was used not only during parsing but when assigning meaning to sentences, determining the scope of quantifiers, and resolving references. When choices could not be made statistically, the natural language processor could query the domain model, the analyzer, or the human user for disambiguation. The final output of the natural language processor are logical representations of the English sentences, which are passed to the analyzer.

3.3 Completeness & Interaction Analyzer

The completeness and interaction analyzer is based on experience with the WATSON research prototype [Kelly & Nonnenmann, 1991]. Originally, WATSON was designed as an automatic programming system to generate executable specifications from episodic descriptions in the telephone switching software domain. This was an extremely ambitious goal and could only be realized in a very limited prototype. To be able to scale to real-world use, the focus has been shifted to merely checking and augmenting given tests and maybe generating related new ones but not to generating the full specification.

Based on the natural language processor output, the input logical forms are grouped as several *episodes*. Each episode defines a stimulus-response-cycle of the switch, which roughly corresponds to the action/verify statements in the original test case. These episodes are the input for the following analysis phases. Each episode is represented as a logical rule, which is checked against the dynamic model. The analyzer uses first-order resolution theorem proving in a temporal logic as its inference mechanism, the same as WATSON.

The analysis consists of several phases that are specifically targeted for this domain and have to be re-targeted for any different application. All phases use the dynamic model extensively. The purpose of each phase is to yield a more detailed understanding of the original test case. Therefore, the structure of a test case is analyzed to recognize or attribute purpose to pieces of the test case. There are four different pieces that might be found: administration of the switch, feature activation or deactivation, feature behavior, and regression testing. The test case is also searched for connections among concepts, *e.g.*, there might be relations between system administration concepts and system signaling that need to be understood. Routine omissions are inserted into the test case as well. Testers often reduce (purposefully or not) test sequences to the essential aspects. However, these omissions might lead to errors on the test equipment and therefore need to be added. Based on the abstract plans in the dynamic model, we can enumerate possible specializations, which yield new test cases from the input example. Finally, plausible generalizations are found for objects and actions as a way to abstract tests into classes of tests.

During the analysis phases, the user might interact with the system. We try to exploit the user's ease at verifying or falsifying examples given by the analyzer. At the same time, the initiative of generating the details of a test lies with the system. The final output of the analyzer is a corrected and augmented test case in temporal logic, which is passed to the translator.

3.4 Translator

To make use of the analyzer's formal representation, the translator needs to convert the test case into an executable test language. This language exercises the switch's capabilities on test equipment with the goal of finding software failures. One goal of the KITSS project was to extend the life of test cases so that they could be used as many times as possible. To accomplish the conversion, it was decided to make the translator support two types of test case independence.

First, a test case must be test machine independent. Each PBX that we run our tests on has a different configuration. The translator loads the configuration setup of a particular switch and uses it to make the test case concrete with respect to equipment used, system administration performed, and permissions granted. Second, a test case must be independent of the automated test language. KITSS generates test cases in an in-house test language. The translator's code is small because much of the translation information is static and can be represented in CLASSIC. If a new test language replaces the current one then the translator can be readily replaced without loss of test cases, with minimal changes to the KITSS code.

4 Related Work

Our work on KITSS is mainly related to three areas of KBSE. The first one is that of acquiring programs from examples. Several approaches have been used to "learn" programs based on sample traces. Most other approaches try to be independent of a specific domain and the result is that their input learning examples must be either numerous or meticulously annotated. The work on NODDY [Andreae, 1985] is the closest in spirit to KITSS by its use of explicit domain knowledge. NODDY, which writes robot programs, uses its solid geometry knowledge to constrain the generality of the programs it writes. In a sense, in both KITSS and NODDY, the domain knowledge serves as the negative examples in an example-based learning system.

The second related field is that of programming from informal specifications such as the SAFE project [Balzer *et al.*, 1977]. SAFE attempts to understand an English prose description of a process and then to infer the relevant entities mentioned in the process and their key relationships. It then generates an algorithm to implement the process.

The third related area is that of automatic specification acquisition as in the WATSON prototype on which KITSS is based. Other examples of such systems are IDeA and Ozym. The Intelligent Design Aid (IDeA) [Lubars & Harandi, 1987] performs knowledge-based refinement of specifications and design. IDeA gives incremental feedback on completeness and consistency using domain-specific abstract design schemas. The idea behind Ozym [Iscoe *et al.*, 1989] is to specify and implement applications programs for non-programmers and non-domain-experts by modeling domain knowledge.

Despite two decades of moderately successful research, there have been few practical demonstrations of the utility of Artificial Intelligence techniques to support Software Engineering activities [Barstow, 1987] other than prototypes as above.

KITSS differentiates itself from these approaches in two antagonistic ways: On the one hand, KITSS addresses a large and complex real-world problem instead of a toy domain as in the above research prototypes. On the other hand, to allow such scaling, we had to relax the ambitious goal of complete *automatic programming*, to the easier task of *automatic testing*. Tests are by definition correct but not exhaustive. KITSS can be seen as performing testing from examples. KITSS' strength lies in its very domain-specific approach and customized reasoning procedures. It will change the software life cycle by modifying the functional tests and then rederiving the system tests. Therefore, maintenance and evolution will be much easier tasks.

5 Status and Future Work

The KITSS project is still a prototype system that has not been deployed for general use on the DEFINITY project. It was built by a team of researchers and developers. Currently, it fully translates 38 test cases (417 sentences) into automated test scripts. While this is a small number, these test cases cover a representative range of the core features. Additionally, each test case yields multiple test scripts after conversion through KITSS. The domain model consists of over 500 concepts, over 1,500 individuals, and more than 80 temporal constraints. The domain model will grow somewhat with the number of test cases covered, however, so far the growth has been less than linear for each feature added.

All of the modules that were described in this paper have been implemented but all need further enhancements. System execution speed doesn't seem to be a bottleneck at this point in time. CLASSIC's fast classification algorithm's complexity is less than linear in the size of the domain model. Even the analyzer's theorem prover, which is computationally the most complex part of KITSS, is currently not a bottleneck due to continued specialization of its inference capability. However, it is not clear how long such optimizations can avoid potential intractability.

The current schedule is to expand KITSS to cover a few hundred test cases in the next couple of months. To achieve this, we will shift our strategy towards more user interaction. The version of KITSS currently under development will intensely question the user to explain unclear passages of test cases. We will then re-target the reasoning capabilities of KITSS to cover those areas. This rapid-prototyping approach is only feasible since we have already developed a robust core system. Knowledge acquisition support is crucial but currently unsolved. Although scaling-up from our prototype to a real-world system remains a hard task, KITSS demonstrates that our KBSE approach chosen for this complex application is feasible.

6 Conclusions

As we have shown, testing is perhaps one of the most expensive and time-consuming steps in product design, development, and maintenance. KITSS uses some novel Artificial Intelligence techniques to achieving several desirable goals in improving Software Engineering tasks. Features will continue to be specified in English. To support this we have incorporated a statistical parser that is linked to the domain model as well as to the analyzer. Additionally, KITSS will interactively give the user feedback on the test cases written and will convert them to a formal representation. To achieve this, we needed to augment the domain model represented in a terminological logic with a dynamic model written in a temporal logic. The temporal logic inference mechanism is customized for the domain. Tests will continue to be specified independent of the test equipment and test environment without the user providing details. Such a testing system as demonstrated in KITSS will ensure both project-wide consistent use of terminology and will allow simple, informal tests to be expanded to formal and complete test scripts. The result is a better testing process with more test automation and reduced maintenance cost.

Acknowledgments

Many thanks go to Van Kelly, Mark Jones, and Bob Hall who also contributed major parts of the KITSS system. Additionally, we would like to thank Ron Brachman for his support throughout the project.

References

[Andreae, 1985] Andreae P.: Justified generalization: Acquiring procedures from examples. In *MIT AI Lab Technical Report 834*, 1985.

[Balzer *et al.*, 1977] Balzer R., Goldman N., and Wile D.: Informality in program specifications. In *Proceedings of the 5th International Joint Conference on Artificial Intelligence*, Cambridge, MA, 1977.

[Barstow, 1987] Barstow, D.R.: Artificial Intelligence and Software Engineering. In *Proceedings of the 9th International Conference on Software Engineering*, Monterey, CA, 1987.

[Brachman *et al.*, 1990] Brachman, R.J., McGuinness, D.L., Patel-Schneider, P.F., Alperin Resnick, L., and Borgida, A.: Living with CLASSIC: When and how to use a KL-ONE-like language. In *Formal Aspects of Semantic Networks*, J. Sowa, ed., Morgan Kaufmann, 1990.

[Brodie *et al.*, 1984] Brodie, M.L., Mylopoulos, J., and Schmidt, J.W.: *On conceptual modeling: Perspectives from Artificial Intelligence*. Springer Verlag, New York, NY, 1984.

[Brooks, 1987] Brooks, F.P.: No silver bullet: Essence and accidents of software engineering. *Computer*, Vol. 20, No. 4, April 1987.

[Howden, 1985] Howden, W.E.: The theory and practice of functional testing. *IEEE Software*, September 1985.

[Iscoe *et al.*, 1989] Iscoe N., Browne J.C., and Werth J.: An object-oriented approach to program specification and generation. *Technical Report, Dept. of Computer Science, University of Texas at Austin*, 1989.

[Jones & Eisner, 1992] Jones, M.A., and Eisner, J.: A probabilistic parser applied to software testing documents. In *Proceedings of the 10th National Conference on Artificial Intelligence*, San Jose, CA, 1992.

[Kelly & Nonnenmann, 1991] Kelly, V.E., and Nonnenmann, U.: Reducing the complexity of formal specification acquisition. In *Automating Software Design*, M. Lowry and R. McCartney, eds., MIT Press, 1991.

[Lowry & Duran, 1989] Lowry, M., Duran, R.: Knowledge-based Software Engeneering. In *Handbook of Artificial Intelligence, Vol. IV, Chapter XX*, Addison Wesley, 1989.

[Lubars & Harandi, 1987] Lubars M.D., and Harandi M.T.: Knowledge-based software design using design schemas. In *Proceedings of the 9th International Conference on Software Engineering*, Monterey, CA, 1987.

[McCartney, 1991] McCartney, R.: Knowledge-based software engineering: Where we are and where we are going. In *Automating Software Design*, M. Lowry and R. McCartney, eds., MIT Press, 1991.

[Myers, 1979] Myers, G.J.: *The Art of Software Testing*. John Wiley & Sons, Inc. New York, NY, 1979.

[Nonnenmann & Eddy, 1991] Nonnenmann, U., and Eddy J.K.: KITSS - Toward software design and testing integration. In *Automating Software Design: Interactive Design - Workshop Notes from the 9th AAAI*, L. Johnson, ed., USC/ISI Technical Report RS-91-287, 1991.

[Nonnenmann & Eddy, 1992] Nonnenmann, U., and Eddy J.K.: KITSS - A functional software testing system using a hybrid domain model. In *Proceedings of the 8th IEEE Conference on Artificial Intelligence for Applications*, Monterey, CA, 1992.

COMPLEX KNOWLEDGE BASE VERIFICATION USING MATRICES

Nancy Botten
IBM Corporation
5 West Kirkwood Boulevard
Roanoke, Texas 76299-0001

ABSTRACT

The paper extends the techniques presented in Knowledge Base Verification Using Matrices[1] from verification of single, unstructured knowledge bases to multiple knowledge bases which interact. The techniques are based on the representation of the knowledge base as a set of matrices and employ simple matrix operations. The anomalies addressed include unreachable and dead-end rules, missing rules, redundancy, subsumption, inconsistency, circular rule chains, and inconsistent rule chains, that occur across the interacting knowledge bases.

KEYWORDS: expert system, knowledge-based system, production system, rule-based system, verification

Overview

The verification is accomplished by analyzing intra-rule and inter-rule connectivity in order to identify and address anomalies in the knowledge base. The formal definition of each of the anomalies is given in Knowledge Base Verification Using Matrices[1]. In general, redundancy means a duplicate rule. Subsumption involves rule(s) with additional constraint(s), which makes the constrained rules unnecessary. Inconsistent rules have conclusions that are contradictory. Unreachable rules cannot be activated because their conditions may never all be true and dead-end rules do not lead to any useful conclusions. Circular rule chains result in rule cycling.

Verification of knowledge bases appears to be a relatively new topic in expert system development[2][3]. Stachowitz and Chang provide a brief history of verification efforts, starting with the early precursors to automatic knowledge base verifiers such as TEIRESIAS, ONCOCIN, CHECK, EVA and Expert System Checker (ESC)[3]. The bulk of the verification techniques that exist may be classified as either consistency checkers or completeness checkers. Tepandi's study indicates that it is necessary to distinguish between structural and functional aspects of verification[4]. A recent verification tool, MIDST (Mixed Inferencing Dempster Shafer Tool), uses a network search algorithm to verify a knowledge base that has been partitioned[5].

The verification analysis is done in two phases. The first phase is the structure verification, which analyzes the composition of the rules by comparing rule elements to rules. The second phase is the inference verification, which analyzes the interconnections between rules. Three main types of matrices are utilized, two that relate rules to rules (adjacency and reachability matrices) and one that relates rule elements to rules (element-rule incidence matrix). For the purposes of this discussion, each separate knowledge base will be considered an "area" of the overall knowledge base.

Structure Verification

Introduction

Structure verification examines the relationship of rule elements to rules with the creation of an element-rule matrix where an element is *incident* to a rule if it is a condition of the rule or if it is a conclusion of the rule. If an element is a condition of a rule, it is of incidence type "I" to the rule. If an element is a conclusion of a rule, it is of incidence type "T" to the rule. Otherwise the element incidence is "0".

Redundancy, Subsumption, or Inconsistent Rules

Identification of redundancy, subsumption and inconsistent rules that exist across areas requires the identification of the elements within the area that are not primary to the area. Since an element may occur in more than one rule and more than one area, it is necessary to assign each element to the most representative area. A convenient coding scheme can be devised, e. g., element E2.13 is primary to area two; the number 13 is significant only as an identifier. Similarly, rule 5.8 resides in area five. If there are no non-primary elements in the area, no further analysis is necessary. Note, however, that the analysis is not symmetric. It is possible to have elements in one area from a different area (the existence of non-primary elements) without that different area having elements from the original area (all primary elements).

If there are non-primary elements in an area, the first step is to reduce the element-rule matrix I_A for the area by removing any rules that contain only elements primary to the area. Any cases of redundancy, subsumption or inconsistency in these deleted rules have been discovered by the within area analysis. Next, the element-rule matrix is augmented by adding columns for all other rules in the knowledge base containing the non-primary elements, which will also result in the addition of rows containing the new constituent elements of these rules. This step is done to represent all of the cases where the non-primary elements are shared by rules throughout the knowledge base. This captures all cases where across area redundancy, subsumption and inconsistent rules may occur. From this point, the modified matrix is referred to as I_A. The procedure for converting I_A to lexicographic order and for identifying and resolving the across area cases of redundancy, subsumption and inconsistent rules is the same as for within area cases.

The technique is illustrated by two examples. The first example, in Figure 1, shows detection of redundancy and subsumption. The analysis is for Area 2, which contains some elements for which their primary area is Area 1. (The analysis for Area 1 would be similar.)

AREA 1	AREA 2
R1.1 IF E1.1 AND E1.2 THEN R2.3 R1.2 IF E1.3 AND E2.1 THEN E1.4	R2.1 IF E2.1 AND E1.3 THEN E1.4 R2.2 IF E1.4 AND E2.1 THEN E2.2 R2.3 IF E1.2 THEN E2.3 R2.4 IF E2.2 THEN E2.3
Note: Figure 1. Across Area Redundancy and Subsumption: Rule Set	

The element-rule matrix is reduced by eliminating R2.4 since it contains only elements primary to Area 2. Then the matrix is augmented by adding R1.1 and R1.2 since they share the non-primary elements from Area 2. The results of the analysis on the elements displayed in Figure 2, shows that rule R2.3 from Area 2 subsumes rule R1.1 from Area 1. There is a duplication between rule R2.1 from Area 2 and rule R1.2 from Area 1.

The second example of the technique, illustrated on the rule set in Figure 3 is displayed in Figure 4, which shows detection of inconsistent rules. The example contains four rules in two areas.

The results of the analysis on Area 2 of the rule set, shows that rule R1.2 from Area 1 and rule R2.1 from Area 2 are inconsistent since they have the same set of conditions, E1.3, but lead to different conclusions, E1.4 versus E2.1. Since Area 1 contains only primary elements, no analysis is necessary.

Sequential Order					
	R1.1	R1.2	R2.1	R2.2	R2.3
E1.1	I	O	O	O	O
E1.2	I	I	O	O	I
E1.3	O	I	I	O	O
E1.4	O	T	T	I	O
E2.1	O	I	I	I	O
E2.2	O	O	O	T	O
E2.3	T	O	O	O	T
Sorted by Element					
	R1.1	R1.2	R2.1	R2.2	R2.3
E1.4	O	T	T	I	O
E2.1	O	I	I	I	O
E1.2	I	O	O	O	I
E1.3	O	I	I	O	O
E2.3	T	O	O	O	T
E1.1	I	O	O	O	O
E2.2	O	O	O	T	O
Lexicographic by rule					
	R1.1	R1.2	R2.1	R2.2	R2.3
E1.4	T	T	I	O	O
E2.1	I	I	I	O	O
E1.2	O	O	O	I	I
E1.3	I	I	O	O	O
E2.3	O	O	O	T	T
E1.1	O	O	O	I	O
E2.2	O	O	T	O	O
Note: Figure 2. Across Area Redundancy and Subsumption: Analysis					

AREA 1	AREA 2
R1.1 IF E1.1 AND E1.2 THEN E1.3 R1.2 IF E1.3 THEN E1.4	R2.1 IF E1.3 THEN E2.1 R2.2 IF E1.4 AND E2.1 THEN E2.2
Note: : Figure 3. Across Area Inconsistent Rules: Rule Set	

Inference Verification

Introduction

Inference verification analyzes the reasoning process by examining the relationship of rules to rules with the creation of two types of matrices, an adjacency matrix and a reachability matrix. Two rules are *adjacent* if the conclusion of one rule is a condition of the other rule. Rules that are adjacent are related by a *direct connection*, which is weighted by the number of direct connections. A direct connection may be either an intra-area connection or an inter-area connection. An *intra-area connection* exists between a rule in one area with a conclusion element that matches a condition element of another rule in the same area. An *inter-area con-*

nection exists between a rule in one area with a conclusion element that matches a condition element(s) of one or more rules in a different area. An inter-area connection may assume a weight up to a maximum equaling the total number of rules in the area; however, in practice the number will be significantly smaller since most knowledge bases will not have that high a degree of interconnectivity. By definition, an intra-area connection will have a weight of one. Since there are N_R rules, an adjacency matrix Λ is defined as the $N_R \times N_R$ matrix with elements (i,j) equal to the weight of the direct connection between rule i and rule j, a_{ij}.

Sequential Order	RI.1	R1.2	R2.1	R2.2	
EI.1	1	O	O	O	
EI.2	1	O	O	O	
EI.3	T	1	1	O	
EI.4	O	T	O	1	
R2.1	O	O	T	1	
R2.2	O	O	O	T	
Sorted by Element	RI.1	R1.2	R2.1	R2.2	
EI.3	T	1	1	O	
RI.4	O	T	O	1	
R2.1	O	O	T	1	
EI.1	1	O	O	O	
EI.2	1	O	O	O	
E2.2	O	O	O	T	
Lexicographic by rule	RI.1	R1.2	R2.1	R2.2	
EI.3	1	1	T		O
EI.4	T	O	O		1
R2.1	O	T	O		1
EI.1	O	O	1		O
EI.2	O	O	1		O
E2.2	O	O	O		T

Note: Figure 4. Across Area Inconsistent Rules: Analysis

A rule is *reachable* from another rule if there is a rule chain linking the two rules; in effect, the condition of the rule is reachable from the conclusion of the other rule via the rule chain. Rules that are reachable are related by an *indirect connection*. An indirect connection between two rules consists of a sequence of direct connections, i.e., intra-rule and/or inter-rule connections, between the rules. The indirect connection is weighted by the number of rule sequences through which rule j is reachable form rule i. Since there are N_R rules, a reachability matrix R is defined as the $N_R \times N_R$ matrix with elements (i,j) equal to the weight of the indirect connection between rule i and rule j, r_{ij}. The reachability matrix may be obtained from the adjacency matrix by taking successive powers of Λ, then adding the transition matrices. The stopping

criteria occurs when all of the matrix elements equal zero, indicating a steady state, or when the pattern of any previous transition matrix appears, indicating a cyclical state.

Unreachable, Dead-End, or Disconnected Rules

Detection of unreachable, dead-end or disconnected rules starts with the creation of an adjacency matrix Λ_A for each area of the knowledge base, where:

$a_{ij} = 1$ if rule i is adjacent to rule j, i.e. if a conclusion element of rule i is a condition element of rule j.

0 otherwise.

The adjacency matrix captures the relationships between the rules within an area. However, we are also interested in discovering the relationships between the rules in other areas, as well as the relationships between the rules and the user interface. Therefore, we need to create an area adjacency matrix $\Lambda\Lambda_A$ by adding a series of columns and rows. One of the columns represents the reportable conclusion which can be reached by a rule in the area. The remaining columns, equal to the number of areas minus one (N_{A-1}) represent the other areas containing rules to which a rule in the area being analyzed are connected. One of the additional rows represents inputs from the user, which will be matched up with rule conditions in the area. The remaining rows represent the other areas containing rules adjacent to a rule in the area being analyzed. In the first added column,

$a_{ij} = 1$ if rule i is adjacent to reportable conclusion j, i.e., if the reportable conclusion element j is a conclusion element of rule i.

0 otherwise.

In the next (N_{A-1}) columns,

$a_{ij} =$ weight of the inter-area connection between rule i and area j rules, i.e., the sum of the connections between rule i in the area being analyzed and area j rules 1,2,...,b, where b is the number of rules in area j.

In the first added row,

$a_{ij} = 1$ if a set of input elements i match the set of condition elements of rule j

0 otherwise

In the next (N_A-1) rows,

a_{ij} = weight of the inter-area connection between area i rules and rule j, i.e., the sum of the connections between area i rules 1,2,...,b, where b is the number of rules in area i, and rule j is in the area being analyzed.

Each of the rows and columns of the area adjacency matrix AA_A are summed. If the sum of any row is zero, that rule is a dead-end rule since it neither contains a reportable conclusion, nor leads to another area that may contain a reportable conclusion. If the sum of any column is zero, that rule is an unreachable rule since there is no way to activate the rule either by using input from the user or by matching conclusions from rules in other areas. If the sum of both the column and row for a rule is zero, the rule is totally disconnected. The unreachable, dead-end or disconnected rules may be true errors and should be deleted. Or they may indicate missing input or missing rules, in which case the missing structures should be generated to tie the unreachable, dead-end or disconnected rule into the existing structure. This requires the knowledge engineer's attention to determine what structures should be removed or added. The following algorithm is applied to the area adjacency matrix AA_A for each area:

[ONE] Initialization:

1. b: number of rules in the area

2. N_A: number of areas in the knowledge base

Begin

Step 1: Sum the rows and columns:

if

$$\sum_{j=1}^{b+N_A} a_{ij} = 0$$

then rule i is a dead-end rule

if

$$\sum_{i=1}^{b+N_A} a_{ij} = 0$$

then rule i is an unreachable rule

If

$$\sum_{j=1}^{b+N_A} a_{ij} = 0$$

and

$$\sum_{i=1}^{b+N_A} a_{ij} = 0$$

and i = j then the rule is disconnected

Otherwise, there are no dead-end, unreachable or disconnected rules in the area.

End

The technique can be illustrated in two examples. In Figure 5, no disconnected rules exist since there are no zero sums in comparable rows and columns of AA_1. However, there is one dead-end rule, R1.3, indicated by the zero sum in row 2. In Figure 6, since rule R1.2 is both an unreachable and a dead-end rule, indicated by the zero sum in row 2 and column 2 of AA_1, it is also a disconnected rule.

R1.1 IF A THEN B
R1.2 IF B THEN C
R1.3 IF D THEN E

Area Adjacency Matrix

From-To	R1.1	R1.2	R1.3	OUT	A2
R1.1	0	1	0	0	0
R1.2	0	0	1	1	1
R1.3	0	0	0	0	0
IN	1	0	0	0	0
A2	1	0	0	0	0

Note: Figure 5. Dead-End Rules

R1.1 IF B THEN D
R1.2 IF A THEN C
R1.3 IF B THEN E

Area Adjacency Matrix

From-To	R1.1	R1.2	R1.3	OUT	A2	
R1.1	0	0	1	0	1	0
R1.2	0	0	0	0	0	0
R1.3	1	0	0	1	0	1
IN	1	0	0	0	0	0
A2	1	0	0	0	0	0
A3	0	0	1	0	0	0

Note: Figure 6. Disconnected Rules

Within Area Circular Rule Chains

Identification of within area circular rule chains requires the adjacency matrix A_A. If a rule cycles with itself, a one will appear on the diagonal for that rule. For this to happen, the rule must be of the general form IF A THEN A, which would not occur very often. To discover all of the circular rule chains, the reachability matrix may be obtained form the adjacency matrix by taking successive powers of Λ until all the elements of the matrix are 0 (a steady state) or until an even multiple of a previous matrix appears (a cyclical state), then adding the transition matrices. As the adjacency matrix is taken to increasing powers, each nth step in the transition represents the number of indirect connections of length n by which rule j is reachable from rule i. The stopping criteria occurs when all of the matrix elements equal zero, indicating a steady state, or when the pattern of any previous transition matrix appears, indicating a cyclical state. When all of the step transitions are added together to form the reachability matrix, all of the indirect connections of any length by which rule j is reachable from rule i are represented. The reachability matrix, by summing all of the powers of Λ, shows all of the rule chains. An important property of both the transition matrices and the reachability matrices is that the trace (the sum of the diagonal elements) is equal to the number of circular chains in the area. A circular chain may be defined as an edge sequence:

$(a_0, a_1) (a_1, a_2),....,(a_{k-1}, a_k)$ where $a_k = a_0$.

This means, for example, in a two-step transition, if the trace of $\Lambda^2 = 1$, there is one circular rule chain of length two. It also means, if the trace of R = 2 that there is a total of two circular rule chains. An example illustrating these concepts will appear later.

The identification of the circular rule chains is accomplished by applying the following algorithm to the adjacency matrix for each area, forming a series of new matrices, each of which is obtained by taking matrix Λ to increasing powers up to a maximum power of b (the number of rules in the area). At each transition step, the new transition matrix is checked to determine if a steady state or a cyclical state has been reached.

[TWO] Initialization:

1 - $n = 1$

2 - b: rule size - 1

Do While n < b

   Step 1: Multiply Λ_n by Λ_1 and set equal to Λ_n

      $\Lambda_n = \Lambda_1 \Lambda_n$

   Step 2: Increment n by 1

      $n = n + 1$

   Step 3: Save $\Lambda_i = \Lambda_n$

   Step 4: Check for steady state or cycling state

      If $\Lambda_n = 0$ (steady state) then proceed to

      step 4.

      If $\Lambda_n = k \cdot \Lambda_i$, where $i = 1, 2,...,n$

      and $k = 1,2....,n$ (cycling state)

      then proceed to step 4.

      Otherwise, go to step 1.

   Step 5: Determine the reachability matrix

$$R = \sum_{n=1}^{b} \Lambda^n$$

   Step 6: Determine the trace (R)

      $b = b + 1$

$$Trace(R) = \sum_{j=1}^{b} r_{jj}$$

End Do While

If the trace of R at steady state = 0, then there are no circular chains within the area; otherwise the identified circular chains must be broken. This requires the knowledge engineer's attention to determine which rule in the chain should be removed or modified. All the rules comprising the circular chain(s) may be read off the diagonal, and the total number of circular chains through each rule will appear on the diagonal for that rule.

The technique can be illustrated in two examples. In Figure 7, no circular chains exist; however, in Figure 8, there are three circular chains: Λ -> B -> C; B -> C -> Λ, and C -> Λ -> B. Note that the algorithm identifies all the cycles, since the chain may begin at any rule in the chain.

Across Area Circular Rule Chains

Identification of circular rule chains across areas is more complex than identifying rule chains within areas. In addition to identifying across area rule chains that contain a single rule from within an area, we must be able to identify across area rule chains that contain a rule chain from within an area. In the latter case, we must determine the boundary rule(s) of each area that delimit the intra-area rule chains, any of which may form a segment of an across area rule chain. There need be no loss of generality, since a single rule may be considered a rule chain of length zero and therefore a boundary rule as well. An across area circular rule chain is an across area rule chain that begins and ends at the same rule.

R1.1 IF A THEN B				
R1.2 IF B THEN C				
R1.3 IF C THEN D				
R1.4 IF B THEN D				
Area Adjacency Matrix				
From-To	R1.1	R1.2	R1.3	R1.4
R1.1	0	1	0	1
R1.2	0	0	1	0
R1.3	0	0	0	0
R1.4	0	0	0	0
Two-Step Transition				
From-To	R1.1	R1.2	R1.3	R1.4
R1.1	0	0	1	0
R1.2	0	0	0	0
R1.3	0	0	0	0
R1.4	0	0	0	0
Three-Step Transition				
From-To	R1.1	R1.2	R1.3	R1.4
R1.1	0	0	0	0
R1.2	0	0	0	0
R1.3	0	0	0	0
R1.4	0	0	0	0
Reachability Matrix				
From-To	R1.1	R1.2	R1.3	R1.4
R1.1	0	1	1	1
R1.2	0	0	1	0
R1.3	0	0	0	0
R1.4	0	0	0	0

Note: Figure 7. Rule Set Without Circular Chains

We start with the reachability matrix R_A generated above, which captures the end rules of any intra-area rule chains. Then we augment the reachability matrix by adding a series of rows and columns to create a boundary/adjacency matrix BA_A. This is done to show the adjacency of the area's boundary rule(s) to the other areas in the knowledge base. A column is added for each of the other areas, where:

a_{ij} = weight of the inter-area connection between boundary rule i and area j rules, i.e., the sum of the connections between rule i in the area being analyzed and area j rules 1,2,...,b, where b is the number of rules in area j.

A row is added for each of the other areas, where:

a_{ij} = weight of the inter-area connection between area i rules and boundary rule j, i.e., the sum of the connections between area i rules 1,2,...,b, where b is the number of rules in area i, and rule j is in the area being analyzed.

R2.1 IF A THEN B			
R2.2 IF B THEN C			
R2.3 IF C THEN A			
Adjacency Matrix		
From-To	R2.1	R2.2	R2.3
R2.1	0	1	0
R2.2	0	0	1
R2.3	1	0	0
Two-Step Transition		
From-To	R2.1	R2.2	R2.3
R2.1	0	0	1
R2.2	1	0	0
R2.3	0	1	0
Three-Step Transition		
From-To	R2.1	R2.2	R2.3
R2.1	1	0	0
R2.2	0	1	0
R2.3	0	0	1
Four-Step Transition (Cyclical State)		
From-To	R2.1	R2.2	R2.3
R2.1	0	1	0
R2.2	0	0	1
R2.3	1	0	0
Reachability Matrix		
From-To	R2.1	R2.2	R2.3
R2.1	1	1	1
R2.2	1	1	1
R2.3	1	1	1

Note: Figure 8. Rule Set with Circular Chains

Now that we have captured the adjacency relationship between the area boundary rule(s) and the other areas, we may apply the algorithm [TWO] to the boundary/adjacency matrix BA_A in order to capture the reachability relationship between the area boundary rules and the other areas. The result will be a boundary/reachability matrix BR_A generated by taking successive powers of BA_A and summing them.

Now we are ready to identify the rules that are candidates for being part of across area circular rule chains. We will identify more potential circular rule chains than actually exist because a level of detail is intentionally missing. Since we are weighting the connection between areas, we are disregarding, for the time being, the individual connections between rules in the areas. We will only look at the individual connections when the possibility of a circular rule chain across those areas has been identified. This has the potential to reduce the complexity of the verification procedure since our initial analysis is performed at a more granular level, areas rather than rules. There is, however, no loss of information regarding the existence of circular rule chains. This approach should prove particularly valuable in terms of computational savings where the number of expected circular rule chains is small, since many areas will not need to be examined at the level of individual rule connections.

To identify the candidate rule(s) for across area rule chains, a list of all the rules in BR_A that either have an entry in the augmented columns or the augmented rows is created. This means that the boundary rules may respectively "reach to" or be "reached from" rules in those areas. When the rule lists from all the areas are combined, an adjacency matrix for this reduced rule set can be created and denoted RA, which is analyzed to determine if there are any circular rule chains across areas. Algorithm [TWO] is applied to the reduced rule adjacency matrix RA. The result will be a reduced rule reachability matrix RR generated by taking successive powers of RA and summing them. The procedure for identifying and dealing with across area circular chains is the same as for within area cases.

The technique can be illustrated in an example consisting of eight rules in four areas, displayed in Figure 9. This example identifies an across area rule chain that contains a single rule from within an area and identifies an across area rule chain that contains an intra-area rule chain. The first can be illustrated by the sequence: rule R1.1 --> rule R2.1 --> rule R3.1 --> rule R4.1 --> R1.1 (which represents four separate circular chains). The latter type can be illustrated by the sequence: rule R2.1 --> rule R2.2 --> rule R3.2 --> rule R2.1 (which represents three separate circular chains), containing the intra-area rule chain: rule R2.1 --> rule R2.2. Although there are seven unique circular chains, the algorithm will only discover six. This will not adversely affect the outcome since all the unique chains will be discovered.

AREA 1	AREA 2	AREA 3	AREA 4
R1.1 IF A THEN D	R2.1 IF D THEN E	R3.1 IF E THEN A	R4.1 IF F THEN C
R1.2 IF B THEN C	R2.2 IF E THEN F	R3.2 IF F THEN D	R4.2 IF G THEN H

Note: Figure 9. Across Area Circular Chains: Four Area Rule Set

We start with the boundary/reachability matrix RA for Area 1, to which the transition steps are performed until a steady state is reached, displayed in Figure 10.

The transition matrices are added to create a reachability matrix that shows which rules are candidates for forming circular chains. In this example, shown in Figure 11, the entry in r_{13} indi-

cates that rule R1.1 is a candidate since Area 2 can be reached from rule R1.1. Similarly, the entry in r_{31} indicates that rule R1.1 is a candidate since rule R1.1 can be reached from Area 1. Therefore, rule R1.1 is added to our reduced rule set list.

A similar process is followed for each of the areas. The results are displayed in Figures 12 through 17. In Figure 12 the transitions end because $n = b$, i.e., no more noncycling chaining is possible.

In figure 13, the entries in r_{14}; r_{15}; r_{24}; and r_{25} of the boundary/reachability matrix indicate that rules R2.1 and R2.2 are candidates for inclusion in the reduced rule set since Areas 3 and 4 can be reached from them. Similarly, the entries in r_{31}; r_{32}; r_{41} and r_{42} indicate that rules R2.1 and R2.2 are candidates since they can be reached from Areas 1 and 3. Therefore, rules R2.1 and R2.2 are added to our reduced rule set list.

Area1 - Boundary/Adjacency Matrix - Start					
From-To	R1.1	R1.2	A2	A3	A4
R1.1	0	0	1	0	0
R1.2	0	0	0	0	0
A2	0	0	0	0	0
A3	1	0	0	0	0
A4	0	0	0	0	0
Area 1 - Two-Step Transition					
From-To	R1.1	R1.2	A2	A3	A4
R1.1	0	0	0	0	0
R1.2	0	0	0	0	0
A2	0	0	0	0	0
A3	0	0	1	0	0
A4	0	0	0	0	0
Area1 - Three-Step Transition (Steady State)					
From-To	R1.1	R1.2	A2	A3	A4
R1.1	0	0	0	0	0
R1.2	0	0	0	0	0
A2	0	0	0	0	0
A3	0	0	0	0	0
A4	0	0	0	0	0

Note: Figure 10. Across Area Circular chains: Area 1 Step Transitions

Area 1 - Boundary/Reachability Matrix					
From-To	R1.1	R1.2	A2	A3	A4
R1.1	0	0	1	0	0
R1.2	0	0	0	0	0
A2	0	0	0	0	0
A3	1	0	1	0	0
A4	0	0	0	0	0

Note: Figure 11. Across Area Circular Chains: Area 1 Reachability Matrix

In Figure 14 the state is cyclical in nature, i.e., a multiple of a previous transition matrix appears ($A_4 = (1)^* A_2$). In Figure 15, the entries in r_{13}; r_{23}; and r_{24} of the boundary/reachability matrix indicate that rules R3.1 and R3.2 are candidates for inclusion in the reduced rule set since Area 1 can be reached from rule 3.1 and Areas 1 and 2 can be reached from rule R3.2. Similarly, the entries in r_{41} and r_{42} indicate that rules R3.1 and R3.2 are candidates since they can be reached from Area 2. Therefore, rules R3.1 and R3.2 are added to our reduced rule set list.

In figure 17, the entries in r_{41} and r_{42} of the reachability matrix indicate that rules R4.1 and R4.2 are candidates for inclusion in the reduced rule set since they can be reached from Area 2. Therefore, rules R4.1 and R4.2 are added to our reduced rule set list.

The complete set of rules in the reduced rule set is represented in Figure 18. The transition steps are performed on this reduced rule adjacency matrix until a steady state is reached. The steady state is cyclical in nature, i.e., a multiple of the three-step transition matrix pattern reappears in the six-step transition matrix.

Area2 - Boundary/Adjacency Matrix - Start

From-To	R2.1	R2.2	A1	A3	A4
R2.1	0	1	0	1	0
R2.2	0	0	0	1	1
A1	1	0	0	0	0
A3	1	0	0	0	0
A4	0	0	0	0	0

Area 2 - Two-Step Transition

From-To	R2.1	R2.2	A1	A3	A4
R2.1	1	0	0	1	1
R2.2	1	0	0	0	0
A1	0	1	0	1	0
A3	0	1	0	1	0
A4	0	0	0	0	0

Area 2 - Three-Step Transition

From-To	R2.1	R2.2	A1	A3	A4
R2.1	1	1	0	1	0
R2.2	0	1	0	1	0
A1	1	0	0	1	1
A3	1	0	0	1	1
A4	0	0	0	0	0

Area 2 - Four-Step Transition (n = b)

From-To	R2.1	R2.2	A1	A3	A4
R2.1	1	1	0	2	1
R2.2	1	0	0	1	1
A1	1	1	0	1	0
A3	1	1	0	1	0
A4	0	0	0	0	0

Note: Figure 12. Across Area Circular Chains: Area 2 Step Transitions

The reduced rule reachability matrix displayed in Figure 19 represents the possible circular chains

across areas, including those that utilize within area chains as part of the across area chains. Note that both circular chains pass through rule R2.1. Although a total of six circular chains have been found, only two are unique:

- rule R2.1 --> rule R2.2 --> rule R3.2 --> rule R2.1
- rule R1.1 --> rule R2.1 --> rule R3.1 --> rule R1.1

Area 2 - Boundary/Reachability Matrix

From-To	R2.1	R2.2	A1	A3	A4
R2.1	3	3	0	5	2
R2.2	2	1	0	3	2
A1	2	2	0	3	1
A3	3	2	0	3	1
A4	0	0	0	0	0

Note: Figure 13. Across Area Circular Chains: Area 2 Reachability Matrix

Area 3 - Boundary/Adjacency Matrix - Start

From-To	R3.1	R3.2	A1	A2	A4
R3.1	0	0	1	0	0
R3.2	0	0	0	1	0
A1	0	0	0	0	0
A2	1	1	0	0	0
A4	0	0	0	0	0

Area 3 - Two-Step Transition

From-To	R3.1	R3.2	A1	A2	A4
R3.1	0	0	0	0	0
R3.2	1	1	0	0	0
A1	0	0	0	0	0
A2	0	0	1	1	0
A4	0	0	0	0	0

Area 3 - Three-Step Transition

From-To	R3.1	R3.2	A1	A2	A4
R3.1	0	0	0	0	0
R3.2	0	0	1	1	0
A1	0	0	0	0	0
A2	1	1	0	0	0
A4	0	0	0	0	0

Area 3 - Four-Step Transition (Cyclical State)

From-To	R3.1	R3.2	A1	A2	A4
R3.1	0	0	0	0	0
R3.2	1	1	0	0	0
A1	0	0	0	0	0
A2	0	0	1	1	0
A4	0	0	0	0	0

Note: Figure 14. Across Area Circular Chains: Area 3 Step Transitions

Area 3 - Boundary/Reachability Matrix

From-To	R3.1	R3.2	A1	A2	A4
R3.1	0	0	1	0	0
R3.2	1	1	1	2	0
A1	0	0	0	0	0
A2	2	2	1	1	0
A4	0	0	0	0	0

Note: Figure 15. Across Area Circular Chains: Area 3 Reachability Matrix

Area 4 - Boundary/Adjacency Matrix - Start					
From-To	R4.1	R4.2	A1	A2	A3
R4.1	0	1	0	0	0
R4.1	0	0	0	0	0
A1	0	0	0	0	0
A2	1	0	0	0	0
A3	0	0	0	0	0

Area 4 - Two-Step Transition					
From-To	R4.1	R4.2	A1	A2	A3
R4.1	0	0	0	0	0
R4.1	0	0	0	0	0
A1	0	0	0	0	0
A2	0	1	0	0	0
A3	0	0	0	0	0

Area 4 - Three-Step Transition (Steady State)					
From-To	R4.1	R4.2	A1	A2	A3
R4.1	0	0	0	0	0
R4.1	0	0	0	0	0
A1	0	0	0	0	0
A2	0	0	0	0	0
A3	0	0	0	0	0

Note: Figure 16. Across Area Circular Chains: Area 4 Step Transitions

Area 4 - Boundary/Reachability Matrix					
From-To	R4.1	R4.2	A1	A2	A3
R4.1	0	1	0	0	0
R4.1	0	0	0	0	0
A1	0	0	0	0	0
A2	1	1	0	0	0
A3	0	0	0	0	0

Note: Figure 17. Across Area Circular Chains: Area 4 Reachability Matrix

Reduced Rules Adjacency Matrix - Start							
From-To	R1.1	R2.1	R2.2	R3.1	R3.2	R4.1	R4.2
R1.1	0	1	0	0	0	0	0
R2.1	0	0	1	1	0	0	0
R2.2	0	0	0	0	1	1	0
R3.1	1	0	0	0	0	0	0
R3.2	0	1	0	0	0	0	0
R4.1	0	0	0	0	0	0	1
R4.2	0	0	0	0	0	0	0

Reduced Rules Two-Step Transition							
From-To	R1.1	R2.1	R2.2	R3.1	R3.2	R4.1	R4.2
R1.1	0	0	1	1	0	0	0
R2.1	1	0	0	0	1	1	0
R2.2	0	1	0	0	0	0	1
R3.1	0	1	0	0	0	0	0
R3.2	0	0	1	1	0	0	0
R4.1	0	0	0	0	0	0	0
R4.2	0	0	0	0	0	0	0

Reduced Rules Three-Step Transition							
From-To	R1.1	R2.1	R2.2	R3.1	R3.2	R4.1	R4.2
R1.1	1	0	0	0	1	1	0
R2.1	0	2	0	0	0	0	1
R2.2	0	0	1	1	0	0	0
R3.1	0	0	1	1	0	0	0
R3.2	1	0	0	0	1	1	0
R4.1	0	0	0	0	0	0	0
R4.2	0	0	0	0	0	0	0

Reduced Rules Four-Step Transition							
From-To	R1.1	R2.1	R2.2	R3.1	R3.2	R4.1	R4.2
R1.1	0	2	0	0	0	0	1
R2.1	0	0	2	2	0	0	0
R2.2	1	0	0	0	1	1	0
R3.1	1	0	0	0	1	1	0
R3.2	0	2	0	0	0	0	1
R4.1	0	0	0	0	0	0	0
R4.2	0	0	0	0	0	0	0

Reduced Rules Five-Step Transition							
From-To	R1.1	R2.1	R2.2	R3.1	R3.2	R4.1	R4.2
R1.1	0	0	2	2	0	0	0
R2.1	2	0	0	0	2	2	0
R2.2	0	2	0	0	0	0	1
R3.1	0	2	0	0	0	0	1
R3.2	0	0	2	2	0	0	0
R4.1	0	0	0	0	0	0	0
R4.2	0	0	0	0	0	0	0

Reduced Rules Six Step Transition (Cycling State)							
From-To	R1.1	R2.1	R2.2	R3.1	R3.2	R4.1	R4.2
R1.1	2	0	0	0	2	2	0
R2.1	0	4	0	0	0	0	2
R2.2	0	0	2	2	0	0	0
R3.1	0	0	2	2	0	0	0
R3.2	2	0	0	0	2	2	0
R4.1	0	0	0	0	0	0	0
R4.2	0	0	0	0	0	0	0

Note: Figure 18. Across Area Circular Chains: Reduced Rules Step Transitions

Reduced Rules/Reachability Matrix							
From-To	R1.1	R2.1	R2.2	R3.1	R3.2	R4.1	R4.2
R1.1	1	3	3	3	1	1	1
R2.1	3	2	3	3	3	3	1
R2.2	1	3	1	2	2	2	2
R3.1	2	3	1	1	1	1	1
R3.2	1	3	3	3	1	1	1
R4.1	0	0	0	0	0	0	1
R4.2	0	0	0	0	0	0	0

Note: Figure 19. Across Area Circular Chains: Reduced Rules Reachability

Inconsistent Rule Chains

Identification of inconsistent rule chains uses a combination of two previous methods, one that identifies rule chains across areas and one that identifies when those rule chains will succeed in the same situation, but with different conclusions. The analysis starts with the reduced rule set list, which contains all the rules that are reachable across areas. i.e., those rules that may succeed as the result of inference chaining. Now we want to determine if any of these reachable rules are inconsistent. To accomplish this, we create an inconsistency adjacency matrix IA for each of the rules in the reduced rule set and all of the rules that are reachable from the rules in the reduced rule set, where:

$a_{ij} = 1$ if rule i is adjacent to rule j, i.e., if a conclusion of rule i is a condition of rule j.

0 otherwise.

Now that we have captured the adjacency relationship between the reduced rule set and their reachable rules, we may apply the algorithm [TWO] to to the inconsistency adjacency matrix IA. At each transition, a consistency check is performed to check the conclusions of rule chains of various lengths that start with a rule in the reduced rule set. If two rule chains succeed given the same condition element set, but lead to different conclusions, there is a rule chain inconsistency. The procedure for the consistency check involves creating a modified element-rule matrix MI_n at each nth transition of the adjacency matrix IA. The element-rule matrix is reduced by eliminating all elements except conclusion elements. This is possible since we have already determined at each transition step that the rules are reachable given the same condition element set, and we are only concerned now with determining if the conclusions are different. If the conclusions are different, we can resolve the inconsistency using the approaches described for within area redundancy, subsumption or inconsistency.

The technique can be illustrated in an example using the knowledge base displayed in Figure 20. We will assume that the reduced rule set has been determined to be (R1.1, R1.2, R1.3, R2.1, R2.2) and that we have the step transitions displayed in Figure 21. From the modified element-rule matrix MI_1, displayed in Figure 22, it is easy to see that there is an inconsistency between rules R1.3 and R2.1. Both rules will succeed when B and C are true, but lead to different conclusions, F and D, respectively. From examination of MI_2 and MI_3, we see that there are no new inconsistencies.

AREA 1	AREA 2
R1.1 IF A THEN B R1.2 IF B THEN C R1.3 IF C THEN F	R2.1 IF C THEN D R2.2 IF D THEN E
Note: Figure 20. Inconsistent Rule Chains: Rule Set	

Evaluation

The techniques described here were applied to the verification of a knowledge base consisting of 190 production rules divided into 15 interacting areas. The case study data were drawn from a knowledge base developed for the Iowa Civil Rights Commission to help investigators process complaints of alleged violations of civil rights laws. On the average, the 15 areas contained 87% primary elements and 13% non-primary elements, which shows the areas have a high degree of internal consistency. A total of 33 anomalies were identified, with the following breakdown: within an area redundancy (1); within an area subsumption (1); within an area inconsistency (2); potential within an area circular rule chain (1); actual within an area circular rule chain (1); potential across area circular rule chain (2); actual across area circular rule chain (0); across area subsumption (2); within or across area unreachable rules (11); within or across area dead-end rules (10); within or across area disconnected chains (5).

Overall, 238 matrices were created, with the number of cells ranging from 39 to 1131. The matrices were created on a standard spreadsheet package, which also facilitated the implementation of the matrix operations. We feel that the effort involved was modest relative to the benefits obtained in improving the quality of the knowledge base. In addition, the anomalies were removed early in the development process, where the cost is likely to be lower.

Inconsistency Adjacency Matrix - Start

From-To	R1.1	R1.2	R1.3	R2.1	R2.2
R1.1	0	1	0	0	0
R1.2	0	0	1	1	0
R1.3	0	0	0	0	0
R2.1	0	0	0	0	1
R2.2	0	0	0	0	0

Two-Step Transition

From-To	R1.1	R1.2	R1.3	R2.1	R2.2
R1.1	0	0	1	1	0
R1.2	0	0	0	0	1
R1.3	0	0	0	0	0
R2.1	0	0	0	0	0
R2.2	0	0	0	0	0

Three-Step Transition

From-To	R1.1	R1.2	R1.3	R2.1	R2.2
R1.1	0	0	0	0	1
R1.2	0	0	0	0	0
R1.3	0	0	0	0	0
R2.1	0	0	0	0	0
R2.2	0	0	0	0	0

Four-Step Transition (Ready State)

From-To	R1.1	R1.2	R1.3	R2.1	R2.2
R1.1	0	0	0	0	0
R1.2	0	0	0	0	0
R1.3	0	0	0	0	0
R2.1	0	0	0	0	0
R2.2	0	0	0	0	0

Note: Figure 21. Inconsistent Rule Chains: Step Transitions

MI$_1$

	R1.1	R1.2	R2.1	R2.2
C	1	0	0	0
D	0	1	0	0
E	0	0	0	1
F	0	1	0	0

MI$_2$

	R1.1	R1.2	R1.3	R2.1
R2.2				
C	0	0	0	0
D	1	0	0	0
E	0	1	0	0
F	1	0	0	0

MI$_3$

	R1.1	R1.2	R2.1	R2.2
C	0	0	0	0
D	0	0	0	0
E	1	0	0	0
F	0	0	0	0

Note: Figure 22. Inconsistent Rule Chains: Modified Element-Rule Matrices

References

1. Botten, N. and T. Raz, "Knowledge Base Verification Using Matrices," Proceedings of The Fourth International Conference on Industrial & Engineering Applications of Artificial Intelligence & Expert Systems, Hawaii, June 1991, pp. 91-98.

2. Green, C. and M. Keyes, "Verification and Validation of Expert Systems," IEEE Knowledge-Based Engineering and Expert Systems, (WESTEX-87), IEEE 87CH2463-8, pp. 38-43. pp. 19-24.

3. Tepandi, J. "Comparison of Expert System Verification Criteria: Redundancy," Validation, Verification and Test of Knowledge-Based Systems, Eds. Ayel, M. and J. Laurent, Wiley, 1991, pp. 49-62.

4. Stachowitz, R. and C. Chang, AAAI 88 Verification and Validation of Expert Systems, Tutorial from Seventh National Conference on Artificial Intelligence, Minneapolis, 1988.

5. Yu, X. and G. Biswas, "CHECKER: An Efficient Algorithm for Knowledge Base Verification," Proceedings of the Third International Conference on Industrial and Engineering Applications of Artificial Intelligence and Expert Systems, Vol. II, 1990, pp. 735-744.

STRUCTURAL TESTING STRATEGIES
APPLIED TO KNOWLEDGE-BASED SYSTEMS

F. Saglietti

Gesellschaft für Reaktorsicherheit (GRS) mbH
Forschungsgelände D-8046 Garching Germany

Abstract. The aim of this work is to extend the existing (though so far still incomplete) testing theories for conventional programs, such as to allow their application also to artificial intelligent systems. The ultimate goal of the present investigations is thus represented by a unified problem representation permitting to define adequate V&V procedures by adapting them to the specific software category considered. In particular, special criteria for AI testing will be included to the usual coverage requirements in order to capture also those faults, which seem to be typical for the expert reasoning.

1 Introduction

The increasing diffusion of artificial intelligence in a variety of application areas has emphasised the urgency for a sound dependability theory. This situation demands for extending the current investigations on adequate V&V procedures for conventional programs; thus the (by far yet unsolved) question on assessing procedural software evolves to a still more challenging problem including declarative languages.

The reasoning principles in declarative programming as opposed to those adopted by procedural software were analysed during past work (s. [9]). These considerations are summarized at the beginning of the paper. In particular, errors affecting knowledge-based rules are classified according to syntactical and semantical features; for each class the existing checking methods are shortly overviewed.

In order to increase the detectability of different fault categories, test adequacy measures have been defined and compared with alternative checking techniques both in terms of expected efficiency and of coverage achievement.

To permit a structural interpretation of such testing criteria, a graph-theoretical representation of knowledge based systems is introduced. It is based on the concept of *logical path* (as a sequence of rules with premises forming an implicative chain), extending the original notion of *execution path* for conventional software structure to a more general view of program functionality.

The final verification strategy proposed consists of integrating a *systematic vertical* phase covering the single rule-paths with a *random horizontal* phase checking to some extent also the interconnections among different rule-paths activated by the same input.

The commonality of the validation process suggested with conventional V&V procedures ultimately consists in considering the *facts* of expert systems analogously to the procedural functions and of verifying in both cases:
- each of them separately by means of *systematic* means and
- their possible combinations by means of *probabilistic* techniques stratifying them according to their *frequency of demand* and their *risk of damage*.

2 A KBS Model

The KBS model chosen for the following notation and terminology is essentially based on the one proposed in [12]. A system E will be regarded as consisting of facts, rules and an interpreter, as briefly summarized in table 1.

System E = (F,R,I) consists of:	Facts (set F) consists of:	initial facts intermediate facts goals
	Rules (set R); each element consists of:	Premises (facts) Conclusion (fact) Uncertainty measure ([0,1])
	Interpreter (set I) consists of:	Operations with uncertainty measures Inference control (procedures to select rules and evaluate premises) decision selection (goal acceptance)

Table 1. KBS model

3 Composition and Decomposition

In [9] we analysed the main differences in the philosophies of procedural and declarative programming, studying the impact they have on the process of dependability evaluation. The fundamental difference between the development of conventional programs and of AI software regards the connection of the information to be elaborated. While software procedures consist of well-determined sequences of commands conceived to be executed in a given order and in a pre-established context, artificial intelligent programs result from the inverse attitude, i. e. from trying to disconnect an overall knowledge or experience into atomic pieces believed to be still valid on their own.

Thus, on the one side conventional programs are obtained by *composition* of single, in themselves meaningless steps into a whole structure giving them significance and logic content. In this sense, the underlying design is essential to permit to achieve the required connection between the pieces of information. On the other side, AI systems result by *decomposition* of theory or reality into micro-cosmos, whose intrinsic consistency and correctness can be either abstractly proved or inferred by observation. This decomposition can result both in simplifications and in difficulties during the evaluation phase, depending on the particular system, as analysed in the following table.

Software Type Language Class	Reasoning type Error Sources	Benefits	Limitations
"Conventional programs", procedural languages	Reasoning based on composition, design errors in logic structure	verification of testing results, failure detection, structural testing	applicability, error localization, error correction, maintenance
"Artificial intelligence" declarative languages	Reasoning based on decomposition, design errors in and among rules	applicability, error localization, error correction, maintenance	verification of testing results, failure detection rule interactions

Table 2. Principles of composition vs. decomposition

As already pointed out, rule-bases are meant to reflect both abstract theories or experienced realities: the former by *deduction* from a set of axiomatic laws, the latter by *induction* from observed reactions. In the first case, each rule represents either an axiom or a proved theorem, so that the overall system produces the statements resulting from these by classical logic inference. In the second case, each rule stands for a correlation between events, which was frequently observed in reality and may therefore be considered to be generally valid. Thus, the underlying knowledge of an AI system may correspond to an abstract *axiom- & theorem-* base or be rather interpreted as a set of empirical laws founded on experimental evidence. These aspects and the following comments are summarized in table 3.

System Type	Inference	Verification	Typical Structure
Deductive	Axiomatic systems universal laws full knowledge	Proof of correctness of universal laws	Star-shaped, multiple recursions, iterative concatenation of theorems
Inductive	Empirical process observed correlation partial experience	Structural test simulation observation	Tree-shaped, multiple cascades, from cursory to fine filtering of features

Table 3. Principles of induction vs. deduction

While deductive systems are meant to model an absolute and self-contained theoretical discipline, inductive rule-bases are intended to approximate reality by an extensive, but nonetheless purely empirical expertise. This fundamental distinction results in dissimilar consequences on the advantages or problems encountered during the testing phase of the respective products:

Within a well-established abstract theory each rule can be validated on its own by *proofs* of correctness (e.g. mathematical demonstration of differentiation theorems). This simplifies also the *correction* procedure of errors detected, in that each incorrectness affects one rule exactly and can thus be uniquely localized and isolated; whereas in conventional software faults may be removed in many different ways, possibly having crucial consequences on adjacent execution paths.

Unfortunately, these potential gains for the V&V procedures of declarative programs are severely counterbalanced by the draw-backs originating from testing inductive systems. Due to the nature of the tasks to be solved, the specifications of expert systems are often nonexistent or imprecise. Thus the principal error sources of expert systems are not a cause of the AI method itself, but rather of the wide application area it allows (or at least encourages) to approach.

As a consequence of inductive reasoning, the validity of tested results often cannot be generally proved, but just confirmed either by past *observations* or by experimental *simulation*. Both checking methods are by far not obvious, requiring an exact and extensive recording of observational settings or (what may be still more crucial) the (at least approximative) reproduction of the marginal conditions to be verified. In particular, this frequently leads to the mere confirmation of an overall combination of parameters, so that rules can be checked in a high-level context, but not singularly. The resulting coverage limitations may have fatal repercussions on testing efficiency, as will be analyzed in the following chapters.

4 Error Classification

We distinguish between errors affecting only the semantical content of single clauses or also their context.

4.1 Syntactical AI Errors

In the following we will define an AI *syntactical fault* to be any inappropriateness in the interrelationships among rules, which may be observable by any analyser, not necessarily provided with any specific expertise on the technical field to be modelled. In this sense the primary syntactical errors concern the *consistency* and the *completeness* of rule-bases (s. [11]).

Consistency. An expert system is consistent, if its rules obey the laws of an axiomatic theory: each one should be essential and should not contradict any other. Thus, the primary failure sources affecting *consistency* are:
- redundancy (i.e. one clause following from others),
- subsumption (i.e. redundancy under given circumstances),
- conflict (i.e. clauses leading to contradictory conclusions).

Completeness. An expert system is complete, if any goal can be achieved, i.e. if any fact belonging to the goal subset can be confirmed or rejected by the rule-base; therefore, *completeness* is prevented by:
- missing rules (i.e. absence of rules required to achieve a given goal).

4.2 Semantical AI Errors

Complementary to the *syntactical* classes of incorrectness studied so far are those errors, which relate to a specific misbehaviour of the expert system due to the wrong modelling of reality in a single non-contradictory clause. Being therefore only detectable by means of a profound knowledge and/or experience in terms of the inherent meaning of each single rule, we will indicate this error class as consisting of *semantical faults*. The testing techniques proposed so far in [12] and [9] are based on:
- statement adequacy,
- contribution adequacy,
- separation adequacy.

Statement Adequacy. The former testing strategy, *statement adequacy*, requires at least one activation of each rule. This condition is meant to detect all those inferential clauses affected by fundamental errors, which are not compensated in combination with further rules.

Contribution Adequacy. The second testing criterion, *contribution adequacy*, asks to verify that each component of a rule-base has a crucial impact on the inference process. Obviously, contribution adequate tests are automatically statement adequate, as they also require the execution of each rule, but under particular conditions, namely proving each of them to be necessary to achieve specific goals.

Separation Adequacy. Finally, *separation adequacy* aims at detecting errors usually masked by failure compensation. This is the case of rules only valid in specific contexts expressed by the pre-conditions of other rules. To increase the detectability of this error class, this testing strategy isolates each rule-premise at least and requires a dynamic verification of any other rule with different pre-conditions: this reduces the effect of rule clusters. In particular, each separation adequate test is also statement adequate.

To compare AI testing criteria with structural coverage measures for conventional programs (s. [1]), we may interpret each rule as an 'if-then'-branch: the branch is executed if both alternative Boolean values of the 'if'-statement are executed: the 'then'-conclusion in case of 'true' and a jump-command 'goto' in case of 'false'. On this basis we can state: Statement adequacy corresponds to complete *statement coverage* for procedural software; separation adequacy is a stronger criterion than complete *branch coverage*, but a weaker criterion than complete *path coverage* for procedural software.

4.3 Summary

The fault categories identified are summarized in table 4 together with their corresponding checking methods.

Error Source to be identified	System Feature to be verified	Error Class to be detected	Detection Strategy to be applied
Syntactic (fault in interaction among rules)	Consistency	Redundancy	Static Analysis
		Subsumption	
		Conflict	Petri nets
	Completeness	Missing rules	
Semantic (fault in rule meaning)	Contribution	Negligible	Contribution Adequacy
	Correctness	No Compensation	Statement Adequacy
		Compensation	Separation Adequacy

Table 4. Overview of fault categories and checking methods

5 Structural Testing Strategies

5.1 A Graph-Theoretical KBS Representation

We propose a graph-theoretical representation of expert systems with rules being interpreted as nodes of a diagram with two sets of arcs:
- *type 1-arcs* are directed (undotted) arcs denoted by " $\rightarrow$ " and
- *type 2-arcs* are undirected (dotted) arcs denoted by " --- ".

Type 1. Type 1-arcs are defined by:
Node r_b is immediate successor of node r_a w.r.t. a type 1-arc iff:
I. (PREMISES(r_b) arc fulfilled) $\Rightarrow$ (PREMISES(r_a) are fulfilled) and
II. There is no further rule r' with:
  [(PREMISES(r_b) are fulfilled) $\Rightarrow$ (PREMISES(r') are fulfilled) and
  (PREMISES(r') are fulfilled) $\Rightarrow$ (PREMISES(r_a) are fulfilled)];

Type 2. Type-2 arcs are defined by:
Two different nodes r_a and node r_b are linked by a type 2-arc iff:
I. None of them is reachable from the other by a directed path of type 1-arcs,
II. PREMISES(r_a) $\cup$ PREMISES(r_b) is a subset of facts,
  which may be *simultaneously valid* (i.e. they do not exclude each other) and
III. There are no different predecessors of r_a and r_b with this property.

Example. A simple instance is shown in figure 1. The previous definitions mainly result in a structure consisting of vertical *logical paths* of type 1-arcs leading from the root to the different goals, e.g.: (r1 - r11 - r13) $\rightarrow$ G1 or (r4 - r42 - r43) $\rightarrow$ G5). The main difference with *execution paths* in conventional control flow structures lies in the possibility of different paths for the same input configuration: namely paths linked by horizontal type 2-arcs, e.g.: (r3 - r31) $\rightarrow$ G3 and (r4 - r41 - r43) $\rightarrow$ G5).

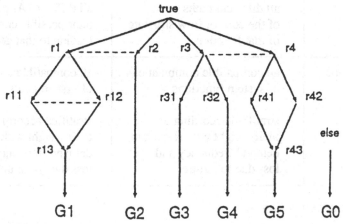

Fig. 1. Example of graphical KBS representation

5.2 Combination of Systematic and Probabilistic Techniques

As adequate dynamic V&V procedures we suggest the following combination of systematic and probabilistic techniques:

1. test each type 1-path systematically, i.e. by means of deterministic techniques aiming at verifying *separately* every rule sequence forming an implicative chain,
2. test type 2-related paths probabilistically, i.e. by means of statistical techniques aiming at verifying *combinations* of rule sequences which may be activated by the same input,
3. distribute the effort of both testing techniques according to a *risk analysis* of expected loss w.r.t. the operational profile.

This strategy shows a considerable analogy with an already existing one for procedural software, s. [4], consisting of the following three phases:

1. Identify the basic system *requirements*:
   originally procedural functions, here: expert system goals.
2. Evaluate their *importance* in terms of:
   frequency of demand and loss due to potential damage.
3. Verify the resulting *strata* accordingly:
   each single (function resp. goal) systematically and their combinations statistically.

This parallelism is summarized in Table 5.

Testing technique	Procedural software	Expert system
Systematic strategy	each individual system function taking into account all different paths of the control flow structure of that function	each individual system goal taking into account all different AI paths and their possible combinations leading to that goal
Probabilistic strategy	all compatible combinations of system functions stratified according to their weight with respect to demand frequency and loss due to failure	all compatible combinations of system goals stratified according to their weight with respect to demand frequency and loss due to failure

Table 5. Unified V&V procedures

6 Conclusion

We started with a brief evaluation of AI as compared to conventional software, especially in the light of the expected error sources and of the resulting impact on the verification process. Successively, AI systems were classified into two main categories according to the nature of the task intended to be solved. Finally, we analysed the error classes to be expected by expert systems, investigating for each identified category the most adequate testing procedure expected to detect the irregularity.

A graph-theoretical representation of knowledge based systems was then introduced: it allows a structural interpretation of this new testing criterion. This results in a new concept of "*logical path*" (as a sequence of rules with premises forming an implicative chain), extending the original notion of "*execution path*" for conventional software structure to a more general view of program functionality. The final verification strategy proposed consists of integrating a *systematic* "vertical" phase covering the single rule-paths with a *random* "horizontal" phase checking to some extent also the interconnections among different rule-paths activated by the same input.

Acknowledgements. The author is indebted to Prof. J. Tepandi, Tallinn Technical University (Estonia) and to Prof. W. Ehrenberger, GRS Garching (Germany) for stimulating discussions on this topic as well as for improving and encouraging remarks during the development of the present work.

References

1. B. Beizer: Software Testing Techniques New York, Van Nostrand Reinhold 1990

2. R.E. Bloomfield, W.D. Ehrenberger:The Validation and Licensing of Intelligent Software, IAEA/OECD/CEC Int. Conf. on Man-Machine Interface in the Nuclear Industry, Tokyo, Japan, 1988

3. H.T. Daughtrey, S.H. Levinson: Standardizing Expert System Verification and Validation, Proc. AI91, Frontiers in Innovative Computing for the Nuclear Industry, Jackson, Wyoming 1991

4. W. Ehrenberger and K. Plögert: Einsatz statistischer Methoden zur Gewinnung von Zuverlässigkeitskenngrößen von Prozeßrechnerprogrammen, Rep. KfK-PDV 151 1978

5. C. J. R. Green and M. M. Keyes: Verification and Validation of Experts Systems, Proc. Western Conference on Expert Systems, IEEE 1987

6. R. M. O'Keefe, O. Balci and E. P. Smith: Validating Expert System Performance, IEEE Expert 1987

7. O' Leary, M. Goul, K.E. Moffitt, A.E. Radwan: Validating Expert Systems, IEEE Expert 1990

8. E. Ness: Verification & Validation of Knowledge Based Systems for Process Control Applications, OECD Halden Reactor Project 1991

9. F. Saglietti: A Comparative Evaluation of V&V Procedures for Conventional Software and Expert Systems, Proc. EPRI Workshop on Methodologies, Tools and Standards for Cost-Effective, Reliable Software Verification and Validation, Chicago 1991

10. G.P. Singh, D. Cadena, J. Burgess: Practical Requirements for Software Tools to Assist in the Validation and Verification of Hybrid Expert Systems, Proc. EPRI Workshop on Methodologies, Tools and Standards for Cost-Effective, Reliable Software Verification and Validation, Chicago 1991

11. M. Suwa, A. C. Scott and E. H. Shortliffe: Completeness and Consistency in a Rule Based System, in "Rule-Based Expert Systems", B. G. Buchanan, E. H. Shortliffe eds. London: Addison-Wesley 1984

12. J. Tepandi: Verification, Testing and Validation of Rule-Based Expert Systems, Proc. 11th IFAC World Congress, Tallinn, Pergamon Press 1990

13. J. Tepandi: What is (Expert System) Redundancy? Dependability of Artificial Intelligence Systems (DAISY_91), Elsevier Science Publishers (North-Holland), IFIP 1991

14. J.P. Vaudet, P.F. Kerdiles: VALID: a Proposition to Advance Capabilities for Knowledge-Based Systems Validation, Proc. of the ESPRIT Conference Week 1989

15. ESPRIT program, VALID project: Expert Systems with Applications journal 1991, Special Issue on Expert Systems V&V, Pergamon Press, New York

Retraining and Redundancy Elimination for a Condensed Nearest Neighbour Network

Dieter Barschdorff, Achim Bothe, Ulrich Gärtner, Andreas Jäger

Institute of Electrical Measurement, University of Paderborn
Pohlweg 47-49, P.O.Box 1621, D-4790 Paderborn

Abstract. A new method of supervised retraining for an artificial neural network model based on the Condensed-Nearest-Neighbour classification principle is presented. Adaptation to time variable features and the classification of new occured objects is possible now. Furthermore different approaches for the elimination of redundant training patterns to reduce the size of the training set and the time required for training and classification is achieved. A speech recognition system is presented as an example of application.

Keywords: Neural network, retraining, adaptive structure, speech recognition

1 Introduction

Artificial neural networks have demonstrated their great potential in manifold classification problems such as image recognition, speech recognition and fault detection. The essential features are: dense interconnection of simple computational nodes, high degree of robustness, only few necessary assumptions about decision regions. In recognition applications neural networks are used to identify the class which corresponds best to the actual input pattern. Many publications cover these areas [3, 4, 5, 9, 10, 13, 14, 15, 18].

As a typical application speech recognition is regarded. These recognition systems can be divided into two major groups: speaker *inde*pendent and speaker *de*pendent systems. In speaker dependent systems each new speaker has to go through a complete training phase with the whole vocabulary which shall be recognizable. The words have to be repeated two or three times, making this training a tedious task for the user, especially if a large vocabulary must be learned.

After a first initial training, speaker independent systems are assumed to recognize speach of speakers, which have not been used for initial training. In cases of misclassification, due to unlearned dialects for example, retraining has to be performed. In such a system solely the misclassified word has to be retrained and only once.

2 Condensed-Nearest-Neighbour-Network

Based on the condensed sample principle the Condensed Nearest Neighbour Network (CNNN) is able to describe and reclassify complex cluster structures. No restrictions of the statistical properties of features have to be imposed.

The network simulation is realized on a PC-AT in the programming language C. Working with our CNN-network requires two steps:

1. Net training with the patterns of a training set until all patterns are classified correctly (reclassification rate of 100%). Storage of training results (net structure and weights) for later use without training.

2. Processing of an input pattern (feature vector). Display of the classification result from the net output.

To make these steps clearer, network structure, network training and pattern classification are described in more detail:

2.1 Network Structure

The network comprises three layers: an input, a hidden and an output layer. Input and output layer are of fixed size due to the number of features J and the number of classes K, respectively. The hidden layer size is variable during training phase and obtains a neuron for each subclass created during the learning phase. The resulting number of hidden neurons L should be smaller than the number of presented patterns. Otherwise the network only maps each training pattern, which should be avoided.

After completion of the training phase there is full interconnection between input and hidden layer, each input neuron has a connection to each hidden neuron. A hidden neuron is connected to one output neuron, but not each output neuron is connected to each hidden neuron (Fig. 1.).

Hidden neurons are processing elements requiring only simple computations. According to a hyperellipsoid approach the output value o_i of the i-th neuron in the hidden layer is given by:

$$o_i = \Theta \left(\sum_{j-1}^{J} \frac{(m_j - w_{ij})^2}{r_{ij}^2} \right) \quad \text{with} \quad \Theta(d) = \begin{cases} 1 & \text{for } d \le \alpha \\ 0 & \text{for } d > \alpha \end{cases} \tag{1}$$

In Eq. (1) $\underline{m}$ is the feature vector to be classified and $\underline{w}$ represents the mean vector of the hyperellipsoid. The radius vector of the hyperellipsoid is named $\underline{r}$. The threshold function Θ depends on the parameter $\alpha > 0$ (initial value $\alpha = 1$).

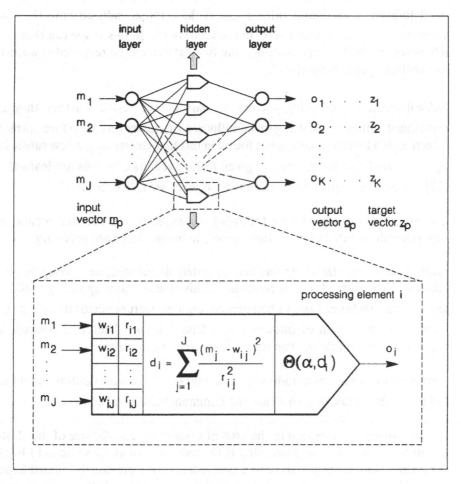

Fig. 1. Topology and hidden layer processing element of the CNNN model

2.2 Network Training

An untrained net contains no neurons in its hidden layer. Therefore no connection between input and output exists, consequently no input vector whatsoever will activate any output neuron. A classification is impossible, hidden neurons have to be created, which occurs during training.

Learning of a set of patterns, inclusive addition of new hidden neurons takes place according to the following rules. The first two steps represent an extension of the "Leader Clustering Algorithm" [12]:

1. For the first feature vector of a class Ω_k, a dispersion zone u_k is defined in the J-dimensional feature space. In this first step all radii r_j (j=1...J) of this zone are fixed to a given maximum value $r_{j,max}$. If these initial radii all have the same value, the result is a hypersphere. In the following steps this shape can change to a hyperellipsoid. The region around the first pattern can be regarded as a cluster or subclass u_k of pattern class Ω_k.

2. Subsequently, all following vectors of class Ω_k are tested whether they are positioned within the first region of influence or cluster and therefore correctly attached. If a feature vector is not found in this first cluster $u_{k,1}$, a new subcluster $u_{k,2}$ is created and so on, until all given feature vectors of a class are learned. In [12] the patterns representing the cluster centres are called "leaders".

3. The process is the same for all following classes. If the class related regions are separately distributed in the feature space, no further steps are necessary.

4. Even if the class related regions are separately distributed, an overlap between clusters of different classes is possible. In this case the corresponding radii r_j of the related subclasses have to be reduced, until no feature vector of a class Ω_k is situated in the region of influence of a false subclass. Therefore overlapping regions of different classes cannot contain any learning pattern.

   Furthermore, a minimum radius $r_{j,min}$ is defined. The current radius should not fall below this value, in such a case the minimum value is taken instead.

   If a feature vector is located in the area of a false subclass, the size of the cluster has to be decreased. One possibility is the reduction of all or some radii by the same ratio, another way is to reduce one radius of the previously defined hyperellipsoid. In this case, it is a question of the semiaxis with the smallest angle between the connecting line of the subclass origin and the misclassified pattern.

5. If a feature vector is positioned in a cluster belonging to a false class and if the radii of the concerned subclusters could not be further reduced ($r_{j,min}$), this area has to be especially marked. This occurs through the introduction of two different types of subclasses. One type represents the normal case with non-overlapping regions, the other marks the cluster overlapping regions.

6. As a result of the radius reduction described in point 4 a previously correctly classified pattern may be no longer in the relevant subclass. For these patterns new subclasses have to be created. Due to this reason points 1 to 5 have to be repeated until all patterns to be classified are attached to correct subclasses and no further radius reduction occurs.

After these computational steps the learning process is completed. The presentation of patterns must not follow a class sequential order. This was only assumed for simplification purpose. A 100%-rate of reclassification for all training patterns is achieved by this learning algorithm. Transfering this algorithm to a neural network structure leads to the definition of the three layer model (Fig. 1).

The number of neurons in the output layer corresponds to the number of classes of the examined problem. Here the neurons are representing a logical OR-operation of the hidden layer outputs, they provide the union of the concerning subclusters. With L_k clusters of a class Ω_k follows:

$$\forall \, \underline{m} \in \Omega_k: \; \underline{m} \in \left(\bigcup_{l-1}^{L_k} u_{kl} \right) \tag{2}$$

The network topology and a processing element of the hidden layer is shown in Figure 1.

2.3 Pattern Classification

During training relevant parts of the feature space are mapped by hidden neurons in subclasses u_k. If a pattern to be classified is positioned in the region of influence of a single subclass u_k, only one component of the output vector is active. A definite class decision is possible. If a previously unlearned pattern is found in regions of two or more classes, a definite class assessment using the output vector is impossible. Now the nearest subclass is sought for using a distance decision (Euclidean distance) and the pattern is attached to this class. A distance-based decision is also necessary if the pattern to be classified is not located in any cluster region.

Another procedure is possible, if there was no definite class assessment. Those uncertain classified patterns can be learned by supervised retraining as described in chapter 3. Next time the same pattern will be classified definite. Application of the algorithm is demonstrated with a speech recognition system.

2.4 Speech Recognition System with CNNN

The speech recognition system was realized on a PC-AT with a digital signal processor board (DSP). Data acquistion of the speech signal is realized with a microphone, analog circuitry and at last an AD-conversion. Word boundaries and different features are calculated in realtime using an TMS320C30 floating point DSP. The essential features are signal energy, variance, zero-crossing rate and spectral features. Non-overlapping time frames with 5 ms duration are used in feature extraction.

Due to the fixed size of input vectors a time normalization is mandatory. Further-more, to minimize differences in loudness between actual pattern and learned patterns a normalization of amplitudes succeeds. Classification of these features is achieved by the CNN network providing the recognized word at its output. If a mis-classification has occured, the actual pattern has to be retrained.

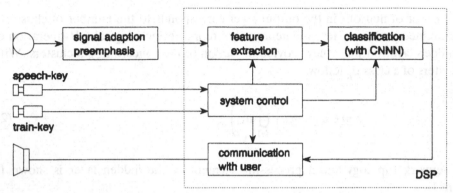

Fig. 2. Speech recognition system

3 New Method of Retraining

In speech recognition the ability to adapt and continue learning is essential, because the set of training patterns is limited and new speakers, new environments and new dialects are encountered. Hence a method of retraining for the system is necessary. This training phase may cause a time delay up to some minutes. The retraining process must be interruptable by user request, so that further word recognitions can be executed, otherwise an inacceptable delay may arise. During training a CNNN cannot be used for classifition. Therefore two networks must be present at one instance; the first one for classification of actual inputs, the second one is trained.

Each time a word was misclassified there is the possibility to retrain the correspond-ing pattern. The following steps are performed:

1.   Detection of the misclassification by the user.

2.   Choice of the correct word (target vector for the network). For this purpose, the four most probable words are presented by a "N-best-algorithm".

3.   Pattern and target are added to the current set of training patterns. Retraining of the whole training set follows.

4.   After retraining the previous net is replaced by the actually trained net, the previous net is discarded.

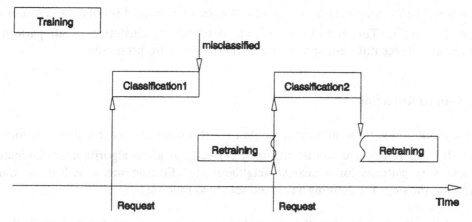

Fig. 3. Interrupted retraining

Adaptation to previously unkown speakers with different pronunciation, other emphasis of words or new dialects is achieved. Figure 4 shows the increase of recognition rates for untrained speakers using the retraining algorithm during three passes. Only one speaker was learned in the initial training phase.

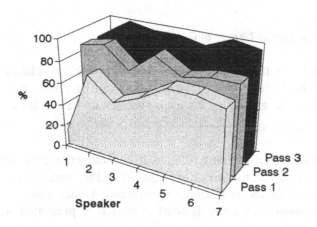

Fig. 4. Recognition rates during adaptation for new speakers

4 Redundancy Elimination

As mentioned earlier, a new pattern is added to the training set each time the network is retrained. The greater the training set, the longer the training times and the greater the needed amount of memory. A method of elimination of redundant pattern or information in the training set is necessary and will thus increase the performance of the system.

Hence aim of the following process is the creation of a reduced set of data ω_r out of a total data set ω. This reduced set should describe the clustering of all patterns sufficiently. Three different approaches to this problem are presented:

4.1 Neuron Reduction

For a given problem the minimum number L_0 of subclusters is defined in a manner, that all training patterns can be classified correctly. A given algorithm to eliminate unnecessary patterns for a nearest neighbour classification was modified for our problem, the steps for defining a reduced set are as follows:

1. Using the training set the network is trained according to chapter 2. The resulting set U_0 of all created subclasses has L clusters or hidden neurons.

2. For l = 1 .. L: Classify all patterns of ω, considering only the clusters $\{U_0 \backslash u_l\}$. If all patterns of the training ensemble ω can be classified correctly, eliminate the cluster u_l in the set ($U_0 \rightarrow U_0 \backslash u_l$), otherwise U_0 is unchanged .

4.2 Retraining starting at Center of Gravity

If the condition holds true, that all subclasses of a class are placed in a single restricted area of the feature space, many of them can be found near their center of gravity. The center of gravity c_k for each subclass u_k has to be calculated before training.

As can be seen from the learning rules, it is convenient to start training with the pattern positioned nearest to the center of gravity and to cancel all patterns inside the region of influence of the corresponding subclass. Under normal conditions this cannot be presupposed. A more general approach is presented in the following chapter.

4.3 Combination of Joined Training Patterns

Let a distribution of subclasses ω, all of the same class, as shown in Figure 5 or similar be given. Only a confined region of feature space is occupied, many overlapping subclasses exist. Then it seems possible to integrate all these overlapping or joined subclasses into one greater subclass. A degree of overlapping v can be defined as follows:

$$v = 1 - \frac{|w_2 - w_1|}{r_2 + r_1} \quad , \quad \begin{cases} v < 0 : \text{non-overlapping} \\ v > 0 : \text{overlapping} \end{cases} \quad (3)$$

This new created subclass is of the same kind as any other subclass, it contains a mean vector $\underline{w}$ and a radius vector $\underline{r}$. The mean vector is placed at the center of gravity of all mean vectors in ω:

$$w_i(\omega) = \frac{1}{J}\sum_{j-1}^{J} w_i(u_j) \qquad (4)$$

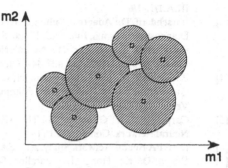

Initially each component of the radius vector is set to a value calculated as shown in Eq. 5:

$$r_i(\omega) = \max_j(r_i(u_j)) \qquad (5)$$

Fig. 5. Example subclass distribution

Thus one new subclass substitutes several subclasses in the set of training patterns. Its region of influence is approximately equal to the union of regions of the subclasses. If any subclass of another class is positioned inside this new region a radii reduction as described in chapter 2.4 has to be performed.

Using the degree of overlapping v as a parameter which subclasses may be integrated into a single subclass has an important advantage in comparision with the method presented in chapter 4.2. If the subclasses of one class are divided into two or more unjoined regions with some overlapping subclusters just the overlapping ones are taken into account.

5 Conclusions

The methods of retraining and redundant pattern elimination have significant advantages in pattern recognition applications.

The results of the presented investigation are:

- adaption to time variable features is possible;
- previously unknown objects, which can be described by the current set of features are trained without further actions necessary;
- redundant patterns in the training set can be eliminated by a cancelling algorithm;
- the run-time for classification and training are minimized;
- the amount of required memory space is minimized.

References

[1]   Barschdorff, D.: Case Studies in Adaptive Fault Diagnosis using Neural Networks. Proc. of the IMACS Annals on Computing and Applied Mathematics, MIM-S2, 3.-7. Sept. 1990, Brüssels, pp. III.A.1/1 - 1/6

[2]   Barschdorff, D.: Adaptive Failure Diagnosis using Neural Networks. 7th Symposium on Technical Diagnostics, Helsinki, Finnland, 17.-20. Sept. 1990

[3]   Barschdorff, D., Becker, D.: Neural Networks as Signal and Pattern Classifiers. Technisches Messen tm 57, 11/1990, pp. 437-444, Oldenbourg Verlag, München

[4]   Barschdorff, D., Cai J., Wöstenkühler, G.: A Learning Pattern Recognition Method for Acoustical Quality Control of Fan Motors. Automatisierungstechnik at 39, 2/1991, pp. 43-48, Oldenbourg Verlag, München

[5]   Carpenter, G.A., Grossberg, S.: The ART of Adaptive Pattern Recognition by a Self-Organizing Neural Network. Computer 21 (1988), pp. 77-88

[6]   DARPA Neural Network Study. AFCEA International Press, Fairfax (Virginia), USA, 1988

[7]   Ersoy, O. K.; Hong, D.: Parallel, Self-Organizing, Hierarchical Neural Networks. IEEE Transactions on Neural Networks, Vol. 1, No. 2, June 1990

[8]   Fukushima, K.: Neocognitron: A Self-organizing Neural Network Model for a Mechanism of Pattern Recognition Unaffected by Shift in Position. Biological Cybernetics 36, 1980, pp. 193-202

[9]   Gramss, T.; Strube, H.W.: Recognition of isolated words by extracting invariant features, NSMS-Symposium, 1990

[10]   Gramss, T.; Strube, H.W.:Recognition of isolated words based on psychoacustics and neurobiology, Speech Communication, Elsevier, North-Holland, 1990, pp. 35-40

[11]   Hart, P.E.: The Condensed Nearest Neighbour Rule. IEEE Transactions on Information Theory 14, 1968, pp. 515-516

[12]   Hartigan, J.A.: Clustering Algorithms. John Wiley & Sons, New York, 1975

[13]   Krause, A.; Hackbarth, H.: Scaly Artificial Neural Networks for Speaker-Independent Recognition of Isolated Words, IEEE ICASSP 1989

[14]   Kohonen, T.; Riittinen, H.; Reuhkala, E.; Haltsonen, S.: On-Line Recognition of Spoken Words from a Large Vocabulary, Information Sciences 33, Elsevier, 1984, pp. 3-30

[15]   Lippmann, R.P.; Gold, B.: Neural-Net classifiers useful for speech recognition, Proc. IEEE First Intern. Conference on Neural Networks, San Diego, 1987

[16]   Lippmann, R.P.: An Introduction to Computing with Neural Nets. IEEE ASSP Magazine, Vol. 4, No. 2, April 1987

[17]   McClelland, J.L., Rumelhart, D.E.: Explorations in Parallel Distributed Processing. MIT Press, Cambridge 1986

[18]   Murphy, O.J.: Nearest Neighbour Pattern Classification Perceptrons. Proceedings of the IEEE, Vol. 78, No. 10, October 1990

A Multiple Paradigm Diagnostic System for Wide Area Communication Networks

B.Pagurek, N.Dawes, R. Kaye

Dept. of Systems and Computer Engineering
Carleton University, Ottawa, Canada. K1S5B6

Abstract. In modern large heterogeneous wide area enterprise networks, fault mamagement requires a great deal of intelligence. This is because of the large amount of monitoring information usually available, and because of the presence of uncertain and missing messages. This paper describes a knowledge-based system that employs multiple diagnostic paradigms in parallel to achieve high accuracy in automated diagnosis. The results achieved in exhaustive testing on a large world scale network are presented.

Keywords: diagnosis, network fault management

1 Introduction

In a complex domain like computer communication networks, human network diagnosticians employ many modes of reasoning to detect and locate faults. It is clear that several reasoning methods will be needed for any automated system to be sufficiently robust and comprehensive. This paper presents a multiparadigm approach to fault management with several cooperating intelligent diagnostic agents. The overall system (Pegasus) was organized along Blackboard architectural lines with the blackboard itself being written in 'C' using UNIX facilities. The cooperating agents used were quite diverse in their approach and capabilities ranging from more conventional graph oriented tools, to quite speculative and experimental tools. The agents cooperating through blackboard management include:

1. An innovative probabilistic belief propagation system that has been extremely useful and successful in diagnosing most of the standard system faults,
2. A rule-based agent using a forward chaining diagnostic reasoning system,
3. A full scale network simulator for system development, testing and performance analysis,
4. Use of the network simulation and a more classical graph based approach in a model-based reasoning component for path and candidate generation and hypothesis testing,
5. A speculative machine learning agent currently using a neural net for pattern recognition and anomaly detection.

All of these agents are in place and have been interfaced to the blackboard. The network simulator is used not only for diagnostic tasks but as a general purpose backbone tool. It simulates the sample network and is used as a surrogate when it is impossible to experiment with the actual network for experimental purposes. Among the other agents, the probabilistic system appears to be the most generally useful and successful tool. The rule-based system implements specialized operator specified diagnostic tasks and currently is the only agent that can request test results. The diagnostic agents do not communicate directly with each other but operate by being triggered into action by blackboard activity and by posting hypotheses to the blackboard after suitable deliberation.

2 The Sample Network

The example network that the prototype was applied to consists of a large corporate wide area network spanning several continents and consisting of many thousands of components. The major components of interest are primarily routers and bridges interconnecting many LANs and workstations. Workstations are connected to bridges which fan into ethernet LANs. Routers connect LAN to LAN by serial links up to T1 speeds. Built in redundancy for fault tolerance coupled with the fact that the routers used dynamic routing sometimes complicated the diagnostic task considerably. Fig. 1 indicates a simplified view of the network connectivity. The width and shading of the links are indicators of the bandwidth while the size of a rectangular node indicates the internal complexity of the node. For example, a large node could represent a metropolitan network with several routers and redundant serial or LAN links.

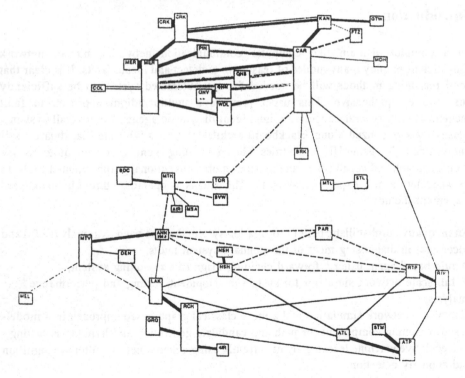

Figure 1. Sample Wide Area Network

The network was organized as an IP/TCP network supporting datagrams and packets, and was managed with SNMP whereby network devices being explored could be queried using special packets. The techniques discussed in this paper however apply to WAN's organized along a wide variety of lines. A hierarchical monitoring network was built on this. In particular four data collector workstations each monitors its own non-overlapping segment of the network, and forwards aggregated status messages to a centralized control station which displays the current status as it sees it using a colour graphical display. The status messages vary depending on how many requests for response failed. The data collectors monitor the router traffic and error rates and are also capable of performing specialized tests where deemed necessary by a network operator.

3 The Diagnostic Problem

Like the Internet, the sample network is an IP/TCP network. However, because of the built in redundancy and multitude of interconnections the diagnosis problem differs somewhat from what has been handled for the internet in for example the "Connectivity Tool"[1]. There the network view consisted mainly of a backbone network with unique gateways into subnetworks and a major diagnostic task appears to be the determination of the appropriate gateways used by packets when a fault is evident. In our sample network, there are many gateways to almost all of the subnets except perhaps the most distant subnet which in the sample network was in Australia. In general therefore, there are very many possible paths or routes between any two nodes (perhaps as many as a hundred or more depending on the nodes). In a recent paper [2], model-based reasoning was explored for a national X.25 network with the conclusion that the technique was not particularly meaningful except perhaps where dynamic routing was a employed. In our case dynamic routing manifests itself somewhat like the problem of reconvergent fanout in model-based diagnosis, where, because of the multiple parallel paths fault masking can occur. Because the system can dynamically route packets around apparent faults or congestion, the fault may seem to disappear until it raises its ugly head during periods of high congestion. Heavy traffic rerouted due to faults can cause overloading elsewhere in the network and make the fault appear to have a different origin. Moreover because of fault masking multiple interacting faults can occur. Multiple faults alone are not unusual because of the size of the network but interacting faults are much less common but much more difficult to diagnose. These complexity problems were addressed in part by using a WAN simulator as a model and for path generation.

Using SNMP and PING the data collectors poll the network devices within their domain. The status messages are not conclusive in pinpointing faults because in most cases the devices involved are not smart enough to announce their exact status. Many status messages indicate only that a device could not be reached for some period of time. This could be due to any number of faults along the route to the device. Moreover status messages are often lost because the faulty network is used to transmit its own management information. The resulting uncertainty motivated the development of the Banes probabilistic reasoner.

A third major problem has to do with the volume of information output by the data collectors, and other sources. This is a typical problem in large systems like nuclear rectors and networks where it is difficult to present a concise picture of the current situation. Error message can cascade and there can be so much "noise" that the network operators have trouble determining when the network is starting to behave improperly. For example there are typically "flaky devices" that are periodically failing and recovering and it is not unusual for lines to exhibit periodic trouble due to temporary environment disturbances or perhaps even uncooperative users of the facility. In such cases, there may not be sufficient evidence in the form of error messages hence it can be difficult for operators to recognize anomalous patterns and to act proactively before the problem becomes very obvious and costly. It is for this reason that the neural network pattern recognition tool was introduced.

SNMP and CMIP for OSI networks make it possible to gather a vast amount of information about the day to day operation of the network. Much of this information could be used to aid in diagnosis but the cost of acquiring and managing this information easily becomes excessive. The requirement for continuous operation also contributes to the buildup of data. The management philosophy of the sample network severely limited the volume of management traffic allowed because it was getting out of hand. This makes the diagnostic task more difficult because an operator often has to try to reconstruct what occurred after the fact with a limited amount of logged data. Nevertheless the volume is still considerable. While the initial operational data was acquired from experts in rule form thereby motivating the use of an expert system, the necessity for handling thousands of interactions per second mitigated against using a pure rule based system. Our rule based component is used therefore to handle special problems and to verify hypotheses suggested by the other diagnostic paradigms.

4 Related Work

Multi-agent sytems for network management have been investigated as for example in the work of Rehali and Gaiti[8]. Their design for managing a campus network involves a set of agents which cooperate in distributed problem solving. While their description of the agents is limited it is clear that the emphasis is on the geographically distributed AI aspects of the problem solving. This has been a component of our investigation because machine intelligence can be distributed throughout the network to reduce overhead traffic and to possibly handle a greater variety of faults. To this end, and to investigate distributed system-level diagnosis[7], we have developed an experimental distributed reasoning tool written in prolog and interfaced to the blackboard. The remote intelligent agents are capable of communicating directly with each other using Unix facilities but they are controlled from a centralized information manager attached to the blackboard. At the time of writing the remote intelligent agents have not yet been fully exploited but are used to gather network statistics from the routers.

Blackboard systems have also been used for fault management in wide area networks. Frontini et al[9] developed an interesting demonstration system using the ART AI environment. A major difference from the implementation point of view is our use of the C language rather than an AI tool. Our system is designed to scale up to real world networks handling hundreds of interactions per second. It is not clear that an ART based system can achieve this. Another major difference is that their knowledge systems (KSs) are organized more as a time progression af agents each being invoked as its turn in the process of solving a given problem occurs. Our KSs represent different problem paradigms and are meant to operate in a more parallel way.

An uncertainty reasoning technique for network fault management has been suggested by Hong and Sen[10] using a belief method inspired by the works of Shafer and Zadeh. The method uses a probability of positive support, a probability of negative support and a calculus for combining probabilities. Their approach reqires a complete causal model of the network and uses a pattern matching procedure applied to network events. Their approach appears to suffer from the computational complexity problems typical of this type of reasoning. It also seems to be still only an idea as no results of any type are reported. BANES on the other hand uses a simpler topological network model, propagates only simple beliefs and is designed to handle the large volumes occurring in a large corporate enterprise network. Moreover BANES is successfully operational as is described later.

5 The Blackboard Architecture

The Pegasus network diagnostic system is organized around a fairly conventional blackboard architecture. Fig. 2 shows the general organization and communication paths of the system. The box labelled KSs represents the knowledge sources listed all of which interact with the blackboard in the same way. The blackboard framework was used to provide centralized communication, a global database, and control for multiple cooperating knowledge systems. In order to ensure real-time speed and the possibility of full integration with the working industrial environment from which the sample network came, the blackboard was written in the 'C' language using the Unix facilities of a SUN workstation.

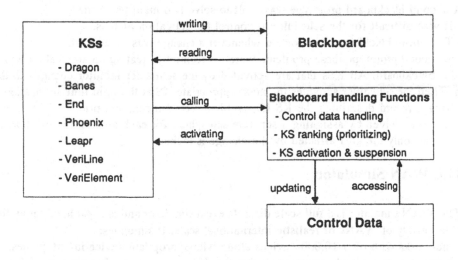

Figure 2. Pegasus' Blackboard Architecture

In the first prototype each knowledge system was implemented as a separate process. The blackboard served as a global database and mailbox using shared memory segments. Communication and access control was managed using semaphores offered in that environment. With this organization KS processes could start or suspend at any time without necessitating a blackboard restart. Unix pipes were particularly avoided in the implementation because of the speed considerations.

The blackboard was divided into several read/write protected areas (e.g. events, problems, hypotheses,.. etc.). Each area consisted of an array of records where processes could post the information they wanted to share with other processes along with some control data structures such as a record of empty records and last changes used to manage the array. Most importantly, a record was kept of knowledge sources interested in the array data so that they could be asynchronously activated when changes of import occurred.

A Knowledge source (KS) can be written in any language as long as it can call C interface functions which provide controlled access to the blackboard data structures. Current KSs include C and Pascal processes but also include one Prolog KS and a forward chaining rule system written in Cosmic Clips. Each KS was assigned a private semaphore on which it could wait for indication of blackboard changes. When such a change occurred, depending on its nature and location, the list of interested KSs was used to selectively unblock KSs by signalling their corresponding private semaphores. The KSs could then look up the change by accessing the last changes list on the blackboard.

In the initial prototype, KSs were being simultaneously developed and operated independently on a test basis. Conflict resolution and scheduling were maintained in only a minimal way and the equivalent of knowledge activation records [3] for KS invocation was not initially implemented. The blackboard was used mainly for communication and its database and not for intricate control. With several KSs in place and full and continuous interleaved operation desirable, reasoning at the blackboard level is being implemented to provide intelligent scheduling of KS resources.

A second blackboard prototype was built to solve two main problems:

1. It was difficult for the scheduler to control the activation of KSs.
2. The entire blackboard was locked whenever a change was made to it.

In the second prototype these problems were overcome by creating an area called KSAR on the blackboard. All KSs that are activated by the scheduler monitor changes to this area. Then the scheduler activates the most appropriate KS at the right time by posting to that area a record describing the KS id., problem record index, and priority. The second problem was solved by assigning a separate semaphore for each area on the blackboard. As a result only the area affected by the change is locked.

6 The WAN Simulator

The WAN simulator is a full scale discrete event simulator and is capable of simulating a wide variety of WANs of realistic international scale. It simulates:
   -networks with over 10,000 devices along with appropriate device in/out queues.
   -workstations as load generators and network servers with packet movement fron device to device including packet drop on timeout,
   -bridges as fan in and fan out devices to workstations,
   -LAN functionality interconnecting bridges and routers,
   -router dynamic routing via adaptive router table updating based on line loads and error rates, and table propagation to neighbours,
   -serial lines of up to T3 speeds with transmission delay and statistical error rates for packet retransmission.

Written in Pascal, it is optimized for efficiency and simulates the large sample network at up to ten times faster than the real world. For the sample network it generates and handles packets in Internet Class Protocols. Data collection is done with SNMP and PING as in the sample network but OSI management standards can be accomodated. All status changes are reported to the blackboard through a program interface.

The simulator is used in two major ways. First of all, since it is impossible to perform many experiments on the real target network the simulation serves to represent the network during the process of development. For this function provision has been made to allow for the dynamic introduction of device malfunction and repair through the blackboard interface as well as for the testing of device functionality. Simulation device data and statistics are available as in the real world.

A second major use of the simulator is as a network model for model based reasoning. In model based diagnosis the discrepancy between observed system behaviour and the model behaviour is generally explained as much as possible by exploring possible faults introduced into the model. It also can be used to enhance the reasoning capabilities of the other KSs through route and candidate generation as well as hypothesis testing. The model is instantiated with current traffic data, and current known faults from the real network so that it follows the real world as closely as possible. Since the proprietary routing strategy of the real network is only simulated and the simulation is not exact in this regard, the simulation model is not perfect as a route generator. What it generates is the most likely current candidate path between a given source and destination given the available network status and recent history. This is very useful information because a classical path generator implemented using classical graph algorithms produces very many alternate routes in this highly redundant network environment. The multiplicity of candidate routes makes it very difficult for a diagnostician to chase down possible faults when there isn't enough concrete information to pinpoint the route in use when a reported access failure occurs. This is not an uncommon problem in diagnosis, so a hierarchical model based approach to reasoning is also being examined to reduce the combinatorial explosion inherent in theses types of problems. This together with the "likely" path generator from the simulation model are under investigation for several classes of problem and should generate fault candidates in a computationaly reasonable way.

7 The Probabilistic Reasoner

The probabilistic reasoner called BANES [4] collects evidence of network faults from the status messages emitted at regular intervals from the data collectors. From the point of view of a data collector the part of the network within its domain appears as an AND/OR tree rooted at the data collector. Using the status message evidence about devices BANES builds up beliefs as to the current device status and propagates these beliefs up the tree toward the data collector accumulating evidence as it goes and combining beliefs from multiple branches at the nodes. Formally speaking, BANES reasons in what is called a nested evidence space. BANES is an effective simplification of much more computationally expensive Dempster- Shafer reasoner operating in the same nested evidence space [5].

Every three minutes BANES issues a report on its current beliefs including an almost certainly broken list and an intermittently failing list. This information is posted to the blackboard as diagnostic hypothesis and can be used as additional information by the other KSs. BANES has proved to be enormously successful as a passive monitor and hypothesis generator. Its main limitations are that it is passive, acting only on status messages and not additional tests, and that it was not designed to respond to end to end faults as would typically be reported in a user's complaint phone call.

8 The Rule-based Expert Network Diagnostician

The rule-based expert network diagnostic KS, called END, is a classical expert system in that it implements human expert reasoning acquired from true expert network operators in the form of rules. To facilitate knowledge acquisition, rule prototyping, and rule addition and editing the rule base was implemented in the shell Cosmic Clips which provides a classical forward chaining inference engine similar to that provided in OPS5. Cosmic Clips itself was written in C and provides excellent facilities for interfacing with other Unix processes. A cosmic clips process can access the blackboard and invoke other processes through prescribed functions or the cosmic clips process can be invoked from blackboard processes. The expert system is meant to be an active KS. For example at the present time it is the only KS that can order tests and respond to the results. It is also the only KS that can respond to specific complaints such as a user calling to say that he cannot reach another network node from his workstation.

In the initial prototype, the expert system implemented several expert diagnostic scenarios exactly as the expert performed them. It accomplished this task through classical sectionalizing techniques which are much like a binary search for a fault on a candidate path. Candidate path generation was accomplished through use of the network simulation acting as a model and within this path components could be examined for operability using for example out of band dialup querying as needed. The expert system could order a test, and then suspend activity until awakened later by the semaphore when the results arrived. The reasoning used in the first prototype was independent of the activity of other KSs. In the target system, the expert system is slotted to operate outside of a purely passive (but still intelligent) monitoring mode. For example END responds to telephone complaints for problems that result from higher level protocol errors that do not cause low level status error messages and are thus hidden from BANES. In its active capacity it is meant to operate and reason in the gaps left by the other passive KSs and to go beyond human expert reasoning which didn't have available the capabilities of the other tools.

9 The Neural Network Analyzer

One of the goals of the target system is that it should be more proactive than is currently possible. That is the system should be able to recognize certain problems early and to react before the problem reaches a critical stage. This is a difficult task because the environment is very complex and there is a great deal of data including status messages, error rates, and performance data to absorb. To be proactive the system must be able to recognise subtle changes in network performance. It is very difficult to acquire such knowledge from expert operators because of the vast amount of experience and intuitive feel that the expert would need to have. Operators can often tell you that the current network operation feels OK but they may not be able to explain exactly why. Moreover, they may say all is well even though there are a number of faults not yet eliminated from the system. In a very large network there may never be a time when the entire network is fault free. Sometimes there can be intermittent nuisance problems that have been around for so long that they are almost considered normal, or there can be other faults considered too minor to respond to at the present time.

The neural network analyzer (LEAPR) is our first attempt at using machine learning to provide intelligent monitoring in the network domain. The basic idea is to train the system what the operator considers to be anomalous or normal behaviour without having to explain why he thinks so, and to use the neural network to recognize such patterns in the future.

Maxion[6] describes an Ethernet anomaly detector based on LAN traffic data as the only input. He used traffic statistics for different key parts of the day to characterize normal activity, and he used a two week moving average to update the statistics. Deviations outside what was considered a normal fluctuation were used to trigger an investigation and to locate faults. Here we are attempting to generalize the idea beyond Ethernet activity alone and to use serial traffic data gathered at routers as the input to the neural nets. We are also attempting to recognise more complex patterns hence the use of the neural net pattern recognizer instead of more simple threshold detectors. Work is currently under way exploring the choice of training data and the recognition effectiveness.

10 Experimental Results

The BANES reasoner has been tested against the simulator and on the live WAN. The simulated data represented six months of network activity with 2653 randomly introduced device break faults. BANES correctly recognized all but 1 (99.96% accuracy) and made 6 false positive diagnoses of devices not broken. As many as 25 faults were present at the same time. It was installed live on the WAN in late December 1991 and since then its performance has been monitored and checked by the human operators. In these first two months of operational use it has been found to make no mistakes in diagnosing broken devices.

The END reasoner has been tested extensively against the simulator and proven to be acceptable and efficient in replicating expert action in several fault scenarios. It has been extended to provide the hypothesis verifiers VeriElement and VeriLine and these two agents now operate as checkers of the BANES diagnosis. When tested on the simulator they correctly remove all the false positives that BANES suggests, but remove none of its correct diagnoses. In other words the concatenation of BANES and the verifiers produces an exceptional level of performance, better than either could produce in isolation. However this performance is restricted to the diagnosis of broken devices.

The neural network tool is now being tested on the simulator at detecting traffic problems such as ARP storms on routers. Often no status message evidence is produced for such faults, only the traffic patterns on the serial lines are abnormal. We used the simulator to generate periods when such faults were active and periods when with no such faults. We then used the traffic patterns corresponding to normal and abnormal conditions to train the neural network to discriminate between the them. This is in contrast to

Maxion's work where no prior knowledge of faulty behaviour was used. Preliminary results indicate the neural network can accurately and efficiently recognize abnormal conditions with a low rate of false positives.

The results so far have been sufficiently encouraging to have motivated further work. Aside from further development of the blackboard control and evaluation of the current components, research will continue in several other thrusts including: an extension to the blackboard with dynamic data structures to allow data area data structures to be extensible while the system is running thereby facilitating the addition of new types of components and the addition of new KSs dynamically, and the feasibility of extending the diagnostic technique to much broader classes of networks e.g. B-ISDN/ATM networks has been examined and we expect to demonstrate this.

11 Acknowledgements:

The authors wish to acknowledge the contribution of J. Altoft, R. Dussault, S. Iqneibi, F. Viens and B. Yuan. This work was supported by The Telecommunications Research Institute of Ontario and NSERC of Canada.

12 References:

1. M. Feridun: Diagnosis of Connectivity Problems in the Internet. Proceedings of the Second International Symposium on Integrated Network Management, Washington D.C. April 1991. pp 691-701, North Holland Publishers
2. B. Lippolt, H. Velthuijsen, E. Kwast: Evaluation of Applying Model-based Diagnosis to Telecommunication Networks. Proceedings of Avignon 91, The Eleventh International Conference on Expert Systems and their Applications, Avignon, June 1991
3. L. Erman, F. Hayes-Roth, F. Lesser, D. Reddy: The HEARSAY II Speach Understanding System. Computing Surveys 12(2):213-253,1980
4. N.W. Dawes, J. Altoft, B. Pagurek: Effective network diagnosis by reasoning in uncertain nested evidence spaces. To appear in Proceedings of Avignon 92, June 1992.
5. J. Altoft, N.W. Dawes: RUNES: Dempster-Shafer Reasoning in uncertain nested evidence spaces. Carleton University Technical Report SCE 91-09
6. R.A. Maxion: Anomaly Detection for Diagnosis. 20th IEEE International Conference on Fault Tolerant Computing, pp 20-27, 1990
7. R. Bianchini, K. Goodwin, D. Nydick: Practical Application and Implementation of Distributed System-Level Diagnosis THeory. Proceedings of the 20th Symposium on Fault Tolerant Computing, pp 332-339, June 1990
8. I. Rahali, D. Gaitl: A multi-agent system for network management. Integrated Network Management II, Proceedings of the 2nd International Conference on Integrated Network Management, North-Holland 1991, pp 469-479.
9. M. Frontini, J. Griffin, S. Towers: A knowledge-based system for fault localisation in wide area networks. Proceedings of the 2nd International Conference on Integrated Network Management, North-Holland 1991, pp 519-530.
10 P. Hong, P. Sen: Incorporating non-deterministic reasoning in managing heterogeneous network faults. Proceedings of the 2nd International Conference on Integrated Network Management, North-Holland 1991, pp 481-492.

Recursive Neural Net Modeling of a Multi-Zone Tenter Frame Dryer

G. Lackner[*], S. S. Melsheimer, and J.N. Beard

Chemical Engineering Department, Clemson University, Clemson, SC, USA

Abstract. A neural net was trained to simulate a single-zone tenter frame dryer over a wide range of operating conditions. The dryer fabric inputs and first zone air temperature were used to compute the first zone outputs of a multi-zone dryer. These outputs (fabric temperature and moisture content) along with the second zone air temperature were then input to the same single-zone model to predict the second zone outputs. This process was repeated until the final dryer outputs were obtained. This recursive modeling scheme provided an excellent simulation of a six-zone dryer over a range of fabric inlet conditions, fabric velocity, and zone air temperatures. This recursive modeling approach should be applicable to other distributed parameter systems, and also to dynamic systems simulation.

1 Introduction

Modern control and optimization methods require accurate mathematical models of the process under consideration. However, the formulation of accurate models can be difficult and requires a high level of technical skill, especially in the case of distributed parameter processes. Moreover, most identification methods are based on linear models, whereas many processes are highly nonlinear. Neural networks provide a new solution to this problem. Neural networks are well suited to modeling nonlinear systems, and are especially useful in the case where the relationship between the input and output variables is not known a priori. Several successful applications in the process industries have been reported [3, 5, 6, 7, 8, 9].

The textile tenter frame dryer is a process of considerable interest in South Carolina and elsewhere due to its high energy consumption [4], and its effects on textile product quality. It is a distributed parameter system, and is quite nonlinear. While differential equation based models are available [4, 10], determination of the empirical parameters involved in the models requires a level of technical expertise that is often unavailable in the textile plant.

This work investigates the application of neural networks to the modeling of textile tenter frame dryers. The specific objective was the prediction of the fabric temperature and moisture profile along the length of the dryer. However, the dryer also serves as a prototype for the study of neural net modeling methods for similar nonlinear, distributed parameter systems.

[*]Institut fuer Systemdynamik und Regelungstechnik, Universitaet Stuttgart, Stuttgart, Germany

2 Tenter Frame Dryers

Figure (1) depicts a typical tenter frame dryer. The fabric first passes through a bath that ensures that the fabric entering the dryer will have uniform moisture content (and may also apply finishing chemicals). The vacuum slot then removes excess moisture from the fabric. The typical dryer is direct fired with natural gas, and has several sections (zones) that are each about 10 feet in length. The temperature in each zone can be controlled independently. Air is exhausted from the dryer by fans located in the exhaust stacks.

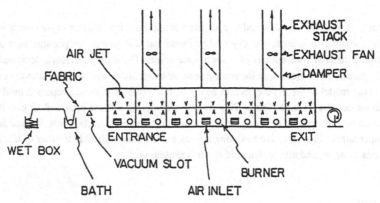

Figure 1. Sketch of a Typical Tenter Frame Dryer.

Viewed rigorously, the fabric drying process involves heat conduction and moisture diffusion through the thickness of the cloth, coupled with transport of the cloth down the length of the dryer. A mathematical model developed by Beard [4] approximates the fabric as a moist interior layer sandwiched between two dry exterior layers. This allows the process to be described by three nonlinear, ordinary differential equations (see Appendix). These equations can be integrated numerically to determine the profiles of fabric surface temperature, interior temperature, and moisture content along the length of the dryer. In addition to physical properties of the fabric and water, the model involves four empirical parameters which are fit using data for a particular fabric and dryer.

This model has been shown by Beard to accurately describe textile dryers, provided that the empirical parameters are properly determined for a given dryer and fabric. Thus, data generated using this model were used as the basis for evaluation of the ability of the neural net to model tenter frame dryers.

3 Neural Network Architecture

The neural net structure used in this work is the feedforward network [2], with training via the backpropagation learning rule [1]. As is common practice, it will be referred to as a "backpropagation network" for convenience. Figure 2 shows a typical neuron with inputs a_1, a_2,....., a_n, and a fixed bias input a_0.

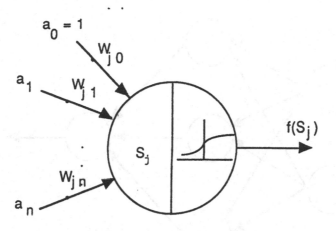

Figure 2. Schematic of a neural net processing unit (neuron).

Equations (1) and (2) describe the calculation performed by the net, given a set of interconnection strengths or weights w_{ji}.

$$S_j = \sum_{i=1}^{n} a_i w_{ji} + w_{j0} \tag{1}$$

$$f(S_j) = \frac{1}{1 + \exp(-S_j)} \tag{2}$$

Note that the neuron output is normalized to a 0-1 range. Thus, the input and output values of the system to be modeled must be scaled within this range. In this work, each input and output was scaled between limits of 0.1 and 0.9.

The structure of a typical backpropagation (feedforward) network is illustrated in Figure 3. This particular net shows three layers; an input layer with two nodes, a hidden layer of three nodes, and an output layer with a single node. The number of hidden layers, and the number of nodes in each hidden layer, may vary. The number of input nodes equals the number of inputs, and the number of output nodes equals the number of outputs. The output of each node is, in general, connected to all nodes in the succeeding layer. The input nodes simply distribute the input values to the second layer (i.e., the outputs from an input node equal its input). In the hidden and output layers, each node is a neuron as described in Figure 2.

In the backpropagation learning algorithm, a training set of input/output vector pairs (or patterns), is presented to the network many times. Each pair consists of an input vector and a corresponding output or target vector, representing the output the system to be modeled would produce if it were excited by the same input vector. After training is stopped, the performance of the network, i.e. the success of the training period, is tested.

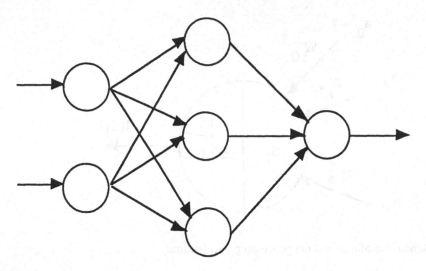

Figure 3. Architecture of a typical backpropagation neural network.

For each presentation of an input/output pair, the backpropagation algorithm performs a forward calculation step followed by a backward propagation of the error. In the forward step the input vector is fed into the input layer and the net computes an output vector using the existing set of interconnection weights. The backpropagation step begins with the comparison of the network's output pattern to the target vector. An error value for each output neuron is calculated based on this difference, and the weights are adjusted for all the interconnections to the output layer. Error values for the hidden layer preceding the output layer are then calculated. The weights of the hidden layer are updated, and the process is repeated until the first hidden layer is reached. Rumelhart and McClelland [1] give a good description of the backpropagation algorithm, and specific details of the algorithm as implemented in this work are given by Lackner [11]. However, it it pertinent to note that the weight adjustment as implemented includes two parameters, a learning rate parameter (η) which scales the response to the current error, and a momentum factor (α) intended to dampen oscillations in the weight adaptation. For step i,

$$\Delta W^i_{kl} = \eta \delta_l f(s_k) \tag{3}$$

$$W^{i+1}_{kl} = W^i_{kl} + \Delta W^i_{kl} + \alpha \Delta W^{i-1}_{kl} \tag{4}$$

These parameters are "tuned" to give rapid convergence [1].

4 Neural Net Modeling of a Tenter Frame Dryer

A large number of variables influence the fabric temperature and moisture profiles of a multi-zone dryer, including

- initial fabric moisture content
- initial fabric temperature
- fabric velocity

- fabric weight
- air temperature in each zone
- air humidity in each zone

Thus, for a six zone dryer there are 16 input variables. Outputs include at least the exiting fabric temperature and moisture content, and quite possibly fabric temperature and moisture at one or more points within the dryer. Thus, a neural network for this system would be large and complex, with many parameters (weights) to be determined. A very large training data set would be needed to cover the possible range of inputs. With the large number of weights, overparameterization (resulting in poor ability to generalize from the training data) would also be of concern. To reduce the net complexity, it was proposed that the problem be reduced to the modeling of a single dryer zone. Simulation of the multi-zone dryer would then be effected by recursive application of the single-zone model.

Initial studies of dryer modeling using neural nets considered, a simple single-zone dryer with only two inputs. The first input variable was the air temperature T_{air}, which is constant over the entire length of the dryer (60 feet). The second input variable was the velocity of the fabric, v. All other variables and parameters were kept constant. The net was trained for all combinations of the following input values:

$$T_{air} \in \{250.,275.,300.,...,400.\} \quad [F]$$
$$v \in \{30.,40.,50.,...,100.\} \quad [ypm]$$

The differential equation model of Beard was used to generate the output fabric temperature and moisture for each of the 56 combinations. A neural net with one hidden layer of five neurons and two output neurons was trained until convergence, which took about one million iterations. The learning rate η was decreased linearly from 2.0 to 0.1 during training and a momentum factor $\alpha=0.4$ was used. The performance of the net was evaluated on the following test data set:

$$T_{air} \in \{260.,280.,300.,...,380.\} \quad [F]$$
$$v \in \{35.,45.,55.,...,95.\} \quad [ypm]$$

The deviation between the actual value of the fabric temperature and the value predicted by the neural net, was always smaller than 3%. However, the prediction of the fabric moisture content at the end of the dryer proved to be poor. Inspection of the results revealed that a broad range of input variable values produces almost the same low fabric moisture output values. Taking the (natural) logarithm of the fabric moisture (x) as target variable instead of x itself makes small changes in low moisture values "more exciting" to the net. With this modification the fit to the moisture content was well within the accuracy of experimental measurements in a textile plant, although the relative error in predicting the fabric moisture remained as high as 20% at very low moisture contents.

The net was then expanded to include the initial moisture and temperature of the fabric as inputs in addition to the air temperature and the velocity. The net predictions of the moisture and temperature of the fabric at the end of a constant air temperature dryer with length equal to the first zone of a multi-zone dryer could then be used as input to a neural net predicting moisture and temperature of the fabric at the end of the second zone, and so on until the last zone of the dryer was reached.

The multi-zone dryer may have zone lengths that are different from the length used to train the net. This does not really present a problem, because the influence of the fabric velocity and the length of one zone of the dryer on the output of the system can be expressed by just one variable, the residence time. The net can either be trained with residence time as the input, or, if fabric velocity is used, the net can be trained for a standard length and the velocity adjusted to give the desired residence time. The latter procedure was used in this study.

The following training data set was chosen to cover the possible range of input variables:

$$T_{air} \in \{250., 275., 300., ..., 400.\} \quad [F]$$
$$v \in \{30., 40., 50., ..., 100.\} \quad [ypm]$$
$$x_0 \in \{0.1, 0.2, 0.3, ..., 1.0\}$$
$$T_0 \in \{70., 95., 120., ..., T_{air}\} \quad [F]$$

The net was trained for two million iterations with a learning rate η decreasing linearly from 2.0 to 0.2 and a momentum factor $\alpha=0.4$. The net structure was 4-20-10-2, the numbers indicating the number of nodes in each layer, beginning with the input layer. Figures (4) to (6) show the results in terms of one dimensional plots of the four dimensional surface learned by the net. Note that all input variables except the one indicated on the x-axis were kept constant at the values

$$v=50.0 \text{ ypm}, \quad T_{air}=250.0 \text{ F}, \quad x_0=0.4, \quad T_0=70.0 \text{ F}$$

The neural net predictions of the fabric moisture and surface temperature match the model output almost exactly. Only the net predictions of the effect of initial fabric moisture content on fabric surface temperature show significant deviations.

As noted earlier, the tenter frame dryer model of Beard consists of three differential equations for the fabric moisture, the fabric surface temperature and the temperature of the interior layer, respectively. The temperature of the interior layer is an important variable, permitting the description of a system which is actually two-dimensional in terms of a one-dimensional model. Using only the the fabric moisture and surface temperature as inputs to the neural net is sufficient for a single-zone dryer because the interior temperature is assumed to be equal to the surface temperature at the dryer entrance. However, for all following zones the initial temperature of the surface layer and the interior layer will be different. As might be expected, a net trained with no information about this difference was not successful in modeling a multi-zone dryer by the recursive approach.

The solution to this problem was to present to the net the interior layer temperature as both an additional input and output during training. To avoid having the training data set become too large, the range of the additional input variable, T_i, was confined to $T_c > T_i > T_e - 50$ F. This restriction was based on inspecting the results of simulations using the differential equation model. Exceptions to this rule are not impossible, but are unlikely to be encountered under normal dryer operating conditions. Therefore, the training data set was expanded to include three or four values of interior layer inlet temperature within the indicated range.

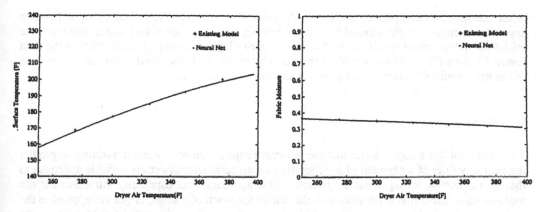

Figure 4: The effect of dryer air temperature on fabric temperature and moisture (single-zone).

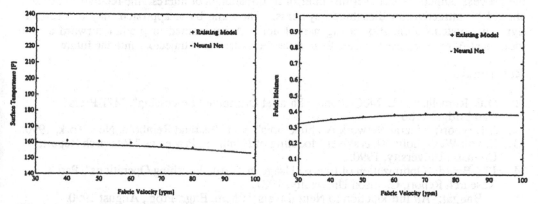

Figure 5: The effect of fabric velocity on fabric temperature and moisture (single-zone).

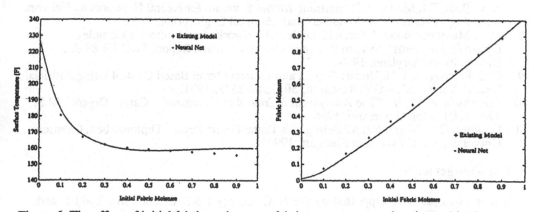

Figure 6: The effect of initial fabric moisture on fabric temperature and moisture (single-zone).

After training was completed, the net (5-20-10-3 structure) was used to simulate multi-zone dryers by recursive application of the single-zone net for a number of test cases. Note that the initial temperatures of the fabric surface and interior layer were equal at the inlet to the first zone. Figures (7) to (9) show typical results. The match of the net prediction with the output of the differential equation model proved to be quite satisfactory.

5 Conclusions

The results of this study indicate that the recursive application of a neural net trained to predict the output values of a segment of a larger distributed parameter system is a viable approach to the neural net modeling of such systems. This approach was especially attractive for the multi-zone tenter-frame dryer since the alternative approach of a single larger net applied to the entire system would have considerably increased the number of net inputs. Moreover, the recursive approach provides interior values of the variables of interest (fabric temperature and moisture in this case), thus giving a profile of the variable along the process rather than only the process outputs. Whenever this interior information is of interest, the recursive approach should be attractive for distributed systems. The same basic approach may be useful in dynamic systems simulation using neural nets. A net trained to predict forward a time increment Δt could be used repeatedly to trace the system time trajectory into the future.

References

1. D.E. Rumelhart, J.L. McClelland: "Parallel Distributed Processing". MIT Press, Cambridge, 1986.
2. J. Dayhoff: "Neural Network Architectures". Van Nostrand Reinhold, New York, 1990.
3. E. von Westerholt: "Neural Net Modeling of Nonlinear Processes." Research Report, Clemson University, 1990.
4. J.N. Beard: "Optimization of Energy Usage in Textile Finishing Operations", Part 2. Research Report, Clemson University, 1981.
5. P. Bhagat: "An Introduction to Neural Nets". Chem. Engr. Prog., August 1990.
6. J.F. Pollard, M.R. Broussard: "Process Identification Using Neural Networks". Presented at the 1990 AIChE Annual Meeting.
7. N.V. Bhat, T.J. McAvoy: "Determing Model Structure for Neural Networks by Network Stripping". Submitted to Computers and Chemical Engineering, 1990.
8. M.L. Mavrovouniotis: "Hierarchical Neural Networks for Monitoring Complex Dynamical Systems". System Research Center Technical Report, SRC TR 89-95, University of Maryland, 1989.
9. D.C. Psichogios, L.H. Ungar: "Direct and Indirect Model Based Control Using Artificial Neural Networks". I&EC Research, 30, 2564-2573, 1991.
10. Robertson, A. E., Jr.: "The Analysis and Control of Commercial Carpet Dryers". M.S. Thesis, Clemson University, 1988.
11. Lackner, G.: "Neural Net Modeling of a Tenter Frame Dryer". Diplomarbeit, Clemson University and Universitaet Stuttgart, 1991.

Acknowledgements

This work was partially supported by the S. C. Energy R&D Center. Mr. Todd Brandes, Ph.D. student at Clemson, wrote the neural net computer code used in this study.

274

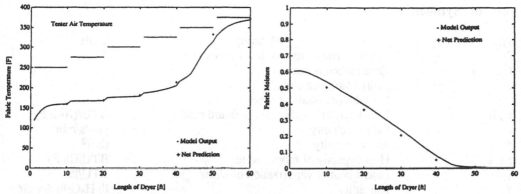

Figure 7: Comparison of net prediction and model output for fabric temperature and fabric moisture (v=30.0 ypm, x_0=0.6, T_0=80.0 F).

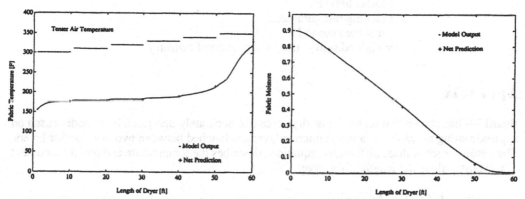

Figure 8: Comparison of net prediction and model output for fabric temperature and fabric moisture (v=50.0 ypm, x_0=0.9, T_0=100.0 F).

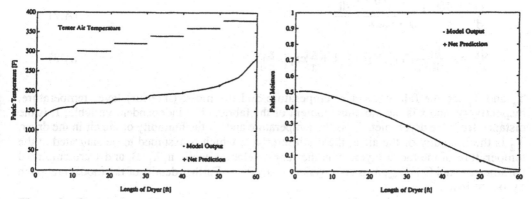

Figure 9: Comparison of net prediction and model output for fabric temperature and fabric moisture (v=70.0 ypm, x_0=0.5, T_0=75.0 F).

List of Symbols

Symbol	Meaning	Units
T_e, T_i	Fabric surface, interior layer temperatures	F
T_{air} or T_a	Zone air temperature	F
x	Fabric moisture content	
l	Length coordinate of dryer	ft
h , h_1	Heat transfer coefficients in Beard model	BTU/(ft^2 h F)
v	Fabric velocity	yards/min
ρ	Fabric density	lb/ft^2
c_{pf}, c_{pw}	Heat capacity of fabric, water	BTU/(lb F)
λ	Latent heat of vaporization of water	BTU/lb
Y	Humidity	lb H$_2$O/lb dry air
τ, B	Parameters in Beard model	
w_{ij}	Weights between neurons i and j	
a_i	Input to neuron i	
α	Momentum factor	
η	Learning rate parameter	
δ_l	Error for neuron l	
S_j	Weighted sum - state of activation of neuron j	

Appendix

Beard [4] has shown that tenter frame dryers can be accurately described by a model based on approximating the fabric as a moist interior layer sandwiched between two dry exterior layers. The resulting set of three differential equations describes the temperature and moisture content of the fabric along the length of the dryer:

$$\frac{dT_e}{dl} = \frac{h(T_a - T_e) - h_1(T_e - T_i)}{180 v \rho c_{pf}} \tag{A.1}$$

$$\frac{dT_i}{dl} = \frac{\frac{h_1}{180 v \rho}(T_e - T_i) + \lambda \frac{dx}{dl}}{c_{pf} + c_{pw} x} \tag{A.2}$$

$$\frac{dx}{dl} = \frac{h_1}{180 B v \rho}[Y_1 - Y_2[1 - \left(1 + \frac{x}{\tau}\right)\exp\left(-\frac{x}{\tau}\right)] \tag{A.3}$$

T_e and T_i are the fabric surface temperature and the moist (interior) layer temperature, respectively, and x is the moisture content of the fabric. The independent variable, l, is the distance from the dryer inlet. T_a is the temperature and Y_1 the humidity of the air in the dryer. Y_2 is the humidity of the air at the cloth interface, which is assumed to be saturated at the temperature of the moist layer. v is the fabric velocity, while h, h_1, B, and τ are empirical parameters, and C_{pf}, C_{pw}, λ, ρ are physical properties. Further details of the model are given by Beard [4].

Representation of the Design to Manufacture Methodology of the Armour Protection Element of the Fibre Optic Submarine Cable Within an Expert System

Darren Bayliss, Sharon Berry, David Curtis, Bernard Cox

STC Submarine Systems & Portsmouth Polytechnic

Abstract. A collaborative program of work was set up between STC Submarine Systems Ltd. (now part of the Northern Telecom group), and the School of Systems Engineering at Portsmouth Polytechnic. The aim of the work has been to establish methodologies (not necessarily new ones) which can be used to formalise methods of design, and to apply these to current design problems within the company. This also includes the representation of these methodologies through practical applications within an expert system tool. We also wish to show how formal techniques can be used to uncover areas of missing knowledge.

1 Product Background

The application of undersea cable for the purpose of establishing telegraph links has existed since the mid-nineteenth century with the first undersea cable being installed in 1850. Since that day, the industry has gone through many changes, notably with the advance of fibre optic technology. One important factor is that although the way in which the signal transmitted has varied, specifically due to material technology, the basic design concepts still exist. That is, the cable must have certain strengths and protection properties, which have been evident from the very early designs. A revolution in design philosophy has coincided with a change in the medium of transmission. The fundamental characteristics of the product follow this evolution modified by problems encountered in service and by customer expectations. Optical cables for transoceanic cable systems have been under development for over a decade with many successful installations. Cable for submarine systems must be designed for the appropriate seabed conditions [1]. The majority of a typical long haul system is in deep water, greater than 2 km. Here the cable is subjected to high hydrostatic pressures, but enjoys a generally benign environment. There is little risk of damage from external agencies. The depth of deployment does mean, however, that tensions in the cable during laying and recovery can be high. It is important that the cable is designed so that the strains in the fibre are sufficiently low that fibre failure will not occur. For the shallow water continental shelf regions of submarine cable routes the major hazard to cables is from attack by fishing vessels and other shipping. In particular, trawl gear which uses a large steel beam dragged across the seabed, presents a major hazard to cables. It is therefore necessary to protect optical cables from breakage or excessive strain caused by damage from such trawlers. Where possible, cable burial is increasingly favoured as a method of protection. However 100 percent burial is not normally possible and it is necessary to design for trawler resistant cable for those regions.

1.1 Requirements of Submarine Cable

The cable is designed to be used (once deployed and tested) for twenty five years. The cable must be deployed without damage, and if damaged whilst in operation, recovered without excessive strain to the fibres. As discussed above, it must be engineered to its operating environment, although not over engineered. i.e. there is no economic or feasible way to design a totally damage resistant cable. The cable must be designed and manufactured to known requirements, physically and functionally and problems, if possible, must be designed out prior to test. Beyond this it must be delivered on budget, and operate under that budget (minimum faults whilst system is in operation).

2.0 Methodologies

The approach that is used to develop practical models for a device has consequence for any future work on that device. There are two methodological theories which may be adopted: either prototyping or structured. Prototyping is the development of a working model of the whole problem area or some subset of it, this model then forms the basis for all future design and implementation activities. The other approach formal through structure. This is characterised by the production of models utilising the concepts of functional decomposition,[2] (the splitting of a complex problem into a number of smaller sub-problems, and those into simpler sub-problems still until the resultant low level problems are of a straightforward and tractable nature).

There are undoubtedly many merits to both, and the approach adopted is based on the latter. As with most successful techniques it is how they are applied to practical problems rather than what capabilities they enhance that makes them used, [3].

3.0 Description of Technique Used

The representation of a device involves theories about languages for representing structure, commitments for representing behaviour and kinds of causation needed to represent behaviours. This can be expressed within knowledge of purpose and knowledge of structure. For this refer to an abstract model for the levels of knowledge representation (Fig1), Milne[4]. From each knowledge representation it is possible to derive the next higher level of representation. From the lowest level, the physical level, which consists of the component parts of the actual device it is possible to develop a behavioral model. This uses the physical simulations to produce causal sequences of how the system should work. The next level of abstraction involves a functional model which is used to organise the causal sequences to produce mental simulations. The intentional behaviour can then be compared to the intended behaviour. The behavioral model, predicted by an envisionment graph is used to relate behaviour states and their transitions. This is a model in which the nodes represent change in state. When a value is generated for a node, the device is allowed to change state. This can update components which are connected and update the systems of which they are a part.

4.0 Design Practices

It is important to understand the engineer's approach of applying experience to the solving of problems. This can be in the form of well understood and documented techniques or heuristic knowledge. The role of the knowledge engineer is to acquire this knowledge, to interpret and improve methods and to resolve regions of missing data. Standard knowledge acquisition techniques have been employed but will not be discussed further in this paper. It is worth briefly elaborating on the subject of missing data. There are two areas in the design-to-manufacture process, (excluding quality which is in-built). The development which involves the design of cable for in-service operation, and manufacture with the role of optimising these designs for production. It is often the case that certain areas in this process are missed. It is important to fully grasp this process when considering solving problems. There will always be sectors of missing knowledge but by applying formal techniques it is possible to eliminate many of them. It is necessary to recognise the process generically and to develop models that represent this concept. An example of this approach is discussed later.

5.0 The Application Field.

As discussed earlier, it is important that the domain of application is of current issue. It is not enough to update technical capability. The domain, the design-to-manufacture of the armour protection element of the submarine cable, is of current on-going interest, (Fig2). There is a rich body of knowledge in the area, but no structure exists for applying this knowledge due to the fact that it is held by a wide range of people.

5.1 The Problem

One of the main design sectors of undersea cable is the protection from the environment and seabed hazards such as shipping. The form of protection consists of steel wires wound in a helical fashion,[5]. The wires form a barrier from natural hazards such as falling rocks and hazards caused by fishing activities, specifically beam trawlers.
There are constraints which fall upon the design-to-manufacture process which limit the protection in use. There is no allowance for analyzing the process wholly and hence certain areas in the process are never evaluated. It has been our intention, by using formal techniques, to bring these areas to light.

5.2 Design-to-Manufacture

For organisational reasons departments within industry are given specific roles. Development engineers design cables and test their designs for operation, meeting well documented and understood prerequisites which are primarily based on geometric and mechanical constraints. The process engineer optimises the plant for given designs. What should be recognised is that the problems of design are the problems of both development and manufacture and hence a common approach to their solution must be sought. The problem is one of total purpose, with the process having a common purpose with design. By using the methodology discussed we have developed models which represent physical and functional structure. These models take on board the concepts of development and of manufacture.

The existing process hides information from the engineer. Data is sent in fixed form with the receiver using this data to optimise their particular section. The data received is processed optimally, but the hidden data, the possible variation is never received. This means that although the sender and receiver may be optimising their particular set of attributes, the common goal is never considered and hence can never be solved. By applying the techniques discussed it is possible to arrive at an integral solution, thus highlighting key regions where there is possible room for improvement.

A problem region identified through using the above formal approach is the nature of the link between development and manufacture. Traditionally, the developed design has been used by the process engineer as a fixed item with the process using look up tables to find machine set up details. The problem with this approach is that design attributes considered are not particular to development or to process but are global. It is this problem which has restricted the possible solutions.

6.0 Representation

A practical approach was required for the representation of the methodology. The pro's and con's of expert system tools are not discussed in this paper but the tools used to develop the application are. The whole solution involved the integration of a number of software tools, working in harmony under the MS-Windows environment.

6.1 Expert System Toolkit

The tool used, Kappa-PC, provides an object based representation system, forward and backward chaining and an interface that lets the developer browse through rules and data in graphical form. It also provides facilities for the development of a user interface, which can be isolated from the development side.

The tool provides an object representation of data. Each object has a name and a number of name/value pairs called slots and slot values (attributes). For example, an object to represent a Double Armour, might be A65F65 and have slots material, type and lay length, with attributes Grade 65, AF and α. Common properties (slots) shared by a number of objects can be gathered into a class, and members of the class inherit all properties of the class. For example, material, type and lay length can be gathered into a class called Double Armour, and A65F65 be declared a member of that class.

Like most expert system tools, decision knowledge is represented as If ... Then rules (Fig3). If the three conditions are true, the Depth is in the region of 0_300 (qualitative), there are no shelf break areas and the fishing activity in the area is light, then the set of consequences set, A65, OneArmour, OneArmourA65 and A. The Object:Slot (NL_Cable:ArmourTypeA65), is where the value i.e. A65 is sent to. The tool has global object which allows similar rules to be grouped. This aids the system, if inferencing within a particular set of rules. The global object facility builds in structure when considering sub-goals (purpose of sub-models).

A useful feature of the tool is a method (demon). The developer defines a series of actions to take place when the value of a property changes (within a slot), either as a result of inferencing or from user input. This is useful for forcing actions when a conclusion has been determined.

As discussed the tool is windows-based with a graphical mouse orientated interface. Most of the windows within the tool represent knowledge graphically. Of particular interest is the rule relations which presents a picture of the IF ... Then dependencies of any rule chosen. The tool also provides verification for the knowledge base by studying its structure. Other interface tools include mouse based forms for creating functions, rules and goals and an index window for selection. Another useful feature is that the user interface can be isolated from the development side. In that, the developer menu's can be removed and the shrink and icon options can also be removed, leaving the end user solely in the interface domain. This prevents the user entering the underlying code. This also ensures the integrity of the developed application. Generalising, the tool has adapted to the overall solution within the windows environment and has proved an effective tool for representing the methodologies discussed.

7.0 Solution

Most expert system tools, including Kappa, are not stand-alone applications. They are embedded in other systems, drawing data from various sources and interacting with other software tools. Commercial expert system tools must be embeddable. The embeddability is achieved through a library of functions, that can be linked into the user program. Users can control the inference engine, manipulate the knowledge base and invoke the graphical interface. This feature can be used to bring new data into the expert system.

In the application area a number of software tools have been used, all compatible and integrated into the final system. The object orientated approach has continued with using Visual Basic, Guide (hypermedia), graphics tools and databases. This version of basic is easily integrated into the expert system, and of great importance was the ability to work in the background while the user remained inside the expert system. An important area, not normally fully realised is the Help facility. The structure of the user interface screens has been simulated within a hypermedia package. It has been possible to use an object approach to represent text and graphics. Objects are used to link information together, thus data can be accessed in a similar manner to the use of contents and index of a book.

A common approach has been applied to the development of the user interface screens. The windowing system and Kappa interface tools have helped provide consistency. Interactive areas have been given a common feel through the use of standard features such as radio buttons and the positioning of commonly used functions. The entire solution aimed to achieve three things, ease of use, efficiency and consistency, [6].

8.0 Conclusion

It has been established that by using a formal approach to design methodology, it is possible to apply a holistic solution to the representation of design problems. This has helped us uncover areas in the design-to-manufacture process which provide greater flexibility in design analysis. It has also been established that through the use of object orientated tools at the core the expert system toolkit, it is possible to represent these techniques in a practical fashion. The expert system Kappa (working under the umbrella of the windows environment) has proved to be a suitable tool for implementing the methodology discussed, although, being personal computer based, it is restricted in power. The object orientated approach and the hierarchical connection of objects within Kappa have provided adequate scope for representing the physical and functional structures. The use of objects has been successful for representing knowledge both in terms of the expert system and ancillary tools. The importance of the user interface has also been touched upon in terms of the usability of the developed application.

References

[1]. P.Worthington, "Cable design for optical submarine systems", J. IEEE, Selected areas in communications, vol.sac-2, no.6, pp. 873-878, Nov. 1984.

[2]. J.Britt, "Structured knowledge engineering", Technical report:91/8, Portsmouth Polytechnic, pp. 2-3, Feb. 1991.

[3]. R.Milne, "Model based reasoning: The application gap", J. IEEE Expert, Dec. 1991.

[4]. R.Milne, "Strategies for diagnosis", IEEE Transactions on Systems Management, May/June 1987.

[5]. M.Takada, K.Sanjo, M.Kameda, "Studies on submarine cables with high resistance to external damage", Sumitomo electric technical review, no.21, pp. 52-68, Jan. 1982.

[6]. K.Taylor, "IBM systems application architecture:common user access from first principles", J. Computing and control engineering, pp. 123-127, May. 1990.

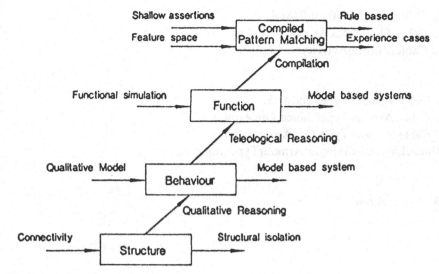

Fig.1. Levels of reasoning

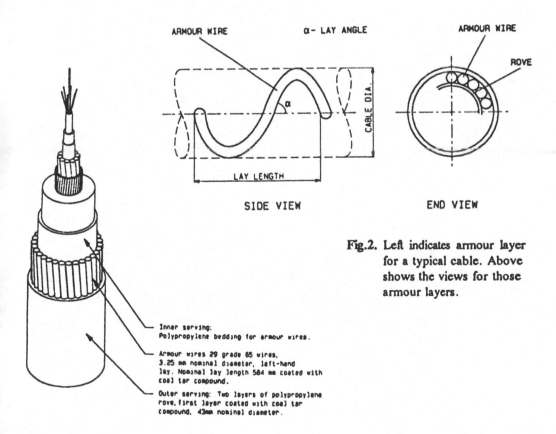

ARMOUR WIRE

α- LAY ANGLE

ARMOUR WIRE

ROVE

CABLE DIA.

LAY LENGTH

SIDE VIEW

END VIEW

Fig.2. Left indicates armour layer for a typical cable. Above shows the views for those armour layers.

Inner serving:
Polypropylene bedding for armour wires.

Armour wires 29 grade 65 wires,
3.25 mm nominal diameter, left-hand
lay. Nominal lay length 584 mm coated with
coal tar compound.

Outer serving: Two layers of polypropylene
rove, first layer coated with coal tar
compound, 43mm nominal diameter.

If   (NL_Cable:Depth#=0_300)And
            (NL_Cable:ShelfBreakAreas#=No)And
            (NL_Cable:FishingActivity#=LightFish);

Then   {
            NL_Cable:ArmourTypeA65,A65;
            NL_Cable:ArmourTypeChoice,OneArmour;
            NL_Cable:ArmourType,OneArmourA65;
            OptimiseLayLengthSingle:ArmourTypeSingle,A;
            };

        Fig.3. A typical rule.

PIM:
Planning In Manufacturing - CAPP using Skeletal Plans

Ansgar Bernardi, Christoph Klauck, Ralf Legleitner[†] ,

ARC-TEC Project
DFKI GmbH, Postfach 2080, D-6750 Kaiserslautern, Germany
Telefon: +49631/205-3477, 205-4068
Fax: +49631/205-3210

Abstract. In order to create a production plan from product model data, a human expert thinks in a special terminology with respect to the given work piece and its production plan: He recognizes certain features and associates fragments of a production plan. By combining these skeletal plans he generates the complete production plan.

We present a set of representation formalisms suitable for the modelling of this approach. When an expert's knowledge has been represented using these formalisms, the generation of a production plan can be achieved by a sequence of abstraction, selection and refinement. This is demonstrated in the CAPP-system PIM, which is currently developed as a prototype.

The close modelling of the knowledge of the concrete expert (or the accumulated know-how of a concrete factory) facilitate the development of planning systems which are especially tailored to the concrete manufacturing environment and optimally use the expert's knowledge and should also lead to improved acceptance of the system.

1 Advantages of knowledge-based systems - Problems in CAPP

Systems to support CAPP encounter a particular problem which arises out of the structure of this domain: In order to create a successful process plan it is not sufficient to rely on theoretically founded rules - which normally can be coded into a program quite easily - but it is also necessary to consider particularities of the production environment and special experience of human process planners. The latter is very often difficult to formalize and consequently hard to be coded, the former may change radically between different working environments and leads to a high maintenance workload.

Under these circumstances the knowledge-based programming methodology may be used: If it is possible to define domain-oriented languages facilitating the explicit representation of the knowledge in question and to provide adequate execution mechanisms for these languages the problem becomes treatable. Suitable representation languages lead to easier representation of the experience and environment-dependent knowledge. The execution mechanisms are independent of the environment and probably useful in a broader domain, thereby reducing the general maintenance workload.

In summary, knowledge-based programming may lead to higher flexibility of the system, easier adaptation to changing environments, reduced maintenance costs, and successful solutions of some problems which can't be tackled with standard techniques.

[†] Cooperation with CCK (CIM-Center Kaiserslautern)

Several projects currently try to use knowledge-based methodologies in this context (cf. the ESPRIT projects IMPACT, PLATO, FLEXPLAN [6]). In the project ARC-TEC at the DFKI Kaiserslautern, these general ideas where exemplarily applied to the automatic generation of work plans for manufacturing by turning, resulting in the prototypical system PIM (Planning In Manufacturing). The developed methodology is of interest for CAPP in general.

2 CAPP from the viewpoint of knowledge-based systems

In order to create a knowledge-based CAPP system, we examine the way a human expert solves the task in question. This leads to the following general model of a process-planning method:

The expert is given the description of the work piece. This description consists of all geometrical and technological data which are necessary for the generation of the process plan.

In this description the expert identifies characteristic parts or areas of interest, the so-called application features. The exact form of these features is influenced by his manufacturing environment (e.g. available tools) and by his personal experience. The expert associates the application features with generalized fragments of a production plan (the so-called skeletal plans). The combination of application features and associated skeletal plans represent the experience of the expert. (An expert may be defined as someone who knows "what to look for" and, having found this, what to do about it). It is important to realize that this observation implies that the application features and the skeletal plans depend on the concrete expert as well as on the concrete working environment and may be different for another expert or another environment.

Having found the features of a given work piece, the expert selects the associated skeletal plans out of his memory (or out of an existing plan library). By combining these partial plans, bearing in mind some general principles, and by adapting them to the concrete work piece, he creates the complete production plan. [14]

We present a set of representation formalisms which allow to model this approach very closely. Based on TEC-REP (TEChnological REPresentation), a general representation formalism for geometrical and technological information about the work piece, an expert's application features are defined. They are described using the language FEAT-REP (FEATure REPresentation) and represent his personal terminology. Skeletal plans (abstracted plans or fragments of plans), represented in the hierarchical formalism SKEP-REP (SKEletal Plan REPresentation), are associated with the features.

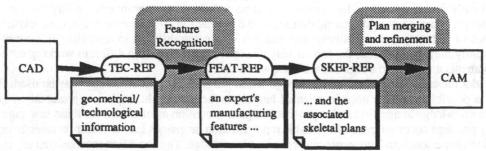

Fig.1. Basic idea and representation languages

Together with a execution mechanism which mimics the sketched method of the human expert, a flexible and reliable CAPP system is possible. [3] The following chapters describe the different representation languages which were developed in the ARC-TEC project according to figure 1.

2.1 TEC-REP

The representation formalism TEC-REP [4] provides the necessary constructs to describe the geometrical and technological information of the work piece. The description must contain all information about the work piece which an expert may use to create the process plan. In order to facilitate the necessary deductions the representation formalism should use symbolic expressions. As shown in a later chapter the TEC-REP formalism is well connected to the CAD-world.

In TEC-REP, the geometry of a work piece is described by surface primitives. These predefined primitives include simple expressions for rotational symmetric parts, which are of special interest in the domain of manufacturing by turning. To be as universal as possible, primitives for non-symmetric surfaces also exist. The extensions of any surface are specified using a cartesian coordinate system. The concrete dimensions are specified for each work piece. Every surface of a work piece description can be identified by a unique identification number.

TEC-REP realizes a simplified boundary representation of the geometry of a work piece. While some surfaces like free formed surfaces which aren't considered in the ARC-TEC project currently can't be expressed, it is possible to expand TEC-REP to cover such cases e. g. by including primitives for surfaces described by B-spline-functions.

The reason to use a surface-based description method is simple: Every manufacturing operation always influences surfaces of an object, therefore the representation of these surfaces seems reasonable.

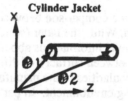

Cylinder Jacket

CJ(reference_point1: <vector>,
    reference_point2: <vector>,
    radius: <num>,
    direction_of_material: [+,-])

Shape Tolerance DIN 7184

ST(surface_number: <num>,
    tolerance_type: [SN, EN, RN, C, LD, PD],
    tolerance_size: <num>)

Circular surface

C(reference_point: <vector>,
    direction_vector: <vector>,
    radius: <num>,
    direction_of_material: [+,-])

Surface Finish

SF(surface_number: <num>,
    surface_finish: [Rt, Rz, Rp, Ra,...],
    value: <num>)

Fig 2. Some TEC-REP Primitives

By using surface primitives, TEC-REP connects the basic data level with the world of symbolic reasoning. With respect to the intended deductions the surface primitives are augmented by information about the direction of material. A circular surface can represent the outer surface of a cylinder as well as a hole in some material, depending on the direction of material.

To facilitate reasoning about topological connections, the neighbourhood of surfaces is expressed in a separate primitive.

The technological information is expressed by attributes to the surfaces. Some attributes (like surface finish or ISO-tolerances) belong to one surface, others connect different surfaces (e. g. length tolerances between different surfaces).

In summary, TEC-REP allows the description of every work piece we currently deal with by using a symbolic and attributed boundary representation.

2.2 Features and FEAT-REP - a closer look

The goal of the feature representation formalism is to allow the expression of a concrete expert's knowledge about characteristic aggregations of surfaces. When CAD/CAM experts view a workpiece, they perceive it in terms of their own expertise: Complex aggregations of surfaces are described by simple names, because the expert associates knowledge with these aggregations. Consequently, we define a *feature* as a description element based on geometrical and technological data of a product which an expert in a domain associates with certain informations (see also [9] or [10]). Features provide an abstraction mechanism to facilitate the creation, manufacturing and analysis of workpieces. They are firstly distinguished by their kind as

- functional features, e.g. *seat of the rolling bearing* or *O-ring groove*,
- qualitative features, e.g. *bars* or *solid workpiece*,
- geometrical (form-) features, e.g. *shoulder, groove* or *drilled hole*,
- atomic features, e.g. *toroidal shell, ring, shape tolerance* or *surface finish*.

and they are secondly distinguished by their application as

- design features, e.g. *crank* or *coupler*,
- manufacturing features:
    - turning features, e.g. *shoulder* or *neck*,
    - milling features, e.g. *step* or *pocket*,
    - drilling features e.g. *stepped-hole* or *lowering*,
    - ...
- ...

Features are domain-dependent. This can be illustrated by a comparison between features in the domain of CAPP and features in the domain of design. While the latter (which are investigated in many modern CAD-systems) represent higher-level informations about the ideas of a designer and describe things like functional relations between surfaces of different workpieces which touch each other, the former depend on particularities of the manufacturing process and are influenced e.g. by the available manufacturing environment. In particular every feature e.g. in manufacturing will be defined by a respective expert because his area, like machines, tools or the characteristics of them, and his ideas, creativity and experience, like special tricks, are included in this definition. A prominent example of such a feature in manufacturing is a insertion of certain dimensions which shall be manufactured using a certain insertion-tool. In a manufacturing environment where this tool is not available this particular insertion may be recognized as a groove or may be not recognized as a feature (figure 3).

In this sense the features can been seen as a *language* of an expert in a domain. Like every language, it is build upon a syntax (geometry and technology) and a semantics (e.g. skeletal plans in manufacturing, functional relations in design). It is important to note that this language represents the *know-how* of the expert respectively the machine shop and that this language is an individual one ("expert in a domain" dependent).

In the domain of CAPP, each feature is associated with knowledge about how the

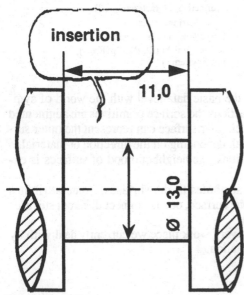

Fig 3. Insertion

feature should be manufactured. If the features embedded in the geometrical/technological description of a given work piece can be recognized, the retrieval of the associated knowledge is a big step towards the generation of the process plan. From this point of view feature recognition forms a major component of the CAD/CAM interface for CAPP [5]. We concentrate on the recognition of geometrical and qualitative features; the functional features are important for design only. Working with manufacturing features means to recognize these features from the CAD data to generate a working plan. Working with design features means to construct by means of these features and to expand them to the CAD data. (cf. [9])

In our paper [9] we have shown that it is possible to describe features by means of formal languages via attributed node-label-controlled graph grammars (ANLCGG's). The area of formal languages is a well established field of research and provides a powerful set of methods like parsing and knowledge about problems, their complexity and how they could be solved efficiently. The use of formal languages for feature descriptions facilitates the application of these results to the area of feature recognition and CAPP. As result ANLCGG's enables a user to define his own feature-language containing complex features and makes feature recognition a parsing process for workpiece interpretation.

To show the usability of our high-level-representation language FEAT-REP the FEAT-PATR-System was implemented as a prototypical part of PIM, by adopting a chart parser for our application in mechanical engineering. Input of our FEAT-PATR-System is a workpiece description in TEC-REP. Input is also the expert's feature knowledge about the workpiece, represented in a grammar. Output of the system is a feature-structure as shown in figure 4.

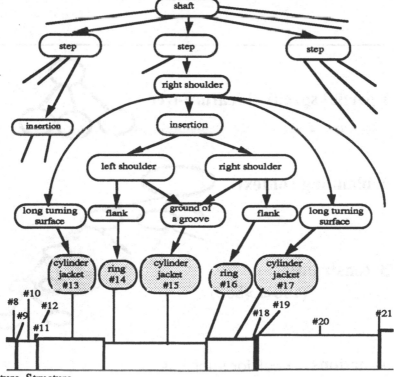

Fig 4. Feature Structure

2.3 Skeletal Plans

To combine the expert's knowledge about the manufacturing process with feature structures as shown above we use skeletal plans [12]. The skeletal plan representation formalism SKEP-REP allows the expert to write down his knowledge about the process necessary for the manufacturing of his workpieces and for special parts of this workpieces (features). It also allows the expert to define how partial (skeletal) plans for special parts of workpieces (features) should be merged to complete manufacturing plans.

The idea of skeletal plans is straight forward: Represent knowledge about the intended solution in some partial (skeletal) plans. Find a first skeletal plan for the given goal. The plan contains some steps which are subgoals to be reached in order to solve the whole problem.

For every subgoal repeat the process until only elementary steps remain. The sequence of elementary steps is the intended plan. [7]

In our domain the skeletal plans represent the expert's way of work and his knowledge. To every feature the expert associates knowledge about the manufacturing process. This may be some rather global information related to the whole workpiece, e. g. "given a workpiece of less than a certain length, use a chuck as the clamping method", as well as detailed instructions concerning small parts, e. g. "manufacture this groove using the special tool X". Such knowledge is represented in the skeletal plans. Every skeletal plan contains concrete and/or abstract information about parts of the manufacturing process or about the whole process. Every skeletal plan is associated to some feature. Feature definition and skeletal plan together represent the expert's knowledge about a certain planning task

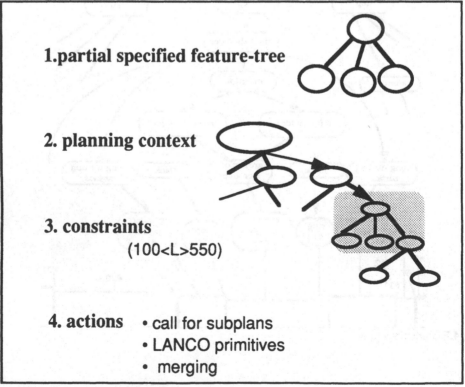

1.partial specified feature-tree

2. planning context

3. constraints
(100<L>550)

4. actions • call for subplans
• LANCO primitives
• merging

Fig. 5. Skeletal Plan Representation

In order to provide a suitable formalism for the representation of skeletal plans, the language SKEP-REP is developed. We currently use a formalism which describes skeletal plans as shown in figure 5. A skeletal plan contains the feature or feature structure it is associated to. It then contains some context information which relates to other skeletal plans which form preconditions for the application of this particular plan. It may also contain some applicability constraints which are not expressed by the features or the context of skeletal plans. Then it contains a sequence of operations, which may result in the subroutine-like call of other skeletal plans or in the generation of concrete planning steps. In the domain of manufacturing by turning, concrete planning steps are chucking commands, cut instructions and tool selections. These steps are represented as primitives of our abstract NC-program representation language LANCO (LAnguage for Abstract Numerical Control). Every operation may access the concrete technological and geometrical informations, especially measurements, which are represented in the TEC-REP of the surfaces which form the features associated with the skeletal plan or with the plans of the context (see above). Beside this, no information about the workpiece can be accessed. This realizes the concepts of modularity and information hiding for the skeletal plans and makes it possible to create skeletal plans for for a large bandwidth of workpieces.
Some of the operations can result in the subroutine-call of special programs for particular tasks. In our prototype, the tool selection operation uses a constraint system to find the suitable group of tools for the intended operation.

To perform the selection and merging of the skeletal plans, a prototypical skeletal planning system was implemented as a part of our PIM system, which uses the following algorithm:
When a given workpiece description has been transformed in a feature structure according to the methods outlined in the chapter above, the skeletal plans associated with these features are found and selected according to the constraints embedded in the plans. The resulting set of skeletal plans is then merged to one final plan, and abstract variables are replaced by the concrete data of the workpiece in question. The merging of the skeletal plans is oriented on several topics: Operations using the same tool should be performed consecutively (minimalization of tool change operations). Operations in one chucking context must be performed together, minimizing the changes of chucking. Different tools belonging to a common group may be exchanged against a more general tool of this group, such that several operations using slightly different tools can be merged to one operation using only one tool. Different surfaces of a workpiece which are treated with similar operations should be grouped together.
These merging operations are supported by a hierarchical ordering of the available tools and a hierarchical grouping of the possible operations. Some heuristical approaches to skeletal plan merging are under investigation.

3 An abstract view on planning using PIM:

Typical AI planning systems usually follow the methodology of STRIPS: The system searches a sequence of operators which transform a given start situation into the goal situation. Every operator is characterized by preconditions which describe the applicability of the operator and postconditions which indicate the changes of the situation resulting from the application of the operator. By standard search methods (e. g. backward chaining) it is possible to find the intended sequence of operators.
Comparing the methodology used in PIM to this, we note interesting similarities and fundamental differences: The features can be seen as aggregations of preconditions relating to geometrical/technological data of the workpiece. The workpiece description contains the description of the goal situation. The feature recognition may be seen as an efficient search method to narrow down the search space to the applicable skeletal plans. The skeletal plans

then serve as indicators for the future search process. Ideally there isn't nearly any real search, as the skeletal plans determine nearly all planning steps.

In summary, the expert's knowledge, represented in features and skeletal plans, is employed to drastically reduce the search space of the planning task.

The planning task itself - to bridge the gap from CAD to NC - is exchanged against some transformation steps from between representation languages. On the level of features and skeletal plans, the gap is bridged by associations representing expert's knowledge. The found skeletal plans are then refined using the concrete data, such that the intended code can be generated. So the planning task boils down to an sequence of abstraction, association and refinement, as illustrated in figure 6, hereby eliminating the need for undirected and inefficient search.

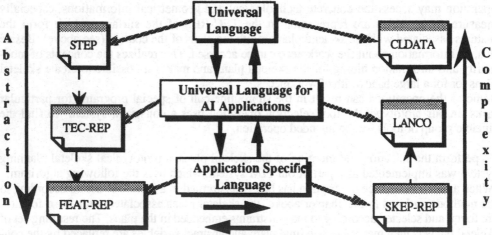

Fig. 6: Abstraction levels of PIM (Planning In Manufacturing)

Beside this it must be remembered that the main advantage of the PIM approach lies in the explicit representation of the special knowledge of an expert in his manufacturing environment. In contrast to systems which employ only general rules this approach solves the problem of the very special and environment-dependent knowledge necessary to create good plans, as motivated in the first chapter.

Intensive talks with many domain experts proved that this approach in fact resembles the way a human expert thinks, which should lead to a high acceptance of such a system by the expert.

4 Connections to the real world

Every CAPP-System serving any useful purpose must be embedded in the existing CA*-world. It has to accept the workpiece description from CAD and its output will be fed into NC machines which manufacture the workpiece. The PIM system fulfils this requirement using interfaces to CAD and NC programming systems.

The interface to the CAD world transforms the necessary geometrical/technological information about the workpiece from the CAD data into our TEC-REP. The forthcoming ISO-standard STEP [2] promises to provide a data exchange format covering all necessary information and providing a system-independent interface. While currently only some prototypes of CAD systems are able to produce STEP files, this will certainly change in the near future. Consequently, a prototypical processor was developed to create TEC-REP files out of a given STEP data file using the information of the geometry, topology and technology models of STEP.

Another usable interface connects the CAD system "Konstruktionssystem Fertigungs-gerecht" of Prof. Meerkamm, University of Erlangen [11], with TEC-REP. This system is a augmented CAD system based on SIGRAPH. It uses an internal data format which can directly be transformed into TEC-REP. Thanks to the cooperation of Prof. Meerkamm and his team, this connection works satisfying.

To get connected to the NC-machines we rely on components of NC programming systems available on the market today. Existing NC programming systems support the human process planner. The system is given the geometrical information about the work piece. The human expert interactively selects areas to be manufactured and suitable tools. The system creates NC-Code according to these commands. The code generation part of a NC programming system doesn't rely on human experts giving the commands, the output of a planning system like PIM can serve as a command sequence as well. So a commercially available NC programming system is connected to PIM as follows:

The output of the PIM system is compiled into the command language of the NC programming system.Then, the NC programming system can generate CLDATA Code as well as machine specific NC-Code, without any further human interaction.

5 Some tools to support our approach

The basic idea behind the PIM system is the explicit representation of a concrete expert's knowledge in his given manufacturing environment in order to closely mimic the expert's problem solving behavior. This methodology leads to special requirements in the area of knowledge acquisition as well as in the area of AI reasoning methods.

According to the intended use of the PIM system, a lot of knowledge acquisition must be performed in the field. In any given manufacturing environment, the expert's knowledge must be acquired and represented separately. This is especially important for any proprietary knowledge and special manufacturing tricks of a factory. When the manufacturing environment changes during the life time of the system, the adequate knowledge must be acquired and represented again in order to adapt to new possibilities. (New tools can lead to new manufacturing methods and consequently to new manufacturing features). The acquisition group of the ARC-TEC project provides tools to support the necessary knowledge acquisition for every step in the systems life cycle [13].

The compilation group of the ARC-TEC project deals with suitable and efficient reasoning mechanisms for the represented knowledge. Taxonomies are handled using the terminological representation language TAXON [1] This extension of a KL-ONE like language can be used to represent the hierarchical dependencies, e. g. a hierarchy of manufacturing tools. Constraints arising in the area can be tackled using the tool CONTAX, which couples a constraint propagation system with the taxonomies. The integrated forward/backward reasoning system FORWARD/RELFUN is suitable for reasoning about the preconditions of the skeletal plans.

6 Conclusion

The observation of human expert's problem-solving behavior resulted in a model of process planning which supports a knowlegde-based approach to CAPP. Based on technological/geometrical information about the workpiece, higher-level features are defined and associated with skeletal plans.

Special languages for a adequate representation of the necessary knowledge on the different abstraction levels were presented. The transformation and interpretation steps between the different languages have been implemented and form the planning system PIM. This system creates a production plan to a given workpiece by performing a sequence of abstraction, selection and refinement.

The approach pursued in the PIM system closely mimics the expert's problem solving behavior. The resulting system is especially tailored to a concrete manufacturing environment and uses the expert's knowledge optimally. This should lead to good quality of the produced plans and to high acceptance of the system. Positive comments of many domain experts support this claim.

7 References

1. Baader, F. and Hanschke, P.: A Scheme for Integrating Concrete Domains into Concept Languages. In *Proceedings of the 12th International Joint Conference on Artificial Intelligence*, 1991.

2. Bernardi, A., Klauck, C., and Legleitner, R.: *Abschlußbericht des Arbeitspaketes PROD*. Dokument, D-90-03 Deutsches Forschungszentrum für Künstliche Intelligenz GmbH, Postfach 20 80, D-6750 Kaiserslautern, september, 1990.

3. Bernardi, A., Boley, H., Hanschke, P., Hinkelmann, K., Klauck, C., Kühn, O., Legleitner, R., Meyer, M., Richter, M.M., Schmalhofer, F., Schmidt, G., and Sommer, : ARC-TEC: Acquisition, Representation and Compilation of Technical Knowledge. In *Expert Sytems and their Applications:Tools, techniques and Methods*, 1991, pp. 133-145.

4. Bernardi, A., Klauck, C., and Legleitner, R.: *TEC-REP: Repräsentation von Geometrie- und Technologieinformationen*. Dokument, D-91-07 Deutsches Forschungszentrum für Künstliche Intelligenz GmbH, Postfach 20 80, D-6750 Kaiserslautern, june, 1991.

5. Chang, T.C.: *Expert Process Planning for Manufacturing*, Addison-Wesley (1990).

6. Computer Integrated Manufacturing - Results and Progress of Esprit Projects in 1989. Esprit XIII/416/89, november, 1989.

7. Friedland, P.E. and Iwasaki, Y.: *The Concept and Implementation of Skeletal Plans*. Journal of Automated Reasoning (1)(october 1985), 161-208.

8. Grabowski, H., Anderl, R., Schilli, B. and Schmitt, M.: STEP - Entwicklung einer Schnittstelle zum Produktdatenaustausch. VDI-Z 131, september 1989, Nr. 9, pp. 68-76.

9. Klauck, C., Bernardi, A., and Legleitner, R.: FEAT-REP: Representing Features in CAD/CAM. In *IV International Symposium on Artificial Intelligence:Applications in Informatics*, 1991.

10. Klauck, C., Bernardi, A., and Legleitner, R.: *FEAT-REP: Representing Features in CAD/CAM*. Research Report, RR-91-20 Deutsches Forschungszentrum für Künstliche Intelligenz GmbH, Postfach 20 80D-6750 Kaiserslautern, june, 1991.

11. Meerkamm, H. and Weber, A.:Konstruktionssystem mfk - Integration von Bauteilsynthese und -analyse. In *Erfolgreiche Anwendung wissensbasierter Systeme in Entwicklung und Konstruktion*. VDI Berichte 903, 1991, pp. 231-249.

12. Richter, M. M., Bernardi, A. ,Klauck, C. and Legleitner, R. : Akquisition und Repräsentation von technischem Wissen für Planungsaufgaben im Bereich der Fertigungstechnik. In *Erfolgreiche Anwendung wissensbasierter Systeme in Entwicklung und Konstruktion*. VDI Berichte 903, 1991, pp. 125-145.

13. Schmalhofer, F. , Kühn, O., and Schmidt, G.: *Integrated knowledge acquisition from text, previously solved cases and expert memories* . Applied Artificial Intelligence (5)(1991), 311 - 337.

14. Tengvald, E.: *The Design of Expert Planning Systems: An Experimental Operations Planning System for Turning*, Ph.D. dissertation, Lingköping University, Department of Computer and Information Science, S-58183 Linköping, Sweden, 1984.

Visual Modelling: A Knowledge Acquisition Method for Intelligent Process Control Systems[1]

D. Schmidt, J. Haddock, W.A. Wallace and R. Wright[2]

Abstract: Intelligent process control requires theoretical and experimental knowledge of physical processes combined with the judgement and experience of experts in plant operations and management. An effective integration of these components depends on effective communication.

The objective of this paper is to propose visual modelling as a knowledge acquisition tool for eliciting, integrating and formalizing the three different types of knowledge into a knowledge base for intelligent process control. The approach is illustrated by the manufacturing processes in the annealing of copper wire and the galvanizing of steel.

KEYWORDS: Intelligent Processing, Process Control, Expert Systems, Visual Modelling

1 Introduction

For the past 25 years, computer systems have been used for process control (8). The first devices, programmable controllers, had a simple operating system, limited instruction, but no capability for programming in a higher level language. Keeping pace with advances in software and memory, automated process control systems began to add in-site sensors, process models and optimization algorithms. The term "intelligent" was added to describe the incorporation of process heuristics into the control system (7). The result is an integrated knowledge base containing theoretical, empirical and expert knowledge for process control denoted intelligent process control (IPC).

However, the difficulty in acquiring process heuristics or codified expert knowledge, and integrating it with the theoretical and experimental knowledge has yet to be resolved. As Albaharna (1) states: "The importance of combining both qualitative and quantitative analysis techniques cannot be over emphasized." The goal of IPC is the integration of theoretical or book knowledge, laboratory knowledge (the quantitative knowledge), and the heuristics of the experts (the qualitative knowledge). The author continues "...acquisition of the knowledge necessary to power an expert system and structure the knowledge into a useable form is one of the primary bottlenecks to expert systems development."

This paper will present the methodology of object oriented analysis and design for acquiring, visualizing, and formalizing the knowledge needed for intelligent processing.

1. Research partly supported by the High Temperature Project at Rensselaer Polytechnic Institute sponsored by the New York State Energy and Development Authority.

2. The authors are respectively Research Assistant, Associate Professor and Professor of Decision Sciences and Engineering Systems, and Professor of Materials Science and Engineering. Professor Wallace is currently at the Swiss Federal Institute of Technology, Zurich, Switzerland.

The resulting model is presented using object oriented software, in this case, EXTEND. Our applications are the annealing of copper wire and the hot-dip galvanizing of steel.

2 Visual Modelling

Modelling is the process of providing an abstraction of reality, i.e., a model. The form and content of the result of this process depends on how the model is to be used. A theoretical model of a physical process provides guidance on what the process ought to be doing, based on a very idealized representation of reality. Models based on experimental data describe what actually happened in a laboratory setting. The judgment and experience of the expert must also be modelled to capture their view of reality, i.e., what goes on in the plant. These three models are different representations of the same reality, and must be combined and integrated into an intelligent process control system.

The process of modelling can be divided into three procedural components defined as: (1) conceptualization; (2) formalization, and (3) exercising and learning. Conceptualization identifies, defines and formulates the process we are trying to control; formalization translates the conceptualization into a formal structure; and exercising and learning involves interacting with the model to ensure validity and address various scenarios based on the objective of the modelling process.

Conceptualization requires consultation with the experts. In the case of models based on theory or experimentation, the experts are material scientists or process engineers. For the expert system component of the IPC, the experts are plant operators and managers. The method of communicating with these two groups most reflect the language and understanding of the experts.

The knowledge engineer(s) must start with "book" knowledge; documented information on the process. However, the sources of this knowledge will differ. Research papers and books supply the theoretical knowledge, and research and laboratory reports supply the experimental data. Plant knowledge is typically found in operating manuals prepared by company or plant management. Although the content may be similar, the form of communications and the language used will differ. The modeler or knowledge engineer (KE) must be able to capture both types of knowledge and integrate them into a common structure. Then, the KE has to communicate with both groups of experts in order to elicit their knowledge. The result of this process is a set of conceptualized models that represent the theory, experimental results and expert knowledge. These models can take on various characterizations including numeric equations, tables or graphs of data, and process flow charts -- each representing an appropriate means of communicating with experts. However, there must be a common structure and language underlying all these representations. If not, it is likely that miscommunication and misinterpretations will occur between the modeler or KE and the experts (as well as among the experts).

Formalization involves translating the conceptual models into representations that constitute the knowledge base of the IPC. A process model is a combination of theoretical equations and empirical data. The control component is some form of optimizing algorithm. These models are encoded into a numeric programming structure. The expert knowledge component is represented in the knowledge base using symbolic processing or a frame-rule system. Matrices or tables store the data on the process, and the program logic, in both numeric and symbolic form, represents the conceptual models. The resulting knowledge base consists of: (1) experimental and plant data on the process; (2) models for the theoretical relationships, estimation and prediction, and the control algorithms; and (3) a conceptual model for the rules and procedures about the process.

The exercising and learning process first addresses the validity of the models. Typical questions are: does the model contain all the relevant variables; are the relationships in the model verifiable statistically; does the output of the process being modeled match the output of the actual process; and can we predict the output of the actual process with the desired degree of confidence? When these questions have been addressed, the model can be exercised by considering various scenarios or situations faced by the plant operators or engineers. Extreme situations are typically used to test the robustness of the model. Mechanisms for learning or adaptation are then incorporated into the system to help ensure successful implementation; techniques include the use of on-line parameter estimation or neural nets for pattern recognition.

The most difficult and time consuming part of the modelling process is the elicitation and subsequent representation of human judgment and experience - expert knowledge. Visual modelling is a knowledge acquisition approach that capitalizes on the capability to use pictures to facilitate the communications process. These "pictures", graphs, icons, figures, etc., are used to elicit the experts' representation of reality. These mental models are the way humans understand the world (6). A model is constructed when the human makes inferences about reality. These tacit inferences can be explicit and made consciously, or implicit and subconsciously derived. It is postulated that these models can be described in terms of (a) sets of concepts; (b) sets of relationships; (c) statements, where a statement is a relationship between a pair of concepts; and (d) a map, which can be a visual representation of a set of statements (3). These statements can be, for example, facts like "if the temperature is below 0°C, the reaction will not occur," or associations like "that material is rare earth." A map is a graph formed from the interrelated statements, and is a representation of a mental model. The process of knowledge acquisition seeks to develop these maps by identifying the concepts and relationships between them. In order to accomplish this task, we try to make the mental models of the KE or modelers and the expert similar, since shared mental models help ensure effective communications.

Modelling a physical process provides the modeler with a ready-made means of communications, graphical depictions of the process itself. In addition, there are theoretical and empirical relationships that can be shown using icons or graphs. Animation can also be used to portray the dynamics of the process. Concepts and their relationships can be defined in terms of visuals, greatly facilitating communications and the building of common mental models between the modeler and the expert. As importantly, these "pictures" can be computer generated, helping to ensure that the resulting knowledge can be integrated with the theoretical and empirical knowledge into a common structure for formalization.

Visual modelling is important throughout the modelling process. Obviously the conceptualization component is greatly supported by this approach. In addition, visualization, in particular animation, can be very useful in the validation of the models.

Even in formalization of the model, the computer programmer can use flowcharts or other graphical tools as a basis for the structuring and integrating needed to develop the knowledge base. In the sections to follow, we will show how object-oriented programming can be used as a tool for visual modelling, and how it was applied in modelling the annealing of copper wire and the hot-dip galvanization of steel.

3 Illustrative Applications

To illustrate the use of visual modelling in the knowledge acquisition process, we will briefly describe two intelligent processing projects, copper wire annealing and steel galvanization, undertaken as part of the High Temperature Technology Program at Rensselaer Polytechnic Institute. One of the objectives of this research program is to improve high temperature materials processing by utilizing and extending expert system technology. In the knowledge acquisition process, we are employing object-oriented programming to support the visual modelling.

3.1 Object Oriented Modelling

Object-oriented modelling (OOM) has a basic premise to use standard, "off the shelf" sub-components or objects to create the model (9). OOM uses C or Assembly language code to create the objects which will be combined into complete models. OOM requires four characteristics: polymorphism, encapsulation, inheritance, and dynamic binding.
Polymorphism is equivalent to saying that each object is a specific example of some generic group. The group has certain attributes that are unique to that class alone. Encapsulation means that each object is complete in itself. Including the data necessary for that object, ideally, one object cannot access the internal data of another object - they are completely self-contained. In OOM the data and the process that uses that data are created at the same time. Inheritance means that a subclass of objects have all of the attributes of the parent class by default. For example, within the parent class CARS, the subclass FORDS would already have the attributes of a Motor, Transmission, and Wheels. Dynamic binding means that a particular message will result in different actions depending on which object receives the message.
The benefits of these characteristics are that many components of a given application are reusable in other applications so that most of any given model will not have to be written from scratch.
Object-oriented programming (OOP) creates reusable code, simplifying the modelling process. OOP, as implemented in EXTEND, adds the additional capability of portraying an object pictorially as an icon separate from other objects. The connections between two objects are also explicitly shown in the model. This allows communication between the expert and the modeler in representing the relationships between objects. EXTEND uses a library of objects that the programmer can combine to produce a model. The user also can use the Model script language included in EXTEND to produce new objects for the library. This allows both the rapid formulation of new models by using the library of objects and the flexibility to improve the library. Data is also input by the user via dialogue boxes. Input data can be integers, real numbers, or arrays. Since the data are easily modified by the user, various "what-if" scenarios can be tested. Other parameters that can be modified by the user are the duration of the model, the number of models to

be repeated, the random sequence number used, the accuracy of recalculations (continuous modelling requires more accuracy); the number of steps per time unit, and whether the processing of the model should be constrained by the physical location of the objects or not. Output in EXTEND is both tabular data and graphical display with respect to time. The outputs are color coded on the screen.

3.2 Annealing of Copper Wire

This project was undertaken to assist the copper wire manufacturers increase product quality and gain control over the annealing process (11). The final properties of wire can be erratic since the exact relationship between the control parameters and properties of the annealed wire are unknown. In addition, expert advice is scarce and expensive. When the metal is plasmically deformed at temperatures that are low relative to their melting point, it is said to be cold worked. Changes occur in almost all of a metal's physical and mechanical properties during this process. Working increases strength, hardness, and electrical resistance, and decreases ductility. To release the stored energy inside the metal and regain its physical properties, the cold worked metal is heated continuously and the energy release is determined as a function of temperature. In the case of copper wire, this process is accomplished by annealing using electrical current after the wire is drawn.

The wire, as an annealed product, may be evaluated with spiral elongation tests, grain size determination, tensile test, hardness test, and so on. Other features of interest are microstructure, surface finish, and texture. It is generally the case that the process parameters are set and the wire is annealed without sophisticated control over the properties of the wire being produced.

After an examination of the wire annealing process and interviews with an expert, it was decided to focus on predicting how various settings of the process parameters would affect the time/temperature profile of the wire. Furthermore, knowledge on how these time/temperature profiles map into the physical and mechanical properties of the wire (elasticity, texture, softness, etc.) would also be in the domain.

A program was developed using EXTEND that simulates the effects of the wire annealing system and allows for control of the process parameters. The simulation was based on the physical principles of heat transfer and generated time/temperature curves of the wire. These curves were obtained in the laboratory under various scenarios employing different settings of the process parameters. The behavior of a small piece of wire as it moves from the drawing process through the annealing process and finally to the spooling process was simulated and displayed graphically (See Figure 1). The controllable parameters of the process were voltage, wire speed, water level, water temperature, and radius of the pulley. The air and pulley temperatures were assumed to be constant. This continuous process simulation was augmented by the addition of production rules written in KEE that relate the time/temperature curves to the characteristics of the product. This augmentation required incorporating both materials research and expert judgment to complete the knowledge base.

The project demonstrated that one can build a model of an annealing process based on both scientific knowledge and the experience of experts. The resulting system was able to provide a measure of process control and improve product quality.

3.3 Steel Galvanization

According to the American Galvanizers Association (2) metal corrosion cost an estimated $222 billion to the U.S. economy in 1989, and $33.3 billion of this cost could have been saved by using proven corrosion control technologies. The most common method to prevent the corrosion of steel is the galvanizing process. The galvanizing process involves first cleaning the steel of impurities, then coating the steel with zinc, either using molten zinc or by electroplating.

Current production of galvanized sheet steel is a continuous process. Cold rolled sheet metal on reels is welded into one long continuous strip for electrogalvanizing or hot-dip galvanizing. While much progress has been made in the quality of the product from U.S. steel production facilities, as Dunbar (4) states: "Today, our customers expect a surface finish (of galvanized steel) equivalent to cold rolled steel." In fact, also according to Dunbar (4): "None of these defects (currently of galvanized steel) would have been considered a cause for rejection on our regular spangled products of a decade ago." A primary motivation for this increase in the standards imposed on the galvanizing process is the U.S. automotive industry. Industry requires not only the corrosion protection available from galvanizing, but also the surface finish necessary for painting of car panels and other exterior sheet metal parts such as car bumpers. The surface issue is compounded since the shade of color must be the same for all the exterior metal parts of the car. Another issue is the formability of the galvanized sheet steel. The galvanized sheet steel is press-formed into car panels using equipment that was originally designed for uncoated low carbon steels. The addition of the galvanized coating makes it that much more difficult to form the panel. This is due to the different friction coefficients for galvanized sheet steel versus uncoated sheet steel.

The hot-dip galvanizing process was selected as an application area (10). The line uses a hot-dipped galvanizing process with batch annealing and a flux furnace. Special effort was made to visually portray the process as closely as possible with the current visual information already used by the steel plant staff, such as brochures and process flow diagrams.

The theoretical knowledge and the descriptive information from steel plants were studied to become familiar with the process of galvanizing steel. An example of the theoretical knowledge used in the model was identifying that the longer the steel was immersed in the zinc pot, the thicker the alloyed coating would be, but there were diminishing returns to how beneficial that would be. The dependent variables for these relationships were the maximum size of the zinc pot and the line speed. Assumptions were stated and explicitly shown in the model. For example, it was assumed that the loop storage had an optimal size, which was adjusted during the coating of each coil, in addition to adjusting loop storage after coil changes. Some of the judgmental knowledge obtained from the steel plant staff dealt with identifying which specific operations were important to this process. They decided that the actual chemical composition and operating temperatures of the cleaning tanks use of secondary importance in the model. The staff also corrected the assumption of the loop storage operation, i.e., the loop storage actually only changed size during coil changes. Basic control and communications within the process were explained by the staff. Worst case scenarios (such as having the sheet stop) were explained in detail as to why that was undesirable. Emergency actions were also described, such as the frequency of preventive maintenance, line speeds, and product mixes in a typical schedule. The resulting visual model of the entire process is shown in Figure 2.

4 Concluding Remarks

Visually modelling the material processing in the annealing of copper wire and the galvanizing of steel greatly facilitated knowledge acquisition from the experts, plant operators and engineers. Visual objects and their explicit connections reduced the modelers' misconceptions of how the processes actually work. In addition, the level of aggregation was changed in the modelling process. For example, the experts in the steel plants felt that two payoff reels and one welder (three objects) could be combined into one uncoiler (one object). The process of information flow in steel manufacturing, as well as the frequency with which certain data are collected had to be modelled in greater specificity. This requirement has important ramifications for the design of the IPC system. In both projects the use of visual modelling simulated the interest of the plant personnel. The ability to view the model in color on a Macintosh computer in a plant work space involved the operators in the modelling process. This interaction has increased their commitment to our activities, and will be valuable in validating the intelligent process controller.

The object oriented approach, as implemented in EXTEND, was excellent for visual modelling in the knowledge acquisition process. However, EXTEND is limited in its ability to use general purpose languages such as Fortran or PASCAL; languages more suitable for numeric routines used in modelling the theoretical and experimental knowledge. In addition, EXTEND has difficulty in being integrated with simulation language like SIMAN or SLAM, as well as software for artificial neural networks. Advances in hybrid or coupled system technology should result in the complete, seamless integration of these modelling approaches into an intelligent process control systems.

References

1.   Albaharna, O. and S. Argyropoulos (1988), "Artificial Intelligence for Materials Processing and Process Control", *JOM*, (40), 6-10.

2.   American Galvanizers Association (1990), *Galvanizing for Corrosion Protection: A Specifier's Guide*, Alexandria, VA.

3.   Carley, K.M. (1988), "Formalizing the Social Expert's Knowledge", *Sociological Methods and Research*, (17), 165-232.

4.   Dunbar, F. (1988), "Defects of the 80's - A Closer Look at the Critical Requirements of Today's Hot Dip Galvanized", *Proceedings of the Galvanizers Association*, (80), 92-101.

5.   Imagine That, Inc. (1988), *EXTEND*, San Jose, CA.

6.   Johnson-Laird, P. (1983), *Mental Models: Toward a Cognitive Science of Language, Inference and Consciousness*, Cambridge, MA, Harvard University Press.

7.   Parrish, P. and Barker, W. (1990), "The Basics of Intelligent Processing of Materials", *JOM*, (42), 14-16.

8.   Roffel, B. and Chin, P. (1987), *Computer Control in the Process Industries*, Chelsea, MI, Lewis Publishers, Inc.

9.   Schmidt, D., Haddock, J. and Wallace, W.A. (1991), "Modelling the Manufacturing Process: An Assessment of Three Approaches to Simulation", *1991 International Industrial Engineering Conference Proceedings*, Detroit, MI, 401-410.

10.   Schmidt, D., Haddock, J., Wallace, W.A, and Wright, R., (1992), Visual Modelling for Intelligent Process Control of Hot Dip Galvanizing", *JOM*, (44), 28-31.

11.   Singh, A., Wallace, W.A., Haddock, J. and Wright, R. (1991), "Expert Systems and the Direct Resistance Annealing of Copper", *Wire Journal International*, September.

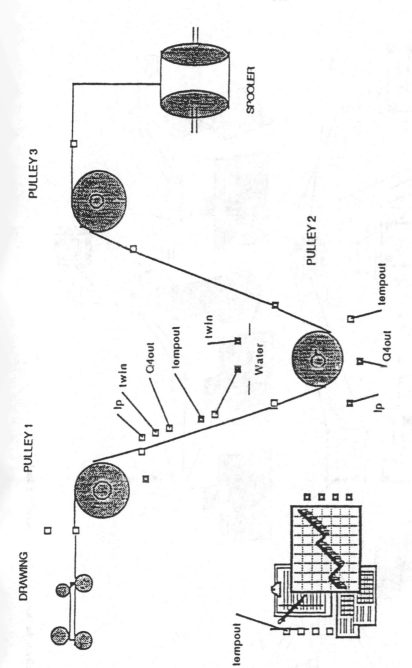

LEGENDS

lp   length of wire touching the pulley.
twin   temperature of water.
Q4out   radial heat loss from wire to pulley.
tempout   temperature of wire at a given time.

Fig. 1 Display of Wire Annealing Process

303

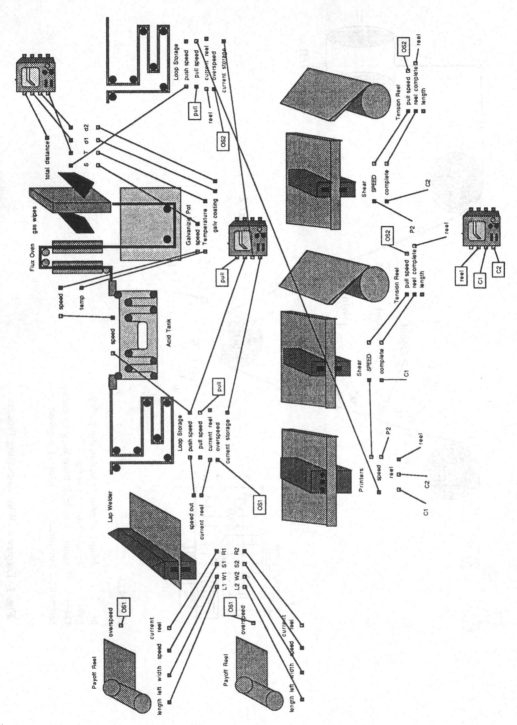

Fig. 2 Display of the Descriptive Model of Galvanizing Steel

Machine Learning in Communication Nets

F. Lehmann, R. Seising, E. Walther–Klaus
Fakultät für Informatik
Universität der Bundeswehr München
D–8014 Neubiberg

Abstract: A concept of learning from observations is presented using a combined theory of subjective and objective probability notions in Bayesian networks with tree structure. The principle of learning is applied to communication nets.

Introduction

A communication network is a very complex system, whose components are interconnected and impact each other. The processes in these systems often are not deterministic but nearly always impaired with uncertainty. The transmission of information is not free of errors, and occasionally random influences cause corruption of the transmitted messages. Hosts, nodes and other components, too, may not work correctly. It is even not necessary that an error is detected at the place where it ocurred. Therefore wrong informations might have been transmitted and become apparent at some other place in the network. This makes the management of the network difficult and expert systems might assist the manager. These expert systems should know about the propagation of errors and deal with the uncertainty in the system. "Uncertainty needs to be correctly represented so that formal deductions may be made that lead to valid conclusions, and the uncertainty can be reduced through formal assimilation of more data" [GAI]. To this end we use a quantitative approach.

Looking carefully at the processes in the systems, one observes that there are basically different types of uncertainty involved. In our example there is the uncertainty concerning the random events in the system, e.g. the occurrence of a transmission error or faulty components. This uncertainty has its origin in the natural randomness of the processes. This type of uncertainty is handled by the concept of probability in the sense of "objective probability".

Another type of uncertainty, however, arises from the limited knowledge of a person about a process. For example, the manager of a system might not know exactly the probability of an error at a certain component of his system. This type of uncertainty has been addressed by the concept of **"subjective probability"**, which is considered to encode the knowledge of a person at a certain moment. While the objective probability is thought to be like other physical values independent of an observer and normally unchanged by observations, the subjective probability, as an encoding of the knowledge of a person, should be changed after the observation of an event to adjust to the new state of knowledge. Therefore it seems to be obvious and necessary to use both notions to describe the different contexts of uncertainty.

A theory, combining both of the notions has been developed by H. Richter [RIC1]. The theory contains a formalism to adapt the subjective distribution to observations of random events which can be used for automatic knowledge acquisition in machine learning.

In complex systems the processes are seldom isolated. Usually there are relationships between some of them, which often can be quantified by conditional probabilities. Bayesian networks have been used to model this type of "causality". In this paper we consider the conditional probabilities to be known, whereas the probabilities of the events are not known exactly and should be learned from observations. This is accomplished by using Bayesian networks combined with the learning theory mentioned above. Proceeding from this theory, a process of learning unknown error probabilities in a communication network can be accomplished using a logical network in form of an acyclic directed graph.

The paper is organized in the following way: At first, the mathematical model of a probabilistic logical network will be presented. In the next section the special learning process and its propagation within this logical learning net is developed and **a formula for learning from observations** is given. This type of learning is different from learning additional constraints as it can be found, for example, in the method of maximum entropy. The mathematical model will then be applied to a communication net with fault management.

The Mathematical Model of a Learning Net

The topology of a learning net is a directed acyclic graph G with tree-structure. Let E = $\{e_1, e_2, ... e_n\}$ denote the set of nodes and K $\subset$ E x E the set of directed arcs. Each directed arc $\alpha = (e_i, e_j)$ is uniquely determined by the associated nodes $e_i, e_j \in$ E, $e_i \neq e_j$. Each node e_i represents a random experiment A_i which may have a finite number of outcomes $A_{i,k}$, k=1,..,m; the arcs are to represent "causal" conditions between random events. The random experiment at the node e_i is described by the probabilities $P(A_{i,k})$, k = 1,..,m.

The arcs α describe "causal" relations between the events in e_i and e_j in form of conditional probabilities $P(A_{i,j}|A_{i,k})$, for l, k =1,..,m. In this logical net an arc $\alpha = (e_i, e_j)$ is only established if the experiments in e_i and e_j are <u>not</u> independent.

In this paper we restrict to $\kappa=2$ for ease of description. Extensions to arbitrary n can be found easily. In the case for the number of nodes $\kappa=2$, we have at each node a and b the event A its complement B and B^c.

Independence of the experiments at a and b with the respective events A, A^c and B, B^c is implied then by

$$P(B|A) = P(B|A^c).$$

For mutual dependence, therefore, it must be valid that

$$P(B|A) \neq P(B|A^c).$$

To conceive the probabilties P(A), P(B), P(B|A) and P(B|A^c) as probabilities in a coupled experiment, they have to satisfy the following consistency condition:

$$
\begin{aligned}
P(B) &= P(B|A) \cdot P(A)) + P(B|A^c) \cdot P(A^c) \\
&= (P(B|A) - P(B|A_c)) \cdot P(A) + P(B|A^c)
\end{aligned}
\tag{1}
$$

Thus given P(A), P(B|A) and P(B|A^c), P(B|A) $\neq$ P(B|A^c) the value of P(B) can be computed using this consistency constraint. Of course we can also compute P(A), if P(B) is given.

In a consistent network (with conditional independence property [PEA]), there exists one unique joint probability distribution containing the given probabilities. Knowing this distribution, we are able to compute all conditional and unconditional probabilities inside the network.

However, as a rule, the probability P(A) that describes the random experiment at a node in the logical net, representing an event at one node (host, component) in the communication net, will not be exactly known. This objective uncertainty is "weigthed" by a subjective probability of the manager, describing his uncertainty about the probability of the random event. If he has reason to put the same confidence in all of the objective probabilities, his subjective distribution over the set of all possible objective distributions may be the equidistribution. With every new observation, this subjective distribution has to be changed according to the new state of knowledge.

The Learning Theory

Consider the following situation: The objective probability p of an event usually is unknown to the observer. For him p belongs to a set M of possible probabilities. Each element of M can be weighted differently by the observer according to his rudimentary knowledge about p. This subjective knowledge is used by the person for decisions. It is represented in the model by a subjective distribution with a probability density function $\phi(p)$ over the set M of all possible objective probabilities. Examples are contained in the next figure:

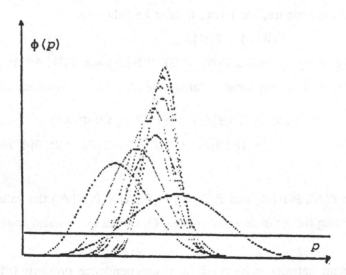

Each subjective distribution includes an estimation system of the unknown "true" objective probabilities. The subjectively expected probability that an event E will happen is given by the following formula:

$$P(E) = \int_M P(E|p)\ \phi(p)\ dp \qquad\qquad (2)$$

where $P(E|p)$ is the objective probability of E, if p is the "true" distribution in **M**, and $\phi(p)$ the subjective density over **M**.

The decisive point in this integrated concept is that a subjective distribution - because it represents the knowledge of a person - has to be accommodated whenever new information is accessible, for example if new events have been observed.

In [RIC2] departing from weak assumptions about the rational behaviour of a person in bets the following formula for the changing of the subjective distribution is derived: having observed an event E, the new subjective distribution with its probability density function $\phi_i^E(p)$ is obtained by the **learning formula**

$$\phi_i^E(p) = c^{-1} \cdot P(E|p)\,\phi_{i-1}(p) \qquad\qquad (3)$$

with $P(E|p)$ being the probability of E assuming that p is the true objective probability, $\phi_{i-1}(p)$ the subjective density before E had happened and c a normalization constant which has, in the case that we have density functions, the following value:

$$c = \int_M P(E|p) \cdot \phi_{i-1}(p)\,dp \qquad\qquad (4)$$

The formula has the appearance of the famous Bayesian formula, but with regard to its contents there are considerable differences:

ϕ_i^E and ϕ_{i-1} are densities subjectively weighting the possible objective probabilities according to the information, whereas the Bayesian formula is a mathematical theorem of a probability theory with a single concept of probability.

The combined concept strictly differentiates between objective and subjective probability and therefore it seems to be appropriately applicable to many problems found in practice as our example shows. It can be shown that, if the experiment is governed by the true probability distribution p' and the experiment will be repeated N times independently, the subjective probabilities $\phi_N(p)$ will converge with objective probability "1" to a distribution which is concentrated in p^t and the subjectively expected distribution p, converges to p^t again with probability "1", if N grows to infinity, if p^t is contained in the support of ϕ_0. This means that a person using this **principle of learning** can find the true distribution, if he can obtain infinitely many observations in the experiments, and if he has not excluded the "true" probability p^t at the very beginning by choosing an unsuitable ϕ_0. The learning process that is described above takes place at one node. It can be propagated throughout the whole logic net structure: ϕ_a and ϕ_b are the subjective distributions at nodes a and b and the arc (a,b) be weigthed by $P(B|A)$ and $P(B|A^c)$. Then, from the transformation rules for densities, the following consistency condition must hold:

$$\phi_b(x) = \phi_a\left(\frac{(x - \beta)}{\alpha}\right) / |\alpha| \tag{5}$$

for $\beta \leq x \leq \alpha+\beta$, if $\alpha > 0$, and $\alpha+\beta \leq x \leq \beta$, if $\alpha < 0$,

with $\alpha = P(B|A) - P(B|A^c)$ and $\beta = P(B|A^c)$.

If an event A is observed at node a, the subjective distribution ϕ_a has to be accomodated with respect to the new information according to the learning formula. In order to make the network consistent again, the subjective distributions at all other nodes must be adjusted by use of the given formulas, or their inverses. In this way the whole network learns from an observation at one node.

The Application of Multipoint Distributions in a Communication Net

In the application of the theory to the management of errors in a communication system we have to consider more than two events at a node; but we can restrict to the case that this number m is the same at each node of the network.

Therefore, the objective probability distributions at node e_i are m-point distributions (p_1, p_2, ..., p_m) with $p_j = P(A_{ij})$, $j = 1,2,...m$, which belong to a set M of possible objective distributions. This set M can be parameterized by m-1 parameters, $p_1,...,p_{m-1}$, because $p_m = 1 - (p_1+...+p_{m-1})$. The subjective distribution ϕ_i encoding the knowledge about this node is then a distribution over M. Observation of an event E leads to an adaption of

ϕ_i to ϕ_i^E according the learning formula

$$\phi_i^E(p_1, \ldots, p_m) = c \cdot P(E|(p_1, \ldots, p_m)) \cdot \phi_i(p_1, \ldots, p_m) \qquad (6)$$

where $P(E|(p_1,...p_m))$ is the objective probability of E, if $(p_1,..p_m)$ is the true distribution and c is a normalizing constant. If the nodes e_i with events $A_1,...,A_m$ and e_j with events $B_1,...,B_m$ are connected by an arc $\alpha = (e_i,e_j)$ the matrix

$$C = \begin{pmatrix} P(B_1|A_1) & \cdots & P(B_1|A_m) \\ \cdots & \cdots & \cdots \\ P(B_m|A_1) & \cdots & P(B_m|A_m) \end{pmatrix} \qquad (7)$$

is associated to α. It is assumed that C be nonsingular, which implies that the random experiments at e_i and e_j are dependent (the converse is not true in general). Let $P_B = (P(B_1),...,P(B_m))^T$ be the column vector of the probabilities at node e_j and $P_A = (P(A_1),...,P(A_m))^T$ the vector of probabilities at node e_i. Then the following consistency condition must hold:

$$P_B = C \cdot P_A \qquad (8)$$

This allows the computation of P_B from P_A and, because C is nonsingular - inversely of P_A from P_B. The transformation formulas for m-dimensional densities lead to consistency conditions for the subjective densities ϕ_i and ϕ_j, which allow the propagation of learning from one node to the whole network.

Application of the theory to communication nets

We consider a communication net consisting of several components, which are connected by transmission lines.

Along a transmission line a sender transmits a stream of bits. This transmission is in general not free from errors. We consider the case that a bit, sent by the sender might be inverted randomly during the transmission and that the transmission of different bits is stochastically independent. The quality of the transmission line is characterized by the probabilities $P(i|j)$, that the receiver receives bit i, if the sender has sent bit j (i,j=0,1). Of course we assume, that $P(0|0) \neq P(0|1)$ (then $P(1|0) \neq P(1|1)$ too). Otherwise, the received bit is independent from the bit sent by the sender. We assume further, that these probabilities be fixed and known.

To recognize and correct transmission errors the data link layer of the sender transmits frames of fixed lenght, say n bits, which fulfill a certain condition that is known to the receiver, so that he can detect a transmission error. Such a condition might be even or odd parity, CRC-conditions etc.

If the sender would work always correctly, he would produce only frames, that fulfill the condition. In reality, however, the sender might malfunction and produce any of the $m = 2^n$ possible frames. The objective probability distribution is therefore a 2^n-point distribution, which might not be known to the manager (or his expert system). So he will have a subjective distribution ϕ_s over the set of all possible 2^n-point distributions for the sender. The erroneous frame might be recognized at the sender. In this case the observed event (error indication or no error indication) is used to change ϕ_s according to the learning formula.

But the error might not be noticed at the sender, so that the erroneous frame is sent to the receiver, possibly disturbed further by transmission errors. For the receiver there is a subjective distribution ϕ_R over the set of all 2^n-point distributions for the probabilities to receive each of the 2^n possible frames. Observations at the receiver lead to appropriate modifications of ϕ_R according to the learning formula.

To propagate the learning through the whole network we need the matrix C associated with the arc from sender to receiver. It is a $(2^n \times 2^n)$- matrix with elements $P((r_1,r_2,...,r_n)|(s_1,...,s_n))$, the probability that the receiver receives frame $(r_1,..r_n)$ under the condition, that frame $(s_1,...,s_n)$ has been sent.

We can apply the consistency formulas, if C is nonsingular. If we set $f := P(0|0)$ and $g := P(0|1)$, then it can be shown by some calculation [L/S/W$_{91}$], that
which is clearly not zero for the system considered, so that the propagation of learning

$$\det C = (f-g)^{n \cdot 2^{n-1}} \tag{9}$$

in the whole network can take place.

The application of the described model implies that there is **one** generic cause of fault manifestation. This manifestation propagates throughout the tree-net according to the learning mechanism with the given conditional probabilities at the arcs between each node. If there is more than one error cause, and the effects can not be seperated distinctly, a Bayesian network has to be introduced for each cause. Several causal conditions at the arcs may be combined, provided that they are independent, only relating the events at each node e_i and and e_j respectively.

As a general property of this learning process it is known ([L/S/W], [L/W]) that in the long run this learning will be successful in the sense that the true objective distribution at each node will be found with probability 1, if it had not been excluded at the very beginning. In special cases mathematical formulas have been developed in [L/S/W$_{vi}$]. In other cases the learning can be simulated and the subjective distribution together with its moments can be estimated.

Conclusion

A theory has been presented, which can be used as a foundation of learning from observations in expert systems.

In this study we gave an example, where the theory is applied to fault managment in communication systems. However, the error pattern has been quite simple. More general cases are under consideration. Problems of efficient implementation are being studied, too.

Literature:

[GAI]   Gaines, B.R., Boose, J.H., Machine Learning and Uncertainty Reasoning, London 1990

[L/S/W]   Lehmann, F., Seising, R., Walther-Klaus, E.: Unterschiedliche Wahrscheinlichkeitsbe—
               griffe in Bayes—Netzwerken, Fortschritte in der Simulationstechnik (ASIM), Bd.1:
               Simulationstechnik, 6. Symposium in Wien, September 1990, Tagungsband, hrsg. v. F.
               Breitenecker, I. Troch, P. Kopacek, Braunschweig 1990, pp. 160—164.

[L/S/W$_{91}$] Lehmann, F., Seising, R., Walther-Klaus, E.: Analysis of Learning in Bayesian Networks
               for Fault Management, Fakultätsbericht der Fakulät für Informatik Universität der
               Bundeswehr München Nr. 9106, Juli 1991

[L/W]   Lehmann, F., Walther-Klaus, E.: Combination of different concepts of probability, in: 6th
               UK Computer and Telecommunications Performance Engineering Workshop, Bradford,
               1990.

[M/C/M]   Mitchell, T.M., Carbonell, J.G., Michalski, R.S., Machine Learning — A Guide To
               Current Research, Boston/Dordrecht/Lancaster 1986

[PEA]   Pearl, J.: Fusion, Propagation, and Structuring in Belief Networks. Artificial
               Intelligence 29, 1986, pp. 241—288.

[RIC1]   Richter, H.: Zur Grundlegung der Wahrscheinlichkeitstheorie. Teil I: Math. Annalen,
               Bd.125, pp. 129—139, 1953, Teil II: pp. 223—234, Teil III: pp. 335—343, Teil IV:
               Bd.126, pp. 362—374, 1953, Teil V: Bd.128, pp. 305—339, 1954,

[RIC2]   Richter, H.: Eine einfache Axiomatik der subjektiven Wahrscheinlichkeit. Inst. Nazionale
               di Alta Matematica Symp. Mathem. Vol IX, 1972, pp. 59—77.

Supporting Model-based Diagnosis with Explanation-based Learning and Analogical Inferences

Priv.-Doz. Dr.-Ing. Dieter Specht

Dipl.-Inform. Sabine Weiß

Fraunhofer Institute for Production Systems and Design Technology

and

Technical University Berlin

Institute for Machine Tools and Manufacturing Technology

Pascalstr. 8-9, W-1000 Berlin 10, Germany

SABINE.WEISS@IPK.FHG.DE

Abstract

This paper introduces two learning approaches used in different phases of a diagnostic system's life cycle. First, an initial knowledge-base is built using an explanation-based learning approach which generates diagnostic rules. A functional model of the object to be diagnosed constitutes the necessary domain knowledge. Later when the system is operational, analogical inferences which utilize taxonomic information continue to improve its diagnostic performance. In this way knowledge which is 'objectivized' by the model can be acquired, greatly improving the performance of a pure model-based diagnosis while preserving the advantages of the model-based approach.

Keywords: analogical inference, explanation-based learning, machine learning, model-based diagnosis

1 Introduction

Until some time ago the prevailing species of diagnostic expert systems solely relied on empirically found associations between symptoms, faults, and their causes, often represented in the form of rules. Beside the successful realizations reported, severe limitations of this approach became known. Frequently, after a painstaking knowledge acquisition process the knowledge base aged with the changes occuring in the system's application area. Moreover, inconsistencies appeared in the empirically determined knowledge base and the diagnostic process abruptly failed as soon as situations appeared which were not considered during knowledge acquisition.

In response to these problems the so-called second generation expert systems or deep knowledge systems were introduced. Unlike the former systems with only 'shallow' knowledge about their application area, these systems are supplied with 'deep' knowledge — usually a model of the object or process under consideration — intended to largely elude the above mentioned limitations. Practical experiences show, though, that using this approach often results in long response times.

In this paper we introduce a hybrid approach which uses a model of the object to be diagnosed to guide the learning of diagnostic rules from single examples and which uses this model for analogical inferences or a model-based diagnostic process in case the learned rule base is not able to cope with a distinct symptom appearence. Following this approach promises to generate more efficient systems than realizing a pure model-based approach while preserving the desired characterisitics of deep knowledge systems.

The system currently under development is based on our former work on model-based diagnosis of a lathe (cf. [Spur & Weiß 90]). The representation formalism allows to model an artifact's construction as well as its functionality. For purposes of this representation, a component-oriented approach similar to the ones known from the literature is employed (cf. [Davis 84, deKleer & Brown 84, Struß 89]). A detailed modeling of the interaction of the affected components is possible by optional connections between component parts. Substructures of components and connections allow for a view of the model to be recurrently refined at different levels of abstraction. The functionality of the artifact's components and its connections is represented in the form of constraints. They describe the correct behavior of machine elements and subsystems involved in the production process, i. e. constraints indicate the required relationship of inputs to outputs of a component or connection in a production process considered to be fault-free. Inputs and outputs are represented internally by variables which are assigned concrete values in the course of the diagnostic process.

Beside this representation formalism for the object to be diagnosed there exists a taxonomic hierarchy of components and parts. While the model mainly describes a system's topology and behavior the hierarchy contains all components and connections modelled so far. Thus, it serves as a library for all kinds of artifacts. By this means, any component description no matter how detailed or general it may be is reusable in different models. Its main purpose from the point of view of learning, though, is to provide a base for the analogical inference process.

The following describes a variant of explanation-based learning relying on the representation outlined above and an approach for learning from analogies. The concluding paragraph shows how these learning mechanisms interact with the model-based diagnostic system that aids in adjusting high voltage circuit-breakers.

2 Acquisition of an Initial Knowledgebase

One possible way to take advantage of experienced diagnostic knowledge while preserving a consistent knowledge base is to 'objectify' empirical rules against a correct and complete domain description. A major characteristic of explanation-based learning approaches is that these methods do not acquire fundamental new knowledge. They transform already known information into a form useful to the problem solver instead. Generally, this transformation of the known information, called 'operationalization', consists of two steps:

- Proceeding from a complete and correct description of the problem domain, an explanation is generated on why an example presented is an example for a concept to be learned.

- Subsequently this explanation is generalized in such a manner that it is applicable to a whole class of examples which contains the original example.

This procedure may be interpreted as objectivizing an exemplary concept description. Applied to the model-based diagnostic approach the first step can be reformulated as follows:

- Proceeding from the model of the object to be diagnosed, an explanation is generated on why an example rule is an example for a diagnostic concept to be learned.

Here, the model serves as the domain knowledge from which to infer explanations for a diagnosis of the symptoms given. These explanations constitute the base for an operational and objective definition of diagnostic rules. In the following, a description of an approach for the derivation of explanations and diagnostic rules is given. This approach follows the proposal of DeJong and Mooney [DeJong & Mooney 86] and adapts it to the representation formalism of the model-based reasoning system depicted above. Production rules with conditions describing symptom appearances and their causal faults as consequences represent the diagnosis examples.

Step 1 All variables of the model occuring in the example are assigned the specified values. These values serve as seeds for the formation of an explanation structure of the fault which the given example is an instance of.

Step 2 The ensuing constraint propagation infers the example's complete causal interpretation. It consists of variable values and their dependencies, i. e. information on which constraint, variables, and assumptions lead to a derived variable value. The constraint propagation uses data and justifications supplied by an assumption-based truth maintenance system which likewise supports the diagnostic process (see [Spur & Weiß 90] and [deKleer & Williams 87]).

Step 3 A first explanation is generated which eliminates all values not supporting the fault, i. e. all values not supporting the variable value given in the example's consequent. Two cases may be distinguished. On the one hand, values not connected to the fault in any way by a chain of values produced in the constraint propagation process must be excluded from the further procedure. This case only arises if an example contains conditions that are not proper symptoms in a narrow sense. All those values which only have one connection to the causal chain, we call 'leaves' because they are terminal nodes (see Fig. 1). If they are not conditions or consequences or if their values cannot be tested on the running manufacturing system they must also be eliminated.

Step 4 The representation formalism supports the description of functionality on different levels of abstraction. In addition, constraints may be defined between different levels of abstraction. Because of this, less specific values may be derived in the course of constraint propagation. Therefore in this step, a more specific explanation is given which excludes all those values of a more abstract level.

This procedure results in a constraint net from which a rule for the given example is constructed. The rule's conditions are formulated using all leaves apart from the one describing the fault itself and all symptoms not derivable from leaves. The

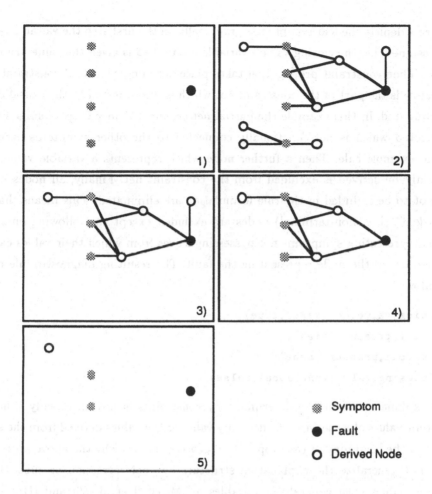

Figure 1: The Explanation-based Learning Procedure

consequence remains the same as in the example. In this way, the acquisition of an initial knowledge base by learning from single diagnostic examples is possible.

An example taken from the system's application to adjusting high voltage circuit-breakers may illustrate the procedure. It is concerned with a circuit-breaker's hydraulic operating system. As a training instance consider the following rule formulated by object-attribute-value-triples:

```
(closing.valve,activated,true),
(red.oil,pressure,high),
(main.valve,open,false),
(blue.oil,pressure,high)
→ (closing.valve,functional,false).
```

Figure 1 depicts the learning process graphically. In the first step the variable values
are assigned, i.e. the closing valve's variable `activated` is given the value `true` and
so on. Then constraint propagation takes place deriving the actual constraint net.
Afterwards the part of the constraint net which is unconnected to the causal chain
is eliminated. In this example the partial net around `(blue.oil,pressure,high)`
is affected which is not functionally connected to the other symptoms occuring
in the example rule. Then a further node which represents a variable value that
can only be derived is excluded from the constraint net. Finally, all nodes which
are not to be included in the rule formulation are eliminated. This means that all
"non-leaf" (i. e. non-terminal) nodes are excluded except the following ones: (a)
nodes representing symptoms not possessing leaves from which their values can be
derived or (b) the node representing the fault. The resulting diagnostic rule reads
as follows:

```
(closing.valve,activated,true),
(red.oil,pressure,high),
(brown.oil,pressure,none)
→ (closing.valve,functional,false).
```

The explanation structure determined by constraints is not universally valid for
different values of its leaves, but it is only valid for the values derived from the seeds
given in the beginning of the propagation process. This is why the above procedure
does not generalize the explanation structure by goal regression or substituting
specific values with generalized variables as [Mitchell et al. 86] and [DeJong &
Mooney 86] respectively propose. Yet, applying the procedure yields a diagnostic
rule which contains the most specific necessary conditions for the occurence of the
exemplary fault. It is efficient in the sense that it does not contain any superfluous
conditions. However, the main achievement is the desired 'objectification' of the
example rule, if the system is able to construct a rule in this way. If the system
fails, this indicates incompleteness or incorrectness of the model.

3   Analogical Inference During Diagnosis

Usually knowledge once generated by learning from examples provided prior to a
knowledge-based system's use is not sufficient to cover all cases actually occuring. To improve the system's diagnostic capabilities and to avoid abrupt failures
it appears suitable to apply a method exploiting analogies between symptoms yet

inexplicable and diagnostic rules already known . Originally, the process of learning from analogies consists of four phases. First, a domain from which analogous inferences may be drawn to solve the actual problem in the problem domain must be identified. The second step consists in finding concepts which possess similarities in the problem domain and the analogous domain chosen. To allow the transformation of solutions from the analogous domain into the problem domain a mapping between these concepts is necessary. Finally the newly derived knowledge must be stored to make it available in the future.

Since in our case the problem domain and the analogous domain are the same the first phase is omitted here. Our approach basically follows the analogical inference process as suggested by [Carbonell 86]. The following describes the procedure's course:

Step 1 Because symptoms are specified by variable values while rules are represented as object-attribute-value-triples, a transformation of the symptoms' representation according to the rule formalism is made.[1] This transformation is done using the model's information about the objects, in which the variable occurs.

Step 2 A list with possible candidates for the construction of an analogous diagnostic rule is created matching the symptom appearance with the conditional part of each rule. The candidate list is ordered with respect to the 'quality' of the match. This quality is determined by the number of complete, partial, and 'non' matches using weighting factors. Here, complete means that objects, attributes, and values of one symptom and one condition matched are the same; partial means that the objects are different while the attributes and values are the same. Only rules for which the match exceeds a threshold are included in the candidate list.

Step 3 Beginning with the qualitatively best rule the candidate list is processed until either (i) a rule is found which possesses only complete matches, or (ii) a rule is found for whose partial matches an analogous pattern can be constructed, or (iii) the end of the list is reached. The first case simply means that the diagnostic rule is more specified than the symptom appearance given, i. e. it contains more conditions than there are symptoms. The

---

[1] A variable stands for a physical quantity which may belong to several different objects of the model at the same time. Thus, a symptom being a variable value does not stipulate its assignment to a distinct object.

interesting case is the second one where the system's taxonomy is used to derive analogous object-attribute-value-triples that constitute the new rule. Finally the third case indicates the failure of the attempt to find an analogous diagnostic rule.

Step 4 If the previous step was successful, before adding it to the knowledge base the newly found rule is presented to the human using the system who is to decide, if this new rule is sensible.

For instance, consider a rule similar to the above example:

`(closing.solenoid,activated,true),`
`(red.oil,pressure,high),`
`(brown.oil,pressure,none)`
`→ (closing.valve,functional,false)`

and assume, in addition, the following symptoms which cannot be explained by this rule:

`(closing.solenoid,activated,true),`
`(brown.oil,pressure,high),`
`(yellow.oil,pressure,none).`

Surely, the rule will be part of the candidate list, because there are one complete and two partial matches between the rule's conditional part and the set of symptoms. Assume further it belongs to the ones processed in Step 3. Since the rule does not possess only complete matches, the system consults the domain model and its taxonomic information. Here it finds that both objects of the rule's partial matches, `red.oil` and `brown.oil`, are connections to the diagnosis object `closing.valve`. This relationship is shown in Figure 2. To construct an analogous pattern the system now searches for an object to which the objects of the symptom set's partial matches, `brown.oil` and `yellow.oil`, are connections. As Figure 2 shows such an object, namely `main.valve`, exists. Because it also belongs to the same class as `closing.valve`, the following rule is formulated analogously:

`(closing.solenoid,activated,true),`
`(brown.oil,pressure,high),`
`(yellow.oil,pressure,none)`
`→ (main.valve,functional,false).`

Beside the case depicted here, there are of course further modelled relations as for

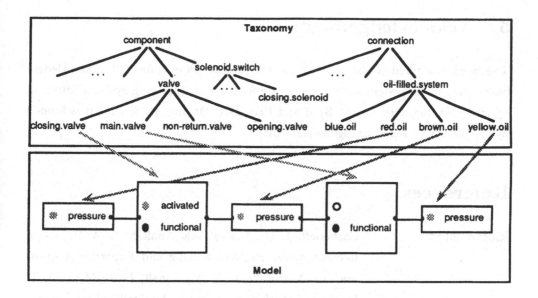

Figure 2: Parts of the System's Knowledge Used for Analogical Inferences

example component–subcomponent relations between different objects which serve as a base for the construction of analogies. By this means, the system is able to react to situations not covered by the set of diagnostic rules already learned thus improving its performance gradually.

4   Concluding Remarks

This paper describes two learning mechanisms applied in two phases of a diagnostic system's life cycle. In the pre-production process phase, an initial knowledge base is built from single training instances of diagnostic rules by means of an explanation-based approach. After setting up the initial knowledge base, the system is applied to a diagnostic task. At first, the system searches the knowledge base for rules diagnosing given symptoms. If this fails, the analogical inference process tries to find new rules. Should this be unsuccessful too, the system falls back on a model-based diagnostic approach. At present, the task for which the modelling is carried out involves adjusting high voltage circuit-breakers and their hydraulic operating systems.

5 Acknowledgements

The work described is part of the research project 'Development of a knowledge-based diagnostic system with learning ability for manufacturing systems' directed by Prof. Dr.-Ing. Drs. h.c. G. Spur and Priv.-Doz. Dr.-Ing. D. Specht. It is funded by the Deutsche Forschungsgemeinschaft.

References

[Carbonell 86]   Carbonell, J. G.: Derivational Analogy — A Theory of Reconstructive Problem Solving and Expertise Acquisition. In: Michalski, R. S.; Carbonell, J. G.; Mitchell, T. M. (eds.) : Machine Learning — An Artificial Intelligence Approach, Vol. 2. Los Altos, Calif.: Morgan Kaufmann, 1986.

[Davis 84]   Davis, R.: Diagnostic Reasoning Based on Structure and Behavior. In: Artificial Intelligence 24(1984), pp. 347–410.

[DeJong & Mooney 86] DeJong, G.; Mooney, R.: Explanation-Based Learning — An Alternative View. In: Machine Learning 1 (1986) 2.

[deKleer & Brown 84] de Kleer, J.; Brown, J. S.: A Qualitative Physics Based on Confluences. In: Artificial Intelligence 24(1984), pp. 7–83.

[deKleer & Williams 87] de Kleer, J.; Williams, B. C.: Diagnosing Multiple Faults. In: Artificial Intelligence 32(1987), pp. 97–130.

[Mitchell et al. 86] Mitchell, T. M.; Keller, R.; Kedar-Cabelli, S.: Explanation-Based Generalisation — A Unifying View. In: Machine Learning 1 (1986) 1.

[Spur & Weiß 90]   Spur, G.; Weiß, S.: Application of Model-based Diagnosis to Machine Tools. Proc. of the International Workshop of Expert Systems in Engineering, September 24-26, 1990, Vienna. Berlin, Heidelberg, New York: Springer, 1990.

[Struß 89]   Struß, P.: Structuring of Models and Reasoning About Quantities in Qualitative Reasoning. Ph. D. Thesis, University of Kaiserslautern, January 1989.

A knowledge-based system * **
for the diagnosis of waste-water treatment plants

Lluís Belanche[1], Miquel Sànchez[1],
Ulises Cortés[1] and Pau Serra[2]

[1] Universitat Politècnica de Catalunya Departament de Llenguatges i Sistemes
Informàtics Pau Gargallo, 5. Barcelona. 08028. Catalonia, SPAIN
[2] Universitat Autònoma de Barcelona Departament de Química Edifici C, 08193
Bellaterra (Barcelona). Catalonia, SPAIN

Abstract. In this work we discuss the development of an expert system with
approximate reasoning which resorts to a new methodology for attribute selection in knowledge-based systems. First, we make a survey of the pu-
rifying process and its problems, as well as those of conventional automatic
control methods applied to industrial processes. Next, we establish a definition of the relevance concept for a given set of attributes, which includes
the special case of non-relevant attributes or **nought attributes**. A new
heuristic is here proposed in such a way that it finds out the more relevant
attributes from those initially selected by the expert, reducing the cost of the
formation & validation of decision rules and helping to clarify the underlying structure of a non well-structured domain as are waste-water treatment
plants.

1 Introduction

A malfunction of a waste water treatment plant is a social and biological hard
problem. A poorly treated waste water outside the plant can achieve dangerous
consequences for human beings as well as the nature itself.

In this paper we show the application of Knowledge Based Systems to the diagnosis and control of waste water treatment plants. The work has been developed in
a wide cooperation among the Chemical Engineering group of the UAB, the Functional Languages and Artificial Intelligence of the UPC and experts from real plants.
The use of an AI tool as the Expert Systems, as a first approach, gives a new vision
to control and diagnose industrial processes.

During the last years the AI research effort in the field of industrial processes
control has been increasing quite a lot ([**MAED89**], [**FLOR89**], [**SANZ89**].) The conventional automatic process control systems (feed-back control, feed-forward control,
optimal control, adaptive control, ...) have to cope with some difficulties:

- The problem's complexity.
- A non well-structured domain.
- Most information is neither numeric nor quantified.

* Partially supported by GRANT TIC-90 801/C02. CICyT. ESPAÑA
** Partially supported by GRANT ROB-89 0479-C03-02. CICyT. ESPAÑA

— Uncertainty or approximate knowledge.

— A dynamic system.

In addition, we can say that the classical methods work on the "normal" state of the plant but not in other "extra-normal" states of working. Bringing to mind the main advantages of expert systems and the above difficulties, one could think that match very well. Hence, we have developed an expert system to the diagnosis with uncertainty reasoning [SANC91], [SANC90].

2 Goals

The main goal, already acquired, is to develop an expert system with approximate reasoning applied to diagnosis in waste water treatment plants in order to make easier the plant control and supervision, to increase the water quality outside the plant and to reduce the cost of the operations. The plants taken as models are based on the main technic usually applied: the activated-sludge process [ROBU90]. In the future, we shall extend the system to build a Supervisor Knowledge Based System (see Fig. 1) applied to diagnosis and control, integrating both heuristic knowledge and numerical processing algorithms (statistical estimations, ...).

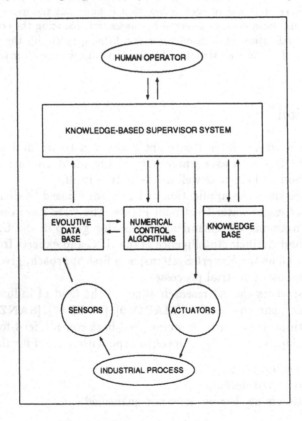

Fig. 1. A Supervisor Knowledge-Based System.

3 Development and Design

We have built up an expert system called DEPUR, over the shell MILORD, in order to support the approximate reasoning. The main problem on a waste water treatment plant is the balance's break between biomass and substrate. This problem may be originated in one of the three treatments that experiments the water until it exits the plant: primary, secondary and tertiary. Each one carries out the elimination of a concrete pollutant. Thus, the diagnosis process tries to determine the causes of the problems detected. An expert system is composed by the Data Base, the Knowledge Base, the Meta-knowledge Module, and other general modules independent of the application domain. Hence, to develop the system it is required to build the Data Base, the Knowledge Base as well as the Meta-knowledge Module (the control strategies). The design process of an intelligent system can be expressed commonly as shown in Fig. 2.

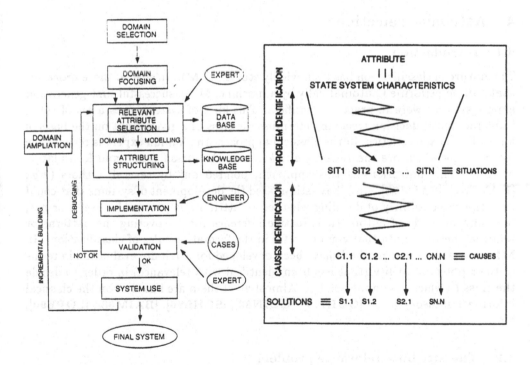

Fig. 2. Design process of an intelligent system. Diagnostic process in an Expert System

A waste water treatment plant is composed usually by two independent but interacting subsystems: sludge line and water line. We have focused on the water line subsystem domain. The *domain modelling phase* includes relevant attribute selection and attribute structuring. The attribute selection process has been developed twofold: firstly, in a typical manner (expert attribute selection) and, secondly, by means of a new automatic attribute selection approach based on the relevance and potential utility concepts [BELA91]. The knowledge engineering process (attribute structuring) relies upon two approaches:

⋆ *Decision trees*: the classical approach derived from the interactions between the expert and the engineer. This approach is governed by the cause-effect criteria. We have used multi-valued decision trees from which inference rules are derived. The rules point out to discover the causes of a certain problem or situation previously detected.

⋆ *Automatic classification techniques*: more recent approach derived from automatic learning in AI. We have used the LINNEO software in this phase, based on fuzzy non well structured domain classifications. From the characterization of the classes obtained more general inference rules are derived. Them lead to know the most probable working situation of the plant.

Decision trees have been used in the causes identification phase and automatic classification techniques in the problem identification process as shown in Fig. 2.

4 Attribute selection

4.1 Introduction

The nature of classification has been widely studied in **ML**. In special those processes that can be performed automatically by a machine. Many successful first-generation expert systems were in practice classifiers. This kind of systems uses a set of rules, implemented as decision trees, to determine the class of a given input on the basis of a set of its characteristics. In the classical approach, the human expert is responsible for deciding which are the *relevant attributes* for the classification and formulation of rules. The limitations of this approach, pointed out by several authors (*e.g.*: [HAYE84], [BAIM88], etc.), have stimulated the development of systems that could take the responsibility of deciding whether an attribute is potentially useful or not, by using methods of automatic inductive inference for discovering the patterns or relations between data that can be useful in the formation rules of classifications. In **ML**, several heuristic methods have been developed for attribute evaluation in terms of their potential utility (that has been identified with relevance) in order to decide the class to which the input belongs. Almost all of them are based upon the classical information theory (*e.g.*: [QUIN79], [QUIN86], [SCHL86], [BAIM88], [LOPE90], etc.).

4.2 The attribute relevance problem

In this section a characterization of the attribute relevance problem is made, and a formal definition of relevance and non-relevant attributes or **nought** attributes is proposed.

Characterization of relevance: Let $\mathcal{U}$ be a universe and $\mathcal{C} = \{C_1, C_2, \ldots, C_m\}$ a classification (partition) of $\mathcal{U}$. Let us assume that the elements of $\mathcal{U}$ can be represented by the ordered n-tuple $(A_1(u), \ldots, A_n(u))$ of attributes (*i.e.* measurable characteristics) of the objects in $\mathcal{U}$, with ranges $X_1, X_2, \ldots, X_n$, respectively. Formally speaking, every A_i is a function with domain $\mathcal{U}$ and range X_i. Notice that, in general, $\{(A_1(u), \ldots, A_n(u)) \mid u \in \mathcal{U}\}$ is a subset of $X_1 \times \ldots \times X_n$ but not necessarily equal. Let us designate r_i as the cardinality of X_i.

Intuitively expressed, the more *information* an attribute supplies to a classification, the more relevant to the classification it is. The natural way to *measure* the information an attribute supplies is by means of its capacity to discriminate the elements of $\mathcal{U}$. The relevance of an attribute is not an absolute property, but is relative to a classification or interpretation. An attribute can be very important for a classification but *irrelevant* in other. For instance, people's color of skin is not important to determine (discriminate) the kind of cardiovascular diseases they have, but it is important to determine their ethnic classification. In practice, the potentially useful initial set of attributes should be selected by the expert [BAIM88].

We state here that the relevance of an attribute depends not only on the proportion of objects that this attribute discriminates, but in *how* it interacts with the rest of the attributes. Any formal definition of relevance must then capture these two concepts. The following definitions are taken from [NUÑE91], and will serve as a theoretical approach to the relevance problem.

Definition The attribute A_i is more relevant than the attribute A_j, if and only if the minimum number of attributes to be added to A_i, to obtain a fully discriminating set, is less than the number required to A_j. If this number is the same for both attributes, then we conclude that they are equally relevant.

This definition could be expressed as: Let $N = \{1, 2, \ldots, n\}$, and, for every $S \subset N$, let

$$A_S = \{A_s \mid s \in S\},$$
$$S_i = \{S \subset N \setminus \{i\} \mid \{A_i\} \cup A_S \text{ is fully discriminating}\}$$
$$S_j = \{S \subset N \setminus \{j\} \mid \{A_j\} \cup A_S \text{ is fully discriminating}\}$$

Then A_i is more relevant than A_j if, and only if,

$$min\{\#S \mid S \in S_i\} < min\{\#S \mid S \in S_j\}.$$

A_i is equally relevant than A_j if, and only if,

$$min\{\#S \mid S \in S_i\} = min\{\#S \mid S \in S_j\}.$$

Definition Let S_1 and S_2 be subsets of N, and let

$$S_{S_1} = \{S \subset (N \setminus S_1) \mid A_{S_1} \cup A_S \text{ is fully discriminating}\}$$
$$S_{S_2} = \{S \subset (N \setminus S_2) \mid A_{S_2} \cup A_S \text{ is fully discriminating}\}$$

A_{S_1} is more relevant than A_{S_2} if, and only if,

$$min\{\#S \mid S \in S_{S_1}\} < min\{\#S \mid S \in S_{S_2}\}.$$

The sets of attributes A_{S_1} and A_{S_2} have the same relevance $\mathcal{R}$ if, and only if,

$$min\{\#S|S \in S_{S_1}\} = min\{\#S|S \in S_{S_2}\}.$$

The intuitive idea that underlies the definition is that two sets of attributes, regardless of their cardinality, are equally relevant if they have the same discrimination capacity, or, equivalently, if *the amount of information that is left for them to be fully discriminating is the same.*

Hence, in order to determine if a given set of attributes is more relevant than another, we should generate, at least, all the fully discriminating (of maximum relevance) sets that contain any of the two given sets. But this is a non-practical and intractable method from the point of view of complexity. Thus the proposed solution has been to establish heuristics to evaluate the potential utility of an attribute, in such a way that they pick up, successively, those which will lead to build decision trees close to the optimum one. The solutions proposed so far have some elements in common, and its discussion lead us to to the heuristic that we propose later on. For a complete survey of these measures and their characteristics, see [NUÑE91].

The definition of relevance for a given set of attributes $\mathcal{A}$ introduces an equivalence relation $\mathcal{R}$ in the power set $\mathcal{P}(\mathcal{A})$, where the equivalence classes are formed by equally relevant sets. Joined with the previous definition, this allows us to introduce a *total order*, $\leq$, in the quotient set of equivalence classes $\mathcal{P}(\mathcal{A})/\mathcal{R}$.

Definition

Let $\mathcal{A}_{S_1}$, $\mathcal{A}_{S_2}$ be two attribute sets and $[\mathcal{A}_{S_1}]$, $[\mathcal{A}_{S_2}]$ their equivalence classes. Then:

$$[\mathcal{A}_{S_1}] \leq [\mathcal{A}_{S_2}]$$

if, and only if,

$$min\{\#S|S \in S_{S_1}\} \leq min\{\#S|S \in S_{S_2}\}.$$

Note that the classes $[\mathcal{A}]$ and $[\emptyset]$ are, respectively, the maximum, and minimum with respect to this order, *i.e.* $[\emptyset] \leq [\mathcal{A}_S] \leq [\mathcal{A}]$, $\forall \mathcal{A}_S \subset \mathcal{A}$. Another important remark is that if $\mathcal{A}_{S_1}$ is more relevant than $\mathcal{A}_{S_2}$, then any subset of $\mathcal{A}_{S_1}$ is more relevant than any other subset of $\mathcal{A}_{S_2}$. The problem, in practice, is to obtain the *optimum representative of [A], which classifies the original set U with the minimum effort.* This is the aim of the above mentioned algorithms.

The nought attributes: It has been pointed out so far that the relevance of a set of attributes is not inherent to the attributes, but it depends on the classification. Frequently we have to deal with situations in which an attribute, or in general, a set of attributes, are not important to a given classification process (though they could be relevant to a different one), i.e., *they have no relevance for a given classification.* This null relevance will be referred to as nought [SANC89] relevance. Hereafter, we will refer to non-relevant attributes as **nought** attributes. In such cases these attributes are yet given and cannot be ignored. What we might want is that such attributes do not disturb the classification considering only the *non*-nought ones.

In accordance to the relevance definition, **nought** attributes are those without discrimination capacity. Therefore, any **nought** set has null relevance and is an element of $[\emptyset]$, the empty set.

Definition A set of attributes $\mathcal{A}_n$ is nought if, and only if, $[\mathcal{A}_n] = [\emptyset]$.

Note that any nought set of attributes $\mathcal{A}_n$, when added to a given set of attributes $\mathcal{A}_S$, does not alter its relevance, that is, the discriminating capacity of class $[\mathcal{A}_S]$ is the same than that of class $[\mathcal{A}_n \cup \mathcal{A}_S]$, $\forall \mathcal{A}_S \subset \mathcal{A}$.

Theorem Let $\mathcal{A}_n$ be a nought set of attributes, then $[\mathcal{A}_n \cup \mathcal{A}_S] = [\mathcal{A}_S]$, $\forall \mathcal{A}_S \in \mathcal{P}(\mathcal{A})$

4.3 A measure of the potential utility of an attribute

Given $i \in N$, for every a_{ij} $(j = 1, 2, \ldots, r_i)$, let $p_{jk} = \#(A_i^{-1}(a_{ij}) \cap C_k)$, where $k = 1, \ldots m$, be the number of objects taking the value a_{ij}, and placed in the class C_k. Since $C_1, C_2, \ldots, C_m$ is a partition of the universe, we have that

$$p_j = \sum_{k=1}^{m} p_{jk} = \#A_i^{-1}(a_{ij}).$$

Let $\overline{p_j}$ be the arithmetic average of objects with value a_{ij} in the classes, that is, $\overline{p_j} = \frac{\#A_i^{-1}(a_{ij})}{m}$. Let $D = \{(p_{j1}, p_{j2}, \ldots, p_{jm}) \mid j = 1, 2, \ldots, r_i\}$, $i \in N$. Then, the potential utility of each attribute A_i may be expressed as a function from set D to $R^+ \cup \{0\}$.

Definition Let $u : \mathcal{A} \to \mathcal{R}^+ \cup \{0\}$ be the function which evaluates the potential utility of an attribute, defined as follows:

$$u(A_i) = \sum_{j=1}^{r_i} \sum_{k=1}^{m} (p_{jk} - \overline{p_j})^2.$$

Note that such a function u is non-negative, and that $u(A_i) = \sum_{j=1}^{r_i} (m-1)S_j^2$, where S_j^2 is the statistical variance of the values $\{p_{jk} \mid k = 1, \ldots, m\}$.

Definition Given a pair of attributes, A_i, A_l, attribute A_i is potentially more useful than attribute A_l if, and only if, $u(A_l) < u(A_i)$.

That is, attribute A_i is more relevant than attribute A_l, if and only if,

$$\sum_{j=1}^{r_l} \sum_{k=1}^{m} (\#(A_l^{-1}(a_{lj}) \cap C_k) - \frac{\#A_l^{-1}(a_{lj})}{m})^2$$
$$< \sum_{j=1}^{r_i} \sum_{k=1}^{m} (\#(A_i^{-1}(a_{ij}) \cap C_k) - \frac{\#A_i^{-1}(a_{ij})}{m})^2.$$

For every $j = 1, 2, \ldots, r_i$, let $u_j((p_{j1}, p_{j2}, \ldots, p_{jm})) = (m-1)S_j^2$. The sum $(m-1)S_j^2$, for every j, stands for the contribution of value a_{ij} to the potential utility of attribute A_i, and can be interpreted as the *potential utility of value a_{ij}*. This assumed, the potential utility of an attribute is the sum of the potential utility of its values.

A useful feature of the function u_j is that, the farther from the average $\overline{p_j}$ the values $p_{j1}, p_{j2}, \ldots, p_{jm}$ are, the greater the function's value is. An increment to some values makes a decrement to some others. In particular, if distribution is altered in such a way that the proportion of objects in one of the classes is incremented —being decremented in the same amount in any other class—, resulting in a distribution not as random as before, then the potential utility of the value is greater. This can be formally expressed as:

If for some positive integer z, $z \leq p_{js}$ we consider $(p_{js} - z)$ and $(p_{jt} + z)$, $s, t \in M = \{1, \ldots, m\}$ and

$$(p_{js} - \overline{p})^2 + (p_{jt} - \overline{p})^2 < ((p_{js} - z) - \overline{p})^2 + ((p_{jt} + z) - \overline{p})^2,$$

then

$$u_j((p_{j1}, p_{j2}, \ldots, p_{jm})) < u_j((p_{j1}, \ldots, p_{js} - z, \ldots, p_{jt} + z, p_{t+1}, \ldots p_m)).$$

Another useful feature of the estimation of an attribute's potential utility is the fact that u_j is very *sensitive* to the number of objects that each value classifies: the greater the number of objects is, the greater u_j is, being *very* significative the difference between a value that classifies a lot of objects and one not classifying many. As a result of this *sensitivity*, the evaluation of the potential utility of each attribute's value —and that of the attribute itself— is better in much cases, since it takes into account the classification ability of each value and, hence, of the attribute.

The function $u(A)$ is a sum of the $u_j((p_{j1}, p_{j2}, \ldots, p_{jm}))$, $j = 1, 2, \ldots, r_i$. From the characterization of functions u_j is possible to figure out the features of the function u. The following are properties of the functions u_j, which are next spread out for the function u. Such properties are useful to estimate the potential utility of an attribute. For simplicity, we do not include the subindex j for p_{jk}. Let $(p_1, p_2, \ldots, p_m) \in D$, arbitrary.

Property 1 (Symmetry)
$$u_j((p_{\sigma(1)}, p_{\sigma(2)}, \ldots, p_{\sigma(m)})) = u_j((p_1, p_2, \ldots, p_m)),$$

for every permutation σ of degree m.

Property 2 (Maximum of the function u_j)

$$u_j((p_1, p_2, \ldots, p_m)) \leq u_j((0, \ldots, 0, p_1 + p_2 + \ldots + p_m, 0, \ldots, 0)),$$

for every position k, $1 \leq k \leq m$.

Property 3 (Minimum of the function u_j)

$$u_j((\overline{p_j}, \overline{p_j}, \ldots, \overline{p_j})) \leq u_j(p_1, p_2, \ldots, p_m).$$

(In fact, $u_j((\overline{p_j}, \overline{p_j}, \ldots, \overline{p_j})) = 0$).

The first property says that what actually matters is the number of objects classified in a given class, rather than which class it is. Moreover, in what concerns to function $u(A)$, no matters which is the attribute's value assigning the biggest number of objects to a given class, but that there exists such a value. In particular (property 2), u_j takes its maximum value when all the objects are concentrated in the same class. Finally, u_j is minimum when we have a uniform distribution on the classes (property 3). Then we say that value a_{ij} has its minimum potential utility.

Another remarkable property of function u is that, if the number of objects that a value assigns to one class equals the sum of the number of objects that two values assign to that class, then the potential utility of the single value is greater or equal than that of the two values. This means that the function u does *not* enhance attributes having a lot of values, fundamental for the sake of efficiency.

In order to obtain an operational characterization of the **nought** attributes, those having potential utility zero could be regarded as potentially nought, or we could even force an attribute to have potential utility zero if we wish to *neutralize* it for a classification.

4.4 Validation of the measure

A model for fuzzy classification: LINNEO⁺. This (LINNEO⁺, [MART91]) is a software for fuzzy classification and rule generation which resorts to analytical and empirical knowledge acquisition techniques. The human expert abstracts a collection of observations and a set of attributes he thinks of as relevant. At the same time, the expert is also able to express what he already knows about the domain. This knowledge is represented as constraints and/or examples on classes already known to exist.

Starting with this knowledge and data, LINNEO⁺ uses induction over the observations to generate a classification. The expert then evaluates the results and eventually modifies some attributes or some aspects of his domain theory in order to start the classification again. When a correct classification is obtained, fuzzy rules are derived from the generated classes.

The relevance module: A relevance module has been attached to LINNEO in order to guide it in the classification process. This is accomplished in a cooperative way: firstly LINNEO produces a classification in accordance with the expert. Next, using the initial domain theory and the resultant classification, the relevance module produces a list of the relative relevance of the attributes to that classification. Now, this list is subject of an eventual modification by the expert. In addition, a threshold is established in such a way that the module generates a final list in which just a few of the initial set of attributes (25%, say), are present. The classifier then restarts using this new list as the final set of attributes and a new classification is obtained. This is expected to be very close to that obtained resorting to the entire initial list. The attributes not considered are then taken as to be nought. A further approach has been developed in order to improve the final relevance list. Given a classification, we can apply the relevance module to each single generated class, then considering only those objects belonging to that class. What we obtain is the relative relevance of the attributes when considered as to characterize a subclass of the classification. An interaction with the expert is also possible here for each of the lists of relevance, and local thresholds can be derived. Next, a global list of relevance is formed, by means of merging all the subclasses' relevance lists. How must one interpret these local lists? The first attributes of each list are those more relevant to discriminate between the objects of the class it refers to. The last ones (in order of relevance) will serve to distinguish that class as a whole from the rest. Not surprisingly, these attributes appear on top of the original global list, and the last attributes of this list appear as very relevant in some (often just one) local list.

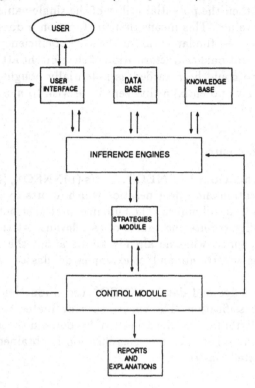

Fig. 3. General architecture of DEPUR.

5 Architecture

The architecture of the prototype system DEPUR (see Fig. 4) is a classic one of an expert system, but we have considered some important aspects as the microbiological and temporal ones. The dictionary of the data base is composed by 167 attributes, wit a 15The application groups 79 rules and 12 modules. Also we have obtained 15 metarules, oriented to control and lead the inference process. The modularity of the application, the self-explanation and the user friendly interface proportioned by the shell MILORD are other outstanding properties.

6 Results

We have obtained 10 classes, each one representing good related prototypical situations, that show the most common situations and problems in the working state of the waste water treatment plants. These classes lead the inference process in the diagnosis. The contraposition of system diagnostics against the answers of human experts, has yielded very good results. It has been tested with several prototypical situations in the working of a plant (bulking, foaming, ...). A diagnostic tool based on Knowledge Based Systems for a waste water treatment plant it is a useful approach in the field of intelligent systems when applied to industrial processes.

The new fuzzy techniques for automatic classification represent another important aspect in order to structure the knowledge domain. The treatment of uncertainty in a Knowledge Based System applied to industrial processes control is not very usual. In doing so, DEPUR supposes a main step towards the building of Ecology Preserving Intelligent Systems.

Another interesting point is the possibility to adapt the system to the specific characteristics of the waste water treatment plant that is under control. All in all, we can say that DEPUR is like a fresh air to the Conventional Automatic Control Processes.

In the future we have to reach a second phase: the development of a Knowledge Based Supervisor System, that incorporates classical control techniques with the diagnosis process. We have to extend the domain to sludge line, investigate in a more accurate way the microbiological aspects and make a better temporal information treatment.

7 Conclusions

Doubtless, the construction of a diagnosis tool for controlling a waste-water plant is an interesting contribution to the field of intelligent systems when applied to industrial processes, and can be of great help to the plant manager. The use of new attribute selection methods are also significant in what concerns to the extraction of the expert's knowledge, together with the treatment of uncertain knowledge, expressed and dealt by means of a fuzzy classifier and a module for attribute relevance. The system represents one step towards the construction of systems capable of preserving a natural resource as is water.

8 Suggested future work

Nowadays, research concentrates on the relationship between the notions of *potential utility* and *relevance*. It has been concluded that an attribute having zero as potential utility does not have to be a nought attribute. What is more, reciprocally, it is not necessary for nought attributes to have null potential utility. Moreover, an operational characterization of the *nought* attributes is needed, based on the proposed heuristic of potential utility, all in all leading to a classification process able to clarify & simplify the domain it is being applied to, as is −in our current development− waste-water treatment. In the future, the work has to advance towards the construction, as mentioned above, of an Supervisor Knowledge Based System including diagnosis and control, and integrating numerical control algorithms and heuristical and qualitative knowledge.

9 References

- [BAIM88] Baim, P. W. *A method for attribute selection in inductive learning systems.* IEEE Transaction on pattern analysis and machine intelligence 10 (6), (1988) pp. 888–896.
- [BELA91] Belanche, Ll. *To be or nought to be. An irrelevant question?.* Master Thesis. Departament de Llenguatges i Sistemes Informàtics. Universitat Politècnica de Catalunya (1991).
- [FLOR89] Floris, V., *Artificial Intelligence in the operation and management of water resources in South Florida.* Personal Comm. (1989).
- [HAYE84] Hayes-Roth, F. *The knowledge-based expert system: A tutorial.* Computer 17 (9), (1984).
- [LÓPE90] López de Mántaras, R., Crespo, J. J., *El problema de la seleción de atributos en el aprendizaje inductivo: nueva propuesta y estudio experimental.* In Memorias IBERAMIA 90, 2º Congreso Iberoamericano de Inteligencia Artificial. Ed. Limusa-Noriega. México, (1990) pp. 259–271.
- [MAED89] Maeda, K. *A knowledge-based system for the wastewater treatment plant.* A future generation computer systems 5, pp. 29-32. North Holland (1989)
- [MART91] Martín, M., Sangüesa, R., Cortés, U. (1991) *Knowledge acquisition combining analytical and empirical techniques.* In Birnbaum, L., and Collins, G. (eds.) Machine Learning: Proceedings of the Eighth International Workshop. San Mateo, CA: Morgan Kaufmann.
- [NUÑE91] Núñez, G., Alvarado, M., Cortés, U., Belanche, Ll. *About the attribute relevance's nature.* In Proceedings of the TECCOM-91. México, D.F.
- [QUIN79] Quinlan, J. R. *Learning efficient classification procedures and their application to chess end games* In Michalski, R.S., Carbonell, J.G. and Mitchell, T.M. (Eds): Machine Learning: An Artificial Intelligence Approach. Tioga, PA, California, (1984) pp. 463–482.
- [QUIN86] Quinlan, J.R. *Induction of decision trees* In Machine Learning 1(1). Kluwer Academic Publishers, Boston MA (1989), p81-106.
- [ROBU90] Robusté, J. *Modelització i identificació del procés de fangs activats.* Ph. D. Thesis. Departament de Química. Universitat Autònoma de Barecelona (1990).
- [SÁNC89] Sánchez, E. *Importance in knowledge systems.* Information System 14 (6) (1989) pp. 455-464.
- [SÁNC90] Sánchez, M. *DEPUR: una aplicació de MILORD al tractament de les aigües residuals en una depuradora.* Research report. (LSI 90-12). Departament de Llenguatges i Sistemes Informàtics. Universitat Politècnica de Catalunya (1990).
- [SÁNC91] Sánchez, M. *DEPUR: Aplicació del sistemes basats en el coneixement al diagnòstic en plantes de tractament d'aigües residuals.* Master Thesis. Departament de Llenguatges i Sistemes Informàtics. Universitat Politècnica de Catalunya (1991).
- [SANZ89] Sanz, R. *et alter. Adaptative control with a supervisor level using a rule-based inference system with approximate reasoning..* In A.I. in scientific computation: towards 2-generation systems. IMACS 1989.

- [SCHL86] Schlimmer, J. C., Fisher, D. *A case study of incremental concept induction*. In Proc. of the fifth national conference on artificial intelligence. Ed. Morgan Kaufmann, (1986) pp. 496–501.

Connectionism for Fuzzy Learning in Rule-Based Expert Systems

LiMin Fu, University of Florida

University of Florida
Department of Computer and Information Sciences
Gainesville, FL 32611

Abstract. [1] A novel approach to rule refinement based upon connectionism is presented. This approach is capable of performing rule deletion, rule addition, changing rule quality, and modification of rule strengths. The fundamental algorithm is referred to as the Consistent-Shift algorithm. Its basis for identifying incorrect connections is that incorrect connections will often undergo larger inconsistent weight shift than correct ones during training with correct samples. By properly adjusting the detection threshold, incorrect connections would be uncovered, which can then be deleted or modified. Deletion of incorrect connections and addition of correct connections then translate into various forms of rule refinement just mentioned.
Keywords: Expert System, Neural Network.

1 Introduction

The neural-network approach may not arrive at an inference structure close to that in expert minds. In recognition of this fact, more initial semantics and knowledge should be instilled into the network before it learns. This consideration leads to a special kind of neural network (or connectionist) called a knowledge-based conceptual neural network (KBCNN) [2], whose topology and initial connection weights are determined by initial knowledge which may just be imprecise and incomplete. We have shown that the KBCNN approach is capable of detecting and removing incorrect rules [3]. Extending the same idea, we will show in this paper that this approach is also able to perform various forms of rule refinement including rule deletion, rule addition, changing rule quality, and modification of rule strengths (the conditional belief values attached to the rule).

2 Knowledge Representation

We construct the KBCNN network in the following manner. First, data attributes or variables are assigned input units (nodes), target concepts or final hypotheses are assigned output units, and intermediate concepts or hypotheses are assigned hidden units. The structure is based on our earlier work [2]. Then, the initial domain knowledge determines how the attributes and concepts link and how the links are weighted.

The knowledge is described in production rules (if-then rules). Each rule has an antecedent (premise) consisting of one or more conditions as well as a single consequent. In the network configuration, the premise is assigned a hidden unit, each condition corresponds to an assigned attribute or concept node, and the consequent

[1] This work is in part supported by FHTIC (Florida High Technology and Industry Councils) and ACS (American Cancer Society).

corresponds to an assigned concept node. Each condition node is connected to the premise node which in turn connected to the consequent node. Under such construction, the rule strength corresponds to the weight associated with the connection from the premise node to the consequent node.

If a variable is binary-valued, then a node performing scalar computation suffices for simulating the variable. If a variable assumes a finite set of discrete values (may be hierarchical), then it is assigned a group of nodes storing a multi-dimensional weight matrix and transmitting vector messages to its neighbors. In fact, a variable with q values can be transformed into q binary-valued variables. In addition, if a variable assumes continuous values, we can either properly discretize it or assign it a node with graded response (the former alternative is more favored). Under such variable-node mapping, the activation level of a node translates into the belief value associated with the corresponding variable value.

A hidden unit is introduced to explicitly represent the conjunction of one or more conditions in a rule's premise part. Such a hidden unit is called a *conjunction unit*. Other hidden or output units are called *disjunction units*. A layer containing conjunction (disjunction) units is called a conjunction (disjunction) layer. The neural network built in this way comprises alternating conjunction and disjunction layers.

Suppose an attribute conjunction involves p positive attributes and q negated attributes. We set the initial threshold of the corresponding conjunction unit to 0 and set the initial weight for each input connection from positive attributes to $1/p$ (no lower than 0.25) and negated attributes to -1. The initial weight corresponding to a rule strength is set to 0.5 for a confirming rule and to -0.5 for a disconfirming rule. Such a weight is associated with a connection pointing from a conjunction unit to a concept node (intermediate or final). In the KBCNN network, we also add potential connections initialized with small random weights. The weight setting is intended to conserve the initial given semantics while leaving flexibility for the neural network to learn.

We propose to use the following incremental updating formula of the CF model [1] to combine activations:

$$f(X, Y) = X + Y - XY$$

This formula is applied to combine positive activations and negative activations separately. Then, the overall activation is the sum of combined positive activation and combined negative activation. Application of this formula results in a nonlinear transfer function in replacement of the sigmoid function normally used in artificial neural networks. Note that this combining function maps [-1, 1] (the interval of -1 and 1) into [-1, 1].

The issue raised is that this combining formula assumes independence between activations, which may not be true in practical scenarios. However, the Consistent-Shift algorithm (described next) has dealt with this issue adequately. If activations (either positive or negative) are dependent (correlated), then the combined activation should be less than is calculated by the above formula. If the activation of a node reaches the bounds (1 or -1) and the error drops to zero, then its incoming weights will not be modified further. It follows that the associated weights would be under-modified for correlated activations. Therefore, the weight shift would be smaller than expected. The Consistent-Shift algorithm handles such undesired consequence of the independence assumption by looking at the direction of post-training weight shift more than the magnitude of post-training weights. Aside from the incremental updating formula, the KBCNN model does not adopt other assumptions or features of the CF model. Hence, further discussion related to the criticisms against the CF model is not necessary here.

The main justification for the use of the CF-updating function is our observation that it can preserve more intended semantics than the sigmoid function. The semantics of a concept node is determined by its input weight pattern. Semantics may be lost because of concept shift, merge, or split during net training. How to minimize this loss is a crucial concern.

3 The Consistent-Shift Algorithm

The basic argument of the Consistent-Shift algorithm is the following. A physical system at an equilibrium will tend to maintain that equilibrium when undergoing small perturbation. Likewise, when a neural network is moved away from an established optimum state, it will tend to restore (relax toward) that state. Suppose in a neural network, most of connection weights are correct. Then, if we train the network with correct samples, the incorrect weights will be modified in the direction of minimizing their effect. What happens is that the incorrect weights will move toward zero and even cross zero during training. Since this weight shift may be small quantitatively, it should be interpreted more qualitatively.

The notion of *consistent shift* for connection weights is introduced as follows. If the absolute magnitude of a weight after training is greater than or equal to that of the weight before training and their signs are the same, then the weight shift is said to be semantically consistent with the weight before training; otherwise the shift is inconsistent. The function *Consistent-Shift* is defined by

$$\text{Consistent-Shift} = \left\{ \begin{array}{ll} w_a - w_b & \text{if } w_b > 0 \\ w_b - w_a & \text{if } w_b < 0 \\ |w_a - w_b| & \text{if } w_b = 0 \end{array} \right.$$

where w_a and w_b denote the weights after and before training respectively. A shift of weight is said to be consistent if its Consistent-Shift value is greater than or equal to zero; else it is inconsistent.

A pragmatic rule for detecting semantically incorrect connections is introduced as follows:

"If the Consistent-Shift value of a weight is less than a predefined negative threshold, then the weight shift is referred to as inconsistent and the pre-training weight is hypothesized as semantically incorrect".

The procedure for revising the neural network is given below:

- *Step 1.* Apply the backpropagation procedure [5] (adapted for the KBCNN model) until the system error converges on an asymptotic value.

- *Step 2.* Compute the Consistent-Shift value for each weight.

- *Step 3.* For each weight, Do

  - if the Consistent-Shift value is less than a selected negative threshold and if the absolute value of the post-training weight is less than a selected positive threshold, then delete the corresponding connection,

  - Else, retain the connection.

The Consistent-Shift algorithm refers to step 2 and 3.

Detected semantically incorrect connections can be totally eliminated from the network or be kept with post-training weights, depending on the magnitude of post-training weights. In case there is doubt, simulation is warranted.

After the neural network is revised by the Consistent-Shift algorithm, the revised network is translated into rules.

4 Related Work

In contrast to other knowledge-based neural networks such as KBANN [6], KBCNN provides bi-directional linkage between neural networks and rule-based systems. On one hand, a rule-based system can be mapped into a neural network. On the other hand, neural-network knowledge can be transferred back to the rule-based system. We expect that this idea can be extended to a rule-based system with variables where a unification mechanism such as the one described in [4] is needed.

5 Results

The KBCNN model was evaluated in the domain of diagnosing clinical icterus. The domain knowledge was mapped into a sparse four-layer neural network. The rule semantics was changed by altering the sign of the rule strength. A set of experiments were designed to demonstrate that the KBCNN model was effective for spotting incorrect rules, and to cross-validate the approach. The approach was able to handle a net with more than one hidden layer.

In addition, the KBCNN approach has been applied to revise the domain theory for recognizing promoter regions in DNA sequences. The initial theory [6] gave only 53/106 accuracy on 106 training instances (53 are positive and the others are negative instances). The domain theory was mapped into a sparse six-layer neural network. The theory revised by KBCNN yielded 103/106 accuracy. 100% accuracy can be achieved if a mechanism for learning new rules (rather than just revising old rules) is added. The KBCNN approach was also cross-validated in this domain. The details will be reported elsewhere.

6 Conclusion

A rule-based system can be mapped into a fine-grained parallel-distributed architecture. Under this construct, significant inconsistent shift of connection weights after training with correct samples suggests semantically incorrect connections. The viability of this approach for rule refinement has been demonstrated empirically. It can be scaled up to accommodate a large problem domain due to the sparse nature of the network.

References

1. Buchanan, B.G. and Shortliffe, E.H., *Rule-Based Expert Systems*, Addison-Wesley, Massachusetts, 1984.
2. Fu, L.M. and Fu, L.C., Mapping rule-based systems into neural architecture, *Knowledge-Based Systems*, Vol. 3, No. 1, 1990, 48-56.
3. Fu, L.M., Knowledge base refinement by backpropagation, to appear in *Data and Knowledge Engineering*, 7, 1992.
4. Holldobler, S., A connectionist unification algorithm, Technical report, ICSI-TR-90-012, International Computer Science Institute, Berkeley, CA., 1990.
5. Rumelhart, D.E., Hinton, G.E. and Williams, R.J., Learning internal representation by error propagation, In *Parallel Distributed Processing: Explorations in the Microstructures of Cognition, Vol. 1*, MIT press: Cambridge, 1986.
6. Towell, G.G., Shavlik, J.W., and Noordewier, M.O., "Refinement of approximate domain theories by knowledge-based neural networks", in *Proceeding of AAAI-90*, Boston,

Extending Constraint Satisfaction Problem Solving in Structural Design

Qi Guan and Gerhard Friedrich
Christian-Doppler Labor für Expertensysteme
Institut für Informationssysteme
Technische Universität Wien
Paniglgasse 16, A - 1040, Vienna, Austria

Abstract

In this article we address the problem of constraint satisfaction in structural design and present a theory of fuzzy constraint satisfaction (FCSP). Constraints based on fuzzy relations can be either hard and soft. Using fuzzy constraint satisfaction we are able to reason about the degree a constraint is satisfied, thus avoiding a static partition into hard and soft constraints as traditionally used. Various functions and comparators are defined for searching the best solution. We exploit object-oriented programming methods to implement FCSP for structural design.

Key words: Fuzzy set, Constraint satisfaction, Structural design.

1 Introduction

A design problem can be described as assignment of values to a set of parameters P. Some of these are pre-defined, e.g. in structural design (height of building, span of bridge etc.). They are not changed within the design process. Other parameters may be changed in the design. They are called "design variables". We define $D = \{d_1, d_2, d_3, \ldots, d_n\}$ as n design variables in a design task and $D \subseteq P$. Each d_i takes a value $V_{d_i} = \{v_{1d_i}, v_{2d_i}, \ldots, v_{md_i}\}$. The set of V_{d_i}, where $i = 0, \ldots, n$, is called *design space*.

For a given design task, not all values in d_i are available. They must meet with the description from design codes as well as specifications and we call them constraints. From this point of view a design problem can be seen as constraint satisfaction problem (CSP) in AI and many articles have addressed this problem. But in design there are special problems. For example, the goal of design can be either searching for an optimum solution under given constraints or for a compromise solution if only some constraints are satisfied. The latter case has been addressed by

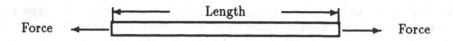

Figure 1: A steel element subjected stress F

many researchers applying relaxing technique of constraints [1, 5, 6]. They classify constraints into *hard* and *soft* [2] or more levels defining a hierarchy [3, 4]. But all of those have their own specialties and cannot cover all application areas. We will discuss those problems starting with a very simple steel structural design problem (see Figure 1). In section 3 we will show how those methods fail to treat this example and so present a theory called *fuzzy constraint satisfaction (FCSP)*. In section 4, we give some definitions about $FCSP$ and section 5 is about how to implement $FCSP$ with object-oriented programming methods.

2 Example

The goal of the example shown in Figure 1 is to find a suitable cross-sectional area A under the given load F. In this case it is needed only to satisfy one constraint:

$$\frac{F}{A} \leq [\sigma]$$

$[\sigma]$ is the allowable tensile stress. In engineering different steel materials have different $[\sigma]$. Usually, the higher the grade of steel is, the higher $[\sigma]$ is. Of course, the higher the quality is, the higher the expenses are. The following table shows one kind of steel types with $[\sigma] = 2800 \ kg/cm^2$, its subtypes, cross-sectional areas, and minimum radius of gyration (I_{min}) of those subtypes.

Subtype	a	b	c	d	$\cdots$
Area(cm^2)	5.942	9.956	15.824	30.392	$\cdots$
$I_{min}(cm^2)$	14.36	38.06	97.580	354.32	$\cdots$

In the normal case there are many solutions to this problem. However, if we add other constraints, there may be no solution. For example, if we say the cross-sectional area can not be larger than 20 cm^2, the steel element is subjected with F = 45000 kg and the allowable tensile stress $[\sigma] = 2800 \ kg/cm^2$, then the constraint above can not be satisfied.

If we examine this case more closely, the smallest value of F/A under the condition of the cross-sectional area should be smaller than 20 cm^2 is 2843.78 kg/cm^2 with type c. It is only a little larger than $[\sigma] = 2800$. Just because of such a little difference the system rejects it. One solution is to use the higher grade steel with a higher $[\sigma]$. While an experiencing engineer will choose another way, i.e. type c and the lower grade of steel, because either there is no difference between 2800 and 2843 in principle or we can use another method to remedy this weakness.

3 Related Works

The example of section 2 is an extreme case, but it does exist and much work has been done on this problem. The common usual way is to relax constraints.

[5] gives a solution, the so called *Partial Constraint Satisfaction (PCSP)*. According to the motto "find me the closest problem that you can solve (or one at least as close as ...)" *PCSP* relaxes *hard* constraints and gives those constraints a partial order. With the comparison of different solutions based on different relaxed constraints a better solution can be found among them.

Another way to relax constraints is found in [3, 4] called *Constraint Hierarchy*. It assigns to each constraint a label *strength*. One of these strengths, *the required strength*, is special. The constraints labeled by this strength must be satisfied. The remaining strengths all indicate preferences of varying degrees. There can be an arbitrary number of different strengths, C_0 indicates the required constraints and C_1 to C_n denote the preferred constraints. With the increasing of the level of n, constraints become weaker. A *solution* to the hierarchy is that all required constraints must be satisfied. The *best* solutions to the hierarchy are those that satisfy the required constraints, and also satisfy the preferred constraints such that no other better solution exists.

[2] gives another way to solve *hard* and *soft* constraint problems. It regards the *hard* constraint as a first-order formula in logic. A solution to those hard constraints becomes an interpretation which satisfies the axiom set, and *soft* constraints can be regarded as providing an order over those interpretations. Most preferred solutions are the most preferred interpretation in that order.

All these methods have a common point: a solution must satisfy hard constraints. Soft constraints represent preference over solutions. The differences between the various methods are the ways they treat the soft constraints. Recall the example above, the constraint $F/A \leq [\sigma]$ is a hard constraint. The problem the example shows, is that a value to hold this constraint is just beyond some boundary: F/A is larger than $[\sigma]$, but just a litter bit too much. We should relax this *hard* constraint and one way is to extend such a one dimensional constraint into two dimensional – with a membership function $\mu_G(C)$ over constraint C. (see Figure 2)

The distribution type of $\mu_G(Stress)$ has many forms. In Figure 2 we use a linear function to represent it. The fuzzy constraints can be described as:

$$C(D) : G$$

Functions $C(D)$ might be in engineering stress, displacement, slenderness and fab-

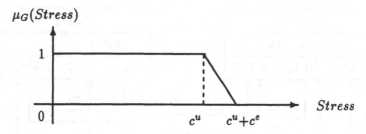

Figure 2: Constraint with membership function

ricational constraints etc. D is a design variable space and G is a fuzzy subset. The constraint in Figure 2 shows that stress must be limited in its allowable range. While G is now a fuzzy set, it describes a not exact *in* and *out* bound of $C(D)$ and thus we can use it as a way to relax *hard* constraints. In Figure 2, c^u is the upper bound of allowable tensile stress and c^e is the extended bound, usually called "tolerance". When c^e is equal to zero, this fuzzy constraint becomes a normal constraint.

Recall that $F = 45000\ kg$, $A = 15.824\ cm^2$ and $[\sigma] = 2800\ kg/cm^2$. We can define $c^u = 2800\ kg/cm^2$ and $c^e = 700\ kg/cm^2$. If $\mu_C(Stress)$ is

$$\mu_G(Stress) = \begin{cases} 1 & when \quad 0 \le Stress \le c^u \\ 1 - (Stress - c^u)/c^e & when \quad c^u \le Stress \le c^u + c^e \end{cases}$$

then $Stress = F/A$ and $\mu_G(Stress) = 0.94$. In design we define $\lambda = \mu_G(Stress)$ a satisfaction index. We can say the satisfaction index of the selected steel type c under a given force to the allowable stress is 0.94.

4   FCSP Definitions

In this section we give formal definitions on FCSP. It is very close to the normal CSP. The main differences are that we compare these solutions to the fuzzy constraints and so we will introduce new definitions about *simple, comprehensive, best* and *best solution* satisfaction indexes. Further we use object-oriented programming (*OOP*) method to describe FCSP and it is also easier to implement. In OOP a class has a set of attributes and a set of constraints on those attributes. The class describes the structure of a set of objects. All such objects are called *instances* of the class. In FCSP those instances act as nodes in the network of a fuzzy satisfaction problem. Some classes in the network are driven from other classes. To satisfy the fuzzy constraints we have to find suitable value tuples for the instances. The suitable value tuples of an instance may be more than one, i.e. all of them satisfy the constraints, and one (or some) of them is(are) the best solution(s) to the FCSP.

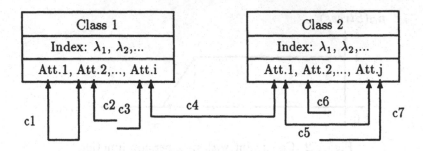

Figure 3: Classes and constraints (normal and fuzzy)

The example of Figure 1 can be defined as a class *tension* with its attributes of cross-sectional area, input force etc. The constraint is $F/A \leq [\sigma]$.

Definition 1 *A class has a set of variables (or attributes) $D' \subseteq D$ (D the set of variables in a design task), a set of constraints, some of which are fuzzy constraints, and a set of satisfaction index λ to every fuzzy constraint. An instance of the class has n possible value tuple assignments to D'.*

Definition 2 *A fuzzy constraint consists of a set of variables to or among classes and a fuzzy relation of these variables.*

Definition 3 *A fuzzy constraint network consists of a set of fuzzy constraints sharing variables.*

Figure 3 shows the relation among classes and constraints (normal or fuzzy). The constraint 4 (C4) is a global constraint, i.e. it has relations among more than one class. In this article we focus on local constraints.

Definition 4 (Simple fuzzy constraint satisfaction) *Let m be the number of variables declared in a fuzzy constraint network for one class and $l \leq m$, $(d_1, \ldots, d_l)$ a tuple of single values of one instance for a subset of those variables, each d_i takes a value $V_{d_i} = \{v_{1d_i}, v_{2d_i}, \ldots\}$, c a fuzzy constraint referring to the i_1th, i_2th, $\ldots$ i_lth variable and λ satisfaction index to c. c is simply satisfied by $(d_1, \ldots, d_l)$ such that $(d_1, \ldots, d_l)$ is in the relation of c and $\lambda > 0$.*

The simple fuzzy satisfaction describes how satisfiable the constraint is under the given data. If there are many constraints in a class, each fuzzy constraint may have different simple satisfaction indexes and together we have a comprehensive description of those indexes:

Definition 5 (Comprehensive fuzzy constraint satisfaction) *Let k be the number of fuzzy constraints in a class, m and $(d_1,\ldots,d_m)$ declared as above, C constraints declared in the class, λ_j a satisfaction index to c_j ($c_j \in C$) and $\vec{\lambda} = (\lambda_1,\ldots,\lambda_k)$. C is comprehensively satisfied and λ' is its index such that c_j is satisfied by $(d_1,\ldots,d_m)$, $j=1,\ldots,k$, $\lambda_j > 0$, $\lambda' \leq 1$ and $\lambda' = g'(k,\vec{\lambda})$. Here g' is a judgement function for comprehensive satisfaction.*

While an instance of a class may have many value tuples, each instance value tuple has its comprehensive satisfaction index. And one of those is the best.

Definition 6 (Best fuzzy constraint satisfaction) *Let $U = (u_1, \ldots, u_n)$ be the set of possible value tuples an instance takes, $(c_1, c_2, \ldots, c_k)$ a set of fuzzy constraints, $\Delta_{nk} = (\lambda_{11},\lambda_{12},\ldots,\lambda_{21}, \ldots,\lambda_{2k},\ldots,\lambda_{nk})$ a matrix of satisfaction indexes associated with the class. C is best satisfied by the tuple $u_i = (d_1,\ldots,d_m)$ and λ'' is its index such that $\lambda_0 \leq \lambda'' \leq 1$ and $\lambda'' = \max(g''(n,k,\Delta_{nk}), u_i \in U)$.*

Here $\lambda_0 > 0$ is a user pre-defined index below which there is no solution. And $g''(i,j,\Delta_{ij})$ is a judgement function or comparator for the best fuzzy constraint satisfaction. If all constraints in the network are *hard* and equally important, we can define following comparator:

$$\begin{cases} \lambda'' = g''(n,k,\Delta_{nk}) = \bigvee_{i=1}^{n}(g'(k,\vec{\lambda}_{ik})) \\ g'(k,\vec{\lambda}_{ik}) = \bigwedge_{j=1}^{k} \lambda_{ij} \end{cases}$$

$\wedge$ and $\vee$ are fuzzy conjunctive and disjunctive operators respectively. Usually, these two operators are considered as operators to take values of minimum and maximum. The meaning of $\wedge$, if it acts as minimum operator, is that an instance with value tuple i satisfies many constraints with different indexes $\lambda_{i1},\ldots, \lambda_{ik}$ and only one, the minimum, is considered as the instance satisfaction index of these constraints. Among all value tuples we take the one with the maximum index as the best candidate:

$$\begin{array}{ccc}
\lambda_{11} \cdots \lambda_{1k} & \Rightarrow & \lambda'_1 = \min(\lambda_{1j}) \quad (j=1,\ldots,k) \\
\lambda_{21} \cdots \lambda_{2k} & \Rightarrow & \lambda'_2 = \min(\lambda_{2j}) \quad (j=1,\ldots,k) \\
\cdots\cdots\cdots\cdots & & \\
\lambda_{n1} \cdots \lambda_{nk} & \Rightarrow & \underline{\lambda'_n = \min(\lambda_{nj}) \quad (j=1,\ldots,k)} \\
& & \lambda'' = \max(\lambda'_i) \quad (i=1,\ldots,n)
\end{array}$$

So far we have introduced some definitions on a class. A structural design problem, for example, may need to choose suitable values for many steel elements, i.e.

search for many *best* satisfactions for many steel elements. They are declared as instances of some classes and independent of each other, i.e. there is no global constraint among them. In the following we define the best solution for a problem which has many different and independent instances declared:

Definition 7 (Best solution for fuzzy constraint satisfaction) *Let x be the number of instances (they are declared by different classes) in a fuzzy constraint network for a design task. The best solution for the system is with an index $\lambda_{solution} = solution(x, \lambda)$.*

if we simply consider all elements be equal, the *solution*() operator can be expressed as

$$\lambda_{solution} = solution_{\lambda''}(x) = (\sum_{l=1}^{x} \lambda_l'')/x$$

again λ_l'' is the best satisfaction index for instance l.

While steel elements in a structural design problem are not all equally important. Some are main elements carrying most of the load and should be more restricted. We can give main elements higher weight than other elements and thus have a new formula of *solution*(). In this case, we need not calculate the best satisfaction index and the *solution*() operator could be changed as follows:

Let l_i be the set of all λ' of an instance i, $L_x = \{(\lambda_1', ..., \lambda_x') \mid \lambda_i' \in l_i$, for $i = 1...x \}$:

$$\lambda_{solution} = solution_{\lambda'}(x) = \max(\frac{\sum_{\lambda_i' \in l_x} (\lambda_i' * w_{element_i})}{\sum_{\lambda_i' \in l_x} \lambda_i'}, l_x \in L_x)$$

5 Implementation

With the help of object-oriented programming the FCSP can be implemented very easily. At the beginning of a design task we should only define instances and the classes they belong to. If we design a steel truss structure only considering tension and compression, then two sets of instances should be declared. A multiple inheritance structure is shown in Figure 4.

At the top of the hierarchy there is a class *type* which has attributes of cross-sectional area, element length and the force on the element etc. It is a base class for all other classes. An element subjected by tension can be declared as an instance of class *Etension*, and an element subjected by compression is declared as an instance of class *Ecompression*. The class *Etension* inherits from two base classes *tension* and *shape*. In the class *shape* there is a constraint *shape_select*() which decides

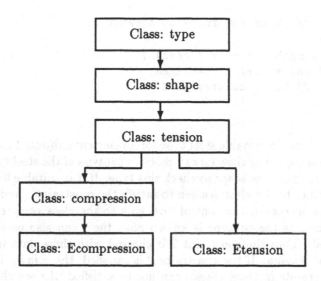

Figure 4: Inheritance of classes

which shape of a steel type should be selected under certain conditions of area, absolute value of force and physical descriptions of the steel structure etc. The class *Ecompression* is derived from the class *compression*. While the class *compression* is driven from the class *tension* it inherits all properties of class *tension*, i.e. data as well as constraints. In addition the class *compression* has its own slenderness constraints to the element subjected compression. The following shows two class definitions.

```
class name: tension
  derived from class shape
            // contains variables force and area
  attributes:
    c_index,   // Comprehensive index for FCSP
    sigma,     // Allowable tensile stress
    ce, cu,    // Upper and  expended bound of lambda
    index_1;   // index for fuzzy constraint 1
  constraint:
    force / area <= sigma;
            // Fuzzy constraint 1
end_of_class

class name: compression
  driven from class tension;
  attributes:
    r,         // Minimum radius of gyration
    phi,       // Reduced coefficient of compressive element
```

```
    index_2;   // index for fuzzy constraint 2
  constraint:
    phi = f( length / sequ(r / area) )
    force / ( phi * area )  <= sigma;
              // Fuzzy constraint 2
end_of_class
```

Now we try to find a suitable steel type for an element subjected to compression. The first step is done in the class *type* to choose a subtype of the steel type. Then the execution goes to the class *shape* to check this type. If it is suitable for the element then it goes further to the class *tension* to satisfy the constraint 1 and calculate the index_1. If it is successful the control flow goes to the class *compression*. In the class *compression*, as the subtype is known now, the *r* can also be found and the *phi* is calculated. Thus the constraint 2 is checked with those data inherited from classes *type* and *tension*. If the constraint 2 is satisfied, the *c_index* is updated. If one of the constraints in these classes can not be satisfied, the search goes back to the class *type* to select the next value. The process continues until all subtypes have been checked.

6   Conclusion

Structural design leads to new demands for constraint satisfaction applicable for this domain. Applying fuzzy set theory to constraint satisfaction shows many advantages. Using fuzzy satisfaction for a constraint does not mean that the constraint is a *weak* one. A fuzzy constraint can be either *hard* or *soft*. If a hard constraint in design can not be satisfied, the usual way is to relax the hard constraint. In a fuzzy constraint, the relaxation gives a range to the constraint and an index to indicate its deviation. We therefore avoid the drawbacks of usual techniques where soft constraints are either considered or completely ignored.

Based on the satisfaction index of the system we can define many judgement functions and comparators to search the best solution. With OOP such kind of constraints as well as comparators can be easily interpreted and implemented.

7   Acknowledgements

We wish to gratefully acknowledge Georg Gottlob for his useful guides as well as helps from colleagues in the Christian-Doppler Labor für Expertensysteme, Technische Universität Wien.

References

[1] Alois Haselböck, Constraint Satisfaction Problems - Ein Überblick über Verfahren und Anwendungen. DBAI-CSP-ST 91/1, Technical University of Vienna, 1991

[2] Ken Satoh. Formalizing Soft Constraints by Interpretation Ordering, In *Proceeding of 9th European Conference on Artificial Intelligence*, pages 585 - 590, Stockholm, Sweden, August 1990

[3] A. Borning, M. Maher, A. Marindale, and M. Wilson. Constraint Hierarchies and Logic Programming. In *Proceedings of the International Conference on Logic Programming*, pages 149-164, Lisbon, Portugal, June 1989.

[4] Bjorn N. Freeman-Benson, John Maloney, and Alan Borning. An Incremental Constraint Solver. *Communications of ACM*, 33(1):54-63, 1990.

[5] Eugene C. Freuder. Partial Constraint Satisfaction. In *Proceedings of the International Joint Conference on Artificial Intelligence*, Pages 278-283, Detroit Michigan USA, August 1989. Morgan kaufmann Publishers, Inc.

[6] Alber Haag. Ein Beitrag zur praktischen Handhabung von Planungs- und Konfigurierungsaufgaben. *Arbeitspapiere der GMD*, 388:83-105, 1989.

[7] Takanori Yokoyama. An Object-Oriented and Constraint-Based Knowledge Representation System for Design Object Modeling. In *Proceedings TOOLS*, Pages 156-152, 1990.

[8] Kefeng Hua, Boi Faltings, and Djamila Haround. Dynamic Constraint Satisfaction in a Bridge Design System. In *Proceedings of international Workshop on Expert Systems in Engineering Principles and Application*. Pages 217-232, Vienna Austria, September, 1990. Springer-Verlag.

[9] Lin Shaopei, Gu Xiuling, Guan Qi, and G.K. Mikroundis. On Development of Expert System for Steel Structural Design. In *Proceedings of the 12-th World Congress on Mathematics and Computer in Simulation*. Pages 447-449, Paris, July, 1988.

[10] Ronald R. Yager. The Representation of Fuzzy Relational Production Rules. In *Journal of Applied Intelligence*. 1. Pages 35-42, 1991.

A Modulation Package Tuning Machine Applying Fuzzy Logic

Akio Ukita and *Toshio Kitagawa

Production Eng. Dev. Lab. and *Microwave Satellite Communication Dep.
NEC Corporation
Tsukagoshi, Saiwaiku, Kawasaki, 210 Japan

Abstract. A tuning machine, developed to tune the analog circuits in a micro-wave modulation circuit board, is presented. The machine's purpose is to tune the circuits automatically. This tuning is conventionally carried out by highly experienced workers. To choose the most appropriate trimmer resistance among the candidates which give good characteristics for the circuits, a method involving fuzzy logic is used.

To develop an appropriate process for choosing the trimmer, at first, the human experts' tuning know-how was utilized.

The prototype machine can tune the circuits within a specific time limit, which is almost equal to that employed by human experts.

1 Introduction

In producing circuit boards, adjusting the circuit's characteristics is often needed. To enable adjusting the circuit characteristics, several trimmers, such as trimmer resistances or trimmer condensers, are usually prepared beforehand in the circuits. They are usually adjusted by hand by highly skilled workers. In the trend toward total factory automation, however, the adjusting process tends to remain un-automated, and it is becoming difficult to gather together a sufficient number of highly skilled workers to handle the workload. That is why automatic adjustment is needed. To realize automatic adjustment, it is not only necessary to develop automatic handling for the drivers which rotate the trimmers, but it is also necessary to realize an effective adjustment procedure. For example, the trimmer rotation rate should be determined reasonably. If the rotation rate is unreasonable, the machine will carry out cut-and-try actions very often and the total adjustment time will become long. Selecting an appropriate trimmer from among the candidates, that seem to be able to improve the characteristics, should also be carried out carefully.

When the tuning process can be defined as "to adjust a specific trimmer among the candidates at a certain rate, until the measured data satisfy the required specific conditions", the process will become more complicated with an increase in the number of candidates and the amount of data which must be included in the specific conditions. Emphasis in this paper is placed on realizing accurate tuning of many trimmers to satisfy certain specific conditions using many data sets.

2 Tuning Objective

The objective circuit for tuning, discussed in this paper, is shown in Fig. 1. This circuit is part of a modulation circuit board, which is used for microwave telecommunications. The circuit action is as follows. Signal P, the output signal from D/A converter D/AX, is an analog signal, whose magnitude is in proportion to digital data P. Similarly, analog signal Q corresponds to digital data Q. The output signal from MIX is a harmonic wave at very high frequency, that is amplitude-modulated by signal P and signal Q. Several trimmer resistances and trimmer condensers are made available to tune the circuit. Among the trimmers, 6 specific trimmers influence each other. That is, when one of them is adjusted, the others may have to be re-adjusted. That makes the adjustment process more difficult.

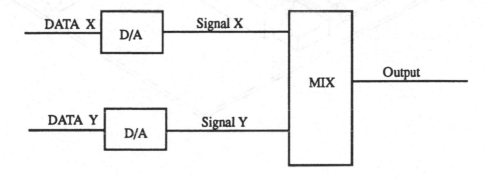

Fig. 1. Objective Circuit

3 Machine Construction

To automatically tune the circuits, a prototype machine, shown in Fig. 2, was constructed. The tuning head, for rotating the trimmer, is moved in X and Y directions by the X-Y robot. The head can also move up and down, to attach the rotators to the trimmers. The head movement, the input data pattern P and Q, and measuring output signal O are all controlled by a personal computer. Since many trimmer categories are used, several tuning heads are prepared to match the different requirements. If need for a substitution occurs during the tuning sequence, the heads are substituted automatically.

Tuning heads for substitution

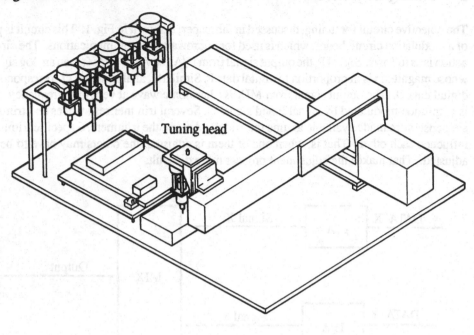

Tuning head

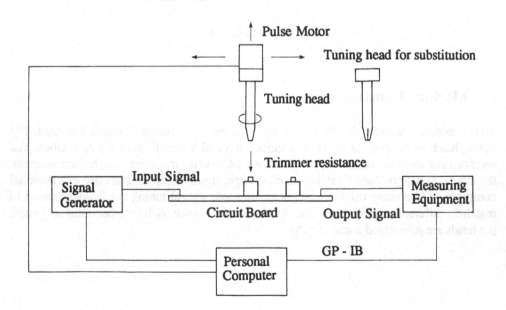

Fig. 2. Mcchine Construction

4 Tuning Process

The tuning process can roughly be divided into two parts. The first part is completely sequential. That is, once a trimmer is adjusted, there will be no need to readjust it later. This process is relatively easily automated. The second part is a cut-and-try process. This is the process mentioned before. The know-how for tuning it was elicited from highly-experienced workers.

The second process can be described as follows. Figure 3 shows output data O, plotted by both a real part and an imaginary part. Individual symbols, A to H, correspond to individual input conditions. Ideally, each output signal must be plotted at the center of each circle. In practice, however, there is some deviation from the center of the circle to the measured data position. The circles indicate the area for "good characteristics" and the second tuning process can be said to be used to get all the measured data into the circles.

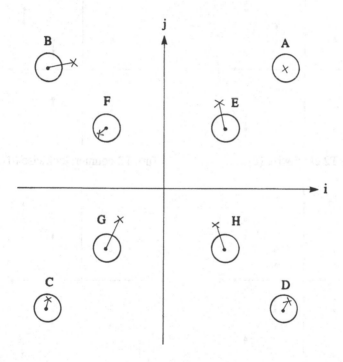

Fig. 3. Output Signal Representation

Consider the case in Fig. 3, where B,E,G and H must be adjusted, while the others need no adjustment. For adjusting B, E, G and H, some trimmer rotation is needed.

Figure 4 shows the tendency for moving measured data, corresponding to each trimmer rotation. For example, Fig. 4 (a) indicates that, if trimmer T1 is rotated clockwise, all the output data will go to the right. The shift values are not equal from A to H and there is no linearity between the rotation rate and the shifts. The non-linearity is mainly caused by trimmer hysterisis.

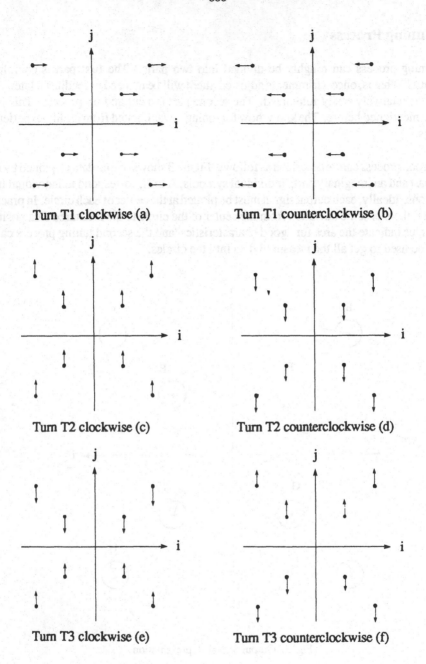

Fig. 4. Individual Trimmer Functions

Considering the example shown in Fig. 3, T2 counterclockwise rotation seems to be appropriate, because E, G and H characteristics will become better by such rotation. However, the F characteristic may become worse, so an appropriate combination of rotations is needed.

When a human expert meets this situation, he rotates trimmer T2 counterclockwise and then measures the data again. Then, he chooses which trimmer to rotate next. By repeating this process, he determines all the data with good characteristics. If the rotation rate is not determined reasonably, the repetition number becomes big and the time required for tuning becomes long.

5 Software Implementation

5.1 Tuning Process

The tuning process, which is taken by this machine, is basically the same as the human experts' process. The process is as follows:

(1) Measure the output signals from the circuit board, under 16 initially scheduled conditions.
(2) Determine whether or not all measured data satisfy the conditions for finishing the tuning.
(3) Select a trimmer and a rotation value.
(4) Move the tuning head to the selected trimmer position and rotate it.
(5) Back to (1)

When all measured data satisfy the conditions at step (2), the machine stops tuning.

5.2 Tuning Know-How Representation

To represent the human experts' tuning know-how, specific production rules are used. By using such production rules, the tuning know-how can expressed as follows.

Rule 1:
   If the measured data would be improved the most by rotating T1 counterclockwise, then, rotate T1 counterclockwise.

Rule 2:
   If the measured data would be improved the most by rotating T1 clockwise, then, rotate T1 clockwise.

Rule 2n-1:
   If the measured data would be improved the most by rotating Tn counterclockwise, then, rotate Tn counterclockwise.

Rule 2n:
   If the measured data would be improved the most by rotating Tn clockwise, then, rotate Tn clockwise.

5.3   Trimmer Selection

For using this machine, two rules are prepared, corresponding to each individual trimmer [1]. When operating the tuning process, first, all the rule condition parts are calculated. By comparing the condition parts' value, it is decided which trimmer to rotate. Condition parts are represented by membership functions as fuzzy logic, which is explained later.

To select which trimmer to rotate, condition parts are used, regarded as a certainty factor. In general cases, the rule with the largest certainty factor value is selected. However in this case, the counter part rule has to be considered. If the certainty factors, concerning whether to turn counterclockwise and to turn clockwise are almost equal, there would be no advantage in rotating the trimmer corresponding to it. Almost equal certainty factor values in opposite directions can be regarded as when "rotating the trimmer may improve characteristics at several points along A to H, but may cause harm at other points." So, some other trimmer may have to be selected.

Instead of using the certainty factor, in order to select an appropriate trimmer, observing the difference between two certainty factors is considered as a pertinent trimmer rule. If the difference is the biggest, the characteristics are most likely to be improved by rotating the corresponding trimmer. Instead of the difference, a division of two certainty factors may be selected.

5.4   Condition Parts Calculation

The membership functions for condition parts are derived as follows. Figure 5 shows an example of membership functions and the derivation of an appropriate rotation rate. Consider expressing the conditions for the rule to rotate T3. The rule to turn T3 clockwise, that is rule 5, can be written in more detail, refering to Figs. 3 and 4:

Rule 5:
    If

                data A shifts to the upper part
      and   data B shifts to the upper part
      and   data C shifts to the lower part
      and   data D shifts to the lower part
      and   data E shifts to the upper part
      and   data F shifts to the upper part
      and   data G shifts to the lower part
      and   data H shifts to the lower part
    Then,
        rotate T3 clockwise.

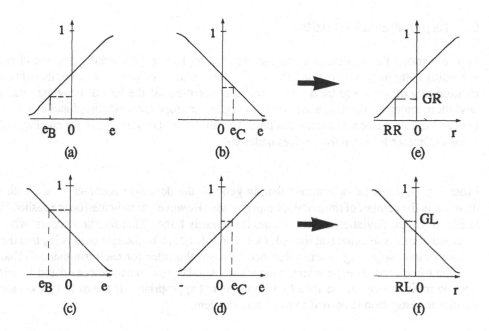

Fig. 5. Rotation Rate Calculation

The membership function for the second item in the conditions is expressed as Fig. 5(a). The horizontal axis is an imaginary part of the data. Figure 5 (a) indicates "the larger the imaginary parts of data become, the greater the membership function becomes". Similarly, the 3rd item in the conditions is expressed as in Fig. 5 (b). The membership function for individual items in the conditions differs in inclination and shape.

By putting the measured data to the horizontal axis, the grades are decided for individual functions. Then, the geometrical average for all the grades, GR, is calculated as the conclusion of the "and" operation. This GR is considered as a certainty factor and is used for trimmer selection.

5.5   Rotation Rate Determination

When a decision is being made as to which trimmer to rotate, its rotation is determined by one of two rules. One is for counterclockwise rotation, the other is for clockwise rotation. The rotation rate derivation from the rule set is as follows [1] [2] [3].

The geometrical average GR, which is previously calculated, is put into the function, shown in Fig. 5 (e). The rule for counterclockwise rotation can also be written as above. Figure 5 (c) shows the membership function for "data B shifts to the lower part" and Fig. 5 (d) shows the membership function for "data C shifts to the upper part". Rotation rate RL is also derived by putting GL into the function shown in Fig. 5 (f). Finally, rotation rate RG is determined as a weighted average for RL and RR:

$$RG = (GL * RL + GR * RR) / (GL + GR)$$

6 Experimental Results

Figure 6 shows the experimental results obtained by letting the machine tune the circuits according to the method explained above. The horizontal axis represents the number of times rotating any trimmer was tried. The vertical axis represents the largest imaginary part of deviation among all the data at each try. In Fig. 6, the slope for weighting functions differs between the solid line and the dashed line. The difference between Fig. 6 (a) and Fig. 6 (b) is the difference between the devices under test.

From Fig. 6, it can be determined that, in general, the deviation becomes small, with an increase in the number of times the attempt is made. However, as indicated on the dashed line in Fig. 6 (b), the deviation often becomes temporarily large. This situation occurs, when a certain trimmer is selected and rotated. One reason for it can be thought of as being that there is insufficient weighting function balance between the rules for each trimmer and that a specific trimmer tends to be selected too often. Another reason can be considered as being that the rotation rate for the specific trimmer is not appropriate. It seems to be too much. Further investigation is needed to solve this problem.

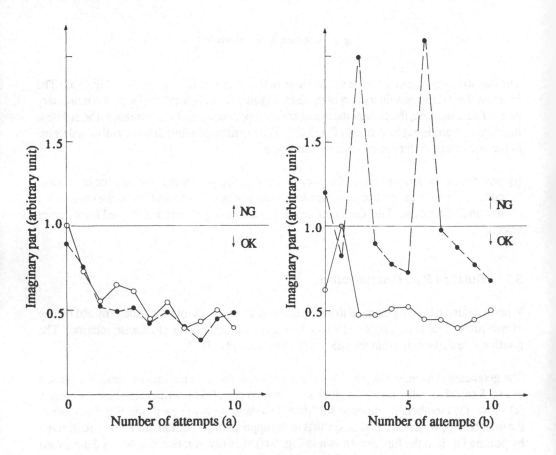

Fig. 6. Tuning Ability

7   Conclusion

The tuning machine, presented in this paper, can automatically tune circuits. This action is conventionally accomplished by highly experienced workers. The most complicated tuning process in this case is realized by applying fuzzy logic. The total tuning time, including the sequential tuning process, is comparable to the time expended by human experts. The machine was proved to be effective for practical use. The next step is a self parameter tuning of fuzzy rules. That is, to automatically determine the shape and declination for weighting functions.

References

[1] M. Sugeno, An Introductory Survey of Fuzzy Control, Information Science, 36, PP59~83
[2] H. Akahori, Y, Egusa, T. Takagi, N. Wakami and S. Kondo, Self-Tuning Method of Fuzzy Reasoning by Optimization Technique and its Application to Vehicle Control, 5th Fuzzy System Symposium (Kobe, 1989) 77-81
[3] M. Maeda, Y. Maeda, and S. Murakami, Fuzzy drive control of an autonomous mobile robot, Fuzzy Sets and Systems 39(1991) 195-204

Towards Intelligent Robotic Assemblies

Carlos Ramos ; Eugénio Oliveira

Department of Electrical Engineering and Computing
Faculty of Engineering of Oporto University
Rua dos Bragas
4099 PORTO CODEX, PORTUGAL

Abstract. The aim of our work is the implementation of an Intelligent Robotic System able to deal with assembly tasks. The efficiency of our approach is mainly due to the choice of good heuristics and the early pruning of unnecessary paths in the planning search tree.

Once generated, the high level plan is translated into robot instructions. At this level, some intelligence is needed to guarantee the feasibility of the robot actions. This is the case of object grasping, regrasping to change from one stable position to another, geometric constraints analysis, collision avoidance and free space management.

Some low level sensing operations have been introduced during task execution. It is possible to recover from very simple errors and deal with unexpected situations. The consideration of uncertainties, incomplete or incorrect descriptions of the current state of the objects is also envisaged.

1 Introduction

Planning involves searching. Unfortunately, for large dimension problems, the combinational explosion during searching process may happen. The approach that we propose bears in mind this problem. A great effort has been done to choose the best heuristics to guide that search. The avoidance of incorrect or non-promising paths is also important to guarantee a great efficiency. The system that we will describe is designed to be integrated in a flexible assembly cell. It involves an interface with a Computer Vision System in order to convert the usual outputs of Image Analysis to the symbolic inputs of the Planning System. Moreover the generated plan is not only displayed at the computer monitor but it is also translated to the appropriate instructions which are executed by a Renault-APRA robot. The full consideration of Geometric Constraints has been introduced to enhance the system flexibility.

Distributed Artificial Intelligence is the framework used to encapsulate the overall system ([6,7]). The modules involved in our robotic testbed are the following: TLP (Task Level Planner), TE (Task Executor), WD (World Descriptor), VISION and MODELS. TLP is an automatic generator of plans able to apply operators in order to convert the initial configuration of objects in a configuration of the same objects corresponding to the assembled objects. TE gets the plan, formulated by TLP, and translates it into orders which are understood by the robot. WD uses spatial reasoning to convert the numeric output of VISION in the symbolic relationships needed as input to TLP. Besides, WD is also concerned with the geometric constraints handling. VISION is a 3D object identifier using camera and laser. MODELS concentrates different object representations to be used by the other modules.

WD, MODELS and TLP are implemented in PROLOG on a UNIX based Workstation. VISION is implemented in C on a Personal Computer. The execution decision support of TE is implemented in PROLOG, on UNIX environment, while the task execution software is implemented in LM, the language controlling the robot, on another Personal Computer. The UNIX Workstations and the Personal Computers are connected by an Ethernet network.

2 Related Work

The first system involving planning ideas was GPS [5]. GPS introduced means-end-analysis where the operators are chosen in order to make the current state more similar to the desired goal state. STRIPS [1] uses three lists for each kind of operation (ADD, DELETE and PRECONDITION lists). Unfortunately, the generation of efficient plans is not certain. Hierarchical Planning appears with ABSTRIPS [12] where planning is achieved as an hierarchy of abstraction spaces. NOAH [13] establishes the preconditions for application of an operator, then the ordering of the operators is chosen. "Critics" were introduced in order to establish the interactions between parts of the plan. NONLIN [14] uses ordering links to treat interactions between parts of the plan.

Unfortunately, real-life assembly tasks introduce many problems which solution is difficult using general-purpose planning methods. For example, it is need to introduce interfaces with the vision system and the robot controller. Therefore, it is necessary to convert the numerical information, given by Vision, into symbolic relationships and to convert the symbolic plan to the robot control language. A great effort must be done to incorporate numerical and geometric constraints such as the decision of how to grasp objects and how to change their stable states by means of regrasping operations.

Nowadays, other approaches are being used to deal with Assembly Robotics. Let us focus our attention in just a few of them. Homem de Mello [4] proposes the use of an AND/OR graph. A relational model is based on quintuples: parts, contacts, attachments, relationships and attribute-functions. A constraint of this approach is that it imposes that no more than two parts are joined at each time. HANDEY [3] includes a sensory system and is able to perform grasping and regrasping operations for pick and place problems. However, HANDEY does not provide reasoning about the actions. SPAR [2] is an interesting system able to perform sensing operations during execution and to generate error-recovery plans. SPAR uses a three-level hierarchy with operational, geometric and uncertainty-reduction goals. The main limitations of SPAR are: there is no subsystem to detect the positions and orientations of objects and the absence of a collision avoidance study.

3 Efficient Plan Generation

TLP is intended to be used for real assembly tasks. The main goal of TLP is the efficient generation of a plan suitable to be handled by TE, which directly commands the robot. There are two phases in TLP activity: goal filtering and plan generation. A detailed description of these phases is found in [9].

The goal filtering module analyses the initial and final states descriptions (both symbolic and geometrical) to discard goals that are already accomplished. Firstly, the symbolic relationships are grouped into a composed state. For example, in figure 1, the states *on(top,top1)* and *on(top,top2)* are transformed into *on(top,[top1,top2])*. A goal is considered as accomplished whenever the intended object locations (positions, orientations and stable states) as well as symbolic relationships have been reached for all objects involved in the goal. Besides, the goals involving the objects where the stacking or insertion are to be performed must be also accomplished. Therefore, the objective is to satisfy grouped, or compound, final goals (those that are not marked as being in the final state description). Figure 1 represents the initial and final (desired) states of the objects. The way how symbolic relationships are extracted from object locations is described in [8].

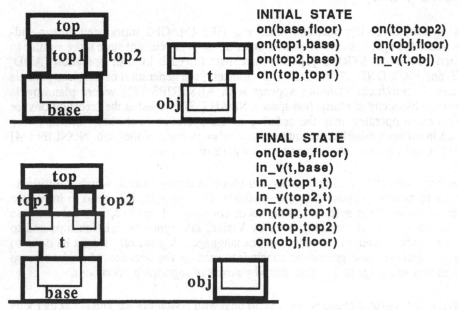

INITIAL STATE

on(base,floor)   on(top,top2)
on(top1,base)   on(obj,floor)
on(top2,base)   in_v(t,obj)
on(top,top1)

FINAL STATE

on(base,floor)
in_v(t,base)
in_v(top1,t)
in_v(top2,t)
on(top,top1)
on(top,top2)
on(obj,floor)

Fig. 1 Initial and Final States of an example

The formulation of compound goals to be achieved, obtained during the goal filtering phase, is the following for the example of figure 1: *in_v(t,[base])* ; *in_v(top1,[t])* ; *in_v(top2,[t])* ; *on(top,[top1,top2])* and *on(obj,[floor])* . The symbolic state *in_v* informs that the object of the first argument is vertically inserted in the ones referred in the second argument.

Best-First is a method of tree searching where, at each step, an heuristic function is used to choose the best path to follow (in the tree). We use the number of objects to be moved, in order to accomplish a specific goal, as the main heuristic on Assembly Robotics. The goal involving the least number of object movements is chosen and a new node is generated. Once a node is chosen it will belong to the solution.

Branch and Bound method is different since the best solution is always envisaged. For each node it is also represented the cost of the path from the root node until that specific node. The next node to be selected is that one with the least cost. This means that we can change from one node to another non-descendant node reformulating the path. The method we are using while similar to Branch and Bound has, nevertheless, some differences: in our case we use a tree building and not a graph search and heuristics are used to establish the costs between one node and the next descendant. Pure Branch and Bound is expensive. However, we also introduced some techniques to avoid incorrect or unnecessary paths. In this way we reduce very much the plan generation time since the number of generated nodes decreases drastically.

Each state is associated with one operator (e.g. *on* with *put_on* and *in_v* with *insert_v*). The generation of the appropriate plan can be seen as a dynamic building of a tree. The state of the world, at each moment, is represented in the nodes and the applied operators to accomplish a goal are the branches. Only those compound goals having in its second argument names of objects marked as being in the correct position (or the *floor*) are considered as candidates to the heuristic selection. This improves the efficiency of the algorithm. If, for example, we try to evaluate the cost to accomplish *on(top,[top1,top2])* and if *top1* and top2 are not at the correct position it must, firstly, insert *top1* and *top2* in *t* . However, there are two goals *in_v(top1,[t])* and *in_v(top2,[t])* having smaller costs than *on(top,[top1,top2])* since the execution of the last one involves also the execution of the previous ones.

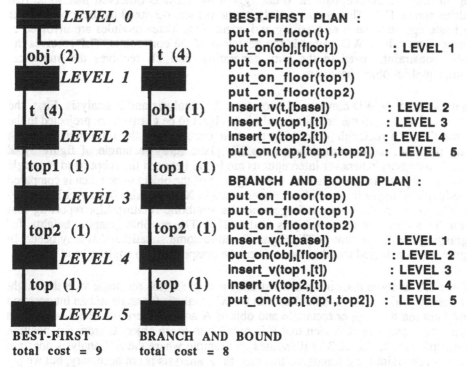

BEST-FIRST PLAN :
put_on_floor(t)
put_on(obj,[floor])   : LEVEL 1
put_on_floor(top)
put_on_floor(top1)
put_on_floor(top2)
Insert_v(t,[base])   : LEVEL 2
Insert_v(top1,[t])   : LEVEL 3
Insert_v(top2,[t])   : LEVEL 4
put_on(top,[top1,top2]) : LEVEL 5

BRANCH AND BOUND PLAN :
put_on_floor(top)
put_on_floor(top1)
put_on_floor(top2)
Insert_v(t,[base])   : LEVEL 1
put_on(obj,[floor])   : LEVEL 2
Insert_v(top1,[t])   : LEVEL 3
Insert_v(top2,[t])   : LEVEL 4
put_on(top,[top1,top2]) : LEVEL 5

BEST-FIRST   BRANCH AND BOUND
total cost = 9   total cost = 8

Fig. 2 Solutions achieved by Best-First and Branch and Bound algorithms

Paths leading to the final state obtained by those two different methods are shown in figure 2. In the figure, branches are labeled only by the name of the objects instead of the operator. The cost of partial goals is embraced by parenthesis. The solution achieved using Best First approach is the one on the left side, with nine objects movements. This plan is derived almost immediately. The solution given by Branch and Bound method is also shown. The cost of the plan is eight and it is generated in almost 1 second.

Our implementation of Branch and Bound is also efficient. As in the Best First approach the unnecessary paths are pruned. On the other hand, pure Branch and Bound use is expensive since it is very common to switch from a node close to the solution to another far from the solution but with a smaller cost than the first one. The values we are comparing are the sum of the known cost from the root node until the current node plus an estimate of the cost from the current node to the solution. The second part of the sum has been introduced to overcome the excessive switching referred before. Note that when we are at the level 1 of Branch and Bound solution, goals related with obj, top1 and top2 are commutative. This implies that there are six possible redundant combinations. The use of minimum costs makes the method much more efficient since only one possible combination will be explored.

4 Geometric Constraints

Once generated the high level plan must be executed. Sometimes, when the robot arm is moving in order to accomplish the orders given by TE it is observed that some real difficulties appear [10]. This is due, mainly, to the volume occupied by the robot gripper during basic operation such as grasping and ungrasping. Three modules are involved in Geometric Constraints: WD proceeds the detection of the constraints, TLP receives the symbolic constraints, used in the plan generation, and TE receives the numerical constraints used for object handling.

The method used by WD consists of two steps: XY analysis and Z analysis. First, the upper faces of objects, in the neighborhood of the object to be grasped, are projected to the XY plane. Then, a rectangle describing the zone occupied by the robot hand, at each grasping position, is also projected to the XY plane (grey rectangle of figure 3). Z Analysis is necessary whenever interceptions are found between the robot hand rectangle and the other objects. In Z analysis the grasping zone of the object to be taken is compared with the height of upper faces of objects intercepted in XY analysis. Whenever it is really impossible to grasp the object it is generated a symbolic relationship, reporting that situation. This symbolic relationship is sent to TLP. The possible grasping heights, for each grasping angle, are sent to TE. Figure 3 shows some situations that may occur. In this figure it is considered that the object G can be grasped by two angles (0 and $\pi/2$).

In 3 a) the XY analysis does not put constraints to angle 0 (the rectangle with the height greater than its width). On the other hand, in the XY analysis for angle $\pi/2$ an interception is found between the gripper rectangle and objects A and B. Therefore, the Z analysis is necessary. In Z analysis, it is seen that it is possible to grasp object G from any position of the grasping zone. Figure 3 b) illustrates an example where the XY analysis does not impose any constraint for grasping. In this case the Z analysis is not necessary, but we put it in the figure in order to show that heights of objects, A and B, are not a real constraint. Figure 3 c) shows an example where the grasping is possible at angle 0 but at angle $\pi/2$ the grasping is only possible for heights greater than the A top. Figure 3 d) is similar to 3 c) in what concerns XY analysis. Meanwhile, it is not possible to grasp G at angle $\pi/2$ since all grasping zone of G is avoided by A in Z analysis. The object needs to be grasped by the angle 0. Finally, in 3 e), it is not possible at all to grasp, directly, object G, since both Z analysis gave impossibilities. Note that in Z analysis only object A and D avoid the access to G. Therefore, the message *obstructed(g,[[a],[d]])* will be sent to TLP. TE receives also messages informing about the impossibility of access (3 e)), or with suggestions for grasp positions.

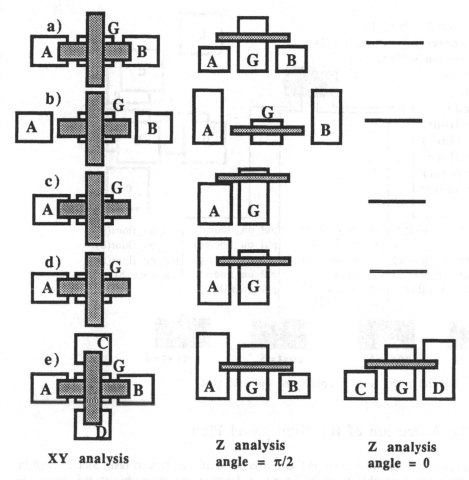

Fig. 3 Examples of XY and Z analysis for Geometric Constraints study

TLP and TE have been modified in order to deal with these geometric constraints. The symbolic relationships "obstructed" are different from compound relationships ("on" or "in_v"). The difference is derived from the interpretation of the second argument. The compound relationships are conjunctive. For example, *on(top,[top1,top2])* means that object *top* is on objects *top1* and *top2*. On the other hand, "obstructed" relationships are disjunctions of conjunctions. For example, "*obstructed(x,[[y,z],[w]])*" means that object x is obstructed by objects y and z or alternatively by object w. This implies that the cost to accomplish a goal is not unique. To solve this problem several paths are permeated for the accomplishment of a goal.

Let us consider figure 4 and suppose that the only intended goal is to put a on h. Four possible solutions are proposed in order to accomplish that goal. The selected node is that one with the smallest cost (only four objects are moved). The different solutions are derived from the different ways to grasp objects. Note that the four generated nodes represent the accomplishment of the same goal. Special attention must be given to those cases for which it is necessary to undo goals already accomplished in order to take a specific object.

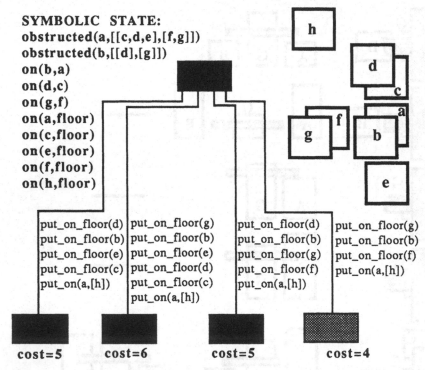

SYMBOLIC STATE:
obstructed(a,[[c,d,e],[f,g]])
obstructed(b,[[d],[g]])
on(b,a)
on(d,c)
on(g,f)
on(a,floor)
on(c,floor)
on(e,floor)
on(f,floor)
on(h,floor)

put_on_floor(d)	put_on_floor(g)	put_on_floor(d)	put_on_floor(g)
put_on_floor(b)	put_on_floor(b)	put_on_floor(b)	put_on_floor(b)
put_on_floor(e)	put_on_floor(e)	put_on_floor(g)	put_on_floor(f)
put_on_floor(c)	put_on_floor(d)	put_on_floor(f)	put_on(a,[h])
put_on(a,[h])	put_on_floor(c)	put_on(a,[h])	
	put_on(a,[h])		

cost=5   cost=6   cost=5   cost=4

Fig. 4   Several paths to accomplish one specific goal

5   The Execution of the High Level Plan

TE is responsible for the correct and safe execution of the planned task. Since it will be necessary to put some objects on the floor in order to access other objects, it is important to define where these objects will be placed. This decision involves the free space management. Note that it is important to have some ideas about the sequence of task execution. For example, it is not a good idea to put an object on a free position that must be occupied in the final state by another object. Another example refers geometric constraints. It is not suitable to put an object on a free position close to another object so that it will avoid the access to the last one. It is often the case that objects are placed in a stable state different from the most convenient one. Therefore, a regrasping procedure is necessary. In order to perform regrasping the robot puts the object on an auxiliary table, with the suitable angle, and grasps it again laterally being then the object ungrasped on the appropriate location.

Handling of objects must be guided in order to the desired action. This means that the ungrasp operation of the object constraints the grasping. Let us consider an object with height equal to 100 mm and another object with height equal to 80 mm and with a vertical hole. Both are on the floor and the goal is to introduce the first one in the hole of the second one. Let us also suppose that the suggested grasping height of the first is 50mm and the grasping zone goes from this height until 90 mm. Since the hole is so small that it is impossible to introduce the gripper two possibilities could be considered. The first object can be grasped at the suggested height (50 mm) and ungrasped at a height superior to the hole top (80 mm) and the object will fall free from at least 30 mm. On the other hand, a graceful solution is to grasp the object between 80 mm and 90 mm and ungrasp it at the same height without the need to fall down.

The robot gripper is supplied with sensors. They are used for the following purposes:
• to avoid collisions during the execution of the basic operations such as grasping of objects
• to map the working area to complete the results of VISION
• to detect the position and orientation of unexpected or not recognized objects
• to recover from simple errors.

Several routines, using proximity sensors, have been written in LM ("Language de Manipulation" that controls our robot). There are routines for object manipulation (eg *GRASP_ON* , *UNGRASP_ON* and *GRASP_BY_SIDE* , being the last one used for regrasping), for checking (eg *VERIFY_IF_FREE* and *CLEAR_DESTINATION*), for the management of the working area (eg *FIND_FREE_AREA*) and for approaching, mapping and handling of unknown objects (eg *GET_CLOSER* , *MAP_UNKNOWN* and *GET_FROM_TOP*). Each high level operation is converted into a set of routine calls. The parameters of these routines define the physical movements of the robot arm. However, the sequence of routine calls as well as the parameter values of these routines may be changed whenever an exception occurs [11].

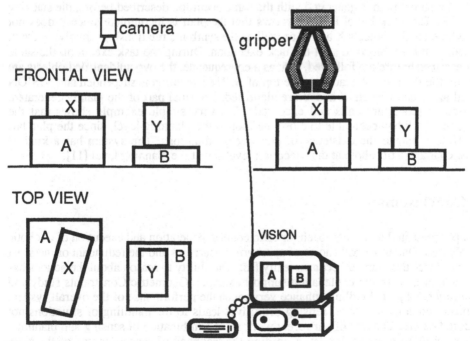

Fig. 5   How to deal with unexpected situations

Let us consider an example of the use of sensors during the execution (see figure 5). VISION identifies two objects, A and B, however two other objects are not identified, X on A and Y on B. We will not detail why they are not identified (eg because only the 2D part of VISION is used or because X and Y have been introduced later). WD detects that A and B are on the floor. The user has said that the goal to be achieved is the stacking of A on B. The plan is easily generated (put_on(a,[b])). TE translates the plan and calls the following procedures: a procedure CLEAR_DESTINATION to see if the position where the ungrasping is to be performed is really clear; another procedure GRASP_ON to grasp the object A and a final procedure call UNGRASP_ON to ungrasp A on the top of B. During the verification, performed by CLEAR_DESTINATION, the existence of an unknown object is detected on B. Then some samples are performed using the installed proximity sensors in order to map the unknown object. Reference [10] explains the methodology used for mapping. The center of area (Xc,Yc) and the orientation ALFAc of the unknown object (Y) are computed (ALFAc extraction is based on the minimum square root method). Therefore, the unknown object, Y, is grasped and ungrasped on the floor. Now, the final destination of A is clear and we will try to grasp this object to put it on B. During the GRASP_ON operation it is detected the existence of another unidentified object. The method used before to map the unknown object is used again. Once more, the center and orientation of the unexpected object are determined being this object (X) grasped and placed on the floor. Therefore, it is possible to grasp A and ungrasp it on B using the UNGRASP_ON basic procedure.

Let us now suppose that both unknown objects are really necessary for the task. The user informs that both objects, X and Y, are to be placed on A. Their positions, orientations and stable states at the final state are also provided by the user. However these objects have not been identified by VISION. The results of VISION and the WD conversion are the same and a partial plan is generated with the same operation described before, the stacking of A on B. The Task Level Planner knows that the plan is incomplete since it does not know where are the objects X and Y as well as the symbolic relationships involving them. Meanwhile, it is intended to start the task execution. During the task execution the same steps described before are followed, and, as a consequence, the two unidentified objects are placed on the floor and A placed on the top of B. The execution is suspended and VISION is called again and objects X and Y are identified. The final part of the plan is generated, translated to procedure calls and executed. This very simple example shows that the sensing feedback has been able to close the loop at the planning level, since the plan has been affected due to the existence of unexpected situations. The system has a kind of reactive characteristics, both at the execution level and at the planning level [11].

6   Conclusions

This paper presented a new approach for efficient plan generation and execution of a robotic assembly task. Due to a good choice of appropriate heuristics and the redundant or incorrect paths avoidance the planner becomes really fast. The ability to reason about the actions has been made clear with the explanation support example. Geometric Constraints study and the Intelligent Object Handling enhance very much the performance of the overall system. The introduction of low-level sensing primitives leads to the handling of situations not considered before. The aim of our recent work is the combination of sensing and planning. The possibility to react and replan according the feedback of sensors is one of the main features desired for intelligent robotic systems. Meanwhile, we give a great importance to the possibility to plan and execute the actions under uncertainties. In fact, the results of

VISION are sometimes incomplete or incorrect being the symbolic descriptions of the initial state also incomplete or incorrect. But this does not avoid the start of the plan formulation and task execution. Once detected the incompleteness or incorrectness, by means of sensors or vision, the system must be able to reformulate the plan and the execution of the task.

References

1. R. E. Fikes, N. J. Nilsson: STRIPS: A New Approach to the application of Theorem Proving to Problem Solving: Artificial Intelligence, Vol. 2, 1972
2. S. H. Hutchinson, A. C. Kak: Spar: A Planner that satisfies Operational and Geometric Goals in Uncertain Environments: AI Magazine, Spring 1990
3. T.Lozano-Perez, J. L. Jones, E. Mazer, P. A. O'Donnell: Task-Level Planning of Pick-and-Place Robot Motions: Computer IEEE, March 1989, pp 21-29
4. L. H. Mello, A. Sanderson: A Correct and Complete Algorithm for the Generation of Mechanical Assembly Sequences: IEEE Trans. on Robotics and Automation, v. 7, n. 2, pp. 228-240, April 1991
5. A. Newell, H. Simon: GPS: A Program that simulates Human Thought: in Computer and Thought: E. Feigenbaum, J. Feldman: N.Y., McGraw-Hill, 1963
6. E. Oliveira, R. Camacho, C. Ramos: A Multi-Agent Environment on Robotics: Robotica, vol. 9, pp. 431-440,Cambridge Univ. Press, 1991
7. C. Ramos, E. Oliveira: Cooperation Between Vision and Planning Agents in a simple Robotic Environment: Intelligent Autonomous Systems 2, Amsterdam, 1989
8. C. Ramos, E. Oliveira: The Generation of Efficient High Level Plans and the Robot World Representation in a Cooperative Community of Robotic Agents: ICAR-91-5th International Conference on Advanced Robotics, Pisa, 1991
9. C. Ramos, E. Oliveira: An Efficient Approach to Planning in Assembly Tasks: Lecture Notes in Artificial Intelligence, n. 541, pp. 210-221, Springer-Verlag, 1991
10. C. Ramos, E. Oliveira: Closing the Loop of Task Planning, Action and Sensing: IEEE/RSJ International Conference on Intelligent Robots and Systems (IROS'92), Raleigh, North Carolina, USA, 1992
11. C. Ramos, E. Oliveira: Sensor-Based Reactive Planning and Execution for Robotic Assembly Tasks: IEEE International Conference on Systems Engineering (ICSE'92), Kobe, Japan, 1992
12. E. D. Sacerdoti: Planning in a Hierarchy of Abstraction Spaces: Artificial Intelligence, Vol. 5, pp 115-135, 1974
13. E. D. Sacerdoti: A Structure for Plans and Behavior: Elsevier, New York, 1977
14. A. Tate: Generating Project Networks: 5th International Joint Conference on AI, 1977

Action Planning for Multiple Robots in Space

U. Kernebeck

Institute of Robotics Research, Otto-Hahn-Str. 8, 4600 Dortmund 50
(Germany)

Abstract. In this paper an action planning system is presented which is directly integrated into a multi robot controller (IRCS) used for a space laboratory. By the application of an action planning system the teleoperation on object level becomes available which allows to improve the safety aspects and the user interface. Furthermore, it becomes possible to let the space laboratory operate in an automatic mode in which the experiment controllers directly generate task descriptions for the IRCS. Some of the action planning system methods are presented in this paper, the action representation, and the coordination of multiple robots on task level.

1 Introduction

Robots in space give the chance to relieve the astronauts from long time missions with all their risks and costs. Space laboratories can be serviced and configured by the robots during the time periods the laboratory is unmanned. While the supervision takes place from the ground the difficulty emerges that not enough sensorial information is available to teleoperate the robots safely. Even automated collision avoidance can not cope with all problems of teleoperation because e.g. for mating operations collisions between objects are explicitly demanded. In these cases the teleoperator has to rely on himself so that teleoperation becomes very time consuming and problematic because of the sensorial information delay. Due to the information run-time and information processing the delay can reach up to a few seconds. The solution to this problem is teleoperation on object level. Especially, the risk that the teleoperator looses his concentration during numerous and repeating standard tasks can be avoided by the teleoperation on object level. The teleoperator on earth describes the task by specifying the goal properties of objects, for example a goal position of a material probe. Then he sends the task description to the robot controller on board of the space laboratory. The adequate robot program is generated on board and can be executed with direct sensorial feedback. The system which generates the robot program on board is a component of the robot controller, the action planning system.

The action planning system (R-Planner), developed in the CIROS (Control of Intelligent Robots in Space) project, is designed to control multiple robots. Naturally, the resulting planner can also be used for a single robot system. But in general it is impossible to adapt a single-robot-action-planning system to the requirements of a multiple robot configuration. To improve the reliability and to gain functional redundancy of the multiple robot system, robots with overlapping workspaces have been chosen. If one robot breaks down the other can be brought into action which is automatically achieved by the R-Planner. Figure 1 shows the structure of the intelligent robot control system (IRCS) the R-Planner is integrated in. The coordinated operation, presented in more detail in [3], has the task to execute the robot program generated form the action planning system. Furthermore, it allows to grip and move large objects with both robots. The collision avoidance, described in [4], increases the safety aspects of the whole system by avoiding collision between the robots and between the robots and their environment.

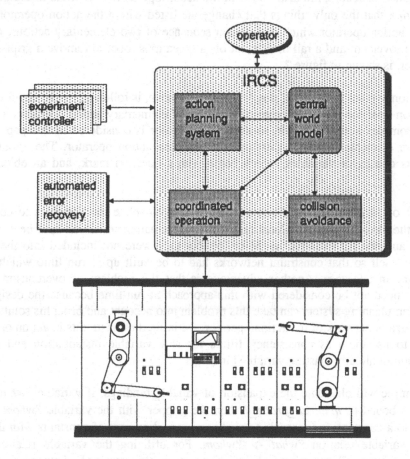

Fig. 1. The logical structure of the intelligent robot control system (IRCS).

In the following the action planning system will be explained in more detail beginning with the action representation in chapter 2. A rough description of the planning process follows in chapter 3. The special features of the action planning system (R-Planner) with respect to a multi robot system are described in chapter 4, 5, and 6. In chapter 5 the goal list modification method is presented to coordinate different robots on task level. The realization is mentioned in chapter 7.

2 Action Representation

Because the action representation influences almost any decision in the planning process, the central concern of an action planning system is a detailed and powerful action model. Actions are represented in form of action operators which are classes of actions containing a list of variables. By instantiating the variables of the action operator during the planning process the individual action can be received. Information about the preconditions of the particular action and about the changes in the world model, when the action is executed, are also coded in the action operator. This leads to the *STRIPS-assumption* that the only things that change are listed within the action operator. A R-Planner action operator which describes a sequence of two elementary actions, namely an arm movement and a rail movement of a seven axis robot to remove a gripper from an object, is shown in figure 2.

The action symbol in figure 2, *simple_gripper_remove*, is followed by the detail level of the action operator which describes the number of elementary action operators involved in the complex action operator, so an action operator is considered in this paper to be either an elementary action operator or a complex action operator. The variable list elements consist of variables, which begin with a question mark, and an object class label.

The list of *variable relations* contains constraints to relate the variables to constants during the planning process. Constraints for variable parameterization have been used by several authors, for example in [5] or [6], but they were not included into the action operator itself so that constraint networks had to be built up at run time which is not necessary in this case. Another advantage is that the problem of overconstraining a variable need not be considered with this approach at run-time because the designer of the action planning system can take this problem into account and bring his solution into the constraint network of the action operator. Defining the constraints in action operators allows to use them as consistency filters for new variable instantiation and for the expansion of already existing variable bindings.

An example will elucidate the expansion of variable bindings. If variable *?what* for an object is bound to *material_probe_1* then the gripper with the variable *?whom* can be related to a constant, for example to *probe_gripper_1* or a set of constants with the help of the variable relation *?what -> ?whom*. For utilizing the variable relations, the constant patterns of instantiated object-gripper-pairs are examined whether a pair exists with an object instantiation *material_probe_1* and a gripper constant. In the example it is assumed that an object-gripper-pair, *material_probe_1 - probe_gripper_1* exists, so that the gripper variable is bound to *probe_gripper_1*. The constant patterns allow implicit geometrical and functional information to be brought into the system. This enhances the performance of the planning process, especially the variable parameterization. Another important function of the variable relations is its ability to work as a consistency filter for variable bindings from the actual situation.

action symbol: simple_gripper_remove   detail level: 2

variable list: ?who   -   robot,
                        ?whom   -   gripper,
                        ?what   -   object,
                        ?deposit   -   deposit,
                        ?path1   -   object-deposit-out-path
                        ?path2   -   object-deposit-in-path,
                        ?pos1   -   endpos-object-deposit-in-path,
                        etc.

variable relations: ?what -> ?whom
                        ?deposit -> ?path1
                        ?deposit -> ?path2
                        etc.

preconditions: begin_1:   fixed_connection(?who,?whom)
                        begin_2:   position(?who,?pos1,?orient1)
                        etc.

intermediate-
conditions: inter_1:   position(?who,?pos2,?orient2)
                        inter_2:   fixed_connection(?who,?whom)
                        etc.

change list: ch_1:   position(?who,?pos3,?orient3)
                        ch_2:   position(?whom,?pos3,?orient3)
                        etc.

action list: op1:   gripper_remove + "variable equivalents"
                        op2:   transfer_move + "variable equivalents"

temporal structure: op1 before op2

Fig. 2. An example for an complex action operator.

The next elements of the action operator in figure 2 are the preconditions. They specify those conditions of the actions, which must be true before the action can be executed. For example, to move a gripper, it must have a fixed connection to a robot. The intermediate conditions allow to decide whether the whole action operator or only a part of it should be used during the planning process. In the example the preconditions of the second action, *transfer_move*, generate the *intermediate condition* of the complex action operator, *simple_gripper_remove*. The *change list* contains the changes of the representation of the R-Planner's space laboratory model that will happen when the action represented by the action operator is executed. Information about the elementary action operators are included in the *action list*. This list consists of the internal action symbols, for example *op1*, and the external action symbol, for example *transfer_move*, of the elementary action operators and their particular list of variable equivalents. The list of variable equivalents assigns the variables from the action operator to the variables of the elementary action operators which can be different. A temporal binary relation, *before*, is used between two elementary actions to order the elementary actions qualitatively in a *temporal structure*. Because simultaneous and sequential actions can be represented in any combination with the same temporal structure, it is also used for action plans .

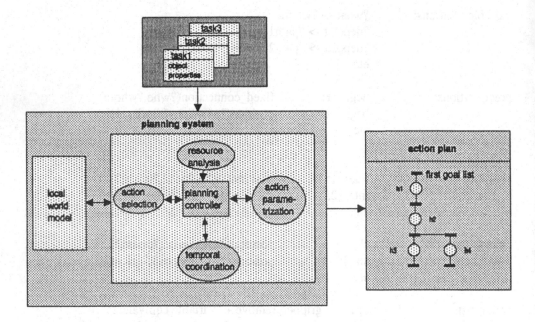

Fig. 3. An abstract model of the action planning process.

3 Planning Process

The planning process, considered from an abstract point of view, consists of the action operator selection process, the variable parametrization process, the qualitative temporal coordination of the action operators, and the planning control, see figure 3. It begins with the specification of the tasks which in the case of the R-Planner results from user inputs via the man machine interface (MMI), from experiment controllers, or from an automatic error recovery component. The task description consists of one or more properties of an object. For example, the goal position of a circuit board is a task description for a complete mating operation of a circuit board. These task descriptions are allowed to be incomplete, so that variables without instantiations can occur in the description of the task and not all objects which are relevant for the task must be mentioned.

After specifying the tasks, see figure 3, the resource requirements of each task are investigated during the resource analysis so that a suitable robot or robot group is related to each task. Then the action selection process begins which selects the appropriate action operators. After each action selection process the action operators are parametrized and temporally ordered. The planning controller has the task to build up a new goal list which is generated out of the preconditions of the selected action operators which are not satisfied by the actual situation or other selected action operators. Like the task description the goal list consists of goal properties of objects which have to be satisfied during the planning process to generate a meaningful and complete action plan. Therefore, the task description specifies the first goal list of the planning process. The action plan is the result of the planning process which consists of several totally parametrized action operators. The action operators build up the action plan incrementally in which sequential and simultaneous actions are allowed. In figure 3 an action plan is presented which consists of several action operators, h1, ..., h4 which are circles in the figure. The small black boxes are the goal lists, one goal list before and after each selected action operator. The planning process begins with the action operator h1 which has to meet the requirements of the first goal list, the task description. The next goal list is build up by the preconditions of h1 which are not satisfied by the actual situation so that the next action operator must be selected and parametrized.

4 Task Orientated Robot Coordination

If two robots with overlapping working areas are available, it is more economic to share tasks between both robots instead of working only with one and making a lot of extra gripper exchanges and additional movements. If several task descriptions are specified by the user, the planning system's task is to treat them separately so that the planning process for one task is not disturbed by the goal list elements of another task. Robot coordination on task level emerges, if one or more subtasks must be generated dynamically by the action planning itself, for example opening a drawer is one subtask generated for a task which describes only the goal position of a material sample, e.g. in the drawer slot. With two robots each robot can execute one subtask, so that subtasks and the original task can be executed simultaneously. Another quality of robot coordination on task level arises if the task demands two additional simultaneous subtasks when only two robots are available. For example, the task is to feed a heater with a material sample which is located in a closed drawer and the flap of the heater is also closed. Examples for the coordination on task level are given in figure 4 and 5.

5 Goal List Modification

During the planning process, several goal lists are generated by the preconditions or intermediate conditions of an action operator. These goal list elements do not match with the model of the actual situation of the space laboratory. Goal list elements, which are satisfied by a selected action operator, disappear form the goal list. In addition, the

R-Planner knows also another kind of goal list modification, called goal list element relations (GLERs) in this paper. A GLER consists of a trigger condition and a goal list modification. Each of them have the form of a goal list element. If the trigger condition of a GLER matches with a subset of the original goal list, these matched goal list elements are replaced by the goal list modification of the GLER. The goal list modification of the GLER has the important feature that it always represents a subtask. An example for a GLER shows how they are used. The original goal list element which will trigger a GLER is:

$$state(endpos\_drawer\_1\_slot\_2\_deposit\_in\_path, accessible)$$

The triggered GLER looks like:

$$state(endpos\_drawer\_1\_slot\_2\_deposit\_in\_path, accessible)$$
$$-->$$
$$state(drawer\_1, auto\_open).$$

That means if the accessibility of a position, *endpos_drawer_1_slot_2_deposit_in_path*, is needed for an action, for example for the deposition of an object, *drawer_1* has to be opened. In addition, it must be distinguished between hard automated and not hard automated drawers in the R-Planner's environment. In the given example *drawer_1* is hard automated which is recognized by the action planning system with help of the symbol *auto_open* instead of *open*. The original goal list element is substituted with:

$$state(drawer\_1, auto\_open).$$

With GLERs the environment specific information, e.g. the special kind of a drawer, need not be kept in the action operator which would demand conditional statements in the preconditions or intermediate conditions of an action operator. The dynamically generated subtask, opening a drawer in this case, can also require the use of a another robot. With the tool of goal list modification by GLERs the coordination of robots on task level can also be realized which is presented in figure 4 in form of an abstract example.

Task *a* was formulated by the user or an experiment controller. The task description constitutes the first goal list for the action planning system. Several action operators, *a1*, *a2* are selected and new goal lists are generated without the use of the GLERs. The third goal list is modified with the help of GLERs, as described above, so that a new subtask is generated, for example for opening a drawer. From this point on two tasks, namely subtask *b* and task *a*, are planned simultaneously until the planning process terminates. For each of the tasks in general one robot is available. Interesting problems for the action planning system arise if more subtasks are generated during the planning process then robots are available or if one robot breaks down. In these cases simultaneous subtasks have to be sequentialized on the planning level.

6 Sequentialization

If not enough robots are available to execute all subtasks and tasks simultaneously they must be executed sequentially until simultaneous actions are possible. Two solutions to these parallel conflicts are possible:

- solving parallel conflicts by modifying the plan,
- solving the conflicts by combining execution and planning.

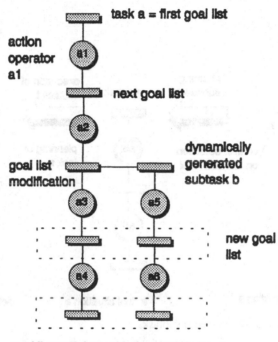

Fig 4. Goal list modification for robot coordination on task level.

An example of the R-Planner's parallel conflict solving method is presented in figure 5. In this case only one robot is available. The action planning begins normally until the goal list modification is made. At this point the R-Planner recognizes that task *a* and subtask *b* should be executed simultaneously with two robots but only one robot is available. To illustrate the example, task *a* may be the charging of a heater with a material sample and subtask *b* should be the opening of the heater flap. Task *a* is planned until its action plan is completed without taking subtask *b* into account. This enables the planning system to get an overview of the subtask and task conflicts involved with the solution of the original task. The conflict solution begins with planning subtask *b* and executing it if there is no other conflict recognized during the planning of subtask *b*. The action plan of task *a* including action operator *a1*, *a2*, and *a3* is stored together with the new goal list. When the execution of subtask *b* is finished, the planning of the rest of task *a* can be restarted but the difference is that additional action operators must be selected to adapt to the new actual situation of the space laboratory. Considering the example for subtask *b*, the opening of the heater flap, different grippers are needed for material probe handling and opening the flap. Also the position of the robot is changed after the execution of subtask *b*.

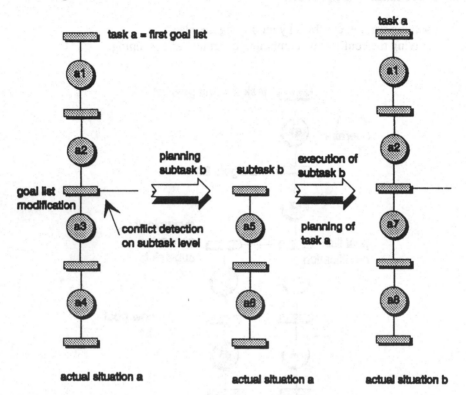

Fig. 5. Sequentialization of simultaneous subtasks.

This kind of conflict solution is a combination of classical planning and reactive planning but used in a deterministic environment. Reactive planning is the way to select actions or action packages according to the behavior of the environment and therefore regards smaller time horizons for the action planning than classical planning does. In reactive planning systems, incomplete plans are executed in contrast to the classical planners like STRIPS, [1] and [2], which only generate complete plans. The similarity between a reactive planner and the R-Planner is the usage of incomplete plans. The difference is that the R-Planner does this only for conflict solutions otherwise it plans incrementally like other classical planning systems. Combining planning and execution on subtask level helps to enhance the performance of an action planning system.

No conflict detection on the action operator level is necessary so that the reason for the conflict can be determined without a big expense at run time. Another feature of this method is, that it helps to reduce a complex planning problem into smaller ones in a recursive way. It might be possible that subtask *b*, in the example opening the heater flap, needs a sequentialization itself because the heater flap is locked. In that case also subtask *b* is stored in a stack until the next subtask, namely unlocking the flap, is planned and executed.

7 Conclusion

In this paper an automatic action planning system (R-Planner) is presented which provides a better user interface for a multi robot system in a space laboratory on one the hand and on the other hand the R-Planner allows to let the robots react on the demands of experiment controllers. This enables an automatic servicing of the space laboratory by means of robots. From the R-Planner's methods the sequentialization of simultaneous subtasks and the goal list modification are stressed in this paper which are the basis of the coordination of multiple robots on task level. The R-Planner's functionality is validated in the test bed of the CIROS project. There are two 7-axis robots in gantry configuration which have an overlapping working area so that each robot can execute the task for the other robot. The set of tasks applied during the validation phase include tasks for experiment servicing, e.g. supplying a heater with material sample, and little repair tasks like mating printed circuit boards into computer slots. To test the sequentialization method a transport task of a material sample was generated which includes three simultaneous tasks or subtasks for two robots: opening the drawer in which the material sample lies, opening a heater flap and moving the material sample into a heater slot. In addition, a mal function of one robot was simulated to test the presented sequentialization method in this case.

Acknowledgements
The R-Planner was realized in the CIROS project (Control of Intelligent RObots in Space) which was supported by a grant of the "Bundesminister für Forschung und Technik (BMFT)" of Federal Republic Germany.

References

1. R.E.,Fikes, N.J.,Nilsson: STRIPS: A New Approach to the Application of Theorem Proving to Problem Solving, Art. Int. 3, 189-208, 1971.
2. R.E.,Fikes, P.E.,Hart, N.J.,Nilsson: Learning and Exectuing Generalized Robot Plans, Art. Int. 3, 251-288, 1972.
3. E.,Freund, P.,Kaever: An Object-orientated Pathplanning System for multiple Robots with Singularity-coping Properties, SYROCO 91, Vienna, 1991.
4. E.,Freund, J.,Rossmann: Teleoperated And Automatic Operation of Two Robots in a Space Laboratory Environment, 41st Congress of the IAF, Dresden, FRG, 1990.
5. M.,Stefik: Planning with Constraints, Art. Int. 16, 111-139, 1981.
6. D.E.,Wilkins: Domain-Independent Planning: Representation and Plan Generation, Art. Int. 22, 269-301,1984.

A Calibration of a Mobile Camera Used for a Camera–Guided Grasping by an Eye–In–Hand Robot

Dipl.–Inform. Anneliese Schrott

Technische Universität München, Institut für Informatik,
Postfach 20 24 20,   W 8000 München 2

Abstract. In order to initialize a camera–guided grasping of a known object, rough 3–D localization data of the object relative to the camera are used. Although the 3–D data need not be more than rough, we need the intrinsic parameters of the camera for calculating these 3–D data. The extrinsic parameters are required to move the camera.

This paper proposes a method for simplified automatic calibration of a mobile camera. The intrinsic parameters are estimated separately in succession with the help of a grid of points. In order to determine the extrinsic parameters, the location of the camera relative to the gripper is needed. This location can be calculated using a certain location of the camera relative to the scene coordinate frame. The exactness of this calibration will suffice for the support a camera–guided grasping.

1 Introduction

In order to make robots more flexible, several types of sensors are used. One possibility is to control or even to guide robot manipulation with the help of a CCD–camera mounted on the gripper. The CCD–camera can be used in different ways.

One method tries to reconstruct 3–D locations of objects as exactly as possible. Therefore, the pinhole model [6] is used as camera model. In this model, the coordinates of an image point, $({}^Ix, {}^Iy)$, are related to its 3–D coordinates, $({}^Cx, {}^Cy, {}^Cz)$, relative to the camera coordinate frame CCF by

$$ {}^Ix = \frac{(-d)\,{}^Cx}{{}^Cz} \qquad\qquad {}^Iy = \frac{(-d)\,{}^Cy}{{}^Cz}. \tag{1}$$

The origin of CCF in the pinhole model is defined by the projection center P (see Fig. 1). The xy–plane is defined by the image plane and the z–axis, which is identical to the optical axis, lies perpendicular to the xy–plane. The image plane has its own 2–D image coordinate frame ICF. The intersecting point I between the optical axis and the image plane is taken as origin of ICF. The axes are identical to the projections of the x– and y–axes of CCF into the image plane. The distance d

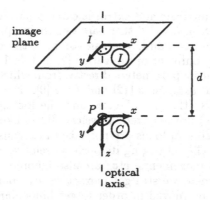

Figure 1: The pinhole model

between the projection centre and the image plane is called *principle distance*. The calculation of the 3–D data can be done using corresponding points of two images taken from distinct projection centres which differ in only one direction, for example in x–direction [5, 7, 8].

The digital imaging process supplies a 2–D matrix of picture elements (pixels) with the coordinate frame *PCF*. The origin of this frame lies in the pixel with the co-ordinates (0,0). The x–axis of *PCF* has the same direction and the y–axis the negative direction of those of *ICF*. As we have to use the image coordinates for the reconstruction of 3–D data, the pixel coordinates have to be transformed into image coordinates. In order to do so, divergences between this model and the reality have to be taken into account.

- The camera chip not being mounted at exactly the ideal position and orientation, *four degrees of freedom* of the chip plane against the optical axis can occur: two translational ones and two rotational ones relative to the x– and y–axes of *CCF*

- In practice, almost every system of lens used for image acquisition causes a *radial symmetric distortion* during the imaging process.

- In addition, the *scale factors in x– and y–directions* are required to transform the unit of a pixel into millimetres.

- Besides these camera parameters, the *principle distance d* is required for the transformation of the coordinates (see equ. 1).

All the parameters mentioned are called *intrinsic parameters*.

In order to move the camera, *position* and *orientation* of the camera relative to a predefined scene coordinate frame *SCF*, called *extrinsic parameters*, are required:

- ${}^S p_x$, ${}^S p_y$, ${}^S p_z$, the coordinates of the projection center representing the origin of *CCF* relative to *SCF*.

- ${}^S \vec{c}_x$, ${}^S \vec{c}_y$, ${}^S \vec{c}_z$, the coordinates of the vectors representing the axes of *CCF* relative to *SCF*. These axes determine the orientation of *CCF*.

The determination of the intrinsic and extrinsic camera parameters is called *calibration*. According to *Puge/Skordas* [10] there are two main approaches of calibrating a moving camera. On the one hand, you can compute the calibration parameters with the help of full scale nonlinear optimization [2, 3, 4, 9, 10]. On the other hand, you can compute the extrinsic parameters directly, from which the rest can be easily calculated as described in *Tsai/Lenz* [12] and *Lenz* [9]. For this calculation, some of the intrinsic parameters (the scale factors and the location of the camera chip relative to the optical axis) are required. Physical effects like the deflection of light at the aperture or not beeing in focus are not taken into consideration. Mechanical problems like the camera chip changing its position relative to the optical axis due to high temperature, fast movements, etc. are also ignored. Although a costly calibration is beeing performed, we still get inexact camera parameters. Moreover, a monocular system has to be moved in order to get binocular images. These movements, which are mostly inexact, together with the inexact camera parameters, lead to inexact 3-D data.

In order to overcome these problems, a "cyclic look-and-move" technique using image features of the object to guide the grasping movement has been developed [11]. The author has minimized the number of movements needed using 3-D data of the object of interest relative to the camera in order to initialize the grasping motion. It will therefore *suffice to estimate the 3-D data relative to the camera roughly*. This means that the intrinsic parameters required for the reconstruction of the 3-D data do not have to be exact; the extrinsic parameters, which are only required to move the camera, can also be approximated. Therefore, no large-scale estimation of the camera parameters is needed.

This paper presents a method of a simplified, automatic estimation of the intrinsic and extrinsic parameters of a camera mounted on a jaw gripper. A simple object, a right parallelepiped, was developed together with the method of calibration. The top of the object consists of a plane, with an aquidistant grid of luminous dots on it. The intrinsic parameters are calculated separately in succession with the help of these dots. In order to estimate the extrinsic parameters of the mobile camera, a certain location of the camera relative to SCF is estimated, also using this object. Then, the location frame matrix ^GF_C of the camera relative to the gripper has to be determined by the help of the certain location estimated.

In the following section, the estimation of the intrinsic parameters will be presented. Chapter 3 describes the reconstruction of the 3-D data. The estimation of the extrinsic parameters is explained in chapter 4. Chapter 5 discusses some results.

2 Estimation of the Intrinsic Camera Parameters

For calibrating the camera, first the object with the grid of luminous dots is arranged in such a way that its axes and those of SCF are parallel to each other.

The automatic determination of the intrinsic camera parameters starts with the estimation of the two *translational degrees of freedom* of the chip plane against the optical axis. They are determined by the *intersecting point* $^PI = (^Pi_x, {}^Pi_y)$ (in pixel coordinates) between the camera chip and the optical axis. This intersecting point is assumed to be the centre of the lens distortion, which causes the image of a straight line lying parallel to the axes of the image plane to become curves. They can be approximated by a polynom of two degrees if the camera chip lies parallel to the plane of the straight lines. For this process, the camera has to be moved in this orientation relative to the grid of luminous dots by help of the symmetric characteristics of the lens distortion. Now, the mean value of the extrema of the curves caused by m rows and n columns of the grid of the luminous dots represents the coordinates of the intersecting point, because this point is the centre of the lens distortion:

$$P_{i_x} = \frac{1}{m}\sum_{k=1}^{m}\frac{-\tilde{a}_{1k}}{2\tilde{a}_{2k}} \qquad P_{i_y} = \frac{1}{n}\sum_{l=1}^{n}\frac{-\tilde{b}_{1l}}{2\tilde{b}_{2l}} \tag{2}$$

where $\tilde{a}_{1k}$, $\tilde{a}_{2k}$ and $\tilde{b}_{1l}$, $\tilde{b}_{2l}$ be the coefficients of the m and n polynoms.

In order to determine the *lens distortion*, first, the resulting polynoms are transformed:

$$\begin{aligned}
{}^P p_y &= \tilde{a}_{2k}{}^P p_x^2 + \tilde{a}_{1k}{}^P p_x + \tilde{a}_{0k} = \hat{a}_{2k}({}^P p_x - {}^P i_x)^2 + \hat{a}_{0k} \\
{}^P p_x &= \tilde{b}_{2l}{}^P p_x^2 + \tilde{b}_{1l}{}^P p_y + \tilde{b}_{0k} = \hat{b}_{2l}({}^P p_y - {}^P i_y)^2 + \hat{b}_{0l}
\end{aligned} \tag{3}$$

Next, the coefficients $\hat{a}_{2k}$ and $\hat{b}_{2l}$ of the different functions for each grid row and column are also approximated by a linear function based on the constant coefficient $\hat{a}_{0k}$ respectively $\hat{b}_{0l}$:

$$\hat{a}_{2k} = a_1\hat{a}_{0k} + a_0 \qquad \hat{b}_{2l} = b_1\hat{b}_{0l} + b_0 \tag{4}$$

In order to get the corrected x- and y-coordinates, which are represented by the constant coefficients, the inversion of equation (3) is needed, replacing $\hat{a}_{2k}$ and $\hat{b}_{2l}$ by the linear functions (see equ. (4)):

$$P_{\tilde{p}_x} = \frac{{}^P p_x - b_0\left({}^P p_y - {}^P i_y\right)^2}{b_1\left({}^P p_y - {}^P i_y\right)^2 + 1} \qquad P_{\tilde{p}_y} = \frac{{}^P p_y - a_0\left({}^P p_x - {}^P i_x\right)^2}{a_1\left({}^P p_x - {}^P i_x\right)^2 + 1}. \tag{5}$$

The distance between the curves and the intersecting point is also affected by the lens distortion. In order to correct this translational part of the lens distortion, a polynom of five degrees with only odd terms is needed:

$$\begin{aligned}
{}^P\hat{p}_x = dc_x({}^P P) &= k_1({}^P\tilde{p}_x - {}^P i_x) + k_3({}^P\tilde{p}_x - {}^P i_x)^3 + k_5({}^P\tilde{p}_x - {}^P i_x)^5 + {}^P i_x \\
{}^P\hat{p}_y = dc_y({}^P P) &= l_1({}^P\tilde{p}_y - {}^P i_y) + l_3({}^P\tilde{p}_y - {}^P i_y)^3 + l_5({}^P\tilde{p}_y - {}^P i_y)^5 + {}^P i_y.
\end{aligned} \tag{6}$$

The addition of ${}^P i_x$ respectively ${}^P i_y$ is needed in order to represent the resulting coordinates relative to PCF.

Since the camera chip lies parallel to the grid plane, the distances between every two neighbouring rows and columns of the image should be identical. Closest to ${}^P I$ there is hardly any distortion. Therefore, the distance closest to ${}^P I$ is taken as measure of the translational part of the lens distortion. The differences between the other distances and this measure is used for the estimation of the coefficients k_1, k_3, k_5 respectively l_1, l_3, l_5.

The *y scale factor* can be taken from the chip's data sheet, because the vertical spacing of the image lines is independent of the digitizing hardware and so the vertical size of a pixel is equal to the size of a sensor element on the camera chip [9].

The horizontal spacing of the image columns, however, depends on the digitizing hardware. Therefore, the scale factor in x-direction is unknown [4, 9]. The x scale factor is estimated by equating the scaled distances between the corrected pixel coordinates $^P\hat{P}_{ij} = (^P\hat{x}_{ij}, ^P\hat{y}_{ij})$ of the image of a dot of the grid and the corrected pixel coordinates $^P\hat{P}_{i+1j}$ and $^P\hat{P}_{ij+1}$ of neighbouring image dots in x- and y directions:

$$\underset{\substack{j=1,\ldots,m-1 \\ i=1,\ldots,n-1}}{\forall} \left|^P\hat{x}_{ij} - ^P\hat{x}_{i+1j}\right| s_x = \left|^P\hat{y}_{ij} - ^P\hat{y}_{ij+1}\right| s_y \quad \Rightarrow$$

$$s_x = \frac{1}{(m-1)(n-1)} \sum_{i=1}^{n-1} \sum_{j=1}^{m-1} \frac{\left|^P\hat{y}_{ij} - ^P\hat{y}_{ij+1}\right| s_y}{\left|^P\hat{x}_{i+1j} - ^P\hat{x}_{ij}\right|}. \tag{7}$$

The *rotational part* of the incorrect location of the camera chip is corrected by defining a new z-axis of the camera frame, which passes through the projection centre and lies perpendicular to the camera chip. Therefore, the intersecting point O between the new z-axis and the camera chip plane has to be determined and ICF has to be redefined by replacing the former origin by the point O. In order to determine O, the *principle distance d* has to be estimated. This distance d can be calculated by help of equ. (1), the distance between two dots of the grid in x-direction, ^Cd_x, and the scaled distances, $^Id_{x_1}$ and $^Id_{x_2}$, between the corrected x- coordinates $^P\hat{x}_1$, $^P\hat{x}_2$ and $^P\hat{x}_3$, $^P\hat{x}_4$ of the images of these two dots, taken by two camera positions which differ in their z-distance to the grid plane by Δz:

$$^Id_{x_1} = \frac{^Cd_x \cdot |d|}{^Cz} \qquad ^Id_{x_2} = \frac{^Cd_x \cdot |d|}{^Cz - \Delta z} \quad \Rightarrow$$

$$|d| = \frac{^Id_{x_1} \cdot ^Id_{x_2} \cdot \Delta z}{^Cd_x(^Id_{x_2} - ^Id_{x_1})} \quad \text{with} \qquad (8)$$

$$^Id_{x_1} = s_x \cdot |^P\hat{x}_1 - ^P\hat{x}_2| \qquad ^Id_{x_2} = s_x \cdot |^P\hat{x}_3 - ^P\hat{x}_4|$$

Now, the x-coordinate of one dot can be represented by help of the equation (1), the corrected x-coordinates of the two images of this dot and the transformation from PCF to the new ICF:

$$^Cx = \frac{s_x \cdot (^P\hat{x}_1 - ^P\hat{o}_x) \cdot ^Cz}{(-d)} = \frac{s_x \cdot (^P\hat{x}_3 - ^P\hat{o}_x) \cdot (^Cz - \Delta z)}{(-d)}$$

$$\text{with} \qquad ^Cz = \frac{^Cd_x \cdot |d|}{^Id_{x_1}}$$

The y-coordinate of this dot can be presented by analogy. Therefore, the new origin of ICF $^P\hat{O} = (^P\hat{o}_x, ^P\hat{o}_y)$ — in pixel coordinates — can be estimated:

$$^P\hat{o}_x = \frac{(^P\hat{x}_1 - ^P\hat{x}_3)^Cz + ^P\hat{x}_3 \cdot \Delta z}{\Delta z} \qquad ^P\hat{o}_y = \frac{(^P\hat{y}_1 - ^P\hat{y}_3)^Cz + ^P\hat{y}_3 \cdot \Delta z}{\Delta z} \tag{9}$$

Subsequently, the relation between the image coordinates $(^Ip_x, ^Ip_y)$ and the pixel coordinates of a pixel point ^PP can be represented by correcting the lens distortion, transforming from PCF into ICF and scaling:

$$^Ip_x = s_x \cdot (dc_x(^PP) - ^P\hat{o}_x) \qquad ^Ip_y = s_y \cdot (^P\hat{o}_y - dc_y(^PP)) \tag{10}$$

3   The 3–D Pose Estimation

For the 3–D pose estimation of a point, the correspondences between the image points have to be found. As the camera moves along the z–axis, the two camera positions differ only in their z–coordinates at a distance Δz (see Fig. 2). Therefore, the 3–D coordinates $^C P_1 = (^C x_1, \, ^C y_1, \, ^C z_1)$ and $^C P_2 = (^C x_2, \, ^C y_2, \, ^C z_2)$ of a real scene point P, which are relative to CCF, differ only in their z–coordinate, just as the camera positions do: $(^C x_1, \, ^C y_1, \, ^C z_1) = (^C x_2, \, ^C y_2, \, ^C z_2 + \Delta z)$. This means that two criteria for correspondence based on polar coordinates can be established.

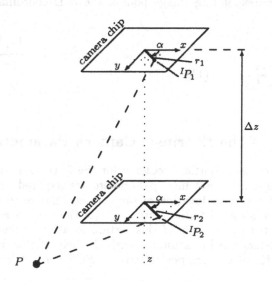

Figure 2: Projective Transformation by two Different Camera Locations

Criteria for Correspondence:

1. *The angles of the polar coordinates of the corresponding image points $^I P_1$ and $^I P_2$ are identical.*

2. *There is a linear interdependence between the radii of the corresponding image points $^I P_1$ and $^I P_2$. Moreover, the linear factor depends on $^C z_1$.*

Using the relative projective transformation (see equ. 1) the ratio of $^I y$ to $^I x$ is identical to the ratio of $^C y$ to $^C x$. Moreover, the identities between $^C x_1$ and $^C x_2$ and between $^C y_1$ and $^C y_2$ can be used to deduce the relations of the correspondence. The image coordinates of the correspondig image points $^I P_1 = (^I x_1, \, ^I y_1)$ and $^I P_2 =$

$(^I x_2, ^I y_2)$ are also necessary:

$$\alpha = \arctan \frac{^I y_1}{^I x_1} = \arctan \frac{^C y_1}{^C x_1} = \arctan \frac{^C y_2}{^C x_2} = \arctan \frac{^I y_2}{^I x_2}$$

$$\frac{r_2}{r_1} = \frac{\sqrt{(^I x_2)^2 + (^I y_2)^2}}{\sqrt{(^I x_1)^2 + (^I y_1)^2}} = \frac{^C z_1}{^C z_1 - \Delta z}$$

After determining the corresponding image points, the 3–D coordinates are estimated:

$$^C z_2 = \frac{^I x_1 \cdot \Delta z}{^I x_2 - ^I x_1} \qquad (11)$$

$$^C x_2 = \frac{^I x_2 \, ^C z_2}{(-d)} \qquad (12)$$

$$^C y_2 = \frac{^I y_2 \, ^C z_2}{(-d)} \qquad (13)$$

4 Determination of the Extrinsic Camera Parameters

In order to perform the camera motion necessary for the 3–D pose estimation and the guidance of the grasping, the extrinsic parameters are required. First, the position and orientation of a certain location of the camera relative to SCF have to be determined. Then, the location frame matrix $^G F_C$ of the camera relative to the gripper has to be estimated by the help of this certain location. Subsequently, the relevant extrinsic parameters can be estimated with the help of this frame and the known location frame $^S F_G$ of the gripper relative to SCF derived from the robot controlling system:

$$^S F_C = ^S F_G \cdot ^G F_C \qquad (14)$$

The location frame matrix $^G F_C$ consists of two parts, the rotational part $^G R_C$, representing the camera orientation, and the translational one $^G \vec{t}_C$, representing the camera position.

In order to simplify the estimation of the *camera orientation* relative to the gripper represented by the rotation matrix $^G R_C$, a certain orientation relative to SCF is chosen in such a way that all axes of CCF and SCF are parallel and the y– and z–axes of CCF have the negative direction of those of SCF. Therefore, the actual rotation matrix $^S R_C$ is determined easily. Now, $^G R_C$ can be calculated by help of the known rotation matrix $^G R_S$:

$$^G R_C = ^G R_S \cdot ^S R_C \qquad (15)$$

After the determination of the intrinsic parameters, the orientation of the camera is identical to the desired orientation.

Next, the distance vector $^S \vec{t}_{GC}$ from the origin of GCF to that of CCF relative to SCF is estimated.

The z–coordinate of the top of the special object $^S o_z$ relative to SCF, the z distance

$\overline{CO_z}$ between the origin of CCF and this object, and the z coordinate ^Sg_z of the position of GCF are used in order to estimate $^{St}t_{GC_z}$:

$$^{St}t_{GC_z} = {}^So_z + \overline{CO_z} - {}^Sg_z \qquad (16)$$

$\overline{CO_z}$ is calculated as described in chapter 3.

In order to estimate the x coordinate $^{St}t_{GC_x}$ of the distance vector, the difference between ^Sg_x and the x coordinate ^Sp_x of that point, the pixel coordinates of which are identical to that of the origin O of ICF, is calculated. The y coordinate $^{St}t_{GC_y}$ of the distance vector is determined by analogy.

In order to determine the *camera position* relative to the gripper $^G\vec{t}_C$, the vector $^S\vec{t}_{GC}$ is transformed by help of the known transformation matrix ^GF_S:

$$^G\vec{t}_C = {}^GF_S \cdot {}^S\vec{t}_{GC} \qquad (17)$$

5   Results

The vision system used is the Panasonic CCD–camera WV–CD50 — with a wide-angle lens (focal length $f = 6.5\,mm$) — and an image processing system which is implemented on a μVAX running the operating system VMS.

The camera was mounted on an electrical servo–gripper SGR–1000 of a PUMA 560 with 6 degrees of freedom. A Cartesian real–time interface developed in our group was used for direct Cartesian control of the robot via the external host processor rt VAX 1000, with the real–time operating system VAXELN and VAXELN–Pascal. The camera parameters are determined by performing the calibration — implemented as a separate VMS–process that communicates with the image processing routines by means of mailboxes — as proposed in this paper for 10 times. Except the coefficients of the translational part of the lens distortion and the coordinates of the new origin of ICF are not estimated. Each time, a 512×512 image of the grid has been acquired and thresholded, and the pixel coordinates of the centroid of the dots needed for the calibration were obtained.

The means and standard deviations of the intrinsic parameters: coordinates of the intersecting point between the optical axis and the camera chip, P_{i_x}, P_{i_y}, the coefficients of the lens distortion function a_0, a_1 and b_0, b_1, the principle distance d and the scale factor, s_x and s_y, are given below. The orientation of the camera is measured with the orientation of the gripper given by the angles φ and ϑ of its x– and z–axes relative to SCF at the certain location. These angles are defined as the ones of the 3–D polar coordinates. The position of the camera relative to the gripper is given by $^S\vec{t}_{GC}$ (see Tab. 1).

The high standard deviation of the intersecting point ^PI is caused by the fluctuation of the repetition of points in the pixel matrix caused by the frame grabber. For a more detailed description of this effect see [1].

The exactness of these parameters are checked for being sufficiency by calculating the distances between points in 3–D space 50 times. The standard deviation of these distances was 2.4 mm. The standard deviation of these distances by translating one image point for 1 pixel was 1.3 mm. For these experiments, a distance $\Delta z = 80\,mm$ was chosen. The distance between the centre of projection and the objects was nearly 145 mm.

	Mean	STD		Mean	STD
P_{i_x} $(pixel)$	274.3	2.1	P_{i_y} $(pixel)$	242.4	1.6
a_0	$4.5\ E^{-6}$	$0.06\ E^{-6}$	a_1	$1.3709\ E^{-6}$	$0.004\ E^{-6}$
b_0	$5.0\ E^{-6}$	$0.05\ E^{-6}$	b_1	$0.6605\ E^{-6}$	$0.087\ E^{-6}$
s_x	0.0158	0.001	s_y	0.0109	0.0
d (mm)	6.749	0.004			
$^S\varphi_z$ $(degree)$	88.589	0.0005	$^S\vartheta_z$ $(degree)$	-5.995	0.00003
$^S\varphi_x$ $(degree)$	-44.497	0.0003	$^S\vartheta_x$ $(degree)$	85.769	0.00003
$^S t_{GC_x}$ (mm)	-0.724	0.7	$^S t_{GC_y}$ (mm)	33.482	0.4
$^S t_{GC_z}$ (mm)	86.368	1.03			

Table 1: Results of the Calibration

6 Conclusion

A method for a simplified, automatic camera calibration is presented. First, the intrinsic parameters are estimated separately in succession with the help of an aquidistant grid of luminous points. Then, a 3–D pose estimation is described. Two images taken at two positions which differ only in their z–coordinates are used to reconstruct the 3–D data. The advantage of the z–distance is the simple determination of the correspondences between the points of two images. The extrinsic parameters at each location are estimated with the help of the camera coordinate frame relative to the gripper frame. In order to estimate this camera frame, a certain location of the camera is used. The advantage of this calibration is the possibility of estimating each parameter separately. No complicated equation system has to be solved by optimization. The exactness of the camera parameters are sufficiently determined for the purpose they are needed, but if the mathematical origin of ICF and the translational part of the lens distortion were also determinated, the results would be more exact.

Acknowledgements

The author would like to thank Prof. Dr. H.-J. Siegert and all colleagues for fruitful discussions.

References

[1] Beyer H.A., *Untersuchungen zur geometrischen Qualität der Datenübertragung bei der Aufnahme mit CCD-Kameras*, 13. DAGM–Symposium München, Springer–Verlag, Berlin, pp. 328–336, Oct 1991

[2] Crowley J.L., Stelmaszyk P., *Measurement and integration of 3-D structures by tracking edge lines*, O. Faugeras (Ed.) First Eur. Conf. on Computer Vision, Antibes, France, Springer–Verlag, Berlin, pp. 269 – 280, Apr 1990

[3] Faugeras O.D., Toscani G.,*Camera Calibration for 3D Computer Vision*, Proc. of Int. Workshop on Machine Vision and Machine Intelligence, Tokyo, Japan, Feb 1987

[4] Föhr R., Ameling W., *Photogrammetric Registration of Spatial Informations Using Standard CCD Video Cameras*, Proceedings SIFIR'89, Int. Workshop on Sensorial Integration for Industrial Robots: Architecture & Applications, Zaragoza, Spane, pp. 194 – 199, Nov 1989

[5] Haralick R.M., Lee C.N., Zhuang X., Vaidya V.G., Kim M.B., *Pose Estimation from Corresponding Point Data*, Workshop on Computer Vision, Miami Beach, Florida, Nov/Dec 1987

[6] Howe H., *Introduction to Physics*, First Edition, McGraw–Hill Book Company, New York, 1942

[7] Kraus K., *Photogrammetrie Band 1: Grundlagen und Standardverfahren*, Dümmlers Verlag, Bonn, 1982

[8] Lee S., Kay Y., *An Accurate Estimation of 3 D Position and Orientation of a Moving Object for Robot Stereo Vision: Kalman Filter Approach*, Int. Conference on Robotics and Automation, Cincinnati, Computer Society Press, pp. 414 – 419, May 1990

[9] Lenz Reimar, *Videometrie mit CCD-Sensoren und ihre Anwendungen in der Robotik*, Habilitationsschrift der Technischen Universität München, Nov 1988

[10] Puget, Skordas *An optimal solution for mobile camera calibration*, O. Faugeras (Ed.) First Eur. Conf. on Computer Vision, Antibes, France, Springer–Verlag, pp. 187 – 198, Apr 1990

[11] Schrott A., *Feature–Based Camera–Guided Grasping by an Eye–in–Hand Robot*, Int. Conference on Robotics and Automation, Nizza, Computer Society Press, May 1992

[12] Tsai R.Y., Lenz R.K., *Techniques for Calibration of the Scale Factor and Image Center for High Accuracy 3-D Machine Vision Metrology*, PAMI-10, 5, pp. 713 – 720, Sep 1988

STRICT: SELECTING THE "RIGHT" ARCHITECTURE

A Blackboard-Based DAI Advisory System Supporting The Design of Distributed Production Planning & Control Applications

Stefan Kirn
Applied Computer Science I
Prof. Dr. G. Schlageter

Jörg Schneider
Dept. of Production Management
Prof. Dr. G. Fandel

FernUniversität Hagen, P.O.B. 940, D-5800 Hagen

ABSTRACT. Modern applications impose an increasing demand for flexible and distributed solutions. This, and the challenge of Open Systems, assigns more and more importance to the architectural decisions during systems design. However, the capabilities of conventional types of distributed systems seem to be restricted to the lower levels of interprocess cooperation. So, the integration of collaborative software systems calls for intelligent distributed architectures. This paper presents the experimental advisory system **STRICT** (Selecting The RIght architeCTure). STRICT serves as an investigation and development environment and provides smart facilities to experiment with (DAI-related) architectural knowledge.

Keywords: CBR; DAI-architecture; production planning and control systems; STRICT

1 Introduction

Problem. Modern applications (CIM, CSCW, Office Information Systems, design, etc.) impose a rapidly increasing demand for very flexible, mostly distributed solutions. This, together with the arising challenge of Open Systems, adds more and more significance to the decisions on architecturs that are made in the design phase of distributed systems. Since the capabilities of conventional approaches seem to be restricted to the lower levels of inter-process cooperation, the integration of collaborative software systems calls for intelligent distributed architectures as they are developed in distributed artificial intelligence (DAI). Unfortunatly, most of the concepts and solutions provided by DAI remain to be well-investigated and -evaluated. That makes it difficult to choose the "right" (DAI-) architecture, and also to assess the most important advantages / disadvantages related to it.

STRICT. The long-term purpose of the STRICT activities is threefold: (1) to develop a tool supporting the selection of the "right" (DAI-) architecture, i.e., that architecture that is most suited to the requirements arising from a specific application, (2) to improve the knowledge on DAI-architectures, to enhance their applicability to distributed systems design, and (3) to develop formal models of generic DAI-architectures and of generic applications with the aim to employ them both in the selection of architectures.

In that sense STRICT is definitly more than just a simple advisory system. It serves as an investigation and development environment by providing smart facilities to experiment with architectural knowledge. Being aware of the large number of open questions, the STRICT development started with the experimental design of a distributed Production Planning and Control Systems (PPC). As PPCs are a well-known domain, they allow us to apply an experimental investigation methodology.

Remainder. Chapter 2 discusses DAI-architectures, with their main characteristics summarized in chapter 2.5. Chapter 3 introduces Production Planning & Control Systems. Chapter 4 presents STRICT and discusses an example, demonstrating the basic ideas of the STRICT problem solving process. Chapter 5 outlines some important research issues.

2 DAI-Architectures: A Problem-Oriented Overview

Among the basic elements that constitute DAI systems we distinguish between the overall architecture, the agents itself, the cooperation concepts and the coordination mechanisms. However, we do not investigate the relationships between DAI-systems and distributed processing platforms. Neither are we concerned with technical issues or even with low-level protocol concerns (e.g., TCP/IP, NFS, etc.).

2.1 DAI ARCHITECTURES

Using the term DAI-architecture we focus on the overall structure and the global characteristics of a DAI-system, addressing its behavior as a whole. To avoid some common confusion, we clearly distinguish between the architecture of the system and the coordination mechanisms. The latter are used to coordinate the interactions within a group of agents. In the following we present four of the most prominent generic DAI-architectures.
Supervisor Systems (SVS). SVSs are to be characterized by a centralized control component. This module schedules all global activities (task allocation, cooperative problem solving, constructing the results). Because none of the agents possesses any global knowledge, the supervisor acts as a master. **Blackboard Systems (BBS).** The main characteristic of a BBS is the blackboard itself (being a centralized data structure, visible from all agents). All iterations of a cooperative problem solving process have to be documented on that board. Any information on the blackboard may be considered as a task. Each agent capable of processing that task generates a bid to be written on the blackboard as well. Processing the set of information on the blackboard (bids, control information, intermediate or final results) is accomplished by an intelligent scheduler. It allocates tasks, recognizes conflicts and decides on how to continue the cooperation process. **Contract Nets (CN).** In contrast to the two concepts above there is no centralization at all in contract nets. The agents in a contract net are completely autonomous. They do not share any resources and do not possess any global knowledge. Using a contract protocol, they cooperate in a flexible, task dependent way. Typically, coordination is based on bilateral negotiations. **Not Explicitly Coordinated Systems (NECS).** NECS typically are build up by very complex autonomous agents. If necessary, they exchange data via a centralized database. NECS do not possess any explicit mechanisms (task decomposition and allocation, synthesis of results, etc.) to support cooperative problem solving further. So, compared to the architectures described above, their cooperative capabilities are limited.
System-features. The most important characteristics of DAI systems (which may be considered as generic attributes) are: *system model* (refers to the number of agents used to solve a problem), *system scale* (number of computing elements to implement the agents), *reliability* (as a measure of the global systems behavior), *degree of coupling the agents* (loose vs. tight), *control* (centralized - decentralized), *selectivity* (as a kind of probability to select the **right** agent in order to solve a certain task), *agents heterogeneity* (knowledge, software, hardware), *mechanisms supporting cooperative problem solving* (task decomposition, planning cooperative processes, partner selection, results synthesis) and *flexibility* (refers to the ease of extensions and modifications of the overall system).

2.2 AGENT MODEL

Intelligent agents of a DAI-system can primarily be described by the following characteristics: *degree of autonomy, cooperation degree* (refers to the number cooperative problem solving processes in relation to pure stand alone problem solving), *agents granularity* (fine - coarse), *agents dynamics* (fixed - programmable - teachable - autodidactic), *resources* (restricted - ample), *global knowledge* and *interactions* (simple - complex).

2.3 COOPERATIVE PROBLEM SOLVING
Distributed Problem Solvers vs. Multiagent Systems. In general, DAI distinguishes between *distributed problem solving (DPS)* and *multiagent systems (MAS)*. In DPS the agents are faced with problems that are beyond their individual capabilities, so that they must cooperate to achieve their goals. In MAS the agents share an environment in which they act collectively to identify and resolve conflicts mutually utilizing each others workings. Consequently, MAS research focuses on coordination [Smit 79], [RoGe 85], [Ferb 89], while DPS research primarily investigates problems like task decomposition / allocation and solution synthesis. **Cooperation Styles.** As different problem types may ask for different types of solution strategies there is some evidence that this holds for cooperative problem solving as well. However, there is no systematic discussion of cooperation styles in DAI. So we only list some of them in order to outline the concept and to give some idea of the complexity involved. First, we distinguish between: distributed, multiagent and federative problem solving, teamwork, problem solving by committee etc. Second, we mention concepts like task sharing, result sharing, knowledge sharing. Other kinds of distinctions may be guided by the interagent relationships (horizontal, vertical, master-slave, etc.).
Summary. Prior characteristics of collaborative problem solving are: *global coherence*; *a priori planning the collaboration* (static - dynamic), (global) *characteristics of problem solving strategies* (hierarchical, decomposable, parallelism) and the question of *how to find the optimal (right) solution*.

2.4 COORDINATION
The Coordination Problem. Coordination is one of the most important DAI topics. Out of the numerous proposals two of them play a major role: Negotiation and Planning. Recently, there are attempts to integrate these two. For this purpose sophisticated collaboration capabilities are to be added into complex intelligent systems [Werk 91], [KiSS 91]. **Negotiation.** Negotiations [Smit 79], [DaSm 83] provide MAS with a powerful technique to coordinate agents towards a common global goal. [CoML 86] introduced multistage negotiations to overcome the problem of incomplete knowledge. They allow agents to bargain on problems with a restricted global significance. They guarantee that the overall system converges towards a global consistent solution. [KuLe 89] applied the multistage negotiation protocol to the resource allocation conflicts in a distributed network. [LâLL 91] introduced cybernetic facilities. Their agents generate goals to be communicated and allocated to achieve a globally consistent solution. Goals not being allocated are reorganized to start a new cycle of goal formulation, communication, allocation and achievement. This idea seems to be a very promising one. However, [LâLL 91] did not discuss how agents can decide on which one of them may process a goal and how they can assess the achievement of a goal after having processed it. Also, they cut all questions related to dynamic goal reorganisation. **Planning.** Closely related to the concept of negotiation a lot of DAI-research deals with planning (for an overview see [BoGa 88]). The basic idea is that agents coordinate their behavior by exchanging their local plans before they move into action. Usually, authors distinguish between multiagent planning (MP) and distributed planning (DP). While MP requires one agent to have all relevant knowledge to generate the global plan, DP requires all agents to act together on one single global plan. To overcome some of the shortcomings of both approaches, [DuLe 87] propose the concept of partial global plans (PGPs) to effectively specify coordinated actions for groups of problem solvers. In this approach coordination is a part of each problem solvers activities and is directly interleaved with local problem solving.

Further Coordination Concepts. There are still a lot of other approaches to the coordination problem. Some of them are directly related to the underlying DAI architecture (blackboard based coordination, shared database). Others can be expected to become useful for different types of architectures (e.g., the focal points approach [RoKr 91] and the concept of arbitrators [Werk 91]).

2.5 SUMMARY: DAI ARCHITECTURES PROFILES

The above considerations have introduced a set of criteria helpful for the description of DAI-systems. Based on them figure 2.5.-1 gives an idea on how to classify the four generic DAI architectures dicussed above. However, it is important to note this is only a first attempt. A lot of research is still needed to get reliable and valid results.

	minor	low	average	higher	high
reliability	SVS	BBS			CN, NECS
decentralization of control	SVS, BBS				CN, NECS
flexibility	SVS			BBS	CN, NECS
selectivity	NECS		CN, BBS		SVS
degree of coupling	NECS, CN	BBS			SVS
system model	NECS		SVS, BBS		CN
agents autonomy	SVS, BBS				CN, NECS
agents granularity	BBS		CN, SVS		NECS
agents global knowledge	SVS, BBS	CN		NECS	
global coherence	NECS		CN	SVS	BBS
a priori planning of collaborations	NECS, CN		BBS		SVS
finding the optimal solution	NECS, CN		BBS		SVS
centrality of coordination	NECS, CN			BBS	SVS
	minor	low	average	higher	high

Figure 2.5.: DAI Architectures Profile

When dealing with the selection of an "appropriate DAI-architecture" the question that immediately arises is: How are these criteria to be "measured"? Obviously, most of them include qualitative knowledge, only few of them (e.g., reliability, system model) can be defined on the basis of a metric. Further more, for different types of applications (e.g., PPC, CSCW, OIS, design, etc.) there may be differences in the interpretation of some of the attributes (e.g., agents granularity). Consequently, using these pieces of architectural knowledge assumes some kind of a (generic) application model. The purpose to deal with such an application model is twofold: First, it supports the system-user dialog by defining the application´s characteristics. Second, it provides context knowledge on the users application definition that is used during the inference process.

3 PRODUCTION PLANNING AND CONTROL SYSTEMS

This section gives a short description of Production Planning and Control Systems (PPC). As all of these systems have the same generic functionality, we introduce the idea of a generic PPC. A generic PPC can be defined as the class *PPC*, from which any specific PPC can be derived via instantiation. It is important to note here, that this instantiation process directly corresponds with the (automatic) configuration of complex software systems, as it has already been described by [Batz 87]. Future PPCs will increasingly be open and distributed ones. So, the long term interest of our work is the design of the overall architecture of distributed PPCs (dPPCs). This approach assumes detailed knowledge about the relationship between the characteristics of a dPPC and those of distributed architectures. As this issue is not yet very well understood, it is the purpose of the following investigations to clarify the basic problems and research issues.

3.1 APPLICATION MODEL

Production can be defined as transforming input materials into outputs. For economical production it is important that this process is planned and controlled with respect to specific goals by an information processing system (PPC). The most relevant issues occurring in this context are the planning of the product mix, requirements analysis, and scheduling. Simultaneous planning of these problems would be necessary to optimize the results. Because this kind of problem solving is too complex in most cases, PPCs work like distributed (hierarchical) planners. That means they usually decompose the overall problem into separate, strictly ordered subtasks. For each of them the agents generate a partial (local) plan which then will be negotiated in order to construct a globally consistent solution . Fig. 3.1. gives a survey of the sequence of planning phases in a PPC.

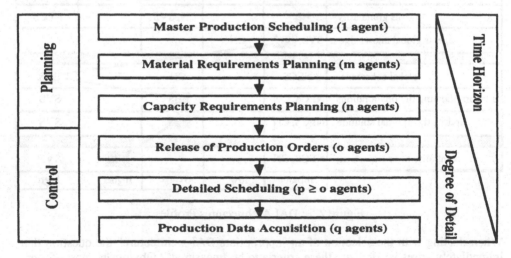

Fig. 3.1.: Model of PPCs [FaFr 88]

Not all details of a plan on a lower level can be anticipated at the higher levels. Thus the previous planning stages must modify their plans when their assumptions cause inconsistencies at subsequent levels. That requires the lower level planners to be able to communicate plan inconsistencies in the plans to the higher level planners. Therefore, top-down problem solving has to be coupled with bottom-up reaction. In consequence, in PPCs planning means an iterative process of initial planning, finding conflicts between the partial plans, restoring feasibility and consistency rechecking. So, tasks and planners can not be accomplished any longer in a fixed order. Instead of this, a coordination mechanism is needed that selects the planners according to the actual state of problem solving.

3.2 EXAMPLE
It follows directly that a PPC is characterized by the concrete algorithms implemented in its modules and the used coordination mechanism. That means instances of the object-class 'PPC' can be built by assigning concrete algorithms and the underlying coordination and communication architecture to its slots. Let us now consider an example. It is not necessary to look at one of the huge PPC-systems available on the market. Instead of this the relevant ideas can be better understood if the example concentrates on a smaller system which already contains the most important aspects.

The system OPIS (opportunistic intelligent scheduler) has been constructed for predictive and reactive scheduling. It follows a dynamical strategy which first generates partial plans and then tries to integrate them. If inconsistencies occur the partial plans are changed. Furthermore the partial plans can be computed by different methods. Their use is coordinated by a control framework based on the principles of blackboard architectures. It seems to be clear that OPIS reflects the problems of a whole PPC which contains more planning tasks than only scheduling. OPIS is implemented as a BBS with a control cycle managing the planning activities. Three alternative methods are available: a machine-based strategy, a job-based strategy and an event-based strategy. The manufacturing environment is a realistic job-shop-system.

4 STRICT: A Blackboard-based DAI Advisory system
4.1 DESIGN
Along with the main requirements (easy to extend, medium selectivity, centralized control, minor agents autonomy, high coherence, centrality of coordination) STRICT has been designed as a blackboard system (figure 4.1.-1). Its main modules are the human computer interface, the blackboard, three different knowledge sources (KSs), the scheduler and the control knowledge base. KSAR stands for "Knowledge Source Activation Record". The DAI-Architecture KS_1 contains the architectural knowledge of STRICT. The PPC-Framework KS_2 provides the generic PPC-model (any other application model can easily be added). Finally, the PPC Case Base KS_3 contains a set of specific PPCs which, as well-described cases, enhances the STRICT´s problem solving capability further.

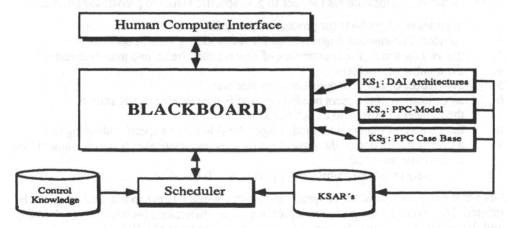

Figure 4.1.-1: The STRICT Architecture

The basic problem solving algorithm works as follows: Once called by a user it asks for a detailed description of the application. Using the knowledge from its knowledge sources STRICT focuses the dialogue on architectural concerns. Following the meta-rule *"prefer specialized against general knowledge"* the PPC Case Base KS3 has a higher priority than the PPC Model KS. The STRICT problem solving strategy is a combination of *hypothesize-and-test* with *establish-refine* and *case-based reasoning*. Applying *hypothesize-and-test*, it uses the relationships between the generic DAI-architecture, the agent model, the problem solving behavior and the coordination mechanism to build up a set of hypothesis mapping the main application characteristics on architectural attributes (application profile). Then, this profile is evaluated (1) against the generic architecture profiles and (2) against the cases in KS3 to *establish* a suggestion on the appropriate generic (and specific) architecture. Please note, that this comparison is done by KS3, applying its similarity knowledge to the profiles on the blackboard and the cases in its own knowledge base. If possible, the suggestion will be *refined* further by the case knowledge included in KS3.

The problem solving process is composed of four interacting layers: (1) generic architecture, (2) agent model, (3) problem solving and (4) coordination. On each layer a configuration decision will be made, which must be consistent with the decisions of all previous layers. If that proves impossible, backtracking over the layers takes place (figure 4.1.-2).

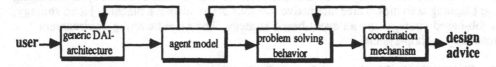

Figure 4.1.-2: The Four-Layer Model of the STRICT Problem Solving Algorithm

At the end of every inference process the user is asked wether he would like to store "his case" just evaluated in the PPC Case Base. This introduces learning into STRICT and improves the STRICT problem solving capabilities over time.

4.2 REASONING PROCESS

This section focuses on the first layer, it shows how an appropriate DAI-architecture can be chosen from the characteristics of a PPC and a manufacturing system. Starting STRICT the user will be asked to define the characteristics of his manufacturing environment and his PPC. This requires the user to give specific values to predefined attributes:

1. the number of products (or production groups)
2. number of production stages needed for completion (product-complexity)
3. the average time for the occurrence of new products (reflecting market dynamics)
4. the number of machines
5. the number of different operations per machine
6. an estimation of the users needing to modify the machine configuration
7. the number of criteria to be optimized
8. the number of planning methods (algorithms) to solve a specific planning task
9. a quality estimation of the algorithms (derived by practical experience, simulations or scientific analysis)
10. the variety of order in which the algorithms will be called

Note that some attributes can be quantified exactly while others can not, and most can be refined. However, the list gives an impression of the difficulties involved when choosing and determining relevant information for the subsequent steps of STRICT.

Given the necessary data the reasoning process will start. The characteristic values must be mapped into aggregates which can be used for the choice of the architecture. This mapping process requires defined aggregates and their measurements. Since the aggregates have to be mapped into DAI-architecture characteristics it is not necessary to define them on the basis of a metric. It seems better to define a set of five values which directly correspond with the values of the architectural attributes. The following structure results:

a. Production complexity (as a consequence of 1,2,4,5)
b. Production flexibility (as a consequence of 3,6)
c. Methodological complexity (as a consequence of 7,8)
d. Methodological flexibility (as a consequence of 8,10)
e. Methodological quality (as a consequence of 8,9)

with possible values from the set $Y = \{$small, moderate, medium, large, immense$\}$.
If you are an expert in PPC STRICT offers the option to set values at this level immediately. In that case you can pass over questions 1 - 10. Otherwise, STRICT uses its domain knowledge to deduce a - e. Making this aggregation will cause a lot of difficulties. For example consider the derivation of e from 8 and 9. E.g., additional algorithms improve the probability of a good one occurring. On the other hand, the question of 'what is a good algorithm' remains to be unanswered. The quality estimation is defined as a measure in how many cases a result near the optimum will be computed. However, now it must be determined what is 'near'. Obviously, in most cases these questions can not be answered exactly. Thus the aggregation problem still requires much more work in detail.
The process of choosing the right architecture can now be divided into two steps:

1. Transforming the characteristics of a PPC into those of DAI-architectures
2. Comparing the profile resulting from step 1 with DAI-architecture- profiles

For the purposes of the first step let $n(x)$ and $n(y)$ be natural numbers denoting the position of the elements x and y in their sets ($X = \{$minor, low, average, higher, high$\}$). Then PPC-characteristics can be changed easily by the following rules:

RULE 1:	IF production complexity	$= y$	THEN system model	$= x$
RULE 2:	IF production flexibility	$= y$	THEN flexibility	$= x$
RULE 3:	IF methodological complexity	$= y$	THEN decentralization	$= x$
RULE 4:	IF methodological quality	$= y$	THEN selectivity	$= x$
RULE 5:	IF methodological flexibility	$= y$	THEN degree of coupling	$= x$

with $n(y) = n(x)$ in each case.

The application of these rules to system OPIS gives the following characterization: system model = low, flexibility = higher, decentralization = minor, selectivity = average, degree of coupling = low. Note that there are still some problems. First, the values of the attributes are not exact. So you can argue about the assignments. Second, there is no valid measure for the methodological quality. You must make an assumption or give the value 'average' if you are uncertain. However, detailed expert knowledge is necessary to transform the system´s features into characteristics of DAI-architectures. Rules 1 - 5 show how such a mapping can be performed in principle.
In the second step a comparison of the profiles of the PPC-system and the DAI-architectures must be carried out. In our example we do not need an algorithm since the profile of OPIS meets obviously the profile of a blackboard-architecture. Thus, BBS seems to be the right architecture for OPIS. In general the application-profile does not meet the DAI-profiles in such an unambiguous way. In this case an algorithm must compute the distances of the profiles and maximize the similarity. As there are still a lot of open questions

real-world application: What are the promises, advantages, disadvantages of the DAI-architectures, compared with each other and with conventional distributed solutions?

6 Summary

This paper presented STRICT, an advisory system supporting dPPC-designers to select the appropriate (DAI-) architecture for their applications. Beside the development of a software engineering tool the STRICT activities aim at a much broader question: The investigation of DAI-architectures to get a sound understanding of their applicability to a given software design problem. The PPC chosen as an example in this paper supports experimental investigations. The domain is well-understood (at least so far as we are concerned with it here), and there are already first examples of DAI-based solutions. However, the blackboard approach provides flexible facilities also to extend STRICT to other applications as well as to other types of (not only DAI-based) distributed architectures. Accomplishing this will enhance the software engineers capabilities by providing them with a powerful tool. In that sense, the prior benefit of the STRICT work described in this paper is to add the DAI-architectures to the software engineers workbench.

7 References

[Batz 87]   Batz, T.: Versionsverwaltung im Datenhaltungssystem PRODAT des Systement-wicklungssystems PROSYT, Softwaretechnik Trends, Oktober 1987.

[BoGa 88]   Bond, A.; Gasser, L.: Readings in Distributed Artificial Intelligence, Morgan Kaufmann Publishers, San Mateo, CA., 1988.

[CoML 86] Conry, S.E.; Meyer, R.A.; Lesser, V.R.: Multistage Negotiation in Distributed Planning, in [BoGa 88], pp. 367.

[DaSm 83] Davis, R.; Smith, R.G.: Negotiation as a Metaphor for Distributed Problem Solving, AI 20(1), 1983, pp. 63.

[DuLe 87]   Durfee, E.H.; Lesser, V.R.: Using Partial Global Plans to Coordinate Distributed Problem Solvers, IJCAI 1987, pp. 875.

[FaDR 88] Fandel, G.; Dyckhoff, H.; Reese, J. (eds.): Essays on Production Theory and Planning, Springer-Verlag Berlin, Heidelberg et.al., 1988.

[FaFr 88]   Fandel, G.; François, P.: Rational Material Flow Planning, MRP and Canban, in [FaDR 88], pp. 43.

[Ferb 89]   Ferber, J.: Eco Problem Solving: How to Solve a Problem by Interactions, 9th DAI-Workshop, Rosario Texas, 1989, pp. 113.

[Kara 90]   Karagiannis, D. (ed.): İnformation Systems and Artificial Intelligence: Integration Aspects, Springer-Verlag Berlin, Heidelberg et.al., 1990.

[KiSS 91]   Kirn, St.; Scherer, A.; Schlageter, G.: The FRESCO Agent Model: Cooperative Behavior in Federative Environments, ICIS `91, Sydney, Australia, August 25, 1991.

[KuLe 89]   Kuwabara, K.; Lesser, V.R.: Extended Protocol for Multi-Stage Negotiation, 9th Workshop on DAI, Bandera, Texas, October 23-27, 1990.

[LâLL 91]   Lâasri, B.; Lâasri, H., Lesser, V.R.: An analysis of negotiations and its role for coordinating cooperative distributed problem solvers, Proc. of the 11th conf. on 2nd generation expert systesm, Avignon, France, 1991, pp. 81.

[RoGe 85] Rosenschein, J.; Genesereth, M.: Deals Among Rational Agents, IJCAI 85, p. 91.

[RoKr 91]   Rosenschein, J.S.; Kraus, S.: The Role of Representation in Interaction: Discovering Focal Points Among Alternative Solutions, MAAMAW 91, Kaiserslautern, Germany, August 5-7, 1991.

[Smit 79]   Smith, R.G.: A Framework for Distributed Problem Solving, IJCAI 1979, pp. 836.

[Werk 91]   Werkman, K.: Using Negotiation and Coordination in Multiple Agent Systems, ICIS `91, Sydney, Australia, August 25, 1991.

related to the comparison of cases in a CBR system we use the method of cluster analysis here. That technique is already well understood. It provides us with interesting facilities. It works as follows: Objects which are similar are grouped together in one cluster while objects which are quite different are grouped in other clusters. Clusters can be built if enough examples are available. That means, generalization of the results assumes that enough cases have been considered before. Since there are not yet enough cases included in the case base KS_3 we apply a basic algorithm to select the architectures and to further build up the case base. The algorithm works as follows: Input are the profiles of the system (s) and the DAI-architectures (SVS, BBS, CN, NECS). The output is given by a value for each DAI-architecture that indicates its suitability in the given case. The main part of the algorithm is presented in the following program:

1. Define the weight of all criteria {condition: sum of all weightings (w_c) is 1}
2. FOR all DAI-architectures a DO
   BEGIN FOR all criteria c DO
              BEGIN compute $d(a.c, s.c) := |n(a.c) - n(s.c)|$;
                         summarize $d(a.c, s.c)*w_c$ {thus getting a value v_a for all a}
              END
          set b(a) := true;
          if any $d(a.c, s.c) > 2$ then b(a) := false;   {KO-criterion}
          if b(a) = true print v_a else print 'not appropriate';
   END;   {The best architecture now is the one with the lowest v_a.}

Note that some problems are simplified. The mapping task is not well-structured. So, the algorithm described above pretends to the reader a degree of exactness that does not exist in reality. The weights reflect individual preferences and can not be objective. The computation of the distance can be implemented in several ways each of them also being only 'right' in a subjective sense. Finally, let us mention the problems of measuring the criteria and bringing them into a comparable form. All of the criteria used in this approach are qualitative from nature. So, we have to deal with the question of how to classify any PPC. The actual STRICT solution (clustering analysis) still bases on a metric, but in future we intend to enhance the clustering analysis by methods of qualitative reasoning.

5   Research Issues

Developing STRICT, some important research issues have already been identified. First, the knowledge on DAI-architectures is still very scarce. There are only few systematic investigations on the applicability of any of these architectures to a concrete distributed application. As far as we know, there is no work on the development of a methodology on how to select an appropriate generic DAI architecture with respect to a given application. Also, we are not aware of any comparative work which applies different architectures to the same application in order to evaluate benefits / disadvantages of the different solutions. In other words: we do not have at our disposal any reliable engineering methodology to design and specify DAI-applications. That is a severe restriction on the work of the software engineers. Some questions of prior interest are: What are the definitional characteristics of DAI-architectures? What are the relationships between DAI-architectures and conventional distributed architectures? What are the characteristics of DAI-architectures which must / should be investigated when applying a DAI approach to a distributed application? And how are they measured and compared? What possibilities are given to substitute any DAI-architecture by any other or even a conventional approach? With respect to a

SOLVING TEMPORAL CONSTRAINTS SATISFACTION PROBLEMS WITH AN OBJECT-ORIENTED MODEL

Laurent CERVONI

• I.T.M.I.
92, Bd du Montparnasse
75682 Paris Cedex 14
France

• Rouen University
L.I.R.
BP 118
76134 Mont Saint Aignan Cedex
France

Francis ROUSSEAUX

• SYSECA
315, Bureaux de la Colline
92213 Saint-Cloud Cedex
France

• Paris VI University
LAFORIA
Place Jussieu, Tour 46-00
75005 Paris
France

ABSTRACT

Object Oriented environments are becoming well accepted as suitable media for implementing applications, because of their qualities of extendibility, reusability, and compatibility.

From an operational point of view, a Control, Command, Communication & Intelligence System (C3I system) can be considered as a set of objects under specific constraints, and particularly under temporal constraints. However, from a design point of view, the object organization has nothing to do with specific constraints: on the contrary, it is very important to be able to design objects and their links, separately from the specific constraint network: then, it allows the Design of the system to be Object Oriented (OOD), objects being reusable and easy to modify.

Our purpose has been to develop an operational tool able to reuse some OOD application and objects represented in PROLOG to design a C3I System, taking in account some Temporal Constraints Satisfaction Problems (TCSP). This paper presents our tool, Objects under Constraints (C/O), on a theoretical and practical point of view, and discusses a temporal model which extends an Object Oriented environment.

Keywords : Temporal Constraints Satisfaction Problems, Object-oriented Design, Prolog

1 INTRODUCTION

Object Oriented environments are becoming well accepted as suitable media for implementing applications, because of their qualities of extendibility, reusability, and compatibility ([Balthazaar & al. 90], [Jones & al. 90], [Meyer 88]). This paper presents our tool C/O on a theoretical and practical point of view, and discusses a temporal model which extends an Object Oriented environment. This extension is justified because in all real world activities, time is an important dimension, as events and actions occur continuously over time.

Our purpose was to develop a tool, named C/O for "Objects under Constraints", dealing with that particular type of problem ([Rousseaux 91]). For practical reasons (most of our operational systems we are working on were modeled by means of PROLOG environments), we wrote it in PROLOG ([Cervoni 90]).

Following this introduction, the second section recall the different formalisms used in the domain.

The third section describes a logical programming environment fitted to that particular problem. We present a way to improve constraints treatments with a particular object representation, building constraints classes.

The fourth section returns to the question "how to reuse an OOD in TCSP environment ?". A simple tool (C/O) to build objects or to reuse them is defined as an overhead of our Prolog environment.

In the conclusion, there is a discussion of the results obtained and those hoped from such an environment, before concluding on our further research.

Remark : to improve clearness of the document, all the examples are typed in a font different from the rest of the text.

2 C3IS AS AN EXAMPLE OF HOW TO DEAL WITH OOD & TEMPORAL CONSTRAINTS SATISFACTION PROBLEMS (TCSP)

In high level software industry or military applications, Object Oriented Technology is a masterpiece, and most Object-Oriented systems deal with reusable objects ; however, to be reusable, objects have to be sufficiently generic, and if they are expected to treat temporal events, to be safe to time variations, being always easy to recognize.

Let us first have a view on a typical example of C3IS, considered itself as a typical system dealing with "objects", defined here as representations of real world objects.

The conception of a Control, Command, Communication & Intelligence System includes handling important quantities of heterogeneous information; for example, in military action, we have to manage immobile entities (hospitals, officials buildings, strategical buildings, telecommunication stations, ...) or mobile (armies, tanks, planes, soldiers). Some of them are military, others are civilian: their structure and the data they contain are complex (references to speed, geographical data, space images, available departure dates, available arrival dates, available ways, ...).

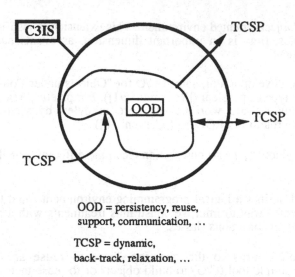

OOD = persistency, reuse,
support, communication, ...

TCSP = dynamic,
back-track, relaxation, ...

Figure 1: C3IS are often dealing with OOD and TCSP.

Such systems are current, and **intrinsically** object oriented designed. But people need also to test dynamically that kind of system. The Object Oriented Design is necessary to describe the actors and the initial situation. But, the **main** aim of a C3IS is to provide a plan simulation using the objects it contains.

The ability to plan actions is an important feature of these tools. Yet, if for example there is an attack on a civilian objective, such a system has to specify who can reach the action place, who has to protect the civilian population, who has to protect such or such important place, given that at the same time one has to settle a surgery, etc... It is absolutely necessary to be able to express that such and such an action is overlapping an other one, starts before or after. This is corresponding to a set of simple constraints, the purpose being to adapt values domains to planned events.

Nevertheless, it is important to take into account some far more complex situations as conditional relations, which are only considered as interesting if a given event should occur in a given time interval. Of course, this has to disappear if the action did not occur during that particular time interval. For instance, if a ship should arrive during time interval I, then the army unit G has to stay in the port during a time interval J containing I ; if the ship arrives during an interval different from I and superior to I, the constraint of G is relaxed.

Then, the main question becomes the following: "How to deal with OOD and TCSP together, from a theoretical, practical, technical point of view ?".

Constraints Satisfaction Problems (CSP) such as resource affectation, planning, optimal sharing, scheduling are being studied more and more and correspond to various needs for industrial tools. Different approaches, languages and developments, different concepts have been explored recently in that field ([Nadel 89], [Puget & al. 91]). The next section presents a way to formalize a particular sub-set of CSP : Temporal Constraints Satisfaction Problems (TCSP). Usual time representation models are described.

A CSP is described by several variables $V_1, \ldots, V_n$ each of them (for instance V_i) getting values in a set D_i, and some relations between these variables called constraints. If domains represent, for instance, real value intervals, the problem becomes optimum research under constraints, and can be solved by some simplex algorithm (if the constraints are linear). When all the domains are discrete sets, solving such a problem means finding one of the Cartesian product sub-sets of the $D_1 \times \ldots \times D_n$ domain. Such a discrete problem is known to be NP-hard (or NP-complete).

3 CSP & TIME REPRESENTATION

With TCSP, several representations are usually proposed (intervals, points, events). The Allen representation is based on an intervals logic, defined as continuous (convex) part of time. Thirteen relations are linked to this model: figure 1 sums up all the different allowed configurations (six relations and their inverse on the one hand, equality on the other hand). One can see that every combination can be defined with the primitive **Meets** only, applying on two intervals: intuitively, **A Meets B** means that interval B is after interval **A**, without any time space between them, the end of the one corresponding to the beginning of the other.

Let us consider now the following expressions, where $\Im$ is the set of time intervals.

Relat.	Inverse	Definition : $\forall\, I, J \in \mathfrak{I}$	Scheme
Meet		This relation is the basic one.	 I J
Before	After	$\exists\, k \in \mathfrak{I} /\, I$ Meets k Meets J	 I J
Equal	Equal	$\exists\, k, l \in \mathfrak{I} /$ k Meets I Meets l k Meets J Meets l	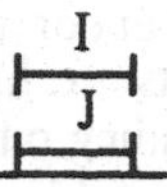 I J
Overlap	Is-overlap.-by	$\exists\, k, l, m, n, o \in \mathfrak{I} /$ k Meets I Meets n Meets o and k Meets l Meets J Meets o and l Meets m Meets n	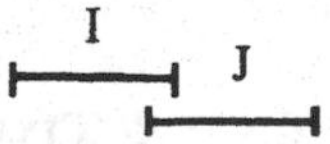 I J
Start	Is-started-by	$\exists\, k, l, m \in \mathfrak{I} /$ k Meets I Meets l Meets m and k Meets J Meets m	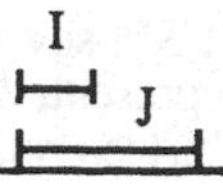 I J
End	Is-achieved-by	$\exists\, k, l, m \in \mathfrak{I} /$ k Meets l Meets I Meets m and k Meets J Meets m	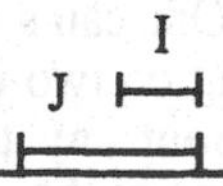 I J
During	Contains	$\exists\, k, l, m, n \in \mathfrak{I} /$ k Meets l Meets I Meets m Meets n and k Meets J Meets n	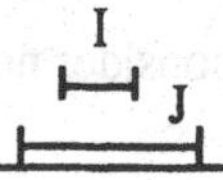 I J

Figure 2 : The Allen relations ([Allen & al. 85], [Allen 83]).

Some other time models are possible ([Vilain & al. 86]): one can find representations by events or some discrete models based on a time definition related with points. In fact, all these models are equivalent ([Tsang 87]). Anyway, the temporal logic based on the intervals (even if it is not the most "up to date") is sufficient and well adapted to the typical problems we want to solve.

If simple constraints can be written from the thirteen relations given in *figure 1*, conditional constraints need a more powerful form and a more **declarative** mode of expression. Prolog provide the required expressiveness. However, an extension is needed to allow Prolog to deal with intervals ([BNR PROLOG 88]). This is done by improving the engine to manage interval "entities" : the usual unification mechanism is modified to deal with intervals; there, to write `I == J` does not mean that I is unified to J, but that I and J are reduced to be unified : it is in fact a kind of "Longer Common Part".

Then, we get predicates (`before, overlap, during, ...`) to test Allen relations between the intervals, without changing these intervals, and predicates (`startc, endc, ...`) to constraint intervals to admit some constraint relation (the constraint aspect of Allen's relation).

This is an example using the two kinds of predicates (range is used to create an interval "object") :

```
?- range(I,[1,3]), range(J,[2,4]), overlap(I,J).
```

verify that I overlaps J (so, generate a success).

```
?- range(I,[1,3]), range(J,[2,4]), startc(I,J).
```

constraints I to start like J does.

```
--> I = [2.0, 3.0]    J = [2.0, 4.0]
```

In order to obtain a complete system dealing with intervals, it is necessary to defined a correct arithmetic on them. Classical mathematic operators are efficient on intervals: the defined arithmetic is monotone, idempotent, commutative and persistent (see [BNR-Prolog 88]) So, every constraint containing operations on intervals is allowed (eg :I = K + J).

4. HOW TO REUSE AN OOD IN A TCSP ENVIRONMENT ?

4.1 External Aspects

A simple tool to build objects or to reuse them is defined as an overhead of our PROLOG environment. For the user (the programmer), the syntax of a Prolog program is preserved. It is possible to create classes, instances and some methods, through some different operators (see also [Gallaire 86]). The main operators are kindOf, isA and with. To keep the compatibility with the Prolog classical behavior, they allow creation or access to objects (if insufficiently instantiated).

A kindOf B
build A as a sub-class of B (if A is a free variable, A can match all the sub-classes of B. If A and B are free, all the classes and their sub-classes are explored).

O isA C
build O as an instance of C: the mechanism, if the variables are free, is the same that for classes definitions; it allows to recognize all the classes and all their instances (other ways of doing that are also provided).

with
is an operator allowing the declaration of the attributes available for each class and each object.

The opportunity to declare objects is not devoted to temporal constraints. However, a specific structure is provided to handle constraints.

The basic instrument of constraint propagation is a specific attribute of the class active_object, connected to the special constraints class. As a matter of fact, this mechanism allows, during the creation of a new object or during the modification of some of its fields, activation of the link constraint, in order to verify the validity or to apply the corresponding procedures.

It exists a class constraint. An object, in order to be considered as an "object under constraint", must contain an attribute which is an instance of the class constraint

> constraint **kindOf** object,
>   **with** list_object,
>   **with** relation,
>   **with** condition,
>   **with** priority.

Figure 3 : Constraint is a class

One can also notice the presence of a **condition** allowing us to define a high level of constraint satisfaction. This condition is a Prolog predicate, able to indicate the necessary conditions to the constraint application, or the modifications which should be carried out on some given objects before the activation of this condition.

4.2 Internal mechanism

An essential aspect of this object model (the first one deals with data structure) is the reduction of the intervals, respecting the given constraints. Usually, it is necessary to adapt an iterative algorithm to cover in deterministic way the elements to be constrained: in such a model, the objects are alone responsible for the constraints propagation. With every object creation, modification or suppression, the link created by the **constraint** attribute actives **relation** checking.

If one considers that the objects are nodes of a graph whose constraints are arcs, this technique provides us with a kind of immediate checking of **arc consistency**: furthermore, this provides a dynamic modification of intervals, to satisfy the consistency. This idea is very important as far as CSP is concerned, because it reduces the real environment by cancelling the improbable values ([Mackworth 77], [Montanari 74], [Dechter & al. 89]).

4.3 Objects, Attributes and Methods

To each attribute is associated either a domain, a value or a default choice. Three methods are provided, by default, to access the instances, their attributes or their values : **modify** (modify the value of an attribute), **add** (add a new attribute to an instance), **delete** (suppress an attribute). The action involved by these methods is executed if and only if the network of constraints stays consistent. The choice we made, in order to stay compatible with the classical Prolog philosophy, is, also, to generate an error when one of the above methods is about to create an inconsistent network.

This choice, combined with a exception handler, leads, in a natural way, to an explanation facility and preserve the behavior of a Prolog program (see [Hwang & al. 90] for an other approach).

5 EXAMPLES

The hierarchy is based on two different kinds of objects. One "world" is composed of "active_objects" that support a constraint attribute, the other one are the "passive_objects" without this attribute (the class constraint is in the second "world"). The following tree is a partial view of these worlds.

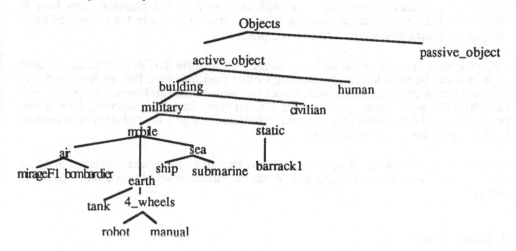

Figure 4: A hierarchy.

In such an environment, it is possible to declare a ship with a constraint on its departure hours :

```
clemenceau isA ship,
            with dep_harbour(paris),
            with arr_harbour(paderborn),
            with dep_hours([1,10]),
            with constraint(unary1),

unary1 isA constraint,
            with list_object([clemenceau]),
            with relation(reduce([1,10], AnOther)).
```

Figure 5 : The Clemenceau.

The predicate "reduce" is the constraint. It can, for example, force the Clemenceau to have its departure hours in a sub-interval of [1,10].

6 CONCLUSIONS AND PERSPECTIVES

In our implementation of C/O, objects support the constraints propagation process. They communicate to each other every time an attribute supports some constraints. This communication is rather "intelligent" and selective because only the sub-graph concerned is informed of what actions are to be taken into account ([Zaniolo 84], [Goldberg 90]). There is no exhaustive training, as it will have been with a classical facts base in PROLOG. Yet, complex cases (cyclic graphs) are dynamically detected and solved by an interval decomposition strategy (a solvec predicate).

C/O provide a quite powerful and easy-to-use way to solve TSCP. Of course, the same approach can be used to manipulate things other than intervals. But we have to notice that, as far as intervals are concerned, we get very well defined arithmetic, compatible with the PROLOG unification. In our model, objects bring structure, methods, information processing, and constraints propagation mechanism through a kind of daemon activation (or messages, links between objects).

By now, we are working on some explanation facilities (in the way described in §4.3), and will provide some optimizations features on particular sets of constraints (integers and strings).

7 BIBLIOGRAPHY

- [Allen 83]; James Allen, "Maintaining Knowledge about temporal interval", Communications of the ACM, November 1983

- [Allen & al. 85]; James Allen & Patrick Hayes, "A common-sense theory of time", IJCAI, Milan Italy, 1985

- [Balthazaar & al. 90]; Camel Balthazaar, Alex Gray & Ted Lawson, "An Object Oriented Temporal Model Embodied as Eiffel Software components", TOOLS 90, 1990

- [BNR-Prolog 88]; BNR-Prolog Reference Manual & user guide, Bell-Northern Research Publishing, 1988

- [Cervoni 90]; Laurent Cervoni, "Planification en construction : utilisation de la programmation logique et de la propagation de contraintes", Xèmes journées d'Avignon, 1990

- [Dechter & al. 89]; Rina Dechter & Itay Meiri, "Experimental Evaluation of preprocessing technics in constraint satisfaction problems", IJCAI 1989

- [Gallaire 86]; Hervé Gallaire, "Merging Objects and Logical programming: relational semantics", AAAI 1986

- [Goldberg 90]; Adèle Goldberg, "Reengineering applications toward Object Oriented Technology", Tutorials TOOLS 90, Paris, 1990

- [Hwang & al. 90]; Soochang Hwang & Sukho Lee, "Modelling Semantic Relationships and Constraints in Object-Oriented Databases", Communication of the ACM 10, 1990, p396-416

- [Jones & al. 90]; C.B. Jones & R.C. Show, "Case Studies in Systematic Software Development", Prentice Hall International, 1990.

- [Nadel 89]; Bernard Nadel, "Constraint satisfaction algorithms", Computer Intelligence Journal 5, 1989

- [Mackworth 77]; A. Mackworth, "Consistency in networks of relations", Artificial Intelligence Journal 8, 1977

- [Meyer 88]; Bertrand Meyer, "Object Oriented Software Construction", Prentice Hall International, 1988

- [Montanari 74]; Ugo Montanari, "Networks of constraints: Fundamental properties and applications to picture processing", Information Science 7, 1974

- [Puget & al. 91]; Jean-François Puget & Patrick Albert, "Pecos : programmation par contrainte orientée objet", ILOG, 1991

- [Rousseaux 91]; Francis Rousseaux, "Les technologies orientées objet chez SYSECA : l'exigence de méthodes adaptées aux grands systèmes temps réel", AFCET Spécial, 1991

- [Tsang 87]; Edward Tsang, "Time Structure for Artificial Intelligence", IJCAI 87, Milan Italy, August 1987

- [Vilain & al. 86]; Marc Vilain & Henry Kaultz, "Constraint Propagation Algorithms for Temporal Reasonning", AAAI, 1986

- [Zaniolo 84]; C. Zaniolo, "Object-oriented programming in Prolog", International Symposium on logical programming, February 1984.

An Integration Mechanism for Design Models in the Design Environment DATE

Hiroshi Arai[1], Takumi Hasegawa[2], Yoshiaki Fukazawa[1], Toshio Kadokura[1]

[1] Waseda University ,Tokyo, JAPAN
[2] NEC Corporation Ltd., Tokyo, JAPAN

Abstract. For designing digital hardware, such as ASICs and PCBs, various kinds of design tools are used. It is very important for hardware designers to integrate these tools into a totally unified design environment. In this paper, we propose an improved mechanism for integrating design models into the design environment DATE (Design Automation Tools integration Environment). By this mechanism, it will become easier to integrate heterogeneous design tools developed by multiple CAD vendors.

Keywords : Design Environment, Framework, Tool Integration, Design Model, Semantic Network

1 Introduction

The more complex digital hardware becomes, the more complicated and difficult to use design tools are getting. To reduce cost and time in using these tools, many kinds of design environments have been developed [1]–[6]. Automatic tool-execution and data-management are two major objectives of these environments. Many of these environments adopt expert system techniques with design knowledge bases. But in these knowledge bases, the knowledge to use each design tool is not clearly separated. The knowledge concerning one design target is distributed over the knowledge base. Therefore, it is difficult to add a new design tool into the environment or to modify the integrated design tools.

In DATE (Design Automation Tools integration Environment), we categorized design knowledge into two types. One is the knowledge on an ideal integrated *design model*, which is the superset of all design models in the environment. The other type is the knowledge specific to each design tool. This type of knowledge includes definitions of each tool's design model, and the correspondences between both types of knowledge. In the knowledge base, these sets of design knowledge are defined separately by the semantic network. To automate tool-execution and data-management, the environment must be able to recognize all information about design targets, design action and status. An inference engine translates this design information dynamically from the design model specific to the tools into the integrated design model, and vice versa. By this mechanism, the design models of each design tool is integrated. Since the design models of each tool is defined separately, it is easy to add new tools into the design environment and to modify existing design tools.

2 Overall architecture of DATE

The architecture of DATE is sketched in **Fig.1**. The environment consists of two sections. The upper half of **Fig.1** shows the mechanism to represent sets of design knowledge and to translate them each other. This part consists of the Knowledge-base and the Knowledge Manager with an Inference Engine. The lower half of **Fig.1** is a section for managing design tools and design data. This consists of five subsystems and the System Monitor to control them.

2.1 Design knowledge and Knowledge Manager

The Knowledge-base consists of following two types of knowledge.

1. *Tool-domain* knowledge
   Design knowledge specific to each design tool is defined. This knowledge is concerned with a design model used in each design tool. The design model consists of a *data-model* which represents the structure and semantics of the design data, and an *operation-model* which represents the design operations supported by the tool. Each tool-designer defines this knowledge. Additionally, the tool integrator defines the correspondence between this data-model and the following data-model.
2. *Common-domain* knowledge
   An integrated design model is defined. This design model is ideal and the superset of all of the tools' design models in the environment. The tool integrator defines this knowledge.

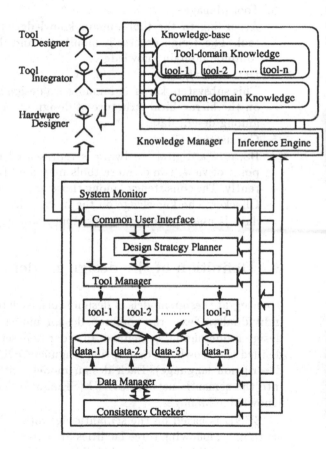

Fig.1. DATE architecture

The Knowledge Manager includes the Inference Engine. This engine translates design information from one design model into another design model according to the correspondence between them.

2.2   Managing design tools and design data

Five subsystems for managing design tools and design data are as follows.

1. Common User Interface
   This subsystem provides a unified human-machine interface for hardware designers. Hardware designers describe design requests in the common-domain knowledge. Inference Engine translates these requests into operation-models of each tool-domain knowledge.
2. Design Strategy Planner
   The requests described by hardware designers may not be simple enough to be implemented using a single design tool. In such cases, the Design Strategy Planner creates a design-plan, which is an execution sequence of multiple design tools.
3. Tool Manager
   According to the tool-domain knowledge, this subsystem invokes each design tool. The design process is monitored and the results are re-translated into the common-domain knowledge.
4. Data Manager
   This subsystem keeps and manages a design history on each data. The knowledge on the hierarchical structure of design objects is used to keep track of relations among design data.
5. Consistency Checker
   Because a digital hardware is represented by many kinds of data from many point of view, two or more tools may modify those data separately and concurrently. The consistency among those data must be guaranteed by the Consistency Checker. The knowledge on the hierarchical structure and relations among each data is used to maintain consistency.

3   Definition of the design model

To integrate design models of design tools, each tool's design model should be defined individually. Next, the common design model and correspondence to each tool's design model should be defined. In order to describe these design models formally, we defined the knowledge description language DOCK (DOmain Concerning Knowledge description language). Each design model written in DOCK is directly translated into the semantic network SEND (SEmantic Network for Design knowledge) by the Knowledge Manager.

SEND is based on the semantic network K-NET [7]. In order to improve representability, following three facilities are extended. First, to define the data-models in detail, *weighted arcs* and *arc bundling* are introduced. Secondly, to define consistency and correspondence between data-models, a rule represented by two *sub-networks* can be described. Lastly, to define design operations, sub-networks are used to represent the pre- and post-conditions of each design operation.

The common-domain knowledge and the set of tool-domain knowledge are represented by sub-networks(**Fig.2**).

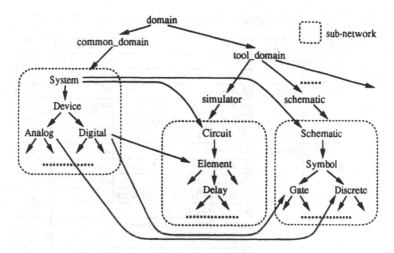

Fig.2. Representation of multiple domains

3.1 Definition of data-models

In SEND, each entity in the data-model is represented by a node. Their relationships
are represented by arcs. All types of nodes and arcs in SEND are listed in **Table 1**.
In this table, *meta-arcs* are used to point to sub network. *Meta-nodes* are incident
to meta-arcs.

DOCK is a formal language to describe SEND. Basically, each statement of
DOCK consists of source node-label, arc-label, arc-weight and destination node-
label. Some of them are omitted if it is clear. Sub-network is parenthesized by square
brackets.

An example of the data-model is shown in **Fig.3**. In this example, two "hs"
arcs are bundled to represent exclusive subsets. Two "he" arcs are also bundled to
represent exclusive elements. The arc "hp" and the arc "Logic_Value" are weighted
to represent the number of "Pin"s and the number of boolean values respectively.
This example represents the following design model: "Device" has two exclusive subclasses, "Gate" and "Terminal". Every "Device" has a "Pin" or up to "Max_Pin"
"Pin"s. Every "Pin" has one "Logic_Value" which has a "Boolean" value "True" or
"False".

To define data-models in detail, SEND has meta-arcs to represent constraints as
rules. Two sub-networks are linked each other using meta-arcs such as "if_and_only_if"
or "if-then". **Fig.4** represents the rule : if the "Type" of an element of "Device" (D)
is "G", then the "Device" (D) is an element of "Gate".

3.2 Definition of operation-models

In the operation-model of the common-domain knowledge, high-level meta-operations
should be defined. Moreover, in the tool-domain knowledge, the primitive design operations of each tool must be defined. SEND has "action" nodes and meta-arcs to
represent these operations. Each action has three kinds of arcs. First, user-defined

Table 1. Nodes and arcs of SEND

object	type	name	short-hand	ext.[*1]
nodes	primitive	set_name	sn	√
		element_name	en	√
	meta	if_and_only_if	iff	O
		imply	imp	√
		user defined actions		O
		is_equivalent_to	eq	O
arcs	primitive	has_part	hp	√
		has_subset	hs	√
		has_element	he	√
		user defined relations		√
		is_equivalent_to	eq	O
		user defined objects		O
	meta	if_and_only_if	iff	O
		if,then	if,then	√
		from, to	from,to	O

[*1] √ — adopted from K-NET
O — extended in SEND

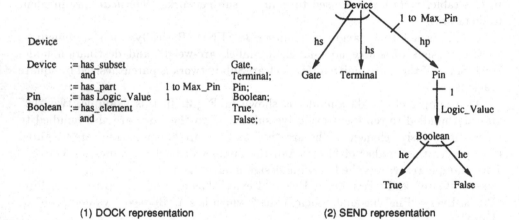

Device

Device   := has_subset   Gate,
          and   Terminal;
          := has_part   1 to Max_Pin   Pin;
Pin   := has Logic_Value 1   Boolean;
Boolean   := has_element   True,
          and   False;

(1) DOCK representation   (2) SEND representation

Fig.3. Definition of data-models

"object" arcs point to the target objects of the action. Secondly, "from" arcs represent the pre-conditions which must be satisfied before the action will be taken. Lastly, "to" arcs represent the post-conditions which will be satisfied after the action is taken. These conditions are represented by a sub-network. **Fig.5** represents the action "Connect". This action has two target objects, "Target_Net" and "Target_Pin". "Target_Pin" (P) is an element of "Pin" and "Target_Net" (N) is an element of "Net". Before this action, "Pin" and "Net" must have their elements (P) and (N), and, after this action, a new element of "Connection" (C) with (P) as "Connected_Pin" and (N) as "Connected_Net" will be created.

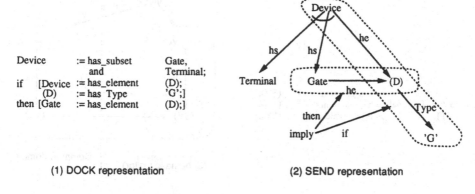

Device	:= has_subset	Gate,
	and	Terminal;
if [Device	:= has_element	(D);
(D)	:= has Type	'G';]
then [Gate	:= has_element	(D);]

(1) DOCK representation

(2) SEND representation

Fig.4. Definition of a rule

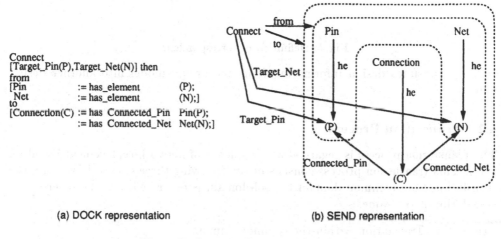

Connect
[Target_Pin(P),Target_Net(N)] then
from
[Pin   := has_element   (P);
 Net   := has_element   (N);]
to
[Connection(C) := has Connected_Pin Pin(P);
              := has Connected_Net Net(N);]

(a) DOCK representation

(b) SEND representation

Fig.5. Definition of an operation-model

3.3 Definition of correspondences between two domains

Each design models of common-domain and tool-domain are defined in sub-networks as illustrated in **Fig.2**. In order to map one data-model to another data-model, correspondences between these design models must be defined. Every element in the tool-domain has an arc pointing to a corresponding element in the common-domain knowledge. This arc is either a primitive arc or an arc defined in a rule. In the example shown in **Fig.6**, the following correspondence is illustrated : If the set "Sym_Vcc" of the tool-domain has an element (SV), then the set "Vcc_Terminal" of the common-domain has a corresponding element (VT).

4 Translation between design models

In DATE, all design information about design targets, design action and status is represented by a semantic-network. In order to automate tool-execution and data-management, the inference engine DOT (DOmain concerning knowledge Translation

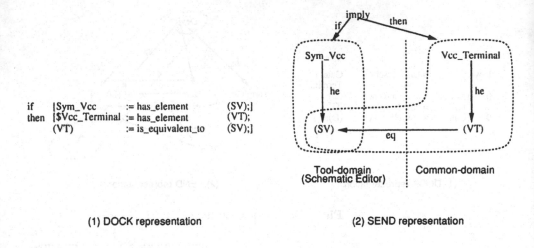


```
if    |Sym_Vcc       := has_element      (SV);]
then  |$Vcc_Terminal := has_element      (VT);
      (VT)           := is_equivalent_to (SV);]
```

(1) DOCK representation

(2) SEND representation

Fig.6. Definition of correspondence

engine) translates design information from one design model into another design model.

4.1 Translation Process

The translation process is regarded as a sequence of network-replacement based on rules. This translation process consists of the following three stages. Here, only the translation from common-domain to tool-domain is described, but the reverse way translation is the same.

Stage 1 : Translation within the common-domain

When the correspondence between the semantic-network and tool-domain is not clearly defined, the sub-network is translated into another sub-network or sub-networks in the common-domain. Every nodes/arcs, which has direct correspondence to the tool-domain, has an arc or rule defining the correspondence. If the correspondence is not defined, DOT translates this sub-network into another sub-networks according to the rules. These rules are categorized into two types. First type is the rules defining design constraints or design knowledges stands in the common-domain. Second type is the rules representing general theorem such as the Set theory. This translation will be repeated until direct correspondences become clear.

Stage 2 : Translation from the common-domain into the tool-domain

In this stage, the design model is actually translated from common-domain into tool-domain. Sub-networks in the common-domain are translated into sub-networks represented in the tool-domain. This translation is performed according to arcs/rules defining correspondences between the two domains.

Stage 3 : Translation to design operations

To realize design requests, the Inference Engine constructs a sequence of design operations. According to the definition of operation-model, the Inference Engine selects design operations, which has pre-conditions satisfiable by current design states, and which can change states as required.

By these three stages, a design request described in the common-domain will be translated into some design operations in the tool-domain.

4.2 Example of a translation

In this section, an example of the translation process will be described. The input design information is **Design Request 1** in the common-domain as illustrated in **Fig.7**. This request will be translated into two tool-domain design operations **Design Action 1** and **Design Action 2** by the following three stages.

Stage 1 : In **Fig.7**, Design request is represented as follows.
Design Request 1 : Set the "Logic_Value" of (P1) which is an element of "Pin".
Here we assume that the direct correspondence of "Logic_Valie" to the tool-domain is not defined. DOT translates this design request into the following design request.
Design Request 2 : Create an element of "Vcc_Terminal" (VT) which has an element of "Pin" (P2). Create an element of "Signal" (S) and create a "Connection" from (S) to (P1) and (P2).
Here the rule **Logic_Value of a Pin (P) is True if the Pin(P) has a "Connection" to a "Pin" of "Vcc_Terminal"** is assumed.
Stage 2 : The **Design Request 2** is translated into the following design request.
Design Request 3 : Create a "Wire" (W) which has a "Connection" to both target "Pin" (P1) and the "Pin" (P2) of "VG_Sym" (VS).

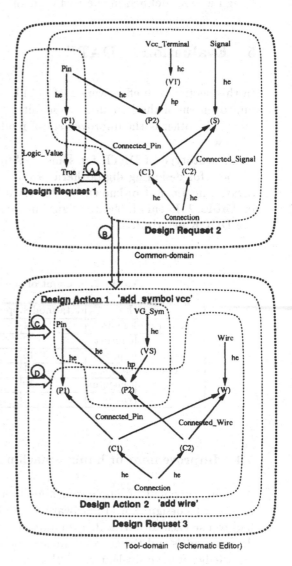

Fig.7. Translation process

Stage 3 : The **Design Request 3** is translated into the following two design
actions.
Design Action 1 : add an element of "VG_Sym" (VS) by the command
"add symbol vcc".
Design Action 2 : add an element of "Connection" (C1) and (C2)
which is connected to "Wire" (W) and "Pin"s (P1) ,(P2) using the
command "add wire".

Thus, the design request described in the common-domain is translated into two
design actions defined in the tool-domain.

5 Evaluation of DATE

In this section, the effectiveness of DATE is evaluated from two aspects. One is the
improvement of the human-machine interface from the hardware designers' point of
view. The other is the improvement of modifiability from the tool integrators' point
of view.

A prototype of DATE for some PLAtools [8] is constructed. PLAtools is a set
of tools for designing digital hardware implemented using the programmable logic
array. The amount of knowledge description for each domain of knowledge is listed
in **Table 2**. In this table, the figure in each entry represents the number of lines in
the DOCK description.

Table 2. Amount of the knowledge description

model		data-model	operation-model	correspondence	total
common-domain		34	90	-	124
tool-domain	eqntott	80	172	39	
	espresso	48	918	17	
	simple	51	184	64	1721
	platypus	30	68	29	
	vi	7	12	2	

5.1 Improvement of human-machine interface

As a measurement of human-machine interface, we adopt the amount of knowledge
required to design digital hardware. In previous environments, a hardware designer
had to use each tool with its own human-interface, and he had to recognize all of the
design models and the correspondences between them. Thus, the amount of required
knowledge was equivalent to 1721 lines in the DOCK description. In DATE, it is
enough for the hardware designer to take care of the common-domain knowledge.
Thus the amount of knowledge required for the hardware designer is only 124 lines
in the DOCK description.

5.2 Improvement of modifiability

In previous environments, relationships between a design model and the other part of the environment are distributed over the knowledge base. Moreover, not only the correspondence of data-models but also the correspondence of operation-models are also described. But in DATE, relationships between the design model in the tool-domain and the other part of environment are encapsulated into the description of correspondence. Thus, it is sufficient for a tool integrator to take care of the correspondence between the data-model of the tool-domain and that of the common-domain. For example, in previous environments, if the data-model of the simulation tool "platypus" is modified, the entire description of common design model(34+90 lines) and all design model of platypus(30+68+29 lines) must be maintained. This is equivalent to 251 lines. But in DATE, a tool integrator only takes care of data-models of common-domain(34 lines) and the data-models of platypus (30 lines), plus, the correspondences(29 lines). This is equal to 93 lines.

6 Conclusions and future works

The knowledge representation and the knowledge translation mechanism in DATE have been described. These mechanisms are used for translating design information from one design model into another. By this integration mechanism of design models, the human-machine interface and the modifiability can be improved. This project of the design environment DATE has not been completed. Research on evaluating the effects of tool modification is now in progress.

References

1. B.McCalla, B.Infante, D.Brzczinski and J. Beyers : CHIPBUSTER VLSI Design System, IEEE Int'l. Conf. on Comput. Aided Des., (1986) 20-27.
2. D.W.Knapp and A.C.Parker : A Design Utility Manager : the ADAM Planning Engine, ACM/IEEE 23rd D.A.Conf., (1986) 48-54.
3. E.F.Girczyc and T.Ly : STEM : An IC Design Environment Based on the Smalltalk Model-View-Controller Construct, 24th ACM/IEEE D.A.Conf., (1987) 757-763.
4. M.A.Breuer, W.Cheng and R.Gupta : Cbase 1.0 : A CAD Database for VLSI Circuits Using Object Oriented Technology, IEEE Int'l Conf. on Comput. Aided Des., (1988) 392-395.
5. M.Bartschi, H.U.Stamer and F.Wunderlich : An Integrated Tool Box for the Electrical Engineer, IEEE 2nd Int'l Conf. on Data & Knowledge Systems for Manufacturing & Engineering, (1989) 144-147.
6. M.L.Bushnell and S.W.Director : Automated Design Tool Execution in the Ulysses Design Environment, IEEE Trans. on CAD, Vol.8, No.3, (1989) 279-287.
7. R.Fikes and G.Hendrix : A Network-Based Knowledge Representation and its Natural Deduction System, 5th IJCAI, (1977) 235-246.
8. User's Manual : PLAtools, Digital Equipment Computer Users Society, Program Library Manual, UX-111 (1987).

This article was processed using the LaTeX macro package with LLNCS style

Configuration of a Passenger Aircraft Cabin based on Conceptual Hierarchy, Constraints and flexible Control

Manfred Kopisch, Andreas Günter

Univ. Hamburg, FB Informatik, Bodenstedtstr. 16, W-2000 Hamburg 50, Germany

Keywords: Configuration, Expert System, Application

Abstract. Starting out from a concrete application - configuring passenger cabins for the aircraft AIRBUS A340 - we discuss the use of a conceptual hierarchy, a constraint system and flexible control in configuration expert systems. In the course of this description we show that these mechanisms allow an adequate representation of the domain knowledge and modular structuring of expert systems Besides, they have advantages over rule-based architectures.
The presented expert system XKL was developed using the expert system kernel PLAKON. XKL proves the useful application of the expert system technology for the configuration of a cabin layout. It shows that PLAKON's formalisms are adequate for representing the domain knowledge and for solving such a task.

1   Introduction

Configuration is one of the fields in expert system technology in which the application of AI-methods has advanced a great deal over the past few years. In this paper, we present an expert system (XKL) that was designed to solve the task of configuring a passenger cabin for the aircraft AIRBUS A340. We show the applicability of some of the latest results in expert system research. The examined application requires the selection of cabin interior components and their arrangement constrained by a large number of restrictions (Sec. 2). After a short description of some methods used in configuration systems (Sec. 3), follows an introduction of PLAKON (Sec. 4). This expert system kernel was used as a basis for the implementation of XKL. The representation of the domain knowledge with the formalisms of PLAKON is described in Sec. 5. The conceptual hierarchy, the constraints and the flexible control mechanism are particularly significant. A short depiction of the graphical user interface (Sec. 6) is followed by the description of future research (Sec. 7) and a summary (Sec. 8).

2   Cabin Layout

A cabin layout defines the passenger cabin of an aircraft by number, arrangement, type and position of the cabin interior components. Cabin interior components are objects like passenger seats, lavatories, galleys or cabin attendant seats which may be installed in the passenger cabin. The entire set of cabin interior components in the passenger cabin defines its capacity, its comfort level, its possible service etc.

Configuring a cabin layout means:
to place a certain number of cabin interior components optimally in the restricted space of the passenger cabin. Optimality may be defined with respect to factors like the number of

seats, the cost, the leadtime or a mix of these factors. In the process of configuring a cabin layout one has to consider the given restrictions and the requests of the ordering airline. The restrictions may be either certification rules of the approving airworthiness authority or technical requirements of either the aircraft builder or the component supplier. The airline requests concern the division of the passenger cabin into passenger classes, the number of seats, the seat pitch etc.

The difficulty in configuring a cabin layout is to consider all restrictions and requirements at the same time. Besides, one has to strive for optimality with regard to criteria like number of seats, comfort level, cost, delivery time etc., which are different for each order. Configuring a cabin layout is getting harder. This is due to the constantly increasing size of aircrafts permitting more and more cabin interior components to be placed and a rising number of certification rules. Especially, the aircraft builder´s representative is no longer able to consider every restriction and thus cannot assure correctness and optimality of his layouts. An expert system might provide assistance by relieving the representative from routine tasks. It could make the necessary tests for correctness and thereby guarantee the consistency of the layout developer´s decisions. Such a system should be able to carry out parts of the configuring and optimizing processes on its own. The system XKL, an expert system for cabin layout, supports configuring a passenger cabin for an AIRBUS A340.

3 Configuration Expert Systems

The application described is a configuration task with special consideration paid to the spatial arrangement of the objects. To solve a configuration task means to compose a system (configuration) from single components which meets all requirements (s. [Mittal89], [Puppe90]). The spatial arrangement of the components is of great importance for the configuration of a passenger cabin and also has consequences for the objects. Thus this application also includes some aspects of design problems. Configuration tasks have the following characteristics [Günter91]:
- large solution space
- objects are composed of components
- dependencies between the objects
- heuristic decisions
- consequences of the decisions are not totally predictable

Expert system technology must provide suitable formalisms and mechanisms to handle typical problems of a configuration task. First we describe some methods shortly, before we introduce the mechanisms offered by the tool system PLAKON which was used for the implementation of XKL.

For a long time, the development of configuration expert systems was influenced by the rule-based paradigm. The best-known representative is the system XCON [McDermott82]. Although XCON is considered a success and it is one of the most-cited expert systems, the concepts of XCON are not applicable in many domains. The rule-based paradigm has been criticized in the latest research [Günter90], [Harmon89], [Soloway87]. The criticism refers to the following aspects: serious problems with knowledge acquisition, consistency checking and in particular maintenance; missing modularity and adaptability. The necessity to reverse decisions leads to problems; and the integration of user instructions and case based approaches is inadequate.

Some promising suggestions already exist in the field of configuration problems and there are also some expert system applications. For example, the systems COSSACK [Frayman87], PLATYPUS [Havens89], PRIDE [Mittal86], VEXED [Steinberg87], VT [Marcus88]. Apart from rule-based systems, the following concepts of AI-research have been employed for configuration systems [Cunis91], [Puppe90]:

- *object oriented representation of the configuration objects*
  AI provides various tools for representing objects. Configuration in technical domains requires classification of objects and description of these objects and their properties. Common demands like expressiveness, maintenance and structure have to be met as well.

- *administration of the dependencies with a constraint-system*
  Constraint-systems are a good choice for representing restrictions and their evaluation during the configuration process. Constraints have been successfully employed in some configuration systems [Frayman87], [Havens89], [Steinberg87].

- *top-down-design*
  An important characteristic of a configuration is its structure of components. This compositional structure may also be employed for controlling the configuration process. Thus the control can refer to the compositional structure of the components to reduce the goal space. The hierarchical structure can often be defined domain dependently in advance.

- *"intelligent" backtracking*
  Heuristic decisions may be recognized as wrong during the configuration process and it becomes impossible to check all consequences of the decisions. That is why the reversal of decisions is necessary. Due to the inefficiency of chronologic backtracking, dependency-directed or knowledge-based backtracking are increasingly used (see [Dhar85], [Markus88]).

The methods listed above are in a stage of development where mature and tested solutions are exceptions. This paper shall, among other things, demonstrate the applicability of some of these methods in a concrete application.

4 The Expert System Kernel PLAKON

The expert system kernel PLAKON is a tool system for planning and configuration tasks in technical domains. PLAKON has been developed at the University of Hamburg together with four industrial partners [Cunis89]. It provides a knowledge engineer with representation and inference methods which are well-suited for planning and configuring tasks. To introduce PLAKON in detail is beyond the scope of this paper. We restrict this introduction to the parts which are relevant for XKL. For a full description see [Cunis91].

Configuration with PLAKON is based on a stepwise refinement of partial solutions. Starting with an initial partial construction which has been derived from a task description a partial construction is refined by a sequence of construction steps until a complete and correct construction is achieved. Each construction step determines a value in the partial construction such as the number of galleys in a passenger class or the position of a lavatory in the passenger cabin (Sec. 5.3). One of the goals during the development of

PLAKON was the strict separation of different kinds of knowledge needed for configuration tasks. Among other things, this serves to avoid the well-known problems of rule-based systems. PLAKON distinguishes the following types of knowledge:

- *configuration objects*
  Knowledge about the objects and their properties is represented in an object-centered fashion in a concept hierarchy. One property which is important for the configuration is the has-parts relationship. This property describes the compositional structure of the objects. The conceptual hierarchy contains a generic description of all configuration objects and thus of all configurations (Sec. 5.1).

- *configuration restrictions*
  Knowledge about the relationships between the objects or their properties are represented by constraints and managed by a constraint net. Thus a declarative representation of relationships is possible without the need to know how and when the constraints have to be evaluated (Sec. 5.2).

- *configuration strategies*
  Knowledge about the course of action during a configuration process (control knowledge) is represented in strategies. Thus several different ways of behavior can be realized. For example top-down, opportunistic, case-based or interactive behavior. The control refers to the conceptual hierarchy [Günter90] (Sec. 5.3).

PLAKON is a tool system and thus provides alternative and optional modules for the different functions. There are optional modules for case-based configuration and simulation. A knowledge engineer determines the modules he needs for his application. Thus PLAKON can be used as a modular building kit. PLAKON is implemented on several workstations under different CommonLisps and CLOS.

5 Cabin Layout with PLAKON

The developer of a cabin layout needs knowledge of the following fields:
- available cabin interior components and their properties
- technical details of the aircraft (measures, capacity of supply for water and electricity, mounting points for the cabin interior components etc.)
- technical restrictions and technical dependencies between objects
- certification rules
- airline requirements
- proceeding during layout development

The following examples show how these kinds of knowledge are represented in PLAKON.

5.1 Conceptual Hierarchy

All objects and their properties are described in the conceptual hierarchy. This conceptual hierarchy consists of a taxonomic and a compositional hierarchy. The taxonomic hierarchy is formed by concepts which describe all cabin interior components and their properties. The possible values for each property are also given in this hierarchy. Diverse object descriptors are available in PLAKON to describe the possible values for the properties of an object. There are, among others, ranges [10 20], choice sets {1 2 3}, concepts (a lavatory), booleans and sets. The following definitions of the concepts *cabin interior component* and *galley* show some of the possibilities of PLAKON for building a taxonomic hierarchy:

```
(is-a    ( a cabin interior component )
         ( a construction object
                  ( part-of         ( a passenger class )
                  ( ref-nr          [ 2500000 2599999 ] )
                  ( x-coordinate    [ 405inch 2408inch] )
                  ( x-expansion     [ 10inch 200inch ] )
                  ( y-coordinate    [ -105inch 105inch ] )
                  ( y-expansion     [ 10inch 70inch ] )
                  ( orientation     [ 0 360 ]
                                    ( default 270 ) )
                  ( weight          [10kg 2000kg ] )
                  ( leadtime        [ 9month 2years ] ) ) )
```

A cabin interior component is described as a specialization of the concept *construction object* which is the given root of the taxonomic hierarchy. The concept *cabin interior component* itself is the root of further specializations like the concept *galley*. A specialization happens either through the restriction of the possible values or through the addition of further properties.

```
(is-a    ( a galley )
         ( a cabin interior component
                  ( ref-nr                    [ 2530000 2533999 ] )
                  ( orientation               { 0 90 180 270 } )
                  ( trolleys                  { 2 3 4 5 6 7 8 9 10 } )
                  ( half-size trolleys        { 0 1 } )
                  ( meals                     [ 28 140 ]
                                              ( computation function meals-fcn ) )
                  (height                     { full half}
                                              ( default full ) )
                  ( water/waste connection    ( a boolean ))
                  ( cooling                   { none dry-ice air-chiller } ) ) )
```

In addition to an object descriptor, PLAKON allows the use of *facets*. In the examples, the facets *default* and *computation function* are used. The set descriptor allows the description of n to m relationships. Thus it is possible to describe the composition of objects. The following description of the general concept *passenger cabin* shows its use:

```
( is-a    ( a passenger cabin )
          ( a construction object
                  ( has-parts   ( :set ( a passenger class [1 3] ) :=
                                        ( :some ( first-class ) [0 1] )
                                        ( :some ( business-class ) [0 1] )
                                            ( a tourist-class ))))
```

A passenger cabin is described as a specialization of the concept construction object. It consists of the optional classes first-class and business-class and an obligatory tourist-class. The has-parts relationship forms a compositional hierarchy which is embedded in the taxonomic hierarchy. This compositional hierarchy describes the decomposition of construction objects into components and thus is of great importance to the control. The information in the structure can be used to control the configuring. The taxonomic and the compositional hierarchy together describe all admissable configurations of a cabin layout. Figure 1 illustrates these hierachies for a small part of the cabin layout domain.

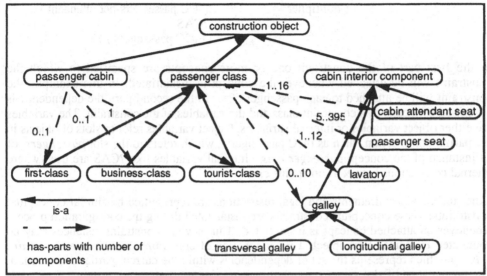

Figure 1: Taxonomic and Compositional Hierarchy of the Cabin Layout Domain

These examples clearly show the object centered representation of knowledge in PLAKON. This representation enables a user of XKL to inspect all information about an object by a single mouse-click on the graphical representation of the object.

A representation of the dependencies between objects is needed in addition to the representation of these objects and their properties. This is done in a separate part of the knowledge base.

5.2 Constraints

The possibilities to choose cabin interior components for the cabin layout are restricted by technical requirements and certification rules. The values for their properties are also restricted. PLAKON allows a declarative description of these restrictions. For this, a three-level constraint model is used. On the first level PLAKON provides several *constraint-classes* like *adder, multiplier, number-restrictiom, sum, equal, etc.* On the second level these constraint-classes are used to define *conceptual constraints*, which describe the dependencies between objects in the conceptual hierarchy or between their properties. On the third level all instantiated constraints form a *constraint net*, which represents all existing dependencies in a partial construction.

In our representation, the airline requests for equipment are described as attributes of the concerned passenger classes. Dependencies between the properties of a passenger class are described declaratively in a conceptual constraint. The example describes the dependency:

The airline request for passengers and for the level of service provided by the cabin attendants results in the necessary number of cabin attendant seats in the concerned class.

as a compound constraint by the use of two constraint-classes.

```
( Constrain        ( ( #?C ( a passenger class ) ) )
                   ( number-restriction        ( #?C has-parts )
                                               ( a cabin attendant seat )
                                               ?CAS )
                   ( multiplier                ( #?C passengers-per-attendant )
                                               ?CAS
                                               ( #?C passengers ) ) )
```

In the first part of the constraint one or more concepts are specified to which the constraint will be attached whenever the concept is instantiated. In this example, the constraint will be attached to each passenger class. In the second part, the dependency is described by means of a constraint-class and the variables of the constraint. The variables are either object variables or internal variables. Object variables refer to slots of objects in the partial construction, such as (#?C passengers), which refers to the slot *passengers* of an instance of the concept *passenger class*. Internal variables like ?CAS are used where internal connections between constraints exist.

Other technical requirements and legal restrictions are represented by similar conceptual constraints. These conceptual constraints are instantiated during the configuration process whenever an attached concept is instantiated. Thus several constraint instances may be connected to a slot of an object. This connection of constraints results in a *constraint network* which represents the set of dependencies within the current partial construction. The propagation of slot values through this constraint net enables the system:
- to check specified values for consistency under consideration of all inter-dependencies and
- to determine new slot values from existing ones or to restrict sets of values.

The knowledge engineer is relieved from thinking about evaluation of a constraint. Even the dependencies between several restrictions are considered by the constraint net. This is a great advantage over rule-based systems.

5.3 Control Knowledge

The control uses the conceptual hierarchy as a guideline. By comparing the instances of the partial construction with the concepts in the conceptual hierarchy the control finds all construction steps which have to be carried out to specify incomplete instances. Possible construction steps are:
- *specialization* of a construction object according to the conceptual hierarchy
- *decomposition* of a construction object along has-parts relations into its parts (top-down approach)
- *aggregation* of components along part-of relations (bottom-up approach)
- *parametrization* of an attribute of a construction object by restricting possible values or determining a concrete value for this attribute.

The configuration process in XKL is divided into three phases (passenger cabin decomposition, passenger class parametrization, specification of the cabin interior components). PLAKON supports such a division by means of *strategies*. Each strategy in PLAKON contains control knowledge about:
- selection of the next construction step from the set of all admissible steps,

- execution of a construction step,
- reduction of the search space by setting a focus on parts of the partial construction,
- resolution of conflicts and
- activation of the constraint net.

XKL does not use all possibilities provided by PLAKON. Often the user wants to make decisions by himself. Thus XKL proceeds automatically until a heuristic decision has to be taken by the user. But everything else is done by the control. PLAKON provides a default strategy which enables a knowledge engineer to create and test the conceptual hierarchy and the constraints without the need to define domain-dependent strategies. By providing more and more domain-dependent strategies he incrementally improves the control. This feature together with the possibility of interactive control proved useful for the implementation of a working prototype of XKL.

6   User Interface

One result of our work on XKL was that a good quality of the constructed cabin layouts does not necessarily get the acceptance of the user. User guidance and a graphical representation of the developing layout are of crucial importance for the use of such an expert system. Thus we developed a user interface, which enables the user to proceed similarly to the previous method of using paper tokens. He is able to touch the cabin interior components on the screen and to place them in the passenger cabin. The available cabin interior components are visualized in a way which shows all possible locations. Figure 2 shows how XKL presents all possible locations for longitudinal galleys.

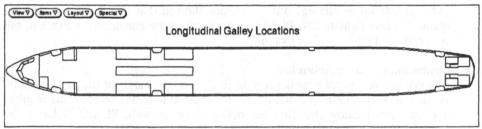

Figure 2: All possible locations for a longitudinal galley

In addition, the user is able to ask for information about the cabin interior components by clicking on an item.
XKL checks the consistency and informs the user of inconsistencies. The result of the configuration is represented on the screen as in figure 3. Thus the user sees for each decision he takes the developing layout on the screen.

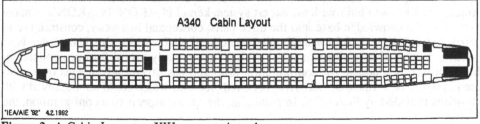

Figure 3: A Cabin Layout as XKL presents it to the user

7 Future Developments

PLAKON proved useful for the implementation of XKL. Nevertheless, during the development some problems occurred which demand an expansion of PLAKON.

- *Consideration of space in PLAKON*
  Layout problems are a special class of configuration problems with a main focus on spatial aspects. The representation of space and the treatment of spatial dependencies should be provided domain-independently by PLAKON. We intend to extend PLAKON by implementing special constraint-classes for spatial dependencies [Kopisch91]. A standardized representation of space is necessary for a domain-independant use of these constraint-classes. For the moment, we limit the representation to 2D-space. We intend to consider the findings in existing systems such as EXIST [Kloth87], with its representation of areas, or WRIGHT [Baykan90], with its great number of spatial relations.

- *Truth-Maintenance for efficient modifications of a configuration*
  Tests with XKL showed that the modification of a configuration is a permanently occurring problem. This problem arises with appearing conflicts or with customer wishes. The TMS of PLAKON saves all construction steps which are not dependent on the construction step which has to be taken back for the modification. The present implementation does not use this TMS but forces taking back all construction steps which occurred after the changed one.

- *Constraint relaxation*
  Some conflicts may be solved by constraint relaxation. Until now we considered every constraint as *strong*. But experience showed that in some cases the user wants to relax constraints. PLAKON does not know constraint relaxation, but concepts exist which will be examined.

- *Optimization of a configuration*
  The interactive control puts the responsibility into the hands of the user. A future version of XKL should provide the user with an automatic control. This requires not only consistency checking but optimization as well. PLAKON has to be extended to allow optimization-based configuration [Günter91].

8 Summary

We introduced XKL, an expert system which supports the configuration of a cabin layout. In the course of this, the use of AI-methods like object-oriented representation, constraint-system and top-down configuration with interactive control were shown in a concrete application. XKL was built with the expert system kernel PLAKON. PLAKON´s concept of dividing the knowledge base into the three parts: conceptual hierarchy, constraints and control, proved to be useful for the implementation of XKL. Beside this knowledge base division, several of its concepts made PLAKON particularly useful for our application. This approach is adequate for configuration tasks and has several advantages over rule-based systems. Nevertheless, there have occurred problems which could not be solved with the means provided by PLAKON. In particular, the spatial aspects of a configuration, the optimization and the modification of a configuration are important for an expert system which is fit for use in the real world of cabin layout configuration. We therefore intend to extend PLAKON for a future version of XKL. We are convinced that the use of an expert system in a domain like cabin layout with its manifold restrictions is possible and useful.

References

[Baykan90] Baykan, C. A., Fox, M. S.: *WRIGHT: A Constraint Based Spatial Layout System*, in: Tong, C. Sriram, D. (eds.): AI in Engineering Design

[Cunis89] Cunis, R.; Günter, A.; Syska,I. u.a.: *PLAKON -- An Approach to Domain-Independent Construction*, in Proc. 2. IEA/AIE,, Tennesse, S. 866-874, 1989

[Cunis91] Cunis, R.; Günter, A.; Strecker, H.: *The PLAKON-Book* (in German), Springer, Informatik Fachberichte N° 266, 1991

[Dhar85] Dhar, V.: *An Approach to Dependence Directed Backtracking Using Domain Specific Knowledge*, in: Proc. IJCAI 85, S. 188-190, 1985

[Frayman87] Frayman, F.; Mittal, S.: *COSSACK: A Constraint-Based Expert System for Configuration Tasks*, in: Sriram, D.; Adey, R.A. (Ed.): Knowledge Based Expert Systems in Engineering: Planning and Design, Computational Mechanics Publications, S. 143-166, 1987

[Günter90] Günter, A.; Cunis, R.; Syska, I.: *Separating Control from Structural Knowledge in Construction Expert Systems*, in: Proc. 3. IEA/AIE-90, Charleston, USA, S. 601-610, 1990

[Günter91] Günter, A.; u.a.: The *Project PROKON -- Problemspecific Tools for Knowledge Based Configuration* (in German), PROKON-Report No. 1, 1991

[Harmon89] Harmon, P.: *How DEC is Living with XCON*, in: Expert System Strategies, Vol. 5, N° 12, S. 1-5, 1989

[Havens89] Havens, W.S.; Rehfuss, P.S.: *PLATYPUS: A Constraint-Based Reasoning System*, in: Proc. IJCAI-89, S. 48-53, 1989

[Kloth87] Kloth, M.; Herrmann, J.: *EXIST — An Expert System for the Innerplant Planning of Locations*, in: Sriram, D.; Adey, R.A. (Ed.): Knowledge Based Expert Systems in Engineering: Planning and Design, Computational Mechanics Publications, S. 429-438, 1987

[Kopisch91] Kopisch, M.: *Configuration of a Passenger Cabin of an Aircraft with an Expert System* (in German), Diploma Thesis, Univ. Hamburg, 1991

[Marcus88] Marcus, S.; Stout, J.; McDermott, J.: *VT: An Expert Elevator Designer That Uses Knowledge-Based Backtracking*, in: AI Magazine, Spring 88, S. 95-112, 1988

[McDermott82] McDermott, J.: *R1: A Rule-Based Configurer of Computer Systems*, in: Artifcial Intelligence 19 (1), S. 39.88, 1982

[Mittal86] Mittal, S.; Araya, A.: *A Knowledge-Based Framework for Design*, in: Proc. AAAI-86, S. 856-865, 1986

[Mittal89] Mittal, S.; Frayman, F.: *Towards a Generic Model of Configuration Tasks*, in: Proc. IJCAI-89, S. 1395-1401, 1989

[Puppe90] Puppe, F.: *Methods for Problem Solving in Expert Systems* (in German), Springer, 1990

[Soloway87] Soloway, E. u.a.: *Assessing the Maintainability of XCON-in RIME: Coping with the Problem of Very Large Rule-Base*, in: Proc. AAAI-87, S. 824-829, 1987

[Steinberg87] Steinberg, L.I.: *Design as Refinement Plus Constraint Propagation: The VEXED Experience*, in: Proc. AAAI-87, S. 830-835, 1987

Configuration of Industrial Mixing-Machines - Development of a Knowledge-Based System

Axel Brinkop, Norbert Laudwein

Fraunhofer-Institute for Information and Data-Processing (IITB)
D-7500 Karlsruhe, FR Germany

Abstract. The paper describes an implemented system designed to assist the sales department.staff in finding a configuration of industrial mixing-machines.

The process of configuration is split into two steps: First process engineering knowledge is used to characterize the mixing-machine by the kind of agitator and the number of revolutions taking into account the mixing task; in a second step all components of the mixing-machine are determined with regard to the laws of mechanics.

The configuration process is driven by a constraint propagation with integrated dependence control which can handle feedback loops. Since often parameters are needed before they can be computed, the configuration process has to contain feedback loops. These parameters have to be preestimated. Later on they will be checked and probably recomputed.

All main decisions in the process of configuration are taken by the user. The system supports the user by exploring alternatives and suggesting standardized and therefore cheap components. Right now the system is tested in the distribution department of the manufacturer and evaluated with respect to its usefulness for the daily work.

1 Domain of Application

Mixing-machines are needed in a wide area of industrial productions with main applications in the chemical industry. The basic structure of a mixing-machine is shown in figure 1. It consists of the following components:

- one or more agitators
- a shaft
- the bearing which is located in the so-called lantern, and
- a drive, probably in combination with a gear.

The basic structure of a mixing-machine is quite simple, but there is a wide variety of possible parts for each component. Besides impellers there are various special forms of agitators; different shafts are available and thousands of drives providing power from 0.5 kW up to 55 kW produced by several manufacturers can be used. There are also some optional components like a sealing or an additional bearing at the bottom of the tank.

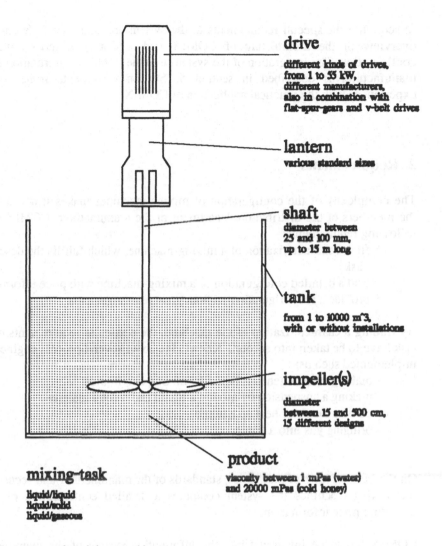

drive
different kinds of drives,
from 1 to 55 kW,
different manufacturers,
also in combination with
flat-spur-gears and v-belt drives

lantern
various standard sizes

shaft
diameter between
25 and 100 mm,
up to 15 m long

tank
from 1 to 10000 m^3,
with or without installations

impeller(s)
diameter
between 15 and 500 cm,
15 different designs

product
viscosity between 1 mPas (water)
and 20000 mPas (cold honey)

mixing-task
liquid/liquid
liquid/solid
liquid/gaseous

Figure 1: Basic structure of a mixing-machine

Usually a customer asks the mixing-machine manufacturer to submit an offer for a mixing-machine for his special purpose. So, a member of the distribution department has to find a configuration which meets the particular requirements of the customer. The information given by the customer consists of the amount of product he wants to mix, i.e. the height and the diameter of the tank, as well as the density and viscosity of the product. COMIX (Configuration of Mixing-Machines) was designed to assist the members of the distribution department in finding a configuration of a mixing-machine which fulfills the requirements of a specific mixing task.

In section 2 the special requirements to the system are introduced. Section 3 gives an overview of the architecture of COMIX and section 4 describes the process of configuration. The integration of the system into the CIM and information system of the manufacturer is described in section 5. Section 6 presents some remarks about experiences with the practical application of COMIX.

2 Requirements

The complexity of the configuration of mixing-machines makes it neccessary to assist the members of the distribution department of the manufacturer. COMIX has to fit the following requirements:
* find a characterization of a mixing-machine, which fulfills the described mixing task
* find a detailed configuration of a mixing-machine with price information
* provide a reconfiguration without great rearrangements

By finding a characterization of the machine, the particular requirements of the mixing task have to be taken into account. Knowledge about several process engineering tasks is implemented such as:
- making a homogeneous liquid by mixing two liquids
- making a suspension by mixing a liquid and a solid material
- cooling down or heat up a liquid.
- bringing gas into a liquid.

On the basis of laws of mechanics, standards of the manufacturer and security guidelines for mixing machines the system computes a detailed configuration of the machine including price information.

COMIX has to be integrated into the information system of the manufacturer. Some printed documents are needed for internal use. Furthermore the system produces as output an offer of the mixing-machine including a drawing of the mixing-machine. In the case the offer is accepted as an order, the data of the configuration of that mixing-machine has to be transferred to an existing CIM-system.

3 Architecture of COMIX

Several different kinds of knowledge have to be implemented. The first type is the so-called structural knowledge about mixing-machines and their components. It is represented in objects (frames and instances) in is-a and part-of hierarchies. The knowledge about available parts and their technical parameters is kept in databases. Process engineering knowledge and component-specific configuration knowledge, often described in an algorithmic way, was coded in functions. Causal dependencies are represented by directed constraints. A more detailed description of the knowledge representation can be found in [1] [3].

The configuration process is done in two steps:

      * with the description of the mixing task and some decisions of the user a characterization of a mixing-machine can be found

      * proceeding from this characterization a detailed configuration with price information can be derived.

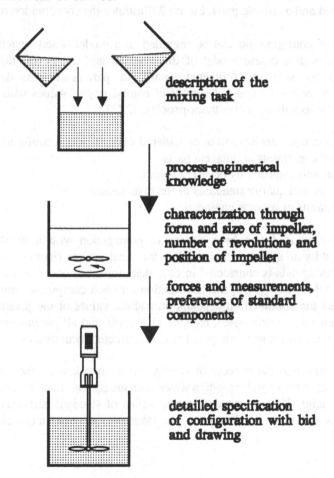

description of the mixing task

process-engineerical knowledge

characterization through form and size of impeller, number of revolutions and position of impeller

forces and measurements, preference of standard components

detailed specification of configuration with bid and drawing

Figure 2: Bi-partition of configuration process

To find a characterization of an optimal mixing-machine reflecting his own opinion and experience, the user can repeat the first step as often he wants to. In this step knowledge about industrial process engineering and experience from the manufacturer is used to determine parameters characterizing a mixing-machine: The type and size of the agitator, its position in the tank and the number of revolutions. All these determinations are valid in intervals and are dependent on each other. The user can vary these suggestions and rerun this step, which not even takes a minute to find another characterization of a mixing-machine which still will be capable of accomplishing the mixing task.

The information derived in the first step is used as input to the second step, that is the detailed configuration of the machine. Here, knowledge of laws of mechanics, industrial standards and quality requirements of the manufacturer are used to find a configuration of standardized and available parts. Figure 2 illustrates the coordination of the two steps.

The process of configuration can be regarded as a model based search problem. The process starts with a coarse model of the machine and transforms this model into a detailed one by stepwise refinement until all parameters are determined. The computation of new values and the proof of consistency of values with the restrictions can be regarded as solving a constraint-problem [2][7].

The following constraints have to be considered when finding a configuration:
- spatial constraints of adjacent parts
- constraints imposed by occuring forces
- security and quality standards of the manufacturer
- constraints of process engineering

The configuration is driven by a constraint propagation with integrated dependence control guided by an agenda which reflects the usual way of finding a configuration by an expert. This agenda is considered in case that more than one constraint can be fired. Simular to an ATMS [4][5][6] the justification for each computed parameter is stored: the parameters the computation depends on and the values of the parameters involved. When a parameter is recomputed, backtracking starts and all parameters which depend on the elder value of the new computed one are retracted recursively .

All main decisions in the process of configuration are taken by the user. The system supports the user by exploring alternatives and suggesting standardized and therefore cheap components. If no consistent configuration of standard parts can be found non-standard parts with the wanted propertiesare taken into account. In this case no price will be computed.

4 Process of Configuration

It showed up that one of the major problems in the process of configuration is the necessity to deal with feedbacks. Often parameters are needed before they are computed; therefore they have to be preestimated. Later on they are checked and probably recomputed.

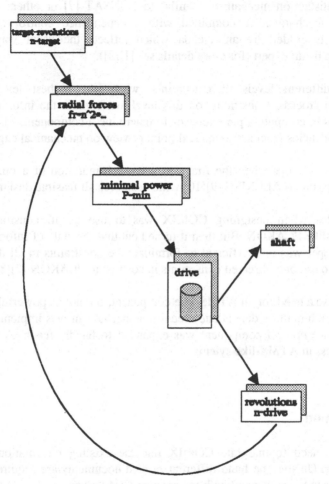

Figure 3: Configuration under feedback

Example: During the process engineering a number of revolutions has been determined which would solve the mixing task (in combination with the chosen design of agitator and so on). This number of revolutions is used to compute the forces which occur at the agitator. These forces determine the minimal power of the drive/gear-combination. One of the process engineering constraints states that the number of revolutions can vary 20 % around the optimal value and still meets the mixing task. Thereafter a database is searched for a drive with the computed minimal power and the wanted number of revolutions. It is highly probable that no drive will be found with the exact number of revolutions. So all parameters depending on the preestimated number of revolutions have to be recomputed.

As mentioned before, the process of configuration is done by a constraint-propagation or constraint satisfaction mechanism, similar to CONSAT [7] or others. In addition the propagation mechanism is combined with a dependence control component. The propagation is guided by an agenda which reflects the usual way of finding a configuration by an expert (for more details see [1][3]).

There are different levels of constraints, where the highest level controls the configuration process. Constraints on this level just contain the information about the preconditions to compute a parameter or to determine a component. They represent the causal dependencies from a pragmatical point of view on mechanical engineering.

The propagation stops when the first consistent configuration of a mixing-machine is found, in opposite to ALLRISE [9][10] which explores all feasible designs.

The first idea when designing COMIX was to use a full constraint propagation mechanism alike PLAKON. But then it turned out that the state of information and skill of the developer was not sufficient to formulate the constraints in all directions. So, it was chosen to use only directed constraints in contrast to PLAKON [2][8].

There was also a need for an ATMS-like component, but not as powerful as described in [4][5][6]. A rudimentary dependency recording mechanism was implemented. Later on, the dependence control component was expanded to handle feedback loops, which is hard to express in ATMS-like sytems.

5 Integration

There was a need to integrate COMIX into the existing information system of the manufacturer. On the one hand different printed documents are required, on the other hand data has to be exchanged with the existing CIM system.

COMIX is implemented in GoldWorks II and is running in a network of PCs under DOS; as LAN NOVELL netware 386 was used. All databases are located on the fileserver. The net is connected through a gateway with the CIM-system on a mainframe. An overview of the network is given in figure 4.

438

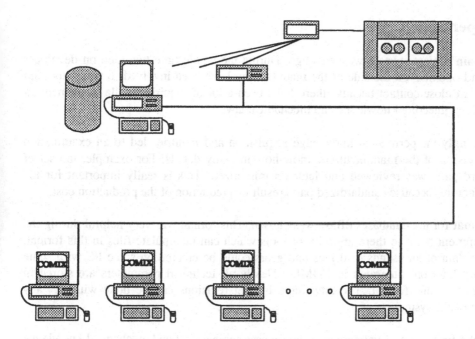

Figure 4: Overview of the network

Several printed documents are needed for the information system. One of them containing details of the configuration including price information is needed to document the configuration in an internal file. Another document containing part identification numbers of the CIM-system and hints about their assembly is used to guide the machine during the whole process of manufacturing.

Additionally a document presenting the final offer including a drawing of the machine is needed. The offer is created by a combination of definite text parts which are depending on the actual configuration of the machine. With an integrated editor the user can change the vocabulary of the offer, add special information or change standardized formulations to get a document ready to be mailed to the customer. The drawing of the mixing-machine is a PostScript-document with symbolic representations of the components of the configuration in scale. It shows the tank as well as the mixing-machine. So it is possible to get an impression of the measurements of the machine and their components.

6 Experiences

The main project started two years ago. During the whole time two men on developer side and one man on the side of the manufacturer have been involved. It was important to stay in close contact because there have been a lot of inquiries while implementing process engineering knowledge and mechanical laws.

Surprisingly the permanent knowledge acquisition and inquiries led to an examination and revision of the manufacturers know-how in many details! For example, the set of standard-parts was reviewed and later on minimized. This is really important for the manufacturer because standardized parts result in a reduction of the production cost.

As format for the databases dBase was chosen. This format was very helpful during the development because there are a lot of tools which can manipulate files in this format. So, the data of thousands of drives and gears could be entered on some PC with some tool and later on integrated in COMIX. Nearly all technical parameters are stored in databases in this format. The advantage is, that revisions can be done without any change of the system.

After the first phase of validation of the process engineerical and mechanical knowledge done by domain experts, the system is now installed on several workstations in the distribution department. The examination of the usefulness of the system for daily work started. This is done by members of the distribution staff by checking the user interface and double-checking configurations found with COMIX.

7 Conclusions

In a preliminary phase of the project the domain of the configuration of mixing-machines was investigated to find out whether it is a possible application for a knowledge-based system. The result of this phase was a first prototype and the choice of GoldWorks II as an appropriate development-tool.

The basic version developed in the main phase of the project was designed first as a stand-alone system. Then the system was extended to run in a network. Six months ago several workstations has been installed in the distribution department. Current work deals with the integration of COMIX with an existing CIM system on a mainframe computer.

It is planned to extend COMIX with a component for case-based reasoning. This component should allow to look for a mixing-machine in a database of already fabricated mixing-machines which is fulfilling a similar mixing task. It has to be investigated, what kind of similarity metric for mixing-machines and mixing tasks can be defined.

8 References

[1]   A. Brinkop, N. Laudwein: A Knowledge-Based System for Configuring Mixing-Machines. Applications of Artificial Intelligence IX, Mohan M. Trivedi, Editor, Proc. SPIE Vol.1468, 227 - 234, (1991)

[2]   R. Cunis, A. Günter, I. Syska, H. Peters, H. Bode: PLAKON - an Approach to Domain-Independent Constuction. Proceedings of the 2nd International Conference on Industrial & Engineering Applications of AI & Expert Systems (IEA / AIE - 89), UTSI, Tullahoma, Tennessee (1989)

[3]   W. Dilger, N. Laudwein, A. Brinkop: Configuration in mechanical engineering. Applications of Artificial Intelligence VIII, Mohan M. Trivedi, Editor, Proc. SPIE 1293, 506 - 514 (1990)

[4]   J. de Kleer: An Assumption Based TMS. Artificial Intelligence 28, 127 - 162 (1986)

[5]   J. de Kleer: Extending the ATMS. Artificial Intelligence 28, 163 - 196 (1986)

[6]   J. de Kleer: Problem Solving with the ATMS. Artificial Intelligence 28, 197 - 224 (1986)

[7]   H. W. Güssgen: CONSAT: Foundations of a System for Constraint Satisfaction. Technische Expertensysteme: Wissensrepräsentation und Schlußfolgerungsverfahren, H. W. Früchtenicht et al., Editor, Oldenbourg Verlag, München, 415 - 440 (1987)

[8]   B. Neumann: Configuration Expert Systems: A Case Study and Tutorial. TEX-K Veröffentlichung Nr. 20, Universität Hamburg, also in Proceedings of SGAICO Conference on Artificial Intelligence in Manufacturing, Assembly and Robotics, Oldenbourg Verlag München

[9]   D. Sriram, M. L. Mahler: The Representation and Use of Constraints in Structural Design.Proceedings of Applications of Artificial Intelligence in Engineering Problems, Springer, Berlin, 355 - 368 (1986)

[10] D. Sriram: ALL-RISE: A Case Study in Constraint-Based Design. Artificial Intelligence in Engineering, 186 - 203 (1987)

A Theoretical Framework for Configuration

O. Najmann   B. Stein

FB 17, Praktische Informatik, Universität-GH-Paderborn

Abstract: We develop a general theory of configuration and give a precise methodology for studying the phenomenon of configuration from a viewpoint which is independent of any knowledge representation formalism. One main result is that we show that the classical skeleton-oriented approach and the resource-oriented approach are in some sense equivalent. We formulate a number of typical configuration problems, like finding a cost-minimal configuration, and show the NP-completeness of one of them.

1   Introduction

Manufacturers of complex technical systems are faced with the problem to satisfy the exceptional demands of their customers in order to stay competitive. Due to the enormous number of possible variants, the process of configuring a technical system became a sophisticated problem. Therefore, many systems have been developed which assist humans in this process. However, only few attempts have been made to formally analyze the nature of typical configuration problems.

The main contribution of this paper is that it provides two models of configuration. A basic idea of these models is that all involved parts of the technical system are solely described by their functionalities. Since we want to describe the configuration problem in a general way, the presented approach does not presuppose any particular knowledge representation formalism.

The paper is arranged as follows: First, a short introduction to configuration problems is given and the idea of a formal description is motivated. Second, two models, M1 and M2, are presented. Model M1 allows to formulate *resource-oriented* configuration problems [Heinrich, 1991; Stein & Weiner 1990], while model M2 allows to formulate *skeleton-oriented* configuration problems [Puppe, 1990]. Several problems that can be formulated under these models are presented. Although model M2 additionally contains restriction rules, it is shown that M1 and M2 are equivalent. This means that it is principally possible to transform explicit knowledge about the structure of a technical system into a set of object functionalities without loss of information. Finally we consider the computational complexity of configuration and show that the general configuration problem is NP-complete.

The operational point of view. Configuration problems are characterized by the fact that the solution (which is the completely configured object) is composed of smaller subobjects. A typical example is the configuration of computers: Particular components such as a harddisk, a processor, etc. are given. The problem is to select components in such a way that their composition fulfills a customer's demand.

A lot of studies in the field of configuration deal with technical aspects of the problem, i.e., description of ideas, concepts, and techniques *how* a certain configuration problem can be solved. Such a pragmatical, teleological point of view is justified. It is caused by the complexity of the general problem "configuration"; most of the configuration systems represent solutions of particular cases.

In this context, PLAKON [Cunis et al., 1991] seems to be the highest developed system dealing with configuration tasks. Also AMOR (cf. [Tank et al., 1989]), a description language for technical systems, should be mentioned here. Brown and Chandrasekaran [1989] developed DSPL, a knowledge representation language for routine design problems.

A theoretical approach. Rather than specifying a particular method of processing or acquiring configuration knowledge, we aim at a formal description of general configuration problems. Such a description could be used to distinguish between different configuration approaches or to compare certain configuration problems with respect to their complexity. One approach, pointing at a similar direction, is the configuration model of Dörner [1991].

In our approach, the central idea is the notion of *functionality*. Objects are described solely by functionalities. When objects are selected, functionalities are composed in order to specify the functionality of the whole system. Because of the functionality-centered approach, our model is especially appropriate to describe problems where the *selection* of components comes to face.

Definition: A configuration problem consists basically of three things, (i) a set of objects, (ii) a set of functionalities which are used to describe certain properties of these objects, and (iii) a set of demands which describe the desired properties of the system to be configured.

2 Model M1

Subsequently, we give a formal definition of the notion *configuration problem*. We will distinguish between the problem itself and the notion *configuration*, which is viewed as the solution of a particular configuration problem.

The system MOKON [Stein & Weiner 1990] operationalizes important parts of model M1. So, the given formalization can be used to distinguish between different classes of configuration problems on the one hand as well as to determine limits of particular domain descriptions on the other (regarding their computer-supported solution).

Definition: A *configuration problem* Π is a tuple $\langle O, F, V, P, A, T, D \rangle$ whose elements are defined as follows:

- O is an arbitrary finite set, it is called the *object set* of Π.

- F is an arbitrary finite set, it is called the *functionality set* of Π.

- For each functionality $f \in F$ there is an arbitrary finite set v_f, called the *value set* of f. $V = \{v_f | f \in F\}$ comprises these value sets.

- For each object o there is a *property set*, p_o, which contains pairs (f, x), where (i) $f \in F$ and $x \in v_f$, and (ii) each functionality $f \in F$ occurs at most once in p_o. A property set specifies the values of certain functionalities of a given object. $P = \{p_o | o \in O\}$ comprises these property sets.

- For each functionality f there is an *addition operator* a_f which is a partial function $a_f : v_f \times v_f \to v_f$. An addition operator specifies how two values of a functionality can be composed to a new value if a new object is *added* to a given collection of objects which themselves describe a part of the system to be configured. $A = \{a_f | f \in F\}$ comprises all addition operators.

- For each functionality f there is a *test* t_f which is a partial function $t_f : v_f \times v_f \rightarrow$ {TRUE, FALSE}. A test t_f specifies under what condition a demand (see below) is fulfilled. $T = \{t_f | f \in F\}$ comprises all tests.

- D is an arbitrary, finite set of *demands*. Each demand d is a pair (f, x), where $f \in F$ and $x \in v_f$. Additionally, the demand set must have the property that no functionality occurs more than once in D. A demand set D describes the desired properties of the system to be configured.

Remarks: O contains objects which can be composed to a system meeting a certain requirement. For example, if we had to configure a computer system, then typical objects of O would be a harddisk, a CPU, a power supply, etc.

Furthermore, functionalities describe certain properties of the objects in O. For example, the harddisk may have the functionalities "capacity," "access time," etc. Then for example, $v_{\text{capacity}} = \{10,20,30,40\}$ [Megabytes].

A typical example of an addition operator is the calculation of the entire capacity of a set of harddisks. For example, a_{capacity} can be defined as follows (where the symbol $\perp$ is used wherever a_{capacity} is undefined):

$$a_{\text{capacity}}(x, y) = \begin{cases} x + y, & \text{if } x + y \leq 40; \\ \perp, & \text{otherwise.} \end{cases}$$

The addition operator in this example can be used to allow multiple use of a harddisk in order to increase the amount of harddisk capacity. An addition operator needs not necessarily specify an *addition* between two numbers, but any kind of operation is possible.

Every object $o \in O$ is characterized by its set of properties, p_o. For example, a particular harddisk can have the following set of properties: $p_{\text{harddisk}} = \{(\text{capacity}, 10), (\text{access-time}, 7)\}$. Note that every functionality can occur at most once in a property set.

Since we have to compose the values of functionalities with respect to a given demand, it is necessary to introduce a test t_f for each functionality $f \in F$. For example, a typical test for "capacity" is the "$\geq$" predicate.

An example of a demand set is $D = \{(\text{capacity}, 30), (\text{mouse}, \text{yes}), (\text{keyboard}, \text{english})\}$.

So far we have only defined the notion "configuration problem"; now we have to define what a solution of such a problem is.

Definition: A solution of a configuration problem must fulfill two conditions: 1. It must be a *configuration*, which is defined below. 2. This configuration must meet all demands of the demand set D.

A configuration contains both, objects and functionalities. Before we give an inductive definition of a configuration, we define how properties can be composed.

Definition: Let (f, x), (g, y) be two properties.

$$\varphi((f, x), (g, y)) = \begin{cases} \{(f, a_f(x, y))\}, & \text{if } f = g; \\ \{(f, x), (g, y)\}, & \text{otherwise.} \end{cases}$$

is called the *composition* of the properties (f, x) and (g, y).

Remarks: The composition of two properties is a set. This set contains either a single property if the functionalities are equal, or it contains these two properties if they are not equal. The rationale of this composition is as follows: If two objects, which have some properties in common, are included in a set containing the configuration objects, then it is necessary to "compute" the values of these properties in some way. The computation is done by the addition operator. If the addition operator is not defined for the given value constellation, these two objects may not both occur in the set of configuration objects. For example, the definition of $a_{capacity}$ does not allow two 40 Megabyte harddisks.

According to our definition, a configuration must specify 1. which objects are parts of the system to be configured, and 2. the entire functionality of the system.

Therefore, a *configuration* is a pair $C = \langle I, Q \rangle$, where I is a set of *items* of the form (k, o) and Q is a set of *qualities* of the form (f, x). An item (k, o) means that object $o \in O$ is used k times in the configured system. A quality (f, x) means that the configured system has the functionality f with value x.

Although qualities and properties are syntactically equal, we distinguish between them since a property is the feature of an object, while a quality (f, x) is the result of the composition of several objects having the functionality f in their property sets.

Based on the above definition of composition, we are now ready to formally introduce the notion *configuration*.

Definition: Let $\Pi = \langle O, F, V, P, A, T, D \rangle$ be a configuration problem. A *configuration* C is inductively defined as follows:

1. $C = \langle \emptyset, \emptyset \rangle$ is a configuration.

2. If $C = \langle I, Q \rangle$ is a configuration and o is an object of O, then $C' = \langle I', Q' \rangle$ is a configuration if the following conditions hold:

    (i) For every $(f, x) \in p_o$ and for every $(g, y) \in Q$, the composition $\varphi((f, x), (g, y))$ is defined or $Q = \emptyset$.

    (ii)
    $$I' = \begin{cases} I \setminus \{(k, o)\} \cup \{(k+1, o)\}, & \exists (k, o) \in I; \\ I \cup \{(1, o)\}, & \text{otherwise.} \end{cases}$$

    (iii)
    $$Q' = \begin{cases} \varphi((f, x), (g, y)), & Q \neq \emptyset; \\ p_o, & Q = \emptyset. \end{cases}$$

3. Nothing else is a configuration.

Remarks: Condition (i) guarantees that only those objects $o \in O$ are added to a given configuration C if all object's properties p_o can be combined with all qualities of C. Condition (ii) specifies how one new object can be added to a given set of items I. Condition (iii) specifies how a new set of qualities can be constructed if a new object is added to the configuration.

Next we give a precise definition of the notion *solution* of a configuration problem.

Definition: A configuration $C = \langle I, Q \rangle$ is a *solution* of a configuration problem $\Pi = \langle O, F, V, P, A, T, D \rangle$ if and only if for each demand $d = (f, x) \in D$ there exists a quality $q = (g, y) \in Q$ such that $f = g$ and $t_f(x, y) = \text{TRUE}$. The set $S(\Pi) = \{C \mid C \text{ is a solution of } \Pi\}$ is called the *solution space* of Π.

Remarks: The above condition guarantees that all demands are fulfilled. Generally there exists more than one solution of a configuration problem Π. Sometimes $S(\Pi)$ is called the "space of variants."

General configuration problems

Based on the above definitions, the following problems may be stated:

Problem CONF
Given: A configuration problem Π.
Question: Does there exist a solution of Π ?

Problem FINDCONF
Given: A configuration problem Π.
Task: Find a solution of Π, if one exists.

Problem COSTCONF
Given: A configuration problem Π, a cost function $c : O \rightarrow \mathbb{Q}$, and maximum cost $c^* \in \mathbb{Q}$.
Question: Does there exist a solution $C = \langle I, Q \rangle$ of Π such that $\sum_{(k,o) \in I} k\, c(o) \leq c^*$?

Note that each of the above problems is essentially a combinatorial problem.

3 Model M2

We will now extend our model M1 in that way that we introduce *rules*. These rules may be interpreted as installation restrictions. For example, one would like to formulate the rule "If harddisk A is used, then either controller B or controller C must be used."

Definition: 1. Let O be a set of objects and $N = \{1, \ldots, k\}$. Let $\Gamma(N, O) = \{[n, o] \mid n \in N, o \in O\}$ denote the set of Boolean variables over N and O. A *configuration restriction rule* r is an implication $[n, o] \rightarrow \psi$, where $[n, o] \in \Gamma(N, O)$ and ψ is a logical formula over $\Gamma(N, O)$ using parentheses, '$\neg$', '$\wedge$', and '$\vee$' in the standard way. A *rule set* R is a finite set of configuration restriction rules over $\Gamma(N, O)$.

2. Let O be given as in 1. Let $C = \langle I, Q \rangle$ be a configuration, where $I \subseteq N \times O$. A *configuration assignment* α_I is a function $\alpha_I : \Gamma(N, O) \rightarrow \{\text{TRUE}, \text{FALSE}\}$ such that for every $[n, o] \in \Gamma(N, O)$:

$$\alpha_I([n, o]) = \begin{cases} \text{TRUE}, & \text{if } (n, o) \in I; \\ \text{FALSE}, & \text{otherwise.} \end{cases}$$

3. A configuration $C = \langle I, Q \rangle$ is called *satisfying for a rule set R* if and only if every rule $r \in R$ is true under the assignment α_I using the known semantics of propositional logic.

Remarks: The semantics of a restriction rule is perhaps best explained by the following example: Let $r = [1, A] \rightarrow ([2, B] \wedge \neg[1, C]) \vee [3, D]$. The meaning of r is: "If a configuration contains exactly one object A, then the configuration must contain either two B's and not one C or three D's.

Definition: A configuration problem under model M2 is a tuple $\Pi_R = \langle O, F, V, P, A, T, D, N, R \rangle$ where all elements but N and R are defined as in model M1, and R is a set of configuration restriction rules over $\Gamma(N, O)$. A configuration $C = \langle I, Q \rangle$ is a solution of Π_R if and only if for every demand $(f, x) \in D$ there exists a quality $(g, y) \in Q$ such that $f = g$ and $t_f(x, y) = \text{TRUE}$ and C is satisfying for the rule set R.

The above problems CONF, etc. can be formulated in a similar way for model M2.

4 Is model M2 more powerful than model M1?

Model M1 is suitable to model a *resource-oriented* configuration problem, while model M2 is suitable to model a *skeleton-oriented* configuration problem. Many systems are built upon the skeleton model. If Π is a problem under model M2, then the skeleton of the configured system can be derived from the rules of Π. The skeleton is the digraph $G = (V, E)$ with $V = O$ and $(o_i, o_j) \in E$ if there is a rule which contains o_i in its left-hand side and o_j in its right-hand side. In typical applications, the digraph is a tree, and a configuration problem is then solved by a top-down strategy.

The model M2, which additionally contains a mechanism to express structural knowledge, seems to be more powerful than the pure resource-oriented model. However, as the following results shows, this is not the case. This is quite surprising since one expects that the rule language enables us to formulate more sophisticated configuration problems.

A central theorem of this work is the following.

Theorem A: Let Π be any instance of problem CONF under model M1 (M2). Then there exists an equivalent instance Π' of problem CONF under model M2 (M1) which can be obtained in polynomial time in the size of Π.

Corollary: Theorem A is also valid for the problems FINDCONF and COSTCONF.

We only prove the theorem, the corollary then follows immediately from the proof. (Readers not interested in the proof may continue with the example of the next section.)

Proof: Part I: Let $\Pi = \langle O, F, V, P, A, T, D \rangle$ be an instance of problem CONF under model M1. Trivially, let $R := \emptyset$, $N := \{1\}$ and let $\Pi' := \langle O, F, V, P, A, T, D, N, R \rangle$.

Part II: Let $\Pi = \langle O, F, V, P, A, T, D, N, R \rangle$ with $N = \{1, \ldots, k\}$ be an instance of problem CONF under model M2. We have to show that there exists an instance Π' of problem CONF under model M1 such that Π has a solution if and only if Π' has a solution.

We construct Π' as follows. The basic idea of the proof is that the rules are replaced by new functionalities and new tests whose behavior is equivalent to these rules.

First, R is transformed into a logically equivalent set $\tilde{R}$ which contains only 3CNF formulas (i.e., propositional formulas in conjunctive normal form where each clause has at most 3 literals).

This transformation is performed in two steps. First, a transformation technique due to Tseitin [1983] is used to transform each rule into a logically equivalent propositional formula in conjunctive normal form. This step requires the introduction of new variables, $\bar{\Gamma} = \{[1, \bar{o}_1], \ldots, [1, \bar{o}_T]\}$. Second, every formula obtained from step one is transformed into an equivalent 3CNF formula by introducing the new variables $\hat{\Gamma} = \{[1, \hat{o}_1], \ldots, [1, \hat{o}_S]\}$. Note that both transformations can be made in quadratic time and linear space. Let $\Gamma = \Gamma(N, O) \cup \bar{\Gamma} \cup \hat{\Gamma}$.

Thus, every rule $r \in R$ is transformed into a set $3\mathrm{CNF}(r) = \{r_1, \ldots, r_S\}$ such that r is satisfiable if and only if every formula $r_i \in 3\mathrm{CNF}(r)$ is satisfiable. Let $\tilde{R} = \bigcup_{r \in R} 3\mathrm{CNF}(r)$.

The introduction of the new variables implies that the new object set O' is defined as $O' := O \cup \bar{O} \cup \hat{O}$, with $\bar{O} = \{\bar{o}_1, \ldots, \bar{o}_T\}$ and $\hat{O} = \{\hat{o}_1, \ldots, \hat{o}_S\}$.

1. For every $r \in \tilde{R}$ we construct a new functionality g_r whose values are sets(!). For each $o \in O'$ which occurs in a rule $r \in \tilde{R}$, we define the property set of o with respect to g_r as $g_r(o) := \{(1, o)\}$.

2. We define the following "union" of an item set X and a singleton $\{(1, o)\}$, $o \in O'$, as follows:

$$X \uplus \{(1, o)\} = \begin{cases} X \setminus \{(n, o)\} \cup \{(n+1, o)\}, & \text{if } o \in O \text{ and } o \text{ occurs in } X \text{ and } n \leq k; \\ X \setminus \{(n, o)\} \cup \{k+1, o)\}, & \text{if } o \in O \text{ and } o \text{ occurs in } X \text{ and } n > k; \\ X, & \text{if } o \in \bar{O} \cup \hat{O} \text{ and } o \text{ occurs in } X; \\ X \cup \{(1, o)\}, & \text{otherwise.} \end{cases}$$

This "union" function will subsequently be used to construct the value sets of the new functionalities.

3. The value set v_{g_r} is inductively defined as follows. Note that Δ is a help variable.

(i) If o occurs in r, then $g_r(o) \in \Delta$.

(ii) If $X \in \Delta$ and o occurs in r, then $X \uplus \{(1, o)\} \in \Delta$.

(iii) Nothing else is in Δ.

(iv) $v_{g_r} := \Delta \cup \{r\}$.

Note that v_{g_r} contains both sets of number/object pairs and the rule r itself. Also note that the computation of v_{g_r} can be made in a finite number of steps since $k+1$ bounds the number n that can occur in a pair (n, o) which itself occurs in a set $X \in v_{g_r}$.

4. As a demand d_r for r we define $d_r := (g_r, r)$.

5. For g_r we define the test t_{g_r} as follows, where $t_{g_r}(X, Y)$ is only defined for $X = r$ and $Y \in v_{g_r} \setminus \{r\}$:

$$t_{g_r}(r, Y) = \begin{cases} \text{TRUE}, & \text{if } r \text{ is true under } \alpha_Y; \\ \text{FALSE}, & \text{otherwise;} \end{cases}$$

where α_Y is restricted to the variable set $\Gamma_r = \{[n, o] \mid n \leq k+1, \text{ and } o \text{ occurs in } r\}$. Note that other variables than those in Γ_r cannot occur because $k+1$ bounds the number n.

6. Next we define the addition operator a_{g_r} for functionality g_r as follows: $a_{g_r}(X, Y)$ is only defined for $X \in v_{g_r} \setminus \{r\}$ and $Y \in \{g_r(o) | o \in r\}$: $a_{g_r}(X, Y) := X \uplus Y$.

7. Henceforth, let $\rho(o) = \{(g_r, g_r(o)) \mid r \in \{r \in \tilde{R} | o \text{ occurs in r}\}$

8. The elements of $\Pi' := \langle O', F', V', P', A', T', D' \rangle$ are now defined as follows:

$$\begin{aligned} O' &:= O \cup \bar{O} \cup \hat{O}, \\ F' &:= F \cup F_R, \text{ where } F_R := \{g_r \mid r \in \tilde{R}\}, \\ V' &:= V \cup V_R, \text{ where } V_R := \{v_{g_r} \mid r \in \tilde{R}\}, \\ P' &:= \{p_o \cup \rho(o) | o \in O\} \cup \{\rho(o) | o \in \bar{O} \cup \hat{O}\} \\ A' &:= A \cup A_R, \text{ where } A_R := \{a_{g_r} \mid r \in \tilde{R}\}, \\ T' &:= T \cup T_R, \text{ where } T_R := \{t_{g_r} \mid r \in \tilde{R}\}, \\ D' &:= D \cup D_R, \text{ where } D_R := \{(g_r, r) \mid r \in \tilde{R}\}. \end{aligned}$$

Case A. We have to show: If $C = \langle I, Q \rangle$ is a solution of Π, then there exists a solution $C' = \langle I', Q' \rangle$ of Π'. We show that there exist an item set ΔI and a quality set ΔQ such that $I' = I \cup \Delta I, Q' = Q \cup \Delta Q$, and $C' = \langle I \cup \Delta I, Q \cup \Delta Q \rangle$ is a solution of Π'. Due to this construction of I' and since $D' = D \cup D_R$, we need only consider the "difference" demands D_R. (The original demands D are satisfied by I.)

We have to construct a ΔI in such a way that I' induces a quality set ΔQ with the following characteristic: For each $d = (g_r, r) \in D_R$ there exists a property $(g_r, Y) \in \Delta Q$ with $t_{g_r}(r, Y) = \text{TRUE}$.

Let $d = (g_r, r)$ be any demand of D_R where $r \in \tilde{R}$ is a 3CNF rule of the form $r = l_1 \vee l_2 \vee l_3$ with $l_i \in \{[n, o] \,|\, [n, o] \in \Gamma'\} \cup \{\neg[n, o] \,|\, [n, o] \in \Gamma'\}$. Note that r is satisfied if some l_i is satisfied. Since C is a solution of Π, it follows that all rules $r \in R$ are satisfied. Hence, all 3CNF rules in $\tilde{R}$ are satisfiable by some truth assignment $\alpha_{I'}$. Note that $\alpha_I \subseteq \alpha'$; this guarantees that an object $o \in O$ occurs with frequency n in I if and only if object o occurs with frequency n in I'. Without loss of generality, we can assume that $l_1 (= [n_1, o_1])$ is satisfied under $\alpha_{I'}$.

Case A1: Let $o_1 \in O$. If $l_1 = [n_1, o_1]$ then $\alpha_{I'}([n_1, o_1]) = \text{TRUE}$, hence (n_1, o_1) must occur in I. The definition of a_{g_r} (cf. the "$\uplus$-operator") guarantees that a $Y \in v_{g_r}$ with $(n_1, o_1) \in Y$ is inevitably constructed as the quality value of g_r. If $l_1 = \neg[n_1, o_1]$ then $\alpha_{I'}([n_1, o_1]) = \text{FALSE}$ (hence $(n_1, o_1) \notin I$). Now, either $(m_1, o_1) \in I$ with $m_1 < n_1$ or $m_1 > n_1$, then $(m_1, o_1) \in Y$, or o_1 does not occur in I at all and $(m_1, o_1) \notin Y$. As before, an appropriate $Y \in v_{g_r}$ is constructed as the quality value of g_r.

Case A2: Let $o_1 \in \bar{O} \cup \hat{O}$. If $l_1 = [n_1, o_1]$ then $\alpha_{I'}([1, o_1]) = \text{TRUE}$ and we put $(1, o_1)$ in ΔI. The quality value Y of g_r will contain $(1, o_1)$. If $l_1 = \neg[1, o_1]$ then $\alpha_{I'}([1, o_1]) = \text{FALSE}$ and o_1 is not allowed to be in ΔI. Hence, the quality value Y of g_r cannot contain $(1, o_1)$. So, ΔI is defined as the collection of all tuples $(1, o_1)$ found in case A2. Furthermore, let the g_r (with $r \in \tilde{R}$) and their corresponding Y form the set ΔQ. As seen above, with these definitions of ΔQ and ΔI it is guaranteed that for each $d = (g_r, r) \in D_R$ there exists the quality $(d_r, Y) \in \Delta Q$ such that $t_{g_r}(r, Y) = \text{TRUE}$.

Case B. We have to show: If $C' = \langle I', Q' \rangle$ is a solution of Π', then there exists a solution $C = \langle I, Q \rangle$ of Π. Since $C' = \langle I', Q' \rangle$ is a solution of Π', all demands in $D' = D \cup D_R$ are satisfied. Let $I = \{[n, o] \,|\, o \text{ occurs in } O\}$ and let $\Delta Q = \{(f, x) \in Q \,|\, f \in p_o, o \in \bar{O} \cup \hat{O}\}$.

1. Clearly, $C = \langle I, Q' \setminus \Delta Q \rangle$ satisfies all demands $d \in D$ since objects which occur in $I' \setminus I$ have no properties for which a demand $d \in D$ exists.

2. To see that $C = \langle I, Q' \setminus \Delta Q \rangle$ satisfies all rules $r \in R$, one need only consider the above transformation which guarantees that α_I is a satisfying truth assignment since $\alpha_I \subseteq \alpha_{I'}$ and $R \iff \tilde{R}$. $\diamond$

5 Skeleton-oriented configuration

In this section we give an example of a skeleton-oriented configuration problem formulated under model M2. Although a configuration problem and its solution strongly depend on particular aspects of the application, this example gives an idea how a hierarchical organized configuration problem can be described.

One characteristic of the skeleton-oriented configuration is that the solution space can be described by a hierarchical graph with two kinds of nodes: AND-nodes and OR-nodes (cf. Puppe [1990]). An AND-node indicates that each direct successor of this node must be selected in the configuration process (more general: to solve the whole problem, each subproblem has to be solved); an OR-node describes mutually exclusive alternatives. The skeleton-oriented configuration approach is appropriate, if we want to configure a system which has always the same basic structure.

In the following example, the task is to configure a tower which has always three planes: An A-plane, a B-plane and a C-plane. For each plane there exists a particular kind of building blocks (A-blocks, B-blocks and C-blocks). Furthermore, the building blocks of the tower have to fulfill the following restrictions: For both plane A and plane B exactly one block of the appropriate kind must be selected. Plane C has to be constructed with at least one C-block where C3 cannot be combined with any of the other C-blocks. If block C2 is used once, block B1 is not allowed to occur once in a configuration. The goal is to build a tower with a given height and minimum cost. The following figure describes the building blocks which can be used to construct a tower.

block	A1	A2	B1	B2	C1	C2	C3		C
height	1	2	1	3	1	2	4		B
cost	2	4	3	5	2	3	6		A

To describe this problem as a hierarchical configuration problem, we introduce particular "dummy blocks" S, A, B, C which have no properties. With the new building blocks, the configuration restrictions are described by the following rules:

[1,S] → [1,A] ∧ [1,B] ∧ [1,C] , [1,A] → [1,A1] ∨ [1,A2] , [1,B] → [1,B1] ∨ [1,B2] , [1,C] → [1,D] ∨ [1,C3] , [1,D] → [1,C1] ∨ [2,C1] ∨ [1,C2] ∨ [2,C2] , [1,C2] → ¬ [1,B1]

The set of objects is $O = \{S,A,B,C,D,A1,A2,B1,B2,C1,C2,C3\}$, and the set of functionalities is $F = \{\text{height}\}$

With the specification of the transformation of model M2 into model M1 (see Section 4), we are now able to reformulate this configuration problem as a problem which solely bases on functionalities and their computation. We do not want to perform this transformation explicitly since we gave examples for single transformation steps in Section 4.

In the reformulated problem, the dependencies between the objects must be derived from their properties. There is an edge between those objects which share at least one functionality.

The following figure shows the original configuration problem and its concrete reformulation. It illustrates in which way the transformation of M1 into M2 comes to effect. The transformation of the above rules yields eight new functionalities $g_1, \ldots, g_8$. In the picture, both functionalities and configuration objects are vertices; an edge (o_i, g_j) indicates that the object o_i has the functionality g_j in its property set.

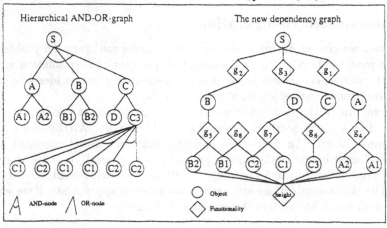

6 A complexity consideration

We present in this section a result regarding the computational complexity of CONF. Note that all other problems are at least as hard as CONF.

Theorem B: Problem CONF is NP-complete.

Lemma C: Problem CONF with restriction rules ($CONF_R$) is NP-complete.

The theorem and the lemma are proven in [Najmann & Stein 1992]. The basic idea of the proof is to transform 3SAT (which is NP-complete, [Cook, 1971]) to $CONF_R$ and to apply the equivalence results of section 4.

Summary: We have developed two models of configuration, M1 and M2. Model M1 allows to formulate typical ressource-oriented configuration problems, while model M2 allows to formulate skeleton-oriented configuration problems. Model M2 was obtained by augmenting model M1 by a restriction rule language. Although model M2 seems to be more powerful than M1, it was shown that both models can be transformed into each other in polynomial time. That means that there is principally no difference between the classical configuration approach and the ressource-oriented approach. Under both models, we have formulated a number of typical configuration problems like the problem of finding a cost-minimal configuration. We have also noted that the fundamental problem, namely to decide whether there exists a configuration for a given problem specification, is NP-complete.

References

Brown, D. C.; Chandrasekaran, B. [1989]   *Design Problem Solving*, Pitman.

Cook, S. [1971]   "The complexity of theorem-proving procedures", *Proc. 3rd Ann. ACM Symp. on Theory of Computing, Association for Computing Machinery*, New York, pp. 151–158

Cunis et al. [1991]   *Das PLAKON-Buch – Ein Expertensystemkern für Planungs- und Konfigurierungsaufgaben in technischen Domänen*, Springer Verlag, Berlin.

Doerner, H. [1991]   "Ein Modell des Konfigurierens" *Beiträge zum 5. Workshop "Planen und Konfigurieren"*, LKI-M-1/91

Heinrich, M. [1991]   "Ressourcenorientierte Modellierung als Basis modularer technischer Systeme," *Beiträge zum 5. Workshop "Planen und Konfigurieren"*, LKI-M-1/91

Najmann, O., Stein, B. [1992]   "Modeling resource-oriented configuration problems." *Internal report, University of Paderborn*.

Puppe, F. [1990]   *Problemlösungsmethoden in Expertensystemen*, Springer Verlag.

Stein, B.; Weiner, J. [1990]   *MOKON*, Interner Report, Universität-Gesamthochschule-Duisburg, SM-DU-178

Tank et al. [1989]   "AMOR: Eine Beschreibungssprache für technische Systeme zur Unterstützung der Wissensakquisition für Konfigurationsprobleme", *Beiträge zum 3. Workshop "Planen und Konfigurieren"*, Arbeitspapiere der GMD 388

Tseitin, G . [1983]   "On the complexity of derivations in propositonal calculus", *Automation of Reasoning 2: Classical Papers on Computational Logic*, pp. 466-483. Springer Verlag, Berlin.

Stock Market Prediction with Backpropagation Networks

Bernd Freisleben

Department of Computer Science, University of Darmstadt
Alexanderstr. 10, D–6100 Darmstadt, Germany

Abstract. In this paper we evaluate the performance of backpropagation neural networks applied to the problem of predicting stock market prices. The neural networks are trained to approximate the mathematical function generating the semi–chaotic timeseries which represents the history of stock market prices in order to predict the values for the future. In contrast to previous investigations, the training data used in our experiments is not exclusively based on stock market prices, but also incorporates a variety of other economical factors. The prediction quality obtained is illustrated by presenting several simulation results.

1 Introduction

The stock market is affected by a large number of highly interrelated economical, political and psychological factors which interact with each other in a complex fashion. Since most of these relationships seem to be probabilistic and therefore cannot be expressed as deterministic rules, financial analysis is one of the most well suited and promising applications of artificial neural networks. Several proposals have been made to use neural network models for prediction and forecasting problems in the financial area, such as locating sources of forecast uncertainty in a recurrent gas market model [11], corporate bond rating [1], mortgage delinquency prediction [3], chaotic timeseries prediction [10], prediction of IBM daily stock prices [12], prediction of three selected German stock prices [9] and prediction of the weekly Standard & Poor 500 index [5]. In some of these proposals, the neural networks performed better than regression techniques [1, 9] or as good as the Box–Jenkins technique [10], while in others the results were disappointing [3, 12].

In this paper, we apply several variants of the backpropagation neural network to the problem of predicting the weekly price of the FAZ–Index, which is calculated on the basis of 100 major German stocks and may be regarded as one of the German equivalents of the Dow–Jones–Index in the USA. The basic idea is to let the network learn an approximation of the mapping between the input and output data in order to discover the implicit rules governing the price movement of the FAZ–Index. The trained network is then used to predict the weekly closing prices for the future.

Our work differs from previous stock price predictions with neural techniques [5, 9, 12] in the data presented to the networks. While in other approaches the input data was exclusively based on stock prices, we also consider other important economical factors, namely a subset of those considered in the *fundamental* and *technical* analysis methods used by human analysts to make their investment decisions. Thus, although we regard a neural network for stock price prediction primarily as a technical analysis tool, elements of the fundamental and the technical analysis are combined in our approach. Similar to a human analyst who is probably more successful if he or she is aware of both methods, we expect the networks to produce high quality predictions in the combined approach. Several simulation results will be presented in order to see if our expectations will be fulfilled.

The paper is organized as follows. In section 2 the variants of the backpropagation network used in the experiments are described and the way in which the stock market data is processed by the networks is presented. The simulation results are described in section 3. Section 4 concludes the paper and discusses areas for future research.

2 Applying Backpropagation to Stock Price Prediction

The backpropagation algorithm [8] has emerged as one of the most widely used learning procedures for multi–layer networks of neuron–like units. The typical back-propagation network always has an input layer, an output layer and at least one hidden layer of units in between, with full feed–forward connections between the layers. The algorithm gives a prescription for modifying the weights of the connec-tions to learn a training set of input–output pairs. It is an example of a *supervised* learning procedure [4] in the sense that it attempts to minimize the (quadratic) difference between the desired and the produced outputs.

There are several variations on the standard backpropagation algorithm which are aimed at speeding up its relatively slow convergence, avoiding local minima or im-proving its generalization ability [4]. The ones we investigated are: a) the use of different activation functions other than the usual sigmoid function [5]; b) the ad-dition of a small positive offset to the derivative of the sigmoid function to avoid saturation at the extremes [2]; and c) the use of a momentum term in the equation for the weight changes [7].

In previous neural approaches to the stock price prediction problem, the training data set typically consisted of input vectors with N components, where the N com-ponents represented N successive prices of a particular stock. The desired output "vector" had exactly 1 component, namely the price of the stock at time $N + 1$. Thus, the training set was processed in a sliding window fashion, with N being the size of the window (typically $N = 10$). The network was supposed to learn the map-ping between the pairs of input/output vectors, and the recall phase started with the current stock price (the last known desired output) together with the last $N - 1$ stock prices in the training set in order to predict the first unknown price at time $N + 1$. The real value of the stock at time $N + 1$ was then used in conjunction with the last $N - 1$ known prices to predict the value for time $N + 2$. This was repeated for the prediction period desired, i.e. the size of the test set. The prediction quality was measured by comparing the network outputs to the known real values of the test set.

This procedure has essentially been adopted in one of our experiments, but in con-trast to the proposals above we also employed another technique which is different in terms of the input presented to the networks. Instead of merely feeding the network with input vectors representing a sequence of stock prices, the basic idea is to use vectors consisting of only one stock price and a number of other factors at time N in order to predict the stock price at time $N + 1$. These factors are a subset of the criteria considered in both the fundamental and technical analysis methods for assessing the market situation. Since they directly affect the stock market, it is hoped that the networks discover the relationships between them in order to improve the prediction quality.

2.1 Input/Output Data

Our study is aimed at predicting the weekly closing price of the FAZ–Index, one of the German equivalents of the American Dow–Jones–Index. The data available consists of the values between January, 23, 1987 and December, 22, 1989. The first 100 weeks were used as the training set and the remaining 53 weeks as the test set. In addition to the FAZ–Index, the following fundamental factors, collected during the same time period, have been selected:

- order index
  The index of received orders, published on a monthly basis, documents the demand for a company's products and therefore reflects the expected profit. Increasing values of the index usually lead to increasing stock prices and vice versa.

- US–Dollar
  From the viewpoint of the German market, the daily published US–Dollar exchange rate is very important for the German stock prices. A decreasing US–Dollar normally implies decreasing stock prices, because the exported products get more expensive and thus will not be ordered as much as when the US–Dollar is strong and vice versa.

- bond market index
  The bond market index, published daily, expresses the average yield to be obtained with bonds. If it is high, then the bond market is attractive for investments, leading to decreasing stock prices and vice versa.

The technical factors, also calculated during the same time period, are different *moving averages*. They are used to smooth prices and reveal their underlying direction or trend in order to avoid unprofitable investment decisions based on "whipsawing" prices or extreme short–term movements. A N–day moving average is equivalent to the simple arithmetic mean of the last N stock prices. Several moving averages are often used in conjunction and their crossovers usually determine buy or sell signals. The moving averages considered are:

- 5–day moving average

- 10–day moving average

- 90–day moving average

Since the above factors have quite different ranges of values and are collected at different times, it is necessary to convert them to a format suitable for presentation to the backpropagation networks. Considering that the activation functions of the output units produce values between 0 and 1 which represent the predicted stock prices, all the data has been scaled to the interval [0.1, 0.9]. For data available on a daily basis, only the weekly closing values were used, whereas for data published monthly, the monthly value was used for all weeks of the corresponding month.

In order to establish a relationship to the data in the past, we also calculated for each value in any data set the absolute difference to the previous value and expressed the sign of the difference (the trend) by different encodings. For example, a positive difference is coded as 0.8, a negative difference as 0.2 and no difference as 0.0. This type of coding is necessary because we restricted the input values presented to the backpropagation networks to the range between 0 and 1.

Consequently, an input vector X has N components $(x_1, \ldots x_N)$, where

x_1: price of FAZ–Index
x_2: absolute difference to previous price of FAZ–Index
x_3: trend of FAZ–Index price movement
x_4: value of factor 1
x_5: absolute difference to previous value of factor 1
x_6: trend of factor 1
...
x_{N-2}: value of factor $N/3 - 1$
x_{N-1}: absolute difference to previous value of factor $N/3 - 1$
x_N: trend of factor $N/3 - 1$

A large number of data sets with input vectors in the format described above has been used in the different experiments. Figure 1 shows the data sets we refer to in the rest of this paper.

	Data set		
	1	2	3
FAZ–Index (+ abs. diff. + trend(0.8,0.2,0.0)	x	x	
5–day mov. avg. (+ abs. diff. + trend(0.8,0.2,0.0)	x	x	
10–day mov.avg. (+ abs. diff. + trend(0.8,0.2,0.0)	x	x	
90–day mov. avg. (+ abs. diff. + trend(0.2,0.1,0.0)		x	
bond market index (+ abs. diff. + trend(0.8,0.2,0.0)	x	x	
order index (+ abs. diff. + trend(0.4,0.2,0.0)		x	
US–Dollar (+ abs. diff. + trend(0.2,0.1,0.0)		x	
10 successive FAZ–Index prices			x

Fig. 1. Data Sets

2.2 Network Architectures

The vectors contained in each data set determine the number of input units of the backpropagation network which processes the data set. Furthermore, the number of

output units is 1 in our application, because the networks should predict exactly 1
future value for each input vector. The other architectural features, such as the number of hidden layers, the number of hidden units per layer, the activation functions
of each unit, the learning parameter α and the momentum parameter are highly
application dependent. Since no sound theoretical foundation for determining these
parameters is available yet, it has become common practice to determine them empirically in a trial and error fashion. We have tested a large number of architectures,
and the ones relevant to the simulation results presented in the present paper are
shown in figure 2.

	Network			
	1	2	3	4
#input units	12	21	10	10
#hidden layers	1	1	2	1
#hidden units	11	20	(1)18/(2)9	9
activation function	sig+off	sig+off	(1)9 sig, 9 sin/(2) sig	sig+off
learning parameter α	0.9	0.7	0.9	0.9
momentum	0.7	0.7	0.7	0.7

Fig. 2. Network Architectures

The backpropagation networks have been implemented in "C" on a SUN Sparcstation. The details of the implementation are described in [6].

3 Performance

After having initialized all weights of a network to small random values between -0.1
and 0.1, the learning set was presented to the networks for some number of learning
cycles, where one cycle is equivalent to one presentation of the whole learning set.
After training, the test set was fed through the network in the manner described
in section 2. The network outputs produced in both the learning and the recall
mode were then compared to the desired outputs, the real values of the FAZ–Index.
The minimum error, the maximum error and the average error, computed as the
absolute difference between the desired and the produced outputs, are used to assess
the quality of the network outputs. In addition, the trend between two successive
values produced was compared to the corresponding pair of desired output values.
If the two trends follow the same direction or if the difference between the produced
and the desired output value is smaller than the previous difference, a count of the
number of correct trends is incremented.

3.1 Performance Results for Data Set 1

In the first experiment, we examined the network behaviour when data set 1 was
used. We performed several simulations in order to determine the most suitable
network parameters. In all simulations where one hidden layer was present, it was
observed that the average error was lowest when the following holds for the number
of hidden units:

$$\#hidden\,units = (k*n) - 1 \quad \text{where} \quad k \geq 1, k \in \mathbf{N}, \quad n = \#input\,units$$

The best results in terms of the average error were obtained when network 1 processed data set 1. The results are shown in figure 3. In the charts, the x-axis is a count of the weeks and the y-axis represents the scaled values of the corresponding FAZ–Index prices (curve) and the predicted prices ("+").

learning set				
#cycles	min. error	max. error	avg. error	trend correct (abs.(out of), %)
3000	0.0004	0.0584	0.0160	90(99), 90
test set				
	0.0000	0.1545	0.0519	38(51), 74

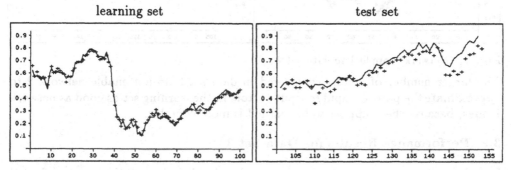

Fig. 3. Network 1 applied to data set 1

We also tested various networks with two hidden layers consisting of different numbers of hidden units in each layer. Although the error values for the training set were nearly as good as for the network with 1 hidden layer, the results for the test set were worse. This also holds for a network with two hidden layers and different activation functions for the units in the hidden layers, such as equipping some units with the sigmoid function and the others with the sine function.

In two further experiments, network 1 was applied to data set 1 with reduced size (learning set: weeks 51–100 (1–70), test set: weeks 100–153 (70–153)), but again the prediction quality for the test set was not as good as for the configuration shown in figure 3.

3.2 Performance Results for Data Set 2

Data set 2, consisting of input vectors with 21 components, was considered next. Similar to the experiment with data set 1, a network with one hidden layer, where the number of hidden units is a multiple of the number of inputs minus 1, gave the best results. The use of different activation functions did not improve the prediction quality. Figure 4 shows the performance of network 2 applied to data set 2.

learning set				
#cycles	min. error	max. error	avg. error	trend correct (abs.(out of), %)
3000	0.0001	0.0656	0.0208	78(99), 78
test set				
	0.0007	0.1722	0.0482	39(51), 76

learning set   test set

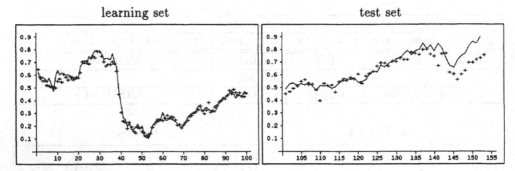

Fig. 4. Network 2 applied to data set 2

The larger number of factors considered in data set 2 do not enable network 2 to approximate the pairs of input/output vectors in the learning set as good as network 1 does, because the mapping to be learned is more complex.

3.3 Performance Results for Data Set 3

In order to compare the approach with purely price–based predictions, we conducted experiments where the data set consisted only of 10 successive prices of the FAZ–index. We first applied network 3 to data set 3, which is exactly the network used in [5]. The results are shown in figure 5.

learning set				
#cycles	min. error	max. error	avg. error	trend correct (abs.(out of), %)
3000	0.0010	0.2070	0.0399	77(90), 85
test set				
	0.0004	0.3337	0.0706	34(51), 66

learning set   test set

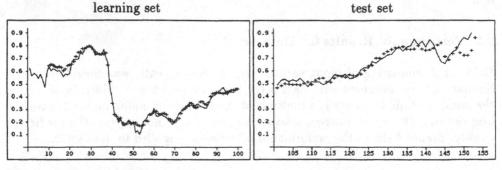

Fig. 5. Network 3 applied to data set 3

In terms of the average error, the results obtained for both the learning and the test set are are not as good as in the previous experiments. The additional knowledge provided within data set 1 and data set 2 seems to be beneficial for the prediction quality.

The previous experiments have indicated that a network with one hidden layer seems superior to a network with more than one layer. In order to see if this is also true for a purely price–based data set, we applied network 4 to data set 3. The results are shown in figure 6.

		learning set		
#cycles	min. error	max. error	avg. error	trend correct (abs.(out of), %)
3000	0.0004	0.1042	0.0327	73(90), 81
		test set		
	0.0001	0.1655	0.0485	37(51), 72

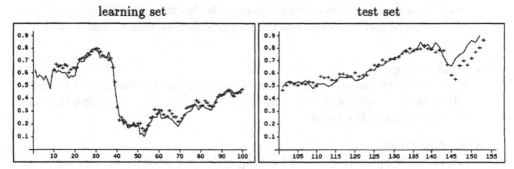

learning set   test set

Fig. 6. Network 4 applied to data set 3

The results confirm our expectations. The average error in the learning and the test set is smaller than that shown in figure 5.

3.4 Discussion

In this section we discuss the results obtained with the different network configurations and data sets.

The following observations have been made for the network configurations:

- number of hidden layers
  The prediction quality of networks with one hidden layer is generally better than that obtained with two hidden layers, and in all trials the best results were achieved when the number of units in the hidden layer is a multiple of the number of inputs minus 1.

- activation functions
  The use of the sigmoid function (plus offset) achieved the best results in the shortest time. The use of the sine function or a combination of sigmoid and sine is inferior to the sigmoid function.

- learning parameter
  If several fundamental factors are present in the input data, the quality of the results is improved by decreasing the learning parameter. In all experiments conducted, the best results were obtained by setting the learning parameter to values between 0.7 and 0.9.

- momentum
  A momentum parameter of 0.7 yields good results in all experiments.

The following observations have been made for the different data sets:

- training set
  The best results in discovering the relationships in the training set are achieved when the number of pairs (input vector, desired output) in the training set is as large as possible. However, if the components of the input vectors reflect contrary developments, it is difficult for the network to learn the mapping. Too many factors do not necessarily improve the learning capabilities of the network, but the presence of several fundamental factors leads to better prediction results.

- input coding
  If the trend of two successive values is included in the data set, it is easy to adjust the influence of the corresponding factor by simply adjusting the values for the coding of the trend.

- learning speed
  On the SUN Sparcstation, 100 learning cycles in a network with 10 inputs and 9 hidden units take approximately 20 seconds. If 29 hidden units are employed, about 60 seconds are required.

The prediction quality of neural approaches to the stock price prediction problem is generally better than the results obtained with conventional statistical techniques. As indicated in [6], regression methods applied to the timeseries used in the present paper predict the trend between two successive values in the test set in 51% to 62% of the cases correctly, whereas the range of correct trend predictions in the test set with the neural networks used lies between 66% and 76%.

4 Conclusions

In this paper we have studied the performance of several backpropagation neural networks applied to the problem of predicting prices of the FAZ stock market index. The neural networks were trained to approximate the mapping between the input and output data which represents the semi–chaotic timeseries of the FAZ–index price movement in order to predict the values for the future. In contrast to previous investigations, the training sets used in our experiments were not exclusively based on stock market prices, but also incorporated a variety of other economical factors. The prediction quality obtained was illustrated by presenting several simulation results. These results are better than those obtained by conventional statistical approaches. In future investigations we plan to confirm our results with other data sets and examine the influence of other economical factors on stock price movements.

References

[1] S. Dutta and S. Shekkar. Bond rating: A non–conservative application of neural networks. In *International Joint Conference on Neural Networks*, volume 2, pages 443–450, 1988.

[2] S.E. Fahlman. Fast-learning variations on back-propagation: An empirical study. In D. Touretzky, G. Hinton, and T. Sejnowski, editors, *Proceedings of the 1988 Connectionist Models Summer School*, pages 38–51, Pittsburg 1988, Morgan Kaufmann.

[3] S. Gosh, E.A. Collins, and C.L. Scofield. Prediction of mortgage loan performance with a multiple neural network learning system. In *Abstracts of the First Annual INNS Meeting*, volume 1, pages 439–440, 1988.

[4] J. Hertz, A. Krogh, and R.G. Palmer. *Introduction to the Theory of Neural Computation*. Addison–Wesley, Reading, Massachusetts, 1991.

[5] A. Lapedes and R. Farber. Nonlinear signal processing using neural networks: Prediction and system modelling. Technical Report LA–UR–87–2662, Los Alamos National Laboratory, Los Alamos, NM, 1987.

[6] G. Leja. Prediction of stock prices with neural networks (in German), Master's Thesis, Dept. of Computer Science, University of Darmstadt, 1991.

[7] D. Plaut, S. Nowlan, and G. Hinton. Experiments on learning by back propagation. Technical Report CMU–CS–86–126, Department of Computer Science, Carnegie Mellon University, Pittsburgh, PA, 1986.

[8] D.E. Rumelhart, G. Hinton and R.E. Williams. Learning internal representations by error propagation. In *Parallel Distributed Processing: Explorations in the Microstructures of Cognition*, Vol. 1, 318–362, MIT Press, 1986.

[9] E. Schöneburg. Stock price prediction using neural networks: An empirical test. *Neurocomputing*, 2, 1, 1991.

[10] R. Sharda and R.B. Patil. Neural networks as forecasting experts: An empirical test. In *International Joint Conference on Neural Networks*, volume 2, pages 491–494, Washington, D.C., 1990.

[11] P.J. Werbos. Generalization of backpropagation with application to a recurrent gas market model. *Neural Networks*, 1:339–356, 1988.

[12] H. White. Economic prediction using neural networks: The case of IBM daily stock returns. In *IEEE International Conference on Neural Networks*, volume 2, pages 451–458, San Diego, 1988.

Forecasting Time Series with Connectionist Nets: Applications in Statistics, Signal Processing and Economics

Claas de Groot and Diethelm Würtz

Interdisziplinäres Projektzentrum für Supercomputing, ETH-Zentrum and Institut für Theoretische Physik, ETH-Hönggerberg CH-8092 Zürich, Switzerland

E-mail: wuertz@ips.id.ethz.ch

1 Introduction

Connectionist networks of the feedforward type [1] have recently been shown to be universal function approximators [2]. The simplest structures for which these theorems hold are nets with a single hidden layer and nonlinear (sigmoid) transfer functions in this hidden layer only. This theoretical result is accompanied by numerical investigations which showed experimentally the capabilities of relatively simple connectionist networks to approximate nonlinear mappings [3]. These findings encourage the application of connectionist networks in the field of (nonlinear) *time series analysis*.

The standard method in linear time series analysis is the autoregressive moving average approach, called ARMA or Box-Jenkins method [4]. For these models we have a complete toolbox for *(i)* model identification and selection, *(ii)* model building and estimation and *(iii)* methods for diagnosis checks on its adequacy. The approach is well established and there are several textbooks available on the *theoretical* as well on the *applied* aspects [5]. However, there is no reason to assume a priori that a particular realization of a given time series was generated by a *linear* process. Therefore, during the last ten years many efforts have been undertaken to achieve more effective signal processing than within linear model building. Prominent examples are the *bilinear* models, the *threshold autoregressive* models, the *exponential autoregressive* models and the *general state dependent* models [6].

In this paper we follow the new and very promising concept of nonlinear function approximation using connectionist networks. Combining the techniques of statistical time series analysis with the ideas inherent in the concepts of connectionist networks we build a feedforward connectionist net approach for modelling and forecasting time series. Our methodology includes all stages of the traditional time series analysis mentioned above. Within this approach we present results for three applications, the first in the field of "statistics" (prediction of sunspots activity data), the second in the field of "signal processing" (noise reduction) and the third in the field of "economics" (currency exchange rate analysis). Our results show that the connectionist approach outperforms traditional methods or yields at least competitive results. For a more detailed description of the theoretical as well as computational aspects of this case study we refer to [7].

2 Statistics: Predictions of Sunspots Activity Data

The analysis of sunspots activity data has a long tradition in the statistical literature. The time series itself is nonlinear, nonstationary and non-Gaussian and thus serves as a benchmark problem in comparing and judging new statistical modelling and forecasting methods [8]. The earliest work in this field goes back to Yule [9] in the year 1922. Since then many other attempts have been undertaken to model the data by a linear ARMA process or closely related models [10]. With the introduction of the bilinear models (BL) [11] and the threshold autoregressive models (TAR) [12] and their application to the sunspots data a major step forward was done in nonlinear model building. Very recently Weigend, Huberman and Rumelhart [13] and independently ourselves [7] applied the concepts of connectionist network modelling to this time series and achieved a promising improvement in comparison to the standard methods.

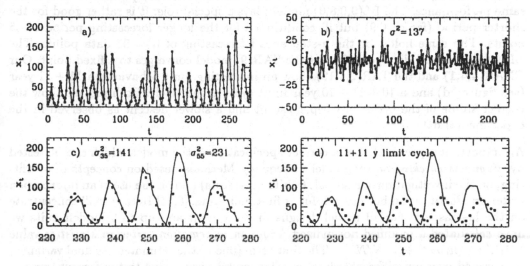

Figure 1: Sunspots activity data analysis. a) trainig set with 221 data points ranging from 1700–1920. b) residuals obtained from the CNAR(4,4,1) estimation. c) 1 step ahead forecasts for 35 and 55 years. d) 55 years ahead forecast (forecasting on the predicted data points).

Figure 1a shows the sunspots activity time series on the basis of 280 data points labeling the years from 1700 to 1979. In Figure 1b we have plotted the residuals of the estimated time series (data points 5 to 221) as obtained from training a feedforward connectionist network with four input, four hidden and one output unit, denoted as CNAR(4,4,1). We used for the transfer function a sigmoid $tanh$ at both hidden and output layers. The original data were shifted by a constant value of 100 and scaled down by a factor 200. The weights and threshold values are given in [7]. Training was achieved by optimizing the least square performance measure for the differences between the target pattern and the obtained pattern. The optimization was performed by the standard backpropagation algorithm or with more advanced optimization algorithms like quasi Newton methods or conjugate gradient algorithms. Usually we used the quasi Newton BFGS method, yielding a speedup in CPU time by a factor of about 100 compared to steepest descent methods like backpropagation. The residual variance of the estimated time series is

$\sigma^2 \approx 137$. The CNAR(4,4,1) outperforms the subset autoregressive model SAR(9) of Subba Rao and Gabr [11] by a factor ≈ 1.5 (ratio of residual variances) and the threshold autoregressive model TAR(2,4,12) of Tong and Lim [6,12] by a factor ≈ 1.1. Compared to the bilinear model BL(9,0,8,6) of Subba Rao and Gabr [6,11] and to the CNAR(12,3,1) of Weigend et al. [13] our CNAR(4,4,1) shows (for the training set) a smaller performance with factors ≈ 0.9 and ≈ 0.8, respectively. Figure 1c illustrates the 1 step ahead forecasts for the next 55 years. The residual variance of the forecasted points yields for the first 35 points (3 cycles) $\sigma^2_{35} \approx 141$, and including additional 20 points (yielding 5 cycles) $\sigma^2_{55} \approx 231$. The second result is of special interest, because the network has to predict an especially intensive cycle, with no "similar" data in the training set available. The forecasting is slightly better compared with the TAR(2,4,12) model and outperforms the simple SAR(9) model by a factor 1.5 and 1.2 for the forecasting period of 35 and 55 data points, respectively. In comparison to the CNAR(12,3,1) it shows almost the same performance. The BL(9,0,8,6) model plays a special role. It is rather good for the shorter period (factor 0.9) but ill conditioned for the longer forecasting period of 55 points. The same holds for the n-step ahead forecasting of $n = 55$ data points. The BL(9,0,8,6) model diverges, whereas the SAR(9) model converges to a fixed point. Our CNAR(4,4,1) and the TAR(2,4,12) show an interesting limit behaviour: a $11 + 11$ year (see figure 1d) and a $10 + 11 + 10$ year limit cycle, respectively. Both models show the correct shape of the cycles, a steeper ascent and a slower descent as observed in the experimental data.

An important aspect is that of a correct specification of the model, which can be tested by *diagnosis checking* on the model's adequacy. Methods based on concepts of overfitting, on the investigation of residuals (white noise tests) and on the use of an *information criterion statistics* can be very helpful as first indicators. [5,14] Here we will concentrate on the last aspect. For all model designs up to 6 input and up to 6 hidden units we have calculated a modified normalized *Bayesian information criterion statistics* value $\text{NBIC}' := \langle \ln \sigma_T^2 \rangle + (\ln N/N) \cdot p$. The first term (due to the estimated residual variance) is averaged over ten different solutions. The second term is due to the free system parameters, where N is the number of training pattern and p the number of weights and thresholds in the net. This expression reflects the idea behind the Bayesian information criterion statistics [14]: Parsimonious models with fewer parameters should be prefered! Since we have not rigorously derived an expression for the NBIC we used it here in a closely related sense indicated by the prime, NBIC'. Introducing more parameters in the net the training error can be more and more reduced, however, on the cost of the second term which increases linearly with the number of net parameters. The models with best forecasting and generalization capabilities should be those where NBIC' becomes a minimum. Within this approach we find, that connectionist nets with three or four hidden units should yield the best forecasting results. It is worth to note, that Weigend et al. [13] using their weight elimimation procedure also observe that about three hidden units may be the best choice. Evaluating the criterion for the BL(9,0,8,6) model (11 parameters), for the TAR(2,4,12) model (19 parameters), for the CNAR(4,4,1) model (25 parameters) and for the CNAR(12,3,1) model (43 parameters) we get in all cases a $\text{NBIC}^{(')}$ value of about 5. Only the SAR(9) model with 4 parameters yields a value which is about 10 percent higher.

3 Signal Processing: Noise Reduction

In order to investigate the capabilities of feedforward connectionist networks in a more systematic manner we also applied them to an artificial problem. The (fully developed) *logistic parabola* ($x_t = 1 - 2 \cdot x_{t-1}$) is known to show two characteristics of white noise: a flat spectrum and δ-autocorrelations. Since time series data are usually corrupted by "real" white noise we tested the nets' capabilities of distinguishing the "deterministic" from the "real" noise by presenting a superposition of both to relatively simple nets. In these experiments we also varied the number of parameters and the strength of "real" noise. The nets behave exactly as expected: the functional part of the data is extracted, while the noise is ignored. Though this experiment was set up to help understanding the complex questions involved in designing a net for time series analysis its results may also be interpreted as a nonstandard method of noise reduction. This is due to the fact that the noise reduction is not based on spectral properties but rather on functional dependencies.

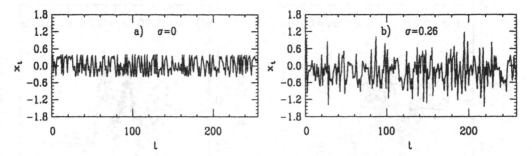

Figure 2: 256 data points obtained from (1) which are used for the training set. a) noise free and b) noise of strength 0.26.

As an example we consider the deterministic time series

$$x_t = f(x_{t-1}) + \sigma \cdot \epsilon_t \quad , \qquad \text{with}$$

$$f(x_{t-1}) = \tanh \left\{ v_0 + \sum_{h=1}^{4} V_h \tanh \left(U_h x_{t-1} + u_h \right) \right\} \quad , \tag{1}$$

where the set of parameters $\{U, V, u, v\}$ is given by { 2.004, 3.984, -14.33, -8.887, $-14.75, 74.297, -1.015, -6.953, -0.8357, -0.8418, -5.387, 13.29, 0.4552$ }. ϵ_t is a white noise signal with mean zero and variance one, σ measures the strength of the noise. This time series, figure 2a, shows on the interval ± 0.4 almost the same behaviour, figure 3a, as the once iterated logistic parabola scaled down to this interval. However, it has the advantage that it can perfectly be modeled by a CNAR(1,4,1). On the other hand it allows to investigate the time series by misspecified models, i.e. underfitted or overfitted connectionist networks. We have investigated this time series by CNAR(1,n,1) models

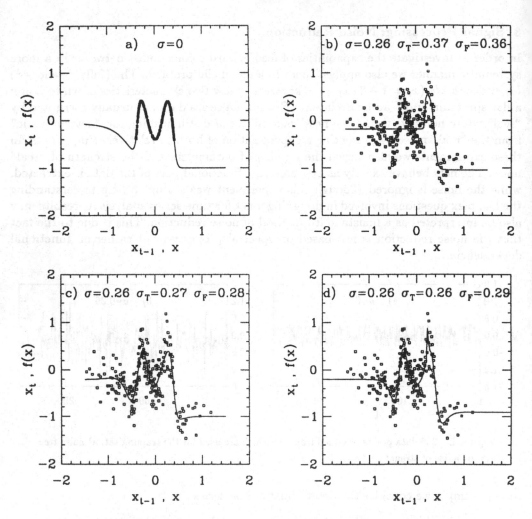

Figure 3: Scatterplots of the pattern x_t vs. x_{t-1}. The squares are the 256 data points and the thin line is the functional relationship $f(x)$ derived from (1), however, with the estimated weights and thresholds of the trained connectionist net. σ denotes the noise strength and $\sigma_{T,F}$ the residual variances of the training and forecasting sets, respectively. a) CNAR(1,4,1) noise free model, b) underfitted CNAR(1,2,1) model, c) correctly fitted CNAR(1,4,1) model, d) overfitted CNAR(1,8,1) model. The three models have noise $\sigma = 0.26$.

with $n = 2, 4, 8$ and different strengths of noise. Figure 2b shows the time series with $\sigma = 0.26$. The training as well as the forecasting set consists of 256 time series points. Figure 3b–d show the scatterplots of the 256 patterns. The "learned" functional relationship may easily be seen even at such a strong noise as $\sigma = 0.26$. For all network structures $n = 2, 4$ and 8 we were able to find solutions with normally distributed residuals, as indicated on a 5% significance level of the Kolmogorov-Smirnov-test. The variance of the estimated residuals σ_T^2 was close to the noise for $n = 4$ and 8. The residual variance of the forecasted series σ_F^2 was closest to that of the training series for all noise strengths when $n = 4$. These results are also in accordance to the model selection based on information criterion statistics.

4. Economics: Currency Exchange Rates Analysis

There are basically two distinct approaches to the problem of forecasting currency exchange rates [15]. The first one is interested in developing economic models, the other one deals with the data available in a statistical environment. There are several reasons why we chose the second approach, amongst them the simple fact that the data base we had to work with only dealt with time series from various stock exchanges. Table 1 gives an overview of the time series we analyzed. We were interested in forecasting time series #8: *IMM Swiss Franc*. We were well aware of the opinion that chances to forecast this relation are low. We leave it to the reader whether he wants to read on!

#	Time Series	#	Time Series
1	Pound Sterling (£/$)	12	SIMEX 3 Months Euro $
2	German Mark (DM/$)	13	LIFFE 3 Months Euro $
3	Swiss Franc (SFR/$)	14	LIFFE 3 Months Sterling
4	Japanese Yen (YEN/$)	15	IMM 3 Months T-Bills
5	US $ Index	16	CBT T-Bond 8% FUT
6	IMM Pound Sterling	17	LIFFE Long Gilt
7	IMM German Mark	18	COMEX Gold FUTURE
8	IMM Swiss Franc	19	COMEX Silver FUTURE
9	IMM Japanese Yen	20	NYMEX Crude Oil
10	FINEX US Dollar Index	21	Euro SFR 3 Months
11	IMM 3 Months Euro $	22	London Gold 50

Table 1: Overview of the time series data for the exchange rate analysis. The first four time series are interbank rates, the other give standard series from various sources. (In most cases *closing, opening, high, low, open interest* and *volume data* were available.)

In a first step we applied several (non-parametric) statistical tests to the time series in order to check for randomness. The vast majority of tests indicated that an univariate approach were useless, since the data represent *random walks*. Thus we reproduced this well-known result for our time series data. — However, we may state a hypothesis in a *multivariate* context: There may exist dependencies between the time series such that some variables may be identified as indicators. We therefore call a value an "indicator", if there is a statistically significant correlation to its "predictant". The first question is, how to identify these indicators. We would like to comment that we do not share the opinion that the connectionist net itself is able to extract these dependencies by some kind of built-in "intelligency". Instead, in our experience on the matter, connectionist nets, as used today, are quite unstable and sensitive to noise, some crucial issues in this highly noisy application. We therefore propose rigorous data preprocessing, leaving the net the "only" task of performing the time series analysis.

In order to find indicators we performed several statistical tests on dependencies between the time series. E.g. we checked the difference of interest rates against the *IMM Swiss Franc* time series. Amongst these tests were the Wilcoxon two sample test, Kendall's τ-test, Spearman's ρ-test etc. All these tests give siginificance levels for the dependencies.

We devised a scheme to combine all these levels in a non-parametric way. Finally, we invented a method to adjust the indicators dynamically. This last method was at least partially a consequence of our lack of (even) more data. (In this study we considered about 250 indicators derived from the 22 time series given in table 1 and performed over the different time periods approximately 2500 statistical tests.)

After the selection of indicators we first developed a *linear* model and performed some trading on that part of the data which we did not include in the search for indicators and parameter estimation process. We will briefly describe the trading scheme. The prediction procedure will tell, how much the exchange rate will raise or fall. If the prediction is "raise" we buy for the amount of one Swiss franc today and sell the dollars we get on the basis of the rate tomorrow. If the system predicts "fall" we sell the dollars today and calculate the return on the basis of the rate of tomorrow. This scheme is very simple, but it has its counterpart in reality. Note that it is possible to make money every single day. The results of our linear method were the following: $(13.1 \pm 4.2)\%$. This figure is an average over three years of trading (1987–1989), and it is based on the maximal possible return. If we compute this return on the basis of the capital involved we get an averaged value of 19.7%.

Having performed the linear prediction we turned to the question whether a non-linear approach allows even larger returns. For this task we chose a CNAR(10,1,1) structure, starting from the linear solutions of the previous paragraph. Since these linear solutions are not unique we get a whole set of solutions. One may judge this as a disadvantage of the method, because unique solutions are somehow related to more stable predictions. However, one may also view this non-uniqueness as a chance to combine several scenarios, based on non-linear prediction. We combined *five* and also *ten* of our connectionist nets. The results are summarized in table 2.

Method	1987	1988	1989	Total
LINEAR	18.7 (29.3)	8.6 (10.7)	12.0 (19.2)	13.1±4.2 (19.7)
5 AVERAGE	10.7 (16.7)	9.4 (11.7)	14.2 (22.8)	11.4±2.0 (17.1)
5 MEAN	14.9 (23.3)	22.7 (28.4)	11.9 (19.1)	16.5±4.0 (23.6)
5 MAJORITY	7.2 (11.2)	15.4 (19.2)	8.3 (13.3)	10.3±4.4 (14.6)
5 CONSENSUS	9.2 (14.4)	3.9 (4.9)	14.7 (23.6)	9.3±5.4 (14.3)
10 MEAN	16.7 (26.1)	13.8 (17.3)	15.4 (24.8)	15.3±1.5 (22.7)

Table 2: Summary of prediction results as explained in the text. E.g. "5 MEAN" indicates that the outcome of 5 nets has been combined.

The first row shows our results from the linear analysis. The number without parantheses is related to the maximal return, the figure in parantheses gives the return based on the capital invested, as described above. "AVERAGE" is based on nets trading on their own and computing their average at the end, while "MEAN" is a strategy that combines the output of the nets *before* trading, i.e. the average prediction is computed. "MAJORITY" considers the sign of the majority of the nets as the final prediction, whereas "CONSENSUS" trades only, if all the nets coincide in their prediction. The conclusion from this table may be that the joint effort of non-linear connectionist nets will lead to a higher return than a linear approach. The return is also more stable than in the linear case.

Very recently, we received a preprint of Weigend et al. [13], which deals with the problem of predicting currency exchange rates. Since they do not give any results on the return that could be compared to our findings, we are almost left with comparing the net structure. Their "weight elimination" procedure indicates that one or two hidden units should do. This agrees with our results remarkably well. This may be surprising, since they forecasted a totally different time series. But it may be seen as a characteristic feature of environments which are extremely noisy. The noisy environment may also be responsible for the fact that Weigend et al. were able to predict only a very small percentage of the exchange rates better than just the simple "stay as it is" strategy.

5 Towards a robust design of connectionist nets

Today, the design of a connectionist nets is a difficult task. The nets have shown surprisingly good results on the sunspots data, but we have to note that there is quite a substantial *lack of robustness* of the method. In nearly every single publication on the subject and also in our work on sunspots and noise reduction, the final solution emerged from a random initial solution. One finds that the results of the optimization, especially the generalization abilities of that solution, depend on the initial configuration. Other problems do exist, but we will not go into detail here.

On the other hand, we think to have established the *principle of parsimony*, which we derived mainly from information criterion statistics. Considering the successful connectionist models for non-linear time series also supports the validity of this principle.

The "weight elimination" procedure [13] is proposed as a relatively new method to cope with the robustness problem. Although it may be a very helpful instrument, we have some objections to proposing this method as an universal problem solver. Skipping the details of a thorough discussion here, we like to mention that the method is based on heuristics, it shows problem dependencies and it introduces quite a handful of dynamically adjustable parameters. This makes the method and its rules rather complicated.

A final observation in this context: It is surprising that nowhere in the literature on connectionist networks one finds rigorous checks for linearity or non-linearity of either the time series or the model given by the net equation. But we have to keep in mind that the net may form a linear model, although there are non-linear transfer functions. If we write down the net equation

$$x_t = w_0 + \sum_{l=1}^{H} w_l \tanh \left(\sum_{k=1}^{I} a_{lk} x_{t-k} + a_{l0} \right) \, , \tag{2}$$

(H: #hidden, I: #input) we observe that $\tanh(x) \approx x$ for small $|x|$. Thus we may rewrite (2) for small arguments of the hyperbolic tangens as follows.

$$x_t = w_0 + \sum_{k=1}^{I} \left(\sum_{l=1}^{H} w_l a_{lk} \right) x_{t-k} + \left(\sum_{l=1}^{H} w_l a_{l0} \right) \tag{3}$$

In (3) we have simply rearranged the terms of the sums and the equation turns out to be the familiar AR-model of order i. In our approach, which will be reported in detail elsewhere [16], we start from this linear net function. We can always guarantee any desired degree of linearity. This introduces one single parameter, however, this parameter has a clear meaning. A unique linear solution may be calculated by imposing more restrictions to the problem ("principal value decomposition"). The learning procedure may now be accompanied by a careful monitoring of the internal net structure. From this monitoring we gain valuable insight in the optimal design of the net and the degree of nonlinearity involved. The basic idea of this approach may be summarized as follows. The initial parameters are a unique solution of the net equation. This linearity allows a fully transparent training procedure. From this information one may gain answers to essential design problems as well as the problem of robustness.

6 Summary

Although these first results are quite surprising they are still not more than preliminary. The quality of the modelling in our case studies motivates further research in this area. We are quite far from a real understanding time series analysis with connectionist networks. Many parameters have to be adjusted—they should be derived from the data to be analysed rather than from empirical analysis of the nets' behaviour. The lack of robustness also is a serious problem.

Acknowledgement

We thank Th. Greminger, B. Huberman, K. Kirchmayr, H. Tong, A. Weigend and D. Wenger for stimulating discussions. This work was partially supported by a PhD grant (CdG) given by the Swiss Bank Corporation.

References

[1] D.E. Rumelhart and J.L. McClelland, *Parallel Distributed Processing, Explorations in the Microstructure of Cognition*, Vols. 1&2, MIT Press, Cambridge 1987.

[2] G. Cybenko, Techn. Rep. No. 856, Urbana Univ. of Illinois, 1988;
K.-I. Funahashi, Neural Networks 2, 189, 1989;
K. Hornik, M. Stinchcombe and H. White, Neural Networks 2, 359, 1989.

[3] A. Lapades, LA-UR87-226 Los Alamos Report, 1987.

[4] G.E.P. Box and G.M. Jenkins, *Time Series Analysis, Forecasting and Control*, Holden-Day, San Francisco 1970.

[5] T.W. Anderson, *The Statistical Analysis of Time Series*, Wiley, New York 1971;
M.B. Priestley, *Spectral Analysis and Time Series*, Academic Press, London 1981;
C.W.J. Granger and P. Newbold, *Forecasting Economic Time Series*, Academic Press, New York ^{2}1986.

[6] C.W.J. Granger and T.W. Anderson, *An Introduction to Bilinear Time Series Models*, Vandenhoeck and Ruprecht, Göttingen 1978;
T. Subba Rao and M.M. Gabr, *An Introduction to Bispectral Analysis and Bilinear Time Series Models*, Lecture Notes in Statistics Vol. 24, Springer, Berlin 1984;
H. Tong, *Non-linear Time Series*, University Press, Oxford 1990;
T. Ozaki, *Nonlinear Time Series Models and Dynamical Systems*, in: Handbook of Statistics, (E.J. Hannan et al., eds.) Vol. 5, North-Holland, Amsterdam 1985;
M.B. Priestley, *Non-linear and Non-stationary Time Series Analysis*, Academic Press, London 1988.

[7] C. de Groot and D. Würtz, *Analysis of Univariate Time Series with Connectionist Nets*, Proceedings of the Munotec Workshop, Dublin 1990 and Neurocomputing (in press), 1991;
C. de Groot and D. Würtz, *Signal Processing and Noise Reduction with Connectionist Networks*, Helvetica Physica Acta (in press), 1991;
Würtz and C. de Groot, *Dollarprognose*, unpublished results, IPS Zürich 1991.

[8] A.J. Izenman, Math. Intelligencer 7, 27, 1985.

[9] G.U. Yule, Phil. Trans. Royal Soc. London A226, 267, 1927.

[10] P.A.P. Moran, J. Royal Stat. Soc. B16, 112, 1954;
M.C. Schaerf, Technical Report, Department of Statistics, Stanford 1964;
M.J. Morris, J. Royal Stat. Soc. A140, 437, 1977;
P. Hokstad, J. Time Series An. 4, 177, 1983.

[11] M.M. Gabr and T. Subba Rao, J. Time Series An. 2, 155, 1981.

[12] H. Tong and K.S. Lim, J. Royal Stat. Soc. B42, 245, 1980.

[13] A.S. Weigend and D.E. Rumelhart, Proceedings of Interface '91, Springer Verlag (in press) 1991;
A.S. Weigend, B.A. Huberman and D.E. Rumelhart, Santa Fe Workshop 1990, NATO Workshop Proceedings, Addison Wesley (in press) 1991.

[14] Y. Sakamoto and M. Ishiguro, *Akaike Information Criterion Statistics*, Reidel Publishing, Dordrecht 1986.

[15] E.W. Heri, Finanzmarkt und Portfoliomanagement 1, 47, 1986/87;
E.W. Heri, Revista Int. di Scienze Economiche e Commerciali 33, 1057, 1986.

[16] C. de Groot and D. Würtz, Proceedings of the IUTAM Workshop, Warwick, 1991, to appear in Physica D.

ILISCE: A System for Learning Control Heuristics in a Scheduling Environment

Thierry Van de Merckt

IRIDIA, Université Libre de Bruxelles
1050 Bruxelles, Belgium
thvdm@is1.vub.ac.be

Abstract. This paper presents a general learning structure called ILISCE which learns meta-rules to guide a scheduling system. The scheduling program, called OPAL, solves real-world problems using a set of local heuristics which incrementaly resolve the whole set of conflicts among the physical resources of the floor. A Selective Inductive learning approach is used to create concepts representing typical states of the job shop which allow a Classifier to "recognize" current situations and to choose the corresponding best operators to apply on. In this paper, we present the general architecture of ILISCE.

Keywords: Scheduling, Learning, Credit Assignment.

1 Introduction

Scheduling is basically concerned with the problem of allocation of resources over time periods to perform a set of tasks. In a job-shop scheduling problem, a number of jobs have to be carried out in a workshop. A *job* is described by a set of related operations whose achievement requires some resources. An *operation* represents an elementary activity and is characterized by its processing time (duration) and its starting time. A *resource* is a physical element (a machine or a work-team). A sequence of operations or *production routing* describes the order in which each operation, allocated to a given resource, has to be done to produce a job and introduces precedence constraints among the involved operations. The predictive scheduling problem is: given a set of production facilities and technological constraints, given production requirements expressed in terms of quantity and time constraints, find a feasible sequencing of processing operations on the various facilities satisfying the production requirements. Pure constraints, such as release dates, operation precedences and durations or resources availability, must be respected. Other constraints such as the respect of due dates, the minimum idle times or the maximum flexibility (slack time), may be view as relaxable preferences. The scheduling problem is NP-complete. Therefore, techniques using heuristic dispatching rules for local decision making were developed [2, 3, 5] to by-pass the complexity problem and to work with realistic models.

AI systems often use a combination between numerical calculus methods (constraint propagation algorithms) adapted to scheduling problems and heuristics based on symbolic knowledge (1, 6, 4). Heuristics, generally represented by production rules, can be separated in two classes: (i) the *strategic* ones which identify and select the sub-problems to focus on; and (ii) the *tactic* ones which actually take the decisions concerning the problem at hand. Non-grey parts of Fig.1 illustrate a general structure of AI scheduling systems. An Expert system selects subsets of strategic and tactic rules which will be used by the Resolution System to control the exploration of the potential solution tree. The Resolution System simulates the physical model of the job-shop and updates its current state when decisions are taken.

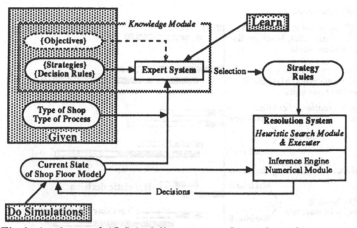

Fig.1 A schema of AI Scheduling systems. ILISCE layer is presented by the greyed items.

It is clear that the selection of heuristics is deeply connected to the type of shop, to the technical constraints of the processes and to the current state of the shop-floor. The efficiency of such a system drastically depends on the knowledge encoded into the Expert System. Two problems can be identified from Fig.1:

(i) *The objectives are implicit* -- The objective the heuristic rules are designed to achieve is most of the time implicit. Therefore, when an objective, such as minimising the batch cycle or maximising the flexibility of the solution is preferred, one has to known about the effects of the selected heuristics.

(ii) *Expert System's knowledge is difficult to define* -- The effects of the heuristics are sensitive to the shop's characteristics [5] and to the context in which they are applied [4]. To specify the relevant conditions under which the implicit objectives will be achieved is therefore a difficult task in a complex domain such as scheduling. It is equally difficult to identify the impact of the heuristics on all the objectives that a user may have in mind.

Our purpose, illustrated by grey parts on Fig.1, is to use simulations and observations of a scheduling system and to apply AI Learning techniques to obtain a set of meta-rules expressing useful relations between (i) the available heuristics and (ii) an explicit objective and the current state of the shop floor.

2 The Scheduling Environment

OPAL[1], a predictive scheduling system for job-shops, is used as the scheduling environment. It is based on a double architecture: (i) a module of analysis & propagation of constraints; (ii) an expert module that guides the search for a solution. Its structure corresponds to the schema presented in Fig.1 where the Expert System selecting the heuristics is a human operator [1].

2.1 Constraint Analysis & Propagator

OPAL uses an object-oriented language to describe the information relevant to build an abstract model of the shop: a description of the resources (type, availability periods, current workload) and of the production routing for each job (sequence of operations, usable resources, durations). From these data and additional information on the required jobs (due dates and quantities), OPAL creates a temporal window for each operation defined

---

[1] OPAL has been provided to us in the framework of a scientific cooperation program between the Univerité Libre de Bruxelles and the DERA-ONERA of the Centre d'Etude et de Recherche de Toulouse (CERT), France.

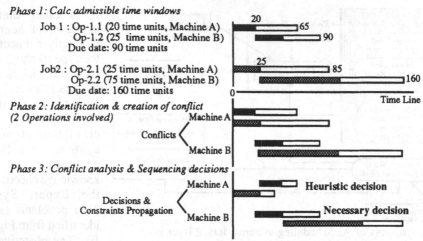

Fig.2. An example of conflict representation and constraint propagation used in OPAL. Dark areas represent operation durations and blank ones are the slack times.

by a *propagation process* based on jobs' routing descriptions: a forward propagation of starting times produces the *earliest starting time* for each operation, a backward propagation of the due dates produces the *last final times*. Therefore, at the beginning of the process, operations are represented by the biggest *admissible* work interval, where only routing precedence constraints are respected (phase 1 in Fig.2). A *conflict*, defined on two operations, is added to a stack when an intersection between the two operation's windows on a same resource is detected (phase 2 in Fig.2). A step towards the solution is taken when a conflict is resolved, i.e. when a sequence between the two involved operations has been decided. Once a decision has been taken, the Constraints Propagator updates the temporal windows of all connected operations, reducing the search space to the admissible solutions. A *Constraint Analyser* takes the trivial decisions, i.e. those where only one sequencing is admissible. In the example, the sequencing "Op2.2 before Op1.2" is not admissible whilst both possible sequencing can resolve the conflict between Op1.1-Op2.1.

2.2 Heuristic Search

Non trivial decisions are taken by a set of heuristics belonging to: (i) *Selection Criteria* which act at a Strategic level by selecting the next subset of conflicts to resolve; and (ii) *Decision Rules* which express weighted preferences for the sequencing of a conflict under the form of "Op_1 before Op_2", "Op_1 after Op_2", "No opinion". The scheduling task , in OPAL, is therefore reduced to taking a set of sequencing decisions concerning the stack of conflicts. The Criteria & Rules (user-defined) reflect this aspect. Here are some examples:
• *Criteria* select a subset of conflicts to focus on:
  - C5: Focus on the conflicts whose earliest starting time of the involved operations are the closest;
  - C1: Focus on the conflicts belonging to the machine with minimum average flexibility;
  - C3: Focus on the the conflicts belonging to the job with minimum average flexibility;
• *Decision rules* choose a sequence between the two operations belonging to a conflict:
  • Priority rules:
    - SPT Operation with the Shorter Processing time is placed first;

- EST   Operation with the Earliest Starting Time is placed first;
- Margin rules:
  - RST   Choose the sequence that maximise the total slack time of both conflicting operations;
  - SRPT Operation that is part of the job with the Shorter Remaining Processing Time is to be placed first.

It should be noted that (i) the preconditions of the heuristics only specify applicability conditions and do not use any information concerning their effects on the objectives; and (ii) the decisions are local since, at each step, only the interaction between two single operations is considered.

The union of one Criteria and one Decision rule defines an *operator*. Each operator may have very different effects and hence may lead to different solutions. Table1 shows the results of three simulations starting from the *same initial job-shop* case, but *applying distinct operators*. It illustrates the differences that may exists among solutions generated by different operators and justify the interest of the learning approach used in ILISCE.

Objective Scores / Solution n° (Operator)	n°1 (C-Asc/R-Asc)	n°2 (C5/Spt)	n°3 (C5/Srpt)
Operations-Avg-Flexibility (to Maximise)	83	46	38
Job-Cycle-Ratio (to Minimise)	32	22	3
Machine-Avg-Idle-Time (to Minimise)	51	57	76
Search-Time (to Minimise)	8"	4"	4"

Table 1. Scores of four objectives issued from three solutions founded by three distinct operators. These solutions were generated from a single starting shop case provided to OPAL.

3   General Structure of ILISCE

ILISCE has two components: a real time one and a batch one. The main goal of the realtime component is to match the current state of the shop model in OPAL to identify

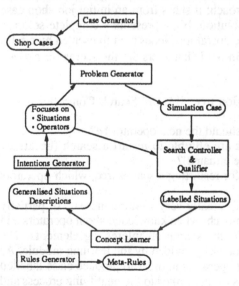

Fig. 3. ILISCE -- General batch structure

important changes that would justify the selection of new heuristics. The batch structure, illustrated by Fig.3, collects data on OPAL's behaviour and produces the meta-rules. To do that, ILISCE uses OPAL as a simulator to produce a number of observations and therefore needs a way to characterise the contexts where heuristics were applied. Consequently, it is composed of : (i) a Language of Representation, used to describe the contexts (Frame-based and user-defined); (ii) a Search Controller which controls the expansion of a solution tree while collecting data on OPAL behaviour; (iii) an Evaluation module which qualifies the effects of the operators on the objectives (the Qualifier) (iii) an Inductive Learning module that creates class descriptions about typical contexts of operator applications. The whole system uses an iterative process. First, the Controller provides a training example to

OPAL and controls it to generate a solution tree using distinct operators. Next, a module qualifies a number of states relatively to each objective. This constitutes the input for the Concept Learning Module which produces the descriptions of "typical" states. Meta-heuristics, in the form of production rules, are derived from these descriptions and used to produce focuses for the Search Controller.

A *state* or a *context* in ILISCE is defined by the set of values taken by the descriptors defined in the Language of Representation. Some examples of such descriptors are: the number of required jobs, the average number of operations per job, the average slack time per job, the number of operations per machine, the average constraint per machine, the total machine idle time, the number of conflicts, the average difficulty of the conflicts, the number of backtracks. A *solution* is a State with zero conflicts. In this sense, any state can be seen as a partial solution. The *meta-rules*, issued from the learning phase, uses the same Language of Representation to express the conditions defining "typical" states or contexts in their precondition part.

4 Search Control & Credit Assignment

The approach used to learn the behaviour of OPAL's heuristics is quite natural: like experts, ILISCE acquires experience from observations and should be able, after a number of observed cases, to identify some regularities of the operators' behaviour when applied to given cases. An operator can be viewed as a function which, when applied to a given state S_i, produces a resulting state S_j that is closer to a final solution S_k. For the sake of simplicity, the following paragraphs will assume that the system focuses on one single implicit objective: the generalisation to multiple objectives is immediate.

4.1 Search Space & Control

Solution Tree. Applying different operators to a root state results in expanding a search tree whose nodes are states or contexts and whose arcs are the operators. ILISCE explicitly creates this tree using a frame-based representation (using KEE3.0™). This search process is done using a top-down approach: it starts from an initial job-shop case and iteratively applies operators to reach a solution. Fig.4 presents a complete solution tree issued from S_0 by applying three available operators. Nodes are represented by circles and solutions (leaves) by rectangles. Figures in bold characters are the scores for a given objective to be maximised.

Search Control. While expending the solution tree, ILISCE' Search Controller has to answer three questions:
1. Each time a new state is generated, what should the next operator be?
2. Among all generated states, which ones are eligible to restart a search (backtrack nodes from which a new search tree can be initiated)?
3. When an eligible state has been chosen for restarting a search tree, which operators will be tested?

The strategy used by the Search Controller will greatly vary with the increase of partially learnt knowledge. ILISCE uses an iterative approach: when knowledge about operators and typical states is available, it is used to focus the search on interesting elements. The Controller answers question 2 & 3 by using focuses provided by the learning module. As the Qualifier compares the actions of different operators in order to produce class-labelled states for the learning module, the Controller is deeply linked to the qualifying process and to the credit assignment problem, as will be explained later.

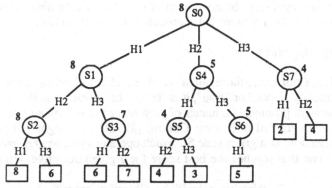

Fig.4. Complete solution tree constructed from an initial job-shop case S_0, using 3 operators: H_1, H_2, H_3.

General Heuristics. When no specific domain knowledge is available, general heuristics are by the Search Controller. Firstly, when a current node is highly constrained, one cannot expect that different operators will generate highly different solutions. Therefore, to analyse the behaviour of the operators, the learning process starts by submitting unconstrained shop cases to OPAL. After some partial knowledge has been acquired, more constrained ones are tested. Secondly, once an operator has been selected, it is applied until a solution is reached. This is justified by the fact that the score obtained for a given objective can only be observed at the solution level. Indeed, analysis done on the dynamic trends of objectives' scores shows that dramatic changes may occur when the last decisions are taken. To assert valid conclusions on the characteristics of an operator, we therefore have to "wait" until the solution is reached. Finally, operators are not always relevant for a given state: it may have 'no opinion'. To take this fact into account, the general search strategy becomes: *apply any chosen operator as long as possible on the way towards the solution and stop it only when it is not relevant any more.* In this case, ILISCE creates a *choice-node*, i.e. a branching point for the selection of a new relevant operator. At the beginning of the process, only those special states are considered as eligible backtracking state points. When a solution is reached, several operators will often have "put their effect" on it, which results in facing the credit assignment problem.

4.2 Credit Assignment Problem

The credit assignment problem is concerned with the following question: when a solution path reflects the action of several operators, how to credit them for the observed result? Two approaches can be applied: a Top-Down one known as *learning while doing*, which credits each operator during the search process, and a Bottom-Up approach known as *learning from solution paths*, which waits until a solution has been founded to propagate the credits back from the solution to the top of the tree. In scheduling, a Top-Down approach is difficult to apply, at least before any learnt knowledge has been made available, because of the unpredictable behaviour of the objective' scores at the end of the search process. A Top-Down approach is only possible when a rich domain knowledge is available which allows, at a node level, to *à priori* evaluate the impact of potential operators. A the beginning of the learning phase, such information is only partially available: the system has some knowledge on the objectives (monotonic behaviour, for example) and has some general heuristics, as will be explained later. But while it acquires meta-knowledge on operators' behaviour, the Controller will be able to use it as domain knowledge to move from a Bottom-Up approach to a Top-Down one.

4.2.1 Bottom-Up Approach

Decision Rule. In our case, the evaluation of the effect of an operator is relative: no operator is absolutely "Good" or "Bad", it is "better" or "worst" than the others. Therefore, the credit assignment problem is a matter of *preference* rather than an *absolute evaluation* as has been widely studied in Process Control [8]. Credit Assignment problem can be stated by this question: at a given state S_i, which operator must be *preferred*? The answer is obvious: the one that reaches the best score for a given objective, represented by the following rule:

DecRule 1 -- IF Objective_score(S_n) > Objective_score(S_m)

AND Reach(S_n,H_i /S_w) $\wedge$ Reach(S_m,H_j /S_w)

THEN Prefer(H_i,H_j /S_w)

where Reach(S_n,H_i /S_w) means 'State S_n is reached by H_i applied from S_w'

Prefer(H_i,H_j /S_w) means 'H_i should be preferred on H_j when applied from S_w'

Score Propagation. On Fig.4, it is clear that H_1 should be preferred to H_3 when applied to S_2 because they both reach a solution whose scores can be directly compared. We can recursively apply the same comparison while going from bottom solutions to upper nodes in the tree: starting from a parent node, a child node located on the path going to the best solution should be preferred on all other children. This iterative process defines a mechanism that makes a parent node inherits the best score of its children. Thanks to this inheritance process, the decision rule 1 can be applied to any node in the solution tree:

PropRule 1 -- IF Type(S_w,SOLUTION)

THEN Objective_score(S_w) = Actual_score(S_w)

PropRule 2 -- IF Type(S_w,NODE)

AND $\Omega = \{S_m: \text{Reach}(S_m,H_j /S_w) \forall H_j\}$

THEN Objective_score(S_w) = (Max[Objective_score($S_m \in \Omega$)])

PropRule 1&2 recursively defines the bottom-up propagation algorithm illustrated in Fig. 4 by the bold values coming up from the leaves to the root. However, it puts a strong constraint on the process: all possible paths from a current node to solution states must be tested before a conclusion concerning this node can be drawn.

4.2.2 Top-Down Approach

General Heuristics. The number of paths to test grows exponentially with the number of operators. However, this combinatorial explosion can be controlled by using simple heuristics. Firstly, restricting eligible backtrack nodes only to choice-nodes is powerful to decrease the branching factor of the expanded tree. Indeed, they are relatively rare at high search tree levels. Secondly, the exhaustive search of the propagation process can be relaxed at a given node when the number of conflicts is small. Indeed, simulations have shown that the range of variation of a given objective becomes narrower as the solution is approached: the last decisions are less significant than the first ones because the degree of freedom is too small for the operators to exhibit their specific impact on the objectives. In these cases, an exhaustive search among all relevant operators is useless.

Therefore, the following rules relaxing the exhaustive search constraint can be defined, while keeping the capacity of asserting reliable conclusions:

PropRule 3 --   IF %Conflicts(S_w) < λ OR   Avg-Constraint(S_w) > β

                        THEN d°Freedom(S_w, NARROW)

PropRule 4 --   IF Type(S_w,NODE)

                        AND d°Freedom(S_w, NARROW)

                        AND $\exists H_j$: Reach(S_m,H_j /S_w)

                        THEN Objective_score(S_w) = Objective_score(S_m)

Background Knowledge can be used to relax the Bottom-Up propagation process. Such an objective as "Total Idle Time" monotonously increases while the solution is approached. This information is very valuable to stop the expansion of a node if it can be seen that its score is higher than all its siblings: the solution score that would be reached from a given node S_j cannot be known, but it can, at least, be said that its actual score is the minimum value that can be reached.

DecRule 2 --   IF Behaviour(objective,MINCREASING) $\wedge$ Target(objective,MIN)

                       AND Actual_score(S_m) > Objective_score(S_n)

                       AND Reach(S_n,H_i /S_w) $\wedge$ Reach(S_m,H_j /S_w)

                       THEN Prefer(H_i,H_j/S_w)

This rule can be defined for all combinations of Decreasing and Increasing behaviour while searching to Maximise or Minimise the objective.

Partially Learnt Heuristics. The major power of ILISCE lies in the fact that it is able to use explicit focuses which identify interesting subregions of the search space to explore for all couple (operator, objective). At the beginning of the learning process, focuses are maximally general: all states and all operators appear to be interesting to explore. While the learning process goes on, focuses will cover less states and operators: well-known states, i.e. those falling within a region of the concept space where meta-rules have a high degree of confidence, will consequently be removed from the explored solution tree. These kinds of cuts restricts the search process to exploring unknown states:

FRule 1 --   IF FMatch(S_n,Focus$_x$)

                    AND Ω = {O_m: $O_m \in$ Focus$_x$}

                    THEN Add (S_n,STACK_NODES_TO_EXPLORE) $\wedge$ Select_op(Ω, S_n)

where FMatch returns true if S_n is included in Focus$_x$, O_m is an operator and Select_op is a function for selecting operators for a node S_n.

These rules illustrate that the whole process, starting as a "blind" search at the beginning, will become a more and more "intentional" by the effects of the incremental acquisition of learnt knowledge.

4   Qualification process

This phase takes place after a solution tree has been completely explored. This is a key process that transforms objective scores to causal links between the action of operators, applied on given states, and the quality of the results for a given objective. The Qualifier produces Labelled Class states under the form of *Class(State, objective, class-*

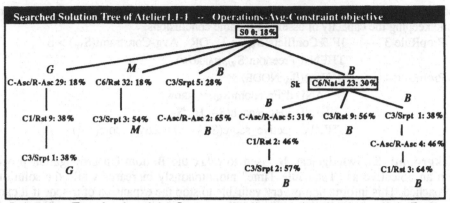

Fig.5. Solution Tree developed by ILISCE on Atelier1.1-1 job-shop case (S$_0$). Each node presents the name of the operator applied on the parent node, the number of its decisions and the actual score of the objective (average constraints on all operations).

Label∈{Good, Medium, Bad}, operators) which means "the class of these operators, applied on this State, for this objective, is Good, Medium or Bad". From these kinds of evaluations, an Inductive Learning algorithm is able to produce generalised descriptions for each class. Three user-defined classes are used to describe the behaviour of each operator relatively to each objective: Good, Medium and Bad. As it has already been said, the evaluation of the quality of one score has to be done by comparison with the others. Therefore, for each evaluated node, a discrete interval is created for each class. Fig.5 presents an example issued from ILISCE based on a small job-shop case. A solution tree, based on a given root S$_0$, has been expanded with five different operators. The standard Bottom-Up propagation algorithm is illustrated by G, M, B. Starting from S_0, minimum and maximum scores are 38 & 65 respectively. The discreet step is (65-38)/3=9 which gives the intervals [38 *Good* 47[[47 *Middle* 56[[56 *Bad* 65]. The qualification process provides then the following data to the Learning Module:

> CLASS(S$_0$ OPERATIONS-AVG-CONSTRAINT GOOD (C-ASC/R-ASC));
> CLASS(S$_0$ OPERATIONS-AVG-CONSTRAINT MEDIUM (C6/RST));
> CLASS(S$_0$ OPERATIONS-AVG-CONSTRAINT BAD (C3/SRPT C6/NAT-D));

where S$_0$ is the frame containing the root node description. If the same process is applied to the state resulting from C6/NAT-D path (node S$_k$ on Fig.5), the discreet step is given by (64-56)/3=2.7, which gives [56 *Good* 58.7[[58.7 *Middle* 61.4[[61.4 *Bad* 64.1] resulting in   CLASS(S$_K$ OPERATIONS-AVG-CONSTRAINT GOOD (C-ASC/R-ASC C3/RST));

> CLASS(S$_K$ OPERATIONS-AVG-CONSTRAINT MEDIUM ());
> CLASS(S$_K$ OPERATIONS-AVG-CONSTRAINT BAD (C3/SRPT)).

5   The Learning Algorithm

The Selective Inductive Learning is an ID3-like algorithms (Top Down Induction of Decisions Trees). Briefly, the reasons for this choice are (i) that it has been proved to be noise resistant; (ii) it provides comprehensible concepts; (iii) it is able to deal with real world applications: (iv) it is robust against irrelevant attributes (which is important in our case since the description language may contain a lot of irrelevant descriptors); and (v) is has been successfully tested on complex concepts [7]. We won't discuss here the technical aspects of the algorithm, we will just expose its principles.

Each instance (a state) is described by a vector containing the values of the descriptors defined in the Language of Representation, plus a label defining its class. The algorithm receives a set of such instances as a training and then builds a decision tree using a Top-Down approach by iteratively partitioning this set using one of the descriptors as a discriminatory test. When a node of the decision tree only contains elements of one single class, it is labelled by this class and is not further expanded. The tests issued from the descriptors constitute a conjunction of predicates that describes each leaf. The basic TDIDT algorithm has been modified in ILISCE to be able (i) to produce rules instead of trees and (ii) to evaluate its concept descriptions [9]. This last modification allows it to produce focuses for the Controller.

6   Limitations and Conclusions

ILISCE is a general structure which aims to learn meta-level rules used to control an AI scheduling system. Nearly all modules are implemented, except part of the score propagation and the Qualifier. This explains why we do not yet have results to evaluate the whole system. Several limitations can be identified: (i) the power of the meta-rules drastically depends on the relevance of the Language of Representation for the learning task; (ii) the system does not create new operating heuristics, it just define utility ranges for the application of existing ones. The first limitation is the most important. The Learning algorithm will not be able to find consistent descriptions if the language do not contains relevant descriptors for the task of identifying states where operators should be applied. As the Language is user-defined and is easy to modify, we guest that the whole process will be iterative: if results are not good enough, the Language will adapted, converging to satisfactory results.

References

[1]   Eric Bensana: Utilisation de techniques d'intelligence artificielle pour l'ordonnancement d'ateliers. Thèse de l'Ecole Nationale Supéricure de l'Aéronautique et de l'Espace. Département Automatique,1987.

[2]   John H. Blackstone , Don T. Phillips, Gary L. Hogg: A State-of-the-Art survey of dispatching rules for manufacturing job shop operations. International Journal of Production Research vol 20 n° 1, 1982.

[3]   W.I. Bullers, S.Y. Nof, A.B.Whinston: Artificial Intelligence in Manufacturing Planning and Control. AIIE Transactions, Dec.1980.

[4]   Anne Collinot, Claude Le Pape: Adapting the behavior of a job-shop scheduling system. Decision Support Systems 7, North Holland, 1991.

[5]   Ali S. Kiran, Milton L. Smith: Simulation studies in Job-shop Scheduling - I: A Survey. Computer & Industrial Engineering Vol 8 n° 2, 1984.

[6]   S.J. Noronha, V.V.S. Sarma: Knowledge-Based Approaches for Scheduling Problems: A Survey. IEEE Transactions on Knowledge and Data Engineering, vol. 3 n°2, 1991.

[7]   J.Ross Quinlan : The Effect of Noise on Concept Learning. Machine Learning, An Artificial Intelligence Approach. Vol II. Ed. Ryszard S. Michalski, Jaime G. Carbonell & Tom M. Mitchell, Springer Verlag, 1986.

[8   R.S. Sutton: Learning to Predict by the method of Temporal Differences. Machine Learning, vol. 3, 1988.

[9]   T. Van de Merckt: NFDT: A Sytem that Learns Flexible Concepts based on Decision Trees for Numerical Attributes. Proceedings of the Ninth International Machine Learning Conference. Morgan Kaufmann, 1992.

KUSET - Knowledge Based User Support for Electron Beam Testing

G. Weichert, R. Lackmann

Fraunhofer Institut für Mikroelektronische Schaltungen und Systeme,
Finkenstr. 61, 4100 Duisburg 1, FRG

Abstract. The implementation of an expert system into the operating environment of an Electron Beam Tester (EBT) is presented. The concept of a knowledge based approach was chosen to meet the specific requirements for supporting the operation and maintenance of highly complex measurement equipment. The expert system KUSET (Knowledge-based User Support for Electron Beam Testing) allows an automatic online diagnosis of the EBT system's performance and offers a variety of tools to assure effectiveness and continuity in the EBT engineers work. A newly designed microprocessor-based hardware, the Modular Control System (MCS) provides a flexible and powerful link between the expert system KUSET and the hardware world of our measurement equipment.

Keywords: expert system, diagnosis, electron beam testing, integrated circuits.

1 Introduction

In the last 10 years Electron Beam Testers have found a wide-spread use in failure analysis and design verification of Very Large Scale Integrated (VLSI) circuits. They allow electrical measurements of chip-internal electrical signals which are not accessible by other techniques [1,2]. Figure 1 gives some examples of typical measurement results.

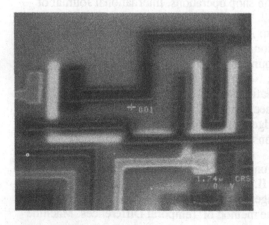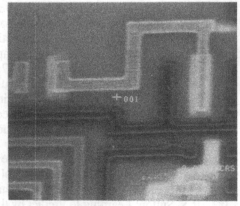

Fig. 1. Example for EBT measurements: Voltage Contrast picture of IC;
bright conductors: 0Volt, dark conductors: 5 Volts; conductors of 2nd layer partly visible

Although the increasing industrial demand for these instruments lead to significant improvements concerning the system performance, problems occure when an EBT has to be operated and maintained under field conditions. These problems arise from the complexity of the instrument (see figure 2) and the great number of possible hardware and adjustment errors. The commercial instruments available today are only partly equipped with automatic adjustment tools and require intensive training for new operators on the instrument. In most cases a specially trained operator is needed to keep the instrument in a normal operating condition. The potential user of the EBT (e.g. a designer of integrated circuits) has only indirect access to the instrument. This leads to

- reduced acceptance of the EBT by possible users
- additional spendings on trained operating personnel
- dependency of measurement results on the user's experience with EBT
- increased costs for repair and maintenance
- poor up/down-time ratio

Users of EBT report that the downtime of their instrument due to maintenance, repairs or adjustments often exceeds the actual uptime of the instrument. A breakdown in the normal operation of an EBT can be caused either by hardware errors or adjustment problems. Ascertaining the problem and localising the root cause requires an intensive knowledge of the hardware and the adjustment procedures of the EBT. Since in many cases this knowledge is incomplete or concentrated on few employees, the application of a knowledge-based software system makes sense [3,4].

KUSET makes distinctions between malfunctions of the EBT and simple operating failures. An automatic error diagnosis and interactive re-adjustment of the EBT is performed by KUSET which works as a 'hidden second operator'.The IC designer or EBT engineer is simultaneously provided with additional background information to improve his knowledge of the physics and hardware of the instrument. For example, if a vacuum problem has been spotted by the expert system, additional information about leak detection or weak spots in the vacuum system of the EBT is offered to the client [5].

It is important to put this approach in contrast to automation tools which simply overtake the control of certain functions of the instrument without including the operator's decision process and experience. In the case of electron beam testing it has been shown that this kind of automation often leads to unreliable or at least insufficient results especially when it comes to the valuation of physical side effects or distortions of measurement quality.

To value this thesis one has to know that the Device Under Test has significant influence on the EBT measurement accuracy and that the consideration of this influence is in many cases essential for a proper adjustment of measurement parameters. Furthermore, the operating conditions of the EBT have to be optimized for the kind of measurements which have to be performed on the actual DUT. The instrument can e.g. be adjusted for maximum signal/noise ratio or maximum temporal or spatial resolution. It is mostly heuristic knowledge which takes these demands into consideration. Therefore the experience of the operator is still the threshold to a successful test session.

The knowledge about the hard- and software of the EBT as well as the experience with physical effects and possible failure sources widens with the number of measurement projects and instrument breakdowns. Our aim is the embedding of this knowledge into the operating environment of our EBT to assure the continuity of research and test activities independent from the actual user's experience. Without great additional effort, the thus collected expert knowledge can be adapted to a tutorial knowledge base for the training of new potential user's or can be made available to other owners of similar EBT systems [5].

2 Implementing KUSET in the EBT Environment

Developing the conceptual structure of the complete system, we watched the following demands:

- simple structure of software and hardware
- automatic measuring and control facilities
- modularity of software and hardware
- portability of the whole system
- links to CAD/CAT computers

These requirements assure good development and maintainance conditions for the expert system and allow the transfer of KUSET e.g. to other beam testing equipment. The realized measurement system is formed by the different components of the EBT, the expert system KUSET and the specially designed, intelligent hardware interface MCS (Modular Control System) which will be described in detail in the following chapter [6].

KUSET was developed on a personal computer using the expert system shell XI PLUS which has proved its worth in a great number of applications. Additional programs for interface control and routines for automatic adjustment of EBT parameters are accessible by KUSET. A simple rule-oriented shell was chosen which runs with minimal computer hardware but puts the essential inference strategies and interfaces at the user's disposal [7]. Consequently, particular interest could be focused onto knowledge acquisition and the adaptation of the expert system to the physics and the hardware world of electron beam testing [8].

Figure 2 reflects the hardware structure of the complete electron-beam test system. The components are linked by databuses. The personal computer plays a central role in providing a platform for the expert-system KUSET and controlling the whole system. The EBT we used for demonstrating our knowledge-based approach consists of a modified scanning electron microscope (Leitz ISI SS130) and a supervisory electron-beam test unit (ABT IL200) which performs voltage measurements and signal-processing. The two probe-stages are controlled by a stage control unit. Different interfaces between the single components had to be installed to prepare the embedding of the expert system KUSET into the operational environment of the EBT. The stage control and the electron-beam test unit are linked to the personal computer by standard RS 232 and IEEE 488 interfaces. An ETHERNET link allows access to the CAD/CAT data and the VAX/VMS computer network. The microprocessor-based Modular Control System acquires analog data of the scanning electron microscope's internal test nodes to give KUSET an online access to the EBT's performance.

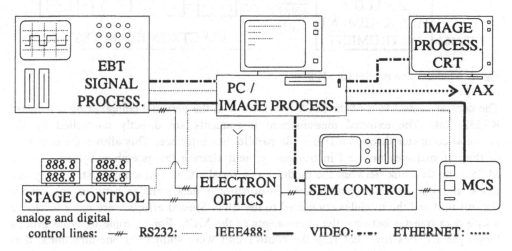

analog and digital
control lines: —⧸⧸— RS232: ········· IEEE488: —— VIDEO: ·—·—· ETHERNET: ·····

Fig. 2. Hardware structure of the complete EBT system

As already mentioned, the whole system was designed with a view to a highly flexible and modular structure to assure its portability to other beam testing equipment. KUSET superimposes the EBT as a separate system. It performs its functions independent of the electron beam tester and can enter data by different data paths. Therefore, even major breakdowns will not affect the performance of KUSET and the MCS.

3 The Modular Control System

The Modular Control System bridges the gap between the hardware world of an EBT and the knowledge-based software of KUSET. It provides a powerful interface to perform analog data acquisition and parameter control. As a stand-alone subsystem it acts on behalf of the expert system controlled by a user defined set of commands.

As the MCS provides the basis for the practical application of our expert system KUSET to our test equipment, we will give some more detailed description of its structure and performance. Without this interface, KUSET would entirely have to rely on interactively obtained data on the EBT's internal parameters. The MCS avoids time-consuming measurements by the operator and allows interactive computer control of different parameters of the EBT such as electron lens or emission currents which are normally hidden from the user. Figure 3 gives an overview of the different components. The MCS consists of a central control unit, a set of modules performing measurement, control and switching functions and one or more affixed commercial measuring instruments.

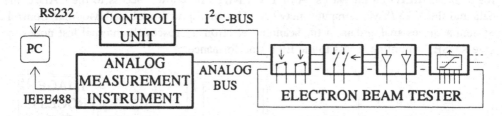

Fig. 3. Modular Control System MCS

The central control unit of the MCS interfaces with the personal computer by a standard RS232 link. The external measurement instruments are directly controlled by the personal computer through a IEEE 488 parallel bus interface. This allows the exchange of the external measurement instrument without altering the assembler software of the MCS. The data link between the modules is a serial I²C bus which consists of only one clock and one data line, but allows bi-directional data transfer between the central control unit and the modules connected to the EBT. Special efforts were made to ensure a safe data transfer between the components of the MCS. For example to avoid ground loops the I²C bus lines entering the control unit were splitted in one-directional data lines to allow the insertion of optocouplers. In general, the modules are designed for only one external supply voltage which is provided by a separate power supply.

Fig. 4. Example for analog data acquisition and D/A modules with I²C interface

The external measurement instrument (e.g. an oscilloscope or a digital multimeter) is linked to the modules by a specially shielded analog bus. Installing the hardware of the MCS in beam testers with complex and decentral hardware structure is easily possible as the control unit of the MCS is connected to the modules only by very few transmission lines. For the same reason, the MCS can easily be adapted to different hardware and measurement requirements. Additional modules are just linked up to the serial analog and data buses. The modules (fig.4) were assembled in a small, modified SMD technology.

They perform different tasks:

- linking the analog bus to internal testpoints of the EBT
- injecting control voltages into the EBT
- switching internal links of the EBT
- readout of digital data from the EBT
- supervision of level-sensitive switches

The range of modules can be adapted to changing requirements since a growing number of integrated circuits with integrated I²C bus interface is commercially available. Supervision of the function and activity of the modules, and communication with the personal computer is performed by the control unit of the MCS. It is based on a 8051 microcontroller with external RAM and ROM and includes different interfaces and self-test abilities. The exchange of commands, error messages and data between the MCS and the personal computer is fixed into easy data formats, and uses ASCII codes to allow a direct operation of the MCS by a computer terminal for debugging purposes.

KUSET can initiate actions of the MCS by calling short C-program modules which control the RS 232 data exchange and send the command sequences to the MCS.

4 KUSET - Structure and Performance

Choosing a knowledge based approach to support our EBT activities we try to take advantage of the separation of problem-specific knowledge from the general processing program which is, of course, typical of expert systems. In this way, we put a highly flexible, modular, and therefore portable tool at our designer's and test engineer's disposal. It is especially the mostly heuristic knowledge and the experience with complex diagnosis and adjustment procedures which can be represented in KUSET in a transparent and well-structured way.

The effectiveness of an expert systems like KUSET, which in a broader sense can be denoted as a diagnostic system [9,10], depends in particular on a well-graded hierarchy to render a fast encircling of the root cause of a problem. KUSET consists of several knowledge bases (KB) which are linked to a hierachical tree-structure. Figure 5 gives an impression of the different main topics. The top level represents the introductory KB. Here, an evaluation of the user's actual interest in the expert system is done, assisted by eventually received critical data from the MCS.

The second level contains KBs which focus on different aspects of operation and maintenance of EBTs such as breakdown diagnosis, tutorial help or parameter optimisation. These KBs are again divided into different sub-KBs (3rd level) containing the concrete expert knowledge for solving the actual problem. A change of sub-KBs or even the KB level can in some cases be necessary to change the focus of the session.

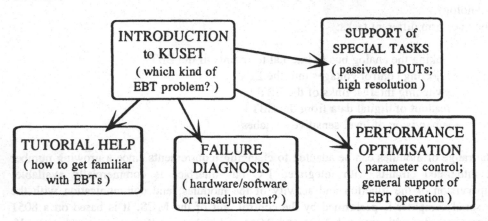

Fig. 5. Main topics of KUSET

The simple hierachical structure of KBs allows a fast introduction of new 'experts' or knowledge engineers to the system's performance and a restructuring or implementation of new knowledge. This is of special importance for us as we intend to transfer KUSET to other test equipment as e.g. our Laser Scanning Microscope (LSM) or different models of EBTs. A close correspondence of the KB's hierachical order and internal structure to the subdivisions of hardware blocks and adjustment or measurement problems has proved to be very useful.

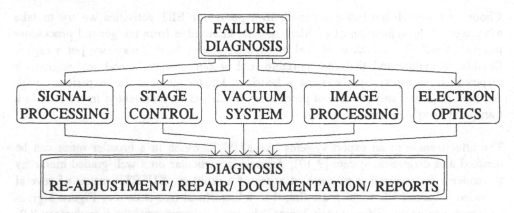

Fig. 6. EBT system failure diagnosis; knowledge bases

As an example, figure 6 reflects the division of the failure diagnosis KB into different sub-KBs, each of them dealing with a separate part or function of the EBT.

The process of Knowledge Engineering (KE) plays a crucial part in building the expert system. The 'natural' organisation of the expert knowledge in question already sets a frame for the structuring of the knowledge bases. For the buildup of the section of

KUSET dealing with diagnosis of system malfunction, e.g. an extraction of structural elements like important symptoms and diagnoses is performed by evaluating diagnostic trees and structural diagrams of the instrument's hardware.

Special efforts were made to optimize the selection of symptoms needed for diagnostic purposes. A minimum number of interactive inputs is very useful to improve the acceptance of an expert system. This can be achieved by including a maximum number of automatically acquired pieces of information. Simultaneously, we tried to optimize the selection of symptoms in respect to their evidential value. For example, if a certain symptom can only be used for one diagnosis in the diagnostic tree, we tried to move it out of the main path of interactive consultation and tried to use it only as a confirmation of an already found hypothesis [11].

On the other hand, a reliable diagnosis can in practice not be made by interpreting a congregation of symptoms which all appear in a vast variety of diagnostic rules. A small uncertainty in the interactive part of data acquisition can easily spoil the reliability of this diagnosis as our experience showed.

We found it very useful to structure the KB and select the symptoms by using schematic diagrams as sketched in figure 7. They give a vivid impression of the transparency and efficiency of the extracted KB to the expert of the focused field of knowledge.

The support of adjustment and control procedures on the EBT orientates itself by the priority sequence given by an expert familiar with the EBT. In parallel, the user's current state of knowledge must be taken into consideration when planning the course of a consultation. When constructing the tutorial part of KUSET, therefore, built-in options are provided that give the chance of changing the run of a consultation in dependence on the results and the user's experience.

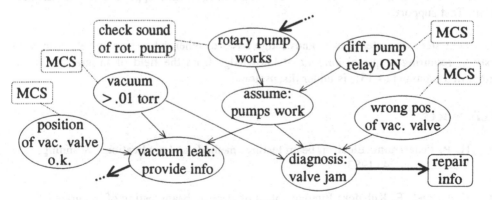

Fig. 7. Diagram of symptoms and diagnoses; section of KB for vacuum-check

The range of KBs accessible by KUSET can easily be extended to meet new demands. For our EBT application a special focus lies on the implementation of additional documentation libraries. Information about difficult repair works like e.g. dismounting of the electron optical column or replacing of spectrometer grids is now saved from getting lost in hidden note books and may be offered instantly to the user's disposal when it is needed.

At the moment, work is under progress embedding IC layout data into the user interface by using a modular CAD/CAT system already introduced [12,13]. We think of a possible application of our expert system to image correction and position adjustment procedures for layout/EBT-image comparisons.

While functionality is in the centre of interest when creating an expert system, we tried not to neglect the design of the user interface when adapting the system to our working environment. Graphic outputs have proved to be very useful in supporting adjustment and diagnosis processes [5,6]. The implementation of symbolic representations of single functional blocks instead of (or better: in combination with) detailed reports enhance the chance of an expert system to be highly accepted by the user.

5 Conclusion

With the support of an expert system, electron beam testers and similarly complex instruments can be operated even by relatively unexperienced users. The expert system KUSET allows, beside the interactive dialogue via user interfaces, the implementation of on-line data from the EBT. The considerable amount of hardware required to gain these data lead to the development of the MCS control system. The modularity of the whole knowledge based support system allows an easy portation to other beam testing equipment.

Beside finishing the knowledge acquisition, a further extension of KUSET and field testing with users is projected. First tests with unexperienced users gave encouraging results and confirmed the efficiency of a knowledge based approach to our Electron Beam Test support.

As a next step, the extension of knowledge bases to other subsections of electron beam testing, requiring a special amount of knowledge from the field of failure analysis in integrated circuits [14,15], is under discussion.

References

1. H. P. Feuerbaum: Electron beam testing: methods and applications. Scanning, Vol. 5, p.14-24, 1983

2. E. Menzel, E. Kubalek: Fundamentals of electron beam testing of integrated circuits. Scanning, Vol. 5, p.103-122, 1983

3. F. Alonso et al.: Knowledge engineering versus software engineering. Data & Knowledge Engineering, 5, p.79-91, 1990

4. G. Buchanan, E. Shortliffe (eds.) et al.: Rule-Based Expert Systems. Addison-Wesley Publishing Company, 1984

5. Navigation through complex diagnostics. Assembly Automation, Feb. 1990

6. M. Ben-Bassat et al.: On the evaluation of real life test expert systems. IEEE, 1989

7. ExperTeam GmbH: Xi Plus Reference Manual. 1990

8. D. Kennett K. Totton: Experience with an expert diagnostic system shell. Br. Telecom Technol. J., Vol. 7 No. 3, July 1989

9. J. Eklund et al.: A knowledge-based system for monitoring and trouble-shooting of production processes. Proc. of Power Plant Dyn., Knoxville, 1989

10. W. R. Simpson: POINTER - an intelligent maintenance aid. IEEE, 1989

11. G. Kahn, J. McDermott: The MUD System. IEEE Expert, Spring 1986

12. J. Fritz, R. Lackmann: Modular CAD environment for contactless testsystems. Proceed. of 2nd EEOBT, 1989

13. J. Fritz, R. Lackmann: A CAD coupled laser beam test system for digital circuit failure analysis. IEEE Proc. of 7th Int. Electr. Manufact. Tech. Symp., San Francisco, Sep. 1989

14. N. Kuji, T. Tamama: An automated e-beam tester with CAD interface, "FINDER": a powerful tool for fault diagnosis of ASICs. Proc. of Intern. Test Conf., paper 23.3, 1986

15. M. Marzouki, B. Courtois: Debugging integrated circuits: A.I. can help!". IEEE, 1989

Knowledge-Based Design of Ergonomic Lighting for Underground Scenarios[1]

Wolfram Burgard and Stefan Lüttringhaus-Kappel and Lutz Plümer

Institut für Informatik III, Universität Bonn, Römerstr. 164, W-5300 Bonn 1

Abstract - In this paper we present BUT[2], a knowledge based CAE-system for the planning of underground illumination in coal mines. The task is to find lighting configurations which are optimal from the ergonomic and economic point of view. The problem is that there is a huge number of possible configurations for lighting satisfying the ergonomic guidelines on underground illumination of coal mines. We present the architecture and main features of BUT, which derives possible solutions to this problem and restricts this set to a manageable set of ergonomically and economically acceptable or even optimal ones.

1 Introduction

Appropriate lighting minimizes the risk of accidents in hardcoal mining. In order to identify objects, sufficient light intensities are required, the value of which depends on the special task to be performed. At the same time, gradual changes of light intensity are desired in order to ease the accommodation of the eyes. Recently the European Coal and Steel Community has adopted guidelines resulting from different research projects and defining the requirements of underground illumination in coal mines [1]. For several reasons, however, it is a technically complex task to fulfill these requirements, and it is even more difficult at minimal or at least reasonable cost.

Until now, there has been no computer support for this kind of planning. The difficulties to be solved are manifold. Different tasks have different requirements on light intensity. With regard to the mine layout, a lot of different scenarios have to be taken into account, which cannot be characterized by simple schemes. Each scenario depends on the timbering, the length of the mine layout, the installed objects and equipment including their dimensions, to mention only a few parameters. A large variety of types of explosion-proof lights is available which can be equipped with different lamps and installed in many different configurations (one or two lines etc.).

2 Planning Underground Illumination

The problem to be solved can be described as follows: given a particular mine layout including its dimensions, installed objects and equipment, identify the tasks to be performed by the miners and determine which minimal light intensities are required in different areas of that mine layout. Then find out which lights with which lamps can be used and where, how and in which distances they have to be fixed. An acceptable solution is one which satisfies the ergonomic requirements in each area, and an optimal solution does this for minimal cost. The overall cost of a lighting configuration depends on the installation cost (lights, lamps, cable, labour cost etc.) and the estimated operating cost (energy, cleaning, maintenance etc.) during the life time of the mine layout.

---

[1] This project carried out by the Ruhrkohle AG, Essen, is supported by the European Coal and Steel Community in the context of the fifth ergonomics program.

[2] But is an acronym for 'Beleuchtung unter Tage', in English: underground lighting.

At present each mine employs a group of experts which plan the lighting of a scenario manually based on their experience with previously planned scenarios. The judgement of ergonomic properties of a proposal is computationally too complex to be performed by the planning experts in advance without any computer support. The only way to verify whether a proposed solution satisfies the ergonomic requirements has been to measure the average lighting intensity and regularity after the installation has been finished.

3 The Planning System

The requirements to a planning system for underground illumination are the ability to represent the necessary information about a mine layout and to implement the guidelines and the heuristics to determine optimal mounting places. Such a system also serves as a tool for the acquisition of the heuristics used by the planning experts. It must provide a great degree of flexibility and user-friendliness, which is a basic precondition for its acceptance.

Therefore, the system must be able to represent complex objects, and to perform inferences with the rules stored in its knowledge base. It must provide a graphics interface to examine proposals for the lighting of mine layouts. To facilitate the implementing and test phase, an explanation tool is required allowing the access to technical documents concerning the domain. A final requirement is the integration of this planning system into the existing environment containing personal computers, mainframes and company's data bases and software systems.

To guarantee a maximum amount of flexibility we decided to implement BUT in Prolog. Logic programming provides both a clear semantics and a practical language so that readable, efficient problem solving programs can be implemented [2, 3, 4]. Knowledge is represented explicitly, and meta-programming can be used to extend the expressiveness of the logic programming language [5]. For pragmatic reasons we had to implement BUT on an MS-DOS PC.

The planning process realized by BUT consists of the following steps. First, the user has to specify the actual mine layout. In order to minimize the amount of information which has to be entered, we have built a knowledge base containing a taxonomic hierarchy of mine layout types. The various types, for example, contain information about possible equipment and possibly installed objects. The inheritance mechanism in the taxonomic hierarchy is top-down, although it is possible to specify exceptions to this rule. According to [6], a taxonomic hierarchy can be represented in Prolog as follows:

```
frame('stone drift', is_a, drift).
frame('stone drift', 'conveyor system', 'band conveyor').
frame(drift, equipment, ramp).
frame(drift, 'conveyor system', 'rail transport').
```

The next step in the planning process is the identification of tasks to be performed by the miners. To identify such tasks, spatial reasoning about the equipment of the mine layout is combined with knowledge extracted out of the guidelines and heuristics acquired from the planning people. In order to implement the guidelines, we formulated them as first order sentences before we transformed these sentences to clausal form. Inspired by [7]

493

we already applied this technique successfully to build up the knowledge base of the diagnostic expert system described in [8].

The guidelines and heuristics, however, often contain vague phrases such as: 'Material is transshipped at the boundary of two transport systems.' The concept 'boundary', however, is vague and not defined precisely. A heuristic to identify a place of transshipment is that there are at least two transport systems close to each other and, additionally, at least one of both systems begins or ends at the particular place. The distance can be computed from geometric data entered by the user [9].

Generally it is possible to deduce several different tasks for a certain point in the mine layout. In this case BUT sorts the tasks according to the priorities implied by the guidelines.

Next BUT determines reasonable places for mounting the lights. For example, the lights are mounted either at the band conveyor or at the stope. For that purpose we store for each object of the equipment a set of points where lights may be fixed.

In the next step BUT generates possible configurations using these mounting places and applies an algorithmic approach, realized in C, to choose lights and lamps and to minimize the number of lights by maximizing their distances. Finally all solutions are judged with respect to their estimated cost.

Graphical simulation of the lighting configuration resulting from a generated plan has been identified as a key component for user friendliness and interactivity. The user can assess directly the pros and cons of a given proposal. He can focus on arbitrary parts of the scenario to analyze critical regions in more detail, and on those regions where only a crucial amount of light is provided. In the actual implementation, BUT allows a top view including the lighting intensity and a sectional view. An interpreter draws all objects, which are represented by vector graphics, in a particular mine layout according to the actual view.

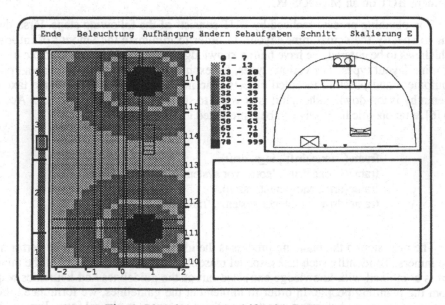

Fig. 1. Planning a stone drift with BUT

Figure 1 shows a screen dump displaying the result of a session. The scenario includes a stone drift containing a ramp, a rail, a band conveyor, explosion barriers and conduits. The illumination resulting from the planning process is shown in the top view.

4 Conclusions

In this paper we presented a planning system called BUT for underground illumination in coal mines. BUT applies the guidelines on underground illumination in coal mines to constrain the set of possible lighting configurations. Furthermore, heuristics acquired from planning experts are used to select reasonable lights and lamps and ways to mount them in the mine layout. To implement BUT, we integrated logic programming techniques, object oriented programming, algorithmic approaches and graphics interfaces. Besides two algorithms written in C, one tackling distance optimization and the other computing a matrix of lighting intensities for the graphical simulation, the whole system has been implemented in Prolog. Despite of the restrictions induced by the MS-DOS operating system, BUT is able to cope with the requirements of a real industrial environment.

Acknowledgements

A. B. Cremers (University of Bonn) and R. Grevé (Ruhrkohle AG) initiated this project. J. Grebe (University of Dortmund), F. Mücher and F. Ende (Ruhrkohle AG) made substantial contributions.

References

[1] Guidelines on the Ergonomics of Underground Illumination in Coal Mines. Tech. Rept. 15, Community Ergonomics Action, European Coal and Steel Community, Luxembourg, 1990.

[2] O'Keefe, R.A. *The Craft of Prolog*, The MIT Press(1990).

[3] Sterling, L. Logical Levels of Problem Solving. *Journal of Logic Programming 1*(1984), 151-163.

[4] Sterling, L. and Shapiro, E.Y. *The Art of Prolog: Advanced Programming Techniques*, The MIT Press(1986).

[5] Sterling, L. and Beer, R.D. Metainterpreters for Expert System Construction. *Journal of Logic Programming 6*(1989), 163-178.

[6] Kowalski, R. *Logic for Problem Solving*, North-Holland, Amsterdam New York Oxford(1979).

[7] Sergot, M.J., Sadri, F., Kowalski, R.A., Kriwaczek, F., Hammond, P., Cory, H.T., The British Nationality Act as a Logic Program, *Communications of the ACM 5* (1986).

[8] Arnold, G., Burgard, W., Cremers, A.B., Lauer, M., and Lüttringhaus, S. Ein Expertensystem zur Prüfung elektrischer Betriebsmittel für explosionsgefährdete Bereiche. In *Proceedings 2. Anwenderforum Expertensysteme*, Cremers, A.B. and Geisselhardt, W., 1988, pp. 57-64, in German.

[9] Grebe, J. *Wissensakquistion für ein Expertensystem zur Planung von Beleuchtung unter Tage*, Diploma thesis, in German, University of Dortmund, 1991.

A Monitoring Approach Supporting Performance Analysis of Expert Systems for the EMC Design of Printed Circuit Boards

R. Brüning[1], W. John[2], W. Hauenschild[1]

[1] University of Paderborn, Department of Computer Science – Cadlab
W-4790 Paderborn, Germany
[2] Siemens Nixdorf Informationssysteme AG – Cadlab
W-4790 Paderborn, Germany

1 Introduction

This paper describes a strategy for the evaluation of the performance behaviour of expert systems, which have been built with the expert system shell Twaice[14]. Rule based expert systems often suffer under a low user acceptance due to the performance behaviour. This applies increasingly for systems built with an expert system shell which often have a slow response time and require a large memory consumption. Solving this performance problem requires the identification of its causes. For this reason the software monitoring approach presented in this paper has been developed. A major accomplishment in this context is that a measurement of the CPU time requirement during a consultation for a problem solution (and all temporary results) can be performed by use of this monitoring approach. The evaluation methods are based on a software monitoring concept using internal Twaice messages for an eventing approach. With this monotoring system the timing behaviour of two expert systems has been analyzed, some of the results are shown in this paper exemplary. The examined systems are intended to solve several tasks of EMC analysis process during the design of printed circuit boards and other electronic subsystems. The consideration of EMC–effects during the design process is a highly complex task, which is a suitable domain to be processed with knowledge based systems [1],[13].
Based on the measurement results a strategy for speeding up existing expert systems has been developed and tested. In a modification–measurement cycle it was often possible to enhance the expert system behaviour, so the modified expert systems were measured again to record the changes in the system behaviour.
After examining the different expert systems, a set of rules for the design of well performed knowledge based systems has been defined; some of these rules are discussed in the last section of this paper.

2 Motivation

Different expert system prototypes were developed in the Cadlab tool environment for the EMC analysis of printed circuit boards (PCB) and for the design of system components under EMC constraints by use of the expert system shell Twaice. This shell is an environment for building expert systems based on IF/Prolog[14]. It provides the knowledge engineer a hybrid knowledge representation mechanism with rules and frames, a modifiable inference engine, an explanation component, interfaces to data bases and programing languages and other features that are neccessary and helpful for building knowledge based systems.

The expert system prototypes are planned to solve several tasks of EMC–analysis. With the application of faster electronic components layout and electronic designers need assistance for the consideration of EMC–problems. During the design process of PCBs the engineer has to account for various effects such as reflection and crosstalk. In many cases errors caused by such noise effects can not be located before the prototype is tested or respectively ascertained by the service technician at the customer. In order to achieve this problem the electronic designers must be supported with suitable EMC–tools. Therefore several expert system prototypes have been designed. Two of this systems were further developed near to a usable status. These systems were the PCB analysis expert system CaPax (Cadlab PCB Analyse Expert) and BaplEx (Backplane Expert), an expert system for the design of computer system backplane components under EMC constraints.

The bad performance behaviour of these systems is considered to be their main disadvantage. To analyse a typical printed circuit board of a computer system the PCB analysis system CaPax requires several hours of computation (up to 20 hours). One reason for this timing behaviour is the high amount of data needed for the internal representation of a printed circuit board (more than 2 MB, e.g. component data, transmission line data, layer data ...). The same effect applies for the backplane design and analysis system BaplEx. The result was, that these systems were not well accepted by the users. Therefore a strategy to overcome such performance problems is needed. In a first step it is important to identify the reasons for this slow performance behaviour.

3 Expert Systems and Performance Measurement

Expert systems are often constructed to solve problems with an unstructed problem space. In general for these problems no good algorithmic solution exists. One characteristic of such problems is that the number of possible solutions, i.e. the number of nodes in a problem tree grows exponentially once the complexity of the problem increases. In rule based systems with most simple search strategies within the rule set, the solution of the problem often does not satisfy real complex applications.

The slow execution speed and in most cases unacceptable memory occupation of rule based systems often restricts their use in domains requiring high performance[8].

One way to overcome this problem is to implement expert systems on more powerful hardware architectures or on dedicated AI systems. But as this strategy is incapable for solving the fundamental insuffiency of most rule based systems, a large number of expert system prototypes were rejected after the evaluation phase and only a small rate of the prototypes — less than 10 % — will be further developed into real working applications[9].

The problem of evaluating the performance of knowledge based systems is that the knowledge within these systems seldom is complete. Due to the complex real applications it is often not possible to build a complete representation of the situation. Therefore it is difficult to decide wether the performance problem is due to the uncomplete knowledge of the systems or to an incompetent system design.

3.1 State of the Art

There exists no widely accepted model to measure the performance behaviour of expert systems in relation to the size and the structure of the included knowledge. In addition, no performance data for the common used expert systems environments is available. For PC based expert system shells some attempts to describe the dependence of the number of rules and the problem solving time exist [12]. These examinations are made with artificial applications and simple rules. The results of the analysis are not appropriate for estimating the performance behaviour of real (often time critical) applications. For this reason, they can only be used as a suggestion for the selection of an expert system tool, but cannot assist a knowledge engineer in constructing an efficient expert system.

As of today, only a small number of publications which deal with the performance of expert systems do exist. One reason is that the term 'performance' in the context of expert systems often differs in meaning of its typical use. In most publications *"expert system performance"* stands for the quality of the solutions found by the system compared to the solutions of a human expert.

Some researches on the acceleration of OPS5–based expert systems were made [2], [3]. Unfortunately this work is very specific to OPS5 and the main interest focusses on the implementation of such systems on parallel hardware architectures.

3.2 Basic Concepts

The structure of typical rule based systems prevents the prediction of the exact system behaviour. This is due to the time dependence of the response on the structure and the arrangement of the objects in the knowledge base and from the order of the system input. Hence it follows that the nondeterministic runtime characteristics cannot be estimated by the knowledge engineer, for they depend on the passing through the rule tree during a consultation. But an accurate system design can help to increase the system performance very effectively.

All rule based expert systems share the same fundamental rule processing mechanism, regardless wether they work with backward or forward chaining. This mechanism is a cyclic process and consists of the following three phases :

1. **Determination of the conflict set** : The conflict set consists of rules which conditions are true. These rules are candidates for the next step of the problem solution process.
2. **Conflict resolution** : At this level of problem resolution the selection of the next firing rule takes place.
3. **Action** : The execution of the chosen rule.

This cycle terminates if the system has computed a problem solution based on its embedded knowledge and of the input data (in Twaice the cycle can be broken by activating Prolog methods during the consultation). In backward chaining systems the conflict set consists of such rules which conclusions compute the current demanded fact. In forward chaining systems the rules with valid conditions build the conflict set. In the rule processing cycle the determination of the conflict set (phases 1) is the most critical step in the inference process. The results of some investigations show, that nearly 80 % of the problem resolution time for rule based expert systems was spent on phases 1 and 2, only 20 % of the time was needed for the rule execution. Hence it follows, that the performance of an expert system can better be accelerated by improving the first two phases than with an optimizing of the rule execution.

3.3 Requirements for a Performance Analysis of Expert Systems

The performance of typical large software systems will be measured by benchmark programs or by software monitors. Benchmarks require the existence of accepted benchmarks programs (like the SPEC benchmark suites for hardware benchmarks). Currently such a set of benchmark programs is in development for database management systems, but for knowledge based systems there exist no known benchmark program. Therefore for the required examinations a monitoring program for expert systems has to be constructed. The required analysis and measurements can be done by :

- executing a measurement program,
- modificating the examined system, or by
- a modification of the operation system.

Because of expert systems represent a special type of software systems, specific features for an expert system monitoring program are needed. Some of these features are shown below :

- Expert systems normally are software products with a high changing rate. Because of the possibility of changing knowledge, these systems have to be a part of a dynamic development process. Therefore it is necessary to measure the performance of expert systems during any phase of the system development process. An update of the knowledge should not prevent further performance analysis.
- Consistency preservation of the system results; the performance analysis should not falsify the solutions computed by the system. The quality of the solutions must be prevented from side effects.
- Consideration of the measurement overhead; the modification of the examined system is necessary for its performance analysis. As every measurement based on a software modification produces an overhead, this overhead has to be analyzed and considered when interpreting the measurement results.

3.4 Performance Measurement on Twaice Expert Systems

Aim of the evaluation of Twaice based expert system's performance data was to obtain information about the time needed for the derivation of facts during a consultation. Facts are the values of the object–attribute slot entries in the object frame hierarchy. A part of the taxonomy schema of CaPax is shown in figure 1.

```
root                             RULE 100
   PCB                           IF Transmission_Line . Critical_Length < 20
      Subnet                     AND Transmission_Line . Prop_Delay > 5
         Transmission_Line       AND Driver . Function = NAND
         Resistor                THEN Resistor . Value_Min = 200
         Driver                  END
         Receiver
               :
```

Fig. 1. Part of the CaPax Taxonomy **Fig. 2.** Rule example with different attributes

In Twaice it is possible to create several 'attributes' for each object by defining them in the taxonomy. An attribute value can be determined from the value of several other attributes.

Because the objects build a hierarchical structure, the tracing time for an attribute may depend on the derivation of other attributes. For the knowledge engineer two time dependent data are of special interest :

- the time total needed for the the derivation of a fact (including the time needed for all depending facts) and
- the *real derivation time* of an attribute.

The 'real derivation time' is a degree for the time–requirement for the rule selection and activation, the search for fact values into the internal data base and the other activities of the system during a problem resolution.

Figure 2 shows an example of a rule with different attributes. The dependency of the attribute values is defined in the rules of the knowledge base. The attribute 'Value_Min' of the object 'Resistor' depends on the attributes 'Critical_Length' and 'Prop_Delay' of the object 'Transmission_Line' and on the value of the attribute 'Function' of the object 'Driver'. As a conclusion, the total derivation time of the fact 'Resistor . Value_Min' includes all the derivation time of this three attributes. Therefore a varied view of the attribute derivation time is needed.

The described data is to be measured for selective areas of the knowledge base. This enables the knowledge engineer to estimate the influence of the internal Twaice processes on the problem solving time.

3.5 Monitoring Expert Systems

There exist two fundamental monitoring strategies, the **eventing approach** and the **sampling approach**. With a sampling–monitor the investigated system is measured in regular intervals. For a close software system like Twaice this can only be realized with a high effort. Therefore an eventing approach was chosen. For the implementation of the monitor the internal eventing concept of Twaice may be used. During a consultation the inference component of Twaice produces a lot of events for special purposes (the start of a rule trace for an unknown variable for example). These events can also be used for the activation of prolog procedures called 'methods'. By means of these methods the knowledge engineer can realize some special computations or other tasks. In this paper we are going to set up some measurement procedures to be called up by a measurement method at the arrival of an event.

The following Twaice events were used for the performance analysis :

- **if_needed** : the value of an attribute is needed, the derivation of the fact starts.
- **if_traced** : the value of an attribute has computed (this event will also be triggered if the fact's value is 'unknown').

The monitor produces selected data for the objects created during a consultation as well as the time needed for computation of the attribute values for the different objects.

Modifying the knowledge base for such measurements is also possible. In this case additional method–calls for all objects and attributes must be inserted into the frame hierarchy. These Prolog methods measure the required time for the calculation of fact values and the measurement methods described above. Therefore Twaice expert systems can also be monitored without changing the code of the inference component. For the analysis of the resulting measurement data, an interface to different graphical postprocessors has been implemented. These postprocessors are AnaRes [6] and Gnuplot. The interface is part of a program which generates the input for the different postprocessors out of the measurement data.

4 Well Performed Expert Systems

A further centerpoint of the presented work was to describe a strategy for speeding up existing expert systems and the definition of a set of heuristical design rules for well performed expert systems. The activities for the optimization of existing systems were tested on different systems, the results of the enhancement of the BaplEx system are also described in this section.

4.1 Activities for a Performance Enhancement

In the following, different possible steps for a speeding up of existing expert systems are presented :

- A modification of the object hierarchy and a reduction of the number of rarely used objects, for this it is advisable to create some additional slots on other objects.
- An optimization of the rule structure. The speed of the inference process depends also on the structure and the arrangement of the rules within the knowledge base.
- A reimplementation or modification of the conflict resolution strategy with Prolog procedures.
- If possible, a splitting of large expert systems into smaller applications.

These and other activities have been carried out to improve the existing systems. An essential result was, that the timing behaviour of most systems could be enhanced. Still for the most problematic system – the PCB analysis system CaPax – the timing behaviour persists a problem. One reason for this is the enormous amount of data needed for an internal representation of a printed circuit board. Each net of the PCB must be transferred into a graphical structure, for large boards the number of nets and so the number of nodes and vertexes of these graphs increases dramatically and causes many problems. In applications with a dynamic generation of more than 3 MB of data, the performance behaviour of Twaice decreases rapidly.

4.2 Acceleration of the System BaplEx

As mentioned above, BaplEx is one of the expert system prototypes which are under development for an EMC–analysis environment. BaplEx is an expert system for the design and analysis of computer backplane boards under EMC constraints. The system applies to the following tasks :

(1) selection and computation of the terminator–resistor–networks of backplanes,
(2) hints for selection of the adequate transmission line type, depending on the clock frequency and other EMC constraints,
(3) assignment of the signals to the the backplane connectors pins,
(4) calculation of the optimal cycle–time for enhancing backplane performance.

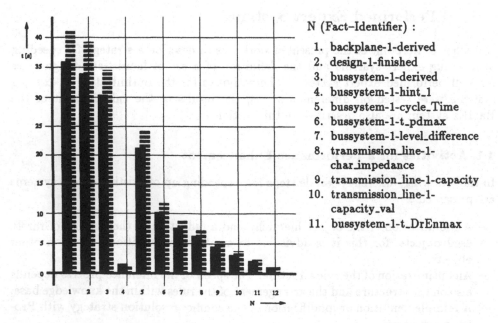

N (Fact–Identifier) :

1. backplane-1-derived
2. design-1-finished
3. bussystem-1-derived
4. bussystem-1-hint_1
5. bussystem-1-cycle_Time
6. bussystem-1-t_pdmax
7. bussystem-1-level_difference
8. transmission_line-1-char_impedance
9. transmission_line-1-capacity
10. transmission_line-1-capacity_val
11. bussystem-1-t_DrEnmax

Fig. 3. Overall–Time of BaplEx and the Sub-System-1

The system BaplEx consists of 146 rules and 27 objects with overall 200 attributes. It uses 5 Prolog procedures for external computations, the Twaice process grows up to 2 MB during a consultation.

For a speeding up of BaplEx, two different methods were chosen: first, the dependency between the different objects has been modified, the total number of objects and attributes was reduced. By this, the timing behaviour of the system improves up to 10 percent.

To overcome the system inefficiency in a different way, BaplEx has been divided into four autonomous subsystems. Each subsystem is designed to solve one of the tasks (1 – 4) explained above. By this strategy, the timing behaviour of each single system for a problem resolution improves about 15 to 30 percent compared to BaplEx. The results of the comparison of the subsystem for the first task with BaplEx are shown in the figures 3 and 4. The figures show the cpu time need for the derivation of selected facts of the dynamic data base. Figure 3 presents the 'Overall–Time', including the time needed for the derivation of all dependent facts. On the other hand figure 4 shows the pure derivation time ('Net–Time') of the facts for the expert systems in comparison. This comparison reveals, that rules which are not activated and objects

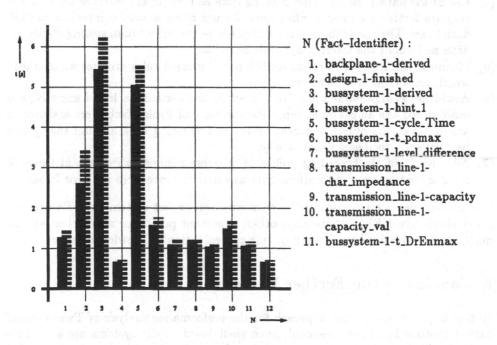

N (Fact–Identifier) :

1. backplane-1-derived
2. design-1-finished
3. bussystem-1-derived
4. bussystem-1-hint_1
5. bussystem-1-cycle_Time
6. bussystem-1-t_pdmax
7. bussystem-1-level_difference
8. transmission_line-1-char_impedance
9. transmission_line-1-capacity
10. transmission_line-1-capacity_val
11. bussystem-1-t_DrEnmax

Fig. 4. Net–Time of BaplEx and the Sub-System-1

which are not referenced are time intensive with respect to the problem resolution. Hence Twaice results in a performance overhead once the knowledge base grows. This is a further reason why building a Twaice expert system should be carried out carefully to avoid such effects.

4.3 Design Rules for Efficient Expert Systems

The results of the analysis and modification of different expert systems allow the definition of a set of design rules for efficient expert systems. Some of these rules are presented below :

(1) Avoid unnecessary object creations and minimize the object hierarchy. An object instance is a very expensive action during the inference process of Twaice. For rarely needed objects it seems to be more efficient to define additional attributes on existing objects than defining additional objects.

(2) Limited use of multivalued attributes. Multivalued attributes are facts with more than one true value. This attribute feature is to be defined by the knowledge engineer during the knowledge base implementation. During a problem resolution, all rules of the knowledge base dealing with a multivalued attribute have to be tested in order to trace the value of this attribute.

(3) Use of backward rules instead of forward rules. Twaice allows the definition of forward rules instead of the common used backward rules. Problem resolution with forward rules is more time expensive than the default backward chaining.

(4) Use of the data base interface. Storing data as facts in the internal data base is very ineffective for most applications. Twaice offers a special interface to SQL data bases. The use of this interface is much more efficient than storing all needed data as facts of the internal dynamic data base.

(5) Minimize the number of rules. Additional or unused rules produce an overhead which costs time and memory.

(6) Avoid knowledge redundancy. In the examined systems, a lot of knowledge is multistored. As an example, input data in form of Prolog facts were assigned to attributes of the object hierarchy. With an increasing data amount the system performance decreases dramatically.

(7) As a result of the preceding points, it is recommended to implement time- or memory extensive computations only as external procedures (in C or Prolog).

The above rules are relevant for the creation of an expert system with Twaice. For other shells like KEE[5] there exist other important points for an efficient system design like knowledge base compiling, rule concentration or knowledge base chaining.

5 Conclusion and Further Research

In this paper, a monitoring approach for the performance analysis of Twaice–based expert systems has been presented. Such shell based expert systems are sometimes unacceptable in their memory occupation and response time. Experience shows, that in many cases these problems base on a defective system design. With the presented monitoring strategy some of the design failures can be identified and corrected.

In the second part of this paper, activities for a performance enhancement of existing expert systems were shown and a set of design rules for efficient Twaice–based expert systems was presented. These results base on the analysis of different expert systems. The design rules for expert system are meant as heuristical hints, which should help the knowledge engineer to build better performing expert systems. The design rules cannot guarantee an efficient system, but they can help to avoid some design failures. For further expert system projects, the results of the examined systems — the presented BaplEx is only one of several expert system prototypes dealing with EMC problems — are to be used as a guideline for building well performed expert systems. Few of the resulting design rules are Twaice specific, others can be adapted to other expert system development tools or can be generalized for expert system design.

References

1. Brancaleoni,C., Cividati, E., Delle Donne, R.: A Design Rule Checker Assistant for Digital Boards. IEEE Workshop on AI–Applications, Munich, October 1987, 69 – 76.
2. Gupta, A., Forgy, C. L.: Measurements on Production Systems. Technical Report CMU-CS-83-167, Carnegie Mellon University, Pittsburgh, USA (1983).
3. Gupta, A.: Parallism in Production Systems. Pitman/Morgan Kaufman Publishers Inc., New York(1987).
4. Hayes–Roth,F.: Towards Benchmarks for Knowledge Systems and Their Implications for Data Engineering. IEEE–Transactions on Knowledge and Data Engineering, Vol. 1, March 1989 : 101–110.
5. KEE 3.1 : Technical Manuals, Vol.1–3, Intellicorp Inc., March 1988.
6. John, W. , Wening, J.: A Graphic–Postprocessor Baesd on X-Window to Analyse the Results of Technical and Scientific Oriented User Programs. DECUS Europe Symposium, Cannes, September 1988.
7. John, W. : An EMC–Analysis–Workbench for Component Design Based on a Framework Approach. Micro System Technologies 91, 2nd Int. Congress and Exhibition, Berlin, Germany,October/November 1991.
8. Laffey, T. J, Cox, P. A, Schmidt, J. L, Kao, S. M, Read, J. Y. : Real Time Knowledge Based Systems. AI Magazine, Number 1 1988, 27 – 45.
9. Mertens, P., Borkowski, V., Geis, W.: Betriebliche Expertensystem-Anwendungen. Springer Verlag, Berlin (1990).
10. McKerrow, Ph.: Performance Measurement of Computer Systems. Addison-Wesley Publishing Company (1988).
11. O'Keefe, R., Balci, O., Smith, E. : Validating Expert System Performance. IEEE-Expert, Number 4, 1987, 81–90.
12. Press, L.: Expert System Benchmarks. IEEE-Expert Spring 1989 37–44.
13. Simoudis,E.: An AI System for Improving the Performance of Digital Circuit Networks. IEEE Workshop on AI–Applications, Munich, October 1987, 104 – 118.
14. Twaice, Produkthandbücher. Siemens Nixdorf Informationssysteme AG, Paderborn, München, Jan. 1991, Rel. 4.0.

This research was sponsored by BMFT (Bundesministerium für Forschung und Technologie) of the Federal Republic of Germany under grant 13AS0097.

Representing Geometric Objects Using Constraint Description Graphs

Borut Žalik, Nikola Guid, and Aleksander Vesel

University of Maribor, Department of Computer Science
62000 Maribor, Smetanova 17, SLOVENIA

Abstract. The paper demonstrates how geometrical constraints can be applied to add a new level of abstraction to description of geometrical objects. Special attention is given to the interactive insertion of constraints. To support incremental design each inserted constraint has to be solved as soon as possible. Because of this requirement a local propagation of known states is used for constraint solving. It is supported by a biconnected constraint description graph structure. The benefits of this structure are insensibility to the order of inserted constraints and ability of replacing constraints with their inverse couples. To override the ambiguities at constraint solving the approximal values of geometrical elements which are inserted through a sketch are used. From the biconnected constraint description graph an acyclic constraint description graph is generated easily. It is suitable for the generation of instances of generic objects.

1 Introduction

In many areas of engineering the information needed in design deals with the shape of artifacts. Geometrical information distinguishes importantly from other information which also describes the artifacts and which is included into design process (e.g. material, price, client). Namely, geometrical information does not define only how a geometrical object looks like, but it determines the majority of the functional properties of the object. During design and manufacturing process geometry takes various roles. Usually, the design is started with a sketch which is needed for breaking the communication barrier among people working at the same project. On this level of design all other aspects of the geometry are insignificant. The exact geometry is determined later during a detailed design [12]. The majority of the actual commercial geometric modeling systems are not able to represent incomplete geometrical information needed in different phases of design. The reason is that almost all geometric problems arising from geometric modeling and computer graphics have been implemented and solved by numerical algorithms exclusively. They require complete information of the used geometric primitives. Therefore the approximal values initially inserted by a sketch are completely useless. In these systems geometrical concepts are immediately converted into numerical parameters, because numerical methods can operate only on numbers. They can not use symbolical data. However, it would be expected that the transformation of information from symbolical to numerical form is reversible [4]. Because of well-known problems of the final arithmetic, a loss of geometrical relations can occur, although these relations have been explicitly given during the creation of the object. This is the reason why algorithms are unstable and why they can make the wrong conclusions from completely correct input data. A good system should store and recall all information which has been given explicitly by usage of the symbolical structures [4]. On the other hand some problems can not be solved without arithmetic operations (e.g. determination of the intersection point between two non parallel lines). Because of this, symbolical and numerical techniques must be coupled together in the most powerful way.

In order to automate the designing process, the designer should be gradually replaced by an intelligent geometric modeling system which should understand the designer's intent [15]. A way to make a geometric modeling system more intelligent is to introduce the concept of geometrical constraints and their automatic solving. In this paper experimental geometric modeling system named FLEXI based on geometrical constraints will be presented. Some benefits of using the constraint description graph for description of the geometry of an geometrical object will be shown.

2 Background

The declarative description of truths that must exist is more natural than specifying the procedures to obtain these truths [19]. In geometric modeling it is also easier to state a set of relations among a set of geometrical elements without thinking of the algorithms used to produce geometrical objects. For example, constructing of a regular hexagon using procedural graphical system (e.g. GKS) forces the user to use trigonometric functions to obtain coordinates of the hexagon. It would be easier for the user to draw an optional hexagon and to require that:

- its vertices lie on a circle with a radius r and a center at point (x, y) and
- all hexagon's edges have the same length.

A relation among one or more geometrical elements (or objects) is called a constraint. Constraints are satisfied by the constraint satisfaction system without help of a human. Some of the methods for constraint solving are [9]: local propagation of know states, relaxation, propagation degrees of freedom, graph transformation, equation solving. It is much easier to state constraints that to satisfy them. This is the reason why constraint-based systems are usually very application specific. It is very hard to extend them even to the very similar domains (for example from 2D to 3D space).

The idea of using constraints in geometric modeling appeared in the early sixties when Sutherland developed his famous Sketchpad [17]. The algebraic approach of variational geometry has been later discussed by Lin et al. where a modified Newton-Raphson method is used to derive new geometry [10]. His 3D system includes the surfaces of the second order, too. Nelson presented Juno, a constraint-based graphics system [13]. Constraints are converted into nonlinear equations. Since the behavior of a numerical solver depends on the initial value, Juno allows the user to give a hint to the solver about where to look for the solution. Aldefeld made a system based on geometric reasoning [1]. The solution is found in two steps. The first step determines the order in which the constraints are used. It is called a construction plan. The data-driven inference method is applied for this task. In the second step construction plan is used for the determination of the position of the characteristic points. Suzuki et al. used a logical framework to represent geometrical objects and constraints. A geometric reasoning mechanism has been used for solving these constraints [18]. Kimura proposed a system based on the first-order predicate logic in the combination with object-oriented techniques for dimensioning and tolerancing in product modeling [7]. Kondo developed a geometric modeling system with parametric design capability based on non-manifold modeling. This system is able to manage 2D and 3D constraints stored in a history graph [8]. Dahshan noticed that designers loose a lot of time on small modifications. He proposed a system named OPAL where modification process has been automated in an object-oriented environment [5]. Shimada also noticed that instead of giving all details of a shape, the designer can give only geometrical constraints. After that the resultant shape is automatically generated by evaluating constraints. He found out that it was more natural for the designer to express his/her shape design intent declaratively in the form of geometrical constraints. Using this tool one is able to change a shape easily and flexibly, because necessary modification and consistency management are automatically done by the system [16]. The practical usage of some systems is often restricted because of their non-interactive behavior and the absence of a good user interface. Van Emmeric proposed an experimental 3D modeling system, GeoNode, based on constraint modeling in CSG where constraints are specified graphically and evaluated in real-time during the manipulation of a model [6]. A survey of several previous works based on geometrical constraints has been made recently by Roller et al. [14]. He divided all methods appearing till now into three groups: primary, algebraic, and artificial intelligence approaches. He concluded that artificial intelligence approaches and algebraic approaches should be incorporated in a common framework to achieve better performances.

3 FLEXI's architecture

The declarative description of a geometrical object should be supported with incremental design. A design community widely accepts incremental design. It has been used for a long time and it is also expected that it is going to be predominant in design for many years to come [15]. Considering modeling with constraints, the designer sets the facts about an object. We can not expect that the user immediately sets enough geometrical constraints which are correct, unambiguous, and without conflicts. The effect of each inserted constraint should be visible immediately, or as soon as possible, giving the user the opportunity correcting himself. The pure declarative approach itself will not inform what has gone wrong if the resulting shape differs from the designer expectation.

The purpose of our system named FLEXI is not to replace the user, but to liberate him/her from boring and time consuming tasks. The architecture of our system is shown in Fig. 1. FLEXI accepts at input a sketch of a geometrical object. A simple 2D drawing system for the insertion of initial geometrical and topological data is used in our case. A topology and the approximal values of an object's geometry have to be determined before a detailed design. The constraints can not change the topology of a geometric object as it is allowed in some other constraint-based geometric modeling systems (e. g. in the system proposed by Ando at al. [3]). In our system the topology is frozen after the sketch is finished. Geometrical objects are represented with the boundary representation (B-rep) which is capable to describe objects of arbitrary shape to any desired level of detail [11].

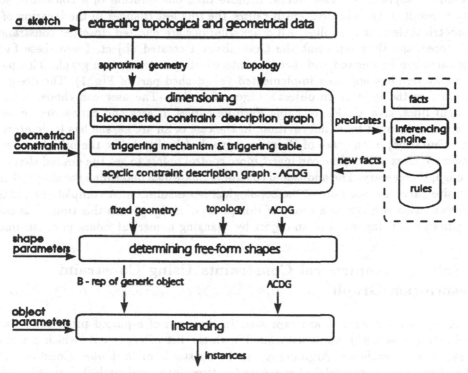

Fig. 1. System configuration

Design is started with a sketch of a geometrical object. Its dimensioning is done by inserting of geometrical constraints. Constraints are set in the form of predicates. Some of them will be briefly discussed later. To support the incremental design, the effect of each inserted constraint should be visible as soon as possible and the appearance of an object has to be immediately updated. A constraint solving method known as local propagation of known states (shortly named just local propagation) satisfies this requirement. The local propagation is the fastest but the least powerful technique for constraint solving. It can be observed as a chain-reaction sequence of computations. Unknown value of one parameter is calculated using other parameters whose values are already known. When values of all input parameters appearing in a constraint are presented, this constraint is solved (or fired) and its output parameters are determined. Output parameters are input for the other constraints. They trigger the solving of these constraints. We call the whole process the constraint propagation. The constraint propagation is managed by a constraint description graph mechanism. Two forms of constraint description graphs are included in our system (biconnected constraint description graph and acyclic constraint description graph). They will be briefly discussed in the next section. During the constraint propagation only exact values of geometrical elements can be used. The selection among alternative solutions is done with the help of approximal values of the geometrical elements. The principle of a minimal disturbance in the position of a particular geometrical element is applied. In other words, if more than one solution of a constraint solving is resulted, the solution which causes the smallest changes in the location of a geometrical element is applied. All other solutions are rejected. Inserted constraints are stored and they represent the facts about a created object. From these facts new facts can be derived and included into constraint description graph. This part of our system has not been implemented yet (dashed part of Fig. 1). The designer can change the type of an object's edges after that. The user can choose among straight lines, arcs, circles, or cubic Hermite curves. If arcs and curves are needed, he/she has to input more information. In the case of an arc he/she has to determine the orientation, in the case of Hermite curve, he/she has to fix the initial and end tangent vectors and he/she can insert more control points to get the desired shape of an edge, if necessary. The additional control points which determine the shape of the Hermite curve are not part of the topological representation. A complete boundary representation is given as a result of this phase of design. From this time, it is easy to generate new instances of an object by changing numerical values in constraints.

4 Solving Geometrical Constraints Using Constraint Description Graph

In our system constraints are expressed in the form of n-placed predicates. Each predicate consists of a name associated with the list of arguments which are enumerated in parentheses. Arguments can be constants or variables. Constants are taken from a set of geometrical elements (vertices, lines, and circles). Variables refer to the values of angles, distances or coordinates. Constraints appearing in geometric modeling have been divided into two groups by Aldefeld [2]:

1. Constraints fixing coordinates, distances, angles, radii etc. are called metric constraints. Constituent elements of their description are numerical variables. Instances of a generic object can be obtained by stating different values to the variables.
2. Structural constraints express spatial relationships, like parallelism, perpendicularity, connectivity, tangency etc. These constraints are rarely changed during object designing. For example, if two lines are parallel they are parallel forever.

A short overview of some constraints used in the rest of the paper is given in Table 1. Similar predicates are used in some other constraint-based geometric modeling systems ([1], [7], [18]).

Different designers work in different ways. Therefore, it can not be expected that constraints can be always given in the most adequate way for triggering the constraint propagation. Choosing a correct order for applying the constraints is one of the most difficult and time-consuming problems in constraint based systems ([1], [9]). We suggest a biconnected graph structure called the biconnected constraint description graph (BCDG) to perform this task. Its benefits are:

- it is flexible enough to support interactive design,
- it does not depend on the order of inserted constraints,
- it is possible to change the inserted constraint by its inverse couple, and
- it directly supports local propagation.

Table 1. Some of the FLEXI's constraints

Constraint	Explanation
On(l/c, v)	a vertex v is situated on the line l (or on the circle c)
Through(v, l/c)	a line l (or a circle c) passes through the vertex v
Perpendicular(l_1, l_2)	lines l_1 and l_2 are perpendicular
Parallel(l_1, l_2)	lines l_1 and l_2 are parallel
Point(v, x, y)	a vertex v has coordinates (x, y)
Angle(l_1, l_2, α)	the angle between lines l_1 and l_2 is α; $0 \leq \alpha < \pi$
Distance(v_1, v_2, d)	the distance between vertices v_1 and v_2 is d
Distance(l_1, l_2, d)	the distance between lines l_1 and l_2 is d
AngleValue(l, α)	the slope of the line l corresponds to the angle α; $0 \leq \alpha < \pi$

Consider A and B to be nodes of the BCDG. The first graph edge proceeds from the node A to the node B, and the second connects these two nodes in the opposite direction. These two connections are represented as a curve with an arrow at each end of the curve (see Fig. 3). A constraint is associated with each BCDG edge. If a constraint with a symmetric effect is presented (e.g. *Parallel, Perpendicular*), only one predicate is drawn. If a relation between two nodes depends on direction (e.g. *On(l_1, v_1)* and *Through(v_1, l_1)*) two predicates are separated by a slash.

A simple example of design in FLEXI is presented in the continuation. Suppose the user has already inserted the sketch of an object (Fig. 2a). For determination of the correct dimensions of the geometric object shown in Fig. 2b the constraints given by means of predicates are set. They are shown in Table 2.

Suppose the first three constraints from our example have already been inserted. They determine the exact position of a vertex v_1 and a line l_1. The line l_1 proceeds through the vertex v_1. Its slope has been explicitly determined with the predicate *AngleValue*. Let us see how the system reacts on the insertion of the fourth constraint *On(l_1, v_2)*. The user chooses the predicate *On* in menu. The line l_1 is selected with a locator device. After that the vertex v_2 is picked by the locator, too. Its position is immediately corrected. The vertex v_2 is moved in the shortest way to the line l_1. The program which controls BCDG building and constraint propagation performs following actions:

Table 2. The constraints used in the example

1. Point(v_1, x, y)	2. Through(v_1, l_1)
3. AngleValue(l_1, α_1)	4. On(l_1, v_2)
5. Distance(v_1, v_2, d_1)	6. Perpendicular(l_1, l_2)
7. Parallel(l_2, l_3)	8. Through(v_1, l_3)
9. On(l_3, v_3)	10. Distance(v_1, v_3, d_2)
11. Through(v_3, l_4)	12. Perperdicular(l_3, l_4)
13. On(l_4, v_4)	14. Through(v_2, l_2)
15. On(l_2, v_4)	16. Distance(l_1, l_5, d_3)
17. Through(v_5, l_5)	18. Angle(l_2, l_6, α_2)
19. Through(v_4, l_6)	20. On(l_6, v_5)

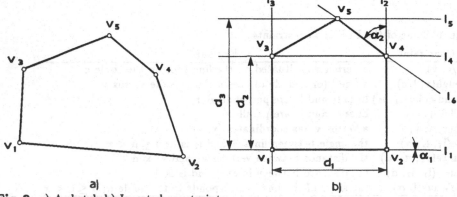

Fig. 2. a) A sketch b) Inserted constraints

1. A new BCDG node containing the vertex v_2 is created because the vertex v_2 has not been a part of BCDG yet.
2. The vertex v_2 is connected with the line l_1 by two links. The first of them starts in the BCDG node storing the line l_1 and ends in the node storing the vertex v_2. The predicate *On* is associated with this link. The second BCDG link binds these two nodes in the opposite directions. It carries the predicate *Through*.
3. The exact position of the vertex v_2 is tried to be calculated. It is concluded that there are not enough constraints to determine the vertex's position exactly. Because of this, only correction in its approximal position is performed (it is moved to the line l_1).

Exact position of the vertex v_2 is achieved by applying next constraint - *Distance(v_1, v_2, d_1)*. Constraint *On/Through* forces the vertex v_2 to be on the line l_1 and constraint *Distance* to be at the distance d_1 from the vertex v_1. Two solutions for the position of the vertex v_2 are possible. Namely, it can lie on the left or on the right side of the vertex v_1. As it has been explained before, the approximal value of the vertex v_2 is used to choose expected solution.

Next constraint in our example - *Perpendicular(l_1, l_2)* - determines the slope of a line l_2. After an appropriate constraint is selected from a menu, the user points at the referenced line (the line l_1 in our case) with the locator. Because the exact position of the line l_2 has not been known yet, the line l_2 appears near to the left screen margin. The exact position of the line l_2 is established no sooner than with the fourteenth constraint. It should be noticed, that the line l_3 is completely defined before the line l_2 although the line l_2 is needed for defining the slope of the line l_3. In Fig. 3 the BCDG of our example is shown.

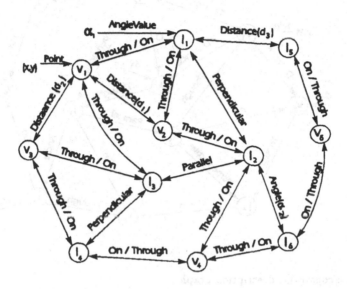

Fig. 3. Biconnected constraint description graph

To define the shape and the position of a geometric object, enough geometrical constraints have to be inserted. For a 2D geometrical object, having n vertices, 2n–3 geometrical constraints which determine angles and distances are necessary. In our example these constraints have the numbers 1, 3, 5, 6, 7, 10, 16, and 18. However, the correct number of the constraints is not a sufficient condition to determine a geometrical object exactly. Namely, while one part of an object can be over-dimensioned, the rest of it can be under-dimensioned. A detection of such occasions is hard in many constraint based geometrical systems. Quite opposite, in FLEXI it is natural and very simple. For example, if the user wants to insert a constraint which has an influence on the already exactly determined geometrical object this is noticed, and a warning is given to the user. Such constraint is immediately removed from the description of an object.

Every BCDG can be transformed into an acyclic graph. The use of this graph has some advantages for the constraint propagation. Here are two examples:

- if one input parameter is changed by the user, it is enough to recalculate and update only a part of the object in many cases;
- one can easily find independent parts of an object for parallel processing.

An acyclic constraint description graph (ACDG) is extracted from the BCDG when the geometry of an object is determined. The links of the BCDG not used in the constraint propagation are eliminated. Resulting ACDG is shown in Fig. 4.

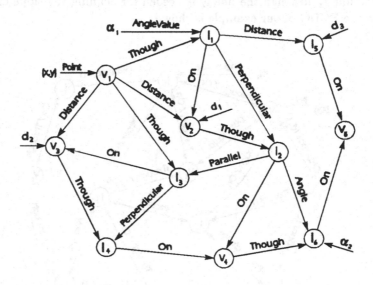

Fig. 4. Acyclic constraint description graph

5 Conclusion

In our paper it has been demonstrated how geometrical constraints can be employed for adding a new level of abstraction to a description of a geometrical object. A lot of attention has been given to the topology of adjacent relations among geometrical elements. For example: has the face f any hole? or which edges touch the edge e? Topological information is explicitly given during designing. It is efficiently stored into data structure and the user (a human or a machine) can have the access to it easily and efficiently without any calculations. Geometrical information is usually directly built in the framework fixed by topology. Geometrical relations are lost and can not be reached without additional computations. Geometrical constraints efficiently capture geometrical relations. Because of this, we are able to recall these geometrical relations among geometric elements without any time-complex calculations, too. At the moment an algorithm for combining different geometrical objects (in fact their ACDGs) is being studied. We do not mean to implement Boolean operations. For our purposes, feature modeling seems to be more promising.

References

1. B. Aldefeld: Variation of geometries based on a geometric-reasoning method. Comput. Aided Design, **20**, 117–126, 1988
2. B.Aldefeld, H.Malberg, H.Richter, and K.Voss: Rule-Based Variational Geometry in Computer-Aided Design. D.T.Pham (Eds.): Artificial Intelligence in Design, Springer-Verlag, 27–46, 1991
3. H.Ando, H.Suzuki, and F.Kimura: A geometric reasoning system for mechanical product design. F.Kimura and A.Rolstadas (Eds.): Computer Applications in Production and Engineering, Elsevier Science Publishers, 131–139, 1989
4. F.Arbab: Examples of Geometric Reasoning in OAR. V.Akman, P.J.W. ten Hagen, and P.J.Veerkamp (Eds.): Intelligent CAD Systems II, Springer-Verlag, 32–57, 1989
5. K.E.Dahshan and J.P.Barthes: Implementing Constraint Propagation in Mechanical CAD System. V.Akman, P.J.W. ten Hagen, and P.J.Veerkamp (Eds.): Intelligent CAD Systems II, Springer-Verlag, 217–227, 1989
6. M.J.G.M. van Emmerik: A System for Interactive Graphical Modeling with Three-Dimensional Constraints. T.S.Chua and T.L.Kunii (Eds.): CG International '90, Springer-Verlag, 361–376, 1990
7. F.Kimura, H.Suzuki, and L.Wingard: A Uniform Approach to dimensioning and tolerancing in product modelling. K.B.Estensen, P.Falster, and E.A.Warman (Eds.): Computer Applications in Production and Engineering, Elsevier Science Publishers, 165–178, 1987
8. K.Kondo: PIGMOD: parametric and interactive geometric modeller for mechanical design. Comput. Aided Design, **22**, 633–644, 1990
9. W.Leler: Constraint Programming Language. Addison-Wesley, 1988
10. V.C.Lin, D.C.Gossard, and R.A.Light: Variational geometry in computer-aided design. Comput. Graphics **15**, 171–177, 1981
11. M.Mäntylä: An Introduction to Solid Modeling. Computer Science Press, 1988
12. M.Mäntylä: Directions for research in product modeling. F.Kimura and A.Rolstadas (Eds.): Computer Applications in Production and Engineering, Elsevier Science Publishers, 71–85, 1989
13. G.Nelson: Juno, a constraint-based graphics system. Computer Graphics **19**, 235–243, 1985
14. D.Roller, F.Schonek, and A.Verroust: Dimension-driven Geometry in CAD: A Survey. W. Strasser, H.-P. Seidel (Eds.): Theory and Practice of Geometric Modeling, Springer-Verlag, 509–523, 1989
15. J.R.Rossignac, P.Borrel, and L.R.Nackman: Interactive Design with Sequences of Parameterized Transformations. V.Akman, P.J.W. ten Hagen, and P.J.Veerkamp (Eds.): Intelligent CAD Systems II, Springer-Verlag, 92–125, 1989
16. K.Shimada, M.Numao, H.Masuda, and S.Kawabe: Constraint-based object description for product modeling. F.Kimura, and A.Rolstadas (Eds.): Computer Applications in Production and Engineering, Elsevier Science Publishers, 95–106, 1989
17. I.E.Sutherland: Sketchpad - A man-machine graphical communication system. Proceedings of the Spring Joint Computer Conference, 329–346 1963
18. H.Suzuki, H.Ando, and F.Kimura: Geometric constraints and reasoning for geometrical CAD systems. Comput. & Graphics, **14**, 211–224, 1990
19. R.F.Woodbury: Variations in solids: A declarative treatment. Comput. & Graphics, **14**, 173–188, 1990

A Terminological Language for Representing Complex Knowledge

Daniela D'Aloisi

Fondazione Ugo Bordoni
Via B.Castiglione 59, I-00142 Rome (Italy)
email: fubdpt5@itcaspur.bitnet

Abstract. This paper describes the current implementation of KRAM, a hybrid system for knowledge representation that is part of a project whose task is to build a realistic model of how agents interact with each other and with the external world: the simulation should be embodied in a cooperative and user-friendly interface for an information system.

The focus of the paper is on the terminological component, TERM-KRAM, that supplies the meaning of all the concepts that are necessary to model man-machine interaction. The assertional component enables the users to make statements about external entities.

TERM-KRAM exhibits some specific features that make it especially suitable for representing actions and complex objects. Moreover, the system can be used for a wide range of applications, in particular those in which it is desirable either to offer explanation to users or to model agents and/or their behaviour, e.g., man-machine interaction, cognitive modelling, planning and help systems.

1 Introduction

The knowledge representation (KR) component is crucial in most Artificial Intelligence (AI) systems, and its capabilities should go beyond the pure representational task: it should offer tools for organizing and retrieving knowledge, and also be competent in reasoning about knowledge, i.e., in performing deductions from implicit knowledge, in maintaining the knowledge base consistent, etc. Moreover it should offer auxiliary services such as support in explanation, problem solving, planning, and diagnosis. As a consequence, the KR component should satisfy some requirements regarding both its representational competence and the classes of aids it can offer.

A lot of different approaches can be enumerated: nevertheless, the hybrid systems take an important role in knowledge representation research. Moreover they seem to be particularly suitable in those applications in which man-machine communication is relevant. The hybrid systems have (or have attempted to) overcome several problems arisen in designing efficient and powerful languages. They have mainly faced the necessity of integrating more than one formalism to adequately describe the different sources of knowledge in a clear and semantically consistent way: the hybrid systems claimed the need of a representational theory that ruled the role of each formalism, and a common semantics that established the interactions between these different formalisms [12]. Brachman and Levesque [1] pointed out the need of two distinct notions of adequacy that a representation tool should address: terminological adequacy and assertional adequacy. The terminological adequacy concerns the capability to describe the *vocabulary* of the domain: it also involves the possibility of stating (or discovering) relations between terms and of offering tools to develop, augment, and maintain the vocabulary. The assertional adequacy concerns the ability of describing knowledge about the world also in incomplete forms, and the consequent need of forming and maintaining a theory. This distinction fixed up the standard architecture of a hybrid system: with very few exceptions, a hybrid system consists of a terminological component, or TBOX, and an assertional component, or ABOX, that reflect the two notions of adequacy stated in [1]. Notwithstanding, some systems insert additional boxes, e.g., LOOM introduces an Universal Box and a Default Box. Each box also involves specialized reasoning capabilities, that strongly depend on the representation language. The research in the area of hybrid systems has produced various and interesting developments, above all regarding the terminological component. A lot of important theoretical results have been obtained with respect to both the complexity of terminological languages (e.g., [14, 13, 8]) and the trade-off between expressiveness and tractability of these languages [3]. Several systems have been also implemented, although none of them completely satisfy the actual representational requirements. (For a complete list, see [12].) The expressive power of terminological languages has been often restricted in order to make them tractable: in fact, the computational costs increase along with the expressiveness of a language, and a powerful language is often an intractable language. At present, many researchers and users of knowledge representation services emphasize the need of fully expressive languages in order to extend their applicability [9, 16]. They stress the point that a knowledge representation component cannot be seen as a separate component, but it must provide an actual service whose utility should be estimated in the context of its application.

KRAM (Knowledge Representation for Agency Modeling) agrees with the point that a KR component should carry out an effective representational service [7]. The system is part of a project whose task is to build a realistic model of how agents interact with each other and with the external world: this model should be embodied in a cooperative and user-friendly interface to be used in an information system devoted to a large class of users. KRAM consists of a terminological component, TERM-KRAM [6], and an assertional component, PROP-KRAM [5]; both the boxes are connected to a TMS (Truth Maintenance System, [11]). Section 5 depicts how KRAM can be applied as a knowledge representation service in a help system.

2 The Terminological Language

TERM-KRAM offers tools for setting up and for reasoning about the terminological network. The nodes of the network are concepts and the edges state relationships between them. A concept can represent any n-ary predicate, overcoming the point that a terminological network describes just a simple taxonomy of objects, as happens in most hybrid systems with very few exceptions, e.g., the Nary system [15]. The edges state *Superc* and *Incorporates* relationships between nodes: the *Superc* link implements a hierarchical nexus between concepts, the *Incorporates* link allows concepts to be embodied in the description of other concepts. The *Definition 1* introduces the syntax of TERM-KRAM: its semantics, given in an extensional form, can be found in [6].

Definition 1: Syntax

`<term-entry>`	::=	*THING* |	
		Top$_n$ |	*with n$\geq$1*
		`<concept>` |	
		`<disjoint-concept>`	
`<concept>`	::=	`<primitive-concept>` |	
		`<complex-concept>`	

\<primitive-concept	::=	*(\<k-predicate-symbol\>*
		((\<var\>$_k$ [\<type-restriction\>])$^+$) NIL)
\<complex-concept\>	::=	*(\<k-predicate-symbol\>*
		((\<var\>$_k$ [\<type-restriction\>])$^+$)
		\<Conceptual Scheme\>)
\<Conceptual Scheme\>	::=	*([\<Concept Conjunction\>] [\<StructualDescription\>])*
\<Concept Conjunction\>	::=	*(AND \<k-predicate-symbol\>*)*
\<StructualDescription\>	::=	*(INCORPORATES \<incorp-concept\>*)*
\<incorp-concept\>	::=	*(\<h-predicate-symbol\> \<arg\>$_1$...\<arg\>$_h$) |*
		(NOT \<h-predicate-symbol\> \<var\>$_1$...\<var\>$_h$))
\<type-restriction\>	::=	*\<1-predicate-symbol\>*
\<k-predicate-symbol\>	::=	**a name of predicate**
\<arg\>	::=	*\<var\> | \<incorp-concept\>*
\<var\>	::=	*x | y | z | ....*
\<disjoint-concept\>	::=	*(DISJOINT \<k-predicate-symbol\>$_1$*
		\<k-predicate-symbol\>$_2$)

THING represents the most general term, the *Universal Concept*, and it is the top of the network; the concepts Top_n represent the most general concepts of arity n , and they are the tops of the sub-networks that gather concepts of the same arity. The partition of the graph makes the computational effort less hard. The concepts may be either primitive or complex: primitive concepts describe entities whose descriptions are not supported by necessary and sufficient conditions; complex concepts refer to partially or completely analyzed entities which own a conceptual scheme. Variables refer to the arguments of the concepts: it is possible to assign a conceptual role to an argument by constraining its domain with a 1-place predicate, i.e., to define a *sort* or *type* for it. Each complex concept *C* holds a *Conceptual Scheme* which explains its meaning: it consists of two parts, a list of concepts whose conjunction makes up *C*, and a list of *terminological statements* that set up its structural content. The AND operator introduces the first list which is denoted by the non-terminal symbol *\<Concept Conjunction\>:* this structure allows the user to explicitly mention a list of concepts that subsume *C*. The *Structural Description* corresponds to a list of statements that the concept being defined involves in its definition. Each statement has a predicative form, i.e., *(k-place-predicate x_1 ... x_k)*, in which x_i can or cannot correspond to the arguments of *C* , and it, in turn, can also assume a predicative form. The INCORPORATION relation allows a concept to be connected to other concepts which have no hierarchical relationship with it, but that concur in denoting its deep meaning. The SUBSUME relation states a hierarchical connection between two concepts C_a and C_b, so that any information on C_a is inherited by C_b. The classification algorithm, that allows a concept to be automatically inserted in the terminological network, is defined on the ground of subsumption.

Definition 2: Subsumption

Let us suppose P be a set of n-ary predicates, and ε an extension function from predicates in P to subsets of the definition domain such that $\varepsilon: P \rightarrow 2^{D*}$ (with $D*$ the power set of D).

A concept C_a subsumes (is a super-concept of) a concept C_b, if the extension of C_a contains the extension of C_b, i.e., $\varepsilon[C_a] \supseteq \varepsilon[C_b]$. The subsumption relationship is transitive, reflexive, and anti-symmetric.

The language can also explicitly declare that two concepts C_a and C_b are disjoint. Each sub-concepts of C_a inherits The propriety of disjointness is transmitted to the chains of the sub-concepts of C_a and C_b.

Definition 3: Disjointness

Given an extension function (defined as in Definition 2), two concepts C_a and C_b are disjoint if the intersection of the extension of C_a and the extension of C_b is the empty set, i.e., $\varepsilon[C_a] \cap \varepsilon[C_b] = \varnothing$.
If $Subc(C_a)$ and $Subc(C_b)$ are respectively the set of the concepts subsumed by C_a and the set of the concepts subsumed by C_b and if C_a and C_b are disjoint, then $\forall C_i \in \{C_a \cup Subc(C_a)\}, \forall C_j \in \{C_b \cup Subc(C_b)\}$ $(\varepsilon[C_i] \cap \varepsilon[C_j] = \varnothing)$.

The TERM-KRAM's language is competent in representing the deep meaning of the terms involved in a communication among human and/or artificial agents: the system's task consists in allowing a formal language to be as expressive as natural language can be. That would make an interface more intelligent and friendly, actually able to capture the deep hints of natural communications. More generally, the system follows the requirements of the *second generation* of expert systems: the representation should be explicit, take care of prime principles, and be able to manage causality relationships. The description of terms is based on their *lexical decomposition* [10], that exploits the internal structure of a concept, i.e., how basic constituents concur in forming its meaning.

3 The INCORPORATES Link

Since the *Incorporates* link is a particular feature of TERM-KRAM, it is pertinent to explain in detail how it works and to show its merits and drawbacks. The terminological statements incorporated in a concept's description make up its Structural Description: the name suggests the old Structural Description (SD) of KL-ONE [2], even if TERM-KRAM's SD resembles KL-ONE's SD only from a semantic point of view, i.e., it structurally connects a concept with its roles (arguments in our case), but the structure is quite different. The KL-ONE's SD has been, in our opinion, a serious attempt to transfer the richness of the natural language in a formalized language.

The *Incorporates* link relates two concepts so that the meaning of the concept being described implies the meaning of the second one: instead of a logical implication, the relationship denotes a type of semantic implication. Each argument in a definition has to be viewed as universally quantified in the scope of its sort domain, and the external variables are existentially quantified. For instance the following (conservative) definition for brother

```
(1)          (define-concept BROTHER ((man x) (person y))
                  (AND
                          Relative)
                  (INCORPORATES
                          (Mother w x)
                          (Mother w y)
                          (Father z x)
                          (Father z y))
```

corresponds to the following interpretation in logical form

$\forall x \forall y$ (brother x y) $\rightarrow$ (man x) $\wedge$ (person y) $\wedge$ (relative x y) $\wedge$
$\qquad\qquad\qquad$ $\exists w \exists z$ (mother w x) $\wedge$ (mother w y) $\wedge$ (father z x) $\wedge$ (father z y)

The assertional component applies the terminological dictionary to assertional entities that belong to the application domain. For instance, if two individuals I_1 and I_2 are brothers then they have the same parents. If one of the incorporated propositions comes to be false as far as the arguments are concerned, it is impossible to assert the predicate on those arguments: the description provides necessary and sufficient conditions for applying that predicate. In some sense, the Structural Description might be seen as the *add list* of a planning operator since it introduces facts whose truth value follows from the assertion of the predicate.

It would seem that TERM-KRAM does not contain a few numeric operators that are typical of standard terminological languages as *all, atleast, atmost*, etc. The *type restriction* substitutes the *all* operator, and more generally it is possible to assign any property to an argument by means of the *Incorporates* link. The number restrictions *atleast* and *atmost* can be translated into predicates in TERM-KRAM: more generally, a predicate substitutes the restriction operators usually used in terminological languages. For example, *(Number-of x n m)* specifies the maximum and minimum values for an object x, and a Conceptual Scheme assigns a meaning to it. Actually every kind of operator can be represented as a predicate, and so every KL-ONE-like definition can be translated in a TERM-KRAM's one.

The Structural Description of a concept in TERM-KRAM is more than a simple list of concepts, but it offers a knowledge structure in which concepts are put in complex relationships. Concepts' arguments can be the same of those of incorporated concepts or different or propositions as well. For instance, the concept TELL is described as follows:

(2) $\qquad$ *(define-concept TELL ((Person x) (Person y) (Fact z))*
$\qquad\qquad$ *(AND*

$\qquad\qquad\qquad$ *Communicate)*
$\qquad\qquad$ *(INCORPORATES*
$\qquad\qquad\qquad$ *(Know x z)*
$\qquad\qquad\qquad$ *(Speak x y)*
$\qquad\qquad\qquad$ *(Cause (Speak x y) (Know y z))*
$\qquad\qquad\qquad$ *(Listen y x))*

The terminological statements that make up the *Incorporates* list are implicitly conjunct (AND), but we are improving the language in order to introduce other logical connectives. The explicit use of the logical connectives allows for a more immediate representation, even if it would be possible to define the corresponding predicates.

4 KRAM: Current Implementation

KRAM is implemented in Common Lisp on a Symbolics Lisp Machine: as the TBOX realizes a stable version, the assertional component is currently under implementation, and it is substituted by a preliminary version, strongly based on a TMS (Truth Maintenance System) drawn from RUP [11] made richer by an interface with the TBOX. The capability of TERM-KRAM in representing n-ary predicates, along with the deep representation of the term meanings, states its specificity with respect to other terminological languages: that makes the language particularly suitable for representing actions and complex objects, and in general applicable for a wide range of applications, particularly for utilizations in which it would be desirable to offer explanation to users and/or to model agents.

TERM-KRAM also offers a collection of TELL and ASK functions whose task is respectively to build and to query the knowledge base. The TELL functions are used to update the knowledge base, i.e., to introduce new concepts, to incrementally define concepts, and to delete concepts. Examples of such functions are DEFINE-CONCEPT, CHANGE-ARGUMENTS, ADD-DESCRIPTION, MODIFY-DESCRIPTION, ADD-SUPERC, and DISJOINT. The ASK functions retrieve information or deduce implicitly stored information from the knowledge base. A first group implements *yes-no-functios*, such as SUPERC?, SUBSUME?, INCORPORATES?, IMPLY?, and DISJOINT?: they verify if a certain relationship exists between two concepts. The second group of ASK functions exploit the concept's description. For instance, DESCRIBE-CONCEPT provides the complete definition of a concept, and LIST-OF-OCCURRENCES gives the list of the concepts that incorporate a given concept.

TERM-KRAM also defines a classification algorithm, whose task is to automatically insert concepts in the terminological network: it takes a concept, and compares it with the concepts previously defined in the network in order to find out if there exists a subsumption relationship between them. The subsumption procedure is sound and incomplete, and its complexity is $O(n^4)$, where n is the number of the nodes in the network. The complexity of the classification is $O(n^5)$, though the average complexity is lower. The incompleteness of the language is a minor point for an application, since logical completeness of subsumption is not always the best criterion for evaluating a system [9].

The assertional component will augment the competences of the whole system by enabling it to assert complex propositions in which n-ary terms are used as predicates and to build complex extensional entities. PROP-KRAM represents extensional (encyclopedic) knowledge by means of a propositional network made up of several kinds of nodes organized in two levels. The first level defines individuals, asserts propositions, deduces new information, and maintains the knowledge base consistent. Its functionalities are currently implemented:
1) To define individuals, a proposition is asserted stating that an individual satisfies a predicate previously described in the TBOX, i.e., *(Man Paul)*.
2) To assert propositions, a TBOX predicate is applied either to individuals defined by the TMS or to other propositions, i.e., *(Run Paul), (Tell Paul Mary (Professor Peter))*.
3) To deduce new information, the system utilizes the ASK functions and other functions defined to better exploit the descriptions of the TBOX in order to deeply understand assertional sentences.

In the second level, complex structures group first level nodes in order to define:
a) *composite entities*, as *classes* and *sets*, that can be used as arguments in propositions;
b) *part-of nodes*, which allow nodes to be grouped to form a whole item, like a story or an object;
c) *clusters*, which allow knowledge about entities or topics to be grouped in an orderly way;
d) *point-of-view nodes*, i.e., belief structures, which group in a context various kinds of knowledge (beliefs, goals, etc.) owned by an agent about itself or about other agents.

Let us show how the components of KRAM interact. The definitions for BROTHER (1) and TELL (2) allow us to assert propositions in the ABOX, for instance that *(Brother Paul Bill)* and *(Tell Mary Peter (Brother Paul Bill))*. The inferential engine of the TMS along with the interface functions (based on the ASK functions) can perform inferences and figure out truth values for the propositions. For example, we know that Paul and Bill have the same father and the same mother, that Peter knows that Paul and Bill are brothers, etc. If we also assert that *(Mother Dory Bill)* the *ad-hoc* functions implemented on the TMS easily deduce that Dory is Paul's mother too. If a new entry denies a previous evidence, all the facts linked to it need to be revised. For example if *(NOT (Brother Paul Bill))* then it will be still true that Paul is a man --because the restrictions on the arguments have been checked before--, but it is not true any more that Paul and Bill have the same parents. Besides, Mary told a false fact to Peter.

5 An Example of Application

An intelligent, friendly, knowledge-based man-machine interaction is a need that is confirmed in many kinds of application, even if it is particularly compelling in computer science and artificial intelligence. In this context, a KR component has to also take into account:
- the user, in particular its goals and expectations;
- the agent's model;
- the model of the system the user has;
- the usual sources of knowledge, i.e., knowledge regarding the domain, general knowledge, common sense knowledge, etc.

As a consequence, it would be utterly useful to have a system able to describe the dictionary (the terms involved in the communication) on the ground of the conceptual schemes that the user itself has about the terms. Obviously, the term descriptions must actually correspond to these schemes, and not simply supply a taxonomy of objects connected by binary relationships. We argue that TERM-KRAM is able to carry out this task, because its language allows for effective description of terms. It can be useful both to understand what the user is asking for and to explain terms to the user itself. When a user queries the system, it can infer, on the ground of the term descriptions, more information with respect to a simple taxonomy. For instance, it can find out connections about the involved entities, generally the arguments of a proposition: this task is difficult to be performed by the classic hybrid system where the roles cannot be explicitly connected. Moreover, the system can classify the expertise of the user on the ground of the kind of terms it uses. In explanation, the system can use the description of the terms (for example technical terms that the user does not know yet) to build a conceptual schema by using structures that the user can understand. The language lets the user free to decide the level of detail of the terminological descriptions: they may be either completely general, e.g., to resemble the use of terms in natural language, or adapted to the application. The first solution is more flexible and largely adoptable since the user will not be compelled in using terms, but it involves a more complex network and a considerable amount of terminological statements to be managed. The second solution would be more specific, since it would allow for a special-purpose description of terms. For the sake of simplicity, in the following example we shall adopt the second option.

In this example we show the use of KRAM as representation service of a help system, called W-HES [4], designed for helping a user to better manage the Microsoft® Word for the Apple® Macintosh™. Fig.1 illustrates an example drawn from this domain. A terminological network is depicted that contains the hierarchies of the terms relevant to the description of some Word commands, i.e., *Change Text, Change Font,* and *Change Style.* The complete descriptions of the three commands are in the following, with the explanations Word associates them:

1) **ChangeText:** *finds and replaces text with replacement text you specify.*

      (ChangeText ((Agent x) (Text y) (Text z))

              (AND   Change)

              (INCORPORATES   (User u)

(Specify u y)
(Specify u z)
(Find x y)
(Replace z y)
(Do x (Replace z y))
(Cause (Replace z y) (Cover y z))))

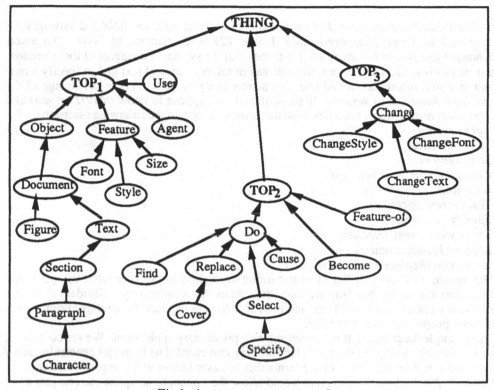

Fig.1. A TERM-KRAM network

2) **ChangeFont:** *changes current font by typing font name and pressing return.*
    (ChangeFont ((Agent x) (Font y) (Font z))
        (AND   Change)
        (INCORPORATES   (User u)
                          (Text w)
                          (Specify u z)
                          (Select u w)
                          (Do x (Replace z y)))

3) **ChangeStyle:** *changes style of the selected paragraphs to style you indicate by typing style name and pressing return.*
    (ChangeStyle ((Agent x) (Style y) (Style z))
        (AND   Change)
        (INCORPORATE   (User u)
                          (Text w)
                          (Specify u z)
                          (Select u w)
                          (Replace y z)

(Do x (Replace y z)))

It is to be remarked that the descriptions are analytical but not operative, i.e., they do not correspond to the sequent of steps needed to perform an operation. It is task of a planning system to specify the actions necessary to accomplish the act: the language can be improved to make it able to represent plans. (We leave for another paper the discussion of that.)

The individuals are *conceived* in the assertional box: for instance, *bold* and *helvetica* are instances of *Font*, *(Centered Bold Times 12)* is an instance of *Style*. To assert *(ChangeFont system helvetica venice)* means that the system is the agent of the operation, and its job is to change the font: the user has to select a piece of text and to specify a new font that will substitute the old one. The action of replacing will cause the change of the text from *helvetica* to *venice*. In particular, if we suppose to know that *(User you)* and *(Text current-selection)*, the assertional database will contain the following assertions:
(Agent system)
(Style helvetica)
(Style venice)
(Change system helvetica venice)
(User you)
(Text current-selection)
(Specify you venice)
(Select you current-selection)
(Replace helvetica venice)
(Do system (Replace helvetica venice))
The system can use this list of propositions to explain to the user what the particular operation means by building explanations about the terminology. Besides, it always knows the current state of affairs, and is so able to help the user, for example pointing out if some proposition is not satisfied.
The example described in this Section shows a preliminary application. We are conceiving a module able to acquire the user's terminology that enables us to model agents: in such a way, we could create a map of the terms that the user knows and compare them with the terms she/he does not know. That would allow the system to adequate the complexity of explanations to the user competence.

6 Conclusions

In this paper a hybrid knowledge representation system has been described, paying particular attention to its terminological component. TERM-KRAM presents some peculiar aspects that differentiate it with respect to other hybrid systems. In particular it naturally describes actions due to the possibility of representing n-ary terms and the ability to define a Conceptual Scheme for any concept. These specific aspects of the language contribute to increase the spectrum of knowledge it can deal with. TERM-KRAM can be placed among the languages that give priority to the expressive aspects even if it means an incomplete classification algorithm. This position is strengthened by the opinion that an expressively rich language is able to better describe the world and as a consequence it allows a larger range of applications.Currently, we are working on the assertional component and also revisiting the terminological box in order to improve its expressive power: another direction of research concerns the extension of TERM-KRAM to make it able to represent complex plans.

Acknowledgements

Most of the ideas developed in KRAM originate from Cristiano Castelfranchi, and Giuseppe Maga played a particular role in the implementation of the KRAM system: my special thanks to both of them. This work has been partially supported by "Progetto Finalizzato Sistemi Informatici e Calcolo Parallelo" of CNR under grant n.104385/69/9107197, and has been carried out in the framework of the agreement between the Italian PT Administration and the Fondazione Ugo Bordoni.

References

[1] Brachman, R.J., Levesque, H.J., Competence in Knowledge Representation. Proceedings of the 2nd AAAI, 1982, 189-292.

[2] Brachman, R.J., Schmolze, J.G., An Overview of the KL-ONE Knowledge Representation System. Fairchild Technical Report No.655, September 1984.

[3] Brachman, R.J., Levesque, H.J., A Fundamental Trade-off in Knowledge Representation and Reasoning. In: Brachman, R.J., Levesque, H.J. (eds.), Reading in Knowledge Representation, Morgan Kaufmann, 1985, 41-70.

[4] Cesta, A., Romano, G., Explanations in an Intelligent Help System. Bullinger, H.J., (ed.), Human Aspects in Computing (Proceedings of the 4th International Conference on Human-Computer Interaction). Elsevier, 1991, 925-929.

[5] D'Aloisi, D., Castelfranchi, C., Tuozzi, A., Structures in an Assertional Box. In: Jaakkola, H., Linnainmaa, S., (eds.), Scandinavian Conference on Artificial Intelligence - 89 (Proceedings of the 2nd SCAI). IOS Press 1989, 131-142.

[6] D'Aloisi, D., Maga, G., Terminological Representation and Reasoning in TERM-KRAM. FUB Technical Report 5T05091, December 1991.

[7] D'Alosi, D., Castelfranchi, C., Propositional and Terminological Knowledge Representations. To appear in: Working Notes of the AAAI Spring Symposium "Propositional Knowledge Representation", Stanford (CA), March 25-27, 1992.

[8] Donini, F.M., Lenzerini, M., Nardi, D., Hollunder, B., Nutt, W., The Complexity of Concept Languages. Proceeding of the Second International Conference on Principles of Knowledge Representation and Reasoning, 1991.

[9] Doyle, J., Patil, R.S., Two theses of knowledge representation: language restrictions, taxonomic classification, and the utility of representation services. Artificial Intelligence, 48, 1991, 261-297.

[10] Lakoff, G., Irregularity in Syntax. New York, Holt, Reinhart, and Winston, 1970.

[11] McAllester, D.A., Reasoning Utility Package: User's Manual. Massachusetts Institute of Technology, Artificial Intelligence Laboratory, AI Memo No.667, 1982.

[12] Nebel, B., Reasoning and Revision in Hybrid Representation Systems, Springer-Verlag, 1990.

[13] Nebel, B., Terminological Reasoning is Inherently Intractable. Artificial Intelligence, 43, 1990, 235-249.

[14] Patel-Schneider, P.F., Undecidability of Subsumption in NIKL. Artificial Intelligence, 39, 2, 1989.

[15] Schmolze, J.G., Terminological Knowledge Representation Systems Supporting n-ary terms. Proceedings of First International Conference on Principles of Knowledge Representation and Reasoning, 1989, 432-443.

[16] Smoliar, S.W., Swartout, W., A Report from the Frontiers of Knowledge Representation. DRAFT, UCS Information Science Institute, Marina del Rey, CA, October 1988.

Knowledge Representation and Decision Making:
A Hybrid Approach[1]

Iain Wallace[a], Simon Goss[b] and Kevin Bluff[a]

(a) Swinburne Institute of Technology   (b) Aeronautical Research Laboratory
    PO Box 218, Hawthorn   506 Lorimer St., Fishermens Bend
    Victoria, 3122   Victoria, 3207
    AUSTRALIA   AUSTRALIA

Abstract

Knowledge representation is a fundamental issue in the simulation of human performance such as the behaviour of a pilot engaged in air combat. Interaction between symbolic and subsymbolic networks is used in a cognitive architecture to represent the effects of emotional and motivational processes on decision making. This enables a shift from the conventional focus of combat simulation on optimal decision making to the modelling of authentic human decision making with its individual variability and imperfection.

Keywords: hybrid systems, knowledge representation, decision making

1 Introduction

The simulation of human performance in interaction with complex technological interfaces highlights some key issues in knowledge representation. A pilot engaged in air combat is involved in detection, classification, and identification which depend on sensori-perceptual processes. Inferring the intentions of other agents and assessing the degree of threat involve the application of general reasoning processes. Generation, selection and execution of tactical options involve the application of planning and executive processes to a range of procedural knowledge, while evaluation of the results hinges on sensori-perceptual monitoring.

This sequence places considerable demands on knowledge representation but still underestimates the challenge of simulating performance. A valid simulation must take account of the prominent involvement of emotional and motivational processes. A pilot makes decisions under constraints of speed and effectiveness in a complex environment that imposes a continuous, yet variable strain upon the information processing limits of cognitive processes. Existing air combat models generally include only a poor representation of individual human decision making. Even where some account has

[1] This research is supported in part by the Aircraft Systems Division of the Aeronautical Research Laboratory, Defence Science and Technology Organisation, Australia.

been taken of human performance in decision making it is confined to the cold cognitive aspects of reasoning, planning and risk assessment. An acceptable approach to knowledge representation for human decision making must accommodate a comprehensive range of cognitive processes and the requirements of representing the influence of emotional and motivational processes.

Minsky (1987, 1991) argues that the degree of versatility required for machine intelligence can only be achieved by capitalizing on specific strengths of symbolic and subsymbolic modes of representation, using each to offset the deficiencies of the other. In the current work the decision to adopt a hybrid architecture reflected the capacity for a better match to human processing than either symbolic or connectionist processing alone. It also reflected the practicalities of system building. Aspects of decision making can be effectively represented as symbolic processes whereas sensori-perceptual processes pose considerable problems for symbolic approaches but are relatively amenable to subsymbolic representation.

2 A General Architecture of Cognition

The general architecture of cognition used has been reported elsewhere [Wallace and Bluff 1991], and is summarised here. It spans the entire range of "hot" and "cold" cognition, providing an explicit framework for the interaction of emotion and motivation with sensori-perception and reasoning, and is presented schematically in Figure 1. The central feature is a core network which contains the current state of the system's world model and is expressed in a hybrid representation. Each of the nodes contains a symbolic representation of procedural knowledge and related declarative knowledge. Symbolic information is passed between these core nodes. Decisions on which nodes occupy the limited number of parallel channels available for semantic processing hinge on both symbolic and subsymbolic inputs to each of the nodes. The symbolic information is derived from other nodes in the core network, or from the sensory networks. The sensory networks register environmental stimuli which are analysed by non-symbolic Sensory Feature Detectors (SFDs). The results are processed to produce symbolic data as input for core nodes. Selective attention is produced by permitting the needs of core nodes for specific symbolic information to provide goals for the sensory networks.

The activation potential necessary for a core node to acquire a processing channel is provided in a non-symbolic form by a motivational network which, in turn, receives input from an emotional network. The emotional network links sensori-perceptual input in both symbolic and non-symbolic form with physiological actions which define primary emotions such as fear, anger and pleasure. Motivational nodes are activated by nonsymbolic inputs representing physiological states and symbolic input representing situational features in the total environment of the system. Core nodes receive activation from one or more sources in the motivational network, as determined by innate or learned associations.

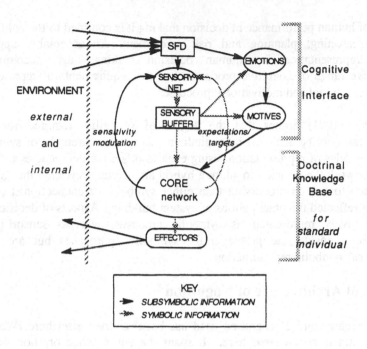

Figure 1

3 Risk Assessment and Decision Making

Pilots are trained to assess a current state of a defensive envelope reflecting their knowledge of the capabilities and current status of enemy ordnance. In practice, this objective assessment of risk is modulated by variations in interpretation arising from individual differences between pilots some of which are attributable to fundamental personality differences. In determining behaviour these differences interact with the effects of variations in experience.

A switch to the defence mode in response to a threat is an example of tactical decision making. It is triggered by registration of specific sequences of spatial relationships between combatants. This occurs in the sensori-perceptual network which transmits symbolic input to the core network and concurrently to the emotional and motivational networks. The modal shift in the core network requires both the symbolic input and a change in the relative levels of activation of the core nodes. This change is brought about by the motivational and emotional networks. Interaction with the motivational and emotional networks also provides the core network with inputs which condition the receptivity of the sensori-perceptual network. Symbolic input enables the setting of attentional targets, endowing the sensori-perceptual network with a "demon flavour". Subsymbolic input determines the sensitivity levels which modulate the transfer of symbolic input from the sensori-perceptual network to the motivational and emotional networks. This determines the rate at which activation is redistributed in the core network and is a source of individual differences in performance.

Figure 2 presents a defensive envelope derived by objective assessment of risk. It includes two subjective variants representing the results of individual differences in the sensitivity levels modulating sensori-perceptual input to the motivational and emotional networks. The practical result would be a significant difference in the timing of the switch to defence mode. Similar variations would result from differences in the missile envelope governing the launching of ordnance in attack.

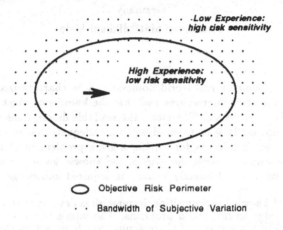

Figure 2

4 Conclusion

Adoption of an emphasis on modelling authentic, fallible human decision making and a modular, hybrid approach to system architecture offers a number of immediate and potential advantages: flexibility in investigating variations in training and/or personality; relatively unpredictable and challenging behaviour of units in combat simulation; hybrid representation reducing the brittleness of rules, providing an operational "fuzzy" effect and producing required results with relatively shallow, less complex rule structures.

References

Minsky, M. (1987). *The Society of Mind.* New York: Simon and Schuster

Minsky, M. (1991). Logical Versus Analogical or Symbolic Versus Connectionist or Neat Versus Scruffy, *AI Magazine*, Summer 1991, pp35-51.

Wallace J.G., and Bluff, K . (1991). A Cognitive Framework for Hybrid Systems, *Proceedings of IJCNN-91*, Singapore , pp491-496.

Knowledge Acquisition from Text in a Complex Domain *

Gabriele Schmidt

German Research Center for Artificial Intelligence

Erwin-Schrödinger-Str. (University Bldg. 57)

W-6750 Kaiserslautern

Germany

e-mail: schmidt@dfki.uni-kl.de

Abstract. Complex real world domains can be characterized by a large amount of data, their interactions and that the knowledge must often be related to concrete problems. Therefore, the available descriptions of real world domains do not easily lend themselves to an adequate representation. The knowledge which is relevant for solving a given problem must be extracted from such descriptions with the help of the knowledge acquisition process. Such a process must adequately relate the acquired knowledge to the given problem.

An integrated knowledge acquisition framework is developed to relate the acquired knowledge to real world problems. The interactive knowledge acquisition tool COKAM+ is one of three acquisition tools within this integrated framework. It extracts the knowledge from text, provides a documentation of the knowledge and structures it with respect to problems. All these preparations can serve to represent the obtained knowledge adequately.

1 Introduction

Although numerous descriptions may be available about some complex real world domain in textbooks, in written documents (e.g. documentations of companies) and through the explanations of various domain experts, these descriptions do not easily lend themselves to an adequate representation. Such descriptions are usually uncommittal, unprecisely stated, at different levels of generality, overall incomplete and at times even contradictory.

This is even true for domains where precise physical laws apply, such as in the real world domain of mechanical engineering. Since the physical laws are idealizations of the real world, unreflected predictions derived from these laws will often crucially deviate from the respective event in the real world. Very often it becomes too complex, or not all data are known to compute a solution from such laws. Therefore, the problem is to find out which data are available, which relations are important, and how to solve the specific problem with the available knowledge. In other words, the complexity of a domain not only consists of the large amount of data, interactions and complex laws. It additionally consists of the difficulties in relating the available knowledge to a specific problem that is to be solved.

---

*This research was funded as part of the ARC-TEC project by grant ITW 8902 C4 from BMFT (German ministry for research and technology).

In the application domain of mechanical engineering, the tool COKAM+ (Case-Oriented Knowledge Acquisition Method from Text) [16, 9] collects the basic taxonomies of materials, and tools, together with the relevant physics knowledge from task-independent descriptions in textbooks. COKAM+ relates the acquired knowledge to concrete problems and their solutions, i.e. cases the solutions of which were already applied to solve real world problems.

First, our application domain, the manufacturing of rotational parts is described. In the next section, an integrated knowledge acquisition framework is briefly presented with respect to one of its tools COKAM+ . Then, it is shown how COKAM+ acquires knowledge from texts and relates it to problems, which are already structured by another tool of the framework. The acquired knowledge is to be formalized so that a domain theory is built. The final discussion describes the applicability of COKAM+ in other domains and compares it to some other knowledge acquisition methods from text.

2 A Complex Application Domain

In this section, the application problems which are in the domain of manufacturing of rotational parts and the problem solutions are introduced. The available knowledge and the necessary knowledge are described.

2.1 Description of the Problem and its Solution

The problem of production planning in mechanical engineering [17, 19] consists of finding an adequate production plan for a given workpiece, which is to be manufactured in some factory. For the manufacturing of a rotational part, the production plan consists of a sequence of chucking and cutting operations by which the workpiece can be manufactured. Drawings (CAD) define the geometric form of the mold, from which the workpiece is to be manufactured and the target workpiece. The chucking fixture, which is rotated with the attached mold with the longitudinal axis of the cylinder as the rotation center, the cuts and their specific order must be planned. For each cut, the cutting tool, the cutting parameters and the cutting path have to be determined. A complete description of the real world problems also includes further technological data of the workpiece (surface roughness, material, etc.) and precise workshop data (CNC machines with their rotation power and number of tools and revolvers, etc.).

2.2 What Knowledge is Documented in Texts?

There is an extremely large spectrum of cutting tools available, which leads to ISO-norms one for toolholders (ISO 5608) and one for tool inserts (ISO 1832) described in catalogues of companies which produce cutting tools. Within these norms about $1, 8 * 10^7$ toolholders and $1, 5 * 10^8$ inserts exist and can be combined, with respect to some restrictions, to more than $3, 5 * 10^{12}$ meaningful cutting tools. From that large spectrum of available tools one medium size company typically uses about 5000 tools for their production. Additionally, some highly specialized cutting tools are constructed and produced on demand. All the tools can be applied differently and can create different effects depending on the cutting parameters. The catalogues give some examples for the application of the tools which graphically show the problem and its solution and provide cases for the knowledge acquisition process.
Textbooks of mechanical engineering often contain very general and theoretical knowledge. For instance, the workpiece materials and the materials of cutting tools are often described as hierarchies. These hierarchies represent the chemical and physical structures of the specific materials. Other relevant properties are the hardness, the persistence of the hardness with increasing temperatures, their sensitivity

to shocks, etc. These properties are only partially correlated with the chemical composition and are thus only to some extent reflected in the taxonomic hierarchy. Although it is hardly possible to derive from this information which cutting material to use in a specific context, this only weakly structured and incomplete information seems nevertheless to play an important role in mechanical engineering. More specific and thereby more useful knowledge has been assembled from systematic experiments and theoretical and approximative calculations. Such results are often presented in table format.

Because of the large amounts of such knowledge and the relatively small usefulness for concrete problems, it does not make sense to bring this knowledge into a knowledge base in an unfiltered way.

2.3 What Knowledge is Necessary?

Think-aloud protocols [15] and the literature in mechanical engineering [6] show that the experts who develop production plans very often overcome the complexity of their task by doing skeletal plan refinement. A general inference structure [3] which we call model of expertise describes this problem solving method for the manufacturing of rotational parts as follows: From the concrete description of the workpiece and the available manufacturing environment more abstract feature descriptions are first constructed. These abstractions are then associated with an appropriate skeletal plan that has been stored in the knowledge base. The skeletal plan is finally refined with the help of the workpiece and the factory description into the concrete production plan.

The model of expertise implies the planning method of reuse of plan schemata [8, 13], which becomes the general structure of the to be developed expert system. Therefore, model of expertise specifies what kind of knowledge has to be acquired for the expert system, namely abstraction rules, refinement rules and skeletal plans which are associated with features of the problem description.

3 Integrated Knowledge Acquisition Method

Within an integrated knowledge acquisition method [14] the knowledge which is described as relevant in the model of expertise is acquired from three different sources of information (texts, cases and the expert's respective memories) with the help of three tools.

First the tool CECoS (Case-Experience Combination System) [2] is applied. CECoS supports the delineation of a hierarchy of problem classes and their description by features. The problem classes are formalized stepwise whereby the formalization is guided by the model of expertise. With the help of CECoS experts also delineate a hierarchy of operator classes – operators are steps of the production plan. The hierarchy of operator classes is utilized for structuring the knowledge acquisition process with the tool COKAM+ .

COKAM+ acquires preconditions and consequences from texts for the operators and the operator classes. Furthermore, abstraction and refinement rules are obtained from texts. The acquired knowledge is also formalized step by step.

The formalized production classes and their feature descriptions obtained through CECoS and the formalized task-related engineering and common sense knowledge, preconditions and consequences supplied by COKAM+ can then be used to automatically construct skeletal plans and associated application conditions through the tool SPGEN (Skeletal Plan Generation Procedure) [13]. SPGEN is based on explanation-based generalization as described by [10].

The interaction among the three tools is determined by this integrated knowledge acquisition method. From COKAM+s point of view, CECoS provides one of its inputs, i.e. an operator class hierarchy and informal feature definitions, and for SPGEN its domain theory consisting of formal preconditions, consequences and abstraction and refinement rules is acquired.

4 Acquisition from Text with COKAM+

In our complex real world domain, the interactive tool COKAM+ is applied to acquire formal descriptions of the operators in the production plan and the appropriate task-related and common sense knowledge, i.e. abstraction and refinement rules. Such descriptions can be found in mechanical engineering textbooks, catalogues and documentations of a company. COKAM+ does not aim to comprehend the text completely, but to extract that knowledge from the text that is needed to obtain a sufficient domain theory and that can be stepwise transformed into a formal representation. In order to support the user as much as possible, information retrieval methods, natural language processing, browsing systems and Hypertext methods can be applied in different ways at different phases of the procedure. In the following, the acquisition process of COKAM+ is described in detail by using examples.

4.1 Input from CECoS

COKAM+ obtains a hierarchy of operator classes from CECoS and defines the concrete operators and their classes within this hierarchy. COKAM+ does not check whether the hierarchy is correct, but uses it as a guidance for the definitions.
Figure 1 shows an example of a part of an operator hierarchy. The hierarchy tree can be read in the following way: The vertical distance is an indication for the similarity. Thus the two operators on the lower left side of figure 1 are more similar than the two on the right side. The expert explains that the two on the left only differ in the tolerances and the type of the insert whereas the two on the right also differ in the cutting path. In order to describe an operator the expert names features (see gray boxes in figure 1), e.g. "cutting material SN80". On the next higher levels the classes are defined by more general features like "bezeling" or "rough turning" which are terms for cutting types. If other single operators are generalized further types of cutting can be identified, e.g. "finishing", "fine finishing", "thread turning" etc..

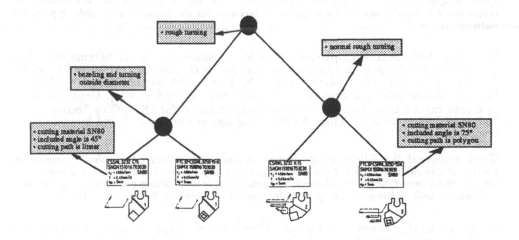

Figure 1: Example a Hierarchy of Concrete and General Operators

To acquire a domain theory with COKAM+ which SPGEN needs to generalize plans into skeletal plans, not only concrete operators and their classes, but also specific sequences of operators (often called macros in mechanical engineering) should be considered. In production plans, some specific types of operators often appear in the same order. Most of such sequences are well-known and the experts describe them by features, e.g. "processing of monotonic contour".

All features which are collected for the operator and macro classes provide cues for the definitions of the latter which are then acquired with COKAM+ .

4.2 Constructing an Informal Knowledge Base

4.2.1 Extracting Relevant Knowledge from Text

At first, texts which contain relevant information for the to-be-developed expert system need to be found. Mechanical engineering textbooks, catalogues and documentations of a company can be used for the collection of presumably relevant knowledge which is subsequently enhanced by the expert's elaborations. In order to find texts segments which can be used to explain the selected operators, the features which were acquired with CECoS can be used as keywords to search a large bibliography of mechanical engineering literature. The information retrieval technique of latent semantic indexing [4] is applied to find relevant texts in bibliographies or text segments which do not directly match any particular keyword. This is very helpful, since a search based on these very specific keywords obtained from CECoS can be too restrictive at that phase. There are various levels of information search for which some examples are given in the following:

Surface Matches In the simplest case, a relevant text segment may be found through a search at the surface level of the text. Such a search is performed by looking for keywords. By searching for the keyword "cutting material" given as the feature description of the concrete operator in figure 1, the following text segment was found:

Text segment 1 *"An important requirement of cutting materials is that the cutting material must be harder than the workpiece material under the temperature resulting from the cutting process."*

Matches between Paraphrases Since the same objects or actions are often referred to by different terms even within one text, not all text segments which describe a certain object or action can be found with surface matches. With a table of aliases, additional relevant text segments can be found [5]. Generalizations of terms can be used such as "silicon-nitrite-ceramic" for "SN80" or "ceramic" for "silicon-nitrite-ceramic". Thus the following text segment was found by looking for the term "SN80":

Text segment 2 *"Silicon-nitrite-ceramic is less shock sensitive than aluminiumoxid-ceramic. It can thus be used not only with higher cutting speeds, but also with the same feeds as coated hardened metal. Its domain of application is the lathing and milling of cast iron workpieces with cutting speeds up to 800m/min or feeds up to 0.8mm/rotation."*

Matches at the conceptual level Even when a paragraph does not contain any of the relevant keywords and aliases, it may contain important information. This occurs when related concepts are discussed in a text. Since the key terms which are used in some text indirectly specify the gist of the text, different clusters of words can be used as indirect indices for the gist of a text segment. By using such latent semantic indices [4], the text can be searched for segments with certain gists. With latent semantic indexing the following text segment which is relevant for the cutting material "ceramic" can be found:

Text segment 3 *"When a shock-sensitive cutting material is used on a mold with a rough surface, bezeling is required."*

"Shock-sensitive" and "cutting material" belong to a word cluster which also contains the word "ceramic". Although the term "ceramic" is not mentioned in the text segment, it can be correctly related to "bezeling" because of the word cluster.

4.2.2 Decontextualization of Text Segments

The text segments which are extracted from texts first must be decontextualized so that they can be understood without the context of the text. However, a link to the original text is stored when they are added as knowledge units to the informal knowledge base. For the support of decontextualization, there are a lot of approaches in natural language analyzing, e.g. anaphora resolution. The following example shows how a decontextualization based on anaphora resolution can be performed. From original text section (from E. Paucksch, Zerspantechnik, p. 14, translated, see text segment 2) the expert selects the last sentences beginning with *"Its domain ..."* as relevant. This sentence must be decontextualized so that the first knowledge unit runs as follows:

Knowledge unit 1 *Application of Silicon-Nitrite-Ceramic:*
"The domain of application for silicon-nitrite-ceramic is the lathing and milling of cast iron workpieces with cutting speeds up to 800m/min or feeds up to 0.8mm/rotation."

4.3 Structuring the Informal Knowledge Base

Now the knowledge unit 1 can be interpreted, but the range of its various interpretations is too large. In a following step, the possible interpretations of the knowledge unit are therefore restricted to the desired interpretation with respect to the task of the future knowledge-based system. For the application of cutting tools in lathing processes, it is of no interest that knowledge unit 1 also is valid in milling processes. The unit is modified according to the concrete application which is termed recontextualization:

Knowledge unit 1: child 1 *Application of Silicon-Nitrite-Ceramic:*
"For lathing cast iron workpieces with silicon-nitrite-ceramic the cutting speeds are up to 800m/min or feeds are up to 0.8mm/rotation."

In COKAM+ there also are two further types of recontextualization. The knowledge units are to be related to one or more categories of the model of expertise. The operators which are given by CECoS and which determine the range of the desired future knowledge-based system describe the specific situations to which the knowledge units are also related.

4.3.1 Structuring through the Model of Expertise

The knowledge units can be assigned to the views of the model of expertise to indicate how they will be used by the knowledge based system. Through the model of expertise the domain can be divided into six views, namely *detailed* and *general product views*, the *detailed* and *general plan views*, and the *detailed* and *general environment views*. The knowledge unit 1: child 1 is assigned to the views *product general* and *environment general* because the material of the workpiece (cast iron) relates to the product and the material of the cutting tool (silicon-nitrite-ceramic) relates to the environment. Additionally, because of the concrete parameters of the plan (cutting feed and cutting speed) it is related to the *plan view*, too. This is a default assumption because the unit specifies a precondition as will be shown in the following section.

4.3.2 Explanation of Concrete Operators

For the explanation of concrete operators each operator is presented to the expert. The expert then searches the informal knowledge base and selects all knowledge units which specify relevant preconditions and consequences of the particular operator. If relevant preconditions cannot be found in the informal knowledge, the expert is to add new knowledge units. This procedure is the best way to really find all the relevant preconditions and consequences. By combining theoretical knowledge from text with the expert experiences, both gaps in the theoretical knowledge as well as gaps in the experts' memories are likely to be discovered.

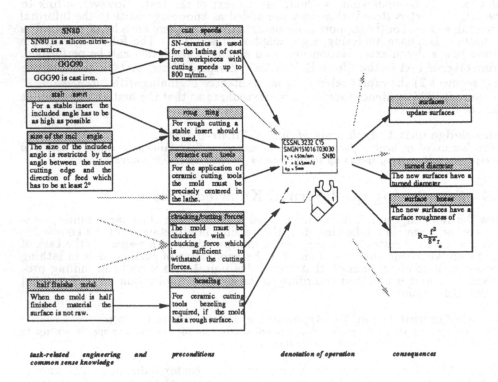

task-related engineering and common sense knowledge   *preconditions*   *denotation of operation*   *consequences*

Figure 2: Explanation for the First Cutting Operator of a Case

Figure 2 shows part of the explanation of a cutting operator. The arrows pointing from the knowledge units to the operator represent preconditions, e.g., the knowledge unit 1: child 1 at the top of figure 2, and the arrows from the operator to the knowledge units represent consequences. Whether the preconditions hold or not, must be checked for the current world state. Since the current world state is described by a number of concrete facts ("SN80") and since the precondition is stated in general terms ("silicon-nitrite-ceramic"), several inference steps which refer to domain knowledge such as the hierarchies of workpiece and cutting materials must be performed. The explanation trees on the left side of the different preconditions represent the knowledge units which are abstraction or refinement rules for testing the preconditions. These knowledge units are also acquired from text and experts. Explanation structures are acquired for the concrete operators as well as for the generalization of operators which are described by different operator classes. The preconditions and the consequences for generalized operators are essential for the construction of somewhat general skeletal plans, which can be applied to different problem types in the hierarchy of problem classes.

4.4 Step-wise Knowledge Formalization

The knowledge units are now well prepared to be incrementally transformed into a formal representation. The knowledge units are attached to the views of the model of expertise and templates corresponding to a specific view or a specific combination of views are provided for the semi-formalization. Additionally, the templates in COKAM+ depend on whether the knowledge units are preconditions, consequences or task-related and common sense knowledge. Several helpful restrictions can be defined because of this information. Examples are the preconditions or consequences which are always related to the views *detailed* or *general plan*.

A very general template for a precondition can be defined:

precondition(operator, expression).

A more specific template for the precondition (knowledge unit 1: child 1) is shown in knowledge unit 1: child 2, where the corresponding knowledge unit is related to the views *general product* and *general environment*.

Furthermore, COKAM+ that allows the construction of a child knowledge unit from any existing knowledge unit and records its source knowledge unit supports the preparation and the formalization of knowledge units. The following two knowledge units show, how the extracted, decontextualized knowledge unit 1 which is specified as a precondition is transformed into a formal representation. Thereby a second generation and a third generation of knowledge units are constructed with the help of COKAM+ . The second generation or child 2 shows the semi-formalization with the help of templates.

Knowledge unit 1: child 2 *Application of Silicon-Nitrite-Ceramic:*
"precondition(operator,
expression(product, environment))"

The last generation consists of the formalization into target language.

Knowledge unit 1: child 3 *Application of Silicon-Nitrite-Ceramic:*
"precondition(cut(speed(Speed), feed(Feed), path(Path), tool(Tool)),
(workpiece_material(Path, cast_iron), cutting_material(Tool, ceramic)) →
(Feed < 0.8, Speed < 800))"

The established child – source – links between the knowledge units help to track down errors or inconsistencies which are later detected in the formal knowledge base. Since the formalization is performed incrementally in small steps, such errors are also less likely to occur.

4.5 Interaction with SPGEN

In SPGEN, the plan execution is simulated on the basis of the available domain theory acquired by COKAM+ . By sequentially executing each operator the preconditions for its application are checked and its effects are determined. A set of Strips-like rules with add- and delete-actions represents the effects of the operators. Thus SPGEN provides the requirement that the domain theory acquired with COKAM+ must be a Strips-like description of the operators.

If the domain theory is sufficient, a complete explanation of the plan will be obtained. Otherwise the domain theory is not complete because of the prerequisite that only such cases are selected which have been successfully used in a real world application. In other words, the simulation of the plan in SPGEN is a formal verification of the domain theory and provides a feedback concerning its completeness with respect to a specific case.

5 Discussion

The prerequisites for an application of COKAM+ in a domain are: there must exist texts of the particular domain, there must exist recorded problem descriptions and solutions, and a model of the problem solving process must be developed. COKAM+ requires the human user to perform a number of operations in a particular sequence to obtain knowledge, which is then related to specific problem solutions within the domain. Furthermore, the application of COKAM+ assists in the structuring and the step-wise formalization of the knowledge and provides a documentation of this process, so that the knowledge acquisition process is also made obvious to other than the involved persons. Thus maintenance can be supported.

In comparison with systems like MEKAS [12] and COGNOSYS II [20] which also acquire knowledge from text COKAM+ can start where the other tools are finished. Whereas MEKAS and COGNOSYS II define the task or goal of the domain just by their application, in COKAM+ the desired competence of the knowledge base is already defined by the model of expertise and the operator classes. By using this additional information the selection of relevant knowledge can be supported. After a complete informal knowledge base is established, the various knowledge units are separately transformed into a formal representation. Such a formalization process can be executed step-wise and guided by templates, which depend on the model of expertise, the developed explanation structure and the desired representations, or some generic structures [20].

COKAM+ also shows a number of similarities to other knowledge acquisition methods from text [1, 11, 18, 7]. The system of Szpakowicz for example supports semi-automated acquisition of structures from technical texts. It presumes an initial general model according to the KADS-method and then acquires conceptual structures from texts. The operator supports the semi-automated system, if it cannot make decisions on its own. The SHELLEY of Anjewierden provides a hypertext-based protocol and concept editor, where the words of the stored text are mouse-sensitive, so that the definitions of the words can be inspected. Also, annotations are stored so that remarks from different people can be read on demand. SHELLY however aims at the development or selection of an interpretation model whereas in COKAM+ the model is a prerequisite for knowledge acquisition.

Acknowledgements

I would like to thank Ralph Bergmann, Otto Kühn, and Franz Schmalhofer for their contributions to this work.

References

[1] A. Anjewierden, J. Wielemaker, and C. Toussaint. Shelly - computer aided knowledge engineering. In B. Wielinga, J. Boose, B. Gaines, G. Schreiber, and M. van Someren, editors, *Current Trends in Knowledge Acquisition*, pages 41 – 59. IOS, May 1990.

[2] R. Bergmann and F. Schmalhofer. CECoS: A case experience combination system for knowledge acquisition for expert systems. *Behavior Research Methods, Instruments and Computers*, 23:142–148, 1991.

[3] Joost Breuker and Bob Wielinga. Models of expertise in knowledge acquisition. In Giovanni Guida and Carlo Tasso, editors, *Topics in Expert System Design, Methodologies and Tools*, Studies in Computer Science and Artificial Intelligence, pages 265 – 295. North Holland, Amsterdam, 1989.

[4] Susan T. Dumais, George W. Furnas, and Thomas K. Landauer. Using latent semantic analysis to improve access to textual information. In *CHI'88 Proceedings*, 1988.

[5] D. E. Egan, J. R. Remde, L. M. Gomez, T. K. Landauer, J. Eberhardt, and C. C. Lochbaum. Formative design-evaluation of superbook. *ACM Transaction on Information Systems*, 7(1):30 – 57, January 1989.

[6] W. Eversheim. *Organisation in der Produktionstechnik, Arbeitsvorbereitung*, volume 4, pages 125–128. VDI Verlag, Düsseldorf, 1989.

[7] R. Jansen-Winkeln. Wastl: An approach to knowledge acquisition in the natural language domain. In J. Boose, B. Gaines, and M. Linster, editors, *Proceedings of the European Knowledge Acquisition Workshop (EKAW'88)*, pages 22-1 – 22-15. Gesellschaft für Mathematik und Datenverarbeitung, June 1988.

[8] Jana Köhler. Approaches to the reuse of plan schemata in planning formalisms. Technical Memo TN-91-01, Deutsches Forschungszentrum für Künstliche Intelligenz (DFKI), January 1991.

[9] O. Kühn, M. Linster, and G. Schmidt. Clamping, COKAM, KADS, and OMOS: The construction and operationalisation of a kads conceptual model. In J. Boose, B. Gaines, M. Linster, and D. Smeed, editors, *Fifth European Knowledge Acquisition for Knowledge-Based Systems Workshop EKAW91*, 1991.

[10] T. M. Mitchell, R. M. Keller, and S. T. Kedar-Cabelli. Explanation-based generalization: A unifying view. *Machine Learning*, 1:47–80, 1986.

[11] Jens-Uwe Möller. Knowledge acquisition from texts. In J. Boose, B. Gaines, and M. Linster, editors, *Proceedings of the European Knowledge Acquisition Workshop (EKAW'88)*, pages 25-1 – 25-16. Gesellschaft für Mathematik und Datenverarbeitung, June 1988.

[12] Hyacinth S. Nwana, Ray C. Paton, Michael J. R. Shave, and Trevor J. M. Bench-Capon. Textual analysis for knowledge acquistion using the mekas approach. In *European Knowledge Acquisition Workshop, Sisyphus Working Papers : Text Analysis*. J. Boose and B. Gaines and M. Linster and D. Smeed and B. Woodward, May 1991.

[13] Franz Schmalhofer, Ralph Bergmann, Otto Kühn, and Gabriele Schmidt. Using integrated knowledge acquisition to prepare sophisticated expert plans for their re-use in novel situations. In Thomas Christaller, editor, *GWAI-91 15th German Workshop on Artificial Intelligence*, pages 62 – 71. Springer-Verlag, 1991.

[14] Franz Schmalhofer, Otto Kühn, and Gabriele Schmidt. Integrated knowledge acquisition from text, previously solved cases and expert memories and expert memories. *Applied Artificial Intelligence*, 5:311 – 337, 1991.

[15] G. Schmidt, R. Legleitner, and F. Schmalhofer. Lautes Denken bei der Erstellung der Schnittaufteilung, der Werkzeugauswahl und Festlegung der Maschineneinstelldaten. Diskussionspapier 90/13, DFKI, Kaiserslautern, Erwin-Schrödinger-Str. (Bau 57), 6750 Kaiserslautern, Juni 1990.

[16] Gabriele Schmidt and Franz Schmalhofer. Case-oriented knowledge acquisition from texts. In B. Wielinga, J. Boose, B. Gaines, G. Schreiber, and M. van Someren, editors, *Current Trends in Knowledge Acquisition*, pages 302–312, Amsterdam, May 1990. IOS Press.

[17] G. Spur. *Produktionstechnik im Wandel*. Carl Hanser Verlag , München, 1979.

[18] Stan Szpakowicz. Semi-automatic acquisition of conceptual structure from technical texts. In *Banff-Proceedings'88*, 1988.

[19] Thien and Chien. *An introduction to automated process planning systems*. Prentice-Hall, 1985.

[20] B. Woodward. Analysing text into general knowledge structures with COGNOSYS II. In *European Knowledge Acquisition Workshop, Sisyphus Working Papers : Text Analysis*. J. Boose, and B. Gaines and M. Linster and D. Smeed and B. Woodward, May 1991.

Parallel Parsing of Ambiguous Languages on Hypercube Architectures

Richard A. Reid and Manton M. Matthews

Department of Computer Science, University of South Carolina
Columbia SC 29208, matthews@cs.scarolina.edu

Abstract. In this paper we describe a generalization of existing LALR techniques to allow the parallel parsing of ambiguous languages on hypercube architectures. In particular we are interested in the parsing of natural languages in parallel. The technique that is used is a "nondeterministic" version of a standard LALR parser, where when a shift/reduce or reduce/reduce conflict is reached the parser branches and pursues them both independently. We have developed a parallel parser generator RACC, which is developed with the use of YACC under Unix. By specifying a YACC like specification file and running it through RACC one obtains a parallel parser that will run on the nodes of the NCUBE/10 hypercube.

1 Introduction

The purpose of this research is to develop an efficient model for the parallel parsing of natural languages that would be well suited to hypercube environments. In addition to a consideration of general parsing issues it is important to consider the specific target environment since there is such a great diversity among parallel architectures.

A distinction can be made between two broad categories of parsing. The deterministic parsing of non-ambiguous languages such as programming languages is inherently very different from the non-deterministic parsing of ambiguous languages such as natural languages. There are efficient sequential algorithms for deterministic parsing which are in widespread use and form the foundation for compiler technology[1]. The situation is quite different for natural languages. The element of ambiguity in natural languages makes parsing a more computationally expensive proposition. Additionally the relationship between syntax and semantics is not a simple one as it is for programming languages. As a result universally accepted approaches to non-deterministic parsing do not exist.

1.1 Existing Parallel Approaches

A widespread approach to natural language parsing is to use recursive transition networks (RTN). RTNs are typically top-down parsers that perform a depth-first search for successful parses. RTNs are normally augmented to allow semantics to interact with the syntactic parsing performed by the RTN. These augmented parsers are referred to as augmented transition networks (ATN)[6]. A drawback to ATNs is that their time complexity can become exponential.

Another approach to non-deterministic parsing is chart parsing [4], [5]. In this approach, dynamic charts are maintained which are similar to the Sets of Items Construction used in deriving LR tables. With a cubic upper bound on time complexity this is a more efficient strategy than that used in ATNs.

Much of the research in parallel non-deterministic parsing has involved mapping chart parsing onto a parallel environment. These approaches require some sort of global data structure to represent the chart. The parsing process is distributed in such a way that each successful parse is distributed across that processing array.

This approach was not at all suited to hypercube architectures. Typically hypercube architectures provide local memory to each processor but do not provide global memory. An approach to non-deterministic parsing was sought that would not require global memory and would use the hypercube interconnection scheme to good advantage. Another consideration was to provide a facility for additional processing during parsing. Although the function of parsing is, strictly speaking, recognition, in practical systems some sort of processing is normally desired during the parse. For instance it is often desired that the parsing phase return a parse tree rather than a single yes or no verdict.

1.2 Our Parallel Approach

The present model involves an adaptation of table driven LR parsing to allow parallel non-deterministic parsing. It is in a sense a combination of the searching strategy fundamental to RTNs with chart parsing. Table driven parsing can be seen as a form of chart parsing where the parsing tables are precomputed and static as opposed to dynamically maintained. A grammar is said to be LR if a LR parsing table can be constructed that contains no multiply defined entries. Such a grammar is non-ambiguous and can be deterministically parsed. In the present model a standard LR parsing table is constructed and multiply defined entires are allowed. With some modifications, the standard LR parsing algorithm can then be used to drive the parsing, thereby allowing 'pseudo LR' parsing of ambiguous languages.

A hypercube array of processors would each be loaded with an identical LR parser and table. Initially all nodes would be operating identically. When a multiply defined table entry was encountered the hypercube would recursively subdivide with one subcube dedicated to each unique table entry. In this way a search for all possible solutions is made in parallel. Since the entire parsing process for each valid solution is performed entirely within a single processor, global memory is not required. Further, if semantic processing has been built into the parsing, the results of such processing is local as well. It would be possible for such a parser to return all possible parses back to the calling process or for the various subcubes containing successful parses to arbitrate among themselves, returning a single 'best' solution.

For many natural language applications where the degree of ambiguity is not great such a parser could be quite effective. If the level of ambiguity inherent in a grammar is great however, one of the branches might exceed the capacity of the hypercube array. The worst case space requirement would be exponential with respect to the length of the input string. The addition of two facilities could greatly reduce this limitation. The first would be to allow a merge facility. In a merge the nodes of a subcube containing a failed parse attempt would be initialized to the state of neighboring nodes that were still pursuing active parses. The result would be the formation of larger dimension subcubes. We have found this to be a difficult problem that requires a good bit of communication overhead. The second (and more difficult) facility would allow for load balancing. This would be useful in the case where a subcube of zero dimension encountered a multiply defined table entry. Load balancing would be an expensive proposition analogous to garbage collection.

2 The Parallel Parser

In the design of the parallel parser there were several design goals that were desired. These include:

1. To be able to parse highly ambiguous languages, especially natural languages.
2. To return all correct parses.
3. To achieve a high degree of parallelism on a hypercube architecture.
4. To minimizing inter-node communication.
5. To make the system easy to use.

These design goals have substantial influence in the direction of the development of the parser. The desire to handle natural languages, meant that some of the more specialized techniques for very fast parallel parsing of restricted languages [3] were not applicable. We considered parallel ATNs, parallel implementations of prolog, and parallel LR parsing techniques. The final decision was to a fair extent influenced by the desire to provide a framework that facilitated the use of the system without having to have a detailed understanding of the system. The decision to pursue a high degree of parallelism is based on the assumption that high parallelism and utilization will lead to the fastest total computation time. In making the system easy to use we chose to follow a framework simliar to YACC (the LALR parser generator under UNIX). While the complexity of a YACC-like framework may be criticized it still allows the user to specify the grammar in a BNF notation as opposed to having to modify the actual code of the parser.

The parser is based on LR parsing schemes following the framework provided by YACC. In YACC when reduce/reduce or shift/reduce conflicts in the grammar are found YACC makes choices for the user (not always the desired ones) about which of the alternatives is placed in the parsing table. In our parallel LR parser all alternatives are placed in the table. When one of these multiply defined entries is encountered during a parse, the parser "forks" and pursues both branches on separate processors. (Fork is used here in the logical sense, and not the Unix system call sense.) When this fork occurs the state of the parse data must be passed to the new processor. To minimize communication costs at this step the following approach was taken: all of the nodes of the hypercube allocated to the parse are initialized and perform the same computations until a branch point is encountered. At that point the cube is subdivided into two equal sized subcubes with each pursuing one of the alternatives. (Assuming for simplicity of discussion that there are only two choices.) The partitioning is determined by the the node number (address) and the AXIS_MAP. The AXIS_MAP is a bit map of the active axes in the current subcube. It has a one in each position that has active neighbors in the subcube and zeroes in

Input: An input string w and a modified LR parsing table where the
'action' function can have multiple entries
Output: If w is in L(G), all bottom-up parses for w;
otherwise an error indication.
Begin:
Initialize
1   let NODE_ADDR be the node number
2   let AXIS_MAP be a bit map of the axes of the cube
           Initialize the stack to state 0 and read the first input token
     repeat forever begin
           let s be the state on top of the stack
                and a the input token
3   if action[s,a] is a multiple entry then begin
4   (if AXIS_MAP is 0 initiate load balancing)
5   let ACTION be one of those entries based on
6   NODE_ADDR and AXIS_MAP
7   update AXIS_MAP
8   end
9   else let ACTION be action [s,a]
           if ACTION = shift then
                shift
           else if ACTION = reduce then
                reduce
           else if ACTION = accept then
                report parse to host
10   else if AXIS_MAP includes all axes
11   report failure to host
12   (else initiate a merge)
     end
end

Fig. 1. Parallel LR Parsing Algorithm

other positions. When the choice point is encountered an active axis is selected and
the subcube is partitioned on that axis, i.e. those nodes with a one in that position of
their node number form a subcube and pursue one action and those with zero form
the other subcube and pursue the alternative action. When there are more than two
choices the partitioning is repeated until there are enough subcubes to pursue all
alternatives. The parallel parsing algorithm is specified in Figure 1. The lines that
are numbered are the deviations from the standard (sequential) LR parser. Also, the
actions indicated in parentheses have not yet been implemented.

3   The Parallel Parser Generator

The parallel parser uses the algorithm in the previous chapter as the skeleton for
the parser running on each node of the hypercube. This parser uses tables, with

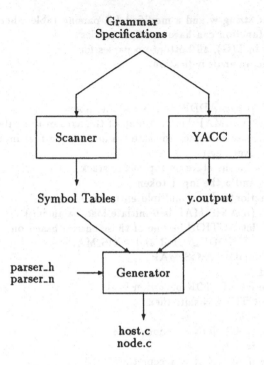

Fig. 2. RACC Parallel Parser Generator

multiply defined entries, that are generated by a "parser generator" like YACC.
The parallel parser generator, called RACC, reads YACC-like specification files and
produces the parallel parsing tables. This is accomplished in several passes and is
illustrated in Figure 2. First the grammar specification is passed through a scanner
that creates symbol tables and also through YACC to generate y.output. The parsing
table information is in the y.output file. It includes a description of the parse table
and also specifies the conflicts that occur. A specialized parser was developed using
YACC, that reads y.output files and generates code for the host and for the nodes
of the hypercube.

4 Performance Improvements of the Parser

As previously discussed, this parsing model has a linear best case execution time. The
worst case execution time is, however, exponential. Although the proposed parser can
follow all the possible parse threads and find all the solutions, its performance can
degrade quickly as the degree of ambiguity inherent in the application increases. This
degradation occurs when the parser is forced into its sequential mode of operation.
The system can be considered to be operating optimally if this sequential mode is
not triggered.

The parser's sequential mode is activated specifically when a zero dimensional
subcube (that is, a single processor node that is following a unique solution) encoun-
ters a multiply defined entry in the parse table. At this point the subcube is forced
into sequential mode since it has no active axis along which to subdivide. There

are basically two reasons why this situation could arise: 1) the number of existing solutions could exceed the number of processors in the array and 2) the number of solutions could be in range but the search could proceed in an unbalanced manner.

In the first case where the actual number of solutions exceeds the number of available processors, there is no alternative to going into a special mode. Since it is always possible that the number of solutions could exceed the size of the processing array, the proposed model must have a facility for mapping multiple parse threads into a single processor. The present implementation provides a sequential mode to address this problem.

It is the second situation which is of primary concern in this section. A natural language application, for instance, might have a manageable degree of ambiguity (that is, the number of possible parse threads might always be fewer than the number of available processors) but might still trigger sequential mode processing due to an imbalanced search for solutions. In this case a merge or load balancing facility could improve performance time.

An unbalanced search is caused by characteristics inherent to a given application. These could be characteristics specific to a particular grammar, input string or both. In an application where there was a high degree of imbalance, a load balancing facility could prevent a triggering of sequential mode (assuming that the number of solutions was in range). Such a load balancing operation would be global in scope and would reallocate the entire array while at the same time interrupting every node. This would be an expensive operation analogous to "garbage collection".

Another method of revitalizing the processing array would be to merge nodes that had completed their work and were idle with nodes that were still working. Such a merge would have the effect of increasing the dimension of active subcubes thereby decreasing the probability that sequential mode would be triggered. As we have seen in the previous section a situation could arise in a given application where the first division of the processing array could yield one succeeding and one failing thread. The failing half of the array could become inactive very early during processing leaving a large and unusable resource. If these nodes could be reactivated it would make the system more robust.

Either facility would significantly enhance the performance of this system. It is difficult to say which might yield the better performance results without actually implementing them. Load balancing would necessarily be a global operation interrupting all processing. It would on the other hand be globally beneficial. Merging, on the other hand, would be more local in scope and wouldn't necessarily interrupt all processing. Also it would address a significant source of imbalance and inefficiency inherent in the proposed system. The synchronization problems, on the other hand, might require more overhead processing and be more difficult to implement. It might be worthwhile to study a wide range of possible applications to determine whether load balancing or merging would be of greater benefit. Alternately, the user could be given the option of specifying one facility or the other if it was found to be counterproductive to implement both concurrently.

5 Conclusions

We have built a parser generator that creates code for both host and node programs for the NCUBE with multiple entries allowed in the parsing tables. The parallel parser handles the ambiguity of natural languages by pursuing choices independently and in parallel. As hypercube multiprocessors enjoy more widespread use, this parsing model could be useful as a foundation for natural language systems. When efficient merging and balancing facilities can be built into our model, this will be a very powerful tool.

References

1. A. H. Aho, R. Sethi and J. Ullman: Compilers: Principles, Techniques, and Tools. Addison-Wesley Publishing Company, Reading, Massachusetts, 1986.
2. Y. Matsumoto, A Parallel Parsing System for Natural Language Analysis. Third International Conference on Logic Programming, pp. 396-409, London, July 14-18, 1986.
3. Y. N. Srikant and P. Shankar: Parallel Parsing of Programming Languages. Information Sciences, vol. 43, no. 1-2, pp. 55-83, 1987.
4. M. Tomita: Efficient Parsing for Natural Language: A Fast Algorithm for Practical Systems. Kluwer Academic Publishers, Boston, 1986.
5. R. Trehan, and P. Wilk: A Parallel Chart Parser for the Committed Choice Non-Deterministic (CCND) Logic Languages. Artificial Intelligence Applications Institute, University of Edinburgh, Edinburgh, United Kingdom, 1988.
6. W. Woods: Transition Network Grammars for Natural Language Understanding. Communications of the ACM, vol. 13, no. 10, pp. 591-606, 1970.

Towards Knowledge Acquisition by Experts

Frank Puppe

Universität Würzburg
Lehrstuhl für Informatik VI
Am Hubland, 8700 Würzburg, Germany

Ute Gappa

Universität Karlsruhe
Institut für Logik, Komplexität und Deduktionssysteme
Postfach 6980, 7500 Karlsruhe, Germany

Abstract: From the three basic knowledge acquisition types for expert systems – indirect knowledge acquisition with a knowledge engineer asking an expert, direct knowledge acquisition mainly by the experts on their own, and automatic knowledge acquisition with machine learning techniques – we currently view direct knowledge acquisition as the most promising approach with respect to total project costs and the technical state of the art. Prerequisites are knowledge representations as well as problem solving methods easily comprehensible for experts and comfortable and easy to learn knowledge acquisition components. In this paper we give an overview on our research to achieve both requirements by "strong" problem solving methods and graphical knowledge acquisition facilities. This is demonstrated with a successful implementation for the well known problem class heuristic classification.

Keywords: Knowledge Acquisition, Expert system, Classification, Graphics

1 Introduction

There are three basic methods how to acquire domain knowledge for expert systems:
- *Indirect knowledge acquisition:* A "knowledge engineer" gets the knowledge from one or more experts and formalizes it for the expert system.
- *Direct knowledge acquisition:* The experts formalize their knowledge by themselves.
- *Automatic knowledge acquisition:* The knowledge is transformed automatically from already existing knowledge, e.g. from the literature or from cases. However no large expert system in routine use has been built with this approach till now.

While indirect knowledge acquisition has shown its practicability in many projects, the approach has two inherent disadvantages:
- It is quite expensive, since at least two highly paid specialists are necessary. The costs remain high because of the often ignored fact, that both specialists are necessary also for the ongoing maintenance of the expert system.
- The experts are often not particularly motivated to communicate their knowledge, since they are only indirectly involved in the project.

Direct knowledge acquisition overcomes these problems, because at best only one person is involved. The responsibility can be compared to the authorship of a book or a paper, which is usually perceived as a rewarding task for the author. However, the author (expert) must understand the language (knowledge representation) in which to express his or her knowledge and needs a comfortable text editing system (knowledge acquisition component).

One of the first knowledge acquisition systems built explicitly for use by experts was OPAL [Musen 87, 89]. OPAL enables physicians to enter their knowledge about cancer therapy protocols in a special graphical programming language, and the variables needed for control can be defined with various kinds of graphical forms. This knowledge is automatically transformed in a form interpretable by the cancer therapy adviser ONCOCIN [Hickam 85].

However, most expert system tools are still difficult to use by experts. Although many graphical facilities like rule forms and frame editors are offered by modern expert system tools, the underlying knowledge representations and problem solving methods like rules with forward or backward chaining and frames with inheritance and attached procedures are quite general (weak). Many researchers including the KADS community [Wielinga 89] feel that stronger knowledge models are necessary to support the knowledge acquisition process. While several tools built on the basis of such stronger models are presented e.g. in [Marcus 88] (MORE, MOLE, SALT, KNACK, SIZZLE), these tools lack sophisticated graphical knowledge acquisition support to be useable by experts. The combination of both requirements in at least a similar complexity than done in OPAL/ONCOCIN, but for a more general problem class still waits for wide-spread practical evaluations.

In the following, we first present MED2/CLASSIKA as an example for such a combination for the well known problem class heuristic classification [Clancey 85] and then generalize our approach. Other problem specific tools for heuristic classification are e.g. AQUINAS [Boose 87], KSSO [Shaw 89] and TEST [Kahn 87], a successor of which is in commercial use. However an important difference is, that MED2/CLASSIKA is based on the hypothesize-and-test strategy with a strong emphasis on separating strategic knowledge about data gathering from structural knowledge about diagnostic inferences, which is less sophisticated or lacking in AQUINAS, KSSO or TEST.

2 Knowledge Model

The application tasks of the classification shell MED2/CLASSIKA are those of *heuristic classification*, in which *diagnoses* (solutions) are identified because of *symptoms* (data) observed. The problem solving method is based on a hypothesize-and-test strategy: starting with the initially given data, the system generates *hypotheses* and pursues them by asking for additional data (see Fig. 1). When new data is available, the *working memory* containing the actual hypotheses is updated again and a new set of data is requested.

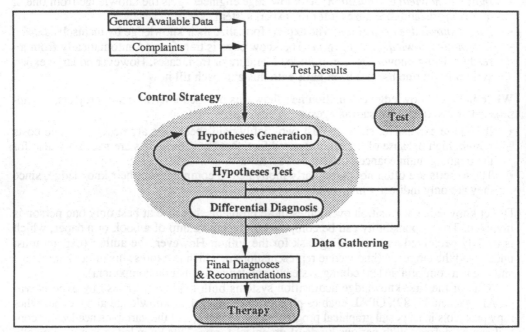

Fig. 1. Overview of MED2/CLASSIKA´s problem solving method

For the purpose of data gathering, questions are grouped into question sets representing basic case information, various *leading symptoms* (complaints) or results of *manual* or *technical examinations*. Once a question set has been selected, a questionnaire of related questions and -

depending on the answers - follow-up questions are displayed on the screen and have to be filled in by the user. Alternatively, they might be set by an automatic data transfer from external data bases or technical sensors. After initially entered complaints, the suspected diagnoses usually have too low evidence and must be confirmed. If the users don´t want to have the initiative, the next question set is either selected directly by dialogue guiding rules or indirectly by a *costs-benefits-analysis* with apriori and case-dependent costs and benefits computed for the currently suspected hypotheses. After the new question set is completed, the interpretation and question asking cycle starts anew until final diagnoses and recommendations are selected.

So, the main object types are *symptoms* (data), *diagnoses* (solutions), *question sets* (symptom classes, tests, questionnaires), *suggestions* (therapies, recommendations) and *rules*. Symptoms represent the *questions* asked to the user (or to a data base or a sensor) and *data abstractions*, which can be infered from the questions by simple database reasoning [Chandrasekaran 83]. Diagnoses represent the diagnostic categories and final diagnoses, which can be structured in hierarchies or heterarchies.

Various kinds of rules represent the relations between the objects. The condition is structured in the main condition, secondary conditions, contexts and exceptions. The latter allows reasoning by default and is implemented with an efficient belief revision system [Puppe 87a]. The most important types of actions are: adding positive or negative evidence to diagnoses (the former are treated as forward rules, the latter as backward rules), infering data abstractions, asking follow-up questions, indicating or contraindicating question sets, assessing the benefits and costs of question sets, and checking the user input for plausibility.

We describe the cognitive model of MED2/CLASSIKA along the intermediating knowledge representations of its knowledge acquisition system. More details on the representation of the heuristic knowledge of MED2/CLASSIKA can be found in [Puppe 87b, 91].

3 Knowledge Acquisition

MED2´s knowledge acquisition environment CLASSIKA (knowledge acquisition for classification) facilitates human experts to represent their knowledge graphically by
- entering their domain vocabulary of symptom names and diagnosis names into hierarchies,
- specifying local information to those terms by filling-in forms and
- establishing their relations by arranging and filling-in tables.

The knowledge model of heuristic classification with its classical phases of *data abstraction, heuristic match* and *refinement* described by Clancey [Clancey 85] is captured in the following graphical representations:
- the data abstraction of the raw data within a *symptom hierarchy* ,
- the heuristic match between data and solution categories by types of *symptom-diagnoses-tables* and
- the refinement of solution categories by a *diagnosis hierarchy* .

The most difficult part of the classification procedure is the heuristic match between the data in the symptom hierarchy and the solutions of the diagnosis hierarchy since one symptom may point to many diagnoses and one diagnosis may be associated with many different symptoms. We found those n:m-relations best arranged in tables (compare fig. 6, fig. 5). An alternative way would be to represent the paths from symptoms to final diagnoses within one single hierarchy or heterarchy. This is done e.g. by the diagnostic shell TEST. The disadvantage of that approach is that the hierarchy tends to contain many redundant nodes if the complexity grows – i.e. "n" and "m" in average are significantly larger than 1.

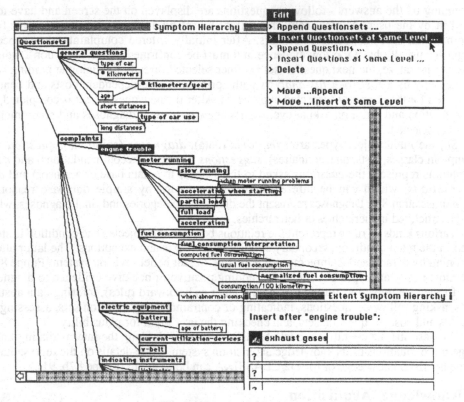

Fig. 2. Symptom hierarchy

Fig. 2 illustrates a symptom hierarchy from a sample knowledge base competent for car engine diagnosis. The data abstractions appear as nodes in bold letters and plain boxes in the graphical hierarchy (e.g. "# kilometers/year") and are calculated from their predecessor nodes. The symptom hierarchy starts with a hierarchical structure of question sets (objects in bold letters surrounded by bold boxes) used to support the end users in the selection of terms concerning their complaints. Each question set is built up by standard questions (nodes in plain letters and boxes), which are asked always when the question set is selected for being filled-in. Those questions may activate follow-up questions depending on the user´s answers (also in plain letters and boxes). Their answer alternatives are hidden at this layer of abstraction. They are entered by opening a form with a double click on an object.

Fig. 3 shows an object form of a question set which asks for attributes like "Name", "Type" of the question set, "Prompt", "Type of Answer" and "Range" of the individual answer alternatives. Standard answer alternatives, "none/something else" and "unknown", are automatically added at run time.

At the abstraction level of the symptom hierarchy, the rules for asking follow-up questions as well as those for deriving data abstractions are represented by lines which can be double-clicked upon in order to enter or modify the preconditions in detail. Fig. 4 shows such a rule for asking a follow-up question and fig. 5 a rule table for the derivation of a data abstraction.

In the table used for the data abstraction "combustion" each column corresponds to one rule. A "+" refers to the row´s label while a "−" would indicate its negation. The rule of the third column for example states that "*If* the color of the exhaust pipe is black sooty *and* the gasoline used is leaded, that the combustion will be not ok". So, marked conditions in a column are "and"-connected, while the statements surrounded by a box are "or"-connected.

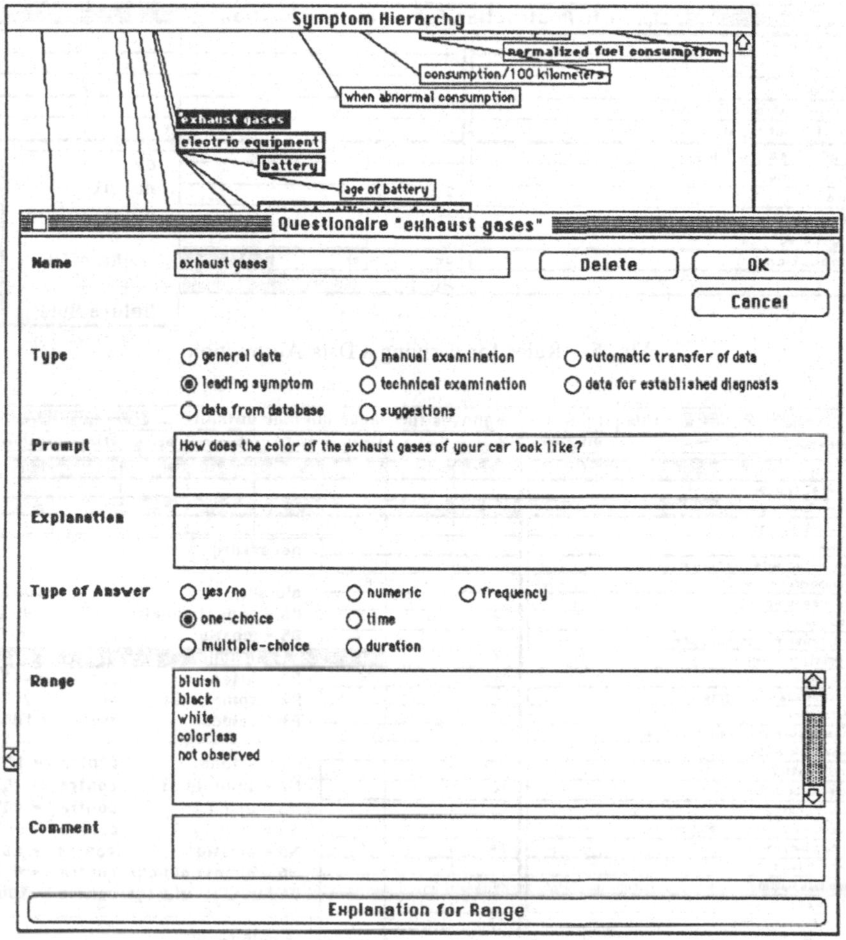

Fig. 3. Attributes of a Question Set

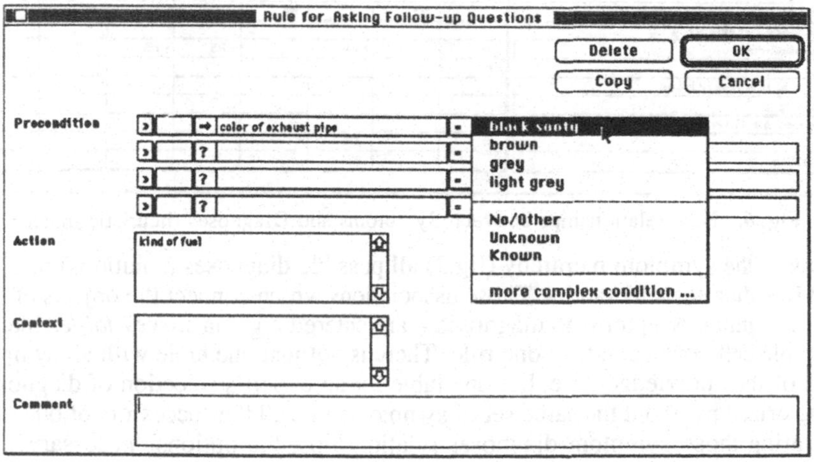

Fig. 4. Rule for asking a Follow-up Question

Data Abstraction Table for "combustion"

kind of fuel					
= unleaded		+			
= leaded			+		
exhaust pipe color known					
= no, I didnot pay attention				+	
color of exhaust pipe					
= black sooty		+	+		
= brown	+				
= grey	+				
= light grey	+				
combustion	ok	ok	not ok		

Popup menu:
```
ok
✓not ok
no attention

-> Rule Form

Delete Rule
```

Fig. 5. Rules for deriving a Data Abstraction

Survey-Table for Diagnoses for "induction pipe untight", ...

Conditions / Diagnoses	induction pipe u...	air filter foul	slow-running sy...	starting system
Predisposition				
General Frequency				
Basic Scoring				
exhaust gases				
• black		+ P5		
pipe color interpretation				
• combustion not ok		+ P5		
fuel consumption				
• too high	N4			
fuel consumption interpretation				
• slightly too high	P3	P4		
• too high	P4	P3		
unusual dias of engine				
= detonating				
when motor running problems				
= when cold engine				
acceleration				
= used to be better	P3	+ N4		
slow running				
= too low number of revolutions				
= irregular running	+ P4	P4		
partial load				
= too less output	P3			
full load				
= no maximum output	+ P3			
start engine				
= hard starting		+ P4		
after engine start				
= low number of revolutions		+ P4		
stop engine				
= problems with unintended engine stop	N5	N4		
not indended engine stop				
= when slow-running		+ P4	+ P5	+ N5
= when braking the speed	+ P5			
motor vehicle inspections				
• regularly		N5		
change of air filter				
> 30000		+ P4		
≥ 50000		P5		

Popup menu:
```
necessary

always              pro    ≈100%
P6 = almost always  pro    ≈ 95%
P5 = mostly         pro    ≈ 80%
✓P4 = usually       pro    ≈ 60%
P3 = often          pro    ≈ 40%
P2 = sometimes      pro    ≈ 20%
P1 = seldom         pro    ≈ 10%

N1 = seldom         contra ≈ 10%
N2 = sometimes      contra ≈ 20%
N3 = often          contra ≈ 40%
N4 = usually        contra ≈ 60%
N5 = mostly         contra ≈ 80%
N6 = almost always  contra ≈ 95%
never = always      contra ≈100%

-> Rule Form
New rule

Delete Rule
```

Fig. 6. N:m-relationships between Symptoms and Diagnoses (heuristic match)

Analogous to the symptom hierarchy (fig.2) all possible diagnoses (solutions) of the domain are listed in a diagnosis hierarchy. Those associations which connect the objects of both hierarchies – assigning symptoms to diagnoses – are entered e.g. via *survey-tables* like fig. 6, in which a table cell corresponds to one rule. There is not just one table with all symptoms and diagnosis of the knowledge base, but one table for an expert´s selection of diagnoses which are characterized by about the same set of symptoms like all the successors of one diagnosis.

For rating those symptom-diagnoses-relationships, the options "necessary", "always" (sufficient) and "never" (exclusion) can be used for categorical conclusions and six positive and six negative categories like in INTERNIST [Miller 82] for expressing probabilistic evidences. In the table of fig. 6, the rule *"If* the fuel consumption is slightly too high, *then* air filter foul can be concluded with the evidence P4" is about to be entered via a pop-up menu. The non-existence of a symptom as row label is represented by the second table cell for each symptom-diagnoses-correlation – according the "+" and "-" of individual object tables like fig. 5 – and is usually used to rule out diagnoses (similar to the frequency value in INTERNIST). The probability of a diagnosis may also be weakened or strengthened by its predisposition appearing in the first section of the survey table (not instantiated in Fig. 6).

Rules with more complex conditions, which cannot be shown in this two-dimensional table, are indicated by a "+" in the table cells and can be viewed in detail when opening the rule´s form to the table´s cell or by opening an individual table for a diagnosis. The individual diagnosis table is the same as the one of fig. 5 except that the pop-up menu in the last line representing the rule´s action offers the rating categories for diagnoses.

So far, we showed how the expert enters the basic diagnostic structure of a knowledge base. We like to demonstrate one more important type of knowledge, the global data-gathering strategy, determining when which question set is be presented to the user. The knowledge for the costs-benefits-analysis of question sets is entered by arranging and filling-in survey-tables for question sets (fig. 7). The question sets label the table´s columns.

Conditions on which a question set should always be activated can be given within the section "Indication" and analogous, conditions which exclude tests in the section "Contra-Indication". One contra-indication represented in the sample table for the test "cylinder head gasket" is for example "water drops are observed at the oil-measuring stick". Besides those categorical conditions, the probabilistic knowledge of how useful a question set is for the evaluation of a diagnosis is entered in the section "Profit". The costs (risk, delay, expenditure) of a test in the section "Costs" are calculated from its "Apriori-Costs" which could be dynamically modified by further symptoms. With this knowledge of the table, the system will at run time indicate that question set which is most useful for testing the actual hypotheses and which costs are the lowest.

Conditions	Questionsets	pressure loss	compression pre..	CO2-test	cylinder measur...	top cylinder cast.	cylinder head g ...
Indication							
Profit							
piston rings defective		50	60				
piston defective		30	50				
cylinders worn-out		50	50		100		
valve/valve spring defective		50	50				
valve guide defective		50	50				
cylinder block defective		50	30	70		100	
cylinder head/gasket defective		50	50	80		100	90
Contra-Indication							
simple cylinder head measurement							
= formation of blowholes in radiator						x	
= water drops at the oil-measuring stick						x	
Costs							
Apriori-Costs		-20	-15	-12	-180	-120	-20

Fig. 7. Knowledge relevant for the costs-benefits-analysis of Questionnaires´ gathering

For additional knowledge types, like plausibility check of input data, therapy selection, further specific forms and tables are available for use by the expert.

The knowledge entered into the graphical knowledge representations is encoded into predefined object types already mentioned (symptom, i.e question and data abstraction, question set, diagnosis, suggestion etc.) with predefined attributes and about 16 types of rules depending on the rule´s purpose. The knowledge is automatically compiled and can directly be tested using MED2´s problem-solver and dialog interface. If the expert disagrees with the systems behavior in solving a case, e.g. in asking questions or drawing conclusions, s/he can locate the knowledge responsible for the misbehavior with the explanation component of MED2, correct the knowledge with CLASSIKA and playback the case with the modified knowledge base. This test-locate-correct-retest-cycle takes a few mouse clicks and may be done within a few minutes, so that thorough testing is encouraged.

4 Evaluation

A convincing success for an expert system can be decided only after several years of routine use, since the ongoing maintenance is the key problem. Since CLASSIKA is fully operational only since one year it is too early for an evaluation of the ongoing projects. However, our expectation that experts will be strongly motivated if they can build expert systems by themselves has not been disappointed. Several experts which have never dealt with expert systems before are very eager to build expert systems and still continue to work as an expert in their environment. Among them are several physicians from different specialities as well as operators responsible for maintaining central computer services and technicians maintaining paper producing machines. The key point is the low budget necessary to start these expert system projects as well as to pass the project mile stones of more and more sophisticated demonstration and field prototypes.

While the experts usually have no difficulties to understand the graphical knowledge representations in CLASSIKA – even without reading the manual [D3 91], but only by playing through an example dialogue to build up a small knowledge base – some experts have difficulties to encode their knowledge in heuristic rules. In particular, they complain about the probabilities and argue either in case examples or in causal models.

5 Generalization

The chances to minimize the conceptual mismatch between the conceptual method of the expert system tool and the expert increase, if more than one problem solving method is offered so that the method best fitting to the mental model of the expert can be selected. While a good fit has clear long term advantages, it poses the new problem of method and tool selection. In difficult cases, the selection even may require an analysis of the domain in some depth. However, graphical knowledge acquisition components (like CLASSIKA) for the various methods resp. tools would greatly facilitate this analysis, since they enable testing the different problem solving methods with a rapid prototyping approach. Fig. 8 overviews the well known methods for classification problem solving.

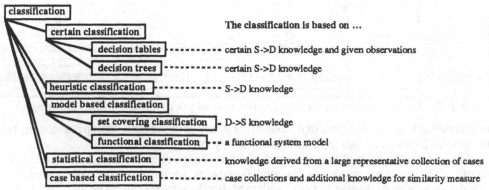

Fig. 8. Overview on well known classification problem solving methods (from [Puppe 90]).
Abbreviations: S = Symptoms; D = Diagnoses.

The simplest problem solving methods for classification are decision tables and decision trees requiring certain knowledge. If the knowledge is uncertain and experts are available, who can give estimations for the heuristic "symptom implies diagnosis" rules, then the techniques of heuristic classification are suitable. Set covering and functional problem solving methods are based on a system model, how the diagnoses cause the symptoms. If large case collections exist, then statistical correlations between symptoms and diagnoses can be computed or the original cases together with a similarity measure can be used for case based classification.

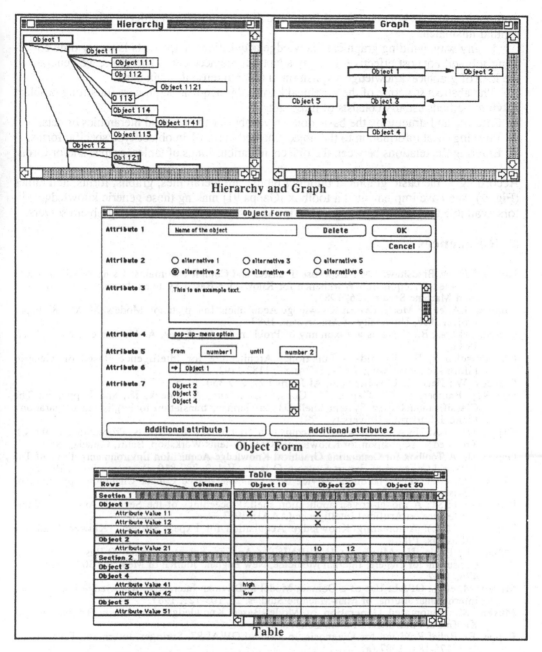

Fig. 9. Generic knowledge editors types (from [Gappa 91])

We are currently experimenting with different versions of corresponding method specific expert system tools. An overview is contained in [Puppe 90]. The idea is to provide a set of ready-to-use specialized tools for strong problem solving methods similar to the tools for role-limiting methods described by John Mc Dermott in [Marcus 88]. However it is an open question, where the compromise lies between the rigidity of strong or role limiting methods and the flexibility of "open" approaches favored by the KADS methodology or by the generic task approach of Chandrasekaran [87], where (ideally) a specific tool is usually not selected from a prestored set but constructed from predefined primitives in order to minimize the conceptual mismatch.

In any way, building graphical knowledge acquisition components is a very time consuming job and not cost effective, if many adaptions are necessary. Therefore, we generalized the idea of graphical knowledge acquisition in order to reuse its main parts.

The abstract structure of the graphical knowledge acquisition process for strong problem solving methods usually consists of three steps:
1. Entering and structuring the basic domain vocabulary in graphical hierarchies or nets.
2. Entering local information to the objects named in step 1 in object-type specific forms.
3. Entering the relations between the objects in various kinds of tables, graphs and/or forms.

Accordingly, the basic graphical building blocks are hierarchies, graphs, forms, and tables (Fig. 9). We have implemented a toolbox [Gappa 91] making these generic knowledge editors available for instantiation in knowledge acquisition systems for other problem solvers.

6 References

Boose, J.H., & Bradshaw, J.M.: Expertise Transfer and Complex Domains: Using AQUINAS as a Knowledge Acquisition Workbench for Knowledge-Based Systems, International Journal of Man-Machine Studies, 26, 1987.

Breuker, J.A. et al.: Model-Driven Knowledge Acquisition: Interpretation Models, MEMO 87 Esprit Project 1098, University of Amsterdam, 1987.

Chandrasekaran, B.: Towards a Taxonomy of Problem Solving Types, AI Magazine 4, No. 1, 9-17, 1983.

Chandrasekaran, B.: Towards a Functional Architecture for Intelligence Based on Generic Information Processing Tasks, IJCAI-87, 1183-1192, 1987.

Clancey, W.: Heuristic Classification, AI Journal 27, 289-350, 1985.

[D3 91]. Bamberger, S., Gappa, U., Goos, K., Meinl., A., Poeck, K., and Puppe, F.: The Classification Expert System Shell D3 (in German, translation to English in preparation). Manual. Karlsruhe University, 1991.

Gappa, U.: CLASSIKA: A Knowledge Acquisition Tool for Use by Experts, Proceedings of the 4th Knowledge Acquisition for Knowledge-Based Systems Workshop, Banff, Canada, 1989.

Gappa, U.: A Toolbox for Generating Graphical Knowledge Acquisition Environments, Proc. of The 1. World Congress of Expert Systems, Orlando, Vol. 2, 787-810, Pergamon Press, 1991.

Hickam, D. et al.: The Treatment Advice of a Computer-Based Cancer Chemotherapy Protocol Adviser, Annals of Internal Medicine, 103, 928-936, 1985.

Kahn, G.: TEST: A model-driven application shell. Proceedings of the Seventh Annual National Conference on Artificial Intelligence (AAAI), 1987.

Marcus, S. (ed.): Automating Knowledge Acquisition for Expert Systems, Kluwer Academic Publishers, 1988.

Miller, R., Pople, H., Myers, J.: INTERNIST1: An Experimental Computer-Based Diagnostic Consultant for General Internal Medicine, New England Journal of Medicine 307, No. 8, 468-476, 1982.

Musen, M. et al.: OPAL: Use of a Domain Model to Drive an Interactive Knowledge Editing Tool, International Journal of Man-Machine Studies, 26, 105-121, 1987.

Musen, M.: Automated Generation of Model-Based Knowledge Acquisition Tools. Morgan Kaufmann Publishers, Pitman, London, 1989.

Puppe, F.: Belief Revision for Diagnosis, in Proc. of GWAI-87, Springer, Informatik-Fachberichte 152, 175-184, 1987 (a).

Puppe, F.: Requirements for a Classification Expert System Shell and Their Realization in MED2, Applied Artificial Intelligence, 1, 163-171, 1987 (b).

Puppe, F.: Problem Solving Methods in Expert Systems (in German; translation to English in preparation) Springer, 1990.

Puppe, F., Legleitner, T. and Huber, K.: DAX/MED2 - A Diagnostic Expert System for Quality Assurance of an Automatic Transmission Control Unit, in Zarri (ed.): Operational Expert Systems in Europe, Pergamon Press, 1991.

Shaw, M. and Gaines, B.: Comparing Conceptual Structures: Consensus, Conflict, Correspondence and Contrast, Knowledge Acquisition 1, No 4, 341-363, 1989.

Wielinga, B., Bredeweg, B. and Breuker, J.: Knowledge Acquisition for Expert Systems, Proceedings of ACAI-88, 1988.

The Rough Sets Approach to Knowledge Analysis for Classification Support in Technical Diagnostics of Mechanical Objects

Jerzy Stefanowski, Roman Słowiński
Institute of Computing Science
Technical University of Poznan
ul. Piotrowo 3A
60-965 Poznań, Poland

Ryszard Nowicki
Institute of Applied Mechanics
Technical University of Poznań
ul. Piotrowo 3
60-965 Poznań, Poland

March 4, 1992

Abstract

Problems of knowledge analysis for decision systems concerning diagnostic classification of mechanical objects are considered in this paper. Knowledge coming from experience is represented in a form of an information system and is analysed by means of the new approach based on the rough sets theory. The application of this approach enables reduction of superfluous data in the information system and generation of classification rules showing relationships between the description of objects and their assignment to classes of a technical state. The use of the rough sets approach is shown on a practical example concerning the evaluation of the technical state of rolling bearings. The bearings are in one of two technical states (good and bad) and are described by a set of symptoms which results from measurements of noise and vibration collected in an industrial environment.

1 Introduction

The paper is devoted to the problems of knowledge analysis for decision support systems. We consider a kind of human knowledge concerning classification of data obtained from observation, measurements etc. which can be represented in a structure called information system. Such data concern a set of objects (examples, states, observations etc.) described by a set of multivalued attributes (features, tests, characteristics, variables etc.). The set of object is classified into disjoint family of classes by an expert according to his experience and knowledge.

Very often, the question about relationships between the description of objects by values of attributes and their assignments to certain classes (e.g. in a form of rules) is very important in the analysis of knowledge represented in information systems. The other essential problem may refer to reduction of all superfluous attributes and objects in the information system. Such problems are considered in the following paper.

The rough sets theory (created by Pawlak in 80's [13]) is a good tool for such an analysis. Using the rough sets theory one can obtain the following results : evaluate importance of attributes for classification, reduce all redundant objects and attributes in the information system so as to get so called minimal subset of attributes ensuring a satisfactory quality of classification. Moreover, it is possible to create "models" of objects in each class on the basis of the analysis of values adopted by the most significant attributes. Then, the reduced information system can be identified with the decision table, and the set of decision rules can be derived from it.

One of the main advantages of the rough sets theory is the possibility of the analysis of information systems containing imprecise descriptions of some objects, i.e. some objects cannot be distinguished because of imprecise information about them.

The obtained results i.e. reduced set of attributes and the set of decision rules represent the knowledge acquired by a specialist on all treated objects (existing in his experience). The results can be helpful directly in decision making concerning the classification of new objects or in construction of decision support system [18]. In the second case, the rough sets approach is one of the methods of knowledge analysis and acquisition for such systems.

The rough sets approach has been already used to analyse some information systems, in particular, medical ones (cf. [16], [4], [8]). It was also initially applied in industry (cf. [10]). This application concerned the analysis of generation of control algorithm for a rotary clinker kiln from observation of a stoker decisions who controlled the kiln. The analysis handled the imprecise description of some observations.

In this paper we discuss possibility of application of the rough sets theory to technical diagnostics of mechanical objects.

One of the main problems in technical diagnostics is evaluation of a technical state of controlled objects (cf. [1],[5],[7],[9]). This evaluation is performed on a base of symptoms of a technical state (cf. [2]). Vibration and noise symptoms are very often used for this aim. Their usefulness results from relatively high reliability of diagnostic information and facility of collecting them.

As the values of symptoms usually change monotonically with deterioration of the technical state, so called symptom limit values are often used in technical diagnostics. The symptom limit values are boundary values of a symptom which divides its domain into intervals corresponding to considered classes of the technical state. For some machines, there are known standards establishing not only the type of measurements but also defining intervals of symptoms. However, practical experience do not confirm these recommendation quite often. So, the problem of estimation of the symptom limit values is important in development of diagnostic procedures for a specified type of machines. In particular it is interesting to have a methodology of comparing different possible methods defining symptom limit values for these machines.

Evaluation of the technical state using the minimal number of symptoms is the most desired in practice because the lower is the number of symptoms used, the lower is usually the cost and time of the diagnostic investigation. However, rarely only one symptom is sufficient for this aim and a subset of symptoms must be used for a reliable evaluation. The problem of a proper choice of symptoms having good diagnostic characteristics has great practical importance. Then, chosen symptoms can be used to create a classifier of the technical state which consists of rules enabling evaluation of the technical state from the values of these symptoms (cf. [1],[11],[12]).

Reduction of a set of symptoms to a minimal subset of symptoms ensuring satisfactory evaluation of the technical state, and creation of the classifier of the technical state are one of the main aims of a majority of technical diagnostic investigations.

In the following paper, the set of identically constructed mechanical objects, i.e. rolling bearings, which are in one of two technical states (good or bad), is considered. The rolling bearings are installed in a band conveyor and considered symptoms are based on their noise and vibration characteristics. The observations were taken in a real industrial environment and are represented in a form of the information system.

The rough sets approach is used to solve the following problems :

- evaluation of different methods of defining symptom limit values,

- evaluation of diagnostic usefulness of particular symptoms of the technical state,

- reduction of a set of considered symptoms in order to get a subset of symptoms ensuring the best evaluation of the technical state,

- derivation of classification rules, i.e. decision rules determining the assignment of the rolling bearing to a given class of the technical state depending on values of symptoms.

The obtained results can be helpful in creation of the decision system concerning classification of the technical state for rolling bearings. In this meaning, the rough sets theory is one of tools of knowledge analysis and knowledge acquisition for such systems.

In the next section, some basic concepts of the rough sets theory are given. Then, in section 3, the analysed data set is described and results of its analysis by means of the rough sets theory are discussed. Conclusions are drawn in the final section.

2 Basic Concepts of the Rough Sets Theory

In this section only basic notions of the rough sets theory created by Pawlak are given. More information can be found in [13],[14],[15],[17].

The observation, that we cannot distinguish objects (patients, states, elements, etc.) on the basis of imprecise information about them is the starting point of the rough sets philosophy. In other words, imprecise information cause indiscernibility of objects. Indiscernibility relation is used to define two main operations on data: lower and upper approximation of sets. Approximation space and approximation of a set or a family of sets (particularly classification) in this space, are two next concepts of the rough sets theory with a great practical importance. Using lower and upper approximation of a set (or a classification) we can define an accuracy and a quality of approximation. These are numbers from interval $[0,1]$, which define how exactly we can describe the examined set (or classification) of objects using available information. The concept of an information system is used to construct the approximation space. It enables to represent data in a useful form of a table.

2.1 Information system

By an information system we understand the 4-tuple $S = \langle U, Q, V, \rho \rangle$, where U is a finite set of *objects*, Q is a finite set of *attributes*, $V = \bigcup_{q \in Q} V_q$ and V_q is a *domain* of the attribute q, and $\rho : U \times Q \rightarrow V$ is a total function such that $\rho(x, q) \in V_q$ for every $q \in Q$, $x \in U$, called *information function*. Any pair (q, v), $q \in Q$, $v \in V_q$ is called *descriptor* in S.

The information system is in fact a finite data table, columns of which are labelled by attributes, rows are labelled by objects and the entry in column q and row x has the value $\rho(x, q)$. Each row in the table represents the information about an object in S.

Let $S = \langle U, Q, V, \rho \rangle$ be an information system and let $P \subseteq Q$; $x, y \in U$. We say that x and y are *indiscernible* by the set of attributes P in S (denotation $x \tilde{P} y$) if $\rho(x, q) = \rho(y, q)$ for every $q \in P$. Equivalence classes of relation P are called *P-elementary sets* in S. Q-elementary sets are called *atoms* in S.

The family of all equivalence classes of relation $\tilde{P}$ on U is denoted by P^*. $Des_p(X)$ denotes the description of equivalence class (P-elementary set) $X \in P^*$, i.e.: $Des_P(X) = \{ (q, v) \mid \rho(x, q) = v, \text{ for every } x \in X, q \in P \}$.

2.2 Approximation of Sets.

Let $P \subseteq Q$ and $Y \subseteq U$. The *P-lower approximation* of Y denoted by $\underline{P}Y$ and the *P-upper approximation* of Y denoted by $\bar{P}Y$ are defined as:

$$\underline{P}Y = \bigcup X : \{X \in P^* \text{ and } X \subseteq Y\}$$

$$\bar{P}Y = \bigcup X : \{X \in P^* \text{ and } X \cap Y \neq \emptyset\}$$

The P-boundary (doubtful region of classification) is defined as $Bn_p(Y) = \bar{P}Y - \underline{P}Y$.

With every subset $Y \subseteq U$, we can associate an *accuracy of approximation* of set Y by P in S, or in short, accuracy of Y, defined as:

$$\mu_p(Y) = \frac{\text{card}(\underline{P}Y)}{\text{card}(\bar{P}Y)}$$

2.3 Rough classification

Let S be an information system, $P \subseteq Q$, and let $\kappa = \{Y_1, Y_2, \ldots, Y_n\}$ be a *clas-sification* of U, i.e. $Y_i \cap Y_j = \emptyset$ for every $i, j \leq n$, $i \neq j$ and $\bigcup_{i=1}^{n} Y_i = U$. Y are called *classes* of κ. By P-lower (P-upper) approximation of κ in S we mean sets $\underline{P}\kappa = \{\underline{P}Y_1, \underline{P}Y_2, \ldots, \underline{P}Y_n\}$ and $\bar{P}\kappa = \{\bar{P}Y_1, \bar{P}Y_2, \ldots, \bar{P}Y_n\}$ respectively. The coefficient

$$\gamma_P(\kappa) = \frac{\sum_{i=1}^{n} \operatorname{card}(\underline{P}Y_i)}{\operatorname{card}(U)}$$

is called the quality of approximation of classification X by set of attributes , or in short, *quality of classification* κ. It expresses the ratio of all P-correctly classified objects to all objects in the system.

2.4 Reduction of attributes

We say that the set of attributes $R \subseteq Q$ *depends* on the set of attributes $P \subseteq Q$ in S (denotation $P \rightarrow R$) if $\tilde{P} \subseteq \tilde{R}$. Discovering dependencies between attributes enables the reduction of the set of attributes. Subset $P \subseteq Q$ is *independent* in S if for every $P' \subset P$, $\tilde{P'} \supset \tilde{P}$; otherwise subset $P \subseteq Q$ is *dependent* in S. In practical applications we are interested in reducing those attributes which are redundant in S (i.e. we are interested in obtaining reducts). Subset $P \subseteq Q$ is a *reduct* of Q in S if P is the greatest independent set in Q and $P^* = Q^*$. As a practical criterion we use the quality of classification Ξ. The reduct gives the same quality as the whole set of attributes in the system. The least independent set which ensures the same quality of classification as the reduct is called the *minimal subset* in S. Let us notice that an information system may have more than one minimal subset. Intersection of all minimal subsets is called the *core*. The core is a collection of the most significant attributes for the classification in the system.

2.5 Decision tables

An information system can be seen as decision table assuming that $Q = C \cup D$ and $C \cap D = \emptyset$, where C are called *condition attributes*, and D *decision attributes*. Decision table $S = \langle U, C \cup D, V, \rho \rangle$ is deterministic if $C \rightarrow D$; otherwise it is non-deterministic. The deterministic decision table uniquely describes the decisions to be made when some conditions are satisfied. In the case of a non-deterministic table, decisions are not uniquely determined by the conditions. Instead, a subset of decisions is defined which could be taken under circumstances determined by conditions.

From the decision table we can derive *decision rules*. Let $\{X_1, X_2, \ldots, X_k\}$ be the C-definable classification of U, and $\{Y_1, Y_2, \ldots, Y_n\}$, the D-definable classification of U. Expression $Des_C(X_i) \Rightarrow Des_D(Y_j)$ is called *(C,D)-decision rule* in S, where $Des_C(X_i)$ and $Des_D(Y_j)$ are unique descriptions of the classes X_i and Y_j, respectively ($i = 1, 2, \ldots, k$; $j = 1, 2, \ldots, n$). The set of decision rules $\{r_{ij}\}$ for all classes Y_j ($j = 1, 2, \ldots, n$) is called the *decision algorithm*. A general procedure for the derivation of an optimal decision algorithm from decision table was presented in [5].

3 Analysis of the Technical State of Rolling Bearings Using the Rough Sets Theory.

We analyse the information system containing data about a set of 55 rolling bearings described by 10 symptoms. The examined set is divided into two classes on the basis of expert's evaluation. The first class consists of 34 bearings recognized to be good, the second class consists of 21 bearings being in different states of failure. Vibration and noise levels of bearing housing were taken as symptoms of the technical state. Two first symptoms (denoted as s_1 and s_2) have the noise nature, others (i.e. from s_3 to s_{10}) have the vibration nature. The analysed data set with detailed description was presented in [12].

The data sets give base to create the original information systems for each diagnostic case. It is known that in the rough sets approach, values of quantitative attributes are translated into some qualitative terms. This translation involves a division of the original domain into some subintervals and an assignment of qualitative codes to these subintervals. In technical diagnostics, attributes are symptoms of the technical state; using symptom limit values one can divide an original domain of a symptom into subintervals corresponding to conventional classes of the technical state (cf. [2], [3]). As a result of this translation, the coded information system can be obtained. i.e. the domain of each symptom has been divided into five subintervals which has been coded by numbers 1,2,3,4,5.

Several methods defining symptom limit values were proposed for this kind of diagnostic problems [2],[11],[12]. In this paper, we consider four of them. They are called L-, W-, P- and C- methods (formulae enabling determination of the limit values by means of these methods are presented in [11] or [12]).

The rough sets approach was used to compare them. The criterion of the quality of classification allowed to rank the considered methods. It was found out that only the L-method was reliable for evaluation of the technical state (quality equal to 0.93). Other methods gave unsatisfactory quality of evaluation, i.e. the W-method - 0.56, the P-method - 0.55 and the C-method 0.44. Hence, only the L-method was chosen to further analysis. Table 1 shows the accuracy of approximation of each particular class by the set of all symptoms using this method.

As the set of all considered symptoms allowed to evaluate, with high quality, the technical state of rolling bearings, we tried to look for minimal subsets of symptoms. We found 5 following subset :

$\{s_5, s_6, s_7, s_8, s_9\}$,
$\{s_2, s_4, s_5, s_6, s_7\}$,
$\{s_2, s_3, s_5, s_6, s_7, s_9\}$,
$\{s_4, s_5, s_6, s_7, s_8\}$,
$\{s_1, s_5, s_6, s_7, s_8\}$,

The core is $\{s_5, s_6, s_7\}$. These symptoms are the most significant for the classification.

Using each of minimal subsets one can classify bearings with the same quality as using all symptoms. Taking into consideration a structure of obtained minimal subsets, it can be noticed that two subsets contain exclusively vibration symptoms and others consist of vibration symptoms with only one noise symptom. As a result

Class	Card. of class	Lower approx.	Upper approx.	Accuracy of class
good	34	32	36	0.89
bad	21	19	23	0.83

Figure 1: Approximation of classification for all symptoms and the L-method

the core is composed of only vibration symptoms and ensures quality of classification equal to 0.65. This value can be increased to the level slightly lower (0.89) than maximum (0.93) by adding symptoms s_8 or s_4 (which are also vibration symptoms). Let us also notice that symptom s_{10} has never occurred in any minimal subset.

We obtained 5 minimal subset of symptoms, so the choice of one of them to create the classifier of the technical state is not univocal. So, other criteria than the quality of classification must be employed. The minimum cardinality of a subset of symptoms and the minimal number of rules belonging to the classifier can be used as the secondary criteria. In practice, it is interesting to take into account also other criteria, e.g. easiness, cost and time of a measurement, or subjective preferences for certain symptoms.

Proceeding in this way, one can choose the best compromise minimal subset to create a classifier.

The subset $\{s_4, s_5, s_6, s_7\}$ was chosen to further analysis in this way and then, the reduced information system has been identified with decision table $S_R = \langle U', C \cup D, V, \rho \rangle$, where $C = \{s_4, s_5, s_6, s_7\}$ is the set of condition symptoms, D is the decision attribute expressing the technical state of the bearing, and U' is the set of $\{s_4, s_5, s_6, s_7\}$-elementary sets.

The set of decision rules derived from the decision table is presented below.

if $(s_7 = 4)$ then (class=good);
if $(s_7 = 5)$ then (class=good);
if $(s_5 = 1)$ and $(s_7 = 1)$ then (class=good);
if $(s_5 = 3)$ and $(s_7 = 1)$ then (class=good);
if $(s_5 = 2)$ and $(s_7 = 1)$ then (class=good);
if $(s_5 = 2)$ and $(s_7 = 2)$ then (class=good);
if $(s_4 = 1)$ and $(s_5 = 5)$ and $(s_7 = 3)$ then (class=good);
if $(s_4 = 4)$ and $(s_5 = 5)$ and $(s_7 = 3)$ then (class=good);
if $(s_4 = 2)$ and $(s_5 = 5)$ and $(s_6 = 4)$ and $(s_7 = 3)$ then (class=good);
if $(s_5 = 5)$ and $(s_7 = 1)$ then (class=bad);
if $(s_4 = 2)$ and $(s_5 = 5)$ and $(s_7 = 3)$ then (class=bad);
if $(s_4 = 1)$ and $(s_5 = 5)$ and $(s_6 = 2)$ and $(s_7 = 2)$ then (class=bad);
if $(s_4 = 2)$ and $(s_5 = 5)$ and $(s_6 = 3)$ and $(s_7 = 2)$ then (class=bad);
if $(s_4 = 2)$ and $(s_5 = 5)$ and $(s_6 = 2)$ and $(s_7 = 2)$ then (class=bad);
if $(s_4 = 1)$ and $(s_5 = 4)$ and $(s_6 = 1)$ and $(s_7 = 1)$ then (class=good or bad);
if $(s_4 = 1)$ and $(s_5 = 4)$ and $(s_6 = 1)$ and $(s_7 = 2)$ then (class=good or bad);
if $(s_4 = 1)$ and $(s_5 = 5)$ and $(s_6 = 1)$ and $(s_7 = 2)$ then (class=good or bad);

values of condition symptoms refer to codes assigned to subintervals in domains of symptoms.

4 Conclusions

The analysis of the diagnostic problem (i.e. evaluation of the technical state of rolling bearings) by means of the rough sets theory leads to the following conclusions :

a) The four discussed methods of defining the symptom limit values give different quality of classification of the rolling bearings from the viewpoint of their technical state. Only, the L-method ensures satisfactory evaluation (quality equal to 0.93).

b) Evaluation of the technical state of rolling bearings can be done using one of the obtained minimal subsets of symptoms instead of considering all symptoms. Considering the structure of minimal subsets and the core, the superiority of vibration symptoms over noise symptoms was noticed. Taking into account additional criteria (e.g. minimal cardinality of the subset, cost and time of measurements and expert's preferences), the subset s_4, s_5, s_6, s_7 was chosen to built the classifier of the technical state.

c). The classifier of the technical state consists of 17 decision rules created using the chosen subset of symptoms. These rules determine an assignment of the rolling bearings to the given class of technical state depending on values of chosen symptoms. Let us notice that the number of descriptors contained in decision rules has been decreased to 10% of descriptors contained in the original information system, without any significant loss of information. These rules are useful for automation of inspection process in production or exploitation diagnostics.

Conclusions a) and b) concerning the comparison of different methods of defining symptom limit values and superiority of vibration symptoms over noise ones were confirmed by similar analysis of rolling bearings (by means of the rough sets theory) performed for data collected in laboratory conditions [11].

The results discussed in this paper and conclusions formulated in [11] and [12] show that the rough sets theory can be used to solve many problems met in the technical diagnostics, in particular when it is necessary to :

- eliminate date which do not provide significant information about the technical state,

- find rules expressing expert's knowledge concerning classification.

These problems are essential in knowledge analysis and knowledge acquisition for classification systems.

It should be noticed that the rough set theory can be used not only to technical diagnostics but also to other problems in industrial engineering, assuming that knowledge about these problems can be expressed in a form of information systems (cf. [10]).

All calculations required by the analysis of reducers were performed using Rough-DAS software [6] which is a very efficient computer program implementing the rough sets approach on a microcomputer IBM PC.

References

[1] . Y. Birger: *Technical diagnostics*. Nauka, Moscow (1978) (in Russian).

[2] . Cz. Cempel: Limit value in the practice of machine vibration diagnostics. *Mechanical System and Signal Processing*, 5 no. 6, 483-493 (1990).

[3] . Cz. Cempel: In Plant Determination Symptom Limit Value for Vibration Condition Monitoring. *Proc. of Conf. on Technical Diagnostics*, Prague, p 79-82 (1989).

[4] . J.Fibak, Z.Pawlak, K.Słowiński and R.Słowiński: Rough sets based decision algorithm for treatment of duodenal ulcer by HSV. *Bull. PAS, Biological Series* 34 (10-12), 227-246 (1986).

[5] . H. Finley: *Principles of optimum maintance*. Course materials. The Howard Finley Corporation (1988).

[6] . G.Gruszecki, R.Słowiński and J.Stefanowski: *RoughDAS-Rough Sets Based Data Analysis Software - User manual*, APRO S.A., Warszawa (1990).

[7] . A.Kelly and M.J.Harris: *Management of industrial maintance*. Newness-Butterworths, London (1978).

[8] . E.Krusińska, R.Słowiński and J.Stefanowski: Rough stes theory versus discriminant analysis. *Applied Stochastic Models and Data Analysis* 8 (2) (1992) (to appear).

[9] .Mitchell J.S.: *An Introduction to Machinery Analysis and Monitoring*, PannWell Books Company, Tulusa, Oklahoma (1981).

[10] . A.Mrozek: Rough sets and dependency analysis among attributes in computer implementation of expert inference models. *International Journal of Man-Machine Studies* 30, 457-473 (1989).

[11] . R.Nowicki, R.Słowinski, J.Stefanowski: Rough sets analysis of diagnostic capacity of vibroacoustic symptoms. *Computers and Mathematics with Applications* (1992) (to appear).

[12] .R.Nowicki, R.Słowiński and J.Stefanowski: Evaluation of Diagnostic Symptoms by Means of The Rough Sets Theory, *Computers in Industry* (1992) (to appear).

[13] . Z.Pawlak: Rough Sets. *International Journal of Information and Computer Sciences* 11 (5), 341-356 (1982).

[14] . Z.Pawlak: Rough classification. *International Journal of Man-Machine Studies* 20, 469-483 (1984).

[15] . Z.Pawlak: *Rough sets. Some aspects of reasoning about knowledge*, Kluwer Academic Publishers, Dordrecht (1991).

[16] . Z.Pawlak, K.Słowinski and R.Słowiński: Rough classification of patients after highly selective vagotomy for duodenal ulcer. *International Journal of Man-Machine Studies*, 24, 413-433 (1986).

[17] . R.Słowiński and J.Stefanowski: Rough classification in incomplete information systems. *Mathematical and Computing Modelling*, 12 (10-11), 1347-1357 (1989).

[18] . J.Stefanowski, Classification support based on the rough sets theory, *Proc. of the IIASA Workshop on the User-Oriented Methodology and Techniques of Decision Analysis and Support*, Serock, September 1991 (to appear).

Bi-Directional Probabilistic Assessment

Brian W. Hagen

Hughes Aircraft Company
Fullerton, California 92634

Abstract. Conditioning has been shown to be one approach for improving subjective probabilistic assessments of a target uncertainty [5]. In practice, conditioning on only one distinction is often employed in an attempt to improve the quality of the assessment while minimizing the level of effort for constructing a distribution for the target uncertainty. Once the actual assessment begins it is not uncommon that the expert realizes that much of his knowledge and beliefs are represented best in the reverse conditioning order. By assessing both conditional orderings (bi-directional assessment) and subsequently "balancing" the two perspectives while reducing assessment errors, a better representation of the expert's knowledge and beliefs is constructed than by a conventional, uni-directional or direct assessment. After the expert has attempted to reduce the assessment errors, total reconciliation of the distributions can be achieved through nonlinear programming.

1 Introduction

Quantifying an expert's knowledge and beliefs in the form of a subjective probability distribution is a necessary step in the development of intelligent systems that perform probabilistic reasoning [4]. Probabilistic conditioning is often used to decompose the expert's complex web of knowledge and beliefs to a partitioning in accord with his intuition at the expense of increasing the magnitude of the assessment. Ravinder, Kleinmuntz, and Dyer [5] have shown that with the proper set of conditioning events, assessment errors can be reduced through decomposition as compared to the direct assessment (i.e., no probabilistic conditioning) of a target uncertainty. In practice, conditioning on only one distinction is often employed in an attempt to improve the quality of the assessment while minimizing the level of effort for constructing a distribution for the target uncertainty. Once the actual assessment begins it is not uncommon that the expert realizes that much of his knowledge and beliefs are represented best in the reverse conditioning order. As an example, this occurs when the expert has both causal and diagnostic knowledge and beliefs relating two uncertainties. The causal knowledge and beliefs suggest one conditional ordering while the diagnostic knowledge and beliefs suggest the opposite ordering. A similar situation occurs when available data relevant to the assessment has been recorded convenient for one direction of assessment; however, the expert is more comfortable assessing in the opposite direction. When a probabilistic assessment of two uncertainties is elicited for both conditional orderings we will say a bi-directional assessment has been made.

At least three measures of goodness are applicable to subjective probability distributions: normative, substantive, and representative. Normative goodness indicates whether or not the distribution satisfies the axioms of probability theory. Substantive goodness is a measure of how consistent the distribution is with reality. Unfortunately, due to the uniqueness of an event this measure is often inappropriate. Representative goodness is a measure of how consistent the distribution is with the knowledge and beliefs of the expert. This is difficult to measure and is itself a subjective assessment. For most assessments we can guarantee normative goodness, attempt to maximize representative goodness, and only hope for substantive goodness.

The intent of this paper is to introduce an approach for constructing discrete marginal distributions for target uncertainties that are derived from bi-directional subjective assessments. The tools discussed throughout this paper are the components of BIPAS (Bi-directional Probabilistic Assessment System), a system that aids experts in identifying assessment errors and reconciles inconsistent bi-directional assessments. The process to be detailed is a direct attempt at improving representative goodness of subjective probability distributions by reducing assessment errors. It is this direct attempt at error reduction that distinguishes this approach from others.

2 An Example of Bi-Directional Assessment

To motivate an example of a bi-directional assessment, consider "Fred the Windsurfer." Fred enjoys windsurfing at a local beach in the afternoons. His level of enjoyment for these outings is determined by the wind conditions at the beach. Fred likes to think in terms of four possible wind states: slight or no wind, mild wind, strong wind, and "blown out" (i.e., severe winds).

The beach wind conditions is Fred's target uncertainty and will be denoted by B. Let B_1 represent the event that there is slight or no wind at the beach this afternoon. Let B_2 represent the event that there is mild wind at the beach this afternoon. Let B_3 represent the event that there is strong wind at the beach this afternoon. Let B_4 represent the event that it is "blown out" at the beach this afternoon. A second, relevant uncertainty is wind conditions at home in the afternoon denoted by H. Let H_1, H_2, H_3, and H_4 be defined similarly to B_1, B_2, B_3, and B_4. Fred has much experience observing the wind conditions at home in the afternoon followed by a trip to the beach for windsurfing. Consequently, Fred is comfortable with the assessment of B given H. Conversely, Fred has much experience driving home from the beach on afternoons when either the beach had no wind or was blown out, both poor conditions for windsurfing. This information is relevant to the assessment of H given B.

It is early in the morning and Fred is contemplating a windsurfing trip for this afternoon. Over the radio he hears a weather forecast that includes a prediction for the beach. As usual, he is not very confident in this forecast. Fred would like to construct his probability distribution for this afternoon's wind conditions at the beach.

To perform the initial bi-directional assessment, Fred would assess his marginal probability distribution for the uncertainty H defined by the probabilities $P(H_1)$, $P(H_2)$, $P(H_3)$, and $P(H_4)$; and then assess the four conditional distributions of B given H defined by the probabilities $P(B_j|H_1)$, $P(B_j|H_2)$, $P(B_j|H_3)$, and $P(B_j|H_4)$ for j = 1, 2, 3, and 4. Keeping in mind the weather forecast, Fred would directly assess B defined by the probabilities $P(B_1)$, $P(B_2)$, $P(B_3)$, and $P(B_4)$; and subsequently assess the conditional distributions H given B defined by the probabilities $P(H_i|B_1)$, $P(H_i|B_2)$, $P(H_i|B_3)$, and

$P(H_i|B_4)$ for i = 1, 2, 3, and 4. Table 1 represents an example of this initial assessment.

$P(H_1) = 0.10$	$P(H_2) = 0.30$	$P(H_3) = 0.45$	$P(H_4) = 0.15$
$P(B_1) = 0.10$	$P(B_2) = 0.287$	$P(B_3) = 0.363$	$P(B_4) = 0.25$
$P(B_1|H_1) = 0.50$	$P(B_2|H_1) = 0.15$	$P(B_3|H_1) = 0.10$	$P(B_4|H_1) = 0.25$
$P(B_1|H_2) = 0.25$	$P(B_2|H_2) = 0.30$	$P(B_3|H_2) = 0.30$	$P(B_4|H_2) = 0.15$
$P(B_1|H_3) = 0.10$	$P(B_2|H_3) = 0.15$	$P(B_3|H_3) = 0.60$	$P(B_4|H_3) = 0.15$
$P(B_1|H_4) = 0.03$	$P(B_2|H_4) = 0.12$	$P(B_3|H_4) = 0.35$	$P(B_4|H_4) = 0.50$
$P(H_1|B_1) = 0.60$	$P(H_2|B_1) = 0.25$	$P(H_3|B_1) = 0.10$	$P(H_4|B_1) = 0.05$
$P(H_1|B_2) = 0.40$	$P(H_2|B_2) = 0.35$	$P(H_3|B_2) = 0.17$	$P(H_4|B_2) = 0.08$
$P(H_1|B_3) = 0.15$	$P(H_2|B_3) = 0.30$	$P(H_3|B_3) = 0.35$	$P(H_4|B_3) = 0.20$
$P(H_1|B_4) = 0.05$	$P(H_2|B_4) = 0.25$	$P(H_3|B_4) = 0.30$	$P(H_4|B_4) = 0.40$

Table 1. Initial Bi-Directional Assessment Of Afternoon Home And Beach Wind Conditions

The result of the bi-directional assessment is two marginal distributions for B. One marginal distribution for beach wind states was assessed directly. A second can be calculated from $P(B_j) = \Sigma P(B_j|H_i)P(H_i)$ for j = 1, 2, 3, and 4. The two marginal distributions for B are almost always probabilistically inconsistent. The two inconsistent probability distributions for afternoon beach wind conditions before Fred attempts to reduce assessment errors are provided in Table 2.

Beach Wind Conditions	From Conditioning	Direct Assessment
B1: Slight or No Wind	0.175	0.100
B2: Mild Wind	0.190	0.287
B3: Strong Wind	0.423	0.363
B4: "Blown Out"	0.212	0.250

Table 2. Two Inconsistent Probability Distributions For Beach Wind Conditions Before Attempt To Reduce Errors

3 Errors in Subjective Probabilistic Assessment[2]

Experience indicates that experts have difficulty in identifying assessment errors when confronted with probabilistic inconsistency in a set of assessments even when the expert is convinced of their existence. Informally we have found that many experts find the task of assessment error identification easier if they are given a simple but complete set of concerns to address. A simple partitioning of the assessment error space provides this set of concerns.

For every subjective assessment the expert has a finite set of relevant knowledge and beliefs. We refer to this set as Ω. During the interviewing process, the expert does not focus on the entirety of Ω. The expert retrieves from memory and focuses on a subset of Ω which we will refer to as the kernel of Ω denoted by ω. The kernel represents all of the knowledge and beliefs the expert uses during the entire probabilistic assessment session for a specific distribution. The process of how the expert retrieves and transforms ω into probabilities is not well understood; however, some theories do exist [3].

We can identify the three sources of error that reduce the representative goodness of an assessed distribution. First of all, the kernel may contain beliefs that are logically inconsistent. Secondly, the kernel may be severely incomplete with respect to the expert's entire set of relevant knowledge and beliefs. By definition, the kernel is incomplete. The concern is the degree of incompleteness. Incompleteness arises from the framing (motivation and structuring) of the assessment and the heuristics used for mentally retrieving relevant knowledge and beliefs. Examples of these heuristics include availability, simulation and representativeness [3]. Finally, the expert or elicitor may make errors during the transformation of the expert's relevant knowledge and beliefs into probabilities. The transformation error results from lack of concentration, recording errors, and biases introduced by some of the employed heuristics.

4 Reducing Assessment Errors

During the error reduction phase the expert will review and adjust distributions. The expert is motivated by two concerns during this phase. The primary concern is the development of probabilistic assessments that represent his relevant knowledge and beliefs about the target uncertainty. Subsumed in this primary concern is the reduction of the three types of error. The secondary concern is that of probabilistic coherence of the entire set of probabilistic assessments. As long as the expert is focused on the primary concern, the expert should continue attempting to reduce assessment errors. Indeed, the expert should only adjust probabilities after he has identified the type of error and subsequently identified the associated biased distribution(s). As soon as the secondary concern of probabilistic coherence becomes the dominant motivation, the expert should stop making adjustments so as not to haphazardly introduce biases into the assessments.

One can design a number of tools to aid the expert in identifying symptoms of the assessment errors. Two such tools are: (1) Comparable Probabilities and (2) a Bug Finder. Comparable probabilities allow the expert to contrast assessments of opposite conditional orderings. In short, the unique comparable probability for a given event is constructed by applying Bayes' rule to the set of assessments for the opposite conditional ordering. Let us return to Fred's initial assessments for an example.

From Table 1 we note that Fred assessed the probability that the beach wind conditions would be strong given the home wind conditions to be mild as $P(B_3|H_2) = 0.30$. The comparable probability for this as derived via Bayes' rule is $P(B_3|H_2)_c = 0.367$. The deviation between this comparable probability and the actual assessment of $P(B_3|H_2)$ is 0.067. If Fred is alarmed by the magnitude of the deviation, Fred should consider the three sources of assessment error. If Fred identifies and locates an assessment error he should subsequently make the appropriate adjustments to the associated distributions.

The Bug Finder is a tool that searches the set of assessments for probabilistic bugs defined in a hierarchical bug library. The library consists of bug checks motivated by the normative laws of probability theory and results from psychological experimentation on individual's ability to construct subjective probabilities. These bugs in their hierarchical order include: overconfidence bias, gross differences, inconsistent marginals/conditionals, equating conditionals bias, conjunction bias, and marginal differences. The position of a bug in the hierarchy is based on two components: the magnitude of the associated probabilistic inconsistency, and the number of associated distributions. This hierarchy is based on the simple heuristic that the larger the magnitude of the probabilistic inconsistency and the larger the set of associated distributions the more probable the associated distributions are evidence of an assessment error. For each bug discovered a set of distributions is identified for the expert to review as to initiate the search for an assessment error. Each of these bugs are rigorously defined by Hagen [1]; consequently only an overview and example of one is provided.

The *inconsistent marginals/conditionals bug* is a severe probabilistic bug. If all of the conditional assessments of an event are less (greater) than the analogous direct assessment from the reverse order assessment then a major probabilistic inconsistency exists. Since the two uncertainties are probabilistically dependent, there should exist a conditional assessment greater than the analogous direct assessment and a conditional assessment less than the direct assessment. This is a simple statement to prove and is provided by Hagen [1].

To illustrate this bug, reconsider the following set of conditional assessments: $P(H_3|B_1)$ = 0.10, $P(H_3|B_2)$ = 0.17, $P(H_3|B_3)$ = 0.35, and $P(H_3|B_4)$ = 0.30. All of these are less than the direct assessment of $H = H_3$, $P(H_3)$ = 0.45. Consequently, the bug finder would identify the conditional probability distributions for H given B and the direct assessment of H as the starting point for an investigation of an assessment error.

Once the expert can no longer identify assessment errors in the set of distributions, the expert should stop making adjustments to the distributions. At this point, the distributions are almost always *still* probabilistically inconsistent.

5 Eliminating Probabilistic Inconsistencies

The ultimate goal is to construct one, best target distribution from the set of inconsistent distributions. The process of constructing one distribution from a set of inconsistent distributions is called reconciliation. After the attempt to reduce errors in the distributions is completed, we reconcile the distributions via a nonlinear programming algorithm.

The best marginal distribution for the target uncertainty will be defined as the one that reconciles the entire set of distributions by minimizing the sum of the squares of the additive perturbations necessary to obtain reconciliation. Additive perturbations, as opposed to multiplicative, for each probability of each assessed distribution insure that the magnitude of the required perturbation for a given probability is independent of the magnitude of the probability. For each assessed probability we define an additive perturbation x_i. The objective is then to minimize Σx_i^2 while achieving probabilistic consistency.

The constraints for the nonlinear programming approach are largely provided by probability theory; that is all probabilities must be no less than zero and no greater than one, and the sum of probabilities over mutually exclusive and collectively exhaustive events must equal one. Probabilistic consistency is imposed by forcing the joint probability distributions for the two conditional orderings to be identical. By removing all linear equality constraints through a redefinition of variables, the general structure of the problem is of the form

$$\min \frac{1}{2} x^T Q x + c^T x \quad \text{subject to}$$

$$A x - b \leq 0$$

$$\frac{1}{2} x^T D_i x + d_i^T x + e_i = 0 \qquad i = 1, 2, \dots, 16.$$

Hagen [1] details a partial duality method employing recursive quadratic programming to solve this nonlinear program.

In some cases the expert may desire to lock an assessed probability so that it is not perturbed during the reconciliation process. As an example, a lock may be motivated by the existence of an "objective" probability for one of the assessed probabilities. Hagen [1] details how this can be easily accomplished through a straightforward modification to the objective function of the nonlinear program.

6 Conclusion

This paper introduced bi-directional probabilistic assessment as an appropriate assessment methodology when the expert has knowledge and beliefs regarding both conditional orderings of two uncertainties. By assessing both conditional orderings and subsequently "balancing" the two perspectives while reducing assessment errors, a better representation of the expert's knowledge and beliefs is constructed than by a conventional, uni-directional or direct assessment. The Bi-Directional Probabilistic Assessment System (BIPAS) was introduced as a tool to aid in the identification of assessment errors and provides the necessary computations for solving the nonlinear programming problem required for total reconciliation of the inconsistent probability distributions.

References

1. B.W. Hagen: Constructing Discrete Marginal Distributions Via Redundant Probabilistic Assessment, PhD dissertation, Department of Engineering-Economic Systems, Stanford University, 1991.

2. B.W. Hagen: "MCPAS: A System for the Acquisition of Knowledge & Beliefs as Subjective Probabilities," in Proceedings of the Fourth International Conference on Industrial & Engineering Applications of Artificial Intelligence & Expert Systems, Koloa, Kauai, Hawaii, June 2-5, pp. 702 - 711, 1991.

3. D. Kahneman, P. Slovic and A. Tversky: Judgement under Uncertainty: Heuristics and Biases, Cambridge University Press, Cambridge, 1982.

4. J. Pearl: Probabilistic Reasoning in Intelligent Systems: Networks of Plausible Inference, Morgan Kaufmann Publishers, San Mateo, Ca., 1988.

5. H.V. Ravinder, D.N. Kleinmuntz, and J.S. Dyer: "The Reliability of Subjective Probabilities Obtained through Decomposition," Management Science, Vol. 34, No. 2, pp. 186 - 199, 1988.

Reasoning under Uncertainty with Temporal Aspects

Detlef Nauck   Frank Klawonn   Rudolf Kruse   Uwe Lohs

Deptartment of Computer Science, Technical University of Braunschweig
W-3300 Braunschweig, Germany

Abstract. The representation and propagation of dynamic knowledge considering temporal changes of attribute values is an important problem in the domain of reasoning under uncertainty. Subject to this paper is an extension of the model of Bayesian networks that is capable of handling temporal constraints. A software prototype has been developed under the expert system shell KEE on a TI micro-explorer.

1 Introduction

The handling of imperfect knowlegde is an important problem in the area of knowledge based systems. We are able to distinct between the two phenomena *uncertainty* and *vagueness.*

Uncertainty corresponds to the valuation of some datum, reflecting the faith or doubt in the respective source. So we have to deal with statements being not just simply true or false but with a validity which is a matter of degree. This is caused by the fact that the actual state of the world is not completely determined and we have to rely on a human experts subjective preferences among different possibilities. The treatment of subjective valuations of evidence requires the use of belief functions measuring the credibility of information.

Vagueness corresponds to the impossibility to observe a crisp event, e.g. to determine the exact state of an object under consideration. An observer or expert is only able to determine a set of possible states and so we have to deal with *imprecise* statements. Imprecision can be handled with set-theoretic concepts. If the elements of the observed set can be more or less possible, fuzzy sets or layered sets are suitable tools for representation.

For a combined treatment of uncertainty and imprecision the belief function theory, and e.g. mass distributions and the concept of specialization [Kruse/Nauck/ Klawonn 91], [Kruse/Schwecke/Heinsohn 91] can be used. If there is no imprecision and we only have to deal with uncertain but precise data, we can use probability theory, and e.g. Bayesian networks.

In addition to the handling of imperfect knowledge the representation of time and temporal relations are important. The changes of an attribute value in time can be more important than the actual value and it is often necessary to consider the delay of the effects of current observations. To deal with different forms of temporal relations there is

a distinction between to main directions of temporal reasoning:
- the temporal odering of a series of events (reasoning about time) and
- the reasoning under the consideration of temporal aspects, where attribute values at certain points of time, or within certain time intervalls respectively, have to be considered (reasoning with temporal aspects).

There already exists a lot of work about temporal logics (see e.g. [McDermott 82], [Allen 83, 91], [Long 89], [Kautz and Ladkin 91]) but a connection of uncertainty and temporal aspects has not been considered to a large extent. To develop a model capable of handling uncertainty and temporal aspects we extended the model of Bayesian networks and included a temporal component. The use of Bayesian networks indicates that we did not consider imprecision. In the following we will give a short description of Bayesian networks, present our mathematical model of temporal reasoning, and give summary of our software tool.

2 Bayesian networks

Subject to this section is a brief introduction to Pearl's model of Bayesian networks [Kim and Pearl 83], [Pearl 86a,b,c], [Pearl 87] as it can be found in [Kruse/Schwecke/ Heinsohn 91]. For proofs and examples the reader is referred to Pearl's original papers.

Let Ω, a finite nonempty set, be our *frame of discernment*. We assume Ω to be a product space $\Omega^M \stackrel{d}{=} \Omega_1 \times ... \times \Omega_m$ with m characteristics $X^{(1)} \in \Omega_1,..., X^{(m)} \in \Omega_m$ where Ω_i $(i = 1,..., m)$ is a finite nonempty set. The knowledge will be represented by a discrete probability distribution over Ω.

Definition 1. Let Ω^M be a product space, $M = \{1, ..., m\}$ its index set and S, T, and C index subsets of M, such that $T = S \cup C, S \cap C = \emptyset$

(i)   the mapping $\qquad \Pi_S^T : 2^{\Omega^T} \to 2^{\Omega^S}$,

$$\Pi_S^T(A) \stackrel{d}{=} \{\omega^S \in \Omega^S \mid \exists\, \omega^C \in \Omega^C : (\omega^S, \omega^C) \in A\}$$

is called the *projection* Ω^T on Ω^S, whereas

(ii) the mapping $\qquad \hat{\Pi}_S^T : 2^{\Omega^T} \to 2^{\Omega^T}$,

$$\hat{\Pi}_S^T(B) \stackrel{d}{=} \{(\omega^S, \omega^C) \in \Omega^T \mid \omega^S \in B\}$$

is called the *cylindrical extension* of Ω^S onto Ω^T.

A Bayesian network is a directed acyclic graph in which the nodes represent propositions and arcs stand for the existence of direct causal influences between nodes. The strengths of these dependencies are quantified by *conditional probabilities*. For instance, the (conditional) dependence of the two characteristics $X^{(i)}$ and $X^{(j)}$ may be given by $P(X^{(i)}|X^{(j)})$. Using the chain-rule representation of joint probability distributions, then, choosing an arbitrary order on the characteristics $X^{(1)}$, ..., $X^{(m)}$, we can write

$$P(X^{(1)},...,X^{(m)}) = P(X^{(m)}|X^{(m-1)},...,X^{(1)}) \cdot ... \cdot P(X^{(3)}|X^{(2)},X^{(1)}) \cdot P(X^{(2)}|X^{(1)}) \cdot P(X^{(1)})$$

Suppose a directed acyclic graph G in which the arrows pointing at each node $X^{(i)}$ emanate from a set S_i of parent nodes directly influencing $X^{(i)}$. Fixing their values would therefore shield $X^{(i)}$ from the influences of all other predecessors of $X^{(i)}$, i.e.

$$P(X^{(i)}|S_i) = P(X^{(i)}|X^{(i-1)},...,X^{(1)}).$$

Now the chain rule representation can be simplified and the joint probability is given by

$$P(X^{(1)},...,X^{(m)}) = \prod_i P(X^{(i)}|S_i).$$

Dependency relations between the nodes can be detected by looking at the *blocking* of paths between the nodes. A *separation criterion* determines conditional independencies between the characteristics (for details see [Kruse/Schwecke/Heinsohn 91]).

For the process of *fusing and propagating* new information and belief through a Bayesian network, we need three entities, which have to be maintained in the nodes:
- a fixed conditional probability matrix $M(X^{(i)}|X^{(j)} ... X^{(r)})$ relating the node $X^{(i)}$ to its immediate parents $X^{(j)}$, ..., $X^{(r)}$. A root node contains a prior probability distribution.
- the current strength of the causal or prospective support attributed to the possible values $X_j^{(i)}$ of node $X^{(i)}$ by the set of ancestors of $X^{(i)}$:

$$\pi(X_1^{(i)}),\pi(X_2^{(i)}),...,\pi(X_n^{(i)}).$$

- the current strength of the diagnostic or retrospective support node $X^{(i)}$ receives from its descendants (i.e. the set of outgoing links):

$$\lambda(X_1^{(i)}),\lambda(X_2^{(i)}),...,\lambda(X_n^{(i)}).$$

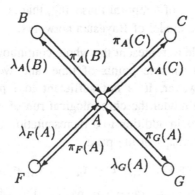

Fig. 1. Updating a singly connected network

Let us consider the node A of the network as it is depicted in Fig. 1. The meaning of the entities $\lambda(A_j)$ and $\pi(A_j)$ can be exemplified by looking at the amount of belief that is attached to a node A. Following Pearl [Pearl 86c], the characteristics D_A^- - the data contained in the network rooted at A - and D_A^+ - the data contained in the rest of the network - are introduced. The amount of belief attached to A induced by the data available in $D_A^+ \cup D_A^-$ is given by the conditional probabilities

$$\text{Bel}(A_j) \overset{d}{=} P(A_j | D_A^+, D_A^-) = \gamma \cdot P(D_B^- | A_j) \cdot P(A_j | D_A^+).$$

This is a result from Bayes' theorem, where γ is a renormalization factor. Looking at the measures $\lambda(A_j)$ and $\pi(A_j)$ introduced above, they can now be identified as

$$\lambda(A_j) = P(D_B^- | A_j),$$

$$\pi(A_j) = P(A_j | D_B^+).$$

The belief vector $(\text{Bel}(A_1), \text{Bel}(A_2), ..., \text{Bel}(A_n))$ reflects the overall belief currently accorded to the propositions "$A = A_j$", $j = 1, ..., n$. This total belief can be computed using the locally available information only:

$$\text{Bel}(A_j) = \gamma \cdot \lambda(A_j) \cdot \pi(A_j).$$

Belief propagation through the network is triggered by changes in the belief parameters of neighboring nodes. The π-messages are sent top-down to *sensory nodes* while λ-messages are propagated bottom-up to the parent nodes (Fig. 1). When a node is activated by receiving a message communicated by its parent or by one of its children, it first updates its parameters λ and π. After that, this node computes messages to be posted to the neighboring nodes.

3 Temporal Reasoning

We will now present our model of temporal reasoning, that is constructed in such a way, that it fits exactly in the static model of Bayesian networks.

Because it is not possible to represent the whole continuous time axis $\tilde{T}$ at once, we have to restrict ourselves to suitable points of time, and we divide the time axis in uniform time intervalls. However, it is not sufficient to represent just a single time intervall, because we want to model the chronological run of events. Therefore we have a *time window*, which includes in addition to a representation of the presence also representations of the future and perhaps of the past:

$$T \overset{d}{=} \{ t_j | t_j \text{ represents } [\tau_j, \tau_{j+1}], \tau_j < \tau_{j+1}, \tau_j, \tau_{j+1} \in \tilde{T}, j, j+1 \in J \}.$$

After the current observation frame has passed, the window can be shifted, by changing to a neighbouring time window.

The frame of discernment Ω has now to be extended, so it is possible to represent states of the world at different points of time. We obtain the new product space

$$\Omega = \Omega^{M \times T} = \underset{t \in T}{\text{X}} \; \underset{i \in M}{\text{X}} \; \Omega^{(i,t)},$$

$$T = \{t_i \, | \, i \in \{1,\dots,n\}\},$$

$$M = \{1,\dots,m\}.$$

We will now distinguish between *statical* and *dynamical* dependencies. The statical dependencies refer to the causal relationships within the representation of a single time interval, and will be modelled by a Bayesian network. The dynamical dependencies refer to the transitions between different time intervals.

To be consistent with the model of Bayesian networks, we cannot model the dynamical dependencies in the same manner as the statical dependencies, because we have to avoid cycles. Therefore we cannot model uncertain temporal relationships through a probabilistic representation. An acceptable restriction is to confine the dynamical dependencies to categorical relations, i.e. to a time dependent propagation of the observations. In this case only the evidence will be propagated in time. Furthermore we will have the same prior probabilities for all time intervalls.

Instead through conditional probabilities, the temporal dependencies will be propagated with the help of categorical constraints or production rules. The handling of the statical dependencies is done in the normal way by propagating conditional probabilities. The advantage of production rules and constraints is the possibility to handle cycles, even if they weaken the result, i.e. the result can contain states, which could already be excluded in the light of the available knowledge. Especially mixed cycles, that are cycles containing causal dependencies interrupted by at least one categorical dependency, can be processed.

The difference between constraints and production rules is that rules point in only one direction whereas constraints always express dependencies in both directions, and therefore their propagation results in a relaxation process. Because production rules are only propagated in one direction, cycles can only appear, if all rules within a loop point in the same direction. If the propagation is restricted to one temporal direction, prognosis or diagnosis, there are no cycles within the system, and relaxation processes can be avoided at all. For the implementation of the software prototype a pure prognosis process was used, by allowing the observation window to be shifted into the future only after a time interval is over.

We will now formalize the above considerations and examine the space of temporal relations at first. For the present we will neglect that the representation of the time zones Ω is a product space. We consider a temporal problem described as follows:

- let $(\Omega, 2^\Omega)$ be a time influenced measurement space with prior probabilities $P^{(t)}$, $t \in T$, with $\forall t \in T: P^{(t)} = P$,
- let $S(t) = \{t\} \cup \{t' \leq t \mid t' \text{ influences } t\}$, $S(t) \subseteq T$, the set of those t', which have a temporal influence on t. Then exists a set $R = \{R_t \subseteq \Omega^{S(t)} \mid t \in S(t)\}$ of production rules, such that R_t is the set representing the rule restricting the space $\Omega^{(t)}$,
- there exists a set of user observations $E = \{E_t \subseteq \Omega^{(t)} \mid t \in T\}$. If there is no explicit user observation for t, then $E_t = \Omega^{(t)}$ is used.

Let $E^* = \{E_t^* \mid t \in T\}$ be the set of evidences resulting through the propagation of observations with respect to the individual time zones.

$$E_t^* \overset{d}{=} \{\omega \in \Omega^{(t)} \mid P^{(t)}(\omega \mid E_t) > 0\}, \quad E_t^* \subseteq E_t.$$

The user observations are at first propagated through the causal dependencies within the current $\Omega^{(t)}$, and after that they are transferred with the help of the production rules into other time zones where they are intersected with the local user observations of this time interval. However, this implies that the dynamical restriction are *independent* from those that are due to the statical, causal relationships. The propagation yields the following posterior probability distribution $\hat{P}$:

$$\hat{P}^{(t)}(.) \overset{d}{=} P^{(t)}(. \mid \Pi_{(t)}^{S(t)}(R_t \cap \underset{t_i \in S(t)}{\times} E_{t_i}^*)).$$

Considering now that Ω is a product space, we derive that the production rules are of the following form:

$$R = \{R_{i,t_j} \subseteq \underset{(m,t_n) \in S(i,t)}{\times} \Omega^{(m,t_n)} \mid (i, t_j) \in S(i,t) \subseteq M \times T\}.$$

For the temporal propagation then the local evidence $E_{(i,t)}^*$ has to be calculated.

We will now introduce the propagation algorithm of our model. Reconsidering the properties of our model, where we have a product space, conditional probabilities for representing statical relations, prognostic production rules for representing dynamical relations, a set T consisting of uniform time zones, the restriction of shifting the time window to the future only, and the same marginal space $\Omega^{(t)}$ for every time zone $t \in T$, we can divide the algorithm in three areas:

- The *vertical knowledge processing* is the propagation of statical knowledge within a single time zone, i.e. the calculation of the posterior probabilities and the change from E to E^*. This is done with the help of Pearl's model, but there are two exceptions:

  - at every node $E_{(i,t)}^*$ is computed and stored,
  - the statical network can have isolated subgraphs, even isolated nodes, that are influenced from other time zones only and are not causally dependent from any node of the current interval.

  The only difference to Pearl's algorithm is that the messages and the belief vector are intersected with another observation in addition to D. This is the evidence E^* from the nodes, which are temporally related to the current node.

- The *horizontal knowledge processing* is the propagation of the dynamical relations within the observation window and is carried out by temporal feed forward production rules. Those rules are described by categorical transition matrices W of the dimension $|S^{(t)}|$. After the new belief vector and the posterior evidence of a node has been computed, it sends E^*-messages to all temporal dependent nodes. There all evidences E^* of the temporal predecessors are gathered and weighted by W. The result is the temporal observation and is intersected with the user observation to compute the evidence. After this the new belief vector of a sucessor node is calculated.

- At the *transition to the next observation window* the states (evidence, belief etc.) of the time zones t_2 to t_n are shifted for one time interval. The information of the time zone t_1 is lost, but its temporal effects are still in the network. The information of the new time zone t_n is calculated as follows:
First the messages from all temporal predecessor nodes are being requested. Then they are being evaluated and the statical propagation process is being started. The calculations are complete after this because nodes of the last time zone do not have any temporal successors.

4 A Software Tool for Temporal Reasoning

The software prototype that was developed in cooperation with Dornier GmbH, Friedrichshafen allows a user to specifiy a Bayesian network including statical and dynamical links with the help of a graphical user interface. The tool was implemented under the expert system shell KEE on a TI micro-explorer integrated in an Apple MacIntosh II Computer.

The user has to specifiy each node of the network together with its attributes and the prior probabilities or the conditional probability matrix respectively. When a dynamical link between two nodes or from one node to itself is specified, the user has to enter a categorical transition matrix. After the network is completed, observations can be entered at each node and the propagation process can be carried out.

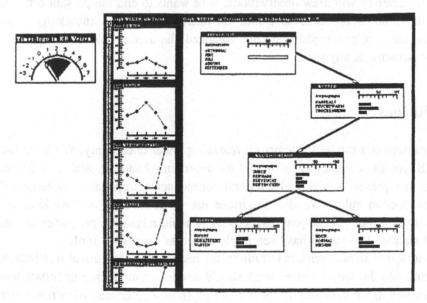

Fig. 2: The user interface of the software tool (statical links)

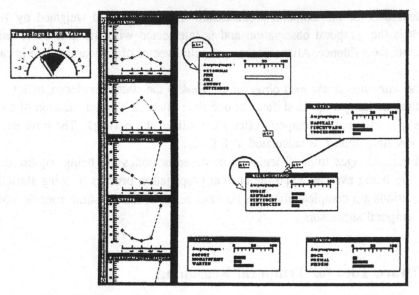

Fig. 3: The user interface of the software tool (dynamical links)

The interface of the tool consists of three windows (see Fig. 2 and 3), where the network is displayed (large window) together with its statical or dynamical links and the belief vector as charts within the nodes. In another window the probabilities of one attribute per node for all time zones is shown (right vertical window). The current time zone of the observation window to which the displayed graph belongs is also shown (small window). The user can freely move between the time zones of the observation window and enter or withdraw observations. If he wants to end the present time zone he can shift the window one time zone into the future. Because the knowledge acquisiton process and the application phase are not separated, the user is able to change the structure of the network at any time.

5 Conclusions

We have presented a model for temporal reasoning under uncertainty. We have used the model of Bayesian networks as the model for uncertainty handling, and have extended it to be able to represent categorical temporal dependencies propagated by temporal feed forward production rules. We did not make use of constraints or two kind of rules pointing into the future or the past respectively to avoid relaxation processes and to keep the model simple. The model has been implemented as a software tool.

The temporal dependecies in our model are restricted to categorical relations and are only directed into the future. Future work should be done on modelling uncertain temporal relations as well as on handling diagnostic and prognostic processes. Also non-monotonic aspects and an extension to evidential reasoning should be considered.

References

[Allen 83] J.F. Allen: Maintaining Knowledge about Temporal Intervals. In: Communications of the ACM, **26**, 1983, pp 832 - 843.

[Allen 91] J.F. Allen: Planning as Temporal Reasoning. In: J. Allen, R. Fikes, E. Sandewall: Principles of Knowledge Representation and Reasoning. Proc. of the 2nd international conference. Morgan Kaufmann, San Mateo, CA, 1991.

[Kautz/Ladkin 91] H.A. Kautz, P.B. Ladkin: Integrating Metric and Qualitative Temporal Reasoning. In: Proc. of the AAAI-91. The MIT Press, Menlo Park, 1991.

[Kruse/Nauck/Klawonn 91] R. Kruse, D. Nauck, F. Klawonn: Reasoning with Mass Distributions. In: B.D. D'Ambrosio, P. Smets, P.P. Bonisonne: Uncertainty in Artificial Intelligence: Proc. of the 7th Conference. Morgan Kaufmann, San Mateo, CA, 1991.

[Kruse/Schwecke/Heinsohn 91] R. Kruse, E. Schwecke, J. Heinsohn: Uncertainty and Vagueness in Knowledge Based Systems. Numerical Methods. Springer, Series Artificial Intelligence, Berlin, 1991.

[Long 89] D. Long: A Review of Temporal Logics. In: The Knowledge Engineering Review **2**, 1989, pp. 141 - 162.

[McDermott 82] D. McDermott: A Temporal Logic for Reasoning about Processes and Plans. In: Cognitive Science **8**, 1982, pp. 101 - 155.

[Pearl 88] J. Pearl: Probabilistic Reasoning in Intelligent Systems: Networks of Plausible Inference. Morgan Kaufmann, San Mateo, CA, 1988.

An Expert System Approach for Power System Diagnosis

Zita A. Vale   A. Machado e Moura

Faculty of Engineering of Oporto University
Rua dos Bragas - 4099 PORTO CODEX - Portugal

Abstract. Power System Control Centers are continuously receiving information about the state of the Power System what allows them to perform efficiently their functions when the Power System is in a normal state. However, in case of an emergency, the quantity of arriving alarms is so great that Control Centers operators are not able to rapidly make the diagnosis of the fault. Decisions must be taken in real time, under great stress and often during the absence of the most skilled operators.
In this paper we present an Expert System that performs an intelligent processing of the alarm lists at Portuguese Control Centers. This system is a good support decision tool to assist operators in Power Systems diagnosis and restoration. It uses an original approach to deal with temporal reasoning and real-time constraints and incorporates explanation facilities enabling its use as a tutor for novice operators.

1 Introduction

During the last decades, we have assisted to a ever increasing complexity of electric Power Systems due to the aim of spreading the deliver of electrical energy with increased quality at moderate costs. The early small isolated electrical networks gave place to internationally interconnected networks that impose a very efficient and accurate real-time control. Control Centers, that are responsible for the decisions taken about Power Systems operation and control, namely in case of disturbances, became much more important and complex than before.

They have access to a great quantity of information about the Power System state through the installation of meter points and large transmission systems that transmit the available information to Control Centers where it is dealt by SCADA (Supervisory Control and Data-Acquisition) systems. The arriving information allows Control Centers to perform efficiently their functions when the Power System is in a normal state. However, in case of an emergency, it may cause serious difficulties because there are so many alarms arriving at Control Centers at each moment that human operators are not able to rapidly understand what is going on. A survey conducted by an IEEE working group [1] about Control Centers alarms revealed that the most serious difficulties found in alarm interpretation are due to the excessive number of alarm messages displayed, specially in case of a disturbance. When a fault occurs in the Power System, operators must make the diagnosis and restore power delivery. Decisions must be taken according to time constraints, under great stress and often during the absence of the most skilled operators. The symbolic and qualitative nature of the reasoning involved in these decisions, strongly dependent on the operators experience, suggests that an Artificial Intelligence approach is suitable to deal with this kind of problems.

2 Related Work

After 1985, having Artificial Intelligence already achieved the state of real applications, Power Systems experts began facing the possibility of implementing Artificial Intelligence based applications. In 1986, Wollenberg [2], Talukdar, Cardozo and Perry [3] and Amelink, Forte and Guberman [4] published the results of developed work related to message handling and diagnosis. The early developed systems had significant limitations but they made clear that Artificial Intelligence approaches could represent a great improvement in Power Systems operation and control.

The most important companies of electricity around the world began to support Artificial Intelligence projects envisaging the integration of Knowledge Based Systems in Control Centers. Systems for alarm processing and fault diagnosis were the basis for a survey conducted by CIGRE [5] where we may find a set of important references.

3 System Overview

The aim of our work is to develop a system that helps Control Center operators in interpreting alarm lists. In order to accomplish this objective, the Expert System reduces the number of displayed messages by making an intelligent synthesis and presents its conclusions in a more flexible and structured way.

The Expert System we are developing runs on a DEC Station 5000/200 and is written in Prolog. It deals with real alarm lists received at the Portuguese National Control Center, making an intelligent processing of these lists. It considers the sixty two plants of the Portuguese electrical network.

The Fact Base contains information concerning the Power System components and topology, the alarm messages arriving at the Control Center and the Expert System conclusions.

The messages arriving are converted into Prolog facts by a pre-processing module, written in C language, and transmitted to the Knowledge Based System that treats them on-line. Let us consider the following message:

01-OCT-90 03:15:12 SSN 103 PALMELA I - BREAKER   BREAKER   CLOSED

that informs operators that Sines Substation (SSN) breaker of the line 1 to Palmela closed at the indicated time. The pre-processor converts this message in the following Prolog fact:

fact(156,message('90/10/01','03:15:12',['SSN','103',['Palmela','1','BREAKER'],
'BREAKER', 'CLOSED']),'23598912').

The last element of this fact encodes the message date and time allowing an easier treatment of temporal problems.

It is important to notice that, while the Expert System is reasoning about the Power

System state, new alarms appear and are transmitted to the Expert System as Prolog facts. This means that the Facts Base is dynamically changed by external events, independent from the conclusions of the Expert System .

The technical and empirical knowledge that allows the Knowledge Based System to reach conclusions is encoded in the form of if...then... rules being the left hand side a disjunction of conjunctions and the right hand side a set of actions to be taken if the left hand side is verified. This knowledge has been gathered from the most experienced operators of Portuguese Control Centers.

A time measure is attached to any fact or rule condition. Facts are true or false during time intervals. When a fact becomes false, it is included in the Fact Base as an "old_fact" what allows a later study of past events. Time conditions included in the rules allow to deal with time and to compensate for the non-chronological receiving of messages.

The inference engine of our Knowledge Based System uses a forward chaining strategy applying the adequate rules when a new fact, external (alarm) or internal (conclusion), arrives, in order to derive new conclusions.

Moreover, the system includes knowledge about the reasoning strategy (metaknowledge) improving its efficiency. The metaknowledge is expressed in metarules and drives the reasoning process triggering the appropriate rules.

The information about Power System state is collected at meter points installed in the network being then transmitted to Control Centers. At these centers, the information is processed and presented to operators on video displays that are continuously updated according to the most recently arrived information. Operators are informed about the Power System most important events through the alarm lists generated. These lists incorporate several kinds of messages providing information about events and the time they occurred. Due to the characteristics of these problems, temporal reasoning is very important in the development of our system.

The time provided with the alarm messages is the time when the corresponding information arrives at the Control Center and not the time when the corresponding fact really happened. This leads to delays that are not important in normal situations but can be of high significance in an emergency state. On the other hand, the messages are grouped in several categories with different priority levels in what concerns its transmission to Control Centers. For these reasons, alarm lists are only apparently chronological and even messages coming from the same plant may be in the wrong order. This kind of temporal delay is taken into account allowing small errors in the time associated to messages.

Moreover, the knowledge used in the interpretation of the alarm lists involves temporal reasoning. A set of predicates defining temporal relationships has been created in order to make easier the expression of that kind of knowledge.

Another important point is that the meaning of a certain message depends on what happened before and, in certain cases, on what may happen next. As an example, let us consider that a message relating a tripping order to a line breaker has arrived. If, associated with this message, a message that reports the breaker opening arrives, we may conclude that the breaker was tripped due to a fault. However, if the last message does not arrive and we know that there are no problems with the transmission of information, we may conclude that the tripping message was due to tests to the protection devices.

Unlike other alarm processors, our system does not wait a specified time interval to process the arrived messages but treats them immediately. When considering a rule that involves information that may arrive later, the system waits the required time to fire it and then derives the appropriate conclusions.

Examples involving temporal reasoning, based on real alarm lists, may be found in [6].

In the development of our system, a special attention was paid to the interface with operators [6]. The graphic interface is based on the X-Window System providing the user with an easy, flexible and well structured way of presenting the required information. It is based on a set of widgets including windows, scrollwindows, buttons, menus and dialogue fields.

A very important feature of a Knowledge Based System is its ability to explain its reasoning. Our system is able to explain how a certain conclusion is achieved presenting the considered set of rules and facts.

4 Conclusions

The alarm processors presently in use in Control Centers make the alarm interpretation a difficult task, specially in case of a disturbance. In this paper, we presented an Expert System that receives and processes the information available at Portuguese Control Centers providing operators a more intelligent and rapid access to the meaningful information they need to base their decisions.

This system uses a forward chaining strategy of reasoning and includes a friendly man-machine interface and an explanation module that is of special importance for the use of the system as a tutor for inexperienced operators.

References

1. William R. Prince, Bruce F. Wollenberg, David B. Bertagnolli: "Survey on Excessive Alarms", *IEEE Transactions on Power Systems*, vol.4, n.3, pp. 950-956, August 1989

2. B. F. Wollenberg: "Feasibility Study for an Energy Management System Intelligent Alarm Processor", *IEEE Transactions on Power Systems*, vol.1, n.2, pp. 241-247, May 1986

3. S. N. Talukdar, E. Cardozo, T. Perry: "The Operator's Assistant - An Expandable Program for Power System Trouble Analysis", *IEEE Transactions on Power Systems*, vol.1, n.3, pp. 182-187, August 1986

4. Herman Amelink, Anthony M. Forte and Robert P. Guberman, "Dispatcher Alarm and Message Processing", *IEEE Transactions on Power Systems*, vol. 1, n. 3, pp. 188-194, August 1986

5. CIGRE Working Group 38.06.02, "Survey on Expert Systems in Alarm Handling", *Electra*, n. 139, pp. 132-151, December 1991

6. Z. Vale, A. Moura: A Knowledge Based System for Power Systems Control. The Practical Application of Prolog International Conference, London, 1-3 April 1992

DELPHI-EXPERT
An Expert System for Error Diagnosis in High Energy Physics Detectors

K.-H. Becks[a], A.B. Cremers[b], A. Hemker[a], R. Meyer[c], J. Ortmann[a], G. Schlageter[c]

a) Department of Physics, University of Wuppertal, Wuppertal
b) Department of Informatics, University of Bonn, Bonn
c) Department of Informatics, University of Hagen, Hagen

Abstract. An expert system for error diagnosis in a large and very complex High Energy Physics experiment has been developed. The system is installed in the experimental environment and has been tested under real experimental conditions during last year's data taking period. From may '92 onwards, the system will cooperate with the experiment's shift crew. We will report on the system's philosophy, the way of inferring symptoms and possible malfunctions, as well as on first experiences on running DELPHI-EXPERT within the data acquisition system of the experiment.

1 Introduction

An expert system[1] has been developed for a particle tracking detector, a so called drift chamber, situated within the DELPHI experiment at the world's biggest electron-positron-collider LEP at the European High Energy Physics Research Center CERN near Geneva. In the collider, electrons and their anti-particles called positrons are accelerated in opposite directions. If an electron collides with a positron, different, new and mainly short living elementary particles are created which are detected with the help of large detector systems. These last generation experiments became so complex that for physicists on shift it is impossible to have all the experience needed to keep all components of such an experiment in good running conditions.

The system is an "intelligent" histogram analyzer which compares particle hit distributions from that detector with a reference distribution and infers from partial differences possible error components. These hit distributions are comparable to a thomographical picture of the detector. They show an "illumination" of the different components created by particles traversing the detector.

---

[1] This work has been supported by the "KI-Verbund NRW", founded by the Ministry for Science and Research of North Rhine Westfalia and by Digital Equipment Corporation, Germany

The kernel of the expert system is implemented in the production rule language OPS5 [1]. As OPS5 interfaces well with other languages, subtasks which are not suitably solved by production systems, are implemented in PASCAL. This increases the performance of the on-line expert system. For communication with the user the Graphical Kernel System GKS is used.

During the development phase of the expert system a possible extension to an error diagnosis shell for nearly all particle detectors had been considered.

2 Structure of a typical Detector

The characteristics of a detector has been studied on a drift chamber built by a group of Wuppertal University for the DELPHI experiment [2]. One of the characteristics of such a detector is the hierarchical structure; the possibility to split a detector in subdetectors, which can be split again in subdetectors etc. up to detector components at the lowest level producing signals, if a particle is detected. In the DELPHI drift chamber, for instance, sense-wires are stretched uniformly. A particle passing a sense-wire produces a signal. The drift chamber system contains about 7600 sense-wires arranged in different detector components with different wire orientations to perform an unambiguous particle track reconstruction in a subsequent analysis step.

Each produced signal is amplified and digitized by a following chain of electronic components and finally recorded in the readout electronics. The drift chamber system contains of several thousands of electronic components, located directly on the detector frames or in the so called counting room near the detector system.

The structure of the detector, as used by the expert system is described by a lot of interconnected objects. Each electronic or detector component is represented by a pair of objects containing the description of a general component type and the description of the special component. These objects are connected in the form of a semantic net and build the model of the specific detector together with objects containing information about the general structure of the detector. By exchanging the part of the knowledge base containing the model of one specific detector, the system is applicable to other detectors with a similar hierarchical structure as desribed before.

3 The Error Diagnosis Task

The error diagnosis process in DELPHI-EXPERT [3] is based on a specific detector model, in our case on the specific model of the drift chamber. The exact knowledge of all available components and all possible connections is sufficient for a first raw error diagnosis task. Knowledge about special faults of these components is not necessary in this initial phase. Although we have to deal with a very high number of components, the number of different types is relatively small. This makes it easy to generalize a special fault to all components of the same type if once a problem has been detected, e.g. in one out of about 1000 preamplifier cards.

3.1 Pattern Recognition

Each signal produced by a sensing device, e.g. a sense wire, is collected by a monitor program in a so called "wire map". A wire map contains the raw data for the error diagnosis. If a component does not work properly, some characteristic pattern will show up in the data. To get the raw data for the diagnosis, DELPHI-EXPERT starts a monitor program running on an equipment computer to collect an actual hit distribution. After a period of time which excludes high statistical fluctuations, an actual hit distribution is sent back to the expert system.

The data are grouped to a distribution suitable for the process of error diagnosis. The first task of the expert system is the comparison of the actual measured wire map with a reference wire map, initially provided by the detector expert for an error free detector. The pattern recognition process is dependent on the detector model. The system searches only for pattern which could be produced by faults in special electronic or detector components. A fault pattern is an area in the wire map with an approximately constant deviation between actual and reference distribution. The result of this task is recorded in objects of the class *Act-pattern*.

Beside the periodically delivered actual wire map there is another possibility to get data from the experiment, usable in the diagnosis. In some cases DELPHI-EXPERT can start so-called hardware tests to prove the functioning of several electronic components. For this task DELPHI-EXPERT determines suspicious components and computes their test addresses dependent on actual fault pattern found in the periodical diagnosis. A test address is sent to a monitor program which sends a constant number of signals to the test input of the electronic component. The result is registered and the collected data are sent back to the system. DELPHI-EXPERT analyzes these data in a pattern recognition process and starts further tests if needed. The result is recorded in objects of the class *Act-hardwaretest*. Hardware tests cannot always be applied, as data taking of the experiment in our case has to be stopped.

By now the system might have inferred two objects, one of class *Act-pattern* and one of class *Act-hardwaretest*, which are related to one special fault area. Both objects are symptoms of one fault. This justifies the introduction of a logically higher class *Symptom*, which we use for collecting all the partial symptoms of one fault and which represents an up to now unknown problem.

In the search process for error pattern in an actual wiremap, two problems occur:
1) A fault area can result from one fault or from a combination of several faults.
2) A fault area can result from a set of electronic faults or from a set of detector faults.

These problems cannot be resolved because of the "incompleteness" of the available data. Nevertheless it is possible to control the different fault possibilities by organizing them in an ordered structure. Consequently, a *Symptom* inferred in the pattern recognition process describes not necessarily the actual problem. The *Symptom* is weighted by a certainty factor to give a measure of belief. To deal with the first problem we introduce a relation called *dominated-by* between two objects of the class *Symptom*. In general the set of dominated *Symptoms* has the same effect in the wiremap as the dominating *Symptom*.

An object *Fault-area* is introduced to manage all *Symptoms* of a fault area and to evaluate both alternatives (detector or electronic fault) by giving a certainty factor for each alternative. After this task we end up with a network of all possible important symptoms ordered in a hierarchical structure and valuated by certainty factors based on heuristics.

3.2 Inferring Known Faults

In the next phase, the knowledge base is searched for faults,which are able to explain the *Symptoms* in the network. For a given *Symptom* the system searches for an error, which can manifest itself by the objects *Act-pattern* and *Act-hardwaretest* of that *Symptom*.

The knowledge about an error is not directly represented by rules, but declaratively by an object of the class *Fault*, containing general knowledge about this error, like fault component, reason of the fault etc. and the connected objects *Pattern* and *Hardwaretest*, containing the general description of this error. The objects *Pattern* and *Hardwaretest* can be interpreted as a heuristic to recognize this fault. This heuristic belongs also to the error description. As the heuristics are completely represented in objects, they can match the condition part of more general rules. This set of rules contains the general relation between the knowledge about detector or electronic errors and their symptoms. By instantiating the rules during run time by the OPS5-interpreter, a general rule is applicable for thc inference of special faults.

As a general strategy the system puts as much knowledge as possible in objects connected together. Each object contains several attributes describing it. The condition part of a rule matches parts of different networks. The action part generates new networks or deletes or modifies existing ones. The advantage of the declarative representation is motivated by an easier extension of the system.

A number of rules is used to explore that a *Symptom* and the related objects, inferred in the pattern recognition process, can be explained by a known fault. One action of these rules is the generation of an object of the class *Possible-fault* as a representative for this fault.

A next problem occurs by the fact that different faults can manifest the same way. In general, the rules do not only generate one *Possible-fault* for one *Symptom* but a number of *Possible-faults*, which are all a more or less plausible explanation of one *Symptom*. Again a management problem occurs, in this case for the different fault possibilities. Therefore the *Possible-faults* of one *Symptom* are ordered in lists with different priorities and the declaration of the class *Possible-fault* is extended by the attribute *Probability*.

Within this list the *Possible-faults* are ordered by their probability factors. The higher the probability factor of a *Possible-fault*, the higher is the certainty for this *Possible-fault* to describe the real fault.

If a *Possible-fault* has been recognized by the system as the real fault, the probability factors of all faults in the list are changed automatically for the benefit of the recognized fault. By this form of "learning by parameter adjustment" the faults which appear most often will get the highest probability values.

After this task the system has generated a network with all the known possible faults ordered in a hierarchical structure and evaluated by probability factors. This shows the abductive strategy of the system. Instead of deducing single faults, which is not possible, the system infers a set of possible faults and orders them in a hierarchical structure for later access.

3.3 Inferring New Faults

A next problem deals with the fact that many errors which might come up in the future are unknown. But one knows from the detector description possible fault components, although one does not know all possible malfunctions which might occur in these components.

So we had to take into consideration that we are working with a not completely filled knowledge base. The possibility exists that a *Symptom* is not explicable by a known fault, i. e. the available heuristics are not applicable or do not give the right result. In this case the system provides help to the detector expert in searching for a new fault. For this purpose the system outputs the knowledge contained in the object *Symptom* and its local objects and by recourse to the detector model it is possible to procure further information concerning possible faulty components. So the output is a network containing all known fault possibilities. Additionally, for each *Symptom* the possibility of a new fault is scheduled.

3.4 User Support

The fault possibilities are hidden in the object network and affected by inheritance with different local certainty factors. Not to waste too much time in searching for the most important fault possibilities, the system transforms them into an ordered list of working proposals. Each element of the list consists of a set of *Possible-fault*s which together are able to explain the complete fault area. By "repairing" this set of *Possible-fault*s the fault area can be removed.

One proposal will be preferred, dependent on the actions done by the user in the preceding diagnoses. For each known fault of a working proposal the user can get information about the fault component, peculiarities of the fault, reasons of the fault and actions to repair the fault. If the system proposes a new fault, the user gets information about the peculiarities of this fault and the possible fault components.

3.5 Verification and Knowledge Acquisition

The user has to announce to the system if he has repaired one of the proposed faults. The system then tries to verify this fault in the next diagnosis. If this succeeds, the verified faults will be recorded and the probabilities of the faults will be changed for the benefit of the verified fault.

If the detector expert has found a new fault with the help of the system, the knowledge base can be extended. The new fault is partly characterized by the knowledge contained in the object *Symptom* and the connected objects *Act-pattern* and *Act-hardwaretest*. To get the new fault description the system generates automatically from the object *Symptom* an object *Fault*, from *Act-pattern* an object *Pattern* and from *Act-hardwaretest* the new object *Hardwaretest*. The detector expert can generalize the values of several attributes and has to input lacking information like reasons for the fault or actions to repair the fault. If, after the acquisition, this fault occurs repeatedly in any other part of the detector, the system infers at least this fault as one *Possible-fault* in the list of the *Possible-faults*. If the repeated occurrence of the fault shows that the description of the new fault was not correct in all details, then this will be detected by the system and the description can be changed.

The heuristic which the system needs in order to recognize a fault is inferred partially by the system itself in the single occurrence of this fault. So one might speak (with restrictions) about a kind of learning system ("learning from examples").

4 Architecture of the System

Although DELPHI-EXPERT is a real-time expert system, the typical problems like time restrictions and fast reaction on new process data do not arise. The data are collected over a relatively large period of time and the time DELHI-EXPERT needs to analyse the data is negligible.

As mentioned before, an easy extension to a diagnosis shell for many different detectors has been taken into account. For this purpose the knowledge base has been divided in a part which contains the general knowledge about detectors and a part in which the knowledge about one specific detector is collected [3,4].

The general part corresponds with the rule base and external routines, whereas the detector dependent part is completely represented in objects. The automatic matching of the special knowledge in the so called Working Memory (objects) with the general knowledge in the Production Memory (rules) by the OPS5 control strategy gives at run time a system which performs a diagnosis for a special detector.

Furthermore the detector specific part of the knowledge base can be divided in permanent knowledge, like the detector model or known detector faults, and temporary knowledge, like inferred fault areas or recognized fault areas. If the shell should be applied to a new detector, only a new detector model has to be entered. The temporal part of the knowledge base will be generated by the expert system in each diagnosis. The part of the knowledge base containing the known detector faults will be filled during run time of the system (each time a fault has been recognized).

The main tasks of error diagnosis, detection, specification, reaction, and verification are performed by DELPHI-EXPERT in four different processes. Two of these tasks build the diagnosis kernel and two of them the interfaces to the different user groups.

In the periodical diagnosis, the analysis is only based on the actual measured hit distribution, which the system receives in periodic intervals. If no fault areas have been inferred, the system starts automatically the next diagnosis. In the other case the fault areas are recorded in the object base for later access. A message is sent to the control monitors to inform people on shift that a problem has occurred and that the diagnosis process has been stopped.

Beside this periodic analysis the user has the possibility to start an additional diagnosis process to run hardware tests. The tests are dependent on actual fault patterns found in the periodic diagnosis step. With this additional information it is possible to reduce the number of fault possibilities significantly.

At the end of each diagnosis the user can get information about the result of the last diagnosis. The system serves two different user groups: for the shift crew, the system plays the rôle of the expert, for the detector expert the system assists in detecting new detector faults and in creating and updating the initial knowledge base. The user interface for the shift crew differs from the expert interface by the restricted view on the temporal part of the knowledge base and the lack of privileges for updating the permanent part of the knowledge base. By going through the list of working proposals, people on shift will either find the real fault or come to a point where the system advises them to call the detector expert because of a high probability for a new, unknown problem in the detector. The periodic diagnosis verifies repair actions performed by the user after the last diagnosis and resumes the monitoring of the detector. If problems occur which are not solvable by the shift crew, then it is possible to engage the detector expert without requiring his presence at the detector location. For this purpose, the temporal part of the know-ledge base is sent via a LAN or WAN to the workstation of the detector expert at his home institute, where he completes the permanent part of the knowledge base. Thus the detector expert can help via remote diagnosis.

The advantage of this architecture makes it possible to place the data taking program, the diagnosis kernel, the object base, and the user interface on different computers.

5 First Experiences

DELPHI-EXPERT was running during last year's data taking time of the DELPHI experiment at CERN. The diagnosis kernel was installed on a special workstation for monitoring purposes of the specific detector. Communication with the data acquisition computer was performed via an ethernet connection; the communication protocol was DECnet. During the whole test phase of about 6 months, the same diagnosis kernel together with the user interface was installed on a workstation at our university in Wuppertal. Every time the diagnosis program at CERN detected a significant mismatch of an actual measured hit distribution with a reference one, the result was automatically sent over a WAN connection to our workstation. In order to debug possible problems within the expert system, the actual measured hit distribution was also sent to our workstation, and the diagnosis task could be repeated for test purposes, if needed. This procedure made it possible to test the expert system without being present at the experiment.

All problems detected by human experts at CERN were also found by the system. Sometimes a problem was detected earlier by the diagnosis system then by the shift crew. For that reason we decided to put the system now into public operation. Experience with this mode of operation is still lacking and must show if the user interface needs some modifications.

The extension of the diagnosis system to an expert system shell for other detectors will no longer be regarded. The reason is that the work which would have to go into the further development for such a general tool is expected to be too much. The effort to build the existing system was about 5 man years (1 for the initial prototype and about 4 for the system as it is now). This long development time caused an decreased interest of the two students who worked on the system. In addition time has changed: The amount of available CPU time grew fantastically so that it is possible now to perform much more of data pre-processing beforehand. This makes such a diagnosis system more reliable especially with respect to fluctuations of the raw data according to accelerator changes (e.g. energy, backgrounds) which sometimes caused artificial detector problems. We had to modify the system in order to handle these conditions properly. Instead of a further development of the existing system, we decided to construct (with our gained experience) a new diagnosis system based on preprocessed and such more reliable input data.

6 Conclusions

We have developed an expert system for error diagnosis and repair suggestions for one specific detector of the High Energy Physics experiment DELPHI at LEP. It is a real time expert system provided with data from on-line monitoring tasks running during data taking of the experiment.

The process of error diagnosis is based on the model of a specific detector. The system uses an abductive inference strategy and organizes all fault possibilities in a semantical network of objects. The representation of the detector model and the detector faults is also object oriented. By separating detector specific knowledge in a object base from the general knowledge in a rule base, the system is adaptable to a whole class of different detectors systems.

References

1. L. Brownston, et. al.: Programming Expert Systems in OPS5 - An Introduction to Rule-based Programming, Addison-Wesley Publishing Company, Inc. (1985)

   R. Krickhahn, B. Radig: Die Wissenrepräsentationssprache OPS5: Sprachbeschreibung und Einführung in die regelorientierte Programmierung, Vieweg (1987)

   T.A. Cooper, N. Wogrin: Rule-based Programming with OPS5, Morgan Kaufmann Publ. Inc. (1988)

2. DELPHI Collaboration: The DELPHI Detector at LEP, Nucl. Inst. Meth. A303 (1991)

3. K.-H. Becks, A. Hemker: An Expert System for Error Diagnosis in High Energy Physics Detectors, Proceedings of the 1988 DECUS Europe Symposium, Cannes (1988)

J. Ortmann, K.-H. Becks, K. Brand, A.B. Cremers, A. Hemker: DELPHI-EXPERT - An Expert System for Error Diagnosis in High Energy Physics Detectors, Proceedings of the International Workshop on New Computing, Techniques in Physics Research, Lyon, Editions du CNRS (1990)

4.   R. Meyer, G. Schlageter: Ankopplung der DELPHI-Expert-Shell an eine relationelle Datenbank , Internal Report, Hagen (1990)

Combining Real-Time with Knowledge Processing Techniques

W. Brockmann

Universität-GH-Paderborn
FB14, FG Datentechnik
Paderborn, FRG D-4790

Abstract. Real-time systems make high demands on the programming style such as timeliness and predictability. The architecture which is presented in this paper uses explicitly given knowledge to meet the requirements of next generation real-time systems. For this purpose a multi-stage architecture for real-time knowledge processing systems was developed. It is based on a simple knowledge model. This can be implemented very efficiently by using the ARON-technique (Alternatives Regularly Organized and Numbered), which is easy to use even by non-programmers. Starting with this elementary system, more complex applications can be handled by decomposing the solution of a problem into interacting subsystems, each working with the ARON-technique. Applicability is demonstrated by two examples of knowledge based process control, e.g. real-time control of an A.C. motor.

1 Introduction

In online-applications, real-time systems interact with an (outer) physical environment typically in a continuously driven mode of operation. This has a strong influence on their architecture for they not only have to provide an appropriate interface but they also must provide a timing that fits real-time requirements. In this context soft real-time systems and hard real-time systems are distinguished. Soft real-time systems can recover from synchronisation or timing faults without external assistance and without loss of information or any damage. In hard real-time environments, however, severe damage such as loss of data or control may be caused by missing real-time constraints, especially timeliness [3].

Timing constraints are especially very strong on lower levels of process supervision or control. Although their nature is less complicated, i.e. operational tasks predominate, large amounts of data have to be processed under (hard) real-time constraints, for instance at the control of technical processes. In this scope most architectural approaches only try to achieve timeliness. That is, real-time systems react timely to external events and respond in time. So their response times have to meet deadlines. But no steps are done to achieve predictability [6]. Predictability additionally means that reponse times can be determined in advance for every particular situation or any data. This makes the temporal behaviour easy to calculate, even in complex applications where multiple real-time tasks interact.

Beyond these timing requirements, next generation real-time systems should be maintainable and flexible in order to simplify changes or extensions to the real-time systems. This is important because they normally operate in a changing environment for a long lifetime. But in conventional programming techniques the application specific knowledge is contained in control and data structures of a real-time program. This makes the design of a real-time program a complicated and expensive job for it normally must be solved by a cooperation of an application expert and a programming expert.

The architecture for next generation real-time systems which is presented here is intended for embedded systems that directly interact with an outer technical environment. We further aimed to achieve response times in the range of milliseconds and below. The basic principle of this approach is to involve an application expert directly in specifying the behaviour of a real-time system with as little support by a programming expert as possible. It therefore makes use of techniques which process explicitly given knowledge. By that a cost effective design is possible which is flexible and easy to maintain because of its abstract level of description.

Due to the field of application, embedded knowledge processing systems directly interact only with a technical process without any direct communication to human operators. This has a strong influence on their architecture. In real-time applications most common approaches to knowledge processing systems try to increase performance in order to meet deadlines for instance by using specialized hardware (e.g. LISP machines), structuring the knowledge base, controlling reasoning depth (and quality) according to time criticalness, focus of attention, compilation of declarational knowledge descriptions into procedural ones, and/or migration of real-time demands to real-time interfaces (e.g. coprocessors). Such systems are primarily applied to higher levels of fabrication or process supervision and control. Here they perform relatively complex tasks on comparatively small amounts of abstract data, e.g. planning or optimizing. But to achieve predictability and to meet hard real-time requirements no approach is known so far which is easy to use by non-programmers.

Because the knowledge is validated by applying it to the application itself, a sophisticated explanatory interface is not necessary to make the reasoning process transparent and also not useful, especially with fast processes under hard real-time constraints. That is why some steps in knowledge acquisition can be omitted by only specifying and performing the desired actions. Thus an interface to higher levels of process supervision is necessary only to handle commands and possible (feedback) messages. Higher levels then perform communications to human operators under soft real-time constraints without explaining why a knowledge processing system acts the way it does.

So a technique was developed exploiting the effect that it only has to match an action, or conclusion respectively, to a given situation. To perform this efficiently, the implementation is done procedurally thus enabling portable run-time systems, which can run on a PC as well as on a process computer.

2 Approach to Real-Time Knowledge Processing Systems

2.1 Basic Principles

The ARON-technique (Alternatives Regularly Organized and Numbered) is designed to process knowledge under hard real-time constraints [1]. To make such knowledge processing systems easy to use, it is based on a simple model of human real-time behaviour. This imitates an application expert who reacts in any given situation by the following scheme:

- He recognizes the system state by considering relevant features. These may be discrete values (like switch positions) and levels or trends of numerical values (like temperature).
- All features are concentrated to a single situational description.
- Due to this situation a suitable reaction is selected and performed.

This knowledge model leads to a three-stage structure consisting of

- feature extraction,
- real-time kernel,
- performing of actions.

The run-time system operates in a sampling mode in such a way that response times are always shorter than the sampling time. This then forms the deadline. Because the length of this sampling period is dictated by the dynamic of the process, it is ensured that the process does not change its state significantly. Therefore no consistency check is necessary when the knowledge processing system gets its results. So the mode of operation is periodic and state processing, in contrast to the wide-spread sporadic and event triggered mode of other knowledge processing systems. The constant sampling rate further leads to a constant rate of data but especially to an implicit notion of ´time´.

Because every feature is extracted and every action is performed procedurally in each sampling period, their (worst case) processing time is easy to calculate. Thus timeliness and predictability only depend on the real-time kernel. Therefore as many operations as possible are performed offline through compilation. Through this the basic bottleneck of search is eliminated. Online only deterministic calculations are necessary to select a suitable action, making operation very fast and easy to predict. This also guarantees hard real-time requirements.

2.2 The ARON-Technique

Knowledge Representation. The whole knowledge is acquired and represented in a structured way on different levels of abstraction. The model of human behaviour leads to the structure of knowledge which is shown in Fig. 1. It is divided into conceptual knowledge (features and their extraction, actions and how they are performed) and operational knowledge (specific behaviour of the real-time system).

knowledge about the problem solution		
conceptual knowledge		operational knowledge
premise	conclusion	
• features	• actions	heuristic
• instances	• action-keys	strategies
• instance-boundaries	• action ranges	rules

Fig. 1. Structure of the knowledge representation

The knowledge representation is based on simple, shallow rules (IF premise THEN conclusion) using stylized natural language like:

IF deviation is positive AND measurement change is very little AND desired value is high THEN add 90% of deviation to the output value.

Determination of the features, which are used, and of the way they are extracted is the first step in extracting conceptual knowledge. Then the instances of each feature are fixed (see example in Fig. 2). In a similar way the scheme is determined under which output values are calculated by using the conclusion.

Deviation	Meas.change	Des.value		Meas. change	Instance
negative (0)	large (0)	high (0)		10 .. 20 ppm	large
positive (1)	medium (1)	med. (1)		5 .. 9 ppm	medium
	small (2)	low (2)		2 .. 3 ppm	small
	very little (3)			0 .. 2 ppm	very little

a) Ordered and numbered (aronized) features   b) Conversion into a qualitative value

Fig. 2. Preparation of features

Because the real-time system has to perform an action on every sampling period, exactly one conclusion must be associated to every possible situation. Consequently one has to ensure that for each situation the conclusion wanted is specified. This can be done offline to minimize run-time. Therefore all possible gaps and conflicts in the knowledge base must be detected and eliminated during compilation. Gaps are situations to which no rule specifies a conclusion. On the other hand conflicts arise if multiple rules fit for the same situation because of 'don't cares' in their premises.

To achieve easy understanding and acceptance, static and automated strategies ought to be used to eliminate gaps and conflicts, e.g. selecting the conclusion of the rule which best fits the appropriate context or preferably selecting simply the rule standing closer to the end or the top of the rule base. Gaps may be filled by explicitly asking for suitable conclusions or using default conclusions which have no effect to outputs of the knowledge processing system. But also analytical strategies are applicable exploiting analogy to determine plausible items out of the context given in feature space.

Compilation. Before compilation the features in the premises of all rules must be sorted. As a next step the premises are completed with 'don't cares'. At run-time all features are extracted from a given situation S(t) at every sample and also arranged according to the given order. Each situation S(t) then is describable as a m-dimensional feature vector $\underline{M}(t)$ (with m = number of features). The order given by that is also used for building the run-time system accordingly.

On a lower level all instances of a feature are arranged in a distinct order in an analogous way, as Fig. 2a also shows. E.g., continuous signals are converted into a qualitative description by classifying them into consecutive linguistic instances. This is done without a gap or overlap according to the distinctions the human expert made in the knowledge base. This ordering allows numbering of instances for each feature. (This process of ordering and definite numbering of ordered instances will be called aronizing, see ARON, hereinafter.) Preferably these instance numbers are succesive integer-numbers starting at 0, as demonstrated in Fig. 2a. Then each situation can be described by an unequivocally aronized feature vector $\underline{M}^*(t)$ having m components and containing the current instance number of each appropriate feature at run-time t, as can be seen in Fig. 3.

$\underline{M}(t) = ($ Deviation(t),
 Meas.change(t),
 Des.value(t),
 Old desired(t)) $\underline{M}^*(t) = (1, 1, 0, 2)$

a) Feature vector at time t   b) Aronized feature vector at time t

Fig. 3. Representing given situations

The totality of all possible aronized feature vectors $\underline{M}^*(t)$ spans a discrete m-dimensional feature space D_m^*. This is completely filled and stored in the run-time system. Thus the premises in the rule base are linked to situational descriptions in feature space D_m^* via compilation. Thereby the conclusion of each rule is related to each point of the feature space D_m^* which is expressible by its premise. I.e., conclusions of rules containing 'don't cares' in their premise are registered at all points which are reachable by filling 'don't cares' with all instances they represent.

The feature space D_m^* is stored preferably as an 1-dimensional array. To do this the features are weighted according to their position in the feature vector and to the number of instances. After that, the position of a conclusion is calculated by means of a weighted addition of instance numbers as shown in (1).

$$G_1 = 1,$$
$$G_i = G_{i-1} \cdot n_{i-1} \qquad \text{with } i = 2 \ldots m$$
$$S^*(t) = \sum_{i=1}^{m} G_i \cdot m_i(t) \tag{1}$$

m   : number of features $m_i(t)$: instance number of i-th feature at time t
G_i : weight of i-th feature n_i : number of instances of i-th feature
$S^*(t)$: situation number at time t

The whole feature space D_m^* as a synonym for the compiled rule base is represented in the memory of the run-time system. There, it only consists of conclusions because premises are unnecessary for they are implicitly comprised in the position of any conclusion in the feature space D_m^* or in memory, respectively.

2.3 Online-Implementation

Online feature extraction determines the values of the feature vector $\underline{M}(t)$ out of a given situation S(t). From this the aronized feature vector $\underline{M}^*(t)$ is built through assigning instance numbers to present values for each feature. Discrete feature values may be encoded directly, whereas quantitative feature values may be converted by continuously checking neighbouring ranges of instances until the suitable one is met. It is faster and of constant processing time to assign the instance number via a table by using the quantized feature value as an index pointing to an entry which contains the appropriate instance number. To determine the desired conclusion, a definite situation number $S^*(t)$ is computed at each sampling period instead of searching or pattern matching. This is done in the same way as it was used to determine the position of any conclusion, namely using also scheme (1). Finally the ARON-technique uses this situation number $S^*(t)$ as an index to access the particular conclusion. For that it exploits hardware support which is given in form of address decoders in memory. To improve speed, multiplications with weights G_i may be included offline in entries of tables used for aronizing of feature values. So at run-time only additions have to be performed and processing-time decreases.

Discrete and numerical output signals must be treated differently. The first ones are aronized and encoded directly while the second type needs some processing because the conclusions normally describe in a shortened way how output values are determined. E.g., conclusions may describe how much an output value is to be increased or decreased. Besides a specific mode of operation this leads to a much smaller range of values and to good memory utilization. Furthermore, specification of offsets to or relative parts of numerically given (outer) values has proven to be useful, e.g. in the example of controlling a biological tank reactor (chapter 3). Rule bases then normally get more simple because they only have to describe coarse behaviour, leaving fine resolution to this way of output processing.

2.4 Discussion

As real-time systems using the ARON-technique work periodically in always the same deterministic scheme, it is ensured that hard real-time requirements are met. Furthermore such systems are maintainable and adaptive [6] concerning operational knowledge. This means that the runtime is independent of the operational knowledge, especially it is independent of the number of rules. This allows variation of strategies in a wide range without changing the program and without the need of checking real-time capabilities again. Moreover, at least the kernel is fully predictable.

Because feature extraction and execution of actions as well as their (worst case) expense (measured in processing time or other costs) are independent from employed reasoning techniques, only used memory space is a measure of cost. In theory, memory requirements may explode combinatorially if numbers of features and instances increase. But practically it is limited to a handy order of magnitude because of limited numbers of features and instances which the human experts use to keep knowledge manageable for themselves. Furthermore the ARON-technique works without producing garbage and explicitly storing premises and other constructs, e.g. pointers in lists or trees, thus also saving memory.

A more stringent restriction is its reflectory behaviour for the ARON-technique basically does not contain a long term memory, in contrast to e.g. expert systems. Because it may be useful in some applications, long term memory must be implemented through one or more additional features representing relevant memory states as instances. These features are updated procedurally or are using fed back outputs of rule based systems. By that, memory demands increase because feature spaces have more dimensions and eventually because of larger conclusions. But the lack of long term memory seems to be no handicap in the intended field of application because it is given implicitly by the outer environment as the following examples demonstrate.

3 Knowledge Based Process Control - First Example

We chose to apply this real-time architecture to knowledge based process control in order to investigate design efforts and applicability. So, control of a simulated biological tank reactor was chosen because biochemical processes gain increasing importance, especially in ecology. But their control still underlies several research efforts because such processes show strong nonlinear characteristics. These are only met by very specialized analytical controllers.

A biological tank reactor consists of a reservoir containing bacteria in nutrient fluid. In our case it is responsible for reducing a (toxic) substrate which is part of the fluid passed through the reactor. The controlled variable is the substrate concentration in the off-flowing fluid while the flow rate is the manipulated variable. As in [2] a nonlinear model according to Monod is used to simulate a continuously driven biological tank reactor. But in this case we used other parameters (according to bacteria culture) and an average sampling time of 8.1 minutes containing an average measurement dead-time of 2 minutes. Thus performance or real-time aspects, respectively, are of no relevance here because sensor elements only allow sampling times of several minutes. This leads to significant changes of the process state between two samples, making controller design a real challenge, especially in conjunction with the multitude of possible disturbances.

Fig. 4a shows control results which are achieved by an experienced human expert under ideal conditions. They are sufficient due to the nature of the controlled process. In comparison Fig. 4b demonstrates the control results of a knowledge based controller using the experience of this human expert. The quality of this controller is much better than manual control by the human expert, especially concerning control speed and deviations from desired values. The reason is that the knowledge based controller always works with its maximum efficiency (no lack of concentration, no stress). Furthermore it can easily be optimized because of its deterministic way of operation.

In order to investigate the robustness of this controller, we examined it by applying disturbances such as noise, variation of measurement dead-time, and variation of bacteria's growth rate. As Fig. 5b shows, the knowledge based controller still performs well. Control results are even comparable to those of an analytically designed and implemented controller (Fig. 5a) which is especially designed for the controlled biological tank reactor. This is remarkable in that designing the knowledge based controller took about 4 weeks, but several months were needed to design (and prove stability of) the analytical controller.

4 Extensions

As mentioned in chapter 2.5, manageable complexity is limited by memory demands as well as by the complexity which can be handled consistently by a human expert. To overcome these drawbacks, the application specific knowledge is divided into separate subtasks representing knowledge accordingly by distinct subsystems. These subsystems cooperate in a static structure and are all evaluated in each sampling period. Through this, the total systems remains capable to satisfy (hard) real-time constraints if the subsystems are able to meet them and are connected non-recurrently (which seems to correspond to the human way of solving real-time problems). Thereby the total real-time problem is reduced to the real-time capabilities of smaller systems which have to perform simpler tasks. They can be realized by an universal and elementary system as described in the chapter 2.

The decomposition into these cooperating subsystems is specific to the application and is related to conceptual knowledge. Its estimation is the first step in knowledge acquisition. In principle the topology of these subsystems may consist of up to three dimensions. The first dimension refers to subsystems arranged in parallel which operate on (partly) the same data or signals, e.g. different evaluations or control of MIMO-processes (Multiple Input, Multiple Qutput). The second dimension concerns the serial structure of subsystems which perform different tasks on a data stream, e.g. plausibility checks and refinement of incoming signals in front of a controller. The decomposition of the application-specific knowledge is intuitively reasonable and intelligible to the human expert in this way. The knowledge is represented clearly and easy to maintain because no interlocked representation schemes are used and because the knowledge base of each subsystem is of independent content and manageable complexity. In addition relatively simple mechanisms can be used to work on these knowledge bases.

The third dimension refers to subsystems which are used to perform a self-organizing process in order to adjust the behaviour of the knowledge processing system to a desired behaviour. This is done by imitating a special learning strategy. By this, a human expert is modelled who makes certain actions depending on each given situation. Afterwards he also reflects his actions by the effects they have on the technical process. If a result is not reasonable the expert modifies that action (in mind) which is responsible for actual deviations from an (implicitly given) ideal case. When this action is triggered again, an altered action will be taken, which he will check again and so forth.

This strategy is modelled by so called hyper-systems which also use the ARON-technique. These hyper-systems do an online-assessment of effects subordinate systems (so-called base-systems) have on the technical process. Therefore they comprise knowledge about which modification has to be done in any particular situation to a base-system's conclusion that was used the latency-time of the process before. To do this, a hyper-system adds a small offset which is depending on the actual situation to the corresponding entry in the feature space of the base-system. According to the approach presented here, such checks and (possible) modifications take place online at each sampling period. So self-tuning and adaptive behaviour can be achieved without human interaction. Furthermore, in some applications machine learning is possible through describing only the frame of a knowledge base on an abstract conceptual level. This leaves the determination of appropriate actions to a hyper-system and the process itself. It may also be used to overcome speed limitations human experts have (see following example).

5 Knowledge Based Process Control - Second Example

In order to demonstrate the usefulness of the decomposition and of the self-learning method described above as well as showing the very high efficiency of the ARON-technique, knowledge based speed control of a three phase A.C. motor is used. A.C. motors are nonlinear processes as well and can be viewed as MIMO-processes because speed is controlled and supply currents are limited by applying an appropriate three phase voltage of variable amplitude and frequency.

The knowledge based speed control system, we used, consists of four subsystems, one base-system for determination of the amplitude and one for the frequency, each having a hyper-system associated to it. Only the knowledge bases of both base-systems are primarily responsible for control because they are directly interacting with the motor. They originally contain coarse knowledge expressed by some rules. These rules are intended to get the motor turning as well as describing how the controlled system is to be protected against overload. Most of these rules are tagged and cannot be modified by a hyper-system while in all other situations process behaviour is rated by the hyper-systems. In that way, knowledge bases of the base-systems are adapted if necessary.

All of the four subsystems are implemented on a single 80386-system working at 20 MHz. A sampling rate of 5 kHz is reached, even though complete feature extraction, four knowledge base evaluations, and four actions are to be performed along with administrative jobs in each sample period. Practical results of controlling a 380W-motor are shown in Fig. 6. Their comparison clearly demonstrates the success of the applied self-learning method. Fig. 6a shows the original behaviour of the control system from which hyper-systems started modifications. After a training period of 2 minutes of applying another randomly changing set point every second, the control precision is significantly improved as can be seen in Fig. 6b. Of course, results do not reach the precision of model based controllers because the knowledge based control only considers outer signals. It therefore has a very limited notion of ongoing processes inside the motor. Possibly results can be improved by giving the expert more insights when he is experimenting with the motor and by using more sophisticated features.

6 Conclusions

In this paper a multi-stage approach to hard real-time knowledge processing systems was presented. The ARON-technique was proven not only to be applicable to such systems but also to be easy to understand and easy to handle. To select an appropriate conclusion, compilation and simple deterministic calculations are used instead of search and pattern matching. Thus the ARON-technique is not only fast and predictable, it also allows great variations of the behaviour of such knowledge processing systems without the need of program changes and without the need of checking timeliness again. But the ARON-technique may find its limit in the memory space needed because it explodes combinatorially by increasing numbers of features. Another limitation is the complexity which a human being can see clearly.

Both drawbacks are to dissolve through decomposition. Thus the problem of processing knowledge under real-time constraints is divided into interacting real-time subsystems which only have to solve simpler tasks and which need less memory. Additionally, automated knowledge acquisition methods or machine learning techniques may be used along with the ARON-technique to aid the human expert.

Practical use was demonstrated in the field of knowledge based process control by the examples of substrate control of a simulated biological tank reactor and of self-tuning speed control of a real A.C. motor. These applications show that a comparatively small amount of memory is required (biological tank reactor some kbytes, A.C. motor some tens of kbyte). So the ARON-technique is suited for next generation real-time systems, especially in conjunction with decomposition. Further applications may be found in the fields of diagnostics, intelligent sensor systems, or pattern recognition.

References

1. Brockmann, W.: Real-Time Architecture for Knowledge Processing Systems. Euromicro '91 Workshop on Real-Time Systems; IEEE Computer Society Press, Los Alamitos, 1991, 52-60
2. Czogala, E., Rawlik, T.: Modelling of a Fuzzy Controller with Application to the Control of Biological Processes. Fuzzy Sets and Systems 31(1989), 13-22

3. Koymans, R., Kuiper, R.: Paradigms for Real-Time Systems. in: Joseph, M. (Ed.): Formal Techniques in Real-Time and Fault-Tolerant Systems. Proc. Symp., Warwick, 1988

4. Leinweber, D.: Expert Systems in Space. IEEE Expert, 1987, 26-36

5. Moore, R.L., Hawkinson, L.B. and others: A Real-Time Expert System for Process Control. IEEE 1984, 569-576

6. Stankovic, J.A., Ramamritham, K.: Hard Real-Time Systems. IEEE Comp.Soc.Press, Washington, 1988

7. Wright, M.L., Green, M.W. and others: An Expert System for Real-Time Control. IEEE Software 3(2), 1986, 16-24

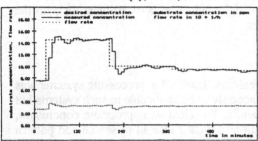

a) Experienced human expert

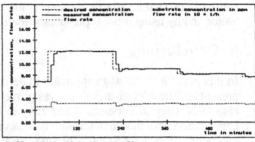

b) Knowledge based controller

Fig. 4. Control of a biological tank reactor under ideal conditions

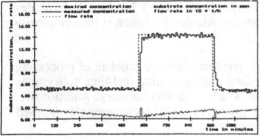

a) Analytical controller

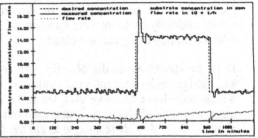

b) Knowledge based controller

Fig. 5: Knowledge based controller of biological tank reactor with disturbances

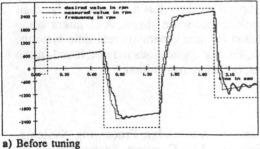

a) Before tuning

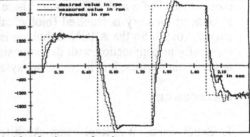

b) After 2 minutes of tuning

Fig. 6. Knowledge based control of a real A.C. motor

Tailoring Advanced Instructional Software for AI

Dr. Dean Allemang[1] and Dr. Robert M. Aiken[2]*

[1] Istituto Dalle Molle di Studi sull'Intelligenza Artificiale
Corso Elvezia 36
6900 Lugano, Switzerland
dean@idsia.ch

[2] CIS Dept. 038-24
Temple University
Philadelphia, PA 19122
AIKEN@TMPLCIS.bitnet

Abstract

A current joint project between three institutions in Switzerland has as its goal to create Artificial Intelligence (AI) software for use in teaching principles of AI at the university level. The modules of this project, the Portable AI Lab PAIL illustrate basic concepts of Artificial Intelligence in a uniform and self-contained manner. This paper discusses the design considerations that were adopted in order to make the presentation of this material effective for students of various backgrounds and interest, particularly intermediate and advanced students, as well as, people in industry wanting to understand better how AI techniques can assist them in problem-solving.

1 Introduction

As Artificial Intelligence (AI) is becoming a more central component of the computer science curriculum, there is a growing need for material which will aid students and instructors in AI education [2]. Three institutions in Switzerland, the Federal Polytechnic of Lausanne, Istituto Dalle Molle di Studi sull'Intelligenza Artificiale (IDSIA) Lugano, and the University of Zürich, received funding from the National Science Foundation of Switzerland, to help address this problem. The goal of this co-operative project, the Portable AI Lab (PAIL), is to create software to assist in the instruction of principles of Artificial Intelligence at the university level. Although the project was initially intended to address the particular educational needs of the AI community in Switzerland, we believe the PAIL will be useful in other countries as well.

The intent of the PAIL is not to write a textbook in AI, but rather to provide a uniform implementation of software. On the one hand, we have tried to construct the modules so that an instructor teaching an Introductory AI course can use them to augment lectures. Thus, we have tried to be sensitive to the needs of beginning AI students. Additional discussion of the design considerations followed in order to construct these modules for easy use by novices can be found in [1]. On the other hand, we believe there will be

* The work reported in this paper was funded in part by the Swiss National Science Foundation Project 23, while both authors were affiliated with the Institut für Informatik, Universität Zürich

interest on the part of more advanced users, students already having a background in AI and Lisp, as well as people working in industry.

Some of the objectives of the PAIL are:

1. The PAIL is portable, that is, it should be possible to run it on more than one machine (At present, the PAIL runs on SUN workstations running UNIX. Possible future targets include the MacIntosh and IBM PCs). This is to guarantee that the PAIL will be usable by a large number of educational institutions in Switzerland.
2. The PAIL is consistent with well-established literature in AI. Whenever possible, the algorithms in the PAIL are based on algorithms defined in AI textbooks, or at least easily accessible conference papers or journal articles. Accompanying every module of the PAIL is a short description that relates the particular module in the PAIL to the conventions in the literature with a few examples to clarify this relationship.
3. One of the problems with using demonstration AI software today is that while many example programs are available in the public domain, they are often written in different dialects of LISP (or even in different languages), so that the interested student must be an expert in LISP in order to investigate different approaches in AI. The PAIL addresses this problem in two ways:
   - Each module is equipped with an interface that will allow the student to explore an AI concept without having first to become an expert in Lisp.
   - Some students might want to delve more deeply into the working details of the implementation, and hence will come to learn Lisp. To faciliatate this process, all the modules have been implemented in the standard Lisp dialect (Common Lisp with CLOS, as specified in [12]), and made to conform to documentation standards that relate them to the literature.
4. The overall PAIL project will cover a wide range of AI topics, including machine learning, knowledge acquisition, natural language, inference engines and theorem proving.
5. The utility of many AI concepts becomes apparent only when they can be seen in the context of other AI research. For this reason, it is important that the modules of the PAIL communicate with one another, i.e., the output of one module can be used as input to another, and so forth. This could, in principle, allow the PAIL to be used for novel experimentation.
6. The PAIL material will be distributed free of charge. The intention is to disseminate it to anyone who might be able to make non-commercial use of them. [3]

Given the breadth of AI research in recent decades it would be a monumental task to incorporate all the mature theories that fall under the AI umbrella. The PAIL includes the following modules for Machine Learning and Knowledge Acquisition:

- Repertory Grid, [6]
- Induction Learning (ID3, see [13])
- Explanation-Based Generalization (EBG) (after Mitchell, see [10])
- Connectionist Models [11]

---

[3] Those interested in acquiring a copy of the modules should contact the Project Director, Mr. Michael Rosner, IDSIA, Corso Elvezia 36, CH-6900 Lugano, Switzerland

- Back-propagation [11]
- Genetic Algorithms [4]

Other modules, for planning, reasoning and natural language processing are also available, but are not the subject of this paper. An overview of how topics in machine learning and knowledge acquisition are supported by PAIL are described in [3].

2 Design Criteria for Different Classes of Users

Three stages of user ability are targeted by the PAIL. The design consideration for the novice user, which includes worked examples, extensive structured help facilities and graphical displays for output (e.g., browsers and spreadsheets), are described in [1]. In this paper we concentrate on the support provided by the PAIL for more advanced users.

2.1 Intermediate and Advanced Users

Intermediate and advanced users of the PAIL share the following characteristics:

- intermediate to advanced knowledge of Computer Science
- solid experience in at least two programming languages
- familiar with computer system on which PAIL modules run
- intermediate knowledge of AI (e.g. knowledge equivalent to that covered in an Introductory AI course, plus some in-depth knowledge of particular areas)
- quite knowledgeable about Lisp

The user's knowledge of Lisp is the most critical feature in determining whether the user belongs to the intermediate or advanced user class, and is the single most important parameter in determining how to design the modules. That is, while both intermediate and advanced users are able to formulate questions about how details of AI algorithms might be worked out, an advanced user would have the Lisp skills necessary to do experimentation on his/her own, while an intermediate user would still need some guidance. For example, an intermediate user could benefit from an exercise in which a pivotal part of an algorithm is left unspecified (say, the conflict resolution strategy for a backward chaining rule engine), and the student is required to do the programming necessary only for that part of the assignment. On the other hand, an advanced user is most likely a "hacker" in that s/he knows what the algorithm will do and is mainly interested in exploring how this particular module does it. This user will want to "play" with the code, see how it works, and then modify, expand or otherwise just experiment with it. For some users the idea will be to see how they can integrate these modules with other "tools" they use in their research or development.

3 Support for the Intermediate user

Considerable attention has been given to intermediate user level support. This includes editors for modifying input to the various algorithms supported by the portable AI lab, graphic tools for output display provided for the beginning user, and a facility for communicating between modules.

3.1 Inter-module Communication

One of the major features of the PAIL modules is that the output from some of them can be used as input to others. This is particularly true of knowledge acquisition techniques, whose output is intended as guidelines for the construction of some knowledge-based system.

As an example of the inter-module communication, consider the relationship between a rule interpreter and explanation based generalization. The rule interpreter takes as input a set of rules, and an initial configuration of known facts (contents of a 'working memory'), in the form shown in figure 1. Figure 1 shows a simple rule set to determine when it is safe to stack one object on another; according to this rule set, lighter objects may always be stacked on heavier ones (rule d). Furthermore, tables always weigh 5 pounds (rule c), while the weight of other objects can be determined by multiplying their density by their volume (rule a). An object is lighter than another if its weight is smaller (rule b). Initially in the working memory are the facts that the density of the *book* is 4, its volume 1, and that *this-table* is a table.

The output from a run of this rule set is the assertion in the working memory that it is safe to stack the book on the table. But beyond this simple output, an explanation of the conclusion, in the form of a rule trace can also be generated. Such a trace is shown in figure 2. This trace can be used as input to an explanation-based generaliztion (EBG) algorithm as described in [9]. The generalized output from the EBG algorithm is also shown in figure 2.

Similar stories can be told for other pairs of modules, e.g., the output of repertory grid can be used in the context of ID3 or of a clustering algorithm, whose output in turn can be used as input to the rule interpreter.

An experienced programmer can, of course, write an interface of this sort using a general purpose language like Lisp, but this requires considerable knowledge of the representation and data structures that are being used. One of the aims of the portable AI lab is to allow students to deal with such data by their conceptual types, e.g., explanation trees or example tables, not as Lisp structures or text.

3.2 Tools

The desire to have users deal with conceptual types requires that there be special-purpose tools for working with data of these types. For this reason, we have provided the following graphical tools for the data types used in the PAIL:

- spreadsheets - ID3 tables, repertory grids, object-attribute-value sets
- browsers - decision trees, Truth Maintenance graphs
- special graphic interface for working with repertory grids

Tools of this sort allow students to work with provided examples of each of these conceptual types, as well as to create new examples of their own. Each of these tools is customized to the particular data type to which it is applied; for example, ID3 works best when the possible values of the attributes are restricted to a small set, so the spreadsheet for ID3 tables includes an option to convert numeric attributes into cluster sets (e.g., low, medium, high). The browser for rule traces shown in figure 2 includes

```
(rule-a
    (IF    ((volume ?p1 ?v1) (density ?p1 ?d1)))
    (THEN (weight ?p1 (* ?v1 ?d1))))

(rule-b
    (IF    ((weight ?p1 ?w1) (weight ?p2 ?w2) (lisp myless ?w1 ?w2)))
    (THEN (lighter ?p1 ?p2)))

(rule-c
    (IF    ((isa ?p1 table)))
    (THEN (weight ?p1 5)))

(rule-d
    (IF    ((lighter ?p1 ?p2)))
    (THEN (safe-to-stack ?p1 ?p2)))

((density book 4)
 (volume book 1)
 (isa this-table table))
```

Fig. 1. Screen dump of rule set and initial working memory.

possibilities for examining the original rule that led to each conclusion. The browser for truth maintenance graphs (figure 3) allows users to set and unset the belief values of TMS nodes, and have the propagated values displayed on the rest of the graph.

3.3 Pool

Defining abstract conceptual types for the inputs and outputs of the modules is only half the job. It is also necessary to provide a means for transmitting data objects from one module to the next. This process is complicated by the fact that often the logical input/output types of two modules may differ, even though the content of the representations is the same. For example, a TMS can be constructed to contain the information present in a rule trace, though the structure of the TMS is different from that of the rule trace. These problems are solved in the PAIL by the use of a pool of strongly typed objects. Each module defines a strongly typed interface to the pool.

The output of the Rules module is labeled with type 'explanation tree', while the output of the TMS is labeled with type 'TMS'. Any output from either module can be put into the pool, and will be of the corresponding conceptual type. The input to the EBG module is labelled with the type 'explanation tree'. When either output (type 'explanation tree' or 'TMS') is requested from the pool as input for EBG, the system automatically searches for a conversion between the type of the pool object and 'explanation tree'. Both conversions are defined in the PAIL (the conversion from explanation tree to itself is trivial), so the conversion is made, with a note to that effect for the user (see figure 4). The user need not keep track of the types him/herself; any data can be used

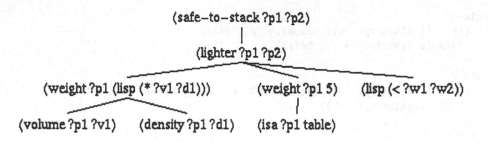

```
(IF ((volume ?var1 ?v1)
     (density ?var1 ?d1)
     (isa ?var2 table)
     (lisp (< (* ?v1 ?d1) 5)))
)
(THEN (safe-to-stack ?var1 ?var2))
```

Fig. 2. (top) Generalized explanation trace of the conclusion that it is safe to stack the book on the table, and the generalized rule from the EBG algorithm. The book and the table have been replaced with variables ?p1 and ?p2. This was deduced from the fact that (lighter ?p1 ?p2), which in turn was deduced from the three facts (weight ?p1 (* ?v1 ?d1)), (weight ?p1 5) and (< (* 1 4) 5). (bottom) The generalized rule states that it is safe to stack anything whose product of density times volume is less than 5 on the table.

in any module, provided such a conversion is possible, and the appropriate conversion is done automatically.

Figure 5 shows how the user can combine modules in different ways to achieve the same result. The same input (rule set) can be run through the rule interpreter to produce an explanation tree, or through a TMS based rule engine to produce a TMS. The content of these two structures is the same, though their form is different. Either can be used as input to EBG, to produce a new rule.

Other conversions are also defined; for example, it is traditional in repertory grids that *constructs* (attributes) be listed down the left side of a table, while *elements* (examples) be listed across the top. For ID3, the tradition is just the opposite. Thus a repertory grid must be transposed before it can be used as input to ID3. Since the repertory grid tables and ID3 tables are strongly typed (i.e., there is a special type of table for repertory grid, and one for ID3), the system can make this conversion automatically.

Finally, the pool is also interfaced with the file system on the supporting machine, so that it is possible to save objects for later use. Of course, the objects are actually saved into files as strings of text, but since the pool is interfaced to the file system both on input and output, the student need only be concerned with the conceptual types of the objects. So when an object is saved to a file, a list of all known data objects is presented,

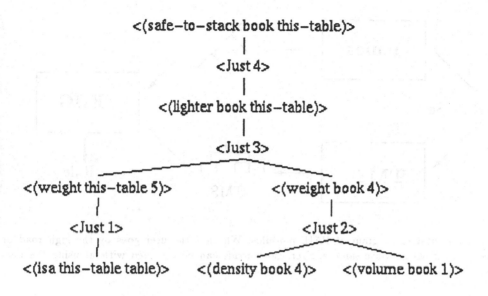

Fig. 3. A browser for the Truth Mainenance Graph produced by running the same rules in figure 1 through a rule engine with TMS. The nodes in the browser are active, and can be specified as IN or OUT by the user.

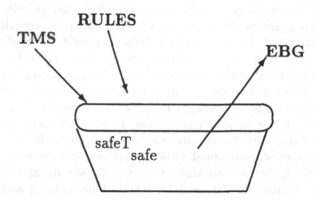

Fig. 4. How the pool mediates between modules. The Rules module places an explanation tree into the pool under the name 'safe'. The TMS module places a TMS into the pool under the name 'safeT'. When the EBG module requests data 'safe' or 'safeT', it is automatically converted into the format of an explanation tree understood by the EBG module.

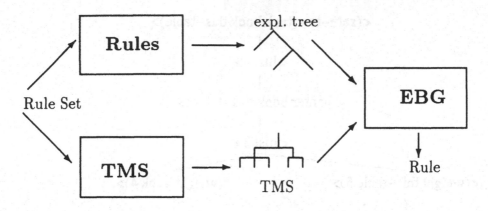

Fig. 5. Commutative diagram of PAIL modules. Whether the user goes by the high road or the low road, EBG gets the same answer. Both roads can be travelled without using the Lisp listener.

from which a selection can be made to enter into the file. When a file is loaded, a table of contents, including the type names, is presented, and these objects become available in the pool. Thus a file is maintained as a set of typed objects, and need never be viewed in its internal form as a lisp source code file.

4 Conclusion

We feel that it has been very useful to have defined the design criteria which should be followed in creating the modules. Since different people are actually programming the various modules then it is particularly helpful to provide each of them with a model to follow. Though it has not always been possible to meet all the guidelines for each module, we were pleased with how many of our design criteria were incorporated into the modules. Figure 6 shows the graphic support for all the modules mentioned in this paper. All of these modules are integrated in the sense described by the examples above, in that output from one module can be used as input to another whenever there is a logical possibility (e.g., using a working memory as a rule set is not supported). We hope that future designers of educational software will benefit from our experience.

The PAIL has undergone an alpha test in a classroom, and has been released for beta testing. In addition to the modules for machine learning and knowledge acquisition described in this paper, the PAIL contains modules for logic programming, natural language, and planning. One of the results we hope to gain from the beta testing is a recommendation for further modules.

5 Acknowledgements

The authors would like to thank the following Graduate Assistants at The University of Zürich who produced the modules for PAIL: Nikolaus Almassy, Stefan Keller, Thomas Rothenfluh, and Thomas Wehrle.

Feature	ID3	EBG	Rules	TMS	Rep Grid
Show abstract concepts in concrete form	Decision trees as browser	explanation browser	rule trace as tree	TMS browser	grid shown as spreadsheet
	tables as spreadsheet		working memory spreadsheet		browser to show clusters
learning by doing	adapt tables to avoid clashes	Can select which point in a tree to generalize from	X	mouse controlled 'in' and 'out' nodes	support for knowledge acquisition lifecycle
'Lispless' capabilities:					
- Edit examples	spreadsheet	NA	working memory spreadsheet	X	Rep Grid interface
- Use Output	change to rules	X	browse rule trace	browse TMS	cluster output with ID3 or other clustering algorithm
- Integrate output to other modules	Trees to rules	X	rule trace to EBG	TMS to EBG	constructs to ID3

Fig. 6. PAIL modules and user support. This table briefly describes the facilities that have been provided for each module (across the top of the figure) to address each requirement (down the left side). 'X' denotes features that have not yet been implemented, NA is 'Not Applicable', features that don't make any sense. References to Rules, EBG and TMS are incorporated as examples in the text. ID3 and Repertory Grid are supported by spreadsheets and a special user interface respectively.

References

1. R. Aiken and D. Allemang. Designing laboratory modules for novices in an undergraduate AI course. In *Proceedings of ACM SIG on Computer Science Education*. ACM, March 1992.
2. Robert Aiken. The New Hurrah! Creating a Fundamental Role for AI in Computing Science Curriculum. *Education and Computing*, 7:119–124, 1991.
3. D. Allemang, R. Aiken, Nikolaus Almássy, Thomas Wehrle, and Thomas Rothenfluh. Teaching machine learning principles with the Portable AI Lab. In *International Conference on Computer Aided Learning and Instruction in Science and Engineering*. EPFL Ecublens, Sept 1991.
4. David Goldberg. *Genetic Algorithms in Search, Optimization, and Machine Learning*. Addison-Wesley, Reading, MA, 1989.
5. J.J. Hopfield. Neural networks and physical systems with emergent collective computational abilities. *Proc. Nat. Acad. Sci.*, 79:2554–2558, 1982.
6. Catherine M. Kitto and John H. Boose. Heuristics for expertise transfer: An implementation of a dialog manager for knowledge acquisition. *Knowledge-based Systems*, 2:175–194, 1988.
7. T. Kohonen. The "neural" phonetic typewriter. *Computer*, 21(3):11–24, 1988.
8. N. Major and H. Reichgelt. Alto: An automated laddering tool. In *Current Trends in Knowledge Acquisition*, pages 222–236. IOS, 1990.
9. Tom Mitchell, Richard Keller, and Smadar Kedar-Cabelli. Explanation-based generalization: A unifying view. *Machine Learning*, 1:47–80, 1986.
10. Tom Mitchell, Paul Utgoff, and Ranan Banerji. Learning by experimentation: Acquiring and refining problem-solving heuristics. In Carbonell Michalski and Mitchell, editors, *Machine Learning*, pages 163–190. Springer-Verlag, Berlin, 1984.
11. D. Rumelhart and D. Zipser. Feature discovery by competitive learning. In D. Rumelhart and J.L. McClelland, editors, *Parallel Distributed Processing*, volume 1, pages 151–193. MIT Press, Cambridge MA, 1986.
12. Guy L. Steele. *Common Lisp, the language*. Digital Press, 1984.
13. Beverly Thompson and William Thompson. Finding rules in data. *Byte*, pages 149–158, 1986.

This article was processed using the LaTeX macro package with LMAMULT style

Interpreting Unexpected User Activity in an Intelligent User Interface

Amir S. Tabandeh

Research Department, Dowty Communications Ltd.
Watford, Herts., WD1 8XH, England.

Abstract. This paper focuses on the interpretation of user activity when a deviation from the projected activity sequence is detected by an intelligent user interface. Following an introduction to the problem, an outline of a proposed interpretation module is described. This module is based on a closed loop control of the interpretation which first describes the difference between the projected plan and the user activity and then using an inference engine attempts to re-adjust its prediction accordingly. The focus of this paper is on the matching process and the appropriate description of its outcome and the construction of an inference engine using this outcome. The proposed architecture is being developed as part of an intelligent user interface to a multimedia catalogue system.

1 Introduction

As the range of users and complexity of computer systems increase, the role of human computer interface is becoming more critical to the overall system performance and usability. One way of coping with this difference in user needs is through incorporation of adaptive user interfaces.

The need for adaptation has pointed towards intelligent user interfaces. An intelligent interface can be seen as an interface which by addressing user's intentions takes actions to dynamically adapt to the user's needs. Adaptive user interfaces have been subject to an increasing attention in the recent years [1, 2].

There are two closely linked categories of uncertainty in the interface between the computer system and the end user:

    i) What action to take (content and form)?

    ii) How to interpret the other agent's actions?

While the former is primarily concerned with the presentation of information from and to the end-user, from the point of view of the computer system agent the latter uncertainty is more complex to resolve since the interface has to establish user's intentions. Knowing user's intention and his/her plans of action allows the system to carry out many actions on behalf of the user without requiring the usual detailed and error-prone commands. The system can also do some error detection and correction. In this way, an intelligent user interface is more than a flexible presentation module.

There are two general ways to address the user's intentions in an intelligent interface:

    i) a top-down approach whereby the user communicates directly with the system concerning concepts, goals and plans which are then mapped into low-level actions necessary to perform the task, for example [3], and

ii) a more popular bottom-up approach in which by observing the user's pattern of activity and based on a model of the world the interface infers user's intentions, forms plans or modifies its models, for example [4].

In this work we are concerned with the latter approach to the design of an experimental intelligent user interface. The remainder of this paper discusses the problem of encountering an unexpected user action when the interface has made hypothesis on the user's plan. Next section will describe the outline of a module of an intelligent user interface, called Chameleon, which addresses this problem. Section 3 describes some experimentation with this module before concluding in section 4. This paper focuses on the matching process and does not cover the presentation engine of the interface.

2 The Chameleon System

Human computer interaction is considered as a trace in state space. At each stage of performing a task there are one or more states which are valid. The sequence of user actions form a user trace of valid states. Even when an intelligent interface has a prediction of the course of action of the user, it should not restrict the user's options at any stage. On this principle, it is possible for a system which has made a hypothesis for user's action plan, to face a user activity which is not compatible with its predicted activity. The causes of this mismatch between actual and expected activities can be:

i) user error; operational (in executing the plan) or conceptual (in understanding the task/plan),

ii) user's change of task or plan of action (a lateral move in the task space), or

iii) system's wrong hypothesis because of misinterpretation of the previous user trace.

These may, to varying degrees, have their roots in the difference between system's model and user's model of the application world. The intelligent interface needs to continuously adjust itself to be in the right context. This section is a brief description of an iterative architecture for interpreting the unexpected user activities focusing on the case where cause of mismatch is due to system's previous misinterpretation of user's intention.

The system, called Chameleon, uses a feedback loop involving the trace of the user's activity and interface's predicted plan of action in order to regulate its prediction, figure 1. Iterations of the loop bring the expected activities in congruence with the actual user activity. Since the interpretation process focuses on the cause of mismatch described above, at the current development stage we take the actual trace as the reference set of data in the closed loop interpretation iteration.

The approach of contrasting the actual and expected user activity has been reported in [5], where it is used to evaluate an interactive system using a neural network approach to pattern recognition.

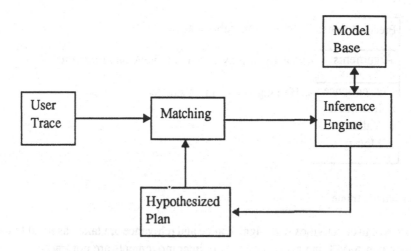

Fig. 1. The closed loop architecture of the interpretation process.

2.1 Representation

In designing a representational scheme for an intelligent user interface there are two prime considerations. Firstly, the representation should be suited to the way it is going to be used in the subsequent stages. Secondly, defining the primitives depends on the adaptation capabilities of the user interface and how fine-grained the analysis is required to be. Representational primitives used in the Chameleon system are "user activity" and "state of interaction", both of which are represented in frame structures. An interaction state is the current state of the interface in terms of the task at hand. Transition in the state space is due to either user or system activities. System activities are transparent to the user and are automatically invoked as soon as their precondition states are met. For example, setting a default automatizes a user activity to a system activity. Some system activities get default parameters from the user model.

It was found to be useful to have both representational primitives since the application interacts with the end user based on both primitives; the system presents to the user the state of interaction, while the user inputs his/her intention through user activities.

The activity's representational frame is depicted in figure 2. The main slots in this frame are Conditions and Consequences of the activity in terms of interaction states. A similar set of representation slots are used in [6] for task analysis in a text editor application. Each activity can have one or more states in its Condition and Consequence segments, in which case those states are connected to each other with logical conjunction or disjunction.

There are two broad classes of activity based on the effect they have on states; an activity could:

    i) invalidate its previous state (Condition) and replace it with a new state, and/or

    ii) modify or supersede the set of its condition state(s).

| Preconditions: <one or more valid states> |
| Arguments: <an activity may or may not have an argument> |
| Consequence: <effecting one or more states>
a. Invalidate
b. Validate
c. Modify |

Fig. 2. An activity frame.

Only those real user activities with significance and relevance are taken as input to the loop, for instance non-modifying actions such as pointer movements are not traced.

Each state indicates what is presented/known to the end user due to system or user activities, and what is known to the system. States may be stable or unstable. A stable state is where neither the system nor the user is expecting an immediate input. It is a logical end/start of a new step. Among the slots in the state representational frame are; its status (leaf, root, non-terminal), presentation (the Graphical User Interface entities managed), sub-tasks that the state can be part of, activities possible due to validity of the state, the activity resulting to its validation, and its stability. Since each state representation includes the list of activities it is connected to, and the activities explicitly contain relevant states in their representation, the activities are linked implicitly through their Condition and Consequence segments.

Finally, in order to be used in the pattern analysis process these representational primitives and classifications are coded using distinct number coding for states and activities.

2.2 Matching

A binary or numeric outcome of comparing the expected and the actual entity is very limited in information that it conveys. In such cases backtracking in the task space is the common use of the matching outcome. In the above architecture, however, the matching process and the nature of its outcome play a central role. At each stage the matching process is based on the set of valid interaction states. The process outcome is a description of similarities and differences between the actual and expected states. In this way the outcome is more indicative of the possible causes of the mismatch between the two comparing data sets, which is vital information for a self-correcting system.

At any stage through the task performance, we have:

a set of valid states due to actual user activities:

$$A : \{A_1, A_2, ..., A_m\}$$

and, a set of expected valid states from the hypothesized plan:

$$E : \{E_1, E_2, ..., E_n\}$$

Describing the matching outcome is divided into two parts of Point Match Description

(PMD) and Line Match Description (LMD). In the continuum of user interaction, PMD may be seen as a transversal match and LMD as a longitudinal match.

The PMD describes the comparison of the two sets based on:

$$PMD : \{A \cap E, A - E, E - A\} \quad \text{where}$$

$A \cap E$:   the set of common states between A and E,

$A - E$:   the set of states present in A and not in E,

$E - A$:   the set of states present in E and not in A.

Hence, both similarities and directed differences between the actual and expected sets of valid states are taken into account in a snapshot comparison of the two sets.

The other match description is the LMD. This is used due to the notion that a pattern of activity is richer in information than a single activity and therefore could assist in gaining a more meaningful indication of the cause of any inconsistency which may have arisen. This extra information is not costless; LMD's appropriate definition and analysis is a complex task.

In the current design, LMD's definition is simplified to be the time series of extracted PMDs together with the hypothesized sub-task (ST) which gave rise to that PMD.

Hence the LMD is:

$$LMD : \{ \{PMD_1 , ST_0\}, \{PMD_2 , ST_1\}, ..., \{PMD_t , ST_{t-1}\} \}$$

Thus, at each stage LMD gives a pattern of PMDs for the states which have been covered. Ideally the LMD should converge to an empty "difference" sub-sets in PMDs.

A design decision regarding LMD is how far back into the trace does an LMD cover. When considering the extent of the user activity trace into the historical data, it should be realised that considering smaller number of activities will reduce the processing effort required while the inference would be based on more unstable input data. On the other hand, user activities over a longer period may be considered which would make the inference a time consuming task. In the current design we have chosen a compromise of the two by using the notion of activity cycle. A cycle is defined as a structured sequence of activity with its start and end nodes being set by the appearance of the top-level menu [7].

2.3 Task Model

A user's actual and desired behaviour is extremely context sensitive. This sensitivity can to some extent be brought into an intelligent user interface through task analysis and modelling. The objective of a task analysis is to describe the activities of the task performer and produce a characterisation of the performer's procedures, related objects and their attributes and actions [8]. In the Chameleon system, direct observation and consultation with system developers are used for task modelling.

The Chameleon system does not consider multi-application tasks. In this system an application may be used to accomplish various tasks, and a given task may be achieved through a number of alternative plans. A plan is an ordered series of valid states.

A task is defined as taking a distinct initial state to a distinct and desired final state using activities for state transition. A task is represented in a finite and explicit manner by a directed graph structure whereby nodes (states) may be shared between two or more paths (plans) of varying lengths. Similarly, a tree structure for representing a task model has been used in [9]. There are limitations with this scheme in that each task/plan has to be thought of at the design level and therefore there are generally insufficient task/plan patterns for the system, and that the size of the search space may be such that this method would become impractical. Also the construction of such a task model is a tedious exercise.

Sub-tasks are defined based on the notion of stable states. A task is thought of as a sequence of sub-tasks. We assume that a given state is not traversed in the state-space of a sub-task more than once.

2.4 Inference Engine

The inference engine takes as its input a match description in terms of PMD and LMD. Based on a framework given by the three sub-sets of PMD and also the frame representation of a state, a set of subjective heuristic rules which are represented in a production system are defined. The definition of these heuristics consolidates the model base (task and user models) and the match description. The main knowledge base used by the inference engine is its task model. At this stage of the work only a coarse user model is incorporated covering attributes on task history and preference of the user regarding the application.

The precedent of the rules are in terms of the slots of the match description frame, and their actions are constraints on the expected plan. Each of these constraint generation rules has a numeric quantifier of certainty associated with it in order to control the effectiveness measure of its action(s). The inference engine optimizes these constraints and makes a fresh hypothesis of the user's intended plan. Essentially the inference engine is minimising the two difference sub-sets in the PMD while maximising the similarity sub-set.

The inference engine initially analyses the match description using some general heuristics and decides how to proceed. For example there are some detected user activities which do not call for a change in the predicted plan but are used to update user model, such as; re-sorting data objects based on an alternative index, or requesting help at any level along the task plan. Other decisions to modify the user model can be taken due to an analysis of the PMD. For example, if a state generated by a system activity is present in the {A - E} subset, the default in the user model concerning the user activity needs to be modified.

The current use of LMD by the inference engine is mainly on influencing the effectiveness of constraints generated by the PMD analysis. For example an increase in the size of {A∩E} relative to {A∪E} strengthens the hypothesized sub-task. Also LMD is used to deal with persistent elements (states) in one of the PMD difference sub-sets. Further, since LMD keeps the track of valid states, it helps determining where on a task continuum the system believes the user is.

Initial hypothesis of a user plan is wholly or partly based on the user model since at the initial stages of traversing through the task, the match description is not rich enough in information to be used conclusively by the inference engine.

If there is no significant difference between the input activity and the predicted one, the plan used in the previous iteration is reinforced and is used in the next iteration as the basis of prediction.

3 Experiments

Preliminary experiments are being carried out to analyse the proposed matching scheme and the inference engine in the feedback loop architecture.

The test application being used is a multimedia catalogue system used by suppliers to give clients remote access to information on their products. The catalogue system is implemented as a client-server model using wide area network. Users of the catalogue have varying degree of familiarity with the system and hence the incorporation of an intelligent user interface is being studied to make the catalogue more efficient to use. The implementation of the test system is under SUN/UNIX/C using OSF/Motif Graphical User Interface environment. The development of the intelligent interface of the catalogue system has been part of the development process of the system.

One experiment is based on the task of a user wishing to keep a copy of a known data object. The root node of this task is the state where the user is presented with the top level menu of the client program without a copy of the target data object on his/her machine. The leaf node of the task is the above but with a copy of the data object on the client machine.

This involves the sub-tasks of choosing a remote catalogue server site, connecting to the site, logging in to the catalogue system, retrieving the object of interest and finally retaining a copy on the user's machine. Each of these comprises three to five activities, and starts and terminates at a stable state. At most states it is possible for the user to take a lateral step from the main task.

At some states along the task a user activity may alter or modify the state. For example when presented with a list of data objects sorted by a certain index, the user may select an object and alter the state or may re-sort the list based on an alternative index and thus modify the set of valid states, while remaining at the same stage of the task.

An example of a system activity in the course of this task is the automatic display of object metadata together with its image once an object is retrieved. These two default outputs of an object retrieval are based on the user model. Alternatives could be any combination of these two and audio and textual outputs.

The results and performance of these experiments are being used to refine the inference engine, in particularly its rule base.

4 Concluding Remarks

This paper has concentrated on the problem of interpreting a deviation between actual user activity and that expected by an intelligent user interface. A feedback scheme involving the predicted and actual user activities has been proposed, based on the belief that there are high potential benefits to be gained by a mechanism of monitoring the hypothesis made by an intelligent interface soon following the prediction.

An important feature of this feedback loop is that the outcome of comparison of the actual and predicted data sets is in a structured and descriptive form. This descriptive outcome provides a rich framework to build a powerful inference engine to analyse the causes of any mismatch between actual and predicted user plan.

Further work and experimentation are concentrating on the enhancement and optimization of representation schemes and the heuristic rules in the inference engine, and also on designing a performance evaluation scheme mainly based on the convergence rate of the LMD to a stable maximum similarity sub-set in PMD.

Acknowledgement

This work is part of the Research and development in Advance Communication in Europe programme (RACE I) under project "CAD/CAM for the Automotive Industry in RACE" (R1079).

References

1. J.W. Sullivan, S.W. Tyler (eds.): Intelligent User Interfaces. ACM Press, 1991
2. M.H. Chignell, P.A. Hancock: Intelligent Interface Design. In: M. Helander (ed.): Handbook of Human-Computer Interaction. Elsevier Science Pub., 1988
3. J. Bonar, B.W. Liffick: Communicating with High-Level Plans. In: J.W. Sullivan, S.W. Tyler (eds.): Intelligent User Interfaces. ACM Press, 1991
4. S. Treu: Recognition of Logical Interface Structures to Enhance Human-Computer Interaction. Int. J. of Pattern Recognition and Artificial Intelligence, Vol.3, No.2, pp.217-236, 1989
5. J. Finlay, M.Harrison: Pattern Recognition and Interaction Models. In: Diaper et al (eds.): Human Computer Interaction - INTERACT'90. Elsevier Science Pub., pp.149-154, 1990
6. R.M. Young, J. Whittington: Using a Knowledge Analysis to Predict Conceptual Errors in Text-Editor Usage. Proc. of Computer Human Interaction'90, April 1990, ACM Press, pp.91-97, 1990
7. M.D. Harrison, C.R. Roast, P.C. Wright: Complementary Methods for the Iterative Design of Interactive Systems. In: Salvendy and Smith (eds.): Designing and Using Human-Computer Interfaces and Knowledge Based Systems. Elsevier Science Pub., pp.651-658, 1989
8. D. Browne, M. Norman, E. Adhami: Methods for Building Adaptive Systems. In: D. Browne, P. Totterdell, M. Norman (eds.): Adaptive User Interfaces. Academic Press, pp.85-130, 1990
9. J. Junger, G. Bouma, P. Letanoux: Intelligent User Interface for a Conventional Program. In: Diaper et al (eds.): Human Computer Interaction - INTERACT'90. Elsevier Science Pub., pp.815-820, 1990

Yet Another Knowledge–Based Front–End–To–Database Project

Jaroslava Mikulecká

Department of Computer Science
Faculty of Electrical Engineering, Slovak Technical University
811 19 Bratislava, Czechoslovakia

Peter Mikulecký

Department of Artificial Intelligence, Comenius University
842 15 Bratislava, Czechoslovakia

Abstract. The aim of the paper is to describe an approach towards *knowledge-based environments to support database design*. The main stress is put towards the role of a restricted natural language facility in *advice-giving systems*, aiming to be expert helpers to the users of database systems. The paper brings also a description of possible solutions in design and implementation of a natural language processing module as a part of a knowledge-based advice-giving environment to database systems.

1 Introduction

Nowadays, commercial database products are being heavily utilized by users who previously belonged to the category of casual or occasional database users. As a result, the practice of creating personal databases is gaining popularity. In this situation the database design process is shifted away from database administrator to the user. Therefore, there is increasing awareness of the value of design tools providing an expert assistance in the specification, usage and maintenance of data-based information systems. This expert assistance needs also the usage of appropriate user interfaces.

As we have stressed earlier [MIK 86], one approach could lie in computer simulation of the appropriate guidance given by an experienced helper, e.g., an expert in effective usage of given database system. Such helper can play the role of an active assistant say, in guiding the novice while using a complex database system for the first time. Some attempts in this direction already have been done [FUT 86].

We can identify six main phases of the *database design process*: requirements collection and analysis, conceptual database design, choice of DBMS, data model mapping (logical database design, or scheme-conversion), physical database design, and database implementation.

In our approach, we shall concentrate on the *phase of data model mapping (scheme-conversion)*, i.e., the replacement of a scheme by another scheme having the same information content. Scheme-conversion is a means of database design: a scheme is first designed in a higher-level database model and then translated into a lower-level model which is supported by the available DBMS). Scheme-conversion is usually done in order to impose implementation restrictions needed because of the database management system, and usually is performed manually by the database designer.

We have designed a tool with a restricted natural language interface as a part of a knowledge-based advice-giving environment which is able of intelligent assistance to the database designer in the scheme-conversion phase of database design. First experience from an experimental implementation will be presented.

2 Natural Language Processing Module

2.1 Issues

There has been a general recognition of the *acceptance of natural language* to successful advisory interfaces. However, a lot of systems have been described in the past which appear to have little or no natural-language capability.

Some results [WIL 84] are encouraging for the possibility of developing empirically adequate but computationally limited natural-language capabilities. There is, however, a *conflict between the natural and command language approaches*. The obvious possibility, how to get a description of user's information needs is to infer the user's skill level from actions and responses. A novice will probably not pose very technical queries, but an experienced user might use relatively general vocabulary and therefore get advice at a level too elementary for him. Some authors bring arguments in favour of specialized command languages, where the main argument is their efficacy. In our opinion, the ideal case could be a combination of natural language and command language approach, resulting in a *restricted problem oriented vocabulary*, which reflects the actual user's skillness and/or the level of user's expertise. Our recent experience is that even a vocabulary of a few hundreds of words can be a basis for a useful and sensible linguistic interface. On the other hand, the people use a surprisingly great variety of words to refer to the same thing. A further linguistic based research in this direction could be useful. Some strong and interesting arguments in favour of restricted natural language interfaces are recently presented in [GUI 91].

In our project, we decided to follow the direction towards effective linguistic user interfaces based on results in natural languages comprehension. Our goal has been to achieve some linguistically interesting results as well, which could be in the case of flexive languages (like the Slovak or Czech language) original and useful. Such results have been achieved as students' projects also in our group at the Comenius University (see [PAL 90], [MAR 91]). The ultimate goal is to develop a prototype of a natural language processing module as a part of more general-purpose knowledge-based user interface.

2.2 English Language Processing Module

The recent development of the natural language processing facility in our project has resulted in a design and experimental implementation of an English language processing module described in [MAR 91]. The underlying user systems are usual available database systems, therefore the discourse area for the module vocabulary had been the area of designing a particular (personal) database. However, the implementation of a module for Slovak language remains as our ultimate goal, mainly because of the challenging difficulty of the problem.

The proposed natural language processing system consists of the following modules: morphologic analysis, syntactic-semantical analysis, inference machine, knowledge-base, and vocabulary.

The designed morphologic analysis reduces partly the size of the vocabulary, speeds up the search of words in the vocabulary, reduces the syntactic ambiguity, and recognizes grammatical features of a word. When a non-determinism appeares during the syntactic analysis, the morphologic process, using the suffix, can unambiguosly determine the type of the successing non-analysed input word. This will eventually decide about the next step of the syntactic analysis. However, the morphology will contribute to the efficiency of the syntactic process only in the case the word belongs to an unique word-class. In the case of a conversion of the unknown word, the syntactic process have to decide itself non-deterministically.

Rewriting rules are used for describing the syntax of an English sentence. The structure of an English sentence consists of *basic (structural) units*. A basic unit is either a *word*, a *surrounding*, or a *clause surrounding*. A surrounding is the structure for describing as well as better understanding a given sequence of words. Well defined sets of rewriting rules are used in order to define all the types of surroundings of particular basic units. These surroundings are hierarchically ordered and in all particular cases used sequentially.

A collection of inference algorithms has been designed and implemented using Pascal with the goal to implement experimentally a system capable of understanding restricted English sentences. The implemented system can be used for dialogues of the type described in the following chapter. Its further development is still in progress and will be published later.

3   Results in a Database Environment

We have concentrated ourselves on the data model mapping or scheme-conversion phase of the database design process. The reason for our choice lies in the fact that this phase is relatively difficult for an ocassional user, but on the other hand it is quite well ellaborated from the algorithmic point of view, and the relevant vocabulary could be considered as sufficiently narrow to demonstrate our approach of a restricted natural language processing. In our experiment, the vocabulary has consisted of words relevant to the scheme-conversion algorithm described in [RIS 88], while the knowledge base included the knowledge expressed by the algorithm as well as some related knowledge on conceptual and relational models. In what follows, we describe a part of possible user-system dialogue, which could be processed by our system.

SYSTEM: Which database management system will you use for implementing your database?

USER: FoxBase.

SYSTEM: O.K. Conceptual model should be mapped into relational model. Follow my next advices.

Step 1. Choose a key for every regular entity type.

USER: I don't understand what is the key.

SYSTEM: A key may be an attribute which is 1:1, time invariant and total.

USER: There isn't such an attribute in every regular entity type... etc.

From the dialogue fragment we can see that the system is actually mimicking a restricted human dialogue on an appropriate professional level. This is enabled by the balance between the vocabulary restriction (because of rather narrow discourse area) and the deepenes of the relevant knowledge accumulated in the knowledge base and used by the system. The knowledge can be organized hierarchically and used in the system answers (explanations) gradually, as the system uses in each case a deeper knowledge about the particular subject. In the dialogue generation, a number of previously prepared templates expressing the knowledge of the system is used. The most answers of the system are actually composed from appropriate pieces of knowledge in the form of such templates. This feature is causing the system to be able of rather fast response.

4 Conclusions

A further development of our knowledge-based advice-giving user interface we see in implementing a working advice-giving system over a widely used commercial database system (e.g., the ORACLE or SYBASE).

References

[FUT 86] Furtado, A. L., Moura, C. M. O.: Expert helpers to data-based information systems. In: *Expert Database Systems* (L. Kerschberg, ed.), Benjamin and Cummings, 1986, pp. 581-596.

[GUI 91] Guindon, R.: Users request help from advisory systems with simple and restricted language: Effects of real-time constraints and limited shared context. *Human-Computer Interaction* 6 (1991), pp. 47-75.

[MAR 91] Martinkovič, M.: *Natural Language Processing in Advice-giving Systems.* Diploma Thesis, Dept. Computer Science, Comenius University, Bratislava, 1991 (in Slovak).

[MIK 86] Mikulecký, P.: On knowledge-based software tools. *Computer Physics Communications* 41 (1986), 397-401.

[PAL 90] Páleš, E.: Co-operation of syntax and semantics in flexive languages. *J. Exp. Theor. Artif. Intelligence* 2 (1990), pp. 1-24

[RIS 88] Rishe, N.: *Database Design Fundamentals.* Prentice-Hall, Englewood Cliffs, N.J., 1988

[WIL 84] Wilensky, R., Arens, Y., Chin, D.: Talking to Unix in English: Overview of UC. *Comm. ACM* 27 (1984), 574-593.

Fault Diagnosis based on Simulation Models

Frank Plaßmeier, Dr. Reimund Küke, Dr. D. Exner, K.F. Lehmann

ERNO Raumfahrttechnik GmbH
Bremen
Federal Republic of Germany

Abstract. This paper describes the SIMEX (Simulation based Expert System Tool) system, which performs fault localization by exploiting the causal and behavioral information contained in simulation models for the generation and verification of fault hypotheses. The system has been developed for space applications in order to improve ground based mission control. It is, however, also applicable for process monitoring purposes. The overall system concept, required model information and the hypotheses generation strategy will be outlined. A sample application will be briefly described. Project status and related research will be sketched.

Keywords: fault localization, expert systems, simulation models, mission control, process monitoring

1 Introduction

Controlling a technical system (e.g. a spacecraft or a power plant) requires knowledge of the structure of that system and the interactions among its components and devices. Process control systems provide relevant data and state information and raise alarms in the case of threshold violations, but the interpretation is left to the operator. It is his task to perform fault localization, i.e. to find the actual cause of alarms or abnormal sensor readings and to perform corrective action.

In the case of mission control, the TM (telemetry) data of a spacecraft usually consists of measurements (e.g. voltages, currents, pressures or temperatures) and state information (e.g. switch states). Some systems also transmit exception messages such as threshold violations which directly support the detection of malfunctions. However, the localization of faulty components or devices is normally not supported, since it requires knowledge of the overall dependencies within a spacecraft. Nowadays, the fault localization task is usually performed by means of paper based contingency procedures, which contain a shallow level representation of this knowledge and describe how to localize and handle preconceived malfunctions.

Due to the importance of fault localization, computer based support for this task is desirable. This support could be achieved by making use of the shallow level knowledge that forms the contents of contingency procedures [16]. However, the utilization of the deeper level knowledge on which these procedures are based is more promising, since in this way the tedious and error prone transformation process between the two levels can be avoided [1, 2]. An even greater advantage could be achieved by a closer integration of the knowledge acquisition task with other design tasks, especially if it were possible to benefit from machine readable system descriptions or models of the spacecraft under consideration and from the software tools for editing and maintaining these models. Similar considerations apply to the development of fault localization support systems for process control applications.

Whilst it cannot be expected that computer integrated support for the spacecraft design process in its entirety will be available in the near future, systems for the development of simulation models can be seen as a suitable starting point. This has lead to the idea of utilizing simulation models for fault localization tasks, since simulation models can be used to predict the behavior of a technical system (e.g. a spacecraft) and therefore support the detection of faults and the validation of fault hypotheses by means of comparing the results of appropriately parameterized simulation runs with observed data. Furthermore, simulation models contain information about the causal dependencies within a technical system, which can be used to generate hypotheses as to the reasons for observed discrepancies by means of upstream tracing.

Whereas the predictive information can be obtained from the execution of a simulation model, the causal dependency information is usually not easy to retrieve since it is hidden in the code of the simulation model. This implies that the common utilization of simulation models for both fault localization and simulation purposes requires either following certain implementation conventions for the simulation model or applying a tool which captures the dependency information automatically at a sufficiently detailed level.

2 Description of the SIMEX System

Within the framework of the Columbus Project the simulation system CSS (Core Simulation Software), which includes the interactive graphical model editing tool MDE (Model Development Environment), is available. Therefore it has been decided to develop the prototype of a simulation based fault localization expert system called SIMEX, that is intended to be integrated with these tools. This system will now be described.

After explaining the overall system concept, the question of how to model a technical system in a way that supports fault localization will be discussed. Then the way in which this information is applied in order to generate fault hypotheses will be outlined.

2.1 System Concept

The basic idea behind the SIMEX system is to acquire a description of the causal dependencies within a spacecraft from MDE models (during the model development phase) and to employ the CSS kernel and the MOCS (Model Observation and Control System) to obtain predictive information describing the expected behavior of that spacecraft (see figure 1).

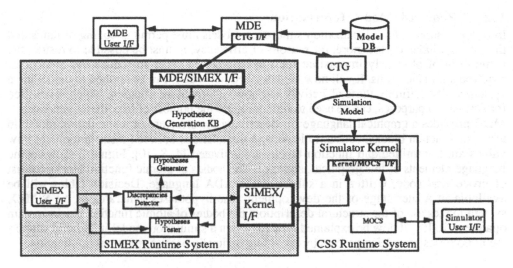

Figure 1: *Integration of the SIMEX system into the MDE/CSS environment*

It is intended to integrate the SIMEX system into the CSS/MDE environment in such a way that a single interface allows for the development of models which serve for both simulation and fault localization purposes. This approach offers considerable advantages since on the one hand a further increase of the diversity of tools applied during the system design process is avoided and on the other hand the simulation models are applicable for other tasks such as FMECA (Failure Modes, Effects and Criticality Analysis) as well [3].

The SIMEX system will perform the fault localization task by following a generate and test strategy consisting of three steps:

- looking for discrepancies between expected and observed spacecraft information (e.g. TM data)
- following the detection of discrepancies, generating hypotheses as to possibly faulty components of the spacecraft, if discrepancies are detected
- testing of hypotheses

The first step is required for detecting the occurrence of faults. It is performed by comparing the spacecraft information with the results of simulation runs. Depending on the onboard capabilities of the spacecraft, this step is sometimes supported by the automatic generation of exception messages which directly pinpoint discrepancies. The second step aims at finding possible reasons which may have caused the detected discrepancies and the third one at finding out the actual reason by comparing the observed spacecraft information with the result of simulation runs, which are this time parameterized according to the fault hypothesis under consideration.

Whereas the first and the third step directly employ the predictive information contained in simulation models (i.e. the transfer function from stimuli to measurements), which can be obtained by executing them, the second step requires additional information describing the causal dependencies between measurements and stimuli. This is usually not easy to retrieve since it is hidden in the code of the simulation model. The situation is different in the case of MDE, since this tool captures a structural description of the simulation model.

2.2 Required Model Information

In order to support fault localization simulation models must permit encoding of faults and their causal links. Furthermore, for reasons of efficiency, it must be possible to restrict the generation of obviously impossible fault hypotheses in order to reduce the number of required simulation runs. It will now be described how MDE can be applied to modelling a system whilst fulfilling these objectives and how additional information, which is not used for simulation purposes, can be used to further reduce the number of hypotheses generated.

MDE provides a graphical language for describing the structure of a simulation model in terms of function blocks, constant blocks and the relations between them in terms of how values are transmitted and function blocks are activated [4, 8, 10]. Figure 2 shows some language elements of this graphical language. The bodies of atomic function blocks consist of procedural code, written in a subset of the ADA language. Decision tables will be available in a later stage of the development. The hypotheses generator of the SIMEX system makes use of the structural description, the bodies of atomic function blocks remain opaque to it. It will now be explained how to design a simulation model in order to support the hypotheses generation process by providing optimally precise structural information.

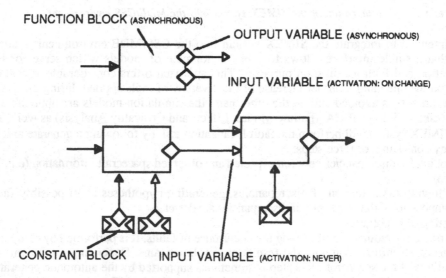

Figure 2:   Sample MDE graphical language elements.

MDE offers a considerable amount of freedom for designing a simulation model. For a model which is intended to be used by the SIMEX system, certain restrictions have to be followed. First of all, a unique way of describing faults is required. Currently, faults have to be encoded as output variables of constant blocks for which write access is permitted (simulation parameters). Furthermore, for efficient hypotheses generation, it is recommended to put as much information as possible into the structure of the model instead of hiding it in the body of atomic function blocks. Three points are particularly important:

- Input and Output Variables should not be combined in one function block, if they are never used simultaneously for a computation, since this way none of the real-world dependencies are introduced into the model.

- The activation mode of input variables of function blocks can be used to prune the search space for hypotheses. An input variable with activation mode "never" needs to be checked only if the execution of the function block is expected.
- Functional Symmetry can be expressed by means of referencing generic function blocks so allowing further restriction of the search space. Furthermore, the use of referenced function blocks reduces the size of the simulation code and facilitates the maintenance of simulation models.

Although taking into account these points increases the efficiency of the hypotheses generation process, one problem remains. Not all dependency information can be obtained from the structural description of the simulation model since dependencies may change dynamically. Therefore it is useful to supply additional information which captures these dynamic changes for guiding the hypotheses generation process. The current implementation of the SIMEX system permits the definition of so called fault exoneration rules which locally restrict the search space for particular function blocks based on information about expected or observed values of its input or output variables. Due to their local character, rules and rule sets are of a concise nature and easy to check for plausibility and to maintain, since they correspond only to the function block for which they are defined.

By means of this approach the number of hypotheses generated can be considerably reduced. The hypotheses generation process will now be outlined in further detail.

2.3 Hypotheses Generation Strategy

The hypotheses generation process is invoked after the initial detection of discrepancies. It returns a list of simulation parameters, each of which is associated with the discrepancies for which the parameter may be responsible. This forms the input for the hypotheses testing process.

The process itself is performed in two steps:

- At the beginning, the search space is restricted by exonerating the input variables of function blocks which either have no parameter as predecessor (neither direct nor indirect) or can be excluded as suspects due to their activation mode or functional symmetry.
- Then a breadth first search backwards from the observed discrepancies is applied in order to find those simulation parameters which may be responsible for the discrepancies. This search is restricted by the result of the previous step, the application of fault exoneration rules and by excluding those candidates which are associated with a set of discrepancies that is a true subset of the set of discrepancies associated with another candidate. A candidate may be either a parameter or the output variable of a function block.

The search terminates if no new candidates can be generated. The hypotheses generation process is based on the so called single failure assumption but does have a limited multiple failure detection capability since it can recognize whether or not there is any parameter that can explain all the observed discrepancies. If no such parameter can be found, more than one parameter must be responsible for the observed discrepancies, hence more than one fault must be present.

3 Sample Application

A prototype of the SIMEX system has been tested against a simulation model of a part a high frequency transmission system, a so called repeater [5, 6]. This device constitutes the payload of a telecommunication satellite.

After a brief description of this device an example of fault hypotheses generation for a part of the model will be given.

3.1 Description of the System under Consideration

The considered KA-band repeater receives high frequency signals at 30 GHz and sends them adequately amplified to the ground at a frequency of 20 GHz. The incoming signals are processed by a preamplification chain and a channel amplification chain. Four different paths can be used for this process, depending on the position of four waveguide switches. For the up/down converters, which are part of the amplification chains, carrier signals are generated by a carrier supply. In order to allow for a continuous tracking of the satellite, a beacon signal is generated by a beacon generator and transmitted. The following figure shows the structure of the KA-band repeater:

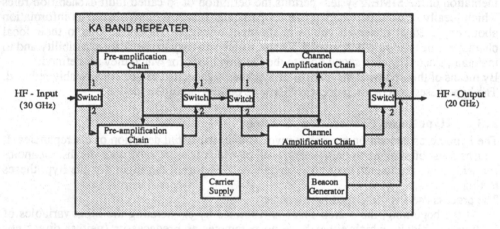

Figure 3:   Repeater KA-Band

The repeater has interfaces with the TTC (telemetry/telecommand) subsystem and with the EPS (electrical power supply) subsystem of the satellite. The TTC subsystem is responsible for interpreting and transmitting commands e.g. the commands for switching amplifiers "on" or "off" or for selecting paths for the high frequency transmission process. The EPS subsystem provides the power for the secondary voltages which are required for the operation of the amplifiers and waveguide switches of the repeater.

3.2 Example: Waveguide Switching

The following figure shows the part of the repeater model which encompasses the mechanism for switching the high frequency transmission between different amplifier chains. Its purpose is to ensure a synchronous switching of two waveguide switches, one at the beginning and one at the end of two parallel amplifier chains, in accordance with the switch position for the related telecommand.

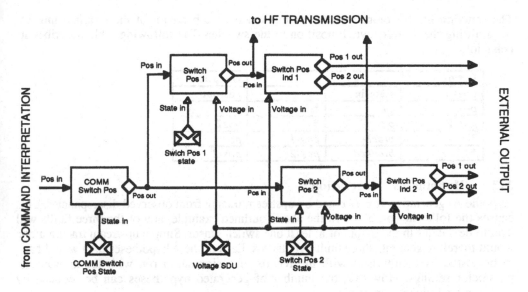

Figure 4: Waveguide Switching Mechanism

Waveguide switching may fail due to blocked switches, malfunction of the terminal unit for transmitting the desired switch position or unavailability of the secondary voltage for operating the switches. Now the function blocks for the waveguide switching mechanism will be described to show the effects of each of these faults.

The function blocks SWITCH-POS_i implement the behavior of the switching relais. Both function blocks have identical behavior and therefore reference a generic function block implementing this behavior, which is described by the following table (* means that the value does not affect the output):

Input				Output
pos-in	state-in	volt-in	pos-out	pos-out
pos-1	ok	available	*	pos-1
pos-2	ok	available	*	pos-2
*	blocked	*	pos-1	pos-1
*	blocked	*	pos-2	pos-2
*	*	not-available	pos-1	pos-1
*	*	not-available	pos-2	pos-2

The function blocks SWITCH-POS-IND_i implement the behavior of the sensors which report the switch positions. Both function blocks have identical behavior and therefore reference a generic function block implementing this behavior, which is described by the following table:

Input				Output	
pos-in	volt-in	pos1-out	pos2-out	pos1-out	pos2-out
pos-1	available	*	*	yes	no
pos-2	available	*	*	no	yes
*	not-available	*	*	no	no

The function block Comm-Switch-Pos implements the behavior of the terminal unit for transmitting the desired switch position to the switches The following table describes its behavior:

Input			Output
pos-in	state-in	pos-out	pos-out
pos-1	*ok*	*	*pos-1*
pos-2	*ok*	*	*pos2*
*	*blocked*	*pos-1*	*pos-1*
*	*blocked*	*pos-2*	*pos-2*

3.3   Hypotheses Generation

Hypotheses generation is performed by upstream tracing from observed discrepancies. This causes the following problem: In the above outlined example, any of the three faults will lead to deviations in the output of at least one switch sensor. Simple upstream tracing alone would therefore generate three fault hypotheses. Each of these hypotheses then would have to be tested by comparison with the results of a simulation run with the appropriate parameter settings. However, the number of generated hypotheses can be reduced by applying the following methods:

- Complete Subsumption: Based on the single failure assumption, blocked switches can be excluded as fault assumptions if the output of both switch sensors show discrepancies since a malfunction of the command transmission or the unavailability of the secondary voltage for the switches can explain all discrepancies itself, whereas a single blocked switch cannot. If, additionally, a discrepancy of the output of the secondary voltage is observed, a malfunction of the command transmission can also be excluded. The required information can be obtained directly from the structural dependencies within the simulation model by tracing the connections of the simulation parameters with the external output.
- Activation mode: Provided that the expected value of the desired switch position has not changed, blocked switches and a malfunction of the command transmission can be excluded since the parameters encoding the state of the command transmission and the switches have only passive influence on the function blocks with which they are related. The activation modes can be directly obtained from the simulation model.
- Functional Symmetry: If for only one switch sensor discrepancies are observed both a malfunction of the command transmission and the unavailability of the secondary power can be excluded as fault hypotheses since a correctly working similar functional chain with similar dependencies exonerates the related parameters. The required information can be obtained from the simulation model, since both the dependency information and the information about the referencing of identical function blocks is contained in the model.
- Exoneration Rules: By specifying a rule for the referenced function block for switch sensor which states that for identical values of the output variables "pos1-out" and "pos2-out" upstream tracing of the input variable "pos-in" can be omitted and another rule which states that in the case of non-identical values of the output variables pos1-out and "pos2-out" the upstream tracing of the input variable "volt-in" can be omitted, the search space for upstream tracing and the number of generated hypotheses can be reduced. This information, however, is not normally contained in the simulation model.

In the example, applying these methods allows for completely avoiding the generation of false hypotheses. For more complex applications this may not be the case, but a considerable reduction is still expected.

4   Project Status

Prototypes of the hypotheses generator and the hypotheses tester have been implemented in Common Lisp and CLOS (Common Lisp Object System). The user interface provides a test environment for the prototypes and is implemented in CLIM (Common Lisp Interface Manager). The system includes a simple simulator for the generation of test scenarios and the detection of discrepancies and the verification of hypotheses. It runs on Symbolics 36xx and Ivory-Machines and on Sun SPARCstations. The system has been successfully tested against the above mentioned KA-band repeater. The next step will be the development of the interface between MDE and SIMEX. A data exchange format has already been defined. The interface between SIMEX and CSS based on MOCS will then be developed.

Mid term goals are investigations concerning the automatic selection and verification of test and recovery actions and a more elaborate representation of faults. Furthermore, the possibility of automatically generating fault exoneration rules from MDE decision tables will be examined.

These features will be evaluated using a more complex MDE/CSS simulation model capturing a subsystem of EURECA (European Retrievable Carrier)

As a long term goal the idea of developing an integrated system development support system should be kept in mind since this way an even better support for the development of both simulation models and fault isolation and recovery knowledge bases can be accomplished. Furthermore the application of the SIMEX system for process control applications will be considered since the underlying technology is not restricted to space applications.

5   Related Research

The SIMEX system takes the approach of combining upstream tracing for hypotheses generation with simulation for discrepancy detection and hypotheses validation. It shares this approach with systems developed by Dvorak and Kuipers on one hand and Pan on the other. [11, 12]. These systems, however, use specifically developed qualitative or semiquantitative simulation models, whereas the SIMEX system is intended to use existing simulation models. The capability of using existing data and utilizing existing environments is considered very important from the application point of view [15]. This is also the reason for choosing a different approach to e.g. Reiter, Böttcher or Scarl, Jamieson and Delaune [7, 13, 14], which are based on constraint satisfaction and require that the inputs of components can be inferred from their outputs.

6   Conclusions

MDE can be seen as a suitable tool for the development of simulation models which can be used as hypotheses generation knowledge bases since it allows the retrieval of structural information about simulation models. In order to ensure sufficient fault localization, sufficiently detailed information has to be contained in the model. This requires certain guidelines to be followed when developing a model. It is especially important to describe as much dependency information as possible by means of the model structure and as few as possible in the code. Some additional information has to be supplied to capture dynamically changing dependencies.

The benefits resulting from the utilization of simulation models are twofold: A further increase of the diversity of tools applied during system design can be avoided and the simulation models can be expected to serve other purposes.

The SIMEX system can be seen as a promising approach for future mission control support systems and may be applied for other purposes (e. g process control).

References

1. Heher D & Pownall P 1989, StarPlan: A Model-Based Diagnostic System for Spacecraft, *Third Annual Workshop on Space Operations Automation and Robotics*, Houston 1989, pp. 117-121
2. Haziza M 1988, An Expert-System Shell for Satellite Fault Isolation based on Structure and Behaviour, *Workshop on AI Applications to Space Projects*, Noordwijk1989
3. Plaßmeier F & Lehmann K F 1991, Utilization of MDE for FDIR Knowledge Acquisition, ERNO internal report
4. Exner D 1990, Core Simulator Software Design Definition Report, COLUMBUS Report
5. Plaßmeier F 1990, System Analysis and Knowledge Representation Requirements for the DFS Repeater, ERNO internal report
6. Lehmann K F 1991, Simulationsmodell DFS-Repeater für SIMEX (Simulation Model DFS Repeater for SIMEX), ERNO internal report
7. Böttcher C 1991, Die Verwendung multipler Modelle in der modellbasierten Diagnose (Use of multiple Models for Model Based Diagnosis), Diploma Thesis, University of Karlsruhe
8. Eiteljörge J 1990, System Fault Simulation for Expert System Verification, *Simulators for European Space Programmes Workshop*, Noordwijk 1990
9. Eiteljörge J 1988, Model Development Environment (MDE) for functional Simulators, *International Symposium on Space Software Engineering*, Turino 1988
10. Eiteljörge J 1988, Graphical Model Development Environment (MDE) for functional Simulators, *MMI Workshop*, Noordwijk 1988
11. Dvorak D & Kuipers B 1991, Process Monitoring and Diagnosis, IEEE Expert, June 1991
12. Pan Y C 1984, Qualitative Reasonings with Deep-Level Mechanism Models for Diagnoses of Dependent Failures, PhD Thesis, University of Illinois at Urbana-Champaign,
13. Reiter R 1985, A Theory of Diagnoses from First Principles, University of Toronto and the Canadian Institute for Advanced Research
14. Scarl E & Jamieson I R & Delaune C I 1985 A Fault Detection and Isolation Method Applied to Liquid Oxygen Loading for the Space Shuttle, *IJCAI-85, Los Angeles 1985*
15. Milne R 1991, Model-Based Reasoning: The Applications Gap, IEEE Expert, December 1991
16. Kellner A & Küke R 1990, Failure Detection, Isolation and Recovery: Current and Future Developments, DGLR Jahrestagung 1990

Real-time Fault Diagnosis - Using Occupancy Grids and Neural Network Techniques

Amit Kumar Ray and R.B.Misra

Reliability Engineering Centre
Indian Institute of Technology, Kharagpur - 721 302, WB, India.

Abstract. This paper presents a methodology for real-time fault diagnosis of manufacturing systems using occupancy grids and neural network techniques. The main advantages of the system over other existing methods are its ability to capture imprecise and time dependent information, ability to accommodate nonlinear relationships, ability to learn and acquire knowledge automatically. A case study related to real-time milling machine fault diagnosis is discussed. The paper also discusses the problems with the proposed method and the future research directions.

Keywords. *Neural networks, Occupancy grids, Information measure, Back-propagation algorithm.*

1 Introduction

In today's highly automated manufacturing plant, on-line fault diagnosis and monitoring is an integral part of the quality control process which ensures that the products coming out of the shopfloor meet design and performance standards. The objective of this work is to automate the diagnostic task by providing computer-based automatic learning schems that are able to code operational knowledge and to use this knowledge for synthesizing optimal strategies for machining operations. In this paper, we proposed a new architecture for real-time manufacturing fault diagnosis using occupancy grids and neural networks.

Neural network consist of a number of simple neuron like processing elements call node or cell. The nodes are inter connected, and strengths of the interconnections are denoted by parameter called weights. The knowledge is internally represented by the values of the weights and topology of the connections. Learning is accomplished by adjusting these weights step by step (by minimizing some objective function). These networks can learn and adapt themselves to inputs from actual processes, thus allowing representation of complex engineering systems, which are difficult to mould either by conventional programming or knowledge-based expert system or by model-based expert systems. The occupancy grid method is proposed to use for estimating the states of the input nodes of neural networks. The node state estimates are obtained by interpreting the incoming sensor readings using probabilistic models. Bayesian estimation procedures are used for incremental updating of the occupancy grids. This occupancy grid framework represents a fudamental departure from traditional approaches. This approach supports incremental discovery procedures, explicit handling of uncertainty.

The paper describes the features of occupancy grids and neural networks that are desirable for knowledge representation for fault diagnosis of manufacturing systems. The paper also discusses the present problems with neural networks and the future research directions.

2 The Proposed Methodology

The objective of this fault diagnosis method is to predict that there is going to be a failure and to find out the causes of the failure based on the values of the manufacturing parameters. In the proposed system, the raw physical data such as speed, current, vibration etc. are obtained through sensors, amplifiers, filters, multiplexers and sampling devices. The collected information is represented internally in the computer in some suitable format. There are many representation techniques, but we wanted an integrated framework that can provide the following facilities :   a) automatic learning capability   b) ability to deal with uncertain and imprecise information   c) ability to deal with time dependent information   d) ability to deal with vast amount of sensor information   e) ability to deal with non-linearity

In order to achieve these, we proposed a occupancy grid based neural network structure. In the proposed method, different features are extracted from the sensor data, using different mapping functions. Numerous possible choices of mapping functions have been proposed [1]. In our model, without going into a specific mapping function, the user may select a set of mapping functions which he feels is relevant to the specific application. After selecting the mapping functions, we propose a Chen's [2] information gain algorithm for evaluating the indices according to their relevance to a specific application. Only highly relevant (i.e. higher information gain) indices, but not redundant are to be considered. Each selected index is classified into several classes by using Devijver's linear optimizing algorithm [3]. The boundaries of each class can be determined by using Devijver's linear optimization algorithm as given below :

$$b_j = \frac{V_j m_{j+1} + V_{j+1} m_j}{V_j + V_{j+1}} \quad j = 2,...,N_c-1. \quad\text{------(1)}$$

where,

b_j are the boundries of the indices belonging to jth class m_j are the cluster centers of the indices belonging jth class $V_{j'}$ are their corresponding varaiances. N_c is the number of classes.

Each class serves as the input node to a feedforward multilayer neural network structure. The occupancy grid method is proposed to use for estimating the states of the input nodes of the neural network structure. The node state estimates are obtained by interpreting the incoming data using probabilistic models. Bayesian estimation procedures are used for incremental updating of the node states. The neural network structure uses a predective back-propagation learning algorithm.

2.1 The Neural Network Structure

Neural network based diagnostic system has been proposed by several authors [4,5,6]. Ray [6] discussed a neural network based expert system for equipment fault diagnosis. Multi-layer neural networks with backpropagation algorithms [7] are the most popular. In this paper the predict backpropagation [8] learning algorithm is used which is an improvement of popular back propagation algorithm [7]. In the original back-propagation algorithm the examples are presented to the network continuously, in a generally random way. The predict back-propagation algorithm determines a dynamic order in which exemples are loaded to the network, depending on the "learning difficulty". The error signal produced by an example is the measure of its learning difficulty. A complete discussion on predict back-propagation algorithm is appeared in [8].

2.2 Occupancy Grids

The basic objective of this occupancy grid methodology is to estimate the state of each input node. The detail discussion regarding occupancy grid is beyond the scope of this article and can be found elsewhere [9]. An occupancy field $O(X)$ is a discrete state stochastic process defined over a set of conitinious coordinates $X = \{x_1, x_2, ...xn\}$. The state variable $s(C)$ associated with a node C of the input layer in the network is defined as a descrete random variable with three states, occupied (+1), empty (0) and unknown (0.5). The occupied and empty states are denoted by OCC and EMP respectively. Since the node states are exclusive and exhaustive, $P[s(C) = OCC] + P[s(C) = EMP] = 1$. The specific steps involved in estimating the occupancy grid from sensor data are as follows:
a) Define the probability density function $p(r|z)$, which relates the reading r to the true parameter space range value z. We have considered Gaussian probability density function

$$p(r|z) = \frac{1}{\sqrt{2\pi}\,\sigma} \, exp\left(\frac{-(r-z)^2}{2\sigma^2}\right) \qquad\qquad (2)$$

b) Use this density function $p(r|z)$ in a Byesian estimation procedure to determine the occupancy grid state probabilities of each node $p[s(x)|r](x)$.
c) Determine the discrete states of the nodes using the maximum a posterior decision rules as follows :
If $P[s(C) = OCC] > P[s(c) = EMP]$ then node C is occupied.
If $P[s(C) = OCC] < P[s(c) = EMP]$ then node C is empty. $\qquad$ (3)
If $P[s(C) = OCC] = P[s(c) = EMP]$ then node C is unknown.
To allow the incrimental composition of sensor information, use sequential updating of Bayes' theorem. Let, $P[s(C_i) = OCC | \{r\}_t$, be the current estimate of a input node C_i based on observations $\{r\}_t = \{r_1,....r_t\}$ and given a new observation r_{t+1}, the improved estimation is :

$$P[s(C_i) = OCC | \{r\}_{t+1}] = \frac{p[r_{t+1}|s(C_i) = OCC] * P[s(C_i) = OCC|\{r\}_t]}{p[r_{t+1}|s(C_i)] * P[s(C_i)|\{r\}_t]} \qquad (4)$$

In this recursive formulation, the previous estimates of the node state, $P[s(Ci) = OCC | \{r\}t]$, serves as the prior and obtained directly from the occupancy grid. The new node state estimate $P[s(Ci) = OCC | \{r\}t+1]$ is subsequently stored again for further processing.

3   A Case Study

The above procedure is experimented for fault diagnosis of a milling machine. The objective of this study was to identify the likely process states from the observable process readings. The possible states in the cutting process is classified into the following four classes : i) normal   ii) worn tool   iii) broken tool iv) tool chatter

These four classes are considered as the output nodes of the neural network structure. The input data are acquired from four channels, one cach for AE signal, current sensor, dynamometer X and Y directions. Data were sampled every 1ms from all four channels. Features extraction from the multiple sensor data is performed over 100 sets of sensor readings, which correspond to a time interval of 100ms each in real time. These features are mapped to the values of the input nodes of the neural network structure using occupancy grid methodology. In our study initialy we have considered 22 different features i.e. indices. In order to assess the relevancy of the indices related to the above four classes of cutting process set of 246 data has been collected, which contain value of the indices and the corresponding cutting states. Using equation (2) the boundray values for all the indicies are estimated, based on these values and using Chen's [2] information measure algorithms, the information measure for all the indices regarding each of the 4 classes has been calculated. After this, the following 12 features were considered :

a) Magnitude of AE signal   b) Change in magnitude of AE signal   c) Frequency of AE signal   d) Average RMS of AE signal   e) Change in average RMS of AE signal from previous interval to the present one.   f) Change in RMS of motor current g) Frequency of X Dynamometer signal   h) Frequency of Y Dynamometer signal   i) Average RMS of X Dynamometer signal j) Average RMS of Y Dynamometer signal   k) Change in average RMS of X Dynamometer signal   l) Change in average RMS of Y Dynamometer signal

Considering these 12 features a three layer feedforward neural network struture has been formed. It consist of (12 x 4) = 48 input nodes, 26 hidden nodes and 4 output nodes. Here, we assume the number of hidden nodes is equal to half of the total input and output nodes. Occupancy grid methods as discussed in section 2.2 are used for determining the state of the nodes. The predict back-propagation based learning algorithm has been used to train the network. A set of 453 data has been clloected cronologically and has been used to train the network. After training the system has been tasted with 228 examples, which shows a success rate of 87 %.

4   Limitations and Future Directions

Neural networks have aptitude for some task but there are many current limitations to the approach. Neural netwoks come in a variety of different architectures, the most popular being the feedforward back-propagation algorithm. However, there is no definite rule exists for specifying the network architecture. There is no theoretical procedure for determining the optimum size of the training examples. The computational time for training the neural network is very high. Deciding the number of hidden nodes in a multilayer neural network is a critical issue. The number of hidden nodes is highly related with the learning parameters. In order to decide this seperate simulation are needed.

5   Conclusions

Neural networks have number of properties that make them attractive for knowledge representation in fault diagnosis area. In this paper we have discussed neural network based system for manufacturing fault diagnosis. We have shon how the occupancy grid representation can be used directly in neural network structure. The main advantages of neural networks in fault diagnosis over other existing methods are its ability to capture uncertainty, ability to accommodate nonlinear relationships, ability to learn and acquire knowledge automatically. We believe that despite some limitations, this approach to real-time fault diagnosis have significant promise.

References

[1]   Monostori, L., 1988, "Signal processing and Decision Making in Machine tool monitoring systems", Proc. of Manufacturing International, 1988, Vol. 1, pp. 277-284.

[2]   Chen, Y.B, Sha,J.L and Wu,S.M, 1990 ,"Diagnosis of the Tapping Process by Information Measure and Probability Voting Approach", ASME journal of En gineering for Industry, Vol. 112, Nov., pp. 319-325.

[3]   Devijver, P.A. and Kittler,J., Patteren Recognition : A statistical Approach, Printice Hall, 1982.

[4]   Hoskins, J.C. and Himmelblau, D.M, 1990, "Fault detection and diagnosis using Artificial neural networks", in the Mavrovouniotis, M.L. (Eds.), "Artificial Intelligence in Process Engineering", Academic Press Inc. , pp. 123-160.

[5]   Peng,Y., and Reggia,J.A.,1989 "A Connectionist model for diagnostic problem solving", IEEE Trans. on Systems man and cybernetics, vol.19,no.2, March/April, pp.285-298.

[6]   Ray, A.K., 1991, "Equipment Fault Diagnosis - A Neural Network Approach", Computers in Industray an International Journal, Vol. 16, No. 2, pp. 169-177,June 1991.

[7]   Rumelhart,D.E., Hinton,G.E., and Williams,R.J., 1986 "Learning internal rep resentation by error propagation " in Rumelhart, D.E, and McClell and, J.L.,(Eds.),"Parallel Distribution Processing", Cambridge, MA : MIT press, chap.8,pp. 318-362.

[8]   Bebis, G.N. and Papadourakis,G.M., "object recognition using invariant object boundary representations and neural network models",vol. 25, no. 1, pp. 25-44, 1992.

[9]   Elfes, A., 1989, "Using occupancy grids for mobile robot preception and navigation", IEEE Computer, June 1989, pp. 46-57.

Expert Systems for Fault Diagnosis on CNC Machines

Dipl.-Ing. Eberhard Kurz

Fraunhofer-Institut für Arbeitswirtschaft und Organisation (IAO)
Holzgartenstr. 17 D-7000 Stuttgart 1, West Germany

Abstract. The development of CNC machines has brought along a rise in complexity and automation. The efficiency of such systems depends strongly on their reliabilty and availability in the production process. Thus, the cause of a break-down of a machine should be detected by the user and in most cases be repaired immediately.

We have developed a diagnosis system prototype which runs on the machine control and reasons about 200 fault possibilities. The architecture of this system is divided into the development environment on a UNIX based, graphic workstation for entering the expert´s knowledge and the target environment on the machine control. The expert system tool on the development environment is written in COMMONLISP and the system on the target environment is written in C.

1   Problem Space

In recent years the market has made an increasing number of demands on Computerized Numerical Control (CNC) machines. The trend is to favour smaller lot sizes and to produce more difficult and complex parts. Nevertheless there should be a reduced time per workpiece and an optimal use of the material should be made. Such highly developed and complex CNC machines guarantee a large degree of automation, flexibility and a high functionality. However, the user interface of these machines, i.e., the programming and controlling of the machine, caused several problems. Here latest developments have shown possible solutions. Techniques as for example graphic programming- or simulation systems, the use of a wide information environment (window technique, colour) and new software tools can help to overcome these difficulties. But this alone will not result in economy. There are two ways to improve the reliability and availability of a machine:

- fault-prevention
- fault-repairing

Fault-prevention has to be considered while constructing the machine, i.e., by the producer. Attention should be paid to the transparency of the system through modularity, the standardization, the reduction of component parts, improvement of the quality of components, the quality and reliability of the software. On the spot, the prevention lies in regular maintenance. This is of great importance, especially if one considers that as the production cycle of any new generation of machines becomes shorter, the improvement of a type of machine has to take place in a shorter time as well.

In the case of breakdown the fault has to be repaired on the spot and as fast as possible. The goal is to give the user sufficient help, so that he will be able to repair the fault in the most of the cases. If that is not possible, the user (or a communication system) should at least be able to transmit a qualified diagnosis. In addition to that, the kind of problems arising in highly development CNC machines has changed:

- Although there is greater complexity there are fewer faults. If there are any, they are usually more complicated and make great demands even on an experienced engineer.

- Frequent changes of machine types or their configuration with devices on the periphery leads to problems concerning the experience and the knowledge of the engineer.

- Due to the wide distribution of the machines, a service technician cannot always be available.

- Knowledge concerning the machine includes complex technologies from different areas, therefore one expert is hardly able to possess the whole existing knowledge.

Thus, todays method of training a panel of specialized service technicians and to have them repair the faults is insufficient and will become obsolete. The development of strongly modularized machines makes the repair work even for a none-expert possible. The remaining problem is the difficulty to find the fault, which is not as easy for a non-expert - in spite of the manuals, construction plans and circuit diagrams. Here conventional data processing has its limitations. Even the most elaborate sensor technique cannot replace the "grey knowledge" of a service technician, i.e. empirical knowledge and overall knowledge, which are of great importance.

Nevertheless, we would like to point out, that due to the complexity of foreseeably detectable faults, an expert system has its limitations too. In the near future we will not be able to do without human experts, but expert systems help the human experts to concentrate upon really important problems. The main advantage of such a system lies in the quick distribution and availability of a basic expert knowledge.

2 An Expert System for Fault Diagnosis on CNC Machines

The Fraunhofer Institute in Stuttgart (IAO) has developed an expert system in cooperation with a CNC machine producer. Special attention was paid to the integration into the work environment and to the user´s problems concerning usability and acceptance of the system. It turned out that a new component, the knowledge compiler had to be added. The purpose of the knowledge compiler is to transfer the knowledge given by the expert from the developing environment to the target environment.

Figure 1 and 2 shows that two expert systems have been developed:

(1)   The LISP version is a knowledge input and testing tool for the expert. In the test phase the expert is identical with the user.

(2)   The C version was designed for the user/service technician and serves solely as a diagnosis system. Here the expert system is integrated into the CNC machine, alterations of the fact and rule definition are not possible.

We are now going to point out some advantages of this special system architecture. These advantages refer to a fully developed expert system which will be explained in the following chapter.

-    The knowledge base is developed at the producer. The development of a knowledge base does not only acquire the construction of a large amount of data but also maintenance and up-dating. The necessary information is only available at the producer of the machine.

-    The service technician/user has no influence on the knowledge base, even if he finds a fault which does not exist in the expert system. For dealing with the knowledge base he has to know the construction, formalism and operation of the expert system. Therefore it is more economic to employ a central expert who will build the new information into the system.

-    The hardware requirements of the expert system consist mainly of supplying storage capacity. Neither a special LISP-processor nor software for special programming languages or graphic environments is necessary.

-    If the expert system is improved or if he CNC machine is altered by periphery tools, local updating is possible via EPROM exchange. This way the expert system can be easily actualized and will come up to the present state of the art.

-    As knowledge about the machine is collected and standardized in the expert system, weak points in machine parts can be more easily detected and possibly improved on. This will be an advantage for the producer.

The system consists of roughly 500 rules. There exist 600 facts with possible faults, 50 of the facts are indicators. Usually, at least one additional fact will be deduced from indicators. The user has to answer 10-20 questions until the expert system will show the result. The system is integrated into the machine control and needs about 300 kByte storage.

This expert system does not require expensive hard- or software, it only uses the existing resources of the CNC machine.

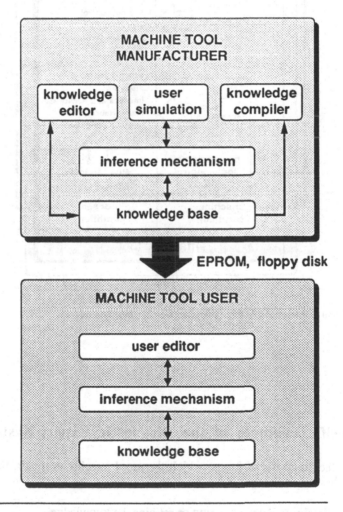

Figure 1: Architecture of the system: Division into
development and runtime version

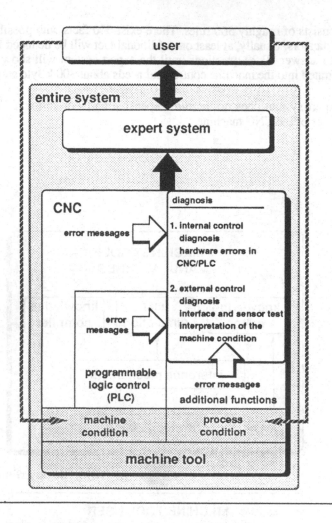

Figure 2: Integration of the runtime version in a machine tool

3 Specific Features of the Diagnostic Expert System

Apart from the diagnosis of faults, which the user himself can repair, the system should meet the following demands too:

-   More flexibility when changing or refining the knowledge.

-   Usability for various machine types. Facts and rules should be easily transferable if the machine components are the same.

- Testing of diagnosis result by comparison with the existing knowledge about the machine, i.e. examination of consistency and eventually following faults.

- Instructions for the repair work of a service technician and tests by changing or removal of machine parts should be possible.

- "Environmental influences", for instances location or age of the machine as well as data about maintenance service should be taken into account too; i.e. storage and processing of data which go beyond one diagnosis session.

These new demands do not only lead to quantitativ changes, but, above all, to qualitative changes of the system. The problems become more complex and lead to an exponential increase of rules. the architecture of the system has therefore to have three essential points:

- Knowledge representation of facts and rules is more detailed and facts are enriched with additional attributes to transfer control information from complex rules into the simpler fact definition. The structuring of the facts supports the extension and recognition of the fact.

- The knowledge editor supports the process of knowledge acquisition by means of windows and menus which increase the clarity of selecting functions.

- The inference engine should be free of control information given by the knowledge engineer. Heuristics, which are applied to the knowledge base and which decide about the next question, can serve this purpose. Apart from that, the inferences are made dependent on uncertainty values. These are helpful for describing correlations more precisely.

Knowledge Editor and Knowledge Base

Very often in existing XPS shells the structure of the facts is not sufficient. Thus several goals should be realized:

- To support the extension and recognition of facts, they should be precisely structured.

- To transfer previous control information from complex rules into the simpler fact defintion, facts should be attributed.

- A knowledge editor, which describes the internal knowledge representation in a formal, but nevertheless easily understandable and uncodified way, should be created.

The definition of facts is subdivided into different windows. The objects are machine components or machine parts and are connected with each other by a subpart relation. By this relation all objects can be shown on the screen in a tree structure. The object tree will only help to structure the facts and has no influence upon the inference process. The knowledge engineer is therefore able to construct the tree in any structure and to change it, depending on how useful the structure of the objects seems to him. Since the properties are left out, the structure remains clear and ist not overloaded with useless information.

The properties are arranged below the objects. By choosing an object they can be depicted and edited. It is possible to add new properties and their values. For concentrating on the essential entry, the display of property attributes can be restrained with the help of global options. If the display of further attributes is restrained with entering a new property, default values will be registered. For example, "certain" ia a default value for the uncertainty value, "up to 10 minutes" is a default value for the answering heuristic function.

There are some functions integrated in the conclusion of facts. You can choose between setting a new fact, making an action or asking the user a question. Making an action is a special feature of this expert system, which allows you to change facts, for instance you can reset a sensor or take off machine parts. An action with a then following question is equivalent to a test. Further rules can follow a question and can then - depending on the result of a question - allow further conclusions.

The data of the knowledge editor and the knowledge base have basically the same structure. The knowledge editor however, additionally offers several demonstrations of the same knowledge; for instance, objects can be shown as alphabetical lists. Windows and menu technique will help the knowledge engineer to control the expert system. The menus will facilitate the choice of the possible functions, the windows will guarantee clarity, because they structure logically related information. Apart from that, the possibility of selecting symbols and choosing from a predefined set of values with the help of the menus will help to avoid syntactical mistakes.

Inference Engine

The whole inference engine has following scheme:

(1) new calculation of the best question.
(2) question to the user editor.
(3) adoption of the answer as fact.
(4) application of the rule.
(5) If a new fact is deduced, go back to (3).
(6) If faults have not been discovered exactly enough, go back to (1). (see figure 3)

The control of the inference engine, will be taken over by a socalled heuristic function. It is advantage of the heuristic function that it does not have to be controlled by the expert or knowledge engineer, but is an automatic process. the goal of the heuristic is to find a good question or rather to achieve a good answer. The heuristic function itself consists of several individual heuristics, which will be evaluated and added and will thus result in the general heuristic function (compare with (Chang/Lee /1/) and (Pearl /2/). This helps to estimate wether a question will be appropriate and sucessful.

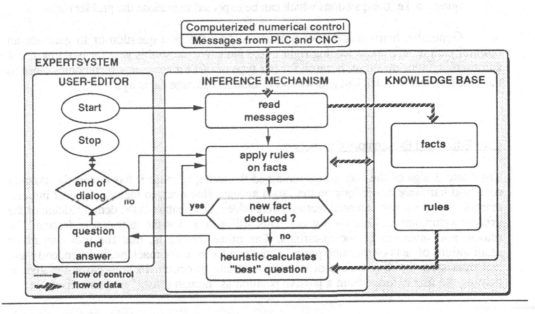

Figure 3: Inference control and working procedure of the expert system

The individual heuristics are:

- *Costs* for answering a question:
  This will mainly concern time. It is for instance usually easy to check the motor-temperature, but it is more difficult to control the wear of the coal brushes inside the motor, because you have to remove some motor parts for checking this factor.

- *Frequency* of faults:
  A fault concerning for instance the motor temperature occurs quite frequently, whereas worn-out coal brushes are rare.

- *Evidence* that a conclusion can be expected:
  This function is based on the idea, that there is a function interdependence between the preconditions. If several preconditions for a rule are fulfilled, one can expect the missing preconditions to be more probabale and thus good to ask.

- *Transitively* obtainable conclusions:
  Answers instantiate a new fact. Thus it is possible to gain indirectly new information by deducing a rule. The new fact itself can be a precondition in other rules and can then set new facts. This transitivity will be measured to find the basic question, i.e. the questions which can be expected to enclose the problem area.

Generally, heuristics do not promise to find the best question or to generate an optimal path of inferences leading right to the fault. We chose this way for building up an automatic control structure, because for our type of tasks a solution which comes close to the problem will be sufficient and will serve time and storage capacity.

<u>User Editor and Explanatory Component</u>

The basic design of the user editor remained the same, because it has to suit the existing soft- and hardware conditions on the target system. However, to include the user into the reasoning process, two essential features were added. One feature is the demonstration of the present symptoms and question goals. This will ensure that the user can understand the reason and intention of the questions. The other feature is, that the user can make assumptions of facts or machine parts, i.e. the condition of the machine, independently and can thus steer the diagnosis process into a certain direction. In both cases the goal is helping the user to move from a passive position as "human sensor" into an active role.

4   Conclusion

The development time for a new tool has been 4 man years and will need probably one more man year. At present the system possesses 150 rules, but the amount of rules does not necessarily imply a valuation. In contrast to the prototype, a rule has become more complex and more accurate, which is due to the more differentiated fact-defintion in conditions and conclusions. We think that about 500 rules should be enough to have a marketable expert system for a lathe with a double turret.

5 References

/1/ Chang, C.L.; Lee, R.C.T. (1973): Symbolic Logic and Mechanical Theorem Proving, Academic Press. New York

/2/ Pearl, J. (1984): Heuristics, Addison-Wesley Publishing Company. Reading, Massachusetts.

Design Efficient Local Search Algorithms

Jun Gu

Department of Electrical Engineering, University of Calgary
Calgary, Canada T2N 1N4
gu@enel.UCalgary.CA

Abstract. Local search is one of the early techniques proposed during the mid-sixties as a practical technique to cope with the overwhelming computational intractability of NP-hard combinatorial optimization problems. In this paper, we give two cases of using local search to solve the n-queens problem and the satisfiability (SAT) problem. We have found that most practical constraint satisfaction problems (CSP) and constraint-based optimization problems can be solved by local search efficiently.

1 Introduction

Given a minimization (maximization) problem with object function f and feasible region F, a typical local search algorithm requires that, with each solution point $x_i \in F$, there is associated a predefined *neighborhood* $N(x_i) \subset F$. Given a current solution point $x_i \in F$, the set $N(x_i)$ is searched for a point x_{i+1} with $f(x_{i+1}) < f(x_i)$ $(f(x_{i+1}) > f(x_i))$. If such a point exists, it becomes the new current solution point, and the process is iterated. Otherwise, x_i is retained as a *local optimum* with respect to $N(x_i)$. Then, a set of feasible solution points is generated, and each of them is "locally" improved within its neighborhood. To apply local search to a particular problem, one need only to specify the neighborhood and the randomized procedure for obtaining a feasible starting solution.

Local search is very efficient in several aspects. First, using a full assignment of all variables in the search space, it reduces an exponential growth search space to a much smaller one. Secondly, it searches for improvement within its local neighborhood using a testing for improvement and, if there is any improvement, taking an action for improvement. Since the object function has a polynomial number of input numbers, both "testing" and "action" are relatively efficient and often perform very well for a search problem. A major weakness of local search is that the algorithm has a tendency of getting stuck at a locally optimum configuration. Many search techniques, such as statistical optimization, simulated annealing, simulated evolution, conflict minimization, and genetic algorithms are either local search or variations of local search.

2 A Local Search Algorithm for the N-Queens Problem

The n-queens problem is to place n queens on an $n \times n$ chessboard so that no two queens attack each other. We now describe briefly some basic ideas of using local search to solve the n-queens problem.

Let n be the size of the board and let each queen be placed in one row only. When n queens are arranged on the board, their column positions are stored in array *queen* of length n. The ith queen is placed on the board at row i and column *queen*[i]. We require that at any moment the array *queen* contains a permutation of

```
procedure Queen_Search ()
begin
    generate_an_initial_permutation_of_queens();
    conflicts := compute_conflicts();
    while conflicts ≠ 0 do
        for queen i := 1 to n do
            for queen j := 1 to n do
                if swap(queen[i],queen[j]) reduces conflicts then
                begin
                    perform_swap(queen[i],queen[j]);
                    conflicts := compute_conflicts();
                end;
            if local then generate_a_new_permutation_of_queens();
        end;
end;
```

Figure 1: An Efficient Local Search Algorithm for the N-Queens Problem.

integers $1, \cdots, n$. This guarantees that no two queens attack each other on the same row or the same column. The problem remains to minimize any conflicts among queens possibly occurring on the diagonal lines. The object function, *conflicts*, in this case is the sum of all conflicts among queens. At the beginning of search (see Figure 1), a random permutation is generated. Conflicts on the diagonal lines are eliminated simply by testing all possible pairs of queens. If a local swap of a pair of queens' positions reduces the object function *conflicts*, then the swap is performed; otherwise no action is taken. This process is repeated until all conflicts among queens are eliminated. If the search meets a local minima, a random search state is assigned.

This local search technique dramatically outperforms the standard backtracking approaches, allowing one to solve million queen problems in seconds, while a standard backtracking approach handles only approximately 97 queens [8]. The basic idea is as follows: while the backtracking explores a space of partial assignments of queens to the board, making sure that there are never any conflicts between the queens already placed, and trying to place all the queens; local search explores a space of complete assignments of all queens to the board, and searches to minimize the number of conflicts within a much smaller search space. Local search algorithms we have developed exhibit polynomial or linear time complexities [7].

3 A Local Search Algorithm for the Satisfiability Problem

One goal of the satisfiability (SAT) problem [1, 2] is to determine whether there exists a truth assignment to a set of variables $(x_1, x_2, ..., x_m)$ that makes Boolean formula $c_1 \wedge c_2 \wedge \cdots \wedge c_n$ satisfiable, where $\wedge$ is a logical *and* connector and c_1, c_2, ..., c_n are n distinct clauses. Each clause consists of only literals combined by just logical *or* ($\vee$) connector. A literal is a variable or a single negation of a variable. We use l to denote the average number of literals in each clause.

```
procedure SAT1.0 ()
begin
    get_a_SAT_instance();
    x̄₀ := select_an_initial_point();
    conflicts := compute_conflicts(x̄₀);
    k := 0;
    while conflicts ≠ 0 do
    begin
        for each variable i := 1 to m do
            if test_swap(xᵢ,x̄ᵢ) then
                x̄ₖ₊₁ := perform_swap(xᵢ,x̄ᵢ);
                conflicts := compute_conflicts(x̄ₖ₊₁);
            if local then local_handler();
            k := k + 1;
    end;
end;
```

Figure 2: **SAT1.0**: A General Local Search Algorithm for the SAT Problems.

A general local search algorithm for the satisfiability (SAT) problem is given in Figure 2. To begin, procedure $get\_a\_SAT\_instance()$ requires a practical or a randomly generated SAT problem instance. Procedure $select\_an\_initial\_point()$ randomly selects an initial starting point: $\vec{x}_0$. The object function, $conflicts$, returns the number of "conflicting" clauses that have a *false* truth value. Function $compute\_conflicts()$ evaluates the number of conflicts in the clauses. Based on such a formulation of the object function, the SAT problem becomes a minimization problem to the object function [3, 4, 5].

The search process is an iterative local optimization process. During a single iteration, for each variable x_i in vector $\vec{x}$, we search through its neighborhood to select a solution point that minimizes the object function. The test to see if a swap of the truth value of x_i with $\overline{x_i}$ will reduce the total number of conflicts in the clauses is performed by function $test\_swap()$. Function $test\_swap()$ returns *true* if the swap operation would reduce the total number of conflicts in the clauses; otherwise, *false*. The actual swap operation is performed by procedure $perform\_swap()$, which produces a new solution point, i.e., vector $\vec{x}_{k+1}$. Function $compute\_conflicts()$ updates the total number of conflicts in the clauses. In practice, before $conflicts$ reduces to zero, the algorithm could be stuck at a locally optimum point. A simple local handler that can negate up to m variables is added to improve its convergence performance. The real execution time of one version of the SAT1.0 algorithm running on a NeXT workstation is illustrated in Table 1.

The running time of the SAT1.0 algorithm can be estimated as follows: Procedure $get\_a\_SAT\_instance()$ and function $compute\_conflicts()$ each take $O(nl)$ time. Procedure $select\_an\_initial\_point()$ takes $O(m)$ time to produce an initial solution point. Accordingly, the total time for initialization is $O(nl)$. The running time of the *while* loop equals the number of *while* loops executed (k) multiplied by the running time of the *for* loop. In the worst case, $perform\_swap()$ takes $O(m)$ time, $test\_swap()$, $compute\_conflicts()$, $local\_handler()$ each will take $O(nl)$ time. Combining $O(m)$ time cyclic scan of m variables, the running time of the *for* loop is

Problems (n,m,l)			Trials		Execution Time			Iterations		
Clause	Variable	Literal	Global	Local	Min	Mean	Max	Min	Mean	Max
100	100	3	10	0	0.000	0.006	0.015	1	1.200	2
200	100	3	10	0	0.000	0.046	0.108	1	3.700	7
300	100	3	10	0	0.046	0.140	0.218	3	7.200	11
400	100	3	10	0	0.156	3.650	10.28	6	130.1	363
1000	300	3	10	0	0.468	1.001	1.634	7	14.10	23
1000	400	3	10	0	0.188	0.472	0.984	3	6.900	14
1000	500	3	10	0	0.156	0.230	0.343	3	3.700	5
1000	600	3	10	0	0.108	0.184	0.265	2	3.000	4
1000	700	3	10	0	0.108	0.179	0.264	2	2.900	4
1000	800	3	10	0	0.109	0.168	0.203	2	2.700	3
1000	900	3	10	0	0.109	0.157	0.203	2	2.500	3
1000	1000	3	10	0	0.046	0.121	0.202	1	2.100	3
2000	1000	3	10	0	0.372	0.564	0.844	3	4.700	7
3000	1000	3	10	0	1.172	2.276	3.954	8	14.10	22

Table 1: Real Execution Performance of one version of the $SAT1.0$ Algorithm on a NeXT Workstation (Time Units: seconds)

$O(m(nl)))$. Summarizing the above, the time complexity of $SAT1.0$ is $O(knml)$.

In the $SAT1.1$ algorithm, if the value of the object function is not zero, the local handler is called. In the $SAT1.2$ algorithm, if the object function does not decrease, the condition of *local* is met. In the $SAT1.3$ algorithm, in addition to the local handler, variables keeping the object function unchanged are negated. For $l \geq \log m - \log\log m - c$ and $n/m \leq 2^{l-1}/l$, where $c \geq 0$ is any constant, we show that $SAT1.1$ exhibits a polynomial average time complexity of $O(m^{O(1)}n)$. For $l \geq 3$ and $n/m = O(2^l/l)$, we show that $SAT1.2$ exhibits a polynomial average time complexity of $O(m^{O(1)}n^2)$. For $l \geq 3$ and $n/m = O(2^l/l)$, we show that $SAT1.3$ exhibits a polynomial time complexity of $O(lmn\log m)$ [6].

References

[1] S.A. Cook. The complexity of theorem-proving procedures. In *Proceedings of the Third ACM Symposium on Theory of Computing*, pages 151–158, 1971.

[2] M.R. Garey and D.S. Johnson. *Computers and Intractability: A Guide to the Theory of NP-Completeness*. W.H. Freeman and Company, New York, 1979.

[3] J. Gu. How to solve Very Large-Scale Satisfiability (VLSS) problems. 1988.

[4] J. Gu. Benchmarking SAT algorithms. Technical Report UCECE-TR-90-002, Oct. 1990.

[5] J. Gu. *Constraint-Based Search*. Cambridge University Press, New York, 1992.

[6] J. Gu and Q.P. Gu. Average time complexities of several local search algorithms. Submitted for publication. Jan. 1992.

[7] R. Sosič and J. Gu. Efficient local search with conflict minimization. *IEEE Trans. on Knowledge and Data Engineering*, 1992.

[8] H. S. Stone and J. M. Stone. Efficient search techniques – an empirical study of the n-queens problem. *IBM J. Res. Develop.*, 31(4):464–474, July 1987.

The Design of Building Parts by Using Knowledge Based Systems

Guardian Ketteler and Mihaly Lenart

University of Kassel, Dept. of Architecture Henschelstr. 2, 3500 Kassel, West-Germany

Abstract With the widespread use of CAD systems in architecture and building construction there is a growing demand for intelligent CAD systems. Such systems have been proposed e.g. for preliminary building design [19], lay-out design [9], [16] or construction management [7]. In this paper we describe a knowledge based framework for the computer aided design of building parts. Our project is divided into two parts: A theoretical one, where methods and their advantages will be identified, and a practical one, where these methods are implemented as independent modules interacting with each other, a data base and a CAD-system. First, a staircase designer will be presented. This expert system is described in details as part of a domain specific shell on one hand and as an independent working module on the other. Based on this prototype expert system, our main objective is to develop a domain specific design shell. Currently, we also investigate specializations of the expert modules in order to tackle particular design and manufacturing problems.
Keywords: Knowledge based applications, domain specific expert system shell, design of building parts.

1 Introduction

A general development in AI of the past 20 years has also influenced architectural design. After a breakthrough in the 60's and the early 70's it was generally expected that AI, in particular expert systems, will have a large impact on industrial design and manufacturing. That the expectations didn't materialize, or at least not in the extent of the expectations, is largely due to the fact that the theories and methods developed earlier didn't provide efficient tools for new applications. It turned out that expert system development is costly and time consuming. It is also widely recognized that expert systems are extremely domain dependent, meaning that the adequacy of formal methods or generic tools is less important than the quality of domain knowledge encapsulated in a particular expert system.

Since most of the planning and design tasks can still be tackled only by expert systems, it is important to resolve this efficiency problem. One possible way of making expert system development efficient is to analyze and organize domain specific knowledge and develop methods for a (well defined) class of domain specific problems. Using this base one can then efficiently develop expert systems for a particular task within the given domain.

The methods and tools developed in this project are intend to support but not to replace the entire design process. The reason for this is twofold: On one hand, design problems in architecture and building construction are so complex that trying to capture the the entire process and to describe a generic design strategy would utterly simplify the problem and throw out the baby with the bath water. On the other hand, certain parts of the design process are readily accessible. Architects do have the (professional) knowledge of solving well defined subtasks for which it is extremely difficult (if not impossible) to find generic rules. Since we do not intend to change the way design is carried out traditionally, some of the decisions are left to the user.

Our aim in this project is twofold: On one hand we would like to tackle well defined tasks within particular areas of the design of building parts, on the other hand we would like to develop generic methods and tools for solving problems in various areas of the design of building parts. This limited, or domain dependent generalization is justified by the failure of large scale generalizations of expert system development tools on one hand and the high development costs of expert systems on the other.

The specific area we have chosen is staircase design, and the generalization is based on STADS, a STAircase Design System developed earlier by the authors (see [6]). A generic shell or design system is derived from domain specific knowledge by combining formal methods with empirical or heuristic ones.

2 Design of Staircases

The first stage of our research was limited to staircase design. The reason we have chosen this scope first lies in the duality of staircase design: it is a well defined task with well defined subtasks and constraints, but also a complex one defying any exact mathematical solution schema. As many famous examples of staircase designs testify, it is also an artistic challenge to the architect. Therefore, staircase design can be considered as a representative for a large number of design tasks. It is also manifold in the sense that designing a staircase requests various design strategies, such as solving configuration problems, drawing analogies or developing solutions iteratively. Design is generally considered as a search process [13].

Typically, the number of solutions for any design problem is rather unlimited. This also holds for staircase design and a major concern is how to find a reasonable number of adequate solutions among the large number of different alternatives. In order to survey solutions, architects classify staircases based on various features. For example, the geometric form of a staircase gives rise to classes of single-flight, two-flights, three-flights, etc. staircases. We can also distinguish staircases with and without landing, with winding steps or spiral geometry. Certainly, any given staircase can belong to more than one class. This classification provides means for decomposing complex design tasks into manageable size problems. An important decomposition is based on building component functions. The functional decomposition is carried out hierarchically so that the structure is partitioned into sets of functional macro elements. These elements can interactively be decomposed into more specific coupled functional subsystems and the decomposition carried out until we arrive at simple building components whose design is a relatively easy task. The decomposition or abstraction hierarchy is represented in Figure 1.

Similar abstraction hierarchy was used by HI--RISE [8], an expert system for preliminary structural design of building systems. Further developments are [18] and [20]. ALL--RISE [20] extends the idea by allowing the system to dynamically decide the order in which to design systems based on heuristics.

However, classification is just one way to reduce the scope of the search space. Another way is by using constraints. Solutions can be eliminated right at the beginning by constraints obtained from e.g. fire or traffic regulations or design rules. Design rules are e.g. step security (riser + tread = 46cm) or step comfort (2 x riser + tread = 63cm). It is advisable to use these constraints as early as possible in order to limit the search space at the beginning of the search process.

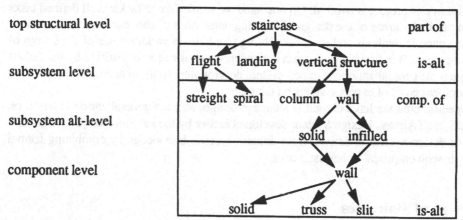

Figure 1: Example for functional decomposition

After preliminary reduction of the search space, it is usually still too large for brute force, but even for other unconstraint search methods. In order to make the search process manageable, we have to use constraints also for guiding the search. We need to develop heuristic rules and evaluate the performance of the system based on provided solutions. In particular, we need heuristic rules describing which constraints and in which order have to be considered. Introducing new constraints later in the design process might modify the design and therefore constraints considered earlier have to be reevaluated with each new constraint. Constraints, search operators and evaluation functions represent design knowledge (expertise), building codes, standards and regulations. Partial designs are also generated by analog reasoning using a case-based approach described in [14] and [15]. Constructive solutions stored in a data base will be applied directly or used for generating new (partial) designs.

In many instances the search provides not a concrete (partial or complete) solution to the given design task, but a generic one. This generic solution is a prototype for a class of solutions that share certain properties, such as given cost limits, area size limits, or any other user defined feature. The decision is left to the user how to modify this prototype according to his/her preferences. Such preferences can be of e.g. aesthetic or economic na ture.

3 Concept of the Expert System

The development and implementation of STADS consists of four major steps:

1. object model,
2. user interface
3. database and
4. efficient search strategy.

In the following we will discuss these steps.

3.1 Object Model

The object model is the representation of a staircase in a design context. The uninitialized object model represents a prototype staircase whose attributes are generic characterizing any staircase or a large class of staircases. This model is implemented as a frame based system whose elements are frames (schemata) connected by (usually binary) relations [11], [4]. Figure 2 shows typical elements (schemata) and relations of the object model.

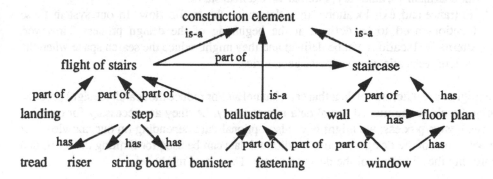

The elements can be objects, class of objects (as in Figure 2), but also relations such that we can easily define new relations within the system. Relational properties, such as transitivity, reflexivity or symmetry are system defined for predefined relations. They can arbitrarily be defined for user defined relations. The objects are either physical objects -as in Figure 2 - or functional descriptions of physical objects, e.g. wall connection. The definition of an object includes various attributes and values such as element properties, requirements and dependencies. A mechanism is provided to define, describe and classify these objects. The object model is organized hierarchically supporting the functional decomposition mentioned earlier. It has also a layer structure representing various design aspects that will be described in section 5.

According to the current abstraction level STADS will consider different composit objects. For example, at the beginning of the design process STADS considers string-walls, flight of stairs and landings, while later in the process these construction elements become more specific and have more specific features such as masonry wall, string of steps, beams, etc.

3.2 User-Interface

The purpose of the user-interface is communication between system and user. In particular, the user-interface enables us to initialize the object model and enter additional information during the design process. It also requests data from the user, displays design evaluations and solutions.

Initializing the object model means entering specific data for a given design task. Construction elements in the uninitialized object model have no, or just default values for their attributes. In the initialization process the user provides concrete values for these attributes in form of graphical or textual data. Characteristic data for initialization and their effects are:

1. Story height: provides the number of steps and landings. (In most of the cases this is not one number but a couple of alternative numbers.) The number of steps and landings gives rise for the least area (length and width) needed in the floor plan for the staircase.
2. Building function: :provides the minimum width of the flights, the angle of the flights, fire requirements, etc.
3. Function of the staircase:provides properties that are relevant in the design context such as main vs. auxiliary, internal vs. external stairs.
4. Entrance and exit locations: they determine the traffic flow. In our system these locations need to be defined at the beginning of the design process. However, alternative locations can be defined and they might reduce the search space when the system seeks adequate staircase geometries.

We might have other initial data that can channel and/or reduce the search (design) process. While the above mentioned initial data is mandatory, i.e. they are necessary for carrying out the design process, we might have also optional data depending on our knowledge at the beginning of the design process. Additional data can be used for limiting the scope and improving the efficiency of the design process. These data might be:

1. Building material: such as wood, concrete, steel, etc.
2. Load bearing concept:such as console, suspended, etc.
3. Construction method: such as prefabrication, casting, etc.
4. Geometry of the staircase: which is a two or three dimensional schematic model of the staircase. It provides geometric data about ceiling, floor, walls and openings.
5. Appearance (aesthetics): can be light vs. heavy, slender vs. robust, transparent vs. closed, etc.
6. Cost limits: which are the relative (compared to the entire building costs) or absolute construction costs anticipated.
7. Area size: can be minimized or constrained depending on the floor plan context.

For all these features we can assign values and define priorities. Priorities can be formulated as constraints, e.g. cost minimizing or resource constraints.

3.3 Database

The database contains information about constraints, context and solutions. In particular, it stores information about:

1. Standard solutions of construction details.
2. Geometric data about the shape of the staircase.
3. Building material.
4. Algorithms representing or implementing design rules.
5. Algorithms solving geometric problems, e.g. calculating measurements for winding

steps.

6. Significant design solutions obtained by case studies or generated earlier by the system.

Information stored in the database is accessible in two ways: by user's or by system's query. Similarly, information can be added to or altered in the database by entries of the user or by the system.

3.4 The Search Process

It is generally accepted that each design activity can be considered as a search process. The reason for this is that on one hand we have goals and evaluation criteria describing how the expected design object should look like and on the other hand we can systematically generate alternative design steps leading to different solutions. Therefore, the design task is to generate alternatives systematically step by step and arrive at solutions satisfying the criteria.

In spite of this fairly simple characterization, the design process as a search process is not easy to implement. The reason is twofold: first, usually we don't have all the necessary information for the entire search process and second, even if we have sufficient information, the search is computationally so expensive that its implementation is either impossible or can not be done efficiently. Problems arising from the incompleteness of information can be solved by carrying out the search partially. This means that our system solves well defined subproblems and let the user decide in which direction the search will go whenever information is insufficient. Problems arising from the size of the search space are solved by using constraints. We apply constraints in two ways: as means for limiting the search space or for guiding the search process. In the first case we talk about constraint bounded, in the second constraint directed search. Constraint directed search is carried out opportunistically. The idea is to identify certainty islands of the search space and use them as starting points or anchors for the search process (see[5] and [2]). This means that in the search process more constraint design decisions will be made first, prior to less constraint ones. In order to identify certainty islands recursively, constraints have to be maintained and their effect has to be evaluated repeatedly in the search process.

Regarding the the relation between constraints and the number of solutions we can distinguish between three major cases:

There is no solution satisfying the constraints and we talk about an overconstraint problem.

There are too many solutions in the sense that either it is computationally intractable to generate all of them or we are unable to evaluate or elaborate the large number of solutions provided by the search process. Intractability might be the consequence of an infinite number of solutions (countable or uncountable) but also of unreasonable computing time. In this case we talk about an underconstraint problem.

There is a reasonable or manageable number of solution to the given design problem.

In the first two cases constraints have to be modified so that we arrive at the third case. Modification means adding new constraints and removing or replacing old ones. The opportunistic strategy determines in which order constraints will be considered in the search process. However, considering stronger constraints at the beginning of the search process might help us to reduce the search space, but it might also lead to loosing relevant solution paths. Therefore, it is of great importance to develop good heuristics for the entire search process. Heuristics is used on two levels: locally, when alternatives are generated by selected search operators and globally, when an appropriate search strategy is chosen for a particular design task. The search process involves one of the following two design methods:

There is a prototype that we modify step by step according to the constraints. We generate solutions i.e.new prototypes by using constraints. In both cases we solve configuration problems by selecting a set of elements and defining a goal [10], [1]. If by the nature of the design task no set of elements is given, we use heuristics to define elements. For example, there is an unlimited number of shapes a step might have. However, architects usually limit their choice to a few number of shapes based on structural, material, delivery, etc. restrictions. They also have personal preferences limiting the number of possible shapes. Goals are generated by rules which often encapsulate some kind of heuristics. If a goal can not be derived from existing rules the system asks for user interaction. The search process itself is supported by domain knowledge that reduces the number of states generated [12]

4 Implementation

As it was mentioned earlier, the system contains three components: user interface, database and object model. The user interface includes a CAD-system and various menus. The CAD-system is AutoCAD and its purpose is to initialize the object model as well as display solutions. In the initialization process the user attaches attributes to the drawing files of the generated construction elements. This means that s/he characterizes the objects and each object with its characterization will be translated and (according to the description above) represented in the object model.

The database links various representation forms of the same object and stores representations of simple as well as complex objects. Complex objects are -among others- complete solutions, i.e. the design of complete staircases. The database is accessible for the object model as well as for the user. It is a relational database, Oracle, with all the functionalities of generating complex data structure and having sophisticated queries.

The object model is implemented in Knowledge Craft, an expert system shell, providing three different representation forms: object oriented, logic programming and production system. Knowledge Craft is based on Common Lisp and its underlying representation language is CRL (Carnegie Representation Language), an object oriented language using frames (called schemata) as basic elements [17]. Logic programming features are provided by (a full fledged) Prolog. It is used mainly for analysis or evaluation purposes. In particular, we use Prolog for analyzing and/or evaluating design solutions w.r.t. aptitude, costs and area demand. This is implemented as a backward chaining system. Production system features are provided by OPS-5, a forward chaining tool. It is used mainly for generating (partial or complete) solutions. Knowledge Craft also has an SQL-interface for interfacing the relational data base.

The implementation of STADS took 4 men/year up to now, and the current prototype can tackle most of the design (sub-)tasks described earlier. However, the system is not complete in the sense that it does not provide complete solutions. The development of a subsystem for designing and manufacturing wooden staircases is underway.

5 Generalization Concept

The underlying idea of generalization is design abstraction [3]. However, the main problem here is to find an appropriate abstraction level, since the level of abstraction has a major impact on implementability and efficiency. As we pointed out earlier, the object model consists of different objects representing data relevant for staircase design. These objects can be classified based on their role in the model. We have e.g. the class "building-function" containing objects that describe building functions. Such an object is e.g. "communal" or "commercial" having information about functional aspects of communal or commercial buildings. We call these classes layers. Each layer represent different aspects of the design task. One object can certainly belong to many layers so that the object "landing" belongs to the class "element-function" as well as to the class "element-type". This means connections between layers that can be extended to or replaced by a network of layers connected by higher level or meta relations. In other words, the layer structure developed for STADS provides a convenient organization of construction objects not only for staircases but also for other kinds of building constructions.

Each layer is selfcontained and initialization of objects is supported by the layer structure. This is demonstrated by the following example: Let's assume we want to design staircases for a building containing apartments, offices and various shops. In STADS we need to represent data (by schemata, rules, clauses and/or procedures) according to the given building function. Since we have here more than one building function, it would be inefficient to develop representations for each function separately. Instead, the system generates new objects at run-time containing just the necessary information for the given design task (see Figure 3).

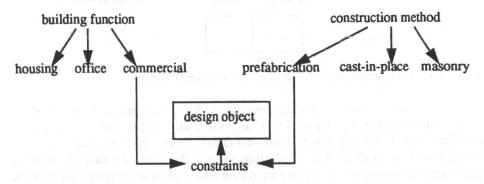

Figure 3: Representation supported by the layer structure

Each layer provides specific as well as generic information. The latter has to be identified and implemented for the design shell. For example, in STADS the layer "building-function" contains information (data, rules, etc.) relevant for staircases. However, some information in this layer can be used for the design of other building constructions. Information about traffic flow is one example that is closely related to building functions and at the same time it is relevant not only for calculating the size of staircases but also of corridors and doorways. This holds for many layers of STADS. Some of the layers, such as e.g."construction-type" consist information relevant for any kind of building construction. Extracting higher level or abstract information about design objects corresponds to well-known and generally practiced construction principles. For example, it is a common expert knowledge that connection details are designed according to the same principles, i.e. wand connections in the attic, ceiling or staircase (or for that matter in any load bearing structure) are constructed based on identical principles.

For STADS we have already established relevant information and how this information will be represented. The next step is to identify generic information and make it available for the design of other building parts. This common, generic information has to be extended in order to represent new building constructions, such as e.g. partition walls or balconies. By analyzing various building parts, we can a) identify building parts that are closely related and have similar representations, and b) develop generic representations, i.e. object model, rule and data base, etc., for each class of closely related building constructions. These steps are represented schematically in Figure 4. The schemata show graphs representing relations between (concrete and abstract) objects. We can obtain generalizations or specializations by extracting subgraphs or generating supergraphs from existing representations (models).

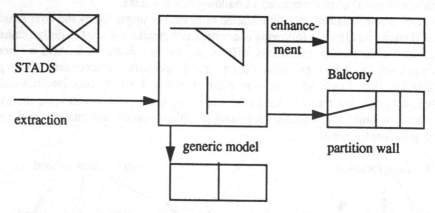

Figure 4: The development of generic and concrete representation

We have here three important development stages. First, one particular system (here STADS) will be analyzed and meta knowledge extracted. Second, based on this meta knowledge, we try to develop new systems, and finally, we go back to evaluate and revise the meta knowledge with the help of existing systems. Although, the first development stage requests additional efforts, we expect to save efforts when we develop new systems. Obviously, there is a trade-off between the generality of the representation and the effort needed to develop a task related representation. In other words, the closer building constructions are related, the easier we can identify common representations and develop task oriented expert systems. However, such representations will be restricted and allow fewer applications. Therefore, a major objective of our research is to define the scope of the shell.

6 Future Extensions

The focus of the current research is a prototype expert system whose domain is staircase

664

design. At the same time we develop methods and tools to handle other kinds of design tasks in the area of building construction. This can be achieved by generalization and specification as it was described in the previous sections. Parallel to this, we also plan to extend STADS in other directions. In particular, we would like to develop an expert system for the design and manufacturing of wooden stair cases. The main purpose of this extension is combining CAD and CIM (Computer Integrated Manufacturing) methods. Parts of wooden staircases are manufactured on CNC woodworking machines. Since these machines are very flexible and capable of manufacturing a wide range of wooden parts (by changing tool sets by themself), the problem we are facing is to control the design according to the machine's parameters on one hand and control the production line on the other hand. The flexibility of these machines and the uniqueness of products simplifies scheduling problems. Figure 5 shows the two extensions of the ongoing research project STADS. One is the generalization aiming at a generic design shell and the other is the specialization within staircase design. Specification and generalization would be an ongoing process as other building construction design systems emerge.

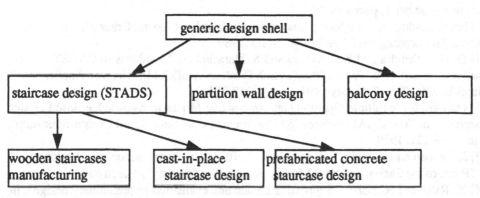

Figure 5: Survey of STADS' extensions

References

[1] V.E. Barker and D.E. O'Connor. "Expert systems for configurations at Digital: XCON and beyond", in: Journal of CACM, 1989, volume 32, number 3, pages 298-318.
[2] C.A. Baykan and M.S. Fox, "An Investigation of Opportunistic Constraint Satisfaction in Space Planning", in: IJCAI, 1987, pages 1035-1038
[3] Coyne, R.D. and Roseman, M.A. and Balachandran, M. and Gero, J.S., "Knowledge Based Design Systems", Addison-Wesley, Reading, MA, 1990.
[4] Fikes, R.E. and Kehler, T "The Role of Frame-based Representation in Reasoning", in: Communication of the ACM, 1985, volume 28, number 9, pages 904-924
[5] Fox, M.S., "Observations on the Role of Constraints in Problem Solving", in: "Proceedings of the Annual Conference of the Canadian Society for Computational Studies of Intelligence", pages 172-187, 1986,
[6] Ketteler, G. and Lenart, M., "Expertensysteme fuer den Entwurf von Baukonstruktionen", editor J. Gauchel, in: "KI Forschung im Baubereich", pages 120-130, Ernst & Sohn, 1990, Berlin,

[7] Levitt, R.E. and Kunz, J.C. , "Using Artificial Intelligence Techniques to Support Project Management", "The Journal of Artificial Intelligence in Engineering, Design, Analysis and Manufacturing",1987,

[8] M.L. Maher, "HI-RISE: A knowledge based expert system for the preliminary structural design of high rise buildings", Carnegie-Mellon University, Dept. of Civil Engineering, 1984, Pittsburgh, PA

[9] McCall, R. and Fischer, G. and Morch, A. "Supporting Reflection-in-Action in the JANUS Design Environment", in: "The CAAD Futures '89 Conference, Computer Aided Design Education", editor : McCullogh and Mitchell and Purcell,1989

[10] J. McDermott, R1: A rule based configurer of computer systems, in: "Artificial Intelligence", September 1989.

[11] Minsky, M.,"A Framework for Representing Knowledge", editor: P. Winston, in: "The Psychology of Computer Vision", pages 211-277, McGraw-Hill, New York, 1975.

[12] S. Mittal and F. Frayman, "Towards a generic model of configuration tasks", in: "Proceedings of IJCAI-89", 1989, volume 2", pages 1395-1401, Detroit,

[13] Mostow, J., "Toward better methods of the design process", in: AI Magazine",1985, volume 6, number 1, pages 44-57,

[14] Navichandra, D., "Exploring Innovative Designs by Relaxing Criteria", Departnment of Civil Engineering, MIT,Cambridge, Mass.,1989

[15] D. Navichandra and K.P. Sycara and S. Narasimhan, "Synthesis in CADET, a case-based design tool", in: "Proc. of the Seventh Conf. on Artificial Intelligence Applications", Miami Beach, FL, = February 1991, pages 217-221.

[16] Oxman, R., "Multiple Operative and Interactive Modes in Knowledge Based Design Systems", in: "The CAAD Futures '89 Conference, Computer Aided Design Education", pages 103-121, 1989.

[17] J. Pappert and G. Kahn", "Knowledge Craft: An Environment for rapid Prototyping", in: "Proc. of the SME Conf. on AI for the Automotive Industry", March 1986.

[18] K. Ryan and N. Harty, "Supporting choice and evaluation in preliminary design", in: "The third Int. Conference on Industrial & Engineering Applications of AI & Expert Systems", 1990, pages 809-818,

[19] Schmitt, G., "Expert Systems for Engineering Design", chapter "ARCHPLAN -An Architectural Front End to Engineering Design Expert Systems", pages 257-278, Academic Press, New York, 1988, editor: Rychener, M.D.

[20]. Sriram", "Knowledge-based approaches for structural design", Carnegie-Mellon University, Dept. of Civil Engineering, Pittsburgh, PA, 1986

Information Fusion in a Knowledge-Based

Classification and Tracking System

Keith P. Mason

Electronics Research Laboratory,
Defence Science and Technology Organisation,
PO Box 1500, Salisbury, S.A. 5108,
AUSTRALIA.

Abstract. KBESM is a laboratory prototype of a shipborne knowledge-based system which interprets data from two types of sensors that provide information about nearby ships and aircraft, with the aim of identifying and tracking such ships and aircraft. The data from one of the two simulated sensors provide information about intercepted radar signals, and the data from the other sensor provide information about positions of nearby ships and aircraft. KBESM uses a blackboard architecture, and the interpretation of data from either sensor alone is achieved by establishing a recognition hierarchy of entities beginning with the sensor data and yielding descriptions of ships and aircraft which have been sensed. In this paper an approach to information fusion which merges the two recognition hierarchies at the earliest point where comparable objects are postulated, and includes steps to assist in the comparison of some features of the different sorts of sensor data is presented. Illustrations of this approach to information fusion in KBESM are provided.

1 Introduction

KBESM is a laboratory prototype of a shipborne knowledge-based system that is designed to interpret data from two types of sensors which provide information about ships and aircraft in KBESM's nearby environment, with the aim of identifying and tracking such ships and aircraft. KBESM provides results via a realtime graphics interface to a human operator.

One of the data streams provide information about radar signals intercepted by KBESM's simulated ESM sensor. This activity of intercepting radar signals is known as Electronic Support Measures (ESM, see Scheler 1986 for details). The other data stream provides information about positions of ships and aircraft in the adjacent geographic region. KBESM uses a blackboard architecture, and the interpretation of data from either sensor alone is achieved by establishing a recognition hierarchy of entities beginning with the sensor data and yielding descriptions of ships and aircraft which have been sensed. KBESM is written in Prolog (Clark 1982).

Characterisations of intercepted radar signals by parameters such as carrier frequency (rf), pulse width (pw) pulse repetition interval (pri) and intercept bearing (but not position), are received by KBESM in near real time. Analysis of signal parameters may provide good identity information about the particular radar (and host vessel or aircraft) which emitted the intercepted signal (Skolnik, 1980), but only poor positional information. Position reports are characterised by good time/location information, however the type/class information may not necessarily be specific, and reports may be historical. Thus in this application, while the different features of the two sensor recognition hierarchies can provide a better basis for making decisions about nearby ships and aircraft, the difficulties of comparison of these features can give rise to ambiguities. The nature of the domain also includes the possibility of inaccurate or even erroneous data which add to the ambiguity.

In this paper the approach to information fusion in KBESM is discussed. This approach merges the two recognition hierarchies associated with the two sensors at the earliest point where comparable objects are postulated, and includes steps to assist in the comparison of some features of the different sorts of data. In this paper only the information fusion task out of the total KBESM system is discussed. Aspects of other KBESM issues have been mentioned in Hood and Mason (1986), Hood, Mason and Penhale (1986) and Hood, Mason and Mildren (1989). The blackboard structure used in KBESM is overviewed in section two. The recognition of radar signals is discussed in section three and an illustration of how radar signal data are fused into a blackboard which also contains items deduced from position reports is presented. Section four looks at the processing of position reports and an illustration of how position reports are fused with blackboard deductions based on intercepted radar signals is given. The elements of the fusion process are identified in section five. Section six summarises this paper Note that the syntax of blackboard entities used in the examples has been chosen to enhance the clarity of the presentation.

2 The Blackboard Model

The Blackboard model (Nii 1986a,1986b) has been used as the architecture of the KBESM system and is described in Mason and Hood (1989). That paper shows how the analogy of a group of specialists gathered around a blackboard to collectively solve a problem is used in KBESM to identify radar emitters from a single sensor measuring signal parameters. In particular the principal control structure of the system, which consists of writing sensor data or a (partial) result on the blackboard to trigger a specialist to solve a particular part of the problem, was shown as a very appropriate paradigm for real-time identification systems. Each specialist contributes to one part of a recognition hierarchy. In the two type sensor system discussed in this paper, although there are two recognition hierarchies, one for intercepted radar signals and one for position reports, the object of information fusion is to merge them into a single structure and to use a unified blackboard to accommodate all observations and deductions.

The KBESM approach to information fusion is to take the recognition hierarchies for both sensors, and at the earliest point where comparable items are produced, to merge the hierarchies. As mentioned in the introduction, entities in these two hierarchies have different features and are difficult to compare. Thus for this approach to fusion to be effective, the recognition process needs to be modified to account for the different features that can be derived from the two sorts of sensor data, and for some effort to be spent fleshing out details that allow for better comparisons. Such instances will be pointed out in the examples of sections 3 and 4.

Other approaches to related problems may be found in Brown, Schoen and Delagi (1986), Evers, Smith and Staros (1984), Lenat, Clarkson and Kiremigjian (1983), Nii, Feigenbaum, Anton and Rockmore (1982), Roe, Cussons and Feltham (1990), and Byrne, Miles and Lakin (1989).

As described in greater detail below, the recognition hierarchy for intercepted radar signals has three levels, *signals*, *emitters* and *platforms*, and the recognition hierarchy for position reports has two levels, *sightings* and *platforms* (*platforms* is a term that includes ships and aircraft, *emitters* means the devices that emit the radar signals which have been intercepted, and *signals* means the symbolic characterisation of radar signals). Thus fusion occurs at the *platforms* level, and processing of *emitters* will therefore also take into account *platforms* postulated from position reports and processing of *sightings* will also take into account *platforms* postulated from radar signals.

The illustrations in this paper are in the context of prior processing which is given by the snapshot of the KBESM blackboard given in table 1. In this snapshot, which was taken at time 1135 minutes, four signals intercepted at times 974, 987, 1018 and 1130 minutes and three position reports received with sightings at times 105, 905 and 1050 minutes have their sensor data listed at the top of table 1. The next block of items pertains to emitters postulated to account for the signal data: 2 emitters have been postulated. The first emitter which accounts for the first 3 signals received has been identified as of type *acme_123*, while the second emitter which accounts for the fourth signal has, at this stage of computation, ambiguous type with two candidates proposed, *flash* and *harbor_guide_2c*. The items, emitter_signal_limits and emitter_has_bearing, conveniently summarise often accessed characteristics of the emitters that would otherwise require continual referencing of the signal data. The next block of items pertains to platform data: 3 platforms have been postulated. Platform 1 accounts for emitter 1 and for sightings 1 and 3, and has been identified as the vessel, *Protector*, of type, *Patrole*, belonging to the *Friendly Islands*. Platform 2 accounts for sighting 2 and has been identified as of type, *Controle*. Platform 3 which accounts for emitter 2 has two type candidates, *Controle* and *Defende*. The item, platform_has_bearing, summarises data similarly as for emitters. The expect_emitter items summarise relevant a priori knowledge and were placed on the blackboard when platform candidates were determined to facilitate the recognition process should a platform start using these radars. The last group of items list the items currently postulated, that is 2 emitters and 3 platforms, and also the current time and the position of the host vessel.

3 Radar Signals

The signal data are interpreted in a three level recognition hierarchy of *signals*, *emitters* and *platforms*. Two specialists are responsible for establishing this hierarchy, *Assign_Signal* and *Assign_Emitter*. The recognition hierarchy is shown in figure 1. That is, an emitter is postulated to account for a sequence of intercepted signals, and a platform is postulated to account for one or more observed emitters. *Assign_Signal* is invoked whenever a signal is intercepted and results in that signal's being associated either with an emitter currently postulated (that is, one listed on the blackboard) or, if no existing emitter could have believably emitted the signal, with an emitter that is postulated to account specifically for that signal. *Assign_Emitter* similarly is invoked whenever a new emitter is postulated and results in the new emitter being associated with either a platform currently postulated or, if no existing platform could believably carry the emitter, with a new platform postulated to specifically account for that emitter.

For example. in the context of the Blackboard shown in table 1 and the database of platform attributes shown in table 2, interception of the signal

signal(time=1150,index=5,rf=3256,pw=300,pri=721,bearing=27,power=-40).

would result in the specialist *Assign_Signal* being invoked, and as the parameters of this signal are dissimilar to those of the currently postulated emitters, execution of this specialist would result in the following items being added to the blackboard to postulate that a new emitter has been observed:

assigned_signal_to_emitter(signal=5,emitter=3).
emitter(index=3).

Note that book keeping specialists triggered by the addition of the item assigned_signal_to_emitter(signal=5,emitter=3) would add to the blackboard:

emitter_signal_limits(index=3,rf=3256..3256,pw=300..300,pri=721..721).
emitter_has_bearing(index=3,bearing=27,time=1150).

and that the emitter identification specialist which is also triggered by the addition of emitter(index=3) would (after consideration of available data) add the following item to the blackboard:

emitter_type(index=3,type=ajax_456).

The scheduling of specialists is arranged so that as much as possible all relevant information is determined before a specialist executes. In this case candidates for emitter identity are determined prior to assigning the emitter to a platform.

Table 1. A Snapshot of the KBESM Blackboard.

Times are in minutes, kilometres for distance, degrees for bearings, megaHertz for rf, nanoseconds for pw, microseconds for pri, knots for speed and decibels for power.

Signal Intercepts

signal(time=974,index=1,rf=9653,pw=500,pri=200,bearing=316,power=-35).
signal(time=987,index=2,rf=9646,pw=500,pri=201,bearing=317, power=-34).
signal(time=1018,index=3,rf=9650,pw=499,pri=199,bearing=338, power=-35).
signal(time=1130,index=4,rf=5855,pw=500,pri=221,bearing=36, power=-40).

Sightings

sighting(time=105,index=1,report_time=110,x=25,y=100,speed=20,course=90,
attributes=[class=surface,type=patrole,hull=protector,country=friendly_islands,kind=military]).
sighting(time=905,index=2,report_time=950,x=370,y=110,speed=20,course=315,
attributes=[class=surface,type=controle,hull=cn_4,country=unknown,kind=military]).
sighting(time=1050,index=3,report_time=1100,x=265,y=100,speed=20,course=90,
attributes=[class=surface,type=patrole,hull=unknown,country=unknown,kind=military]).

Table 1. (continued)

Emitter Associations

```
assigned_signal_to_emitter(signal=1,emitter=1).
assigned_signal_to_emitter(signal=2,emitter=1).
assigned_signal_to_emitter(signal=3,emitter=1).
assigned_signal_to_emitter(signal=4,emitter=2).
emitter_signal_limits(index=1,rf=9646..9653,pw=499..500,pri=199..201).
emitter_signal_limits(index=2,rf=5855..5855,pw=500..500,pri=221..221).
emitter_type(index=1,type=acme_123).
emitter_type(index=2,type=ambiguous).
emitter_type_candidate(index=2,type=flash).
emitter_type_candidate(index=2,type=sky_watch).
emitter_type_candidate(index=2,type=harbor_guide_2c).
emitter_has_bearing(index=1,bearing=338,time=1018).
emitter_has_bearing(index=2,bearing=36,time=1130).
```

Platform Associations

```
assigned_emitter_to_platform(emitter=1,platform=1).
assigned_emitter_to_platform(emitter=2,platform=3).
assigned_sighting_to_platform(sighting=1,platform=1).
assigned_sighting_to_platform(sighting=2,platform=2).
assigned_sighting_to_platform(sighting=3,platform=1).
platform_id(index=1,class=surface,type=patrole,hull=protector,country=friendly_islands).
platform_id(index=2,class=surface,type=controle,hull=unknown,country=unknown).
platform_id(index=3,class=surface,type=ambiguous,hull=unknown,country=unknown).
platform_type_candidate(index=3,type=controle).
platform_type_candidate(index=3,type=defende).
platform_has_bearing(index=3,bearing=36,time=1130).
platform_has_position(index=1,x=265,y=100,time=1050).
platform_has_position(index=2,x=370,y=110,time=905).
expect_emitter(plat=1,emitter=ajax_456).
expect_emitter(plat=2,emitter=harbour_guide_2c).
expect_emitter(plat=2,emitter=ajax_456).
expect_emitter(plat=2,emitter=flash).
expect_emitter(plat=3,emitter=acme_123).
expect_emitter(plat=3,emitter=harbour_guide_2c).
expect_emitter(plat=3,emitter=sky_watch).
expect_emitter(plat=3,emitter=super_flash).
expect_emitter(plat=3,emitter=flash).
```

Entities Postulated

```
emitter(index=1).
emitter(index=2).
platform(index=1).
platform(index=2).
platform(index=3).
own_ship_position(x=265,y=25,time=1135).
```

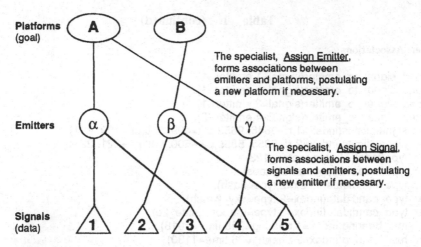

The specialist, <u>Assign Emitter</u>, forms associations between emitters and platforms, postulating a new platform if necessary.

The specialist, <u>Assign Signal</u>, forms associations between signals and emitters, postulating a new emitter if necessary.

Figure 1 Recognition Hierarchy and associated specialists for radar signal data.

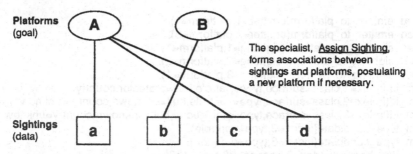

The specialist, <u>Assign Sighting</u>, forms associations between sightings and platforms, postulating a new platform if necessary.

Figure 2 Recognition Hierarchy and associated specialists for position reports.

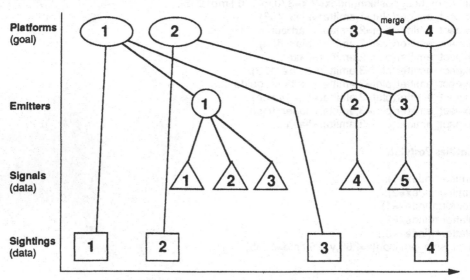

Figure 3 Unified KBESM recognition hierarchy showing associations between data and postulated entities for the Blackboard given in Table 1 and the subsequent data described in Sections 3 and 4.

TABLE 2. A Database of Platform Attributes

```
plat_data(type=patrole,class=surface,kind=military,emitter=[ajax_456,acme_123]).
plat_data(type=defende,class=surface,kind=military,emitter=[harbor_guide_2c,sky_watch).
plat_data(type=controle,class=surface,kind=military,
            emitter=[ajax_456,flash,harbor_guide_2c).
hull_data(country=friendly_islands,type=patrole,hull=[protector]).
hull_data(country=friendly_islands,type=controle,hull=[great_river]).
hull_data(country=southern_empire,type=patrole,hull=[pt_3,pt_4,pt_6,pt_7]).
hull_data(country=southern_empire,type=defende,hull=[admiral_jones,president_smith]).
hull_data(country=northeastern_union,type=patrole,hull=[p_1,p_2,p_3]).
hull_data(country=northeastern_union,type=controle,hull=[cn_4]).
hull_data(country=northeastern_union,type=defende,hull=[president_smith]).
```

The addition of emitter(index=3) to the blackboard invokes the specialist, *Assign_Emitter*, to find or postulate a platform to carry the emitter. *Assign_Emitter* is a rule-based specialist. Incorporating information fusion into the *Assign_Emitter* specialist, requires additional rules to allow matches of platforms with features derived from position data. In particular, these rules need to be able to match an emitter's bearing with a platform position and with allowance for motion since the most recent position report.

Continuing the example, one rule of *Assign_Emitter* will note that platform 2 (a platform deduced from position reports) has an expect_emitter clause for *ajax_456*, which is the same emitter type as has just been determined for the emitter to be assigned, emitter 3. It would also note that it is feasible for platform 2 to have been in a position to emit signal 5 (an angular sector corresponding to resolution of the ESM sensor around the bearing of intercept of signal 5 intersects a circle centred on the position reported in sighting 2 with a radius of the distance that a surface platform could reasonably have travelled in the time since the sighting) and that no other platform is similarly admissible. Hence emitter 3 will be assigned to platform 2 with the following item being added to the blackboard:

```
assigned_emitter_to_platform(emitter=3,platform=2).
```

The book keeping specialist will then update as follows

```
platform_has_bearing(index=2,bearing=27,time=1150).
```

This example has shown how in KBESM radar signals are fused into a recognition hierarchy containing platforms postulated on the basis of positional data.

4 Position Reports

The position reports are interpreted in a two level hierarchy of observations and platforms, and the specialist *Assign_Sighting* is responsible for establishing associations in this hierarchy, as shown in figure 2. That is a platform is postulated to account for one or more position reports. The specialist *Assign_Sighting* is invoked whenever a position report is received and results in that observation being associated with either an existing platform, or if no existing platform could believably account for the report, with a new platform postulated to specifically account for that observation. *Assign_Sighting* is a rule-based specialist, and incorporating information fusion into the this specialist, requires additional rules to allow matches of platforms with features derived from radar data.

For example, subsequent to the context of the processing in section 3, the following sighting is reported:

sighting(time=730,index=4,report_time=1200,x=325,y=175,speed=20,course=270,
attributes=[class=surface,type=unknown,hull=unknown,country=south_empire,kind=military]).

and will cause the *Assign_Sighting* specialist to be invoked, which will consider surface platforms postulated from both intercepted radar signal data and position reports in making its assignment. Due to the distance that the vessel seen in sighting 4 could have travelled since being sighted, either platforms 2 or 3 potentially could have been the source of sighting 4. However, at this stage KBESM is not able to distinguish between these platforms and will postulate a new platform level object and assign the emitter to it. The following items would be added to the blackboard:

platform(index=4).
assigned_sighting_to_platform(sighting=4,platform=4).

The book keeping specialist will then update as follows:

platform_has_position(index=4,x=325,y=175,time=730).

The addition of platform(index=4) will invoke the platform identity specialist which will add the following to the blackboard:

platform_id(index=4,class=surface,type=ambiguous,hull=unknown,country=south_empire).
platform_type_candidate(index=4,type=defende).
platform_type_candidate(index=4,type=patrole).

Also invoked by the addition of a platform level object is the *Details* specialist which fleshes out details implied by any postulated platform type candidates. This specialist is an important part of the fusion process as it provides information about platform features that may assist in subsequent recognition of intercepted radar signal data. As a result of the two candidates above, this specialist would add the following to the blackboard:

expect_emitter(plat=4,emitter=ajax_456).
expect_emitter(plat=4,emitter=harbour_guide_2c).
expect_emitter(plat=4,emitter=acme_123).
expect_emitter(plat=4,emitter=sky_watch).

Indeed, the example of section 3 used an expect_emitter clause created by this method to fuse intercepted radar signal data with a platform postulated from sighting data.

The addition of an ambiguous platform to the blackboard then triggers the *Ambiguity* specialist to undertake further processing to attempt to resolve the ambiguity. This specialist looks at platforms which appear to be in a similar location to see if it is feasible to merge two postulated platforms into a single platform. In this case the specialist will look at platforms 2 and 3. It will note that as platform 2 has been determined to be of type *controle*, which is not a candidate for platform 4, it is an inadmissible candidate for merging. It will also note that platform 4 is compatible with platform 3 (see platform 3 entries in table 1) as they have a common subset of platform type candidates. Thus this specialist will merge platform 4 into platform 3. Also as this subset has only one member, the specialist will deduce that platform 3 is of type *defende*. These processes are depicted in Figure 3. Blackboard items relating to platform 4, and candidates and the identity of platform 3 are deleted from the blackboard during this process and replaced by:

platform_id(index=3,class=surface,type=defende,hull=unknown,country=south_empire).
assigned_sighting_to_platform(sighting=4,platform=3).

When postulating platform 4 the specialist *Assign_Sighting* did have available all the information needed to make the association with platform 3 that was subsequently found. However it would have required *Assign_Sighting* to flesh out the details of (in this case) platform identity. Such an approach compromises the compartmental nature of specialists as it requires the specialist to be omniscient. If taken to extremes such an approach can allow a specialist to become bogged down, possibly interfering with the timeliness of responses.

This example has shown how in KBESM position reports are fused into a recognition hierarchy containing platforms postulated on the basis of intercepted radar signal data.

5 Fusion

The essence of the fusion process in KBESM is the merging of the recognition hierarchies for intercepted radar signal data and for position reports at the *platform* level. The illustration of section 3 showed a process using track features which had been previously sought to facilitate such matches. In that case by determining radars which may be expected to be seen given the postulated type of a sighted platform. Other strategies to assist in this manner include localising a radar emitting platform by using knowledge of where land is and the land or sea based disposition of the postulated platform, or triangulation of observations over time. The illustration of section 4 showed the merging of two tracks after a matching process. Such matches generally are no stronger than to indicate that it is feasible for two postulated tracks to correspond to a single platform. In some cases KBESM may merge tracks that appear to be related but correspond to distinct platforms, however consistency checks will force a reconsideration of the data (possibly splitting a track) if necessary. Such checks are usually based on the subsequent observed velocity/angular velocity of a platform being consistent with its postulated type.

6 Summary

This paper has shown how KBESM fuses data from two types of sensors providing data about intercepted radar signals and data about the position of ships and aircraft in the nearby region. Analysis of signal parameters may provide good type/class identity information about the ship or aircraft which emitted the radar signal, but only poor positional information. Position reports are characterised by good time/location information, however the type/class information may not necessarily be specific, and sightings may be historical. The approach presented in this paper merges the two recognition hierarchies associated with the two sensors at the earliest point where comparable objects are postulated, and includes steps to assist in the comparison of some features of the different sorts of data.

References

Brown, H.D., Schoen, E. and Delagi, B.A. (1986): " An Experiment in Knowledge-based Signal Understanding Using Parallel Architectures", Report STAN-CS-86-1136, Dept of Computer Science, Stanford University.

Byrne, C.D., Miles, J.A.H. and Lakin W.L. (1989): "Towards Knowledge-Based Naval Command Systems". 3rd IEE International Conference on Command, Control, Communications and Management Information Systems", pages 33-42.

Clark, K.L (1982) : "Prolog: A Language for Implementing Expert Systems", in *Machine Intelligence 10*, Hayes J.E., Michie D. and Pao Y-H eds. , Wiley, New York.

Cohen, P.R. (1984): *Heuristic Reasoning About Uncertainty: An Artificial Intelligence Approach*. Pittman.

de Kleer, J. (1986) "An Assumption-based TMS", *Artificial Intelligence* vol 28, no 1.

Evers, D.C., Smithg, D.M., and Staros, C.J. (1984): "Interfacing an Intelligent Decision Maker to a Real-time Control System", SPIE Proceedings 485, pages 60-64.

Hood, S.T. and Mason, K.P. (1986): "Knowledge-based Systems for Real-time Applications", *Proceedings of the Second Australian Conference on the Applications of Expert Systems*, Sydney NSW, 110-137.

Hood, S.T., Mason, K.P. and Mildren, W.J. (1990): A Software Stimulator for a Knowledge-based ESM System, SIMULATION, vol 55 number 3, pages 153-161, September 1990.

Hood, S.T., Mason, K.P. and Penhale, R.G. (1986): "A Laboratory Prototype of a Real-time Shipboard Expert System", *Proceedings of the 10th Australian Computer Conference*, Gold Coast Queensland, 349-357.

Lenat, D.B., Clarkson, A. and Kiremidjian G. (1983): "An Expert System for Indications Warning Analysis", Proc. 8th IJCAI pages 259-262.

Mason, K.P. and Hood, S.T. (1989): KBESM - "A Prolog Blackboard System", Australian Computer Journal, vol. 21, No. 2, August 1989.

Nii, H.P. (1986a): "Blackboard Systems: The Blackboard Model of Problem Solving and the Evolution of Blackboard Architectures", *AI Magazine,* vol 7 no 2, pages 38-53.

Nii, H.P. (1986b): "Blackboard Systems: Blackboard Application Systems", *AI Magazine,* vol 7 no 3, pages 82-106.

Nii, H.P., Feigenbaum, E.A., Anton, J.J., and Rockmore, A.J. (1982): "Signal to Symbol Transformation: Hasp/Siap Case Study", *AI Magazine,* Spring 1982, pages 23-35.

Roe, J., Cussons, S. and Feltham A. (1990): "Knowledge-based Signal Processing for radar ESM Systems". IEE Proceedings, vol 137, Part F, No 5, pages 293-301.

Scheler, D. Curtis (1986): *Introduction to Electronic Warfare*, Artec House.

Skolnik, M.I., 1980, Introduction to Radar, McGraw Hill.

Waterman, D.A. and Hayes-Roth, F. (eds.), (1978): *Pattern-Directed Inference Systems*, Academic Press, New York.

Numerical and Syntactic Tools for Fusion and Diagnosis Provided with the TOPMUSS-System

F. Quante, H. Kirsch, M. Ruckhäberle

Fraunhofer-Institute for Information and Data Processing (IITB),
Fraunhoferstraße 1, D-7500 Karlsruhe 1

Abstract. A toolbox for the development of diagnostic systems is described. The information processing structure chosen for TOPMUSS (*Tools for Processing of Multi-Sensorial Signals*) relies on five building stages from sensing to decision inside which application-specific tools can be exchanged easily. This is achieved by using an object-oriented approach. For scheduling a reasoning mechanism is used and therefore all tools conform to a common functionality definition. Three short examples of numerical and syntactic tools explain the functionality of the tool box. Its usage will reduce the development time needed for a new diagnostic system.

1   Introduction

The TOPMUSS-project (ESPRIT P2255) will provide a tool box for the development of monitoring and diagnosis systems (M&D). For existing tool packages it has been observed, that a lot of them provide tools for processing of individual signals but lack of syntactic tools and of tools for fusion and diagnosis. For the design of M&D systems, developers usually have to adapt incompatible tools from different tool libraries written in different languages which is a time-consuming task. Such design phases are needed for sensor- and signal-based diagnosis tasks and for merging of information contained in signals from multiple sensors. A lot of plants like those for which the TOPMUSS-project will deliver tools, namely FMC's, engines, turbines, rolling mills, are often equipped with a large number of sensors. The tools needed for this system are subdivided into important classes, inside which all tools are made compatible.

This contribution will give an overview on the structure of information processing chosen for the project, then outline important classes of tools and finally give examples of numerical and syntactic tools.

2   Information Processing Structure in TOPMUSS

Information processing in M&D systems is regarded as a closed-loop task (fig. 1) on different stages, for which classes of tools have to be provided.

The selection of an individual tool is based on the application of a formalism for the acquisition of application-specific knowledge described elsewhere /1/. The results of this formalism are the following:

- Identification of items to meet *functional requirements*, such as type and number of sensors, processing speed, faults to be diagnosed.

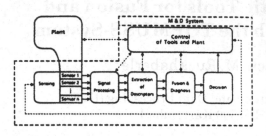

Information Processing Level	Information Processing makes use of	Computational Task
Strategic	Domain specific knowledge	Control and coordination of tools and models, definition of conditions for their activation (TERM)
Tactical	Models	Linking of individual tools to macros and supertools (Linker)
Teleological	Tools of the toolbox, realized as objects	Provision of tools (objects), definition of messages between objects
Functional	Procedures	Generic processing functionalities, provision of basic tools (library)
Programming	Computer language	Implementation

Fig. 1.   Stages of a typical M&D system.
Control tools allow the realization
of closed-loop operation

Fig. 2.   Levels of information processing
based on TOPMUSS toolbox

- Identification of *tools* needed for the stages of sensing, signal processing, descriptor extraction, fusion & diagnosis and control.
- Identification of *strategies* to be followed by the M&D system. Strategy represents the highest level of information processing.

For information processing 5 levels as in /2/ have been identified (fig. 2). On the *strategic level* the coordination and control between various tools is provided by the so-called Tool Executor Reasoning Mechanism (TERM). On the *tactical level* modules to satisfy clearly defined sub-problems are constructed. From the information processing point of view these models often represent models like symbolic and numerical ones for signal or plant behaviour, to be used for the extraction of descriptors. The *teleological level* is the interface to individual tools in the tool box. From the user's perspective the toolbox appears as a set of objects /3/ which are structured into classes corresponding to the stages given in fig. 1. Inside these classes a hierarchy is observed to make efficient use of the inheritance principle in object-oriented programming. A further level of decomposition in TOPMUSS, the *functional level*, is based on identifying generic information processing functionalities that are used to realize the methods which are provided by a tool (object). This decomposition is used to construct individual procedures to be reused in the programming of several tools (objects). The final programming level represents the various procedures and provides access to programming languages.

3   Examples of numerical and syntactic tools

TOPMUSS is still under development, but it has been used in different areas already to test its functionality and to develop new methods.

3.1   Numerical model for machine milling

It has been shown, e.g. /4/, that process signals generated in machining operations like milling, turning and drilling can be predicted fairly well using a numerical model for the cutting process. A model for the prediction of passive forces generated during milling has been introduced into TOPMUSS. It relies on model components shown in fig. 3. The high accurary obtainable with this model is demonstrated in fig. 4.

Such a numerical model can be set up in TOPMUSS by combination of available tools on the tactical level (see above). Ongoing work concentrates on the development and testing of models for other machining operations.

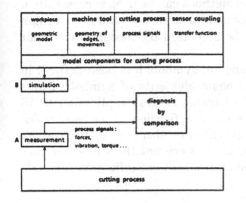

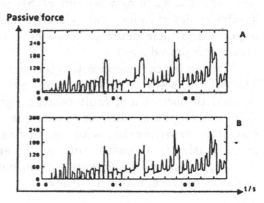

Fig. 3.   Model components for the prediction
          of process signals for milling

Fig. 4.   Example of measured and simulated
          passive forces (tool with 8 edges)

3.2 Classifier tools

For fusion and diagnosis, which make use of descriptors (attributes, features) extracted from heterogeneous sensor signals, methods from cluster analysis, pattern recognition, probability and fuzzy-set theory can be applied. In pattern recognition, classifiers have been developed to allocate an object described by a set of numerical descriptors to one of several possible diagnoses. Such classifier tools have been implemented in the system comprising classical methods like Nearest-Neighbour classifiers, the Mahalanobis classifier, Bayes classifier and Regression classifier. More advanced methods comprise an adaptive Regression classifier and two neural networks, one of them belonging to the type of function networks and the other to the prototype networks. All methods have been applied to the diagnosis of electromotors (using similar data as in /5/) and combustion engines. The performance of classifiers is strongly dependent on the statistical properties of the data and cannot be summarized in short. But there is a strong tendency, that the performance of neural networks is not better in such tasks than that of classical classifiers.

3.3 Syntactic tools

In TOPMUSS syntactic tools are applied to the classification of signals from which symbolic descriptors can be extracted. An example of a process signal for which a human expert also likes to make a symbolic description, is the angular acceleration at the crankshaft of an engine. Normal behaviour could be described as a sequence of similar peaks, which is disturbed under faulty conditions. For the automation of this procedure a mapping of numerical signal values on symbols has to be carried out and the resulting symbol string can be classified using syntactic methods. This procedure has been investigated for biomedical and speech signals in the past (e.g. /6/, /7/) whereelse applications to process signals are very seldom /8/. Syntactic methods can be used successfully if symbolic descriptions and related classifications are easily made by human experts and their heuristic knowledge can be exploited by choosing a similar automatic procedure.

A broad (as compared with literature) range of symbols has been defined in TOPMUSS, see fig. 6. They represent the basic elements of symbol strings related to normal and fault related signal patterns. A first application to 16 different engine conditions resulted in correct classification of 14 conditions. Parallel experiments with numerical classifiers also revealed correct classification in 14 cases, but the two error cases were not the same. Further investigations will show, how a combination of different classification methods can increase recognition accuracy.

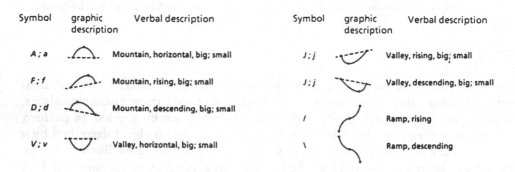

Fig. 6.   Basic symbols for the description of signals in TOPMUSS

References:

1.   D. Seidel, N. Härle, N. Braun: Acquisition of Application Specific Knowledge for Configuring of Monitoring and Diagnosis Systems; IMACS '91 Conference Proceedings, Vol. 3, Dublin, 1991, pp 949-950
2.   R. Leitch, A. Stefanini: Quic, a Development Environment for Knowledge Based Systems in Industrial Automation. Proc. of the 5th ESPRIT Conf., North Holland, Amsterdam 1988, pp 674-696
3.   P. Coad, E. Yourdon: Object-Oriented Analysis. 2nd Ed; Yourdon Press, New Jersey, 1991
4.   N. Höbing: Modellgestützte Signalverarbeitung beim Stirnplanfräsen. Diss. Universität Karlsruhe, Fakultät für Elektrotechnik, 1989
5.   D. Barschdorff, D. Becker: Neuronale Netze als Signal- und Musterklassifikatoren. Technisches Messen tm 57, 1990, 11, pp 437-444
6.   K.S. Fu: Syntactic Pattern Recognition and Applications. Prentice Hall, Inc., Englewood Cliffs, N.J., 1982
7.   E. Pietka: Feature Extraction in Computerized Approach to the ECG Analysis. Pattern Recognition 24, 1991, pp139-144
8.   P.L. Love, M. Simaan: Automatic Recognition of Primitive Changes in Manufacturing Process Signals. Pattern Recognition 21, 1988, pp 333-342

Learning Performance Estimations
in a Knowledge Based CAD-Environment

University of Dortmund, Lehrstuhl Informatik 1
P.O.Box 500 500
D-4600 Dortmund 50, Germany

Abstract. A learning system which is able to learn incrementally analytical knowledge in a knowledge based CAD environment is presented. To speed up the learning process and to cope with the problem of a typically very low example generation rate in interactive design environments different complementary learning techniques are combined. During the normal operation of the CAD-environment learning is based on observation whereas its idle times are exploited by the learning system to perform problem oriented active experimentation. For experiment design meta knowledge about the CAD system's problem solving process is applied.

1 Introduction

In a design process learning is the normal situation, rather than the exception. Design experts are continually reorganizing their knowledge as part of their normal problem-solving activity [1]. The reorganized knowledge improves subsequent problem-solving by avoiding similar failures in the future. In the scope of a fast technical evolution 'intelligent' systems which take over parts of an engineers work should mimic that behaviour. For this reason in recent years considerable effort has been spent to integrate machine learning capabilities into knowledge based design systems (e.g. [2]-[5]). In literature two main subtracks of development can be identified: a) systems which learn procedural design knowledge like design operators, their sequence and applicability (e.g. Mitchell's *LEAP* system, [4]) and b) systems which learn the estimation or classification of certain design constraints or results (e.g. the *LIMES* system [5]).

In this paper a learning system of the latter type is discussed. It represents a novel approach to the integration of machine learning techniques into a cell-based design system. Cell-based design is a specific type of routine design. It is a frequently applied design style in many of the currently available knowledge based design systems especially in the electronic domain. In such a design process for every level of design abstraction a certain set of predefined parameterized templates or *cells* is available, which can be used as building blocks to construct more complex design objects. The prototype of the presented learning system was successfully implemented and tested in the OASE and SILAS analog circuit design environment [6]-[8]. It is able to learn analytical knowledge about worst-case estimations for complex analog cells like operational amplifiers.

Worst-case analysis is an important aspect in the circuit design process. It is used to improve product reliability and minimize redesign costs. With this analysis the effects of instable fabrication process parameters and operating conditions on the performance of a circuit can be explored. Instead of applying a set of exact but time consuming circuit simulations, it will, in practice, often prove to be sufficient to use relatively good estimations of the qualitative and quantitative circuit behaviour. Circuit designers are able to derive intuitively such sensitivity characteristics with a sufficient degree of accuracy on the basis of prior design experience. Nevertheless it is very difficult to elicit this knowledge directly in an acquisition session because of its highly implicit character and the variety of designs to be considered.

2 System Architecture

The integration of machine learning into a knowledge based system for a complex technical application like analog circuit design results in a set of constraints which such an approach has to satisfy. First, the learning system should utilize the circuit instances given incrementally by the normal operation of the CAD system. From the user's point of view this will only be accepted if the learning process is executed in real time and means virtually no additional time effort to her/him. However, the number of these manually generated examples as well as their distribution in the possible design space of the CAD system will normally not be sufficient with respect to the learning efficiency which can be achieved by learning solely from observation. Complementary learning techniques like active experimentation in combination with automated example generation are required to increase learning speed and the reliability of the obtained results.

Although active experimentation is not a novel idea only very few approaches have so far addressed the area of knowledge acquisition support in a complex knowledge based environment [5]. Most of the systems which have been reported in literature are dedicated to more theoretical investigations. Typical application domains for this type of system are the reinvention of physical and chemical laws (e.g. BACON [9][10], KEKADA [11], BLADGEN [12]) or learning solution strategies for mathematical expressions (e.g. LEX [13], PET [14]). Also, not all of these systems apply automated example generation but work with carefully preselected examples or measurement data [9][15][16]. Embedding active experimentation into a knowledge based CAD system therefore is a novel domain, which requires specific procedures and concepts.

2.1 Structural Overview

Figure 1 gives an overview of the presented approach and its integration into the OASE and SILAS design environment.The CAD environment consists of the kernels of the design and the analysis system which are complemented with circuit and technology specific knowledge bases. The kernel systems consist of global task specific knowledge and are implemented in the form of cooperating blackboard systems with distributed, agenda based control [6][8]. The technology specific knowledge mainly describes the parameters of the fabrication process and will not be explained in greater detail. The circuit specific knowledge, however, represents the circuit related expert knowledge which was incorporated into the CAD system. It consists of declarative knowledge like knowledge about circuit structures and procedural knowledge like selection rules, heuristics for calculating device parameters, etc.

The learning system is subdivided into two modules, the experimenter and the estimator. In *learning mode* (default) the *estimator* learns analytical knowledge about the worst-case characteristics of analog circuit blocks by observation of the circuit design cycle. The generated circuit data is evaluated and generalized with statistical functions (see below). The new analytical knowledge derived from this process represents the current circuit specific experience concerning the circuit behaviour. This knowledge is stored in the concept knowledge base and grows with increasing numbers of design in-

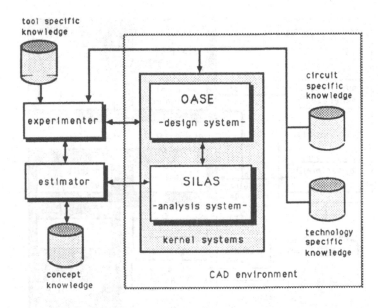

Figure 1: System overview

stances. In subsequent design cycles it can be recalled by the estimator to replace circuit simulations by sensitivity estimations (*estimation mode*).

The second module of the learning system, the *experimenter*, is able to exploit idle times of the design environment for active experimentation. This is necessary to increase the number of available circuit instances and thus to increase the quality of the estimation knowledge as well as to get working results quickly. The experimenter applies tool specific knowledge about the internal organization of OASE and SILAS to stimulate and control the CAD system in remote operation. The tool specific knowledge also comprises meta knowledge about the circuit and technology specific knowledge bases. This enables the experimenter to reuse specific parts of the CAD system's design knowledge in a different context and adapt automatically to any changes in that knowledge.

2.2 Architecture and Basic Functionality

Figure 2 shows a more detailed view of the overall system architecture (without the knowledge bases). The OASE system performs the synthesis part of a design task. On the basis of a functional specification which typically consists of 20-30 circuit specific parameters (e.g. gain > 60 dB, bandwith = 4-6 MHz, etc.) the dimensioned topology of the probably best suited circuit realization is synthesized. In a first step an appropriate circuit configuration is determined and assembled from a library of generic circuit templates and low level cells. In a second step the parameter values of all resulting substructures and devices are calculated using simplified model equations. This design data forms the input for the cooperating SILAS system, which tries to verify the synthesis results by simulation. If the evaluation of the simulation results yields differences with respect to the given specification a redesign cycle is initiated. If the results are acceptable a worst-case analysis of the circuit performance is carried out.

As worst-case analysis is one of the basic functions of SILAS the estimator was also integrated directly into the SILAS kernel. Because of the SILAS blackboard architecture [8] which supports incremental development the implementation was straightforward.

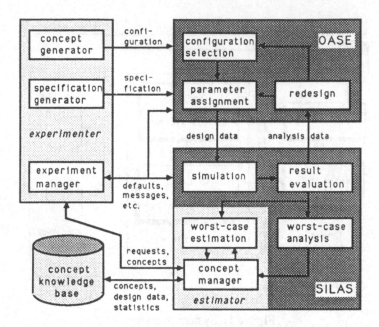

Figure 2: Architecture and information flow

The estimator consists of the *estimation expert* and the *concept manager*. If during a circuit design cycle the user selects worst-case estimation instead of worst-case analysis the estimation expert sends a corresponding request for information to the concept manager. The concept manager exclusively controls the access to the concept knowledge base to ensure its consistency. The requested information is generated by inspecting the analytical knowledge learned so far and, if possible, selecting appropriate circuit specific statistics. This statistical data describe the characteristic worst-case behaviour of similar designs which have been synthesized before. A data set for worst-case estimation comprises information about the typical performance deviations of individual circuit parameters in different worst-case scenarios as well as information about the estimation quality. An example of such a data set for the circuit parameter *open-loop-gain* (A_{V0}) could look like the following (see also Table I):

> worst-case mode: fast
> deviation: average = +4.8%, maximum = +6.6%, minimum = +3.0%
> standard deviation: 0.69.

Circuits of a specific circuit class (e.g. opamp) are considered to be *similar* if they have the same topology but different device geometries. It is expected that similar circuits will also show similar sensitivity concerning variations of operating conditions and fabrication process parameters.

Figure 3 illustrates the structure of the concept knowledge base. It is hierarchically organized in different layers which represent an increasing specialization of a circuit description. Depending on the basic configuration of a circuit and thus the number and type of cells which make up that configuration the topology layer can consist of a variable number of sublayers. On the lowest level, the circuit class layer, the acquired analytical knowledge about the characteristic performance of the individual circuit topologies is stored. These circuit class statistics include information about the ranges of circuit specific parameters which were covered by the instances generated for a specific circuit class as well as information about relative sensitivities of these parameters (see above).

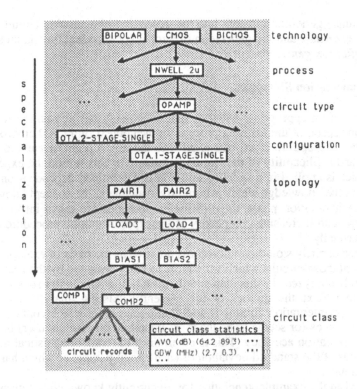

Figure 3: The concept knowledge base

Unlike other approaches which use concept trees for the purpose of learning performance estimations (e.g. [5]) the structure of the concept knowledge base is not static but dynamically generated by the concept manager on the basis of the available circuit instances. If, for example, the knowledge implemented in the CAD system is extended to additional circuit types or cells the concept manager automatically adds corresponding entries to the concept knowledge as soon as instances of those circuits are available. As it is possible to configure the functionality of SILAS to the user's needs (e.g. to perform *no* worst-case analysis during a circuit design cycle) the concept manager is also able to deal with incomplete analysis data which cannot be evaluated directly for knowledge acquisition. These data sets including the circuit specific design data (circuit specification and netlist) are temporarily stored as complete circuit records on the circuit class level of the concept knowledge base. They form the starting point of the next active experimentation phase.

3 Active Experimentation

Active experimentation in a complex CAD environment comprises three basic tasks: a) experiment and CAD tool control, b) problem oriented example generation and c) example evaluation. The last topic has already been discussed so that we will now concentrate on experimentation control and the generation of appropriate circuit instances. These tasks are performed by the experimenter. The experimenter (see Figure 2) consists of three task specific submodules. The *experiment manager* controls the complete experimentation process as well as the set-up and operation of OASE and

SILAS. Example generation is split into the generation of a suitable circuit topology by the *concept generator* and the generation of a topology depedent functional specification by the *specification generator*.

3.1 Experimentation Strategies

The selection of an appropriate experimentation strategy has substantial impact on the efficiency and speed of the learning process. In the context of a CAD environment these aspects have to be optimized at least relative to the user´s point of view to increase the acceptance and applicability of the learning system. For this reason the experimentation strategy which is applied in a specific situation is determined dependent on the amount and quality of the knowledge which is currently available in the concept knowledge base. In each experimentation phase three different strategies are taken into consideration, which represent an increasing complexity of the experimentation procedure and thus are applied sequentially.

First all temporarily stored circuit records are used as the basis of example generation. They represent those circuits which were most recently designed by the user(s) and were not completely analyzed. In this step simply all missing simulation runs are performed and evaluated. Next the quality of the currently existing analytical knowledge is examined. Experimentation continues with that circuit class which contains the highest number of instances (or seems to be the most important one for the user) but still shows insufficient estimation accuracy. This is determined in terms of statistical measures like limit violations of the standard deviation of each circuit parameter which has to be taken into consideration.

Finally, when the example generation for all currently known circuit classes has been completed successfully, the exploration of further circuit types and topologies which can be designed with OASE is started. For this task the experimenter applies meta knowledge about the different types of circuit knowledge which were incorporated into OASE. In this phase the structure of the concept tree (see Figure 3) is dynamically extended.

During the experimentation process the quality of the estimations is periodically checked to avoid unnecessary experimentation effort. If, after having processed a minimum number of circuit instances, three examples in sequence can be estimated correctly in the given limits the example generation for a specific circuit class is aborted. To deal with circuit parameters which turn out to be not suitable for sensitivity estimations the example generation process is also aborted if an upper limit of instances for a circuit class is exceeded.

3.2 Example Generation

The circuit data which is required to stimulate the OASE system for the purpose of example generation consists of a description of a circuit topology and a functional circuit specification. To achieve working designs dependencies between a circuit topology and the corresponding performance requirements have to be taken into consideration. For this reason as the first step of the example generation a circuit topology has to be selected by the *concept generator*. In a second step an individually tailored specification is generated by the *specification generator* which uses that topology information for constraining specification ranges.

As the basic strategies of topology selection have already been discussed above we will now briefly explain the exploration of new circuit classes. For this task the concept generator applies tool specific meta knowledge to directly access the configuration and topology specific knowledge of the OASE system. Figure 4(a) illustrates the example of a corresponding configuration tree for operational amplifiers as it is represented in OASE. This tree describes the available configurations and cells as well as their connectivity. The concept generator simply traverses this tree to identify new circuit topologies which so far were not available from the current concept knowledge base (example: one-stage, single-ended opamp consisting of the low level cells Buffer1, Load3, Pair2, Bias3, Comp3).

To generate an appropriate circuit specification with respect to the selected circuit topology the specification generator has to consider a set of constraints which result from the complex numerical character of this task. The most important ones are:

- the performance limits of the selected circuit realization should not be exceeded but cannot be defined exactly
- the number of possible parameter combinations is virtually infinite due to continous parameter ranges
- nonlinear dependencies between circuit parameters have to be considered
- exact estimations require an even distribution of the circuit instances in the possible design space

Due to the requirements resulting from the problems described above all methods which are based on concept trees like pertubation [5][13][14] or which can vary only one parameter value with each instance [17] are not suitable for this type of problem domain. To deal with this situation a specific procedure which is based on knowledge about relative performances of circuit topologies with respect to all possible topologies was developed (redesign knowledge of OASE, see Figure 4(b)). Starting with the maximum range of each specification parameter (plausibility knowledge of OASE, example: A_{V0} = [40-120] dB) these ranges are, in a first step, restricted dependent on the relative performance of the selected circuit configuration. In subsequent steps the remaining ranges are

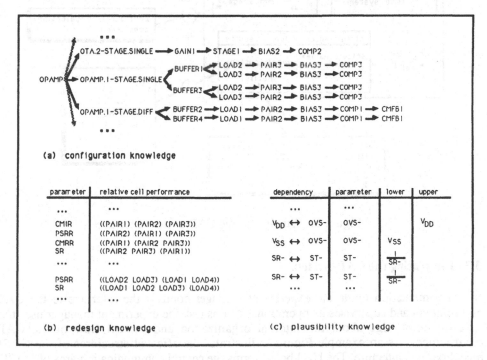

Figure 4: Different types of circuit specific knowledge

further confined in a similar manner with respect to all cells which make up that configuration. In this procedure a certain segment of the specification range of a parameter is cut off for each level of circuit refinement to remove unsuitable parameter ranges and to be on the safe side concerning the specific performance limits of the circuit.

In the case of a one-stage operational amplifier the first step of this procedure would result in an allowed specification range of $A_{V0} = [40\text{-}100]$ dB. This range is calculated by taking into account that one-stage opamps are well suited for low to medium gains while twostage opamps typically are designed for high gain applications. The size of the segment which has to be cut off is determined by multiplying the number of possible design alternatives (2 in the example) with an uncertainty factor (e.g. 2) which expresses the possibility of overlapping subranges and deviding the current overall range by the resulting factor.

Finally the remaining ranges of the specification parameters are partitioned into discrete parameter specific intervals (e.g. 3 dB for A_{V0}). From these intervals suitable values for the specification parameters can be selected by a systematic sample. Different types of parameters like exact numerical values, upper or lower bounds and intervals (see Figure 5) have to be taken into consideration.

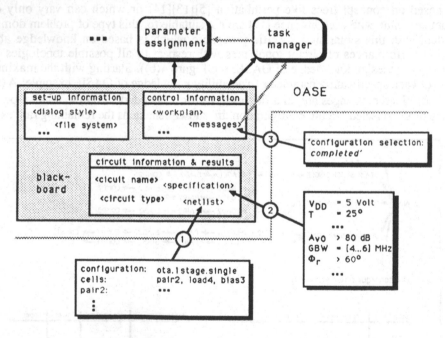

Figure 5: Principle of the CAD system interface

3.3 Interfacing the CAD-System

In experimentation mode the experiment manager controls the interface to the CAD environment and supervises its operation. For this task the experiment manager uses tool specific knowledge about the internal organization and control flow of the CAD environment. As an example Figure 5 illustrates the affected structures of the OASE blackboard architecture. The blackboard forms the central communication area of OASE. All experts (task manager, parameter assignment, etc.) read their task description and input data from the blackboard and, in turn, write their results back. The blackboard comprises three different types of information areas: *set-up information, control information* and *circuit information*. By manipulating this data it is possible to stimulate and control the CAD environment externally. An experimentation phase is started by the experiment manager setting all necessary defaults to specific values (e.g. *no user communication*). Next the concept generator and the specification generator are activated to write a netlist (topology description) and a functional specification to the OASE blackboard. This emulates normal user communication and the operation of the configuration selection of OASE. By sending the message ´*configuration selection completed*´ to the corresponding slot of the control information the task manager of OASE is triggered to start all necessary subsequent activities like parameter assignment and the activation of SILAS.

4 Results

First experiments with the prototype implementation of the described learning system proved the suitability of the approach. Applied to the domain of operational amplifiers the system was able to learn sufficiently exact worst-case estimations (standard deviation < 5) for most of the circuit specific parameters (see Table I). To achieve these results only 15-20 instances for each circuit class taken into consideration were required. A minority of circuit parameters, however, turned out to be too unstable for a sensitivity estimation on the basis of the presented similarity concept. Due to the necessary simulation effort the complete generation and processing of each instance took approximately 10 minutes.

parameter variations

fast: B0 +20%, V_T -20%, T = -20 C, V_{DD} +10%
slow: B0 -20%, V_T +20%, T = +70 C, V_{DD} -10%

parameter	unit	range	fast (%)	slow (%)	
A_{VO}	dB		+4.8	-5.1	1)
		89.3	+6.6	-7.2	2)
		64.2	+3.0	-3.4	3)
			0.82	0.69	4)
GBW	MHz		+25.5	-21.2	
		8.3	+33.0	-25.6	
		2.7	+18.9	-16.7	
			3.71	2.11	
Φ_r	deg		+2.9	+2.7	
		82.1	+7.4	+7.0	
		31.4	+0.2	+0.2	
			2.31	2.17	
. . .		. . .	. . .	. . .	

1) average   2) maximum   3) minimum   4) standard deviation

Table I: Example of a circuit class statistic

Acknowledgements

The research project which this paper is based on is supported by the German Federal Secretary of Research and Technology (subsidy number: 13MV00343). The reponsibility for the contents is by the authors.

References

[1]   Horner, R. and Brown, D.C.: Knowledge Compilation Using Constraint Inheritance, *in:* Gero, J.S. (ed.): *Applications of Artificial Intelligence in Engineering V*, Vol.1 Design, Computational Mechanics Publications, 1990, Boston, pp.161-174

[2]   Richard Forsyth, Roy Rada: *Machine Learning: Applications in Expert Systems and Information Retrieval*, Ellis Horwood Series in Artificial Intelligence, Ellis Horwood Limited, 1986

[3]   Richard Forsyth (ed.): *Expert Systems - Principles and Case Studies*, Chapman and Hall, 1989,

[4]   Mitchell, T.A., Mahadevan, S. and Steinberg, L.I.: LEAP: A learning apprentice system for VLSI design, *Proc. IJCAI 1985*, Los Angeles, CA., pp.573-580

[5]   J. Herrmann: Estimations for High-Level IC Design: A Machine Learning Approach, in: *Artificial Intelligence in Engineering: Diagnosis and Learning*, J.S. Gero (ed.), Elsevier Science Publishers B.V., 1988

[6]   K. Milzner, R. Klinke: Synthesis of Analog Circuits using a Blackboard Approach, *Proc. Third Int. Conf. on Industrial and Engineering Applica-tions of AI and Expert Systems*, 1990, Charleston S.C., Vol.I, pp.114-122

[7]   K. Milzner, W. Brockherde: SILAS: A Knowledge-Based Simulation Assistant, *IEEE Journal of Solid-State Circuits*, Vol.26, No.3, March 1991, pp.310-318

[8]   K. Milzner: An Analog Circuit Design Environment based on Cooperating Blackboard Systems, *Int. Journal of Applied Intelligence*, Vol.1, 1992, pp.179-194

[9]   P. Langley, G.L. Bradshaw, H.A. Simon: Rediscovering Chemistry with the BACON System, in: *Machine Learning I*, R.S. Michalski, T.M. Mitchell, J.G. Carbonell (ed.), Tioga Publishing, 1983

[10]   Pat Langley: Machine Learning as Experimental Science, *Machine Learning*, Vol.3, No.1, August 1988

[11]   D. Kulkarni, H.A. Simon: The Processes of Scientific Discovery: The Strategy of Experimentation, *Cognitive Science*, No.12, 1988, pp.139-175

[12]   D.H. Sleeman, M.K. Stacey, P. Edwards, N.A. Gray: An Architecture for Theorie-Driven Scientific Discovery, *Proc. European Working Session on Learning*, 1989, pp.11-23

[13]   T.M. Mitchell, P.E. Utgoff, R. Banerji: Learning by Experimentation: Acquiring and Refining Problem-Solving Heuristics, in: *Machine Learning - An Artificial Approach*, R.S. Michalski et. al., Tioga Publishing, 1983

[14]   B.W. Porter: Learning Problem Solving, *Ph.D Thesis*, University of California, Irvine, 1984

[15]   J.G. Carbonell, J. Gil: Learning by experimentation, *Proc. 4th International Conference on Machine Learning*; Morgan/Kaufman (ed.), 1987

[16]   Brent J. Krawchuk, Jan H. Witten: On Asking Right Questions, *Proc. 5th International Conference on Machine Learning*, Morgan/Kaufman (Ed.), 1988

[17]   Klaus P. Gross: Incremental Multiple Concept Learning Using Experiments, *Proc. 5th International Conference on Machine Learning*, Morgan/Kaufman (ed.), 1988

Object Oriented Framework for Generating Machine Understanding of a Digital System Design

Prem Malhotra and Rudolph E. Seviora

Department of Electrical & Computer Engineering, University of Waterloo,
Waterloo, Canada, N2L 3G1

Abstract. Machine understanding of a digital system design is needed to perform machine assisted design maintenance activities. This information is usually not available in engineering representation of design. Generation of understanding requires framework for representation and reasoning over design. Object oriented approach has been used to represent design as a model of the digital system, with features to facilitate reasoning over it. A blackboard based reasoning framework integrates with the representation framework to form Digital System Understander.

1 Introduction

This paper presents an object oriented framework for the representation of and reasoning over designs[1] of digital systems. The framework was developed for generating machine understanding, of a design, required for certain design maintenance activities. Machine understanding of a digital system design is a model of the design, which incorporates information that is not available in its standard engineering representation (e.g. circuit schematic). Engineers often generate such information from a schematic diagram for the purposes of design maintenance, design critiquing and fault diagnosis. This information is also needed for automation of these activities.

Digital System Understander (DSU) is a program that generates machine understanding of a digital system design. Given a netlist corresponding to schematic diagram and high level textual summary of a design, DSU would generate a model of the design that includes additional information about it. The designs to which DSU is targeted are composed of components with levels of integration ranging from SSI to VLSI and are dominated by a few semantically rich components. An example of such a design is a microprocessor based single board computer.

Relatively little work has been reported on machine understanding of digital designs. deKleer[8] addressed understanding of circuits consisting of few gates. Lathrop, et.al.[9] involve understanding for purpose of simulation speedup. The understanding comprises of simulation model for a design, obtained using design structure and simulation model of each component. Blaauw, et.al.[2] use understanding to speedup switch level simulation. The understanding consists of functional abstraction of design at four levels in order to identify circuit nodes and transistors that need not be simulated completely, as well as logic gates and clusters of gates for which simulation models are available. In contrast, the understanding in Critter[7] consists of a range of possible behaviors and required specifications of signals internal to a digital circuit. Such information is used to critique the design. Almost all work on understanding of digital designs has dealt with small circuits. Therefore this did not require any extensive framework for representation of and reasoning over designs.

---

[1] The word *design* is often used for both the process of design and the output of such a process. In the paper, the word *design* means the output of a design process.

The three main concerns of design understanding effort are the content of design understanding, its representation and the methodology for its generation. The content is influenced by the intended applications of machine understanding, in our case: redesign and design fault localization. The paper briefly describes the content of machine understanding required for performing these activities through machine assistance. The other two concerns are dealt with by two parts of DSU, the representation framework called SURF and understanding generator called SUDS. Machine understanding generated by DSU is in the form of a model of the digital system. SURF provides object classes whose instances are used to represent the model. SUDS provides object classes to perform reasoning over a design represented using SURF.

The SURF's objects[2] provide partial support for reasoning tasks involving the represented design. Objected oriented approach has proved advantageous in integrating into representation the capability to *perform* some tasks, which are better done by the representational mechanism rather than the user(s) of the representation. Based on analysis of the required content of machine understanding and its use in design maintenance activities, two principles have been formulated. The object oriented paradigm facilitated their incorporation in SURF. The principle of *locality of services* is that objects in the representation should provide services to user(s) of the representation which require extensive access to attributes of objects and/or need application of local constraints. This principle off-loads low level tasks from representation's users, such as SUDS and design maintenance activities. The principle of *breadboardable representation* requires flexibility in interconnection and simulation of objects which have an associated structure. The purpose is to facilitate the extraction of a part of design and *test* it for evaluation, comparison, etc. during the process of reasoning over it.

SURF attends to a number of issues which were not sufficiently addressed by design representation frameworks used in related areas, such as, digital system diagnostics[3, 4, 5], redesign[13, 10], design verification[1], design critiquing[7, 13] and simulation[9, 2]. The issues are: *representation of complex design entities* and the *simplification of reasoning over design*. The prior work usually deals with simple design primitives, such as gates. In contrast, DSU deals with complex primitives which have several external interfaces, such as microprocessor. Therefore a more versatile framework is needed. The earlier representation frameworks also do not factor out simple tasks and move them from representation's users to the representation itself, as is done by SURF.

SUDS is influenced by existing blackboard problem solving schemes[6, 11, 12] which use a data board for evolving problem solution and a second board for control of reasoning. These schemes keep control heuristics and scheduling decisions on the second board. However, SUDS employs the second board differently. It uses a goal board for recording goals pertaining to the problem domain and the knowledge sources which generated these goals. A novel scheduling scheme is employed. It avoids use of commonly employed weighted scheduling criteria, which is difficult to *tune*. The scheduling is performed by *filtering* the set of triggered knowledge sources through a series of domain independent and dependent filters until one knowledge source is left in the set, which is then activated.

The paper is organized as follows. Section 2 describes the selected design maintenance activities and outlines the content of machine understanding. Section 3 describes salient representation objects provided by SURF and their use in representing a digital system design. An example of reasoning over design is presented in section 4 to illustrate how incorporating principles of *locality of services* and *breadboardable*

---

[2] The word *objects* is being used for both class objects as well as their instances.

representation in SURF facilitate reasoning. Section 5 describes the objects of reasoning framework SUDS and how they interact during reasoning. The concluding section, section 6, summarizes key points of the paper, lists advantages and disadvantages of object oriented approach and gives implementation status of the work.

2 Machine Understanding for Design Maintenance Activities

The intended application(s) of machine understanding influence its content. The primary application at which the representation framework presented in this paper is targeted is the design maintenance activities. Such activities require more information about a design than is present in its schematic drawing or its netlist.

2.1 Design Maintenance Activities Considered

The activities targeted are redesign and design fault localization. Redesign is adaptation of an existing design for minor changes in requirement specification. Design fault localization is identification of fault(s) in a structural part of design, assuming that remaining design is correct.

Redesign is restricted to small changes in the existing design. Two common redesign activities are component substitution and design augmentation. During component substitution a component (or a small collection of components) in the given design is replaced with structurally and/or behaviorally different but functionally similar component(s). Design augmentation adds to the design; the changes are typically restricted to a section of design in order to limit the complexity of redesign activity. For example, given a microprocessor based single board computer design, the changed requirements may require addition of a parallel output port.

Design fault localization determines minor mistakes in a design. Such a design fault can be due to incorrect interconnection(s), wrong choice of components with respect to their behavior parameters (such as, propagation delay), etc. For example, in a single board computer system's design, the ROM memory may have access time larger than permitted by microprocessor. The fault is typically described in terms of a structural part of design, *ROM memory* in this case. The fault localization should determine ROM memory section's access time as cause of the fault and further localize it to access time of memory components or bus interface of memory sub-system.

A suitable approach to redesign (or design fault localization) would need detailed information about design. It can have the steps: (a) focus attention on the section of design to be modified (or indicated to have design fault), (b) determine the highest level of abstraction where the design can be changed (or where design fault appears), (c) modify design (or identify entities which are responsible for design fault) at the determined abstraction level and (d) successively propagate the modification (or fault localization) to lower level representations of design. It is relatively easier to modify a design (or localize design fault) at an abstract level because an abstract representation has fewer design entities and structurally simpler external interfaces of the entities.

2.2 Content of Machine Understanding

The content of machine understanding of a digital system design needed for the above maintenance activities would include: (a) representations of design at higher levels of abstraction, where representation at each level consists structure and each structural entity has an associated behavior; (b) constraints on parameters of a design entities, concerning its structure and/or behavior; (c) dependency between parameters of inter-dependent design entities in a representation at a given level of abstraction; (d) mapping (structural and behavioral) between representations at adjacent levels.

Distinction is made between mapping associated with abstract levels and hierarchical levels. Fig. 1 illustrates the difference. In the hierarchical representation of design, passive interconnection(s) between entities at a higher level map as 1:1 to an interconnection at a lower level. However, in an abstraction based representation the mapping 1:(a collection of n) interconnections can occur; because the process of abstraction encompasses structure, associated behavior and trade-off between them. Both kinds of mapping are required because abstract levels described in (a) include hierarchy and abstraction.

The rationale for the content is as follows. Abstract representations are less detailed and therefore facilitate reasoning over sections of design. The mapping information assists in traversing among abstraction levels. Constraints are used to record the assumptions made about design entities. Dependency information records inter-dependence between entities. Constraint and dependency information together yield rationale for some design choices, especially at lower levels of abstraction.

3 Representation Framework: SURF

SURF is an object oriented framework for representing machine understanding of a digital system design. It offers object classes whose instances collectively represent a design. The naming convention for classes and instance objects is as follows: class names begin with an upper case letter while an instance name begins with a lower case letter. For names consisting of more than one word, all words are concatenated and first letter of all following words is made upper case (e.g. aDigitalSimulation).

3.1 Criteria used in Defining Classes

In addition to generalization of entities in the domain, two other principles were applied in class definition. These are *locality of services* and *breadboardable representation*. The nature of digital system domain and the intended reasoning intensive application of machine understanding have led to these principles.

The digital system domain requires that a number of constraints be attached to representational objects. These should be satisfied in order to assure correct operation of the represented entity. Such requirements motivated the principle: *locality of services*. It requires that an object forming a representation should have capability to perform operations (such as constraint satisfaction) that need extensive access to the object's attributes. For example, to enforce constraint such as "a signal source should have worst case current drive capacity", a task with this objective is localized to an interconnection. This approach leads to fewer explicitly represented constraints, the ones imposed during design process.

Reasoning over design during design maintenance activities as well as process of understanding requires comparing behavior of parts of design at same or different levels of abstraction to confirm/reject hypotheses about them and deduce missing pieces (while generating machine understanding). This motivated the principle of *breadboardable representation*. It requires that pieces of design be *executable*, much like executing a design on breadboard with actual components. Breadboardability of a representation requires flexibility in interconnection of design entities and availability of a simulation mechanism.

3.2 Major Classes

The classes responsible for representing structure, behavior, mapping, constraints and dependency are described in this section. Support classes, such as ones which implement discrete event simulation, are not described. The following descriptions often refer to an object as owner of another object or being owned by it. The meaning of owner in this context is that the owner object can perform special operations on the owned object, such as, configuring it and exclusively use it under certain conditions. The owner object also gets informed of changes to the owned object.

Structure and Behavior The classes involved in representing structure are Port, DigitalEntity, ConnectEntity. Instances of Port represent external structural interface of aDigitalEntity or aConnectEntity. Instances of DigitalEntity, or its subclasses, represent active entities at different levels of abstraction, such as, components and their abstractions. Instances of ConnectEntity, or its subclasses, represent interconnections among digitalEntities, such as, a data bus. Fig. 2 illustrates structural representation using these classes.

Instance of Port has a concept of its two ends, internal and external. The internal end refers to the owner of port (e.g. a digitalEntity) and external end refers to external interface of the port's owner. A port permits only one other port to be hooked at each of its ends. Basic operation of a port is to propagate digital signal received at one end to its other end. Attributes of a port are: width and direction. The width refers to maximum number of bits of digital signal it can transfer from one end to other end (width can be undefined too). The direction refers to direction of signal flow through the port with respect to its owner, such as, input, bi-directional and output. A distinction is made between different kinds of output ports, such as, active pullup, open collector, etc. The constraints regarding hookup compatibility of different port directions are applied by a port during hookup to another port[3].

Instances of ConnectEntity are passive in nature. Their operation is to transfer signals between its ports, much like a connecting wire or a bus transfers signals between pins of ICs on a breadboard. A connectEntity has attributes for its width in bits and assignment of its ports to various contiguous bit positions. Therefore it can route signal bits to ports with differing widths. Two additional tasks have been assigned to a connectEntity. These are: checking for insufficient current drive capacity of signal source port and detecting absence of a signal source port. A subclass of ConnectEntity called ConnectEntityWithPullup, provides a variation. Its instances behave as if there exists a pullup resistor from each bit to power supply.

Instances of DigitalEntity and its subclasses are the most complex objects in SURF. A digitalEntity knows its external interface, embedded behavior, associated constraints, typical compositions, and its default self tester. Its complexity is tackled by assigning responsibilities to instances of other classes, such as DigitalBehavior, DigitalComposition, etc.

---

[3] The concept of *port* has been used earlier (e.g. [3]) to represent external interface of an entity. However, no state, attributes or responsibility for port hookup and signal transfer were associated with it.

The provision of separate port and connectEntity objects gives the flexibility in manipulating structure of a design. The external interface of a digitalEntity (embodied in ports) is separate from concerns of interconnection between digitalEntities; therefore its easy to modify interconnections.

Instances of class DigitalBehavior represent the behavior of a digitalEntity. The behavior is represented symbolically and it is directly executable using SURF's simulation mechanism. A digitalBehavior is a collection of finite state machines. It normally keeps an internal state, except when representing combinational behavior. Instances of OperationStep are used to determine state transitions. An operationStep embodies procedural knowledge concerning behavior. It can be triggered by a change in value of state variable of digitalBehavior or that of a port of the associated digitalEntity. Actions of an operationStep modify internal state variables and/or signals at ports.

Mapping Mapping is a means of linking representations of design at adjacent levels of abstraction. It implies both structural and behavioral correspondence. A digitalEntity or connectEntity at higher level can be mapped to a single entity of same kind or a collection of entities at the lower level.

Instances of DigitalComposition represent a collection of interconnected digital-Entities and connectEntities that implement a behavior equivalent to the embedded behavior of the associated digitalEntity. A digitalComposition has an external interface (it uses ports) which can be hooked to inside of its owner digitalEntity (i.e. to internal ends of digitalEntity's ports). Such a hookup replaces the digitalBehavior with digitalComposition for purpose of simulation. This feature gives an important capability: to verify through simulation that a candidate composition is correct, that is, it yields expected behavior.

A digitalComposition, consisting of entities at a level of abstraction lower than that of its owner digitalEntity, represents the composition as well as maps to the digitalEntity. However, to only map to a digitalEntity a digitalComposition should not be owned by the digitalEntity. This illustrates the difference between the concepts of composition and mapping, as used in SURF. A composition means *equivalent to* while mapping all by itself means *corresponds to*.

Instances of ConnEntityCollection represent a collection of connectEntities. A connEntityCollection can be a part of digitalComposition or can map to a connectEntity at higher level of abstraction. Unlike a digitalEntity, a connectEntity does not need composition because its structure and behavior are simple.

Constraints and Dependency Instance of Constraint records constraints applicable to its owner (a digitalEntity or a digitalComposition). A constraint has symbolic parameters, messages to get and/or set values of these parameters and boolean-valued expressions consisting of arithmetic and logical operations on parameter values. A constraint is satisfied if all of its expressions evaluate to true. The messages required to get or set a parameter's value are sent to the owner object or other objects owned by it. A constraint can check that it is satisfied and can attempt to satisfy itself, using satisfaction procedures (optional) attached to it.

Dependency information is available from constraints. If a constraint's parameter values do not use attributes tagged to the constraint's owner, the expressions themselves represent dependency information. Else, the constraints of owner object(s) at a higher level of abstraction have to be scanned to determine ones which affect values of such attributes. Dependency is then deduced through parameters of higher level objects.

3.3 Design Representation Using SURF

A design is represented at successive levels of abstraction. SURF presently supports four levels, though in principle there is no limit on the number of levels. The levels, in increasing order of abstraction, are: component, block, unit and module. Subclasses of DigitalEntity correspond to different kinds of components, blocks, units and modules.

Representation is recorded on an instance of DesignBoard. It keeps objects of different levels separately and provides operations to manage them. In addition, it records the mapping between mapping compatible objects on adjacent levels. The objects explicitly recorded on a designBoard are digitalEntities, connectEntities, digitalCompositions and connEntityCollections. Fig. 3[4] pictorially illustrates representation of ROM section of a microprocessor based design on a designBoard.

4 Reasoning Over Design Represented Using SURF

An example of reasoning over design is presented in this section to illustrate how principles of *locality of services* and *breadboardable representation*, incorporated in SURF, facilitate reasoning.

Consider component substitution activity in the design shown in Fig. 3. The read only memory (ROM) components are to be substituted by structurally identical but faster version (lower access time) components. The 500ns access time 68364 ROM components are to be substituted with similar but faster 250ns components. Purpose of the substitution is to enhance performance of the design.

The given design needs a modification to a part of ROM section, in addition to 1:1 component substitution, to get advantage of the faster ROM components. Data transfer between ROM and processor is asynchronous. The signal *dtack** is required by the processor to indicate that read data is available on *data* bus. This signal is generated by *zfer-ack* generator 500ns after the *enable** signal to ROM. In order to get maximum advantage of 250ns ROM components, the modified design should generate *dtack** 250ns after *enable** signal. This entails taking *dtack** signal from flip-flop output $q2^*$ instead of $q4^*$ of 74LS175 component.

The required design modifications can be achieved through the use of services available with representation objects and application of independent knowledge sources. The services used here are *constraint check* and *constraint satisfy*. These are performed by objects with attached constraints[5]. Referring to Fig. 3, the objects of interest are digitalEntities: *rom block, zfer-ack gen block* and digitalCompositions: *c1, c2, c3*.

A *knowledge source* replaces 68364s in c2 with faster ones as well as changes value of its tagged attribute #accessTime to 250ns. The *rom block* performs constraint check because its composition c2 has changed. The check reports failure. In an attempt to constraint satisfy itself the *rom block* changes access delay of its digitalBehavior to 250ns. The change in rom block prompts *c1* to perform constraint check and subsequently the constraint satisfy procedure. The dtack delay parameter of xfer-ack gen block's digitalBehavior is modified to 250ns. The *zfer-ack gen block* in turn changes value of the tagged attribute #delay of its digitalComposition c3 to 250 ns. Constraint check failure is detected by c3 but it has no constraint satisfaction procedure

---

[4] The figure does not show, due to lack of space, the module level of abstraction and embedded objects, such as digitalBehavior and constraint. The signal labels postfixed with * indicate that active value of the corresponding signal is low.

[5] Constraints record assumptions made about design entities when these are used in a design. See sections 2.2 and 3.2.

since structural change is required. Subsequently, a *knowledge source* moves the dtack output from q4* to q2* of 74LS175. To verify correctness of the structural modification, the knowledge source disconnects enable input and dtack output of c3 from the rest of design and connects them to a digitalTester. The objects c3, digitalTester and clock gen composition are simulated. The digitalTester verifies that delay generated by c3 is 250ns. Lastly, c3 is reconnected to the design.

The above example shows that localizing certain operations (such as constraint check/satisfy) to representation objects reduces the need of knowledge sources to know an object's internal organization. Consequently fewer broadly applicable knowledge sources can solve a problem. The breadboardability of representation is useful in verifying correctness of modifications to design.

5   Reasoning Framework: SUDS

SUDS is an object oriented framework based on blackboard paradigm for reasoning over design represented using SURF. The ideas it uses from blackboard paradigm are: (a) cooperative problem solving by independent agents; contributions of an agent are visible to all others on a common data area, (b) the common data area is organized by levels of abstraction inherent in the domain, (c) the only coupling between agents is through contents of common data area, (d) an agent gets *triggered* if it has potential to contribute to the problem being solved and (e) control of problem solving is done centrally.

5.1   Major Objects

Fig. 4 shows major objects of SUDS. Solution is formed on *design board* through application of instances of knowledge source (KS) classes. The *goal board* records domain related goals and KSs which created these goals. It is organized by levels corresponding to ones on design board. Level of a KS or a goal is determined by the level (on design board) of its principal binding. The goal board has one additional general level for KSs and goals which do not have a principal binding. Both boards are visible to every KS class.

SUDS supports three kinds of KS classes. Instances of a ContributingKS class (or its subclasses) assert new goals on goal board. An example of such a KS is AcquireProblem. It is applicable when goal board is empty. Instance of AcquireProblem, on being activated, interrogates user for the kind of problem (generate understanding, perform redesign, etc.) to be solved and creates appropriate goals. Instances of a DecomposingKS class (or its subclasses) decompose an existing goal on goal board into simpler ones. Instances of a ProceduralKS class (or its subclasses) actually solve a goal by modifying design board.

The *problem solver* is responsible for scheduling applicable knowledge sources and supervision of problem solving activity. It provides a message protocol for user to start/stop problem solving, specify breakpoint conditions to automatically stop problem solving and to modify parameters used for scheduling triggered KS classes. Some of the breakpoint conditions are: a specified KS class is triggered, any KS is unsuccessful, no KS classes triggered for a *ready* goal during an interval, etc.

5.2   Operation

SUDS uses the concept of problem solving cycle. Fig. 4 shows the sequence of messages sent between major objects in a cycle. A cycle begins with both boards performing housekeeping in response to *nextCycle* message from problem solver. Next, all KS classes are interrogated for applicability in this cycle. A KS class caches the result (bindings) from its previous cycle's trigger matching. If none of these bindings changed in previous cycle as well as no changes were made to levels of interest to the KS class, it does not reevaluate trigger and replies the answer (yes/no) it gave in previous cycle. A KS class can interrogate the problem solver for objects changed and board levels modified during previous cycle.

The problem solver selects one KS class out of those triggered by *filtering* them through an ordered collection of filters. Each filter is a method which lets through the KS classes it prefers. The scheduling algorithm has following features: start using filters from beginning of the ordered collection; if a filter does not let through any KS class, ignore it; stop filtering if only one KS class is left, select it for execution; if no unique KS class could be determined after application of all filters, select at random from the filtered set of KS classes. Examples of filters are: prefer KS class(es) triggered for goal(s) that have most complex principal binding, prefer KS class(es) triggered for most recent goal(s), etc. An instance of the selected KS class is created, placed on goal board and activated. A cycle ends when the activated KS has finished processing.

Control strategy is implemented by suitably ordering appropriate filters to be used by problem solver. The strategy cannot be dynamically modified during problem solving (no control knowledge sources are used). The scheduling approach used in SUDS is sufficiently flexible to experiment with different control strategies. Since the selection criteria (the filters) are not weighted, the exhibited scheduling decisions are easier to explain.

6 Conclusion

This paper presented an object oriented framework for generating machine understanding of a digital system design, which is required for selected design maintenance activities. Understanding consists of representation of structure and behavior of design at abstract levels along with additional information, such as, constraints, mapping, etc. The framework has two parts, SURF and SUDS, which are used to represent design and reason over it respectively. SURF is influenced by two principles: *locality of services* and *breadboardable representation*. It provides services and features to facilitate reasoning over a design. SUDS is based on blackboard paradigm. It employs a simple scheduling scheme based on *filtering* competing knowledge source through pre-determined criteria.

Several advantages have accrued from object oriented approach. The modeling related advantages include: (a) objects can be tailored to match structure and behavior of primitives in the domain, (b) semantic hierarchies among primitives of the domain are easily modeled by class hierarchies and (c) since an object has modular and extensible behavior in form of message interface, it can be extended to beyond what is needed for modeling. Frequently performed operations and/or the operations which need extensive access to attributes of an object, can be moved to the object itself.

The implementation related advantages include: (a) compactness of software due to code sharing among classes in a class hierarchy and (b) control over complexity because code is modular, it being distributed over several methods in each class.

SURF and SUDS have been implemented in Smalltalk-80 release 4 on a Sun SPARCstation. The complete implementation comprises of about 50 classes, excluding the subclasses of DigitalEntity corresponding to known components, blocks, etc. The source code size is about 19,000 lines.

References

1. H. G. Barrow. Verify: A program for proving correctness of digital hardware designs. *Artificial Intelligence*, 24:437–491, 1984.
2. D. T. Blaauw and et. al. Snel: A switch-level simulator using multiple levels of functional abstraction. *Proc. ICCAD-90*, pages 66–69, 1990.
3. R. Davis. Diagnostic reasoning based on structure and behavior. *Artificial Intelligence*, 24:347–410, 1984.
4. M. R. Genesereth. The use of design descriptions in automated diagnosis. *Artificial Intelligence*, 24:411–436, 1984.
5. W. Hamscher. Temporally coarse representation of behavior for model-based trouble-chooting of digital circuits. *Proc. IJCAI-89*, pages 887–893, 1989.
6. B. Hayes-Roth. A blackboard architecture for control. *Artificial Intelligence*, 26(3):251–321, 1985.
7. V. E. Kelly and L. I. Steinberg. The critter system: Analyzing digital circuits by propagating behaviors and specifications. *Proc. AAAI-82*, pages 284–289, 1982.
8. J. D. Kleer. How circuits work. *Artificial Intelligence*, 24(1-3):205–280, 1984.
9. R. H. Lathrop, R. J. Hall, and R. S. Kirk. Functional abstraction from structure in vlsi simulation models. *Proc. 24th Design Automation Conference*, pages 822–828, 1987.
10. J. Mostow. Design by derivational analogy: Issues in the automated replay of design plans. *Artificial Intelligence*, 40(1-3):119–184, 1989.
11. W. R. Murray. Control for intelligent tutoring systems: A blackboard-based dynamic instructional planner. *AICOM*, 2(2):41–57, 1989.
12. B. Pomeroy and R. Irwing. A blackboard approach for diagnosis in pilot's associate. *IEEE Expert*, pages 39–46, 1990.
13. L. I. Steinberg and T. M. Mitchell. The redesign system: A knowledge-based approach to vlsi cad. *IEEE Design & Test*, pages 45–54, 1985.

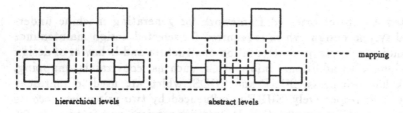

Fig. 1. Hierarchical Levels vs Abstract Levels.

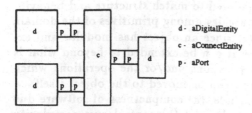

Fig. 2. Conceptual Picture of Structural Representation in SURF.

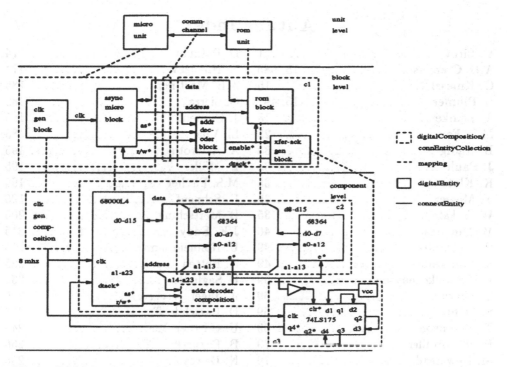

Fig. 3. ROM Section of a Microprocessor Based System.

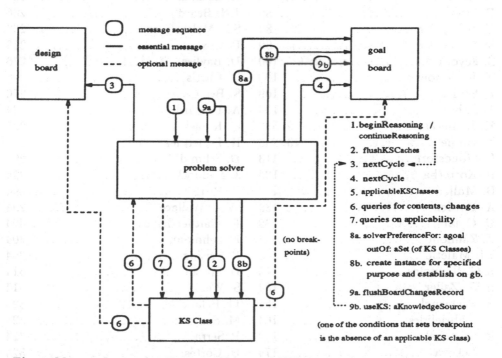

1. beginReasoning /
   continueReasoning
2. flushKSCaches
3. nextCycle
4. nextCycle
5. applicableKSClasses
6. queries for contents, changes
7. queries on applicability
8a. solverPreferenceFor: agoal
    outOf: aSet (of KS Classes)
8b. create instance for specified
    purpose and establish on gb.
9a. flushBoardChangesRecord
9b. useKS: aKnowledgeSource

(one of the conditions that sets breakpoint
is the absence of an applicable KS class)

Fig. 4. Message Interaction Between Objects of SUDS.

Author Index

Lecture Notes in Artificial Intelligence (LNAI)

Lecture Notes in Computer Science